AF782252

Chemical Modeling in Aqueous Systems

Chemical Modeling in Aqueous Systems

Speciation, Sorption, Solubility, and Kinetics

Everett A. Jenne, EDITOR
U.S. Geological Survey

Based on a symposium cosponsored by the Society of Environmental Geochemistry and Health and the ACS Division of Environmental Chemistry at the 176th Meeting of the American Chemical Society, Miami Beach, Florida, September 11–13, 1978.

ACS SYMPOSIUM SERIES **93**

AMERICAN CHEMICAL SOCIETY
WASHINGTON, D. C. 1979

JAN 0 7 1988

Library of Congress CIP Data

Chemical modeling in aqueous systems.
(ACS symposium series; 93 ISSN 0097-6156)

Includes bibliographies and index.

1. Chemistry—Mathematical models—Congresses. 2. Solution (Chemistry)—Congresses.
I. Jenne, Everett A., 1930- . II. Society of Environmental Geochemistry and Health. III. American Chemical Society. Division of Environmental Chemistry. IV. Series: American Chemical Society. ACS symposium series; 93.

QD39.3.M3C49 541'.3422'0184 79-242
ISBN 0-8412-0479-9 ASCMC 8 93 1-914 1979

Copyright © 1979

American Chemical Society

All Rights Reserved. The appearance of the code at the bottom of the first page of each article in this volume indicates the copyright owner's consent that reprographic copies of the article may be made for personal or internal use or for the personal or internal use of specific clients. This consent is given on the condition, however, that the copier pay the stated per copy fee through the Copyright Clearance Center, Inc. for copying beyond that permitted by Sections 107 or 108 of the U.S. Copyright Law. This consent does not extend to copying or transmission by any means—graphic or electronic—for any other purpose, such as for general distribution, for advertising or promotional purposes, for creating new collective works, for resale, or for information storage and retrieval systems.

The citation of trade names and/or names of manufacturers in this publication is not to be construed as an endorsement or as approval by ACS of the commercial products or services referenced herein; nor should the mere reference herein to any drawing, specification, chemical process, or other data be regarded as a license or as a conveyance of any right or permission, to the holder, reader, or any other person or corporation, to manufacture, reproduce, use, or sell any patented invention or copyrighted work that may in any way be related thereto.

PRINTED IN THE UNITED STATES OF AMERICA

ACS Symposium Series

Robert F. Gould, *Editor*

Advisory Board

Kenneth B. Bischoff
Donald G. Crosby
Robert E. Feeney
Jeremiah P. Freeman
E. Desmond Goddard
Jack Halpern
Robert A. Hofstader
James D. Idol, Jr.
James P. Lodge
John L. Margrave
Leon Petrakis
F. Sherwood Rowland
Alan C. Sartorelli
Raymond B. Seymour
Aaron Wold
Gunter Zweig

FOREWORD

The ACS Symposium Series was founded in 1974 to provide a medium for publishing symposia quickly in book form. The format of the Series parallels that of the continuing Advances in Chemistry Series except that in order to save time the papers are not typeset but are reproduced as they are submitted by the authors in camera-ready form. Papers are reviewed under the supervision of the Editors with the assistance of the Series Advisory Board and are selected to maintain the integrity of the symposia; however, verbatim reproductions of previously published papers are not accepted. Both reviews and reports of research are acceptable since symposia may embrace both types of presentation.

CONTENTS

PREFACE

Chemical modeling, particularly as related to environmental chemistry, is a rapidly evolving field. However, because of the wide dispersion of the pertinent literature in scientific publications of many different fields and the absence of suitable review articles, investigators are often unaware of recent advances in disciplines other than their own. Therefore, it appeared that progress in chemical modeling capability could be significantly increased by a symposium which would bring together investigators from the several disciplines. This was facilitated by joint sponsorship of the symposium by both the Society of Environmental Geochemistry and Health and the American Chemical Society.

The introduction to the symposium details the current constraints on research in general and chemical modeling in particular. The 37 technical papers presented fall naturally into the areas of: 1) redox equilibria, 2) organic ligand characterization and stability constant estimation, 3) adsorption processes, 4) evaluation and estimation of thermodynamic and kinetic data, 5) interfacing chemical and biological models, and 6) comprehensive chemical models and their applications.

By arrangement, certain papers are either partially or totally of a critical review nature. For the benefit of the reader it may be noted that these review papers deal with thermodynamic data evaluation and estimation or with bioavailability: D. Langmuir reviews and demonstrates available techniques for the estimation and evaluation of thermodynamic data; J. M. Cleveland reviews the solubility and stability constant data for plutonium compounds and complexes, respectively; J. A. Kittrick reviews the problem of how to treat exchangeable cations in calculating the solubility of clay (layer silicate) minerals; G. H. Nancollas reviews the thermodynamic data for calcium phosphates; L. N. Plummer, T. M. L. Wigley, and D. L. Parkhurst review the kinetics of calcite dissolution–precipitation; J. W. Morse and R. A. Berner review the kinetics and solubility of marine carbonates; S. N. Luoma reviews the general problem of estimating trace element bioavailability from sediments; F. L. Harrison reviews the extensive studies she and others have made of trace element bioavailability and accumulation by macrobenthic organisms; and finally, D. K. Nordstrom, L. N. Plummer, and others compare the trace element speciation and solubility disequilibrium indices obtained with various extant chemical models. The latter authors concluded that differences in the thermodynamic data selected and treatment of redox are the predominant causes of the discordant results obtained.

Acknowledgments

The diligence, industry, and imagination of two of my associates contributed greatly to the quality of the symposium and the proceedings. Jo Buchard style-edited the review copies, proofed the camera-ready copy of the manuscripts, and made some of the necessary contacts with the authors. Janet Wong arranged for much of the peer review of manuscripts, kept track of manuscript receipt and review status, and made corrections to the various "camera-ready" manuscripts as were deemed necessary. The unfailing support and good humor of these two individuals was a major factor in the success of the symposium and the quality of these "rapid publication" proceedings.

Peer reviews were indispensable in maintaining the technical quality of the proceedings. Appreciation is expressed for the assistance of the numerous reviewers who carefully and critically read the papers in this volume. The Editorial Review Committee was of great assistance and made a significant contribution to the quality of these proceedings. This committee consisted of V. C. Kennedy, R. V. James, and R. W. Potter, II, to whom I am indebted.

U.S. Geological Survey
Menlo Park, California
November 14, 1978

EVERETT A. JENNE

Introduction

1

Chemical Modeling—Goals, Problems, Approaches, and Priorities

EVERETT A. JENNE

U.S. Geological Survey, Water Resources Division, Menlo Park, CA 94025

The belief that Science can solve societal problems in acceptable time frames has been discounted by the public in recent years (1, 2). This situation, according to D. S. Saxton (3), is a result of the "...overreaching expectations that research....would visibly contribute to early solutions to difficult social problems." Since the total monetary resources available for studying such problems are limited, the effect of recent policy developments and legislation has been to increase support for applied studies with a resultant decrease in support of fundamental research. The country as a whole has turned, and is turning further, toward the use of computational models utilizing best guesses and readily available parameter information which may have only order-of-magnitude reliability. One result of this trend is an enormous quantity of reports of a highly applied nature coming out of institutes, universities, and governmental agencies. Many of these reports are of dubious value (4). The highly applied research approach produces some short-run benefits, but is virtually certain to produce long-term liabilities. One liability will be that a larger part of the research community will be trained as data gatherers rather than information synthesizers with a resultant decrease in the transfer value of much of the work. In the field of chemical modeling, the principal long-term liability will be that insufficent fundamental knowledge needed for improvement of chemical, biological, and hydrologic models will not be available within the next several years.

Although the fundamental problems being addressed herein are large philosphical in nature, it is appropriate to define what is meant by chemical modeling. As used herein, chemical modeling encompasses the aqueous speciation of dissolved cationic elements among organic and inorganic anionic ligands, of anionic elements among their complexes with cations, and both cationic and anionic elements among their redox states. Chemical modeling also includes calculation of the degree of saturation of an aqueous media with regard both to metastable and to equilibrium solids and calculation of sorption or desorption. Additionally, pre-

0-8412-0479-9/79/47-093-003$05.00/0
This chapter not subject to U.S. copyright
Published 1979 American Chemical Society

dictive chemical modeling must include kinetics. Chemical modeling can be used to describe the chemical characteristics of an aqueous system whether it be a lake water or the fluid in a human digestive system, given the requisite input data for the system to be modeled as well as adequate reference data (thermodynamic and kinetic), and a properly constructed model. Quantitative models are valuable because they increase one's ability to derive information from data, to predict certain effects of anthropogenic inputs, and to do sensitivity analyses. The latter may well be one of the most important and least recognized uses of such models. For example, in recent modeling of the inorganic speciation of dissolved silver in San Francisco Bay, a sensitivity analysis to test the effect of metals and ligands not included in the chemical analyses indicated that hydrogen sulfide at the microgram-per-liter level would complex more silver than would chloride in the low salinity portion of the estuary (5). This finding then led to a preliminary study of dissolved sulfide in San Francisco Bay waters. Sensitivity analyses can also be performed to indicate needed accuracy and precision of particular analyses and parameters. Chemical models can also be used in appropriate situations to estimate biological effects on the chemical part of an ecosystem (6).

Goals

The ultimate goal of all research in general and chemical modeling in particular is additional understanding of processes and events which can, under the proper conditions, facilitate the improvement of human life physically, emotionally, and aesthetically. For example, within the next decade, appropriate chemical- and biological-modeling studies of soil solutions and plant-nutrient uptake can assist in increasing markedly both the quantity and quality of terrestrial food production. Chemical modeling can provide insight into the bioavailability and the variations in the bioavailability of trace elements in man's diet components. Processes such as sulfide complexation, valence reduction, precipitation and organic complexation, may significantly reduce the activity and hence the bioavailability of various trace elements. Linkage to biological models will certainly increase the amount of information which can be derived. In addition to the resultant potential improvement in human nutrition, insight into the bioavailability of trace elements from dietary components may contribute to the control and alleviation of some human diseases related to trace elements.

More immediate goals differ from one scientific discipline to another. One goal of which I am most aware is the prediction of the role of aqueous speciation on the bioavailability and toxicity of trace elements to plankton and estuarine detritus-feeding invertebrates. In the longer range, these studies would be extended to include the various aquatic organisms higher in the food chain.

The overall goal of this symposium is to promote the science of chemical modeling in three ways: 1) by promoting the immediate exchange of information via oral presentations by as many workers in the field as practicable; 2) by fostering the maximum amount of discussion of research goals, problems, approaches, and priorities; and 3) by preparing a state-of-the-art record in the form of symposium proceedings. Further goals are to improve the quality as well as the applicability of chemical models in ensuing years and to help minimize duplication of computerized chemical models by increasing the visibility of those already available. Although new approaches and mathematical frameworks are obviously to be encouraged, it would appear more profitable to the scientific community to upgrade and expand the available models rather than to generate additional ones of similar type and structure. This view comes out of our experience that the computer codes, and indeed the models themselves, can be rendered largely error free only by extended use over time by knowledgeable investigators. It is hoped that the formal (7) and informal discussions of attributes of chemical models will increase the reliability and usefulness of the next generation of chemical models.

Problems

Important restraints to the usefulness of available chemical models are inadequacies in the 1) capability to characterize the organic ligands of natural waters; 2) knowledge of redox status of waters to permit realistic computation of redox-controlled speciation; 3) available thermodynamic data; 4) knowledge of the time dependency of the various processes (such as variation in apparent trace element concentration during a tidal cycle) in general and of kinetic data for chemical and biological processes in particular; and 5) error estimates for the preceding restraints. The field of chemical modeling is also constrained by the: 6) need for comprehensive analyses of the systems to be modeled; 7) limitations of the quality and scope of published data; 8) scarcity of adequate literature reviews; 9) difficulties in organizing and conducting integrated interdisciplinary studies; and to a lesser extent, 10) limitation in quality of graduate training.

Organic Ligands. Investigations of organic compounds dissolved in surface waters have largely dealt with identification of specific compounds or groups (e.g., phenols, polyaromatic hydrocarbons, etc.), determination of complexation capacity, or separation--with variable types of characterization--into fulvic- and humic-acid fractions. The information developed in these studies has not been shown to be particularly useful to quantitative chemical modeling. The analyses of specific compounds or groups of compounds have come about because of taste and odor problems in drinking water and subsequently because of the possible toxicity of these compounds to food chain components and the toxicity and carcinogenic potential of these compounds to man. Complexation-capacity values have served to identify and dramatize the potentially important role of dissolved organic compounds as ligands capable of promoting the solubilization and hydrologic transport of trace elements. However, there seems little point in continuing these determinations since their magnitude is now known and it is not apparent how they can be used in quantitative chemical modeling.

The speciation of cations among organic as well as inorganic ligands requires estimates of their molar quantities and of their stability constants. Two possibilities exist. One is that there are only a few important natural organic ligands for each metal. The other is that organic complexation is the result of a host of specific organic ligands, each of which is present in too small an amount to be individually important. If the former possibility proves to be true, the individual ligands can be identified by techniques such as liquid and thin-layer chromatography. If the latter possibility proves true, then an approach such as that of Leenheer and Huffman (8) which fractionates soluble organic compounds into solubility and functional group classes will be required.

Redox-Dependent Speciation. Valence state determines the toxicity of an element. It greatly affects the complexation strength between metals and various ligands, hence the degree of saturation of water with respect to solid phases. There is almost complete lack of agreement within the soil science and geochemical communities as to the significance and usefulness of platinum-electrode estimations of redox status. The techniques of estimating the concentration of the valence states of individual couples at their environmental levels are only now becoming available. These techniques offer considerable promise for future studies.

Available Thermodynamic Data. There are several problems regarding available thermodynamic data. The appropriate thermodynamic data do not exist for many solids of environmental interest. This is particularly true for the solids which are extensively substituted, such as the cryptocrystalline manganese oxides or the metallic hydroxy sulfates and phosphates. Greatly divergent thermodynamic data exist for numerous aqueous complexes and solids; some of these values differ by orders of magnitude. Thus, there is a great need for thorough evaluation and selective redetermination. The evaluation of thermodynamic data is very time-consuming, requiring extensive literature searches and considerable knowledge of analytical techniques. The expertise and effort required to obtain the required quality of results generally preclude a comprehensive evaluation of numerous complexes, solids, or elements in the course of a single research project or contract. However, scientists frequently are compelled by the nature of their studies to undertake compilation and selection with or without use of evaluation techniques, since adequately evaluated compilations are not available. Hence, the compilations made by single investigators are often not as thorough and comprehensive as is desirable.

The U.S. National Bureau of Standards (NBS) is the governmental agency with primary responsibility for this area although the U.S. Geological Survey shares responsiblity with the NBS in regard to minerals and the U.S. Department of Energy shares responsibility with regard to the transuranic elements. The evaluation and selection of "best" values is a job of such magnitude that most of the NBS publications of the last decade (9, 10, 11, 12, 13) present only selected values. The substantiating references and discussion for these compilations are to follow in subsequent NBS publications. One recent publication (14) on thorium gives the critical evaluation and references. Critical evaluation and documentation efforts need to be greatly accelerated. As one step in this direction, the NBS is funding two studies devoted to the evaluation and selection of "best" solubility studies from the literature for the carbonate, halide, phosphate, and sulfide salts of the alkali and alkaline earths plus V, Cr, Mo, and U (by Allen Clifford of Virginia Polytechnical Institute) and of Cu, Zn, Cd, Sb, Hg, Pb, and As (by Lawrence Clever of Emory University). These two reviews are part of Commission V. 6.1 of the International Union of Pure and Applied Chemistry (IUPAC) program of compilation and critical evaluation of the available literature on solubility. The thermodynamic data for environmentally and geochemically important complexes and solids (chiefly, nonsilicate) of 29 elements (Ag, As, Au, Ba, Bi, Ca, Cd, Co, Cu, Fe, Hg, K, Mg, Mn, Mo, Na, Ni, Pb, Sb, Se, Sn, Sr, Te, Th, Ti, U, V, W, and Zn) are being semi-critically reviewed by Donald Langmuir and Hugh Barnes of Pennsylvannia State University with NBS funding. Polysulfide complexes are to be included in their review. The status of the thermodynamic data on sulfide- and polysulfide-metal complexes will also be discussed in a revised edition of "Hydrothermal Ore Deposits" edited by Hugh Barnes (15).

Critical reviews of thermodynamic data published in the last few years include the four-volume series of Martell and Smith (16, 17, 18, 19), three on organic and one on inorganic ligands; and the single volumes of Christensen et al. (20) for metal-ligand heats, etc., of Baes and Mesmer (21) for cation hydrolysis, of Langmuir (22) for uranium, and of Robie et al. (23) for numerous minerals.

Reliable activity coefficients for marine waters or for waters of even higher salinity continue to be a problem in calculation of aqueous speciation and computation of mineral equilibria. In particular, the appropriate technique to use for uncharged ion pairs remains in dispute (24). However, the chemical-modeling errors attributable to uncertainties in activity coefficients are generally less than the uncertainties due to other causes such as missing or unreliable thermodynamic values, missing analytical data, and incomplete model components (5). Additionally, a major assessment of activity-coefficent estimation is now underway (25).

Another potential problem in quantitative aqueous speciation is polymerization. Various elements (Be^{2+}, Sc^{3+}, Ce^{4+}, Zn^{4+}, Hf^{4+}, Cr^{3+}, Mo^{5+}, U^{6+}, Fe^{3+}, Co^{3+}, Ni^{2+}, Ru^{4+}, Cu^{2+}, Zn^{2+}, Al^{3+}, Ga^{3+}, In^{3+}, Ln^{2+}, Pb^{2+}, Sb^{5+}, and Bi^{3+}) are said to exhibit this behavior (26, 21, 27). To the extent that polymerization occurs and is not accounted for in the aqueous-speciation submodel, the activity of the various solute species and saturation state of solid forms will be overestimated. For example, the apparent oversaturation of San Francisco Bay waters with silver sulfide was decreased by about 10 percent when the polysulfide complexes of Ag and Cu were included in the aqueous speciation submodel (E.A. Jenne and J.W. Ball, unpublished data, 1978).

There is little agreement among the relevant scientific communities as to the priorities for obtaining missing thermodynamic data. Top priorities on my own list are for data on natural organic ligands, polysulfide complexes (data for Ag and Cu exist, 28), low-temperature solid solution of trace elements in iron and in manganese sulfides and in cryptocrystalline substituted (Ba, Li, Sr) manganese oxides; and the solid solutions of Sr, Ba, Mn, and Mg in $CaCO_3$, and various transition metals plus Sr and Ba in $MnCO_3$. It should be noted that because of kinetic inhibitions in the precipitation of various solids, it is highly desirable in solubility and dissolution studies that physical identification of precipitates and residues be made by optical, X-ray diffraction, and electron microscopy and diffraction techniques.

Time Dependency. The time dependency of geochemical and environmental-chemistry processes is a largely unrecognized but often important variable. For example, it is not unusual for 50 percent or more of the total annual sediment transport of a river to occur in the course of a single storm event (29, p. 487). Marked time variations also occur with dissolved constituents. Girvin *et al.* (30) report the concentration of dissolved Ni and Zn at a South San Francisco Bay station to vary by factors of 1.42 and 1.56, respectively, through a tidal cycle. Similarly, concentrations of dissolved trace element varied in a complex manner between main-channel and near-shore (tidal flat) stations. Diurnal variations in the apparent degree of saturation of surface waters with respect to certain solids may result from biological activity and temperature changes. Recent studies (V.C. Kennedy, U.S. Geol. Survey, unpublished data, 1978) have shown pronounced diurnal variation in phosphate concentrations in a short reach of a stream containing considerable attached algae. The apparent saturation of $ZnSiO_3$ in a small stream has been found to vary diurnally (V.C. Kennedy and E.A. Jenne, unpublished data, 1978). These observations and others indicate that realistic application of chemical models must take cognizance of the time scales of biologic, hydrologic, and chemical processes.

Error Estimates. Little attention has been paid to errors associated with either the thermodynamic or the analytical data utilized in chemical models. M.L. Good (31) considers this to be one of the four principal problems of the day with regard to use and misuse of scientific data. It is essential that chemical models compute propagated standard deviations for their analytical, thermodynamic, and kinetic data. This provision exists in the model of Ball et al. (32).

System Characterization. A further limitation to the adequate testing of chemical models and to their application to environmental problems is the extensive system characterization (especially chemical analyses) often required. Adequate characterization often necessitates many difficult and time-consuming analyses, some of which must be done on site, and the collection--with adequate preservation--of numerous subsamples.

The appropriate interfacing of chemical with biologic and hydrologic models is a rather difficult problem. For example, the prediction of trace-element bioaccumulation by phytoplankton may require in some instances that the uptake rates and the compartmentalized loss rates for various solute species of the element present in the system be known. The effect, if any, on compartmentalized loss rates of the particular solute species taken up (e.g. $HgCH_3^+$ vs Hg^{2+}) also needs to be known. The interaction effect of the concentration of one element upon the uptake and loss rates of another element, such as Hg on Se (33, 34, 35), also need to be known. In many instances, hydrodynamic models may have to be linked with, or otherwise incorporated, into the biologic and chemical models to permit predictions of, for example, increased trace-element levels in oysters resulting from increased anthropogenic inputs to an estuary.

Limitations of Published Data. The development, testing, and successful application of comprehensive chemical models requires at least moderate amounts of wisdom, experience, and time. In the first half of this decade, universities, institutes, and government agencies rushed headlong into environmental chemistry. The result has been a marked increase in the number of reports of the occurrence and distribution of trace elements, but this great effort has produced surprisingly few studies of fundamental processes and reactions that would provide an understanding of observed trace element redistribution. Most of the analyses reported in such studies have been on poorly characterized, or even uncharacterized, samples. Generally, trace-element data have been reported for sediment, or for biota, or for water, but only infrequently for all three. Even more rare is the inclusion of useful hydrodynamic data and atmospheric inputs. A trivial but illustrative example of the result of lack of experience plus lack of an adequate literature review is the federally funded contract study of a few trace elements in a surface water wherein the careful handling of sterile bottles--with plastic gloves (to avoid contamination of the bottles with trace elements from the fingers)--is noted with obvious pride but no mention is made of any filtration or preservation of the samples or even of the possible need for these steps! The large number of chemical analyses of so-called "whole-water" samples (i.e., water samples from which the suspended sediment has not been removed by filtration or other means) is an example of the generation of a large amount of data from which little information can be derived. Proper sampling and sample preservation are essential to high-quality analyses. It should be noted that "composite" time, discharge, or load are generally not useful for chemical-modeling studies.

Adequate Literature Reviews. I conclude that factors other than total monies available for research are responsible for limiting the scientific progress in the fields of environmental chemistry in general and of chemical modeling in particular. This conclusion results from the following observations. The increase in the amount of earth-science data published has been tremendous and the rate at which it is being published continues to increase. This lament is of course not new. Barnaby Rich is quoted as writing in 1613 that "one of the diseases of this age is the multiplicity of books: they doth so overcharge the world that it is not able to digest the abundance of idle matter that is every day hatched and brought forth into the world" (36). In the 10 years following 1965, the number of papers, patents, and other published reports cited by Chemical Abstracts was almost as large as in the 58 preceding years (37). The increase in the absolute amount of published products is correlated with the relative lack of productivity of an increasing proportion of available research monies and manpower. Present investigators are increasingly unaware of previously published data, particularly the data which have recently become available or which have been published in journals outside of the investigator's subdiscipline. The large quantity of published data and its dispersion in almost innumerable journals and other publication series is, in part, mitigated by the preparation of review articles. However, the problem still remains because time, in the order of years, passes before synthesizers (reviewers) garner the large quantities of scattered new data and distill information from them. A further problem is that many of the reviews are a compilation of "who found what" instead of a critical evaluation and synthesis.

The rapid rate of data generation and the sophistication of techniques and methods often required for significant research progress can make the products of many small, one-to-two year grant or contract studies of minimal value. These short-term grants often do not permit the buildup of expertise necessary to make significant research contributions. The short-term grants are not conducive to generating information about mechanisms and processes because observations must be terminated on a relatively inflexible time schedule during which completion of a final report is imperative. The time at which it is necessary to terminate observations and measurements often coincides with the time at which further observations have the greatest potential to increase the investigators' characterization and knowledge of the system. (Obviously, the small discontinuous grant or contract may result in a significant contribution where it complements or adds to an ongoing program.) The funding of long-term investigative programs at the national laboratories and within investigative governmental agencies has become increasingly difficult. It must also be noted that there has been a shift in grant and contract funding away from basic research toward applied research.

Interdisciplinary Approach. The integrated interdisciplinary approach to research is difficult. That it should be possible is shown by, for example, the GEOSEX, Manganese Nodule, Deep Sea Drilling, and RANN (Research Applied to National Needs; 38) projects (Pb at the University of Illinois and Missouri, Mo at the University of Colorado), funded by the National Science Foundation. It would be highly desirable, and it should be possible, for funding agencies to promote the common use of selected sites and samples by investigators from the various disciplines. In many cases, a good deal more information would be generated by the use of splits of samples, being collected in other studies, the use of archived samples, and the use of intercalibration and intercomparison samples than by the use of some local or distant set of special samples. Obviously, considerable care must be taken in the selection of such samples where unstable constituents or absolute numbers of trace constituents are to be reported. More generally, use of a limited number of samples collected in previous or other on-going studies would serve to tie the various studies together and to facilitate interpolation and extrapolation. That is, the time has come to concentrate on the generation of information as opposed to the accumulation of data.

The various interdisciplinary groups differ significantly in at least two ways. In some, there is little integration of the substudies wherein each investigator carries out his own investigation more-or-less independently of the other substudies. In others, the emphasis is on interrelationship and priorities are set by consensus. Integrated interdisciplinary groups also differ in the way they form. They may come about via management decisions or they may form voluntarily. The latter type are termed "associative interdisciplinary groups."

Integrated interdisciplinary groups appear to form only where at least part of the funding is in common. The quality of the scientific products from integrated interdisciplinary studies is nearly always above average because of the multidisciplinary view points and wide range of experience of the investigators their broad coverage of journals, and their professional contacts. Such groups consisting of a half dozen or more professional members often have much less elapsed time between availability of funds and significant progress because a base, in terms of experience, and available techniques, supplies and equipment, is normally much broader than that of smaller project groups. These larger groups require less investment in small equipment items than is required for the same number of professionals organized in smaller projects and can frequently afford more expensive equipment.

Associative interdisciplinary groups can be formed by investigators (without common funds) with overlapping interests for the purpose of sharing research-literature results, literature-searching information, experience, equipment, space, support personnel, continuing peer review, and encouragement. This is the common academic approach to interdisciplinary studies. D.R. Carson (3) believes that the shifting of research emphasis towards the applied end of the spectrum will be accompanied by more large-scale interdisciplinary research in universities. However, it should be noted that the language and conceptual barriers encountered by those undertaking interdisciplinary research sometimes take months to years to overcome (39).

Successful interdisciplinary groups require certain common attributes:

1. Both integrated and associative interdisciplinary groups require at least one of the following--a project director, lead investigator, coordinator, or one or more strong individuals to establish and maintain coherence. Associative multidisciplinary groups may function with a committee (which may be informal) of senior investigators when there is a shared research or intellectual commitment.
2. Senior investigators must get along well or the group will break down and cease to function.
3. There must be frequent exchange of information (as opposed to data) via a common coffee group and(or) periodic seminars on problems, preliminary findings, and plans. This provides continuous "peer" review and critiquing of approaches and results. Exchanges of recent literature searches, discoveries, and summaries of "in progress" and unpublished studies then occurs naturally.
4. The group, or individual projects contained therein, should be of a multiyear or continuing nature. In particular, key leadership and analytical personnel should have a high

degree of continuity.

5. One or more of the senior investigators must spend a great deal of time reading, extracting and synthesizing information from literature and experimental data. This task can be shared among several of the participants, as is generally the case with graduate students in associative multidisciplinary academic groups.
6. Establishing time goals for successive phases of study, and reviewing them periodically, is essential to maintaining the necessary sense of purpose and achievement. However, such time goals should be set conservatively and should be recognized as being goals, not as absolutes. The nature of scientific inquiry will often preclude the attainment of such goals within the initial time frame. Their chief value is as a yardstick to measure the relative merit of various promising and interesting substudies.
7. Formal and informal post-doctoral and visiting-scientist appointments provide new areas of needed expertise and provide accomodation for the normal year-to-year accordion-like pattern of funding levels of both contract research and federal agency programs.
8. The administrative chain needs to recognize the existence of these groups, and to encourage, and, to some extent, to deal with them as units.

There are certain problems peculiar to, or at least more prominent with, these interdisciplinary groups than with the more conventional individual-scientist research studies. These include the prepublication sharing of data as well as significant ideas and concepts, which obscures rights to authorship of papers and determination of senior authorship and of senior versus junior authorship. These are important matters because they impinge so heavily on professional standing and advancement, and on access to research funding. Professional and institutional recognition of the efforts of the one or more investigators who contribute a significant or even a major portion of their time to coordination is perhaps the most difficult problem, particularly with associative groups, as V.C. Kennedy points out (U.S. Geol. Survey, written communication, 1978):

> "Most scientists do quality research for one or both of two reasons; first, they wish to satisfy their intense interest or curiosity in a certain field, or second, they see a financial or professional advancement from the successful completion of the work. Success in the second case almost guarantees success in the first. In science, success in the second case commonly is tied to recognition of the scientist as a generator of new ideas or techniques as indicated by first authorship on published talks or papers. The use of "Doe, et al." to cover several authors of a paper is an example of the downgrading of other contributors to a paper. There is no widely

recognized means of distinguishing coauthors from junior authors.

The result is that often each member of a cooperating group gets to publish a short paper with himself as senior author and others as junior authors. Because numbers of papers are important also, we get a repetition, or near repetition, of the same ideas spread widely. Single, comprehensive reports by multiple authors of approximately equal scientific status are rare except where one individual carries much of the load or the report is broken down to identify separate parts by authorship.

One thought is to use some other identification of authorship instead of one person's name. Perhaps the use of an acronym for the authors and an acknowledgements section describing each person's contribution would work in some instances."

An example of the use of an acronym is that of FOAM (Friends of Anoxic Mud, 40).

A significant part of the research effort of interdisciplinary studies is, and will be, done by graduate students and post-doctoral fellows. A major problem in the utilization of graduate students and post-doctoral fellows is the matter of preparation of journal articles from the research work. A completed thesis is in and of itself often a sufficiently traumatic experience that immediately proceeding to the preparation of journal articles is a difficult task for most advanced degree recipients to face. Indeed, thesis completion invariably continues beyond the anticipated date so that there is rarely time to work on follow-up journal articles. It is generally difficult to get post-doctoral fellows to summarize their research at regular intervals. There are inevitably one or two more very important things which they feel must be investigated before termination of the appointment. In this way, the block of time reserved for preparation of journal articles can easily evaporate. However, this difficulty may be minimized by a "contract" to the effect that the thesis or post-graduate study is incomplete until it has been reduced to acceptable drafts of journal articles.

Research scientists involved in interdisicplinary research in the mission-oriented Federal Agencies must face an additional conflict between the "need" to write papers and the "need" at the agency to achieve certain objectives within a fixed time period. Often a complex choice must be made between these two demands on the scientist. On the one hand, it can be argued that the initial phases of a study were at the taxpayer's expense so they must be reported in the archival literature. On the other hand, pressing on to the fundamental goal of an adequate understanding of some process or to the development of some model may in some instances be considered of sufficient importance to override the need for and value of any interim scientific paper(s). It seems clear that if the preliminary or initial findings are such that they can reasonable be expected to affect ongoing or in-the-planning-stage research elsewhere significantly then the finding should be published with the dispatch. If the primary value will be an addition to some author's bibliography and list of project reports, then the public interest may be equally well served by moving as expeditiously as possible to the final objective. This approach is hazardous in the case of a single investigator in that a change of job or an untimely death may cause the loss of the monies which supported man-years of research.

Graduate Training. Another avenue to increased scientific progress includes the improved academic preparation of research scientists. We should, of course, heed the admonition of Hubbert (41)

> "--that thinking is peculiarly an individual enterprise, and that the greatest of all scientific achievements -- those of the great synthesizers from Galileo to Einstein -- have, almost without exception, been the work of individuals."

Training in scientific research is presumed to be of a high quality in this country, but is in fact not invariably so. Such training for the Ph. D. degree should include: 1) a summer in field work (outside of the thesis topic if necessary) and a summer in laboratory work; 2) experience in the preparation of critical literature reviews; 3) at least one technical paper submitted for publication before thesis writing is started; 4) arrangements for adequate technical supervision of thesis research; 5) adequate knowledge plus experience in the behavioral sciences (pyschology, sociology, etc.); and 6) experience in one associative integrated research study. Consideration should be given to dropping non-thesis advanced degrees from both the physical and biological sciences.

Some discussion of the first of the above training items may be in order. In this day of specialization, it may seem strange to require a theoretical chemist to spend a summer in a field ecosystem study. This aspect of graduate education is intended to facilitate interdisciplinary research and to provide the experience to assist in bridging the gap between field and laboratory research. Similarly, the research achievements of many bright young graduates are limited more by their inability to work well with others than by limitation of intelligence or willingness to work hard. Some focused training in behavioral sciences might well increase the intrinsic value of graduates trained in this way.

Approaches and Priorities

How, then, can the progress of Science in general and of chemical modeling in particular be promoted in the following decade? Several avenues exist. It seems clear that one avenue is for research relating to chemical modeling to be carried out within the context of integrated interdisciplinary, multiscientist studies. As Russell Peterson (42) has recently pointed out,

> "...specialization has been necessary for scientific and technological advance; and we have learned much more and learned it quickly by breaking down phenomena into various compartments and studying them from the standpoints of biology, physics, chemistry, and so forth. But we must remember that our world does not exist in compartments; it comes in single, interrelated communities, each part of which affects other parts.... it would do well for scientists to establish the habit of stepping back occasionally from the immediate task to reflect on what kind of world he or she wants for his or her children and grandchildren and consider whether the work he or she is doing is leading in the right direction. If not, the scientist should have the courage to alter his course, even if it means some near-term personal sacrifice. It requires courage to speak and act effectively against the status quo. But we can bring about the essential changes. To do so will require that we look at things comprehensively and work toward worldwide goals that provide for improving the quality of life everywhere."

And as E. Odum (43) has stated so succintly,

> "It is self-evident that science should not only be reductionist in the sense of seeking to understand phenomena by detailed study of smaller and smaller components, but also synthetic and holistic in the sense of seeking to understand large components as functional wholes.--Science and technology during the past half century have been so preoccupied with reductionism that supraindividual systems have suffered benign neglect. We are abysmally ignorant of the ecosystems of which we are dependent parts.--the time has come to give equal time, and equal research and development funding, to the higher levels of biological organization in the hierarchial sequences. It is in the properties of the large-scale, integrated systems that hold solutions to most of the long-range problems of society."

Clearly, integrated, interdisciplinary, multiscientist efforts are required to achieve the goal of predictive chemical modeling of aquatic ecosystems and to make progress toward the goal of understanding the bioavailability of trace elements to the various food chain components within the next decade.

Funding agencies could well develop ways of permitting and even of encouraging individual investigators to continue along given broad topical or technique pathways to enhance the production of significant information--as opposed to gathering more data. This continuity of investigative area or investigative techniques appears to be essential to increase the intrinsic value of research products. The acceleration in rate of data generation by the scientific community and the sophistication of techniques and methods often required for significant research progress require the buildup of expertise and equipment. The rapid changes in societal concerns and governmental programs often result in principal investigators' taking on studies in tangential or even totally new areas this diversion of effort frequently detracts from original goals and from building up of expertise in a given area. The net effect of these conditions is that too many investigators are: 1) insufficiently knowledgeable of published information; 2) unacquainted with the conventional wisdom of the field; and 3) unable to secure adequate funding of sufficient duration to maximize the scientific value of their technical products.

The lack of pertinent reliable thermodynamic data needed for chemical modeling is a problem which could be resolved in the near future for waters to which ion-association models are applicable. This progress would be facilitated by a coordinated effort among the funding agencies, the compilation-evaluation agencies, and professional society groups. Satisfactory modeling of brines and other concentrated waters will require a major focused effort. The priority of various national problems to whose solution chemical modeling could contribute needs to be established. A similar evaluation and prioritization is needed for kinetic data.

Summary

Trust in, and support for, scientific investigation of fundamental processes has weakened in the last decade. Increasing emphasis has been given to the "black-box" type of modeling solutions to environmental problems. The ultimate goal of research in general, and of chemical modeling in particular, is human betterment. I hope this symposium will contribute toward this primary objective. The immediate goals of this symposium are the promoting of the timely exchange of pertinent information; the fostering of extensive discussion of goals, problems, approaches, as well as research priorities; and the preparation of a symposium proceedings. The principal restrictions on useful application of chemical models, to problems ranging from aqueous speciation to bioavailability in the human digestive system, include inadequate knowledge of natural organic ligands; of redox processes; of thermodynamic, time-dependent and kinetic information and anthropogenic inputs; and inadequate characterization of the natural systems being modeled. What many observers consider a low rate of progress being made in solving these problems is due, in part, to the short term of funding available to many investigators, to inadequate cognizance and synthesis of the available technical literatures on the part of some investigators and to the lack of integrated, interdisciplinary, multiscientist studies in many of the systems being modeled.

Acknowledgments

Pleasurable discussions with Ronald James, Jack Feth, Alan Jackman, Ralph Cheng, Ray Wildung, and Harvey Drucker, but most especially with Vance Kennedy, are acknowledged. Donald Langmuir, Harry Leland, Frank Trainer, and particularly David Rickert provided very helpful reviews.

Abstract

The ultimate goal of research in support of chemical modeling, as in most scientific research, is the improvement of human life physically, emotionally, and aesthetically through the understanding and prediction of processes and events. More immediate objectives include reliable aqueous speciation of trace elements between their valence states and among organic and inorganic ligands, and prediction of sorption equilibria and kinetic parameters as well as solubility controls on trace element solute levels. The objective is to predict toxicity and bioaccumulation in aquatic organisms and ultimately in man.

The complex and interactive nature of aqueous speciation (redox states, organic and inorganic complexation) and of sorption-desorption by sediment and biota, and the precipitation-dissolution of solid phases, requires comprehensive models and extensive analyses of an adequate number of appropriately collected and preserved samples. The prediction of chronic toxicity and bioaccumulation of trace elements will require the linking of chemical models to biologic and hydrologic models. This mandates a multiscience, interdisciplinary approach to the development, testing, and application of comprehensive models to assist in solving and preventing environmental problems.

Important restraints on the evolution of superior chemical models are the inadequacies in the: 1) capability to characterize the organic ligands of natural waters; 2) knowledge of redox status of waters, which does not permit realistic computation of redox-controlled speciation; 3) available thermodynamic data; 4) knowledge of the time dependency of the various processes in general (such as variation in apparent trace-element concentration during a tidal cycle) and kinetic data for chemical and biological processes in particular; and 5) error estimates for the preceding restraints. The field of chemical modeling is also constrained by the: 6) need for comprehensive analyses of the systems to be modeled; 7) limitations of published data; 8) scarcity of adequate literature reviews; 9) difficulties in organizing and conducting integrated interdisciplinary studies; and to a lesser extent, 10) limitation in quality of graduate training.

Literature Cited

1. Abelson, P. H. A report from the research community. Science 194, 482 (1976).
2. McElroy, W. D. The global age: Roles of basic and applied research. Science 196, 267-270 (1977).
3. Krieger, J. H. Universities face wide-ranging changes. Chem. Eng. News, Feb. 16, 395 (1976).
4. Schindler, D.W. The impact boondoggle. Science 192,509(1976).
5. Jenne, E. A., Girvin, D. C., Ball, J. W., and Burchard, J.M. Inorganic speciation of silver in natural waters - fresh to marine, Ch. 4, p. 41-61, (Cum. Bib., p. 217-270) in Klein, D. A., ed., "Environmental Impacts of Nucleating Agents Used in Weather Modification Programs." 270 p. Dowder, Hutchinson & Ross, Stroudsberg, Pennsylvania, 1978.
6. Kester, D., Byrne, R. H., Jr., and Liang, Y.-J. Redox reactions and solution complexes of iron in marine systems. Amer. Chem. Soc. Symp. Ser. 8, 56-79 (1975).
7. Nordstrom, D. K., Plummer, L. N., and others. A comparison of computerized chemical models for equilibrium calculations in aqueous systems in Jenne, E. A., ed., "Chemical Modeling in Aqueous Systems. Speciation, Sorption, Solubility, and Kinetics." Amer. Chem. Soc. 1978 (This volume).
8. Leenheer, J. A., and Huffman, E. W. D., Jr. Classification of organic solutes in water using macroreticular resins. U. S. Geol. Survey J. Res. 4(6), 737-751 (1976).
9. Parker, V. B., Wagman, D. D., and Evans, W. H. Selected values of chemical thermodynamic properties. Tables for the alkaline earth elements (Elements 92 through 97 in the Standard Order of Arrangement). Nat'l Bur. Stand. Tech. Note 270-6, 106 p. (1971).
10. Schumm, R. H., Wagman, D. D., Bailey, S., Evans, W. H., and Parker, V. B. Selected values of chemical thermodynamic properties. Tables for the Lanthanide (Rare Earth) elements (Elements 62 through 76 in the Standard Order of Arrangement). Nat'l Bur. Stand. Tech. Note 270-7, 75 p. (1973).
11. Wagman, D. D., Evans, W. H., Parker, V. B., Halow, I., Bailey, S. M., and Schumm, R. H. Selected values of chemical thermodynamic properties. Table for the first thirty-four elements in the Standard Order of Arrangement. Nat'l Bur. Stand. Tech. Note 270-3, 264 p. (1968).
12. Wagman, D. D., Evans, W. H., Parker, V. B., Halow, I., Bailey, S. M., and Schumm, R. H. Selected values of chemical thermodynamic properties. Tables for elements 35 through 53 in the Standard Order of Arrangement. Nat'l Bur. Stand. Tech. Note 270-4, 141 p. (1969).
13. Wagman, D. D., Evans, W. H., Parker, V. B., and Schumm, R. H. Chemical thermodynamic properties of compounds of sodium, potassium, and rubidium: An interim tabulation of selected values. Nat'l Bur. Stand. Interagency Report 76-1034, 76 p. (1976).

14. Wagman, D. D., Schumm, R. H., and Parker, V. B. A computer assisted evaluation of the thermochemical data of the compounds of thorium. Nat'l Bur. Stand. Interagency Report 77-1300, 94 p. (1977).
15. Barnes, H. L., ed., "Geochemistry of Hydrothermal Ore Deposits." John Wiley, New York (in press).
16. Martell, A. E., and Smith, R. M. "Critical Stability Constants, Vol. 1, Amino Acids." 469 p. Plenum Press, New York, 1974.
17. Smith, R. M., and Martell, A. E. "Critical Stability Constants, Vol. 2, Amines." 415 p. Plenum Press, New York, 1975.
18. Smith, R. M., and Martell, A. E., "Critical Stability Constants, Vol. 4, Inorganic Complexes." 257 p. Plenum Press, New York, 1976.
19. Martell, A. E., and Smith, R. M. "Critical Stability Constants, Vol. 3, Other Organic Ligands." 495 p. Plenum Press, New York, 1977.
20. Christensen, J. J., Eatough, D. J., and Izatt, R. M. "Handbook of Metal Ligands Heats and Related Thermodynamic Quantities," 2nd ed. 495 p. Marcel Dekker, New York, 1975.
21. Baes, C. F., Jr., and Mesmer, R. E. "The Hydrolysis of Cations," 489 p. John Wiley, New York, 1976.
22. Langmuir, Donald. "Uranium Solution-Mineral Equilibria at Low Temperatures with Applications to Sedimentary Ore Deposits." 39 p., Penn. State Univ., University Park, Pennsylvania, 1977.
23. Robie, R. A., Hemingway, B. S., and Fish, J. R. Thermodynamic properties of minerals and related substances at 298.15K and one bar (10^5 Pascals) pressure and at higher temperatures. U. S. Geol. Survey Bull. 1452, 456 p. (1978).
24. Reardon, E. J., and Langmuir, Donald. Thermodynamic properties of the ion pairs $MgCO_3^{\circ}$ and $CaCO_3^{\circ}$ from 10 to 50°C. Amer. J. Sci. 274, 599-612 (1974).
25. Pytkowicz, R. M., ed. "Activity Coefficients in Electrolyte Solutions." CRC Press Inc., Palm Spring, Florida (in press).
26. Hunt, J. P. "Metal Ions in Aqueous Solutions." 124 p., W. A. Benjamin, New York, 1963.
27. Langmuir, Donald. Techniques of estimating thermodynamic properties for some aqueous complexes of geochemical interest, in Jenne, E. A., ed., "Chemical Modeling in Aqueous Systems. Speciation, Sorption, Solubility, and Kinetics." Amer. Chem. Soc. 1978 (This volume).
28. Cloke, P. L. The geologic role of polysulfides --Part II. The solubility of acanthite and covellite in sodium polysulfide solutions. Geochim. Cosmochim. Acta 27, 1299-1319 (1963).
29. Vanoni, V. V., ed. "Sediment Engineering." 745 p. Task Committee, Amer. Soc. Civil Eng., New York, 1975.

30. Girvin, D. C., Hodgson, A. T., and Jenne, E. A. Spatial and seasonal variations of silver, cadmium, copper, nickel, lead, and zinc in South San Francisco Bay water during two consecutive drought years. Lawrence Berkeley Lab. UCID-4031. Univ. Calif., Berkeley, California, 1978.
31. Anonymous. Symposium scores misuse of scientific data. Chem. Eng. News, Feb. 10, 17 (1975).
32. Ball, J. W., Jenne, E. A., and Nordstrom, D. K., WATEQ2--a computerized chemical model for trace and major element speciation and mineral equilibria of natural waters, in Jenne, E. A., ed., "Chemical Modeling in Aqueous Systems, Speciation, Sorption, Solubility, and Kinetics." Amer. Chem. Soc. 1978 (This volume).
33. Jenne, E. A., and Luoma, S. N. The forms of trace elements in soils, sediment, and associated waters: An overview of their determination and biological availability, p. 110-143, in Wildung, R. E., and Drucker, H., ed., "Biological Implications of Metals in the Environment." CONF-750929, NTIS Springfield, Virginia, 1977.
34. Kari, Tarja, and Kauranen, Pentti. Mercury and selenium contents of seals from fresh and brackish waters in Finland. Bull. Environ. Contam. Toxicol. 19(3), 273-280 (1978).
35. Freeman, H. C., Shum, G., and Uthe, J. F. The selenium content in swordfish (Xiphias gladins) in relation to total mercury content. J. Environ. Sci. Health A13, 235-240 (1978).
36. Anonymous. Which way now for journals? Nature 262, 731 (1976).
37. Baker, D. E. Recent trends in growth of chemical literature. Chem. Eng. News, May 10, 23-27 (1976).
38. Lomask, Milton. "A Minor Miracle - An Informal History of the National Science Foundation." 286 p. Nat'l Sci. Foundation, Washington, D. C., 1976.
39. Ross, J. E. The Institute for Environmental Studies: Univ. of Wisconsin-Madison. Chemosphere 5, 185-198 (1973).
40. FOAM (Friends of Anoxic Mud): Goldhaber, M. B., Aller, R. C., Cochran, J. K., Rosenfeld, J. K., Martens, C. S., and Berner, R. A. Sulfate reduction, diffusion, and bioturbation in Long Island Sound sediments: Report of the FOAM group. Amer. J. Sci. 277, 193-237 (1977).
41. Hubbert, M. K. Are we retrogressing in Science? Geol. Soc. Amer. Bull. 74, 365-378 (1963).
42. Peterson, Russell. Scientists and the holistic approach. Chem. Eng. News, May 15, 3 (1978).
43. Odum, E. P. The emergence of ecology as a new integrative discipline. Science 195, 1289-1293 (1977).

Disclaimer: The reviews expressed and/ or the products mentioned in this article represent the opinions of the author(s) only and do not necessarily represent the opinions of the U.S. Geological Survey.

RECEIVED November 16, 1978.

Redox

2

Sulfur Speciations and Redox Processes in Reducing Environments

JACQUES BOULEGUE and GIL MICHARD

Laboratoire de Géochimie des Eaux, et Laboratoire de Géochimie et Cosmochime (CNRS LA 196), Université de Paris 7, 2 Jussieu, 75221 Paris Cedex 05, France

Natural waters are generally in a dynamic rather than an equilibrium condition, and the concept of a single redox system characteristic of a given water cannot be maintained. In the most favorable case, measurements of Eh can be related to a particular redox system or systems in partial equilibrium (1). The redox system must be electrochemically reversible at the surface of the platinum electrode at a rate that is rapid compared with the electron drain or supply by way of the measuring electrode (2). In natural waters, only the Fe(II)/Fe(III) and the H_2S/S_n^{2-} systems correspond to these limitations (3,4,5). The electrochemical reactions of hydrogen sulfide and polysulfide ions are known to be rapid at the surface of the platinum electrode (6). Hence, the potentials obtained in the presence of these species should be explainable in terms of equilibrium of the redox couple H_2S/S_n^{2-}; eventually these potentials might be utilized to infer redox processes involving the sulfur species. More information is needed about the relation between measured redox potentials (Eh) and sulfur chemistry in reducing environments.

Reducing environments are frequently characterized by the presence of hydrogen sulfide (even in very low concentration). In these environments several factors, such as slow diffusion of oxygen, the presence of Fe(III) minerals or organic matter, may result in the incomplete oxidation of H_2S, which yields polysulfide ions (S_n^{2-}) and thiosulfate ($S_2O_3^{2-}$), (7,8,9):

$$2\ HS^- + O_2 \rightarrow (1/4)\ S_8 + 2\ OH^-$$

$$HS^- + (n-1)/8\ S_8 \rightleftarrows S_n^{2-} + H^+$$

$$2\ S_n^{2-} + O_2 + H_2O \rightarrow S_2O_3^{2-} + S_{n-1}^{2-} + 2\ H^+$$

$$2\ HS^- + 2\ O_2 \rightarrow S_2O_3^{2-} + H_2O\ .$$

The above reactions proceed easily and rapidly in the physicochemical conditions prevailing in natural waters (5,9). Thus,

0-8412-0479-9/79/47-093-025$06.50/0
© 1979 American Chemical Society

polysulfide ions should be found in reducing environments where the oxidation of H_2S has been incomplete. Polysulfide ions can also be produced via direct reaction between hydrogen sulfide and bacterially produced elemental sulfur (5,10):

$$HS^- + (n-1)/8\ S_8 \rightleftarrows S_n^{2-} + H^+ .$$

The distribution of the sulfur species between sulfide and polysulfide at equilibrium with elemental colloidal sulfur is represented in Figure 1.

In this report we examine the aqueous solutions associated with reducing environments in relation to the composition and redox properties of sulfur species.

Experimental

The study of the electrochemical properties of the sulfide/polysulfide redox couple have been achieved by measurements of pH, Eh (the potential of a platinum electrode) and E_S2- (the potential of a membrane Ag/Ag_2S electrode), the reference electrode being a double junction Ag/AgCl electrode. Potentials given are referred to normal hydrogen electrode (H^+/H_2) potential.

To prevent any interference of oxygen in the chemical processes all the measurements (laboratory and field) were done in an electrochemical cell under purified nitrogen atmosphere. Details are published elsewhere (11). Field sampling was done from a glove box under nitrogen atmosphere.

The technique of study of the electrochemical properties of the sulfur species was to measure pH, Eh and E_S2- on aqueous solutions after imposed pH variations. The pH variations were obtained by HCl or NaOH additions. Eh, E_S2- and pH were measured when the e.m.f. had stabilized. A moderate stirring was maintained during all the measurements. In addition, the concentrations of H_2S, S_n^{2-}, S_8, $S_2O_3^{2-}$, SO_3^{2-} were measured before and after each experiment either by the standard iodine method or following the method of Boulègue *et al.* (12).

Electrochemical study of the H_2S-H_2O and H_2S-S_8-H_2O systems

The results of pH, Eh and E_S2- measurements after imposed pH variations can be utilized to characterize the dissolved sulfur species. This can be achieved by considering successively the pH-E_S2- relations, the Eh-pH relations and the Eh-E_S2- relations.

The pH-E_S2- relations. The pH-E_S2- relations are characteristic of the concentration of the total dissolved hydrogen sulfide ($\Sigma|H_2S|$). In the presence of only dissolved hydrogen sulfide one can write:

$$E_S2- = E_S^\circ 2- - 0.0295\log(S^{2-})$$

and

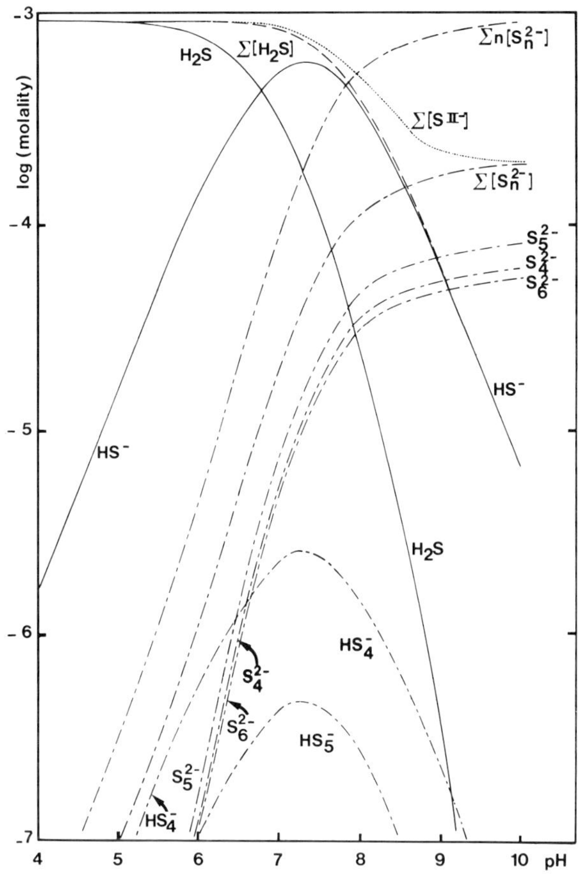

Figure 1. Distribution of sulfur species in the system H_2S–S_8 (colloid)–H_2O–NaCl (0.7M) for $\Sigma[S] = 10^{-3}$ g-atom/kg.Log (molality) vs. pH at 298 K. ΔG_{298} (S_8coll.) = 3.5 kJ/mol S_8 (28).

$$\Sigma|H_2S|/(S^{2-}) = \{1/\gamma_{S^{2-}}\} + \{(H^+)/K_{A2}\cdot\gamma_{HS^-}\} + \{(H^+)^2/K_{A1}\cdot K_{A2}\cdot\gamma_{H_2S}\}$$

with (i) = activity of the ith species, γ_i = activity coefficient of the species i, $K_{A1} = (HS^-)\cdot(H^+)/(H_2S)$, $K_{A2} = (S^{2-})\cdot(H^+)/(HS^-)$. Thus, $E_{S^{2-}}$ is a simple function of pH at constant $\Sigma[H_2S]$. Taking $pK_{A1} = 7.01$ and $pK_{A2} = 14$ (5), one can write:

$$E_{S^{2-}} = E^{\circ}_{S^{2-}} - 0.0295\log(\Sigma[H_2S]) + 0.0295\log(4.854 + 1.455 \times 10^{14-pH} + 8.734 \times 10^{20-2pH})$$

which is valuable in the H_2S-H_2O-NaCl(0.7 M) system. The measurements of $E_{S^{2-}}$ and pH in the H_2S-H_2O-NaCl(0.7 M) system have given results in good agreement with the values predicted from the above equation (see Figure 2), (5,11).

In the presence of hydrogen sulfide, polysulfide ions and elemental sulfur one can write:

$$(HS^-) = \Sigma[S] \Big/ \left(\{1/\gamma_{HS^-}\} + \{(H^+)/K_{A1}\cdot\gamma_{H_2S}\} + \{\textstyle\sum_{n>1} n\cdot K_n/(H^+)\cdot\gamma_{S_n^{2-}}\}\right) \qquad (1)$$

with $\Sigma[S] = \Sigma[H_2S] + \Sigma n[S_n^{2-}]$ and $K_n = (S_n^{2-})\cdot(H^+)/(HS^-)$. Equation 1 can be written alternately $(HS^-) = \Sigma[S]/f(pH)$, where f(pH) is the pH dependent member in 1. Taking into account the expression of K_{A2} one can deduce $(S^{2-}) = \Sigma[S]/g(pH)$. In the H_2S-S_8-H_2O-NaCl (0.7 M) system one can obtain (5):

$$g(pH) = 10^{5.9994} + 10^{14.161-pH} + 10^{20.949-2pH}$$

and

$$E_{S^{2-}} = E^{\circ}_{S^{2-}} - 0.0295\log(\Sigma[S]) + 0.0295\log(g(pH)). \qquad (2)$$

Thus $E_{S^{2-}}$ can be computed in the system H_2S-S_8-H_2O knowing $\Sigma[S]$, which can be done by titration of the dissolved species. In the H_2S-S_8-H_2O-NaCl(0.7 M) system the relations obtained between pH and $E_{S^{2-}}$ are in good agreement with the relations computed from equation 2 as can be seen in Figure 3. At pH > 9, slight deviations were observed owing to polysulfide interaction with the Ag/Ag_2S electrode.

Thus the pH-$E_{S^{2-}}$ relations obtained in the H_2S-H_2O and H_2S-S_8-H_2O systems are characteristic of these systems and they can be employed in natural waters to characterize the presence or the absence of polysulfide ions. However, owing to the lack of sensitivity of the Ag/Ag_2S electrode to small variations of (S^{2-}) the pH-$E_{S^{2-}}$ relations cannot be employed to detect slight departures from equilibrium in the above systems.

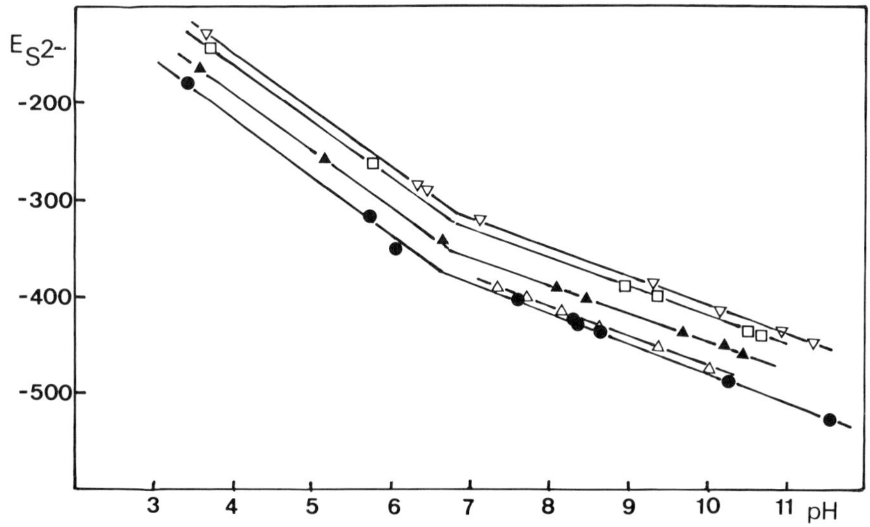

Figure 2. Ag/Ag_2S electrode potentials (E_S2-) vs. pH in H_2S–H_2O–NaCl (0.7M) system at 298 K. The curves correspond to calculated pH–E_S2- relations at constant experimental $\Sigma[H_2S]$. E_S2- are given in mV/(H^+/H_2). $\Sigma[H_2S]$ (mmol/kg): exp. 6 (▽) 0.05, exp. 7 (□) 0.07, exp. 8 (▲) 0.7, exp. 209 (△) 5.8, exp. 203 (●) 6.9.

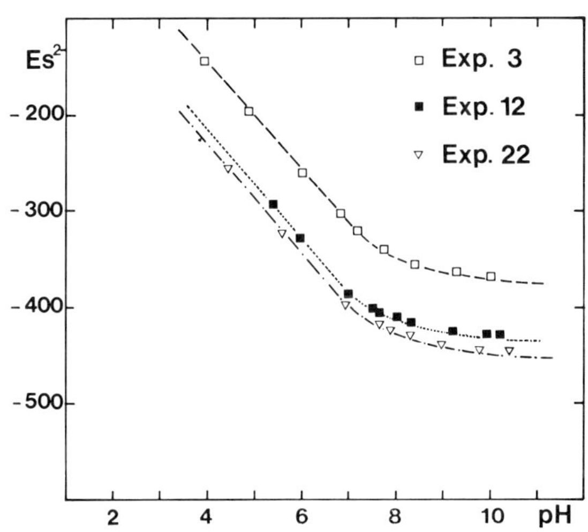

Figure 3. Ag/Ag_2S electrode potentials (E_S2-) vs. pH in H_2S–S_8–H_2O–NaCl (0.7M) system at 298 K. The points correspond to the experimental values and the curves correspond to the calculated values. E_S2- are given in mV/(H^+/H_2). $\Sigma[S^{II-}] = \Sigma[H_2S] + \Sigma[S_n^{2-}]$ (mmol/kg): exp. 3 = 0.024, exp. 12 = 4.17, exp. 22 = 12.8.

The Eh-pH and Eh-ΣS2- relations in the H_2S-H_2O system. The Eh values in the H_2S-H_2O system were found to be slightly dependent upon stirring (±10 mV). At 298°K the potentials of the platinum electrodes were established quite slowly (1 to 2 hr) at pH > 5. At pH < 5 it generally takes less than one hour. The potentials were independent of such electrode pretreatment as cathodic or anodic polarization. In Figure 4 we present the Eh-pH relations obtained in the H_2S-H_2O-NaCl(0.7 M) system.

The Eh values are too high to be explained by the formation of either PtS or PtS_2. The rest potentials of the platinum electrode are dependent upon the concentration of dissolved hydrogen sulfide ($\Sigma[H_2S]$). This can be explained by selective adsorption of H_2S species on the platinum followed by the discharge of the proton mediated via chemisorbed H_2S. The corresponding reactions are (11):

$$H_2S + Pt(e^-) \rightleftarrows (e^-)Pt\text{—}H_2S$$

$$(e^-)Pt\text{—}H_2S \rightleftarrows Pt\text{—}H + HS^-$$

$$Pt\text{—}H + Pt\text{—}H \rightarrow 2\ Pt + H_2$$

$$Pt\text{—}HS^- + H_2S \rightleftarrows Pt\text{—}H_2S + HS^-$$

$$Pt\text{—}HS^- + H_3O^+ \rightleftarrows Pt\text{—}H_2S + H_2O\ .$$

Thus the equivalent process controlling the potential of the Pt electrode is

$$H_3O^+ + e^- \rightarrow H + H_2O$$

where H is hydrogen adsorbed on the platinum surface, and the redox potential is

$$Eh = -0.059(pK_H + pH + \log(H))$$

with $K_H = (H)/(e^-)\cdot(H_3O^+)$, and (H) is the activity of H atoms on the platinum. The surface activity of H atoms can be assumed to follow the law

$$(H) = k\cdot(H_2S)^{1/n}$$

as found in solid-solute interactions (13). In this instance, the experimental data on the Eh - $\Sigma[H_2S]$ relation correspond to n ≃ 2.5 at pH < 6.5 and n ≃ 1.5 at pH > 6.5; and the best fit with the experimental data gives k ≃ 2.5. These values are compatible with an adsorption process (n ≃ 1, (13)). Since n decreases as pH increases for constant $\Sigma[H_2S]$, it is suggested that H_2S, rather than HS^-, is effectively mediating the discharge of the proton on the platinum in agreement with the above processes. With $pK_H = -3$

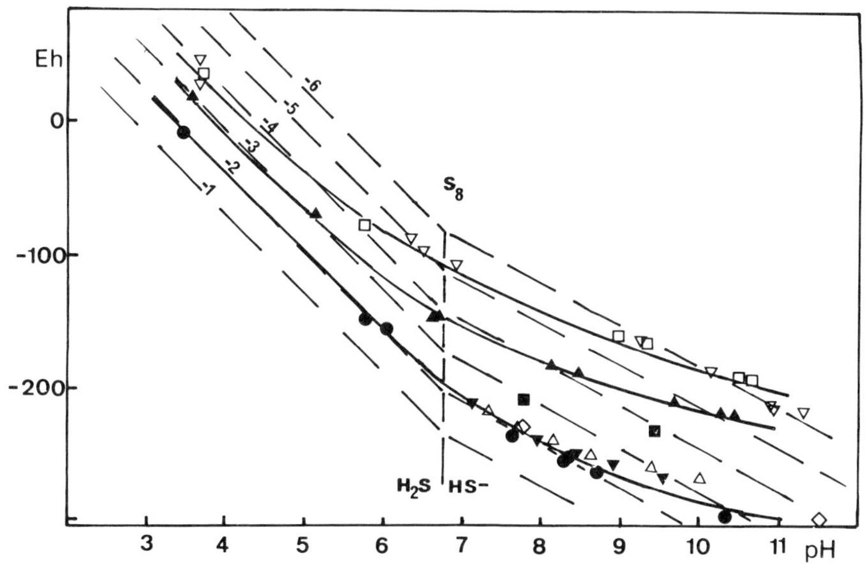

Figure 4. Platinum electrode rest potentials Eh (expressed in $mV/(H^+/H_2)$ vs. pH in H_2S–H_2O–NaCl (0.7M) system for different experimental $\Sigma[H_2S]$ at 298 K. The experimental points corresponding to the same $\Sigma[H_2S]$ have been linked by a continuous curve. The dashed lines correspond to the Eh–pH relationships in the H_2S–S_8–H_2O system for different $\log\Sigma[H_2S]$ = x (5). Experimental $\Sigma[H_2S]$ (mmol/kg): exp. 6 (▽) 0.05, exp. 7 (□) 0.07, exp. 8 (▲) 0.7, exp. 10 (■) 2.4, exp. 11 (◇) 4.6, exp. 212 (▼) 5.7, exp. 209 (△) 5.8, exp. 203 (●) 6.9.

(14), one can write:

$$Eh = 0.153 - 0.059pH - (0.059/n)\cdot\log(H_2S) \quad (3)$$

where n is characteristic of the adsorption-discharge process of the H_2S species. Equation 3 is in good agreement with the Eh values measured in the H_2S-H_2O system. Also the resulting Eh-E_S2- relations obtained in the H_2S-H_2O system are characteristic of this system as can be seen in Figure 5.

Oxidized sulfur species occurring in natural waters (sulfate, sulfite, thiosulfate) do not interact with the platinum electrode when in the presence of H_2S and the pH-Eh-E_S2- relations found were similar to the above relations. Thus, the unambiguous relations found between pH, Eh and E_S2- in aqueous solutions of hydrogen sulfide can be employed to characterize solutions and water samples where hydrogen sulfide is the only reduced sulfur species present.

An application of the above results can be found in the study of the hot sulfurous springs of the French Pyrénées which issue from massive granite (5,11). As seen in Table I, H_2S, SO_4^{2-} together with Na^+ and dissolved carbonate are major components. Polysulfide species were not detected ($\Sigma(n-1)[S_n^{2-}] < 10^{-6}$ g.atS/kg). These sulfurous springs should be equivalent to a H_2S-H_2O system.

Table I. Analyses of two springs from the eastern Pyrénées: H 18 "Beauté" (Thués) and O 5 "Rossignol" (Ax les Thermes).The concentrations are given in 10^{-4} mol/kg.

Spring	Li	Na	K	Mg	Ca	pH	T(°C)
H 18	0.128	28.3	0.659	0.002	0.4	8.89	66.4
O 5	0.131	21.5	0.585	0.005	0.36	8.74	76

Spring	F	Cl	$\Sigma[H_2S]$	$S_2O_3^{2-}$	SO_4^{2-}	$\Sigma[CO_2]$	SiO_2
H 18	3.51	2.09	1.99	0.05	5.57	9.1	1.49
O 5	1.41	3.68	1.68	0.02	1.38	11.2	1.47

The results of the pH, Eh and E_S2- measurements after imposed pH variations are given in Figures 6 and 7. The agreement of the electrochemical measurements with laboratory data is excellent and both selected springs are equivalent to a H_2S-H_2O system. In this instance the chemical analyses of the springs and the electrochemical measurements are also in agreement.

The Eh-pH and Eh-E_S2- relations in the H_2S-S_8-H_2O system. In the H_2S-S_8-H_2O system polysulfide ions are produced following the general process

$$(n-1)/8\ S_8 + HS^- \rightleftarrows S_n^{2-} + H^+ . \quad (4)$$

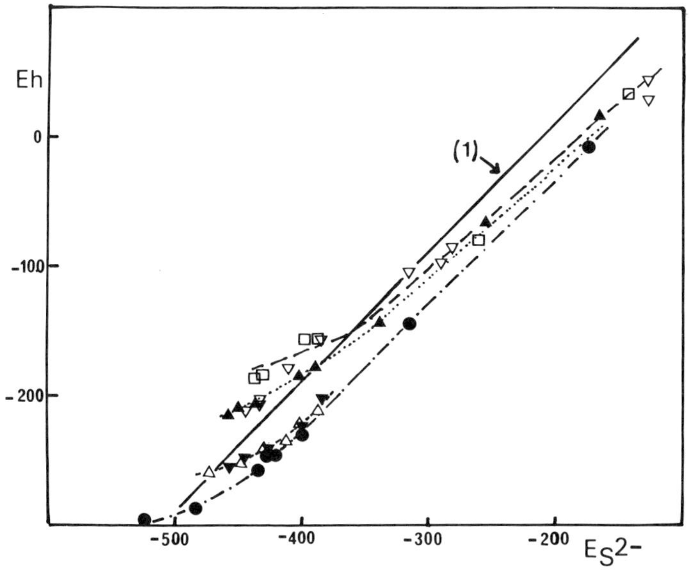

Figure 5. Eh–E_S2– relationships in the H_2S–H_2O–NaCl (0.7M) system for different Σ[H_2S]. The experimental points corresponding to the same Σ[H_2S] have been linked by dashed line curves. The Equation 1: Eh = E_S2– + 210mV corresponds to the Eh–E_S2– relation in the H_2S–S_8–H_2O system (5). Eh and E_S2– are given in mV/(H^+/H_2). The symbols are the same as in Figure 4.

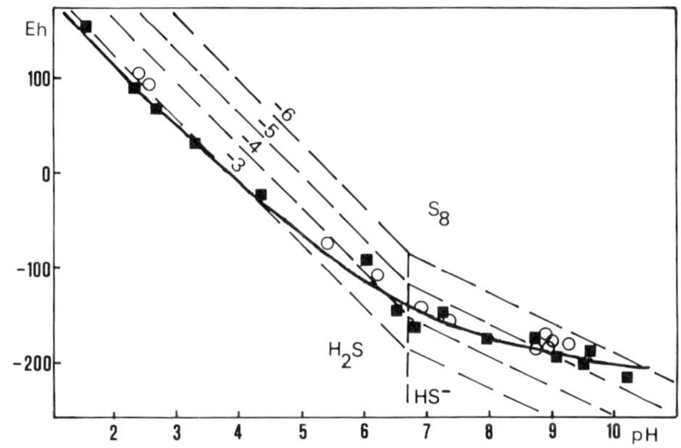

Figure 6. Eh vs. pH in H 18 and O 5 springs. Eh is expressed in mV/(H^+/H_2). The experimental points have been linked by a continuous curve. The dashed lines correspond to the Eh–pH relations in the H_2S–S_8–H_2O system for different logΣ|H_2S| = x at 343 K (5). (○) Spring H 18, (■) spring O 5.

Laboratory experiments (15,16,17) have shown that polysulfide ions are stable in the H_2S-S_8-H_2O system even at very low sulfide concentrations (about 10^{-5} - 10^{-6} M, (17)). The major polysulfide ions are probably S_5^{2-}, S_4^{2-} and S_6^{2-} in a large pH range (17). The following equations can be employed to compute the Eh-pH relation in the H_2S-S_8-H_2O system:

$$H_2S \rightleftarrows HS^- + H^+ \qquad K_{A1} = 10^{-7.01}$$

$$H_2S \rightleftarrows S + 2\,H^+ + 2\,e^- \qquad K_{\alpha 1} = 10^{-4.80}$$

$$HS^- \rightleftarrows S + H^+ + 2\,e^- \qquad K_{\alpha 2} = 10^{2.21}$$

$$5\,HS^- \rightleftarrows S_5^{2-} + 5\,H^+ + 8\,e^- \qquad K_{5/1} = 10^{-0.58}$$

$$4\,HS^- \rightleftarrows S_4^{2-} + 4\,H^+ + 6\,e^- \qquad K_{4/1} = 10^{-2.90}$$

$$S_5^{2-} \rightleftarrows 5\,S + 2\,e^- \qquad K_{5/0} = 10^{11.62}$$

$$5\,S_4^{2-} \rightleftarrows 4\,S_5^{2-} + 2\,e^- \qquad K_{4/5} = 10^{14.37}$$

These equations can be employed to draw the stability diagram of the H_2S-S_8-H_2O system (see Figure 8).

Since the sulfide/polysulfide couple is electroactive and the reaction at the surface of the platinum is rapid (6), the potentials measured in the H_2S-S_8-H_2O system must be characteristic of the equilibria between sulfide and polysulfide species. These equilibria correspond to the redox level within the environment since the reactions between sulfide and polysulfide are rapid (5, 17). In this instance, the redox potential can be directly interpreted in terms of chemical and electrochemical equilibria of the reduced sulfur species. Thus one can employ the notion of pe ($pe = Eh/(2.3 \times R \times T \times F^{-1})$) when in the presence of polysulfide ions.

For the general equation

$$n\,HS^- \rightleftarrows S_n^{2-} + n\,H^+ + 2(n-1)\,e^-$$

the corresponding expression for pe is

$$pe = pe_n^\circ + \{1/2(n-1)\}\log\left((H^+)^n\cdot(S_n^{2-})/(HS^-)^n\right).$$

Taking into account equation 1 and $K_n = (S_n^{2-})\cdot(H^+)/(HS^-)$ corresponding to equation 4, one can write:

$$pe = pe_n^\circ - \{n/2(n-1)\}pH + \{1/2(n-1)\}\log\left(K_n/(H^+)\cdot\gamma_{S_n^{2-}}\right)$$
$$- \{1/2\}\log\left(\Sigma[S]/f(pH)\right). \qquad (5)$$

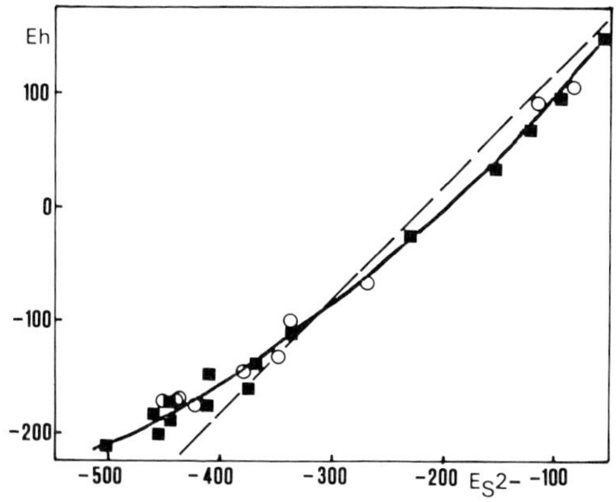

Figure 7. Eh–E_S2– relation in H 18 and O 5 springs. The experimental points have been linked by a continuous curve. The dashed line equation is Eh = E_S2– + 217 mV, i.e., the Eh–E_S2– relation in the H_2S–S_8–H_2O system at 343 K. Eh and E_S2– are given in mV/(H^+/H_2). The symbols are the same as in Figure 6.

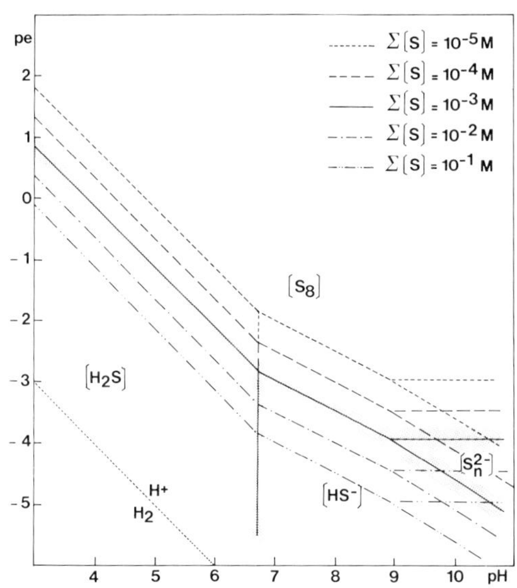

Figure 8. PE–pH stability diagram of the system H_2S–S_8–H_2O–NaCl (0.7M) for Σ[S] = 10^{-x} g-atom/kg. The grey curve on the diagram for Σ[S] = 10^{-3} corresponds to the experimental error on PE computed from the uncertainties on the thermodynamic data (5, 17).

For instance the equation

$$5\ SH^- \rightleftarrows S_5^{2-} + 5\ H^+ + 8\ e^-$$

corresponds to

$$pe = 2.08 - \{5/8\}pH - \{1/2\}\log\left(\Sigma[S]/(1.466 + 10^{6.95-pH} + 10^{-8.0+pH})\right)$$

in the system H_2S-S_8-H_2O-NaCl(0.7 M), (5). The potentials measured in the H_2S-S_8-H_2O system and the H_2S-S_n^{2-}-H_2O system (i.e. when elemental sulfur is not present) can be compared to the potential computed with equation 5. As can be seen in Figure 9, the agreement between measured and computed values is good. Thus in physico-chemical conditions close to those encountered in reducing environments the sulfide and polysulfide species appear to be chemically and electrochemically reversible. The Eh measurements are reliable over a large concentration range since at pH = 8.5, with $\Sigma[H_2S] = 2 \times 10^{-5}$M and $\Sigma[S_n^{2-}] = 5 \times 10^{-6}$M, the exchange current $i_o \simeq 10^{-6}$ amp/cm^2, (5).

Note that at pH > 7 the redox potentials vary only slightly. The H_2S-S_8-H_2O system behaves as a redox buffer. This effect can be of importance in waters where sulfide and polysulfide are found.

Since

$$E_{S^{2-}} - E^{\circ}_{S^{2-}} - 0.0295\log(S^{2-})$$

and

$$Eh = Eh^{\circ} - 0.0295\log(S^{2-})$$

for

$$S^{2-} \rightleftarrows S + 2\ e^-,$$

one can write for the H_2S-S_8-H_2O system:

$$Eh = E_{S^{2-}} + (Eh^{\circ} - E^{\circ}_{S^{2-}}).$$

Thus the Eh-$E_{S^{2-}}$ relation should be linear when in the presence of sulfide and polysulfide ions. The term $(Eh^{\circ} - E^{\circ}_{S^{2-}})$ can be estimated from the thermodynamic data on S^{2-}, S_8, Ag, Ag_2S. For the Ag/Ag_2S electrodes employed in our work $E^{\circ}_{S^{2-}} = -660(\pm 10)$mV, (5); with $pK_{A2} = 14$ one can compute $Eh^{\circ} = -480$mV; thus at 298°K:

$$(Eh^{\circ} - E^{\circ}_{S^{2-}}) = 180\text{mV}$$

and

$$Eh = E_{S^{2-}} + 180\text{mV}. \qquad (6)$$

These expressions can be applied to the H_2S-S_8-H_2O system. As can be seen in Figure 10, the experimental data are in good agreement with equation 6. Equation 6 can be compared with

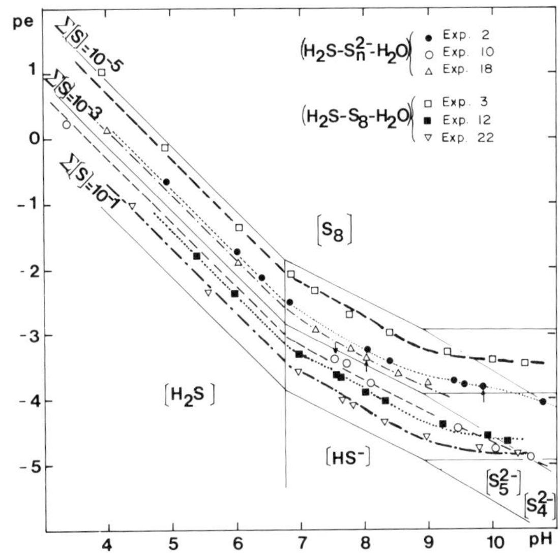

Figure 9. PE–pH relations in the H_2S–S_8–H_2O–NaCl (0.7M) and H_2S–S_n^{2-}–H_2O–NaCl (0.7M) systems. The points correspond to the experimental data. The dashed curves correspond to the computed values. The full lines correspond to the PE–pH stability diagrams in the H_2S–S_8–H_2O–NaCl (0.7M) system for different $\Sigma[S]$. $\Sigma[S^{II-}] = \Sigma[H_2S] + \Sigma[S_n^{2-}]$ (mmol/kg): exp. 2 = 0.18, exp. 18 = 0.255, exp. 10 = 2.28, exp. 3 = 0.024, exp. 12 = 4.17, exp. 22 = 12.8.

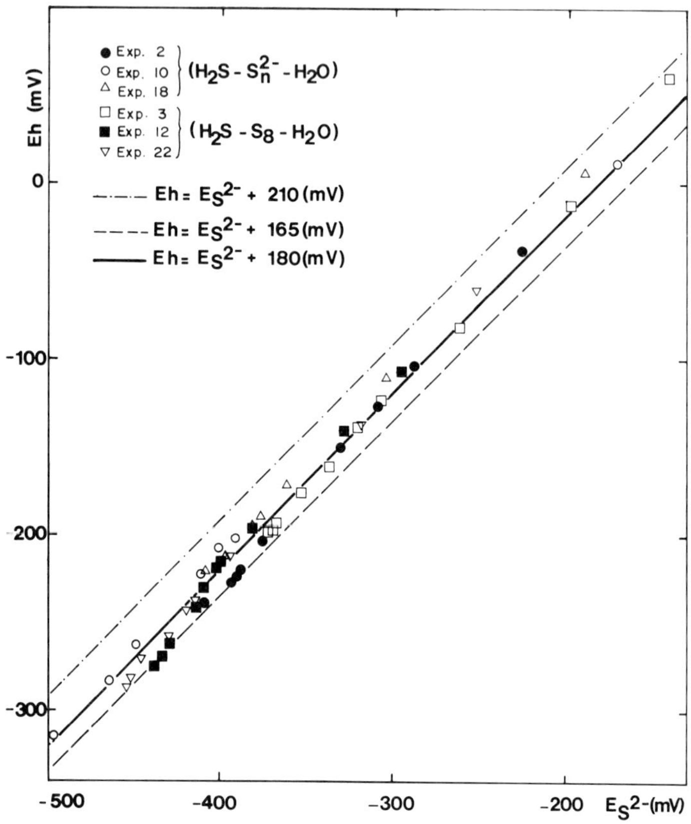

Figure 10. Eh–E_S2– relations in the H_2S–S_8–H_2O–NaCl (0.7M) and H_2S–S_n^{2-}–H_2O–NaCl (0.7M) systems. The points correspond to the experimental data. Eh and E_S2–are given in mV/(H^+/H_2). The symbols are the same as in Figure 9.

$$Eh = E_{S}2- + 165mV$$

as obtained by Berner (4) in various reducing environments. Equation 6 is characteristic of the presence of sulfide and polysulfide ions and it can be employed to distinguish the corresponding systems from environments where only hydrogen sulfide is found.

Thus the pH-Eh-E_S2- measurements can be employed to characterize the state of reducing environments concerning sulfur chemistry and electrochemistry. The redox level deduced from the Eh measurements should be in agreement with the redox level estimated from the chemical analyses of the sulfur species as found in laboratory experiments (5).

Field measurements of the pH-Eh-E_S2- relations in reducing environments where hydrogen sulfide, polysulfides and elemental sulfur were found were in good agreement with the redox level estimated from the concentrations of the sulfur species (10,18). For instance in hot brine from the French Dogger, where reduced sulfur species (hydrogen sulfide, organic and inorganic polysulfides, elemental sulfur, thiosulfate) were found, there was a good agreement between analytical and electrochemical results as can be seen from the comparison of the data of Table II and Figures 11 and 12 (18).

Table II. Results of the analyses of the brine from the Melun drill-hole (PM2). (a) 10^{-1} mol/kg; (b) 10^{-3} mol/kg; (c) 10^{-4} mol/kg.

Li (c)	Na (a)	K (b)	Mg (b)	Ca (b)	Sr (c)	NH_4 (c)	SiO_2 (c)
2.0	1.78	1.9	5.8	13.6	4.0	11.0	6.9
F (c)	Cl (a)	HCO_3(b)	$\Sigma[CO_2]$(b)	B (c)	pH	pe	T(°C)
1.7	2.02	6.3	12.5	7.4	6.40	-2.20	70
SO_4 (b)	S_2O_3 (c)	S_8 (c)	$\Sigma(n-1)[S_n^{2-}]$ (c)		$\Sigma[H_2S] + \Sigma[S_n^{2-}]$ (c)		
4.91	2.0	0.05	10.6		5.90		

Laboratory experiments have shown that the incomplete oxidation of H_2S yields polysulfide and thiosulfate ions (9). In this case also the electrochemical measurements were in agreement with the redox level estimated from the concentrations of the sulfur species.

Thus electrochemical measurements can be employed to characterize reducing environments where reduced sulfur species are present.

Computer simulation of the weathering processes involving reducing environments.

The above data together with data on iron sulfides in reducing environments (3,19,20,21) emphasize the possibly important part played by metastable sulfur species such as polysulfides and

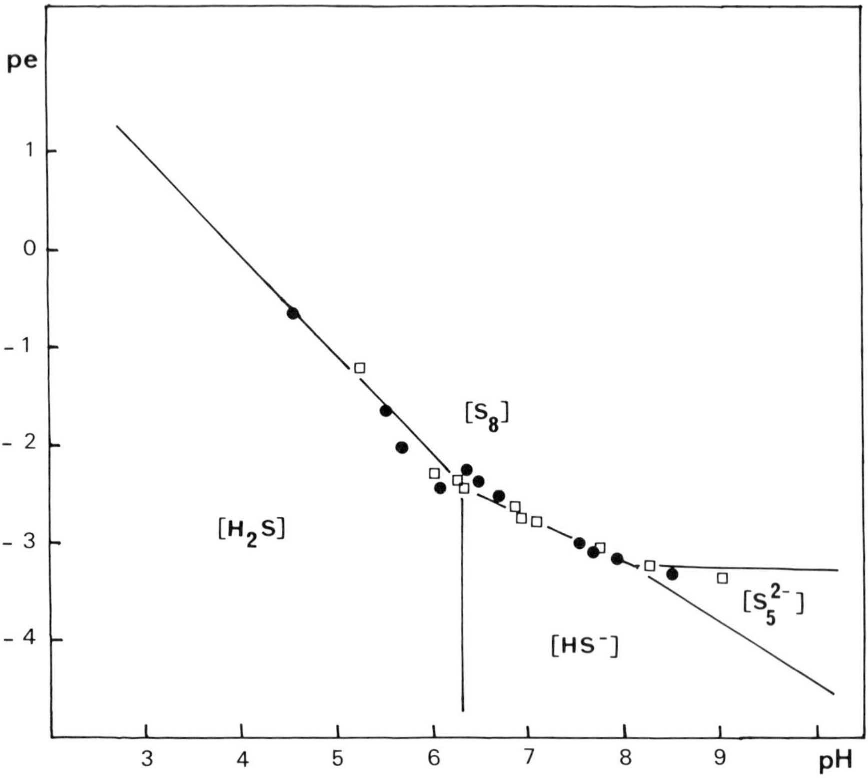

Figure 11. PE–pH diagram of brine PM2; (□) sample PM2b, (●) sample PM2c. The full line corresponds to the PE–pH stability diagram of the H_2S–S_8–H_2O system for $\Sigma[S] = 1.5 \times 10^{-3}$ g-atom/kg at 353 K and 1 bar (5). The dominant species in each domain is put in brackets.

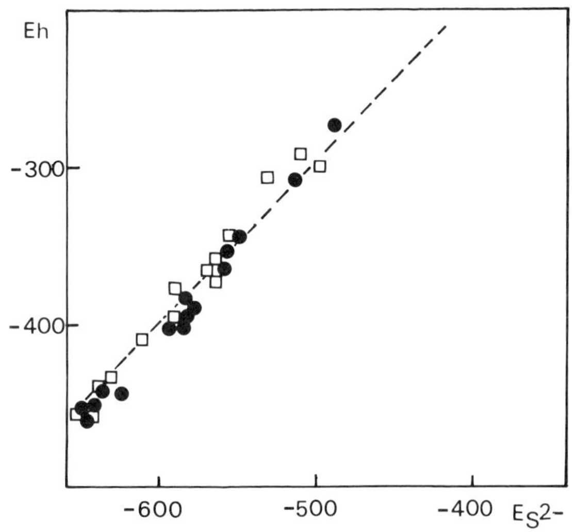

Figure 12. Eh–E_S2- relationship in brine PM2; (□) sample PM2b, (●) sample PM2c. The equation of the dashed line is Eh = E_S2- + 210 mV. All potential values are given vs. the potential of an Ag/AgCl electrode at the temperature of the brine.

thiosulfate in these environments. Since the information from these results is accurate enough we developed a computer program which enables us to describe the processes of the chemistry of dissolved species and minerals in reducing environments (22).

Methods of computer simulation. Our purpose is to calculate the evolution of the water composition in reducing environments. The sulfur system is assumed to be the result of a constant yield of sulfur at a given initial oxidation level included between -2 and 0, plus modifications brought about by possible oxidation.

The mathematical model of Helgeson (23) as modified by Fouillac *et al.* (24) was employed as a basic tool of calculation. It enables us to calculate mass transfer in open aqueous systems and to compute the evolution in space and time of a weathering profile. The following modifications were necessary:
(i) The redox potential was introduced as a master variable. This enables us to take into account disproportionation and oxidation processes. At each step of the computation the oxidation-reduction potential is deduced from the oxidation-reduction level of all electroactive species.
(ii) Since oxidation of sulfide to thiosulfate and sulfate is an irreversible process, we have considered the oxidation processes as the "weathering" of an "imaginary mineral" which yields oxygen and consumes electrons. This "weathering" is linked to a kinetic process as described by Chen and Morris (8):

$$-d\Sigma[H_2S]/dt = K\cdot(\Sigma[H_2S])^{1.34}\cdot([O_2])^{0.56} \qquad (7).$$

Practically this was achieved by fixing the initial value $(-d\Sigma[H_2S]/dt)_{t=0}$. The other kinetic variables are deduced from equation 7 and the stoichiometric coefficients of the reactions of oxidation of sulfur and iron species.
(iii) The solubility of elemental sulfur in water is not very well known. Its appearance as an individual phase is an important event. The precipitation of sulfur was effected at a value of the degree of advancement of the "weathering" reaction (ξ) which is characteristic of the maximum polysulfide concentration in the system H_2S-S_8(colloid)-H_2O (5).

Since pyrite is a quite ubiquitous mineral we have applied this method of simulation to the weathering of FeS_2 in the pH range 5 - 9. In this pH range the activity of the autotrophic iron bacteria *Thiobacillus* and *Ferrobacillus ferrooxidans* is negligible, thus microbiological catalysis of reactions is improbable. Practically we have chosen to study the alteration of a sandstone containing pyrite and calcite to look for the environmental conditions prevailing during the formation of roll-type uranium ores.

The thermodynamic data on polysulfides and thiosulfates are those of Boulègue and Michard (17) and Giggenbach (16). Other data are from Sillèn and Martell (25,26).

Computer simulation of the weathering of a pyrite rich sandstone.

(i) Initial conditions. The initial mineral assemblage represents a pyrite rich sandstone: (in mol %) quartz = 80, orthoclase = 6, kaolin = 2, calcite = 6, dolomite = 3, pyrite = 3. The composition of the pyrite is: (in mol %) S = 66.7, Fe = 33.3; [U/Fe] = 5 × 10^{-6} within the sandstone. The composition of the initial percolating water (IPW) is: (in mmol/kg) Na = 2.6, K = 0.26, Ca = 1.5, Mg = 0.99, Cl = 2.2, SO_4 = 1.4, $\Sigma[CO_2]$ = 1.6, HCO_3 = 1.45, H_4SiO_4 = 0.25, $0 < O_2 < 0.4$; T =298°K.

(ii) Weathering conditions. The pyrite is dissolved yielding Fe^{2+} and S_2^{2-}, and it is also partly oxidized. In this process S_2^{2-} ions are disproportionated to sulfide, polysulfide and thiosulfate ions with a total redox level equal to -1. Elemental sulfur and mainly thiosulfate are the products of the incomplete oxidation of reduced sulfur species; sulfur S_8 reacts with hydrogen sulfide to yield polysulfides (5,9).

In the calculations of Fouillac et al. (24) dξ represents the ratio of renewed water over total content in an open "box" where water and minerals are in interaction. In this condition the parameter V_x is the ratio of the speed of alteration over the speed of the water as explained in details by Helgeson (23), but it has the dimension of a concentration. We have adopted $V_x = 10^{-3}$ which corresponds to a slow speed of the water within the rock.

(iii) Results. The program describes the evolution in space and time of the composition of the percolating water and the neoformed minerals. In Figure 13 we present an "instant view" of the weathering profile with the neoformed minerals obtained in a simulation where the sandstone is divided in 20 equivalent boxes at the beginning of the weathering process (27).

In the downstream part of the weathering profile there is no more oxygen in the percolating water and only congruent dissolution of FeS_2 occurs. The increase of the concentrations of reduced sulfur species results in a steady decrease of pe. Dissolved iron issued from the weathering is mainly represented by Fe^{2+} free ion. The saturation with respect to FeS is reached at log(ξ) = -1.86, and it is soon followed by the precipitation of colloidal elemental sulfur. A steady state is established, corresponding to the global process

$$FeS_2 \rightleftarrows FeS + (1/8)\ S_8.$$

The evolution in time of the weathering conditions and the composition of the percolating water in the downstream part of the weathering profile is given in Figure 14. As shown by Rickard (21) the FeS-S_8 paragenesis is not stable in the presence of polysulfide ions and the reformation of pyrite will occur quite rapidly in the conditions encountered in the percolating water. The process is probably (21):

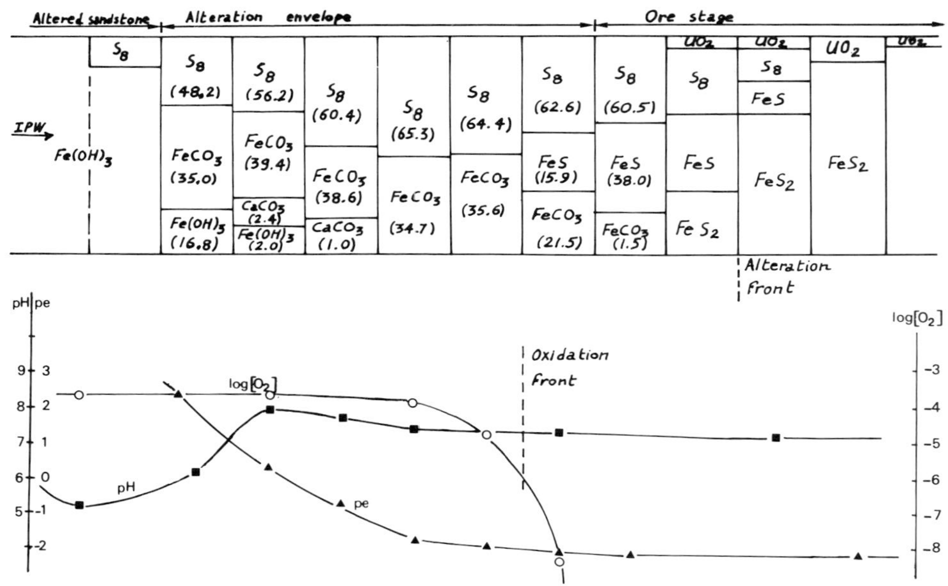

Figure 13. Simulation of the genesis of a roll-type uranium deposit and of the weathering of a pyrite-rich sandstone. Twenty boxes model at logξ = −1 with V_x = 10^{-3}. In the upper part we present an instant view of the alteration profile with neoformed minerals. Numbers in brackets correspond to the molar percentage of the neoformed minerals vs. total neoformed minerals in each box. In the lower part we present the major physicochemical parameters PE, pH, and [O_2] (mol/1) in the percolating water in the above boxes. The scale is the same in the upper part and the lower part. IPW = initial percolating water.

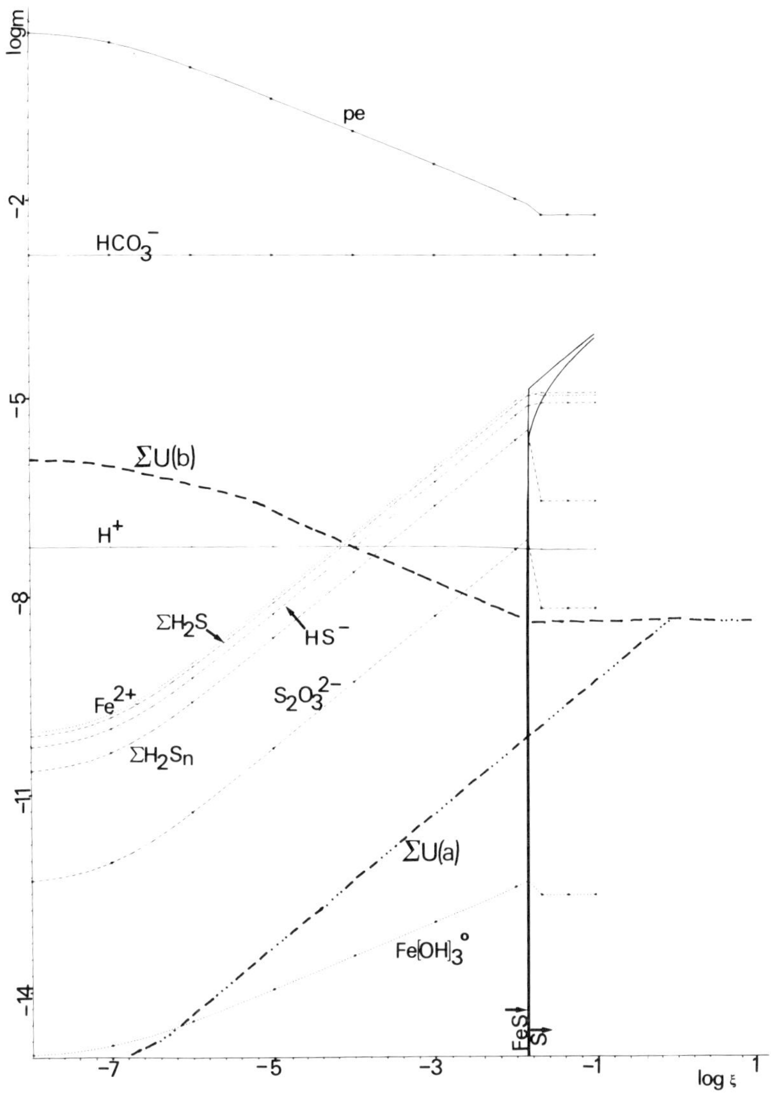

Figure 14. Time evolution of the composition of the percolating water in the downstream part of the alteration profile of a pyrite-rich sandstone. $V_x = 10^{-3}$. *The concentration of the dissolved species are given in mol/kg and the quantities of neoformed minerals are given in mol as a function of the parameter of advancement of the reaction ξ. All data are represented as the logarithm of the molality (log m) vs. log ξ. ΣU(a) corresponds to [U/Fe] = 5 × 10*$^{-6}$ *(molar ratio) leached within the sandstone. ΣU(b) corresponds to the maximum possible dissolved uranium concentration. All the curves are direct Benson plots from the computer.*

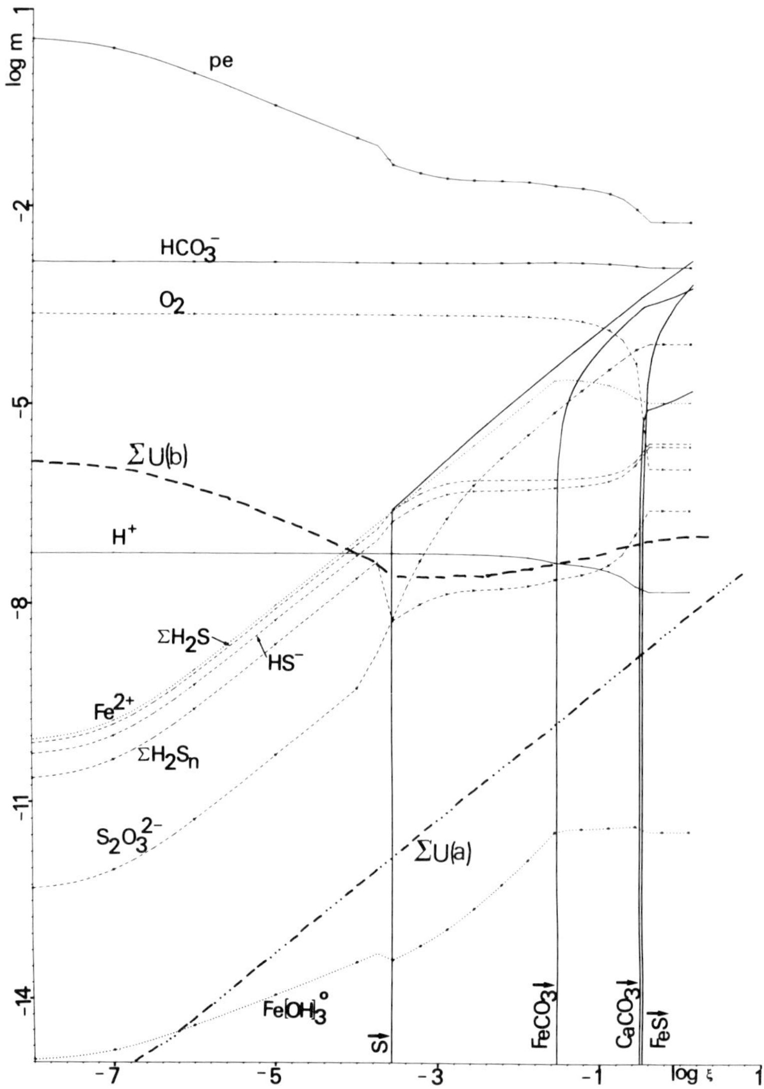

Figure 15. Time evolution of the composition of the percolating water in the alteration envelope upstream of the oxidation front (see Figure 13). $V_x = 10^{-3}$. *$[O_2]$ in IPW = 2.36×10^{-4} mol/kg. Evolution of the molality of characteristic species as a function of ξ. Other definitions are similar to Figure 14.*

$$FeS + S_n^{2-} \rightleftarrows FeS_2 + S_{n-1}^{2-} . \qquad (8)$$

Elemental sulfur, in the form of highly reactive colloid (28), can be the nucleus of additional pyrite reformation following polysulfide formation on the surface of the colloidal particles and reaction with Fe^{2+} as shown by Kribek (29). The stoichiometry of this process is

$$HS^- + S_8{-}S_n^{2-} + Fe^{2+} \rightleftarrows FeS_2 + (9/8)\ S_8 + S_{n-2}^{2-} + H^+ . \qquad (9)$$

Since Fe^{2+} and elemental sulfur are major species in the percolating water it is highly probable that the reaction 9 is competitive with reaction 8. Thus it is possible to write that the weathering of pyrite in the downstream part of the weathering profile results in an ideal composition of the percolating water corresponding to

$$FeS_2 \rightarrow \text{dissolved species} \rightleftarrows FeS + (1/8)\ S_8 \rightarrow FeS_2 .$$

In the upstream part of the weathering profile, oxidizing conditions are prevailing, leading to the accumulation of neoformed iron hydroxide $Fe(OH)_3$, while the sulfur species are completely oxidized to sulfate.

In Figure 15 we have represented the evolution in time of the composition of the percolating water in the intermediary zone (in Figure 12 it corresponds to the alteration envelope upstream of the oxidation front). Dissolved oxygen is still present in the percolating water and oxidizes FeS_2. The main products of the incomplete oxidation of the reduced sulfur species are elemental sulfur and thiosulfate. Elemental sulfur yielded by the oxidation process adds to elemental sulfur coming from the disproportionation of polysulfides, so that colloidal sulfur precipitates at an early stage of the weathering. In agreement with the experiments of Gourmelon et al. (9), slightly oxidizing conditions lead to a pH increase, which results in the formation of $FeCO_3$ and then $CaCO_3$. FeS precipitates in the later stage when the concentration of dissolved oxygen becomes negligible. The conditions following FeS formation result in the breakdown of $CaCO_3$ formation.

The above results on the simulation of the weathering of a pyritic sandstone are in good agreement with the results of the experimental study of pyrite oxidation at pH 5 - 9 (30). They are also in good agreement with the field observations and limitations imposed on the processes of genesis of roll-type uranium deposits (31,32).

Conclusion

The electrochemical measurements of pH, Eh and $E_{S^{2-}}$ after imposed pH variations enable an unambiguous characterization of the sulfur speciation and the redox level in reducing environments. The formation of metastable sulfur species (polysulfides and thiosulfate) should be effective in numerous reducing environments as shown by the results of laboratory and field studies.

The redox processes involved in the chemistry of sulfur and iron can be simulated on a computer. The feasibility of such simulations is supported by the good agreement between computer results and field data on the corresponding system of interest as shown by the example of roll-type uranium deposits.

Acknowledgments

Contribution number 293 from Institut de Physique du Globe de Paris. J.W. Ball and J. Burchard have helped with manuscript editing.

Abstract

In sulfurous waters where hydrogen sulfide is the only reduced sulfur species present in significant concentrations, the redox potential of the environment is difficult to assess. The potentials of the platinum electrode are characteristic of hydrogen discharge mediated via selective chemisorption of H_2S on the platinum electrode. Metastable sulfur species such as polysulfide and thiosulfate ions are stabilized in reducing environments and they play an important part in redox processes. The amounts of sulfur in these forms are related to the oxidation-reduction level of the environment. The simultaneous measurements of pH and of the potentials of platinum and Ag/Ag_2S electrodes after imposed pH variations enable one to characterize the oxidation-reduction level of such reducing environments. The results of laboratory experiments and field data are in good agreement. Considering the above results, we have developed a computer program which enables one to describe the chemistry of sulfur species in aqueous solutions. Kinetic processes as well as metastable species are taken into account. This program also enables one to describe the evolution in space and time of a weathering profile where sulfur species are playing an important part. We have applied this program to the simulation of the weathering of pyrite at neutral pH. The results are in good agreement with the results of laboratory oxidation of pyrite and field data on weathered pyritic sandstones.

Literature Cited

1. Morris J.C., and Stumm W., Redox equilibria and measurements of potentials in the aquatic environment, p. 270-285, in "Equilibrium Concepts in Natural Water Systems", Adv. Chem. Ser. 67, Washington, D.C., 1967.

2. Bockris J.O'M., and Reddy A.K.N., "Modern Electrochemistry", 1432 p., Plenum, New York, 1970.
3. Doyle R.W., The origin of the ferrous ion-ferric oxyde Nernst potential in environments containing dissolved ferrous iron. Am. Jour. Sci. 266, 840-859 (1968).
4. Berner R.A., Electrode studies of hydrogen sulfide in marine sediments. Geochim. Cosmochim. Acta 27, 563-575 (1963).
5. Boulègue J., "Géochimie du soufre dans les milieux réducteurs", 288 p., Ph.D. Thesis, Université de Paris 7, 1978.
6. Allen P., and Hickling A., Electrochemistry of sulfur. I . Over potential in the discharge of the sulfide ion. Trans. Farad. Soc. 53, 1626-1635 (1957).
7. Boulègue J., Mise en évidence et dosage des polysulfures au cours de l'oxydation de l'hydrogène sulfuré dans l'eau de mer. C. R. Acad. Sci. Paris 275(C), 1335-1338 (1972).
8. Chen K.Y., and Morris J.C., Kinetics of oxidation of aqueous sulfides by O_2. Environ. Sci. Technol. 6, 529-537 (1972).
9. Gourmelon C., Boulègue J., and Michard G., Oxydation partielle de l'hydrogène sulfuré en phase aqueuse. C. R. Acad. Sci. Paris 284(C), 269-272 (1977).
10. Boulègue J., Equilibria in a sulfide rich water from Enghien les Bains (France). Geochim. Cosmochim. Acta 41, 1751-1758 (1977).
11. Boulègue J., Electrochemistry of reduced sulfur species in natural waters- I. The H_2S-H_2O system. Geochim. Cosmochim. Acta 42, 1439-1445 (1978).
12. Boulègue J., Ciabrini J.P., Fouillac C., Michard G., and Ouzounian G., Field titrations of dissolved sulfur species in anoxic environments. Geochemistry of Puzzichello waters (Corsica, France). Chem. Geol. (1978), in press.
13. Defay R., Prigogine I., Bellemans A., and Everett D.H., "Surface Tension and Adsorption," Longmans Green, London, 1966.
14. Schuldiner S., Piersma B.J., and Warner T.B., Potential of a platinum electrode at low partial pressures of hydrogen and oxygen. II. An improved gas-tight system with a negligible oxygen leak. J. Electrochim. Soc. 113, 573-577 (1966).
15. Teder A., The equilibrium between elementary sulfur and aqueous polysulfide solutions. Acta Chem. Scandinavia 25, 1722-1728 (1971).
16. Giggenbach W., Optical spectra and equilibrium distribution of polysulfide ions in aqueous solutions at 20°. Inorg. Chem. 11, 1201-1207 (1972).
17. Boulègue J., and Michard G., Constantes de formation des ions polysulfurés S_6^{2-}, S_5^{2-} et S_4^{2-} en phase aqueuse. Jour. Fr. Hydrologie 9, 27-34 (1978).
18. Boulègue J., Metastable sulfur species and trace metals (Mn, Fe, Cu, Zn, Cd, Pb) in hot brines from the French Dogger. Am. Jour. Sci. 278 (1978), in press.
19. Berner R.A., Thermodynamic stability of sedimentary iron sulfides. Am. Jour. Sci. 265, 773-785 (1967).

20. Rickard D.T., The chemistry of iron sulfides formation at low temperatures. Stockholm Contrib. Geol. 20, 67-95 (1969).
21. Rickard D.T., Kinetics and mechanism of pyrite formation at low temperatures. Am. Jour. Sci. 275, 636-652 (1975).
22. Gourmelon C., Michard G., and Boulègue J., Metastable equilibria in sulfide rich waters. Application to a computer simulation of the weathering of pyrite rich sandstones and sulfide ore formation. p. 56-65, in "Proceedings of 2nd Intern. Symp. on Water-Rock Interaction", vol. 4, Sciences Géologiques, Université de Strasbourg, 1977.
23. Helgeson H.C., Evaluation of irreversible reaction in geochemical processes involving minerals and aqueous solutions - I - Thermodynamic relations. Geochim. Cosmochim. Acta 32, 853-877 (1968).
24. Fouillac C., Michard G., and Bocquier G., Une méthode de simulation de l'évolution des profils d'altération. Geochim. Cosmochim. Acta 40, 207-213 (1977).
25. Sillèn L.G., and Martell A.E., "Stability Constants of Metal - Ion Complexes", The Chem. Soc., Spec. Publ. 17, London, 1964.
26. Sillèn L.G., and Martell A.E., "Stability Constants of Metal - Ion Complexes. Suppl. 1", The Chem. Soc., Spec. Publ. 25, London, 1971.
27. Gourmelon C., "Simulation des interactions eaux-minéraux sulfurés des roches", Ph.D. Thesis, Université de Paris 7, 1977.
28. Boulègue J., Equilibres dans le système H_2S-S_8(colloïde)-H_2O. C. R. Acad. Sci. Paris 283(D), 591-594 (1976).
29. Kribek B. The origin of framboidal pyrite as a surface effect of sulfur grains. Mineral. Deposita 10, 389-396 (1975).
30. Goldhaber M.B., and Reynolds R.L., Experimental study of pyrite oxidation at pH 5 - 9.5; implications for formation of roll - type uranium deposits. p. 59, in "Amer. Assoc. Pet. Geol. Rocky Mount. Sect., Abstract," 1977.
31. Granger H.C., and Warren C.G., Zoning in the altered tongue associated with roll-type uranium deposits. in "Formation of Uranium Ore Deposits", I.A.E.A., Vienna, 1974.
32. Harshman E.N., Distribution of elements in some roll-type uranium deposits. in "Formation of Uranium Ore Deposits", I.A.E.A., Vienna, 1974.

Received November 16, 1978.

Redox Equilibria of Iron in Acid Mine Waters

DARRELL KIRK NORDSTROM
Department of Environmental Sciences, University of Virginia, Charlottesville, VA 22903
EVERETT A. JENNE and JAMES W. BALL
U.S. Geological Survey, Water Resources Division, Menlo Park, CA 94025

As the fourth most abundant element in the earth's crust, no other metal is as important as iron in geochemistry (1), natural water systems (2), microbial metabolism (3), the evolution of the earth's past and the formation of ore deposits (4,5). The chemical behavior of iron is known to be a key to the interpetation of processes involving trace elements, nutrients and oxidation-reduction reactions in natural waters, soils, sediments and groundwaters (6-13). An appropriate starting point for the cycling of iron through the lithosphere is the weathering of iron minerals, especially pyrite because it commonly occurs in many rock types and it provides a major source of sulfate and acidity as well as iron to natural water systems. Pyrite weathering also leads to the formation of acid mine drainage, a major cause of water pollution in many freshwater rivers and lakes.

Unlike iron transformations in marine systems or freshwaters of neutral to alkaline pH, acid mine waters contain very high concentrations of iron, which are therefore more easily determined. There are also minimal problems in these waters with non-equilibrium polymerization and colloid formation. Attempts at chemical modeling, however, are complicated by the large quantities of aqueous complexes that must be considered and the inadequacy of individual ion activity coefficients at moderate ionic strengths (up to 0.6 molal) when the ion association method is used. Calculations which are based on ion association and assume mineral-water equilibrium have been applied to problems of aqueous geochemistry with encouraging results (14-19). Although we consider our initial efforts as estimates, our calculations can be checked against field obser-

0-8412-0479-9/79/47-093-051$07.25/0
This chapter not subject to U.S. copyright
Published 1979 American Chemical Society

vations. Our results should indicate where the major weaknesses of this approach remain. We believe that any chemical model must be continually tested against many different field situations so that it can be improved and modified and thereby enhance our understanding of natural systems. In this paper we shall focus on three important aspects of acid mine water equilibria: 1) the measurement and interpretation of redox reactions, 2) the formation of aqueous complexes and 3) solution-mineral reactions.

Formation of Acid Mine Waters

Numerous investigations of pyrite oxidation and the production of acid mine drainage have partially unravelled the mechanisms of this complex process. The initial step in pyrite oxidation was demonstrated by Sato (20) to be the release of ferrous ions and elemental sulfur:

$$FeS_2 \rightarrow Fe^{2+} + S_2^{\circ} + 2e^- \qquad (1)$$

Likewise, the initial oxidation step of other metals sulfides (chalcocite, covellite, galena and sphalerite) appeared to produce the aqueous divalent metal ion and elemental sulfur.

In the next step of the oxidation, sulfur is oxidized to sulfate and protons are produced

$$S_2^{\circ} + 8H_2O \rightarrow 2SO_4^{2-} + 16H^+ + 12e^- \qquad (2)$$

Iron will remain in the reduced ferrous state for quite some time as long as the solution is acid (pH $\leq$ 4). Although the overall oxidant which drives this reaction is oxygen from the atmosphere, several investigators have demonstrated that dissolved ferric iron is the primary oxidant which directly attacks the pyrite surface (21,22). Since the reaction rate depends upon the availability of Fe^{3+}, the oxidation half-reaction

$$Fe^{2+} \rightarrow Fe^{3+} + e^- \qquad (3)$$

has been called the rate-determining step in the formation of acid mine drainage (23). Reaction 3 proceeds so slowly under acid conditions that acid mine waters would not commonly occur were it not for the acidophilic iron-oxidizing bacterium *Thiobacillus ferrooxidans*. This bacterium is such an effective catalyst that the oxidation rate is increased by 5 or 6 orders of magnitude over the abiotic rate (24,2). The bacterium can grow in the absence of light and it is found deep inside mines as long as some minimal amount of oxygen is available. The production of acid mine drainage is thus a rapid, self-perpetuating process catalyzed by bacteria which continues as long as air, water and pyrite are available. The geochemical formulation of this process is summarized in Figure 1. Pyrite is attacked by ferric iron, an acid ferrous sulfate solution is produced, and the ferrous iron is catalytically reoxidized to ferric by *T. ferroxidans*. This cycle continues until the pyrite is gone or the mine water leaves the sulfide surfaces (issues from the mine) where it will fully oxidize and hydrolyse to form amorphous ferric hydroxide (yellow-boy), goethite or jarosite depending upon the acid, iron, sulfate content of the water and the degree of ageing. Oxygen enters this reaction scheme by accepting the electron donated from the iron oxidation of reaction 3 through the metabolic pathways of the bacterium. The reduction half-reaction is:

$$\tfrac{1}{2}O_2 + 2H^+ + 2e^- \rightarrow H_2O \qquad (4)$$

Since these bacteria are aerobic, this scheme implies that an important function of oxygen is to provide oxic conditions for adequate respiration. These half-reactions (3 and 4) need not be in equilibrium with each other; only a small amount of oxygen, enough for respiration, is necessary to drive this process. In this study, we are chiefly concerned with the reactions occurring in the effluent water after it has left the mines and has entered natural streams.

Field Site, Sampling and Analytical Methods

In Shasta County, California, inactive copper mines containing several millions of tons of sulfides are being weathered to produce highly acidic mine waters. The watershed surrounding Iron Mountain (Figure 2) was chosen for detailed investigations because it was the largest source of acid drainage in the region (25,26). Approximately 14 km of stream waters are affected by acid effluent issuing from the mines at Iron Mountain. These streams show a large gradient in redox state, pH and total dissolved solids during downstream transport due to mixing and dilution as well as rapid oxida-

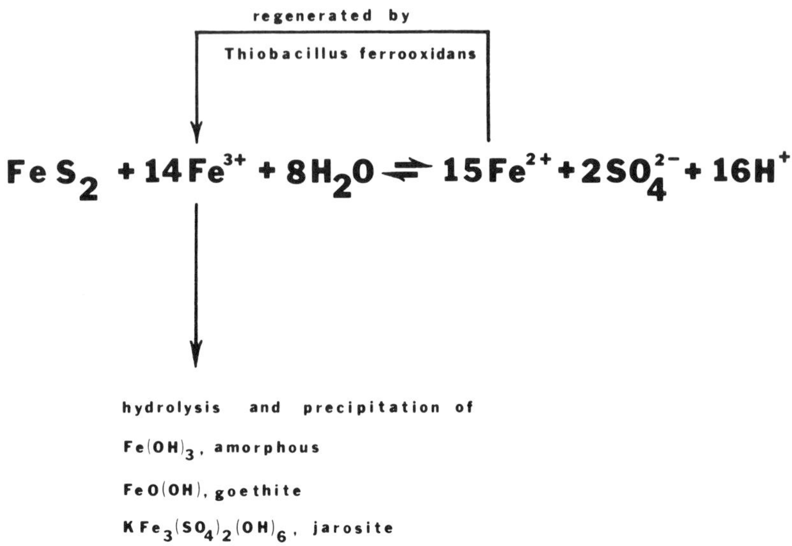

Figure 1. Mechanism of acid mine water formation

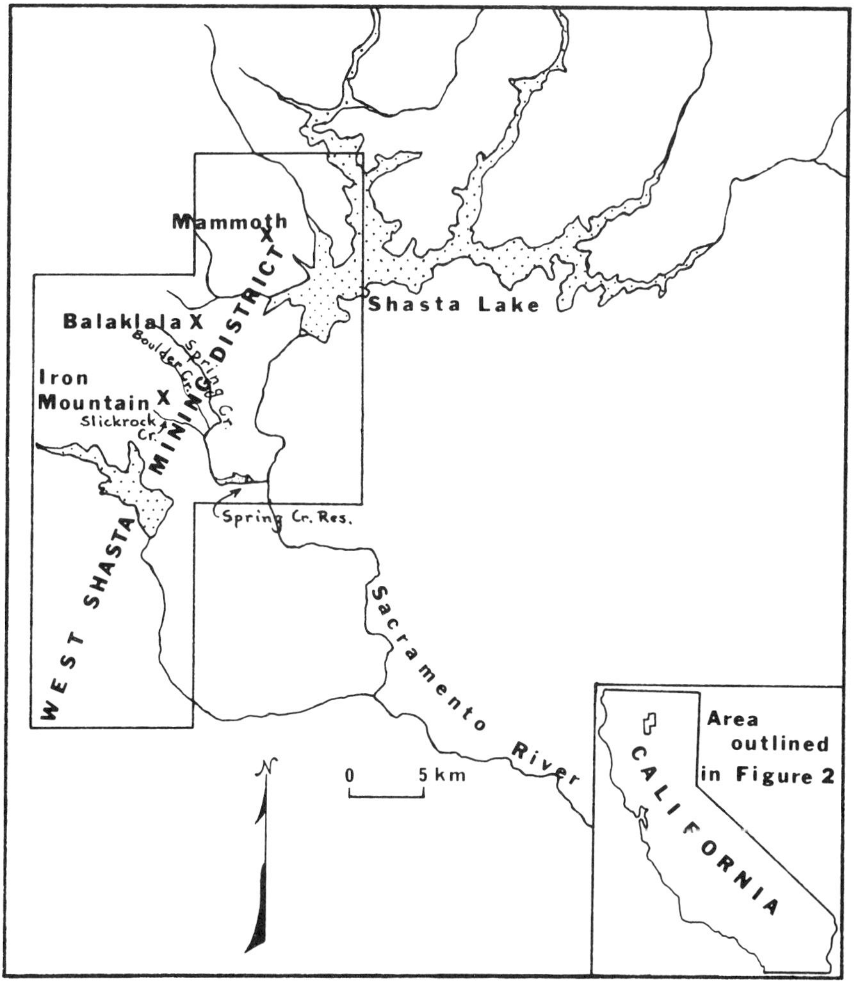

Figure 2. Location of Spring Creek and its two tributaries, Boulder and Slickrock Creeks, which contain acid mine drainage and drain the Iron Mountain watershed

tion rates. The weathering ore bodies are typical massive sulfide deposits, and they contain fine-grained pyrite with variable amounts of chalcopyrite and sphalerite occurring as pods or lenses within a rhyolite of Devonian age. Both air and groundwaters easily infiltrate the ore bodies and stoped out areas where the acid waters develop. The acid waters eventually find their way to a portal or adit and are finally discharged into either Boulder Creek or Slickrock Creek, the principal tributaries of Spring Creek (Figure 2). During an 18-month period over 100 water samples were collected at several different locations on these streams in order to interpret the hydrogeochemical processes controlling the heavy metal chemistry (26).

Temperature, conductivity, pH, Eh and dissolved oxygen (D.O.) were measured on site. Water samples collected for analysis were filtered on site through 0.10 μm membranes in an acid-cleaned plastic filtering apparatus (27) connected to a portable peristaltic pump via acid cleaned silicone tubing. Samples were preserved by acidification to a pH of about 1 with nitric acid for heavy metal analysis and with hydrochloric acid for ferrous iron, aluminum, alkali metals and alkaline earth metals. Filtration through 0.10 μm membranes remove iron-oxidizing bacteria and prevents biological oxidation while acidification minimizes inorganic oxidation. Anion analyses were carried out on samples which were filtered but otherwise were not preserved. Eh measurements were made with a combination polished platinum electrode which was checked at least once in a sampling day with ZoBell's solution (28). The precision on emf readings was ± 5 mv; the precision on pH readings was ± 0.02, although the accuracy cannot be considered better than ± 0.05. Measurement of pH in the most concentrated acid mine effluent (pH ≃ 1.0) were quite slow to come to a steady value but were always reproducible. Measurements of D.O. were made by a commercial oxygen electrode which was calibrated by a Winkler titration on a clean oxygen-saturated stream water.

Aluminum, silicon and zinc were determined by d.c.-argon plasma jet emission spectrophotometry. The remaining cations were analyzed by atomic absorption spectrophotometry (AAS) except ferrous iron which was done by a modification of the Ferrozine method (29,30,31). Total iron was determined by AAS and Fe^{3+} by difference. Sulfate was analyzed by the Thorin method after removal of interfering cations by ion exchange. Fluoride and chloride were extremely difficult to analyze because of the high concentration of interfering ions, and they were present at very low concentrations (about 0.1 mg/L or less). For chemical equilibrium computations they were assumed to be 0.1 mg/L for all samples.

Water samples collected from the Iron Mountain watershed provided a large variation in pH (1 to 7), Eh (350 to 900 mv), total dissolved iron (10 to 12,000 mg/L), temperature (4° to 31°C) and the ionic strength varied up to 0.6 molal.

Calculations of Activities and Saturation Indices

The water analyses were coded and then processed with the computer program WATEQ2. This program was modified in several ways to handle acid mine waters: (a) the Eh could be calculated from the Fe^{2+}/Fe^{3+} activity ratio or vice versa, (b) several sulfate minerals were added, (c) metal sulfate and hydroxide complex constants were carefully evaluated and included, and (d) Mn, Cu, Zn and Cd species were added since they are major constituents for several of the water samples. These modifications and the evaluated thermodynamic data are described by Ball, Jenne and Nordstrom in this symposium (32).

Redox Status of Acid Mine Waters: Equilibrium or Disequilibrium?

What we mean in this report by equilibrium and disequilibrium requires a brief discussion of definitions. Natural physicochemical systems contain gases, liquids and solids with interfaces forming the boundary between phases and with some solubility of the components from one phase in another depending on the chemical potential of each component. When equilibrium is reached by a heterogeneous system, the rate of transfer of any component between phases is equal in both directions across every interface. This definition demands that all solution reactions in the liquid phase be simultaneously in equilibrium with both gas and solid phases which make contact with that liquid. Homogeneous solution phase reactions, however, are commonly much faster than gas phase or solid phase reactions and faster than gas-liquid, gas-solid and liquid-solid transfer reactions. Therefore it seems appropriate to assume that natural waters would be in equilibrium with respect to dissolved species in the aqueous phase but might be in varying stages of disequilibrium with respect to gaseous and solid phases. Such a system is in partial or local equilibrium (33). In fact, there may be multiple and diverse local equilibrium states in such a heterogeneous environment. Equilibrium thermodynamics defines the extent to which a natural water has reached chemical equilibrium even though it may be undergoing dynamic changes in fluid flow, heat flow, biologic activity or mass transfer. This assumption of local equilibrium is a fundamental postulate of irreversible thermodynamics which justifies the calculations of equilibrium properties for a restricted spatial or temporal element in an irreversible process (33,34,35). With these concepts in mind, we can appreciate that a natural water may have one fixed, representative Eh but it may not be in equilibrium with dissolved gases or the solid surface of a platinum electrode. We now consider Eh measurements and chemical equilibrium in natural water systems.

The Eh of natural waters has been calculated theoretically (2,36), measured with inert metal electrodes, calculated from analyses of individual redox species (37) and measured by equilibration with known redox couples (38,39). Eh measurements have been used qualitatively as an operational parameter and quantitatively as an indication of a dominant redox couple. The qualitative use of Eh, advocated by ZoBell (40), has resulted in a great many measurements (41). As a quantitative tool, the use of Eh has not enjoyed widespread success. Several criteria must be met before the Eh can be related to a specific redox couple or mechanism: 1) the net exchange current at the electrode-solution interface, i.e., the difference in the rates of electron transfer to and from the platinum surface, must be negligible (an equilibrium criterion required for the application of the Nernst equation), 2) the individual exchange current, i_o, should be greater than about 10^{-7} ampere/cm^2 (43), 3) all aqueous electroactive redox species should be in homogeneous equilibrium, 4) the electrode surface must be free of electroactive surface coatings and adsorbed impurities and 5) activities of the redox species must be either measured or calculated by considering the effects of complexing and ionic strength on activity coefficients, since the activities, rather than the concentrations of the participating species are the quantities that determine the potential due to the redox couple. These criteria are difficult to meet and have led to the skeptical outlook that most Eh measurements are not amenable to quantitative interpretation (2,42,43). Contamination of platinum electrode surfaces by oxygen in aerated waters (43,44), by sulfur in anaerobic waters (44) and by iron in surface sediments (45) may cause errors in the measured values. Furthermore, many Eh measurements are thought to be mixed potentials (43). For these reasons, most Eh measurements have been used only in a qualitative sense.

In contrast to this general pessimism, several investigators have found good agreement between measured Eh and a dominant redox couple. Sato (20) found a relation between the measured Eh and pH of subsurface waters and the peroxide-oxygen redox couple. Berner (46) demonstrated a Nernstian relationship between measured Eh and the S^{2-}/S° couple in anoxic marine sediments. This equilibrium has also been substantiated by Kryukov, et al. (47), Skopintsev, et al. (48) and Whitfield (49). Thorstenson (37) found good agreement between the redox couples of sulfur and nitrogen species for five different reducing environments, implying that mixed potentials may not be a major problem. In laboratory experiments, Natarajan and Iwasaki (50) found Nernstian behavior of platium electrodes in the presence of varying dissolved oxygen and ferrous-ferric ratios. Bricker (51) achieved reversible equilibrium behavior with a platinum electrode in solubility studies of manganese oxides and hydroxides. Eh measurements of sulfur-rich waters can deviate from calculated sulfide redox values due to the presence of colloidal sulfur, but Boulegue (52) has shown that these interferences can be identified by acid-base titrations of the sample. Thus, amid the skepticism, there is already evidence from several different environments indicating that Eh measurements can be amenable to quantitative, thermodynamic interpretations.

Acid mine waters should be particularly well-suited to reliable Eh measurements because the high concentrations of electroactive iron species would clearly satisfy criteria 1) and 2) above (42,43). In addition, the acid conditions inhibits surface coatings of iron oxides on the electrode. The measured Eh values in this study were compared with calculated values based on ferrous-ferric analyses of 60 samples all of which had cation-anion balances better than 30 percent. Calculated values were derived from the Fe^{2+}/Fe^{3+} activity ratios after species distribution and temperature corrections to the standard ferrous-ferric couple had been computed with WATEQ2. The Nernst equation used for the calculation is

$$Eh = E^\circ\ (T,\ Fe^{2+}/Fe^{3+}) - \frac{RT}{F} \ln \frac{[Fe^{2+}]}{[Fe^{3+}]}$$

where the square brackets indicate activities, R is the ideal gas constant, T is the Kelvin temperature, F is the Faraday constant and E° is the temperature-dependent standard electrode potential for the ferrous-ferric couple. The results are shown in Figure 3, where the solid line represents perfect agreement. Seventy-seven percent of the values fall within ± 30 mv, which would be an anticipated uncertainty for most Eh measurements. No value is greater than 80 mv from the other and over half of the values agree to within ± 10 mv, which is twice the precision of the emf readings. This excellent agreement suggests a simultaneous validation of the equilibrium condition of acid mine waters, of the chemical model and its ability to represent that equilibrium, and of the accuracy of the Eh measurement.

Several features of this study should be pointed out. First, the reliability of the Eh measurement depends greatly on the technique. We used a closed line to pump water from the stream to a flow-through cell where a continuous supply of fresh sample water made contact with the electrode. No streaming potential effects were noticed because there was little or no change in potential when the pumping was stopped. The lack of streaming potential was probably due to the slow flow velocities and the high iron concentrations. The method of using a closed flow-through cell supplying a large volume of water to the surface of the Eh and pH electrodes has been found to be quite successful for a wide range of natural waters (53). We also would commonly remove the platinum electrode and gently buff the polished surface with a soft cloth and repeat the measurement if we noticed any drift. Second, we felt it was valuable to see the amount of change in the calculated Eh when activity coefficients and complexes were not considered, that is, when measured total concentrations of ferrous and ferric iron, rather than their respective calculated activities, were used to compute the Eh. We chose two measured Eh values, one which agreed to within 1 mv of the calculated value and another which differed from the calculated value by 79 mv. Without activity and complexing corrections, the discrepancies increased to 30 and 180 mv, respectively, a large error in each case.

It appears that the largest source of error in these comparisons is the analytical data. The next largest source of error seems to be the adequacy of activity coefficients and stability constants used in the model and last is the reliability of the field Eh measurement. Close inspection of Figure 3 shows a slight bias of calculated Eh values towards more oxidizing potentials. Fe(III) complexes are quite strong and it is likely that some important complexes, possibly $FeHSO_4^{2+}$ (54,55), should be included in the chemical model, but the thermodynamic data are not reliable enough to justify its use.

Finally, we return to a clarification of what parts of the system are in equilibrium. The comparison of calculated with measured Eh values gives strong evidence for homogeneous solution equilibrium and for equilibrium conditions at the surface of a platinum electrode in these waters. To determine whether the D.O. content of the water related to these equilibria and whether the oxidation of iron consumed oxygen from the stream water, we made measurements at the mouth of Boulder Creek and downstream in Spring Creek where most of the iron has been oxidized. We compared our D.O. measurements with the saturation oxygen solubility for the appropriate temperature and barometric pressure and found each part of the stream very close to saturation. The streams should contain saturated amounts of D.O. since they are commonly turbulent and well-mixed. The results in Table I, however, show that the Eh calculated using the O_2/H_2O couple is considerably higher than the measured value. The redox state of the water is thus determined by the ferrous-ferric ratio; oxygen dissolved in the water, while in equilibrium with the atmosphere, is not in equilibrium with the ferrous-ferric couple. This situation also reflects the fact that the exchange current for the O_2/H_2O couple is far less than for

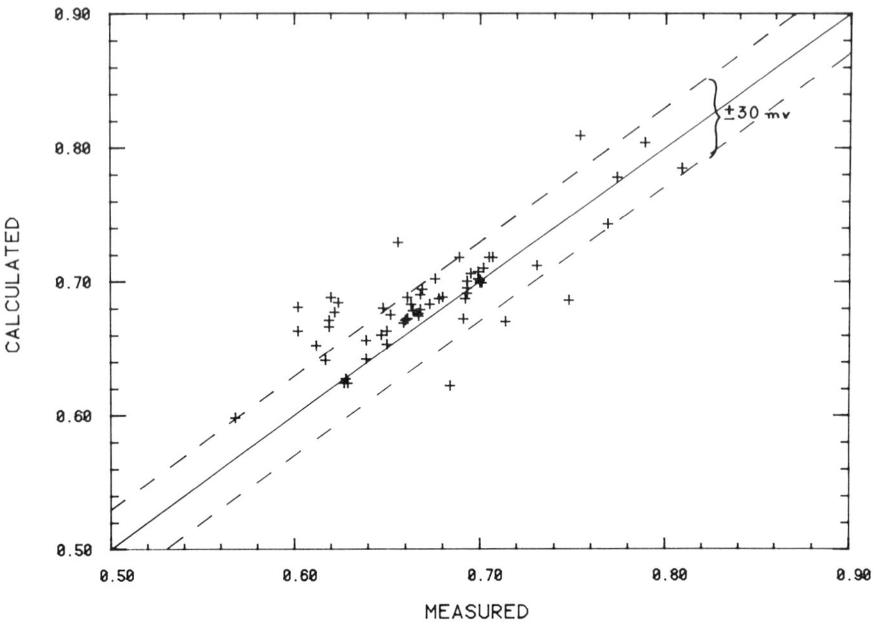

Figure 3. Comparison of measured and calculated Eh values. Errors of ±30 mV are given for the uncertainty in the Eh measurements.

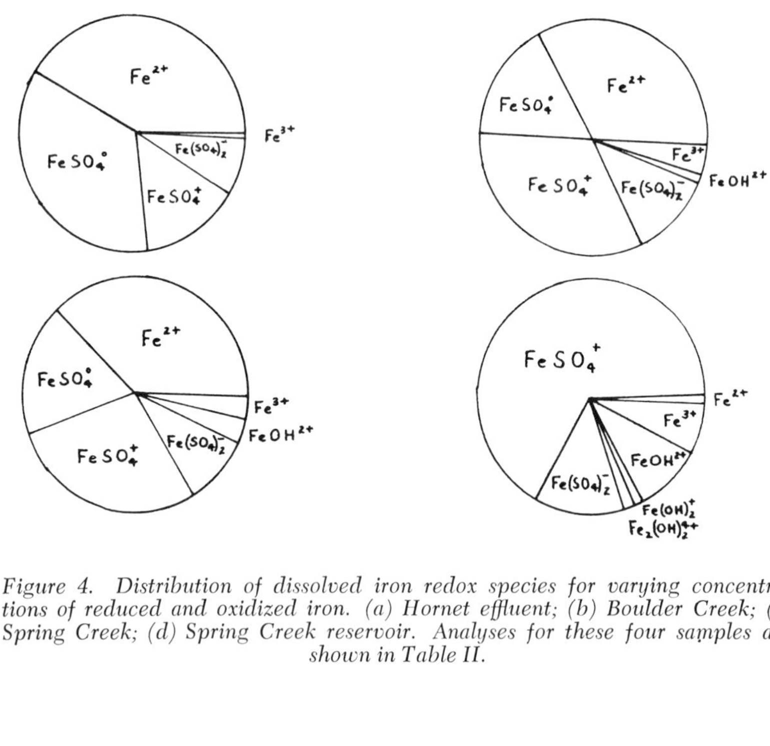

Figure 4. Distribution of dissolved iron redox species for varying concentrations of reduced and oxidized iron. (a) Hornet effluent; (b) Boulder Creek; (c) Spring Creek; (d) Spring Creek reservoir. Analyses for these four samples are shown in Table II.

Table I
Comparison of measured Eh values with those calculated from dissolved oxygen and the O_2/H_2O redox couple

	(H)Spring Cr.	(C)Boulder Cr.	(G)Boulder Cr.	(J) Spring Cr.
pH	6.40	3.02	1.85	2.32
D.O.(mg/L)	10.3	10.5	11.0	11.8
Eh, measured (V)	.385	.694	.625	.640
Eh, O_2/H_2O (V)	0.948	1.156	1.227	1.197

Table II
(Concentrations in mg/L except where noted)

Site	B*	G	J	N
T°C	25.5	6.0	5.5	5.5
pH	1.10	2.00	2.50	2.90
Eh (V)	.622	.378	.650	.770
Spec. Cond.**	65000	9900	2970	1810
Ca	173	45	17	18
Mg	685	107	27.5	30.8
Na	92.5	14.2	6.00	5.40
K	128	13.8	3.75	2.40
Fe (II)	9050	-	260	-
Fe (III)	2650	1280	354	208
Al	1400	188	47	47
SiO_2	130	55	26	32
SO_4	60000	5310	1280	1080

*B - Hornet Effluent, G - Boulder Creek, J - Spring Creek, N - Spring Creek.
**Units for specific conductance are microSiemens.

the Fe^{2+}/Fe^{3+} couple (43) and does not affect the measured potential.

Solution Complexes of Iron in Acid Mine Waters

The wide range of Eh, pH and dissolved solids for the acid mine waters in the Iron Mountain watershed provides an opportunity to survey the dominant complexes which are calculated from an ion association model. Rather than choosing some "average" composition for acid mine water, we have used naturally-occurring waters which represent the range of conditions found in the study site. The analyses shown in Table II are a series of water samples containing the most ferrous-rich to the most ferric-rich in iron content. These analyses show progressive downstream dilution and oxidation beginning with the source water coming from the Hornet tunnel which discharges into Boulder Creek (see Figure 2). The two Spring Creek samples were taken at 1) a point just below Boulder Creek confluence and 2) a point 4 km below the Boulder Creek confluence at the mouth to the Spring Creek Reservoir. The list of iron complexes which were used in the aqueous modeling computations are shown in Table III. Of these complexes, the major ones are expected to be hydroxide and sulfate. The distribution of species is shown in Figure 4 as percentages of the total dissolved iron. Only species which contributed 0.2 percent or more of the total dissolved iron were included. From Figure 4, it is clear that only the $FeSO_4^{\circ}$ contributes to the complexing of ferrous iron but that this complex alone can constitute up to nearly 50 percent of the total ferrous iron. Ferric iron is much more highly complexed than ferrous and under most conditions the free ferric iron is never more than about 8 percent of the total iron. $FeSO_4^+$ is always the dominant ferric complex, followed by $Fe(SO_4)_2^-$ > $Fe(OH)^{2+}$ > $Fe_2(OH)_2^{4+}$ > $Fe(OH)_2^+$ in order of decreasing abundance. An objection might be raised that organic complexing plays an important role in the species distribution of these waters. This possibility is certainly very real, and there is some evidence that the dissolved organic carbon in these waters is very high, but the good agreement on the Eh calculations and the low pH values of these waters suggest that any organics present are probably fully protonated and have only a minor effect on complexing.

Any chemical modeling of acid mine waters

Table III
Aqueous complexes and solid phases of iron considered for computation

a) Complexes:

$FeOH^{+}$, $Fe(OH)_2^{\circ}$, $Fe(OH)_3^{-}$, $FeSO_4^{\circ}$

$FeOH^{2+}$, $Fe(OH)_2^{+}$, $Fe(OH)_3^{\circ}$, $Fe(OH)_4^{-}$, $Fe_2(OH)_2^{4+}$, $Fe_3(OH)_4^{5+}$ $FeSO_4^{+}$, $Fe(SO_4)_2^{-}$

$FeCl^{2+}$, $FeCl_2^{+}$, $FeCl_3^{\circ}$, FeF^{2+}, FeF_2^{+}, FeF_3°

b) Solids:

Melanterite	$FeSO_4 \cdot 7H_2O$
Copiapite	$Fe^{2+}Fe_4^{3+}(SO_4)_6(OH)_2 \cdot 20H_2O$
Coquimbite	$Fe_2(SO_4)_3 \cdot 9H_2O$
Jarosite	$KFe_3(SO_4)_2(OH)_6$
Ferric hydroxide, amorphous	$Fe(OH)_3$

relies heavily on accurate thermodynamic data for iron sulfate complexes. Unfortunately, data for ion triplets such as $Fe(SO_4)_2^-$ are not accurately known and possible complexes such as $FeHSO_4^{2+}$ and $Fe(HSO_4)_2^0$ have not been properly identified. Better data are needed for these species in order to enhance the capability of generalized chemical models.

Water-Mineral Equilibrium Relationships

A suite of both oxidized and reduced iron minerals has been found as efflorescences and precipitates in or near the acid mine water of Iron Mountain. The dominant minerals tend to be melanterite (or one of its dehydration products), copiapite, jarosite and iron hydroxide. These minerals and their chemical formulae are listed in Table III from the most ferrous-rich at the top to the most ferric-rich at the bottom. These minerals were collected in air-tight containers and identified by X-ray diffractometry. It was also possible to check the mineral saturation indices ($\log_{10}(AP/K)$, where AP = activity product and K = solubility product constant) of the mine waters with the field occurrences of the same minerals. By continual checking of the saturation index (S.I.) with actual mineralogic occurrences, inaccuracies in chemical models such as WATEQ2 can be discovered, evaluated and corrected (19), provided that these occurrences can be assumed to be an approach towards equilibrium.

We have chosen an equilibrium zone around the S.I. equal to the estimated uncertainty of the solubility product constant. Outside of these limits, the solution is either supersaturated (positive direction) or undersaturated (negative direction). The S.I. values were then plotted vs. the log conductivity to show approaches to mineral saturation as an approximate function of the total dissolved solids. Conductivity was chosen as a convenient parameter because 1) it points out dilution trends or undersaturation and corresponds to a spatial trend for downstream sites, 2) it changes more consistently than other parameters, and 3) it covers a larger range of values than Eh, pH or an individual ion. Activity diagrams (53) could have been used but then certain assumptions have to be made about the mineral stability boundaries such as constant activity of water or of pH or of sulfate.

Melanterite. The most concentrated acid mine

waters, such as those flowing from the Hornet tunnel, commonly precipitate melanterite, a pure ferrous sulfate, along their edges. Water which had infiltrated a massive ore body would also precitipate melanterite along the surface of an open cut exposed to dry outside air. But melanterite was found only on sulfide surfaces which were regularly washed by rainfall or running water. In sheltered areas, such as the entrance to the Richmond mine, the dominant mineral occurring on open cuts is copiapite or a mixture of copiapite and coquimbite.

In Figure 5 we have plotted S.I. values for melanterite indicating a trend towards saturation for the Hornet effluent (labeled B). All of the other waters (collected at downstream sites) have been diluted and oxidized and therefore appear undersaturated. The results of these calculations compare quite favorably with field observations. Unfortunately, there is a large uncertainty associated with the thermodynamic solubility constant for melanterite. Although its solubility is well-known, the thermodynamic equilibrium constant is difficult to obtain because the compound is highly soluble and therefore becomes saturated only at high ionic strenths.

More importantly, it should be noted that melanterite commonly forms on rock surfaces, especially along fractures where water can be pulled up from a water-saturated zone to the surface by capillary forces. Thus, melanterite saturation is occurring in a microenvironment in many instances which makes sampling and interpretation difficult.

Copiapite. Theoretically, melanterite can oxidize to form copiapite by the reaction:

$$5FeSO_4 \cdot 7H_2O + O_2 + H_2SO_4 \rightarrow$$
$$Fe^{2+}Fe_4^{3+}(SO_4)_6(OH)_2 \cdot 20H_2O + 15H_2O.$$

As mentioned earlier, large amounts of copiapite have been found accumulating on ore surfaces in areas protected for direct rainfall. The mechanism of its formation is not at all clear but from field observations it appears that acid mine water is being drawn by capillary forces to an exposed surface where it quickly evaporates to melanterite and/or copiapite. Coquimbite is intimately associated with copiapite and these two minerals appear to be very stable as long as they are protected from rainfall or running water. No thermodynamic data

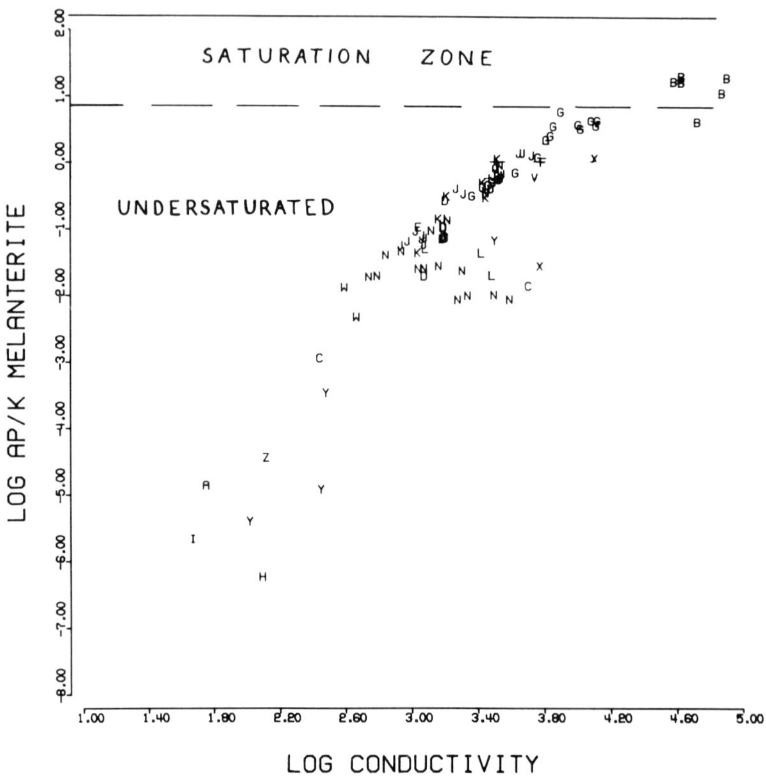

Figure 5. Saturation indices for melanterite plotted as a function of the log conductivity showing an approach to saturation for the most concentrated ferrous-rich waters of the Hornet effluent (B)

are available on either copiapite or coquimbite so that S.I. values cannot be calculated. Copiapite contains both ferrous and ferric iron but a form called ferricopiapite containing only iron in the oxidized ferric state may also occur. There is little information on the relative abundance of these minerals nor whether copiapite is a necessary precursor to the formation of ferricopiapite and coquimbite. A further complication to the interpretation of mineral reactions is the solid substitution properties of copiapite. It not only can have a change in the ferrous/ferric ratio but the divalent and trivalent sites can accept a large number of divalent and trivalent ions such as Mg^{2+}, Cu^{2+}, Zn^{2+}, Cd^{2+} and Al^{3+}.

Jarosite. The waters which are associated with such soluble salts as melanterite and copiapite commonly have pH values of 0.8 to 1.5. As the pH rises to the range of 1.5 to 2.5, jarosite will commonly precipitate if the dissolved iron content is high enough. A fine-grained yellow precipitate forms in Boulder Creek every summer and fall, which when washed and filtered, gives a good X-ray pattern for potassium jarosite. This mineral is precipitating directly from the water and it appears to have good crystallinity. Both copiapite and jarosite are a bright yellow color and may easily be confused with sulfur.

In Figure 6, the S.I. values for potassium jarosite are given for all the samples. The values show supersaturation for nearly every sample and prompts us to ask several questions: (a) are these waters truly supersaturated and is this a kinetic requirement for precipitation? (b) are organic complexes important for these waters? (c) are the thermodynamic data reliable? Some of these questions can be answered. First, the amount of supersaturation is more than the $\pm$ 1.1 log K unit uncertainty in the thermodynamic data. Second, the model does not appear to be inadequate for the Eh comparison and other calculations, and if organic complexing were introducing a significant error it would have shown up in the cation-anion balance or other calculations. Downstream in Spring Creek where no jarosite has been found the S.I. is often more supersaturated than in Boulder Creek. This supersaturation suggests that jarosite solubility does not control the water chemistry of any of the streams. The remaining possibility is that jarosite is precipitating only in microenvironments where the

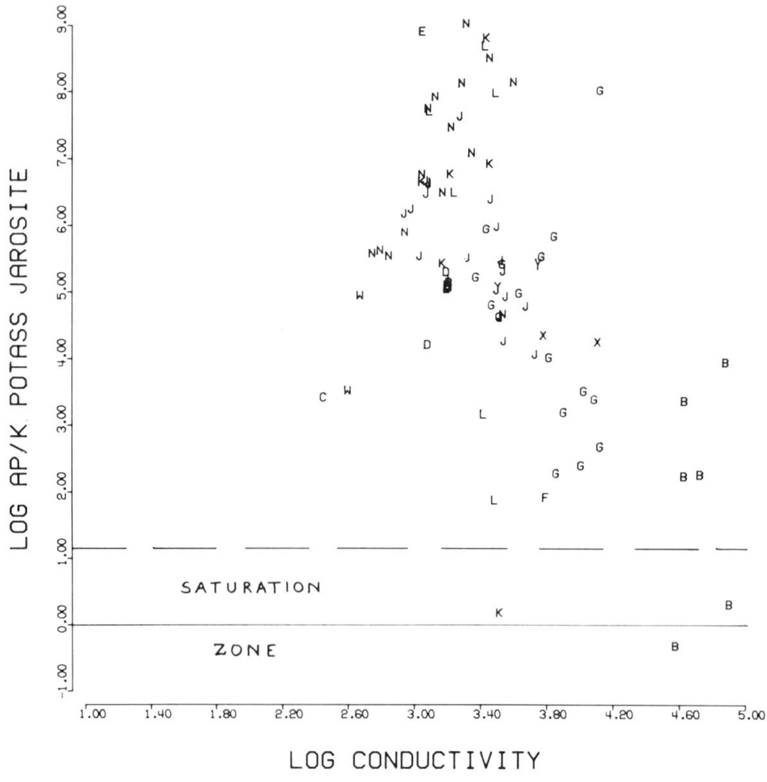

Figure 6. Saturation indices for jarosite plotted as a function of the log conductivity showing supersaturation for nearly all values

water chemistry is different than in the bulk solution. The evidence for this is that jarosite precipitation is always found on stream bed surfaces where bacteria have been growing, and is also found inside of acid slime streamers produced by these bacteria. Beyond this, there may be important particle size effects (12). There could still be local equilibrium conditions existing here, but they are probably below the resolution of the sampling techniques. Clearly, more work is needed before the jarosite geochemistry can be adequately interpreted.

Amorphous ferric hydroxide. Most of Spring Creek below the Boulder Creek confluence is iron stained and actively precipitates iron hydroxide. The pH values are commonly in the range of 2.5 to 3.5 depending on the season. The iron precipitates are amorphous to X-rays and appear to be amorphous ferric hydroxide from their color and texture. S.I. values calculated for amorphous ferric hydroxide show a trend towards saturation with downstream dilution and oxidation (Figure 7). We have used the upper solubility limit of pK = 37.1, where pK is the apparent stability constant for ferric oxyhydroxides (12) and most of the downstream sites (labeled N) fall neatly into the region of pK = 37 to 39 as would be expected for a freshly precipitated material. Very little supersaturation appears in our computations. Again, the S.I. calculations match very well with the field observations and with previous studies of ferric hydroxide equilibrium in natural waters (12,15,17). A noticeable hiatus occurs at about log K = 41. Below this value the samples plotted in Figure 7 are very high in ferrous iron, low in pH and show no sign of ferric hydroxide precipitation.

Conclusions

We have utilized a chemical model in this investigation to interpret the equilibrium behavior of iron in acid mine waters. A successful correlation between calculated and measured Eh values has been found, using WATEQ2, the computerized ion association model. This correlation supports the basic assumption of homogeneous solution equilibrium in these waters and simultaneously corroborates both the validity of the aqueous model and the quantitative interpretation of Eh measurements in these waters. This interpretation makes it possible to calculate the distribution of iron

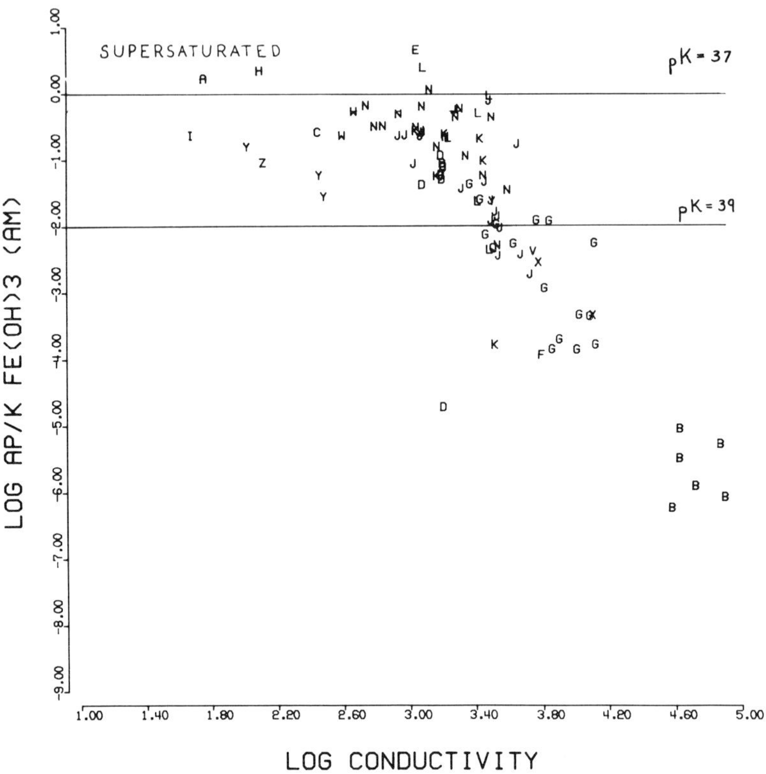

Figure 7. Saturation indices for amorphous ferric hydroxide plotted as a function of the log conductivity showing an approach to saturation for the more dilute and more oxidized downstream waters

species based on Eh measurements when ferrous-ferric analyses are not available. Although the stream waters show equilibrium saturation with dissolved oxygen, there is no equilibrium between the dissolved oxygen and ferrous-ferric redox couple.

Mineral saturation indices for melanterite and amorphous iron hydroxide agree quite well with field occurrences of the same minerals. Jarosite, however, appears to be supersaturated for nearly all of the samples regardless of the presence or absence of the mineral in these streams. Field observations indicate that jarosite precipitation occurs in the microenvironment of bacterial colonies where the chemical conditions may be quite different from the bulk solution. These considerations lead us to suggest that there is a kinetic barrier which hinders jarosite precipitation but does not hinder ferric hydroxide precipitation and that this barrier is overcome by the surfaces of bacterial colonies (both iron-oxidizers and unidentified non-oxidizers).

The chemical equilibria approach used in this study have enabled us to identify which parts of an acid mine drainage system are in equilibrium and which parts are not. Our results have provided greater insight into the chemical processes of acid mine waters in particular and the redox relations of iron in general.

Acknowledgements

The research in this paper was done in partial fulfillment of a Ph.D. dissertation completed by the senior author. Support from the U.S. Geological Survey, especially Bob Averett, Rich Fuller, Jo Burchard, Doris York, Dave Vivit and the staff of the Redding Field Office, are gratefully acknowledged. Critical comments by G. A. Parks, K. B. Krauskopf, J. O. Leckie and B. Hallet are greatly appreciated. We also wish to acknowledge the careful review of this manuscript by L. N. Plummer and D. C. Thorstenson, both of the U.S. Geological Survey. We also thank Janet Batten for typing the manuscript.

Abstract

A detailed hydrogeochemical investigation of 14 km of stream waters contaminated by acid mine drainage in Shasta County, California provides insight into the equilibrium transformations of iron during the oxidative weathering of pyritic ore. Over 60 water analyses covering a range of pH, redox state, total iron concentration, temperature and ionic strength were processed with the computerized chemical model WATEQ2 so that activities of free ions and complexes and satutation indices could be determined. The results demonstrate (a) measured Eh values agree quite well with Eh values calculated from the ferrous/ferric activity ratio, (b) the dominant complexes in acid mine waters are $FeSO_4^o$, $FeSO_4^+$, $Fe(SO_4)_2^-$, $Fe(OH)^{2+}$, $Fe_2(OH)_2^{4+}$ and $Fe(OH)_2^+$ and (c) acid mine waters low in pH (1 to 2) and high in reduced iron approach saturation with respect to melanterite whereas progressive downstream oxidation and dilution (pH = 2 to 3.5) of these waters forces these waters to become saturated with respect to amorphous iron hydroxide. For these two minerals, the field observations match the saturation calculations very well. However, all of the waters show supersaturation with respect to jarosite regardless of the presence or absence of jarosite in the stream waters. These results suggest a strong kinetic hindrance to the precipitation of jarosite.

Literature Cited

1. Lepp, H., ed. "Geochemistry of Iron," Benchmark Papers in Geol. 18, Halstead Press, 464 p. (1975).
2. Stumm, W. and Morgan, J. J., "Aquatic Chemistry: An introduction emphasizing chemical equilibrium in natural waters," 583 p., Wiley-Interscience, New York, 1970.
3. Neilands, J. B., ed. "Microbial Iron Metabolism: A Comprehensive Treatise," 597 p., Academic Press, New York, 1974.
4. James H. L. Chemistry of the iron-rich sedimentary rocks, U. S. Geol. Survey Prof. Paper 440-W, 61 p. (1966).
5. Klemic, H., James, H. L. and Eberlein, G. D. United States mineral resources, U.S. Geol. Survey Prof. Paper 820, 291-306 (1973).
6. Jenne, E. A. Controls on Mn, Fe, Co, Ni, Cu and Zn concentrations in soils and waters: the significant role of hydrous Mn and Fe oxides, p. 337-387, in "Trace Inorganics in Water," Adv. Chem. Ser. 73, 1968.

7. Jenne, E. A. Trace element sorption by sediments and soils--sites and processes, p. 425-533, in "Symposium on Molybdenum in the Environment, M. Dekker, New York, 1977.

8. Krauskopf, K. B. "Introduction to Geochemistry," 721 p. McGraw Hill , New York, 1967.

9. Krauskopf, K. B. Geochemistry of micronutrients, p. 7-40 in "Micronutrients in Agriculture," Soil Sci. Soc. Am., Madison, Wisc. 1972.

10. Berner, R. A. Low temperature geochemistry of iron, section 26, in "Handbook of Geochemistry," vol. II-I, Springer-Verlag, Berlin.

11. Berner, R. A. "Principles of Chemical Sedimentology," 240 p. McGraw-Hill, New York, 1971.

12. Langmuir, D. and Whittemore, D. O. Variations in the stability of precipitated ferric oxyhydroxides, p. 209-234, in Hem, J. D., ed., "Nonequilibrium Systems in Natural Water Chemistry," Adv. Chem. Ser. 106, 1971.

13. Kester, D. R., Byrne, R. H., Jr. and Liang, Y.-J. Redox reactions and solution complexes of iron in marine systems, p. 56-79, in Church, T. M. ed., "Marine Chemistry in the Coastal Environment," Amer. Chem. Soc. Ser. 18, 1975.

14. Paces, T. Chemical characteristics and equilibration in natural water-felsic rock-CO_2 system, Geochim. Cosmochim. Acta 36, 217-240 (1972).

15. Jones, B. F., Kennedy, V. C. and Zellweger, G. W. Comparison of observed and calculated concentrations of dissolved Al and Fe in stream water, Water Resour. Res. 10, 791-793 (1974).

16. Helz, G. R. and Sinex, S. A. Chemical equilibrium in the thermal spring waters of Virginia, Geochim. Cosmochim. Acta 38, 1807-1820 (1974).

17. Gang, M. W. and Langmuir, D. Controls on heavy metals in surface and ground waters affected by coal mine drainage; Clarion River - Redbank Creek watershed, Pennsylvania, p. 39-69, in Fifth Symp. Coal Mine Drainage Res., N.C.A., B.C.R., Louisville, Kentucky, 1974.

18. Bricker, O. P., III and Troup, B. N. Sediment-water exchange in Chesapeake Bay, p. 3-27, in Cronin, L., ed., "Estuarine Research, vol. I, Chemistry, Biology and the Estuarine System," Academic Press, New York, 1975.

19. Nordstrom, D. K. and Jenne, E. A. Fluorite solubility equilibria in selected geothermal waters. Geochim. Cosmochim. Acta 41, 175-188 (1977).
20. Sato, M. Oxidation of sulfide ore bodies. I. Oxidation mechanisms of sulfide minerals at 25°C. Econ. Geol. 55, 1202-1231 (1960).
21. Garrels, R. M. and Thompson, M. E. Oxidation of pyrite in ferric sulfate solution, Am. J. Sci. 258, 57-67 (1960).
22. Smith, E. E. and Shumate, K. S. Sulfide to sulfate reaction mechanism, Fed. Water Qual. Admin. Rept. 14010 FPS 01/70 (1970).
23. Singer, P. C. and Stumm, W. Acid mine drainage: The rate determining step. Science 167, 1121-1123 (1970).
24. Lacey, D. T. and Lawson, F. Kinetics of the liquid-phase oxidation of acid ferrous sulfate by the bacterium Thiobacillus ferrooxidans. Biotech. Bioeng. 12, 29-50 (1970).
25. Nordstrom, D. K., Jenne, E. A. and Averett, R. C. Heavy metal discharges into Shasta Lake and Keswick Reservoirs on the Upper Sacramento River, California: a reconnaissance during low flow. U. S. Geol. Survey Open-File Report 76-49, 25 p. (1977).
26. Nordstrom, D. K. "Hydrogeochemical and Microbiological Factors Affecting the Heavy Metal Chemistry of an Acid Mine Drainage System," Ph.D. Thesis, Stanford University, Stanford, California, 1977.
27. Kennedy, V. C., Jenne, E. A. and Burchard, J. M. Backflushing filters for field processing of water samples prior to trace-element analyses. U. S. Geol. Survey Water Resour. Inv. Open-File Report 76-126, 12 p. (1976).
28. Nordstrom, D. K. Thermochemical redox equilibria of ZoBell's solution. Geochim. Cosmochim. Acta 41, 1835-1841 (1977).
29. Stookey, L. L. Ferrozine -- A new spectophotometric reagent for iron, Anal. Chem. 42, 779-781 (1970).
30. Gibbs, C. R. Characterization and application of Ferrozine iron reagent as a ferrous iron indicator. Anal. Chem. 48, 1197-1200 (1976).
31. Nelson, M. B., Nordstrom, D. K. and Leckie, J. O. Modifications of the ferrozine method for ferrous iron in the presence of ferric with

applications to natural waters, (submittted for publication).

32. Ball, J. W., Jenne, E. A., and Nordstrom, D. K. WATEQ2--a computerized chemical model for trace and major element speciation and mineral equilibria of natural waters, in Jenne, E. A., ed., "Chemical Modeling in Aqueous Systems--Speciation, Sorption, Solubility, and Kinetics," Amer. Chem. Soc., 1978 (This volume).
33. Thompson, J. B., Jr. Local equilibrium in metasomatic processes, p. 427-457, in Abelson, P. H., ed., "Researches in Geochemistry." John Wiley, New York, 1959.
34. Onsager, L. Reciprocal relations in irreversible processes. I. Phys. Rev. 37, 405-426 (1931).
35. Helgeson, H. C., Brown, T. H., Nigrini, A.. and Jones, T. A. Calculations of mass transfer in geochemical processes involving aqueous solutions. Geochim. Cosmochim. Acta 34, 569-592 (1970).
36. Garrels, R. M. and Christ, C. L. "Solutions, Minerals, and Equilibria." 450 p. Harper and Row, New York, 1965.
37. Thorstenson, D. C. Equilibrium distribution of small organic molecules in natural waters. Geochim. Cosmochim. Acta 34, 745-770 (1970).
38. Lamm, C. G. Reduction-oxidation level of soils. Nature 177, 620-621 (1956).
39. Breck, W. G. Redox potentials by equilibration. J. Mar. Res. 30, 121-139 (1972).
40. ZoBell, C. E. Studies on redox potential of marine sediments. Bull. Am. Assoc. Petrol. Geol. 30, 477-509 (1946).
41. Baas-Becking, L. G. M., Kaplan, I. R. and Moore, D. Limits of natural environment in terms of pH and oxidation-reduction potentials. J. Geol. 68, 243-284 (1960).
42. Stumm, W. Redox potential as an environmental parameter; conceptual significance and operational limitation, p. 1-16, in 3rd Intern. Conf. on Water Pollut. Res., Proc. Munich, 1966.
43. Morris, J. C. and Stumm, W. Redox equilibria and measurements of potentials in the aquatic environment, p. 270-285, in "Equilibrium Concepts in Natural Water Systems." Adv. Chem. Ser. 67, 1967.

44. Whitfield, M. Thermodynamic limitations on the use of the platinum electrode in Eh measurements. Limnol. Oceanogr. 19, 857-865 (1974).
45. Doyle, R. W. The origin of the ferrous ion-ferric oxide Nernst potential in environments containing dissolved ferrous iron. Am. J. Sci. 266, 840-859 (1968).
46. Berner, R. A. Electrode studies in hydrogen sulfide in marine sediments. Geochim. Cosmochim. Acta 27, 563-575, (1963).
47. Kryukov, P. A., Zadodnov, S. S. and Goremykin, F. E. Sulphide and carbonate equilibrium and the oxidation-reduction state of sulfur in the mineral regions of the Crimean mineral waters. Dokl. Akad. Nauk. SSSR 142(1) (1962).
48. Skopintsev, B. A., Fomenskaya, N. N. and Smirnov, E. V. New determinations of the oxidaton-reduction potential in Black Sea water. Oceanology (USSR) 6, 653-659 (1969).
49. Whitfield, M. Eh as an operational parameter in estuarine studies. Limnol. Oceanogr. 14, 547-558 (1969).
50. Natarajan, K. A. and Iwasaki, I. Eh measurements in hydro-metallurgical systems. Minerals Sci. Eng. 6, 35-44 (1974).
51. Bricker, O. P. Some stability relations in the system $Mn-O_2-H_2O$ at 25°C and one atmosphere total pressure. Am. Mineral. 50, 1296-1354 (1965).
52. Boulegue, J. Equilibria in a sulfide rich water from Enghien-les-Bains, France. Geochim. Cosmochim. Acta 4, 1751-1758 (1977).
53. Ball, J. W., Jenne, E. A. and Burchard, J. M. Sampling and preservation techniques for waters in geysers and hot springs, p. 219-234, Proc. First Workshop of Sampling Geothermal Effluents, October 20-21, 1975, Las Vegas, Nevada, EPA-600/9-76-0111, 1975.
54. Sykes, K. W. The structure and reactivity of the complexes of ferric ions with some simple anions. Chem. Soc. London Spec. Pub. 1, 64-70 (1957).
55. Sapieszko, R. S., Patel, R. C. and Matijevic, E. Ferric hydrous oxide sols. 2. Thermodynamics of aqueous hydroxo and sulfato ferric complexes. J. Phys. Chem. 81, 1061-1068 (1977).

56. Helgeson, H. C., Brown, T. H. and Leeper, R. H., "Handbook of Theoretical Activity Diagrams Depicting Chemical Equilibria in Geologic Systems Involving an Aqueous Phase at one Atm. and 0-300°C." 253 p. Freeman, Cooper and Company, San Francisco, 1969.

Disclaimer: The reviews expressed and/ or the products mentioned in this article represent the opinions of the author(s) only and do not necessarily represent the opinions of the U.S. Geological Survey.

RECEIVED November 16, 1978.

Influence of Redox Environments on the Mobility of Arsenic in Ground Water

J. GULENS and D. R. CHAMP
Atomic Energy of Canada Ltd., Chalk River Nuclear Laboratories, Chalk River, Ontario, Canada K0J 1J0

R. E. JACKSON
Inland Waters Directorate, Fisheries and Environment Canada, Ottawa, Ontario, Canada K1A 0E7

The concept of a sequence of redox reactions in ground water flow systems has been developed (1). This concept is based on a modified version of the theory presented by Stumm and Morgan (2, p. 326-339), and uses the flow system as the hydrogeological framework on which a thermodynamically based sequence of redox reactions is superimposed. In a confined aquifer containing excess dissolved organic carbon (DOC) and some solid-phase Fe(III) and Mn(IV) minerals, redox reactions can occur between the DOC and the oxidized species present in the ground water. As the water flows from recharge to discharge within this confined aquifer, the oxidized species will be reduced in the following sequence: dissolved oxygen, nitrate, (solid) manganese oxides, (solid) ferric hydroxides, sulfate, dissolved carbon dioxide and finally dissolved nitrogen (1).

As a consequence of this concept of redox sequences, it can be concluded (1) that three sequential zones or environments may exist in confined aquifer systems: an oxidizing zone (in the recharge area), a "neutral" zone (in the transition area), and a reducing zone (in the discharge area), Figure 1. The mobility and concentration of multivalent transition metals and nonmetals varies in each of these zones, a principle that can be usefully applied to the abatement and prevention of ground water pollution. As an example, Fe is immobilized as the Fe(III) hydrous oxide in the oxidizing zone; further downstream in a less oxidizing environment, the "neutral" zone, Fe(II) is stable and as it is also more soluble than the Fe(III) hydrous oxide, the latter becomes reduced and mobilized as Fe(II). In the reducing zone, sulfate reducing bacteria may exist and Fe may again be immobilized as the insoluble sulfide.

Hydrous oxide surfaces of sand immobilize As by adsorption processes (3). The results of our studies show that the extent of adsorption varies with the oxidation state of the As, the redox environment and/or the pH of the eluting water. The influence of these parameters on the mobility of As was studied by eluting As through sand columns: waters of different redox

0-8412-0479-9/79/47-093-081$05.00/0
© 1979 American Chemical Society

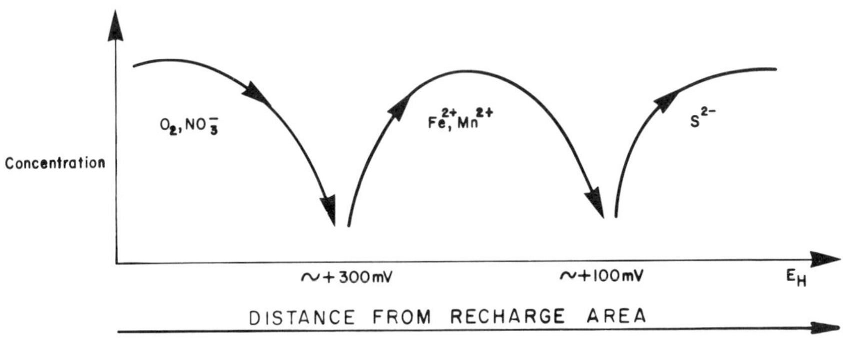

Journal of Earth Sciences

Figure 1. Sequential redox zones within a confined aquifer: oxidizing (O_2, NO_3^-); "neutral" (Fe^{2+}, Mn^{2+}); reducing (S^{2-})

characteristics wereused for elution, and the elution behaviour of As(V) was compared to that of As(III).

Studies in buffered solutions showed that an Fe-As complex was formed, the solubility of which was also dependent on the oxidation state of the As and the solution pH.

Experimental

Elution Studies. The elution profiles of As(V) and As(III) through sand columns were obtained using radioactive arsenic. ^{76}As was prepared by irradiating reagent grade solid As_2O_3 and As_2O_5 in the Chalk River NRU reactor. The irradiated solids were dissolved and loaded immediately. ^{74}As was obtained from Amersham/Searle Corp. as a solution of arsenic acid in 0.04 M HCl; a portion of this solution was converted to As(III) by heating in the presence of HSO_3^- (4) (estimated conversion was 70%).

The elution studies were conducted in plexiglas columns (2.5 cm I.D. and 28 cm long) packed with sand (60-230 mesh). The Fe and Mn content of the sand was 0.6% and 0.01% (w/w) respectively, determined by atomic absorption spectrometry following an HCl extraction. The columns were hand packed with the sand, using techniques to minimize density and size segregation of the sands (5) and thereby minimizing flow inhomogeneities (6), and then pre-equilibrated by flushing with the eluate for 3 days prior to loading. An aliquot of radioactive arsenic (^{76}As or ^{74}As) was loaded onto the base of the column and eluted with upward flow. Eluant fractions were collected in an automated fraction collector and analyzed for As by γ-ray spectrometry. The As(V) and As(III) elution profiles were obtained from separate columns, eluted in parallel. The results have been corrected for the decay rate (^{76}As, $t_{\frac{1}{2}}$ = 26.4 h, ^{74}As, $t_{\frac{1}{2}}$ = 17.7 days).

The columns were eluted at 0.5 mL min^{-1} both in laboratory and field studies. Elution studies in the laboratory were conducted at room temperature 298 K, using air-saturated distilled water, pH 5.7. Elutions in the field were carried out at ambient temperatures ranging from 278 K to 298 K, using ground water at a temperature of 279 K. Ground water was pumped continuously from piezometers by a Masterflex peristaltic pump at 50 mL min^{-1} into a 500 mL plexiglas sampling cell where E_H and pH values were measured; water was then pumped from this cell into the columns. The measured potential of a platinum electrode, E_H, was determined relative to a saturated calomel reference electrode, but values are quoted relative to a normal hydrogen electrode.

Ground Water Geochemistry. Ground water from the lower aquifer in the lower Perch Lake Basin at the Chalk River Nuclear Laboratories, Figure 2, was used for field elution studies to provide water of different redox characteristics. Water from piezometer "O" in the transition area, and from KNEW in the discharge area was used as being representative of "neutral" and

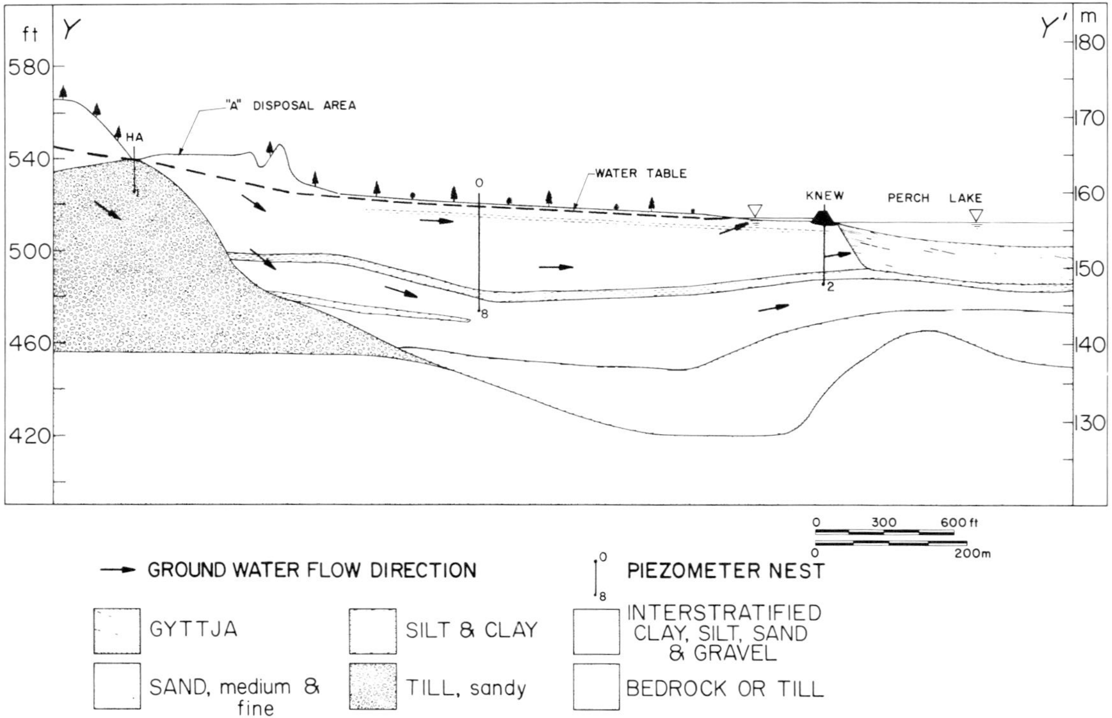

Journal of Earth Sciences

Figure 2. The lower Perch Lake basin, Chalk River Nuclear Laboratories, showing the location of piezometers HA, "O," and KNEW. The screens of these piezometers are 60 cm long and are located at the bottom of the piezometer, adjacent to the numbers on the figure.

reducing environments respectively. Piezometer HA in the recharge area represents an oxidizing environment but inaccessibility to electricity precluded its use in the present situation. Thus air-saturated distilled water was used in the laboratory elution studies to simulate an oxidizing environment (it is readily admitted that the absence of DOC and major and minor elements from the distilled water may alter the elution profile). The geochemical parameters characterizing the ground water at these locations are presented in Table I. It should be noted that at piezometer "O", while the dissolved Fe concentration is low and is entirely due to Fe(II) (polarographic determination), the ground water here also contains large amounts of suspended hydrated ferric oxide.

Solution Studies. Reactions between Fe and As in solution were also studied. An amount of $FeCl_3$ solution was added to a buffered solution containing As so that the total arsenic and iron concentrations were equal at 5×10^{-3} M. The solutions were stirred for 16-24 hours, allowed to settle and aliquots of the supernatant analyzed polarographically for Fe and As, using a Princeton Applied Research Model 174 Polarograph in the sampled DC mode (2 mV s^{-1} scan rate, ΔE = 50 mV, 1 second drop time). A 0.2 M sodium oxalate (pH 4) solution was used to determine Fe(II) and Fe(III) (7), while As(V) was determined in a 2 M perchloric acid - 0.5 M pyrogallol solution (8) and As(III) in a 2 M sodium hydroxide - 0.2 M sodium tartrate solution (7).

For heterogeneous studies, As(III) was adsorbed onto silica gel impregnated with ferric hydroxide, according to the batch method of Yoshida et al. (9). The initial supernatant solution, and the supernatant solution resulting from an acid rinse (1 M HCl) of the silica gel were analyzed polarographically for As and Fe.

Results and Discussion

Column Elutions. The elution behaviour of As(III) is significantly different from that of As(V), in terms of both its time of initial appearance and the quantity of As eluted. These parameters vary for each species with the redox characteristics of the water used, Figure 3. In an oxidizing environment, As(III) is detected in the column eluate 5-6 times sooner than As(V), and the amount of As(III) eluted ($\sim$60% of loading) is $\sim$8 times larger than that of As(V). In the "neutral" environment, the relative amounts of both species eluted are unchanged; however, the As(V) moves through the column much more rapidly than before but still is retarded with respect to the As(III). In the reducing zone, the mobility of As(V) is accelerated: both species now appear in the column effluent after less than one column volume is displaced. Both species are also eluted almost quantitatively ($\sim$100% for As(III), $\sim$80% for As(V)).

TABLE I

Ground Water Geochemical Data[a]

	HA	"O"	KNEW
pH	5.4	6.9	8.3
E_H	580	140	75
DO	2.4	<0.2	<0.2
NO_3^-	3.5	0.4	0.5
SO_4^{2-}	14.0	17.3	5.7
S^{2-}_{TOT}[b]	n.d.[c]	0.02	0.06
Fe_{TOT}	n.d.	6.2	0.2
Mn_{TOT}	0.03	0.05	0.06
Ca^{2+}	5.2	4.3	7.1
Mg^{2+}	2.0	6.9	5.5
K^+	1.8	3.2	1.4
Na^+	91.3	11.9	3.5
Cl^-	170	22.5	<1.0
DIC[d]	10.4	11.2	17.4
DOC[e]	2.0	0.8	1.8

[a]All data in mg/L, except E_H (mV)

[b]The subscript TOT indicates sum total of species

[c]n.d. = not detectable

[d]DIC = dissolved inorganic carbon

[e]DOC = dissolved organic carbon

Journal of Earth Sciences

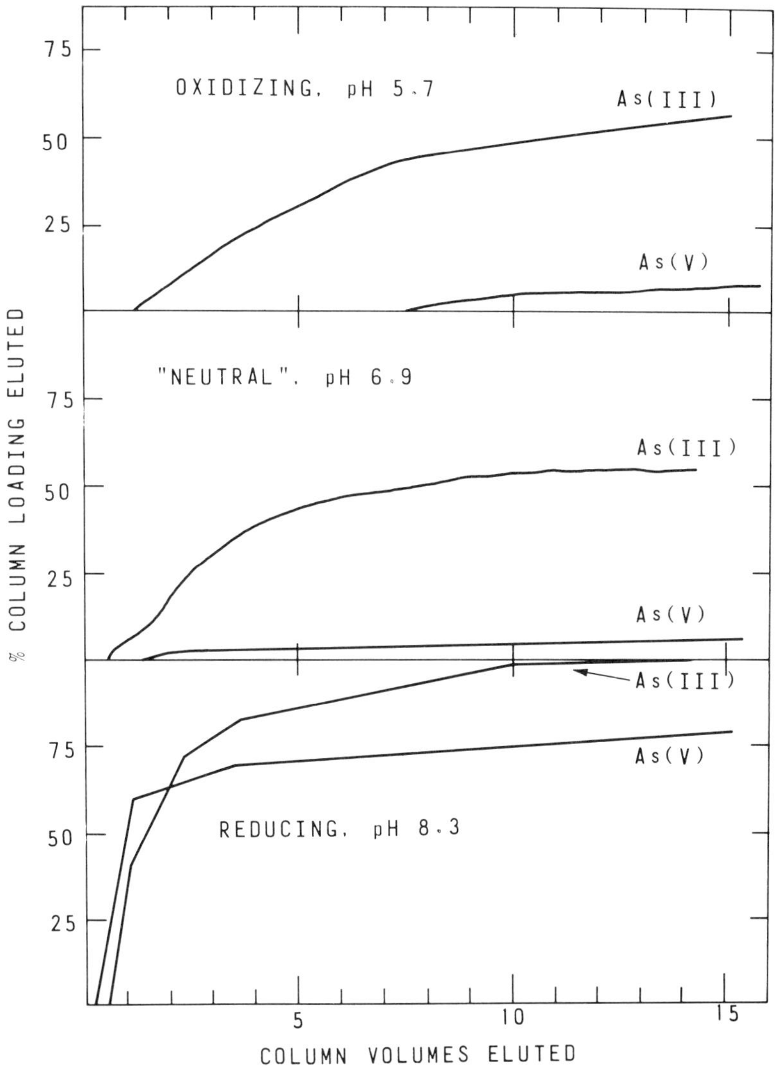

Figure 3. Elution profiles of 180 μg of ^{76}As(V) *and* ^{76}As(III) *on fine sands as a function of the redox characteristics of the water used for elution*

The adsorption and retention characteristics of arsenic are also influenced by the amount of As loaded on the column and by the nature of the column materials. Much stronger retention of both arsenic species is observed, Figure 4, when the amount loaded is decreased from 180 μg to 0.01 μg, but the influence of the initial oxidation state of the arsenic and of the redox environment on mobility are still evident. The change in mobility of both arsenic species as a result of the decrease in the amount loaded reflects the limited adsorption capacity of these sands. The influence of the column material on mobility is illustrated in Figure 5: increased adsorption of both As(III) and As(V) species is observed when medium-grained sands (5-35 mesh size and with Fe and Mn content 0.8% and 0.013% (w/w) respectively) are used as compared to the fine-grained sands (both sands were loaded with 180 μg As).

pH Effects on Adsorption. Yoshida *et al.* (9) have studied the equilibrium adsorption of As(V) and As(III) ions on silica gel impregnated with ferric hydroxide as a function of solution pH. These workers observed that both arsenic ions had not only the same adsorption-pH profile but also the same value of the adsorption coefficient (at a given pH); on this basis, it would be expected that the elution profiles for As(III) and As(V) from sand columns would be very similar. A later, more recent publication by the same authors (10) dealt with the selective separation of As(III) and As(V) ions using the ferric complex of the chelating ion-exchange resin, Uniselec UR-10 (whose functional group is a quadridentate O-hydroxybenzyl-nitrilodiacetic acid). In this latter study, the maximum adsorption for As(V) was observed at pH 5.5, while for As(III) it was at pH 9.2; the reasons for the differences in the pH-adsorption profiles were not discussed.

Frost and Griffin (11) studied the effect of pH on the adsorption of As by kaolinite and montmorillonite and observed that the maximum adsorption of As(V) on both clays occurred at pH 5, while the adsorption of As(III) increased with increasing pH. At pH 5, both clays removed more As(V) than As(III), while at pH 8 more As(III) was removed than As(V) (the authors presented the adsorption isotherm at pH 5; the isotherm at pH 8 can be constructed from the data presented).

Gupta and Chen (12), observed that As(V) was effectively adsorbed in the pH range 4 to 7 by activated alumina and bauxite, but that adsorption decreased sharply above pH 7; activated carbon had maximum adsorption for As(V) in the pH range 3 to 5, with sharply decreased adsorption at both lower and higher pH values. Only slight variations in As(III) adsorption were observed over the pH range 4 to 9 on activated alumina and bauxite, but adsorption decreased markedly above pH 9.

Other Controls on Mobility. The mechanism of adsorption of As onto hydrated ferric oxide surfaces has not yet been elucidated.

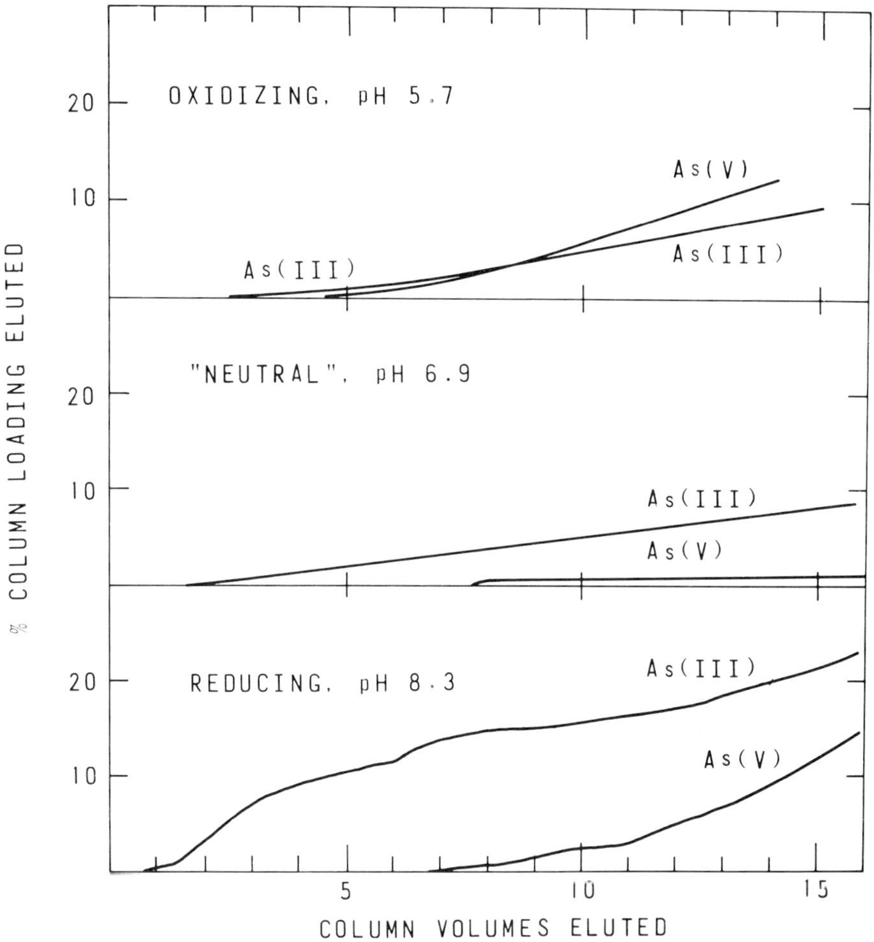

Figure 4. Elution profiles of 0.01 μg of ⁷⁴As(V) and ⁷⁴As(III) on fine sands as a function of the redox characteristics of the water used for elution. Vertical exaggeration, ~ 7.3 ×.

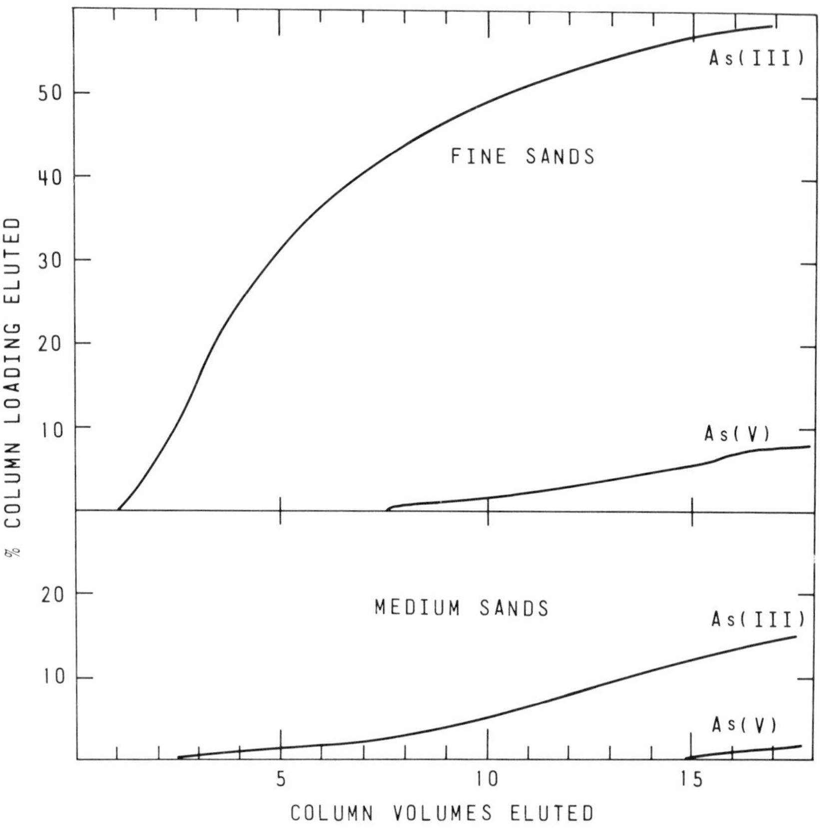

Figure 5. Elution behavior under oxidizing conditions of 180 μg of $^{76}As(V)$ and $^{76}As(III)$ on fine sands vs. medium sands

However, the nature of the adsorbent affects the adsorption-pH profiles (9,10,11,12). Although the difference in the Fe content of the sand fractions used in this study is small, the large change in the mobility of both As(III) and As(V) between the fine and medium fractions may be due to the difference in the nature of the sand i.e. due to a higher amount of discrete iron oxides in the coarse fraction (13).

Arsenic mobility is also affected by the amount of As loaded onto the columns, as is evident from this study, Figure 4, and from the equilibrium adsorption studies of Gupta and Chen (12). These workers observed that the percentage of As(III) removed was greatly affected by the initial As(III) concentration, while for As(V) the removal was only slightly affected.

pH versus Redox Control on As Mobility. The strong retention of As(V) observed in the oxidizing environment, Figure 3, is attributed to its adsorption by the hydrated ferric oxide in the sand (3). The increase in mobility of As(V) as more reducing ground waters are used may be due to the increase in pH of the water, or to the reduction on the column of not only Fe(III) to Fe(II) but also As(V) to As(III). The concentration of sulfide in the discharge zone apparently is not high enough to immobilize the arsenic as a sulfide precipitate.

The weak retention by the sand of As(III), particularly in the oxidizing environment, indicates that the interaction between As(III) and Fe(III), the hydrated ferric oxide, is much weaker than that between As(V) and Fe(III). On the basis of other studies (10,11,12) increased adsorption of As(III) would be expected as the pH of the eluting water is increased (as more reducing ground waters are used). However, decreased adsorption of As(III) is observed, and this is attributed to the reduction of Fe(III) to Fe(II) on the column.

On the basis of this study, it is not possible to decide whether pH or the redox environment control the mobility; each has an important role. Nonetheless, a marked difference is observed in the mobility between As(V) and As(III), and the relative concentration of each species is governed primarily by the redox environment.

Solution Studies

Redox Studies. A possible explanation for the similar values of the adsorption coefficient and the remarkable similarity in the adsorption-pH profiles for As(V) and As(III) initially reported by Yoshida *et al.* (9), is a redox reaction between As(III) and excess Fe(III) to produce As(V) and Fe(II):

$$3Fe(OH)_2^+ + HAsO_2 + H^+ \rightleftharpoons 2Fe^{2+} + FeAsO_4 + 4H_2O$$

This reaction is thermodynamically favoured ($\Delta G°rxn = -39.2$ kcal mol^{-1}). Redox reactions of this kind, occurring at the surface of hydrous iron oxide and affecting the concentration of certain metal ions, have recently been discussed by Hem (14), although he has not considered this particular reaction.

A redox reaction between Fe(III) and As(III) has not been observed, despite its thermodynamic favourability. In a heterogeneous study, As(III) was reacted with a slurry of silica gel impregnated with ferric hydroxide. Polarographic analysis showed that while As(III) was removed from solution, Fe(III) was released into solution; however, As(V) and Fe(II) were not produced. A redox reaction was also not observed when solutions of Fe(III) (or freshly precipitated $Fe(OH)_3$) were mixed with a solution of As(III) at pH 4 or pH 9. However at pH 4 interactions between Fe(III) and As(III) were observed that were significantly different from those between Fe(III) and As(V).

Fe-As Complexes. When a solution of $FeCl_3$ was added to an acetate buffer solution (pH 4.3, [HOAc] + [NaOAc] = 0.5 M) containing As(V) (the amount of As present being equal to the amount of Fe added), a pale yellow-brown precipitate formed immediately. Analysis of the resultant supernatant solution showed that 97% of the Fe and 70% of the As had been removed from solution. Increasing the Fe:As ratio to three resulted in quantitative removal of the As from solution while the amount of Fe removed was approximately 85% of that added. If, however, the buffer solution initially contained As(III) instead of As(V), a precipitate was not visible when the $FeCl_3$ solution was added, even when the Fe:As ratio was increased to three. Less than 10% of the Fe or As(III) were removed from either of these solutions.

These solution reactions were also studied in an ammoniacal buffer solution (pH 8-9, $[NH_3] + [NH_4^+] = 0.1$ M), to determine the effect of increased pH. When a solution of $FeCl_3$ was added to the ammoniacal buffer solution, $Fe(OH)_3$ was precipitated quantitatively. However, if the buffer solution initially also contained As(V) or As(III) (the amount of As present being equal to the amount of Fe to be added), only a small amount of precipitate formed on adding the $FeCl_3$ solution; the concentration of both the Fe and As in the supernatant solution had decreased by only 15-20%. Different results were obtained if the order of addition was reversed (the $FeCl_3$ was first quantitatively precipitated from the buffer solution as $Fe(OH)_3$, and then the As(III) or As(V) solutions were added and allowed to mix with the $Fe(OH)_3$ precipitate): the concentration of As in the resultant supernatant solution had decreased by approximately 35%, but the concentration of Fe had increased by approximately the same amount.

These results demonstrate that an Fe-As complex is formed in solution, and that the solubility of this complex depends both on the As oxidation state and on the solution pH. Since very little

Fe (or As) is lost from an ammoniacal buffer solution containing As(V) or As(III) when $FeCl_3$ is added to this solution, the Fe(III) must be stabilized in solution by the formation of an Fe-As complex. The solubility of the complex formed between As(V) and Fe(III) decreases with decreasing pH, while the solubility of the complex formed between As(III) and Fe(III) decreases with increasing pH. In addition, both As(V) and As(III) interact heterogeneously with precipitated $Fe(OH)_3$: some As is removed from solution but some Fe also redissolves into solution.

The fact that the solubility of the As(III)-Fe(III) complex is higher than that of the As(V)-Fe(III) complex is consistent with the results from the column elution studies, where the mobility of As(III) was greater than that of As(V). These facts are also in agreement with the results of Shen (15), who studied the removal of arsenic from drinking water and observed improved arsenic removal if the water was oxidized with chlorine prior to coagulation with $FeCl_3$, as compared to coagulation alone. The water studied was either well-water (with no dissolved oxygen but containing sulfide and ammonia) or synthetic water with added sodium arsenite. The improved removal of arsenic is due to its oxidation from As(III) to As(V), a conclusion also arrived at by Gupta and Chen (12).

Summary

The mobility of As(III) through sand columns is significantly different from that of As(V). The mobility of both species varies with the redox characteristics of the ground water used for elution, and this change in mobility may be due to oxidation-reduction reactions occurring on the column and/or due to changes in the pH of the ground water. Both As(V) and As(III) form a complex with Fe(III) in solution, with the Fe(III)-As(III) complex being more soluble than the Fe(III)-As(V) complex.

This study has not considered the effects of organics on the mobility of arsenic; they may play a role in determining the arsenic mobility in ground water or surface waters (3).

When dealing with problems of disposing As-containing wastes or removing As from ground water, the oxidation state of the As in the sample is of major importance in determining methods of immobilizing and/or removing the As. If the arsenic wastes are to be disposed to the subsurface, the redox characteristics of the infiltrating ground water must also be determined and considered; a geochemical barrier (or sequence of barriers) capable of altering the redox characteristics of the infiltrating ground water may be required to ensure that the As remains immobilized.

Subsurface storage of radioactive wastes in geologic formations such as plutons or salt deposits is being considered (16); the concepts presented here are also applicable to this waste management problem.

Acknowledgement

We gratefully acknowledge J. Young and K. Inch for technical assistance, and J.F. Pickens for helpful discussions. We particularly thank W.F. Merritt for his invaluable assistance.

Table I and Figures 1 and 2 are reproduced by permission of the National Research Council of Canada from the Canadian Journal of Earth Sciences, 1978, In Press.

Abstract

The influence of redox environments and/or pH on the mobility of arsenic through sand columns was studied by using waters of different redox characteristics for elution and by comparing the elution profiles of As(V) to that of As(III). The mobility of arsenic was affected by each of the above parameters, by the amount of arsenic loads onto the columns and the nature of the column materials. Solution studies have shown that both As(III) and As(V) from complexes with Fe(III); the solubility of the complexes being dependent upon the oxidation state of the arsenic and the solution pH.

Literature Cited

1. Champ, D.R., Gulens, J. and Jackson, R.E., Oxidation-reduction sequences in ground water flow systems, Can. J. Earth Sci. (in press).
2. Stumm, W. and Morgan, J.J., p. 326-339, "Aquatic Chemistry," Wiley-Interscience, New York, 1970.
3. Ferguson, J.F. and Davis, J., A review of the arsenic cycle in natural waters, Water Res. 6, 1259-1274 (1972).
4. Henry, F.T., Kirch, T.O. and Thorpe, T.M., Speciation of trace level arsenic (III) and arsenic (V) by differential pulse polarography, Abstract No. 180, Pittsburgh Conference on Analytical Chemistry and Applied Spectroscopy, 1978.
5. Gray, W.A., "The Packing of Solid Particles," 128 p., Chapman and Hall Ltd., London, 1968.
6. Ripple, C.D., James, R.V. and Rubin, J., Packing-induced radial particle-size segregation: influence on hydrodynamic dispersion and water transfer measurements, Soil Sci. Soc. Am. Proc. 38, 219-222 (1974).
7. Meites, L., p. 654-662, in "Polarographic Techniques" 2nd Ed., Wiley Interscience, New York, 1965.
8. White, M.C. and Bard, A.J., Polarography of metal-pyrogallol complexes, Anal. Chem. 38, 61-63 (1966).
9. Yoshida, I., Kobayashi, H. and Ueno, K., Selective adsorption of arsenic ions on silica gel impregnated with ferric hydroxide, Anal. Lett. 9, 1125-1133 (1976).
10. Yoshida, I., Ueno, K. and Kobayashi, H., Selective separation of arsenic (III) and (V) ions with ferric complex of chelating ion-exchange resin, Sep. Sci. 13, 173-184 (1978).

11. Frost, R.R. and Griffin, R.A., Effect of pH on adsorption of arsenic and selenium from landfill leachate by clay minerals, Soil Sci. Soc. Am. J. 41, 53-57 (1977).
12. Gupta, S.K. and Chen, K.Y., Arsenic removal by adsorption, J. Water Poll. Control Fed. 50, 493-506 (1978).
13. Jenne, E.A., Controls on Mn, Fe, Co, Ni, Cu, and Zn concentrations in soils and water: the significant role of hydrous Mn and Fe oxides, p. 337-387, in "Trace Inorganics in Water," Adv. Chem. Ser., 73, American Chemical Society, Washington, D. C. 1968.
14. Hem, J.D., Reactions of metal ions at surfaces of hydrous iron oxide, Geochim. Cosmochim. Acta 41, 527-538 (1977).
15. Shen, Y.S., Study of arsenic removal from drinking water, J. Am. Water Works Assoc. 65, 543-547 (1973).
16. Tammemagi, Y.T., Geological disposal of radioactive wastes - the Canadian development program, A. Energy Can. Ltd., AECL Rep., AECL-5392 20 p. (1976).

RECEIVED November 16, 1978.

Organic Ligands

5

Solution Thermochemistry of Humic Substances

Acid–Base Equilibria of River Water Humic Substances[1]

E. M. PERDUE

Environmental Science/Chemistry Department, Portland State University, Portland, OR 97207

Current concern over the potential ecological impact of trace metals in the aquatic environment has led to extensive research in two main areas: 1) the coordination chemistry of trace metals in natural waters and 2) the role of trace metals in the growth of aquatic organisms. Recent studies have shown that natural waters possess some capacity to complex metals and that this "complexation capacity" can strongly influence the biological activity of a metal (1, 2, 3). Due to the presence of a diverse array of inorganic and organic ligands in natural waters, a very large number of metal-ligand complexes could conceivably contribute to the "complexation capacity" of natural waters. The availability of sophisticated computer programs such as MINEQL (4) has made it possible to calculate the distribution of free and complexed trace metals in well-defined media (i.e., aqueous solutions in which the analytical concentrations of all trace metals and complexing ligands are known and where stability constants for all possible metal-ligand complexes are available). While stability constants have been compiled for a large number of trace metal complexes with inorganic ligands and with simple synthetic organic ligands such as NTA and EDTA (5), no such data are available for interactions between trace metals and naturally occurring organic ligands. Until these data are obtained, equilibrium model calculations will at best provide an approximate description of the distribution of trace metals in natural waters.

The composition of naturally occurring organic matter in river water is determined mainly by the input of allochthonous organic matter (6) which includes biopolymers (proteins, carbohydrates, and lignins) and geopolymers (humic substances). Since the chemical character of the geopolymers requires specialized, less common enzyme systems for breakdown, microbial attack introduces a strong bias favoring digestion of the biopolymers. Consequently, humic substances usually constitute the major fraction of organic matter in river water (7-13). Humic substances are generally considered to be composed of three operationally distinct fractions: 1) fulvic acid, which is soluble in both

[1] Part II of a series.

0-8412-0479-9/79/47-093-099$05.00/0
© 1979 American Chemical Society

acidic and basic solutions, 2) humic acid, which is soluble in basic solution but insoluble in acidic solutions, and 3) humin, which is insoluble in both acidic and basic solutions. The bulk of river water humic substances generally resembles the more readily solubilized fulvic acid, with the relative amount of the less soluble humic acid probably being dependent on the pH of the natural water (10).

The actual structural features of humic substances in natural waters remain unresolved. It is generally accepted that humic substances are complex polymers which possess both carboxyl and phenolic hydroxyl acidic functional groups. These functional groups are thought to be involved in metal complexation reactions since protons are released during such reactions (for review, see Schnitzer and Khan, (14); Flaig et al., (15). The extent of complexation of a metal will thus depend not only on the stability constant of the metal-ligand complex but also on the acid dissociation constant of the acidic functional group which serves as the complexing ligand. Consequently, many efforts have been made to determine both the concentrations and chemical characteristics of the acidic functional groups of humic substances (for review, see Reuter and Perdue, (10)).

In general, while the potentiometric titration methods used to characterize acidic functional groups in humic substances have been useful in determining the concentrations of carboxyl and phenolic hydroxyl groups, these methods have not been very useful in determining the pK_a values associated with each type of acidic functional group. Some of the more successful studies have utilized various mathematical techniques to try to take into account the effect of accumulating negative charge on the acid dissociation constants of progressively weaker acidic functional groups (16, 17, 18). While direct titrations utilizing Gran functions for endpoint location are sometimes used (16, 17), the concentrations of carboxyl and phenolic hydroxyl groups in humic substances are most often determined using the calcium acetate exchange reaction and barium hydroxide total acidity reaction, respectively (19). The specificity of these reactions has, however, been questioned (20, 21). In summary, the need for knowledge of the concentrations and acid dissociation constants of naturally occurring organic ligands has not been adequately met by the use of potentiometric titration methods.

The technique of titration calorimetry has been successfully used to determine the nature and abundances of a variety of acidic functional groups in proteins (22). Several investigators have made rather limited efforts to use titration calorimetry to study humic substances, usually as a method to determine the cation exchange capacity or titratable acidity of humic substances (23, 24). Choppin and Kullberg (25) have recently used titration calorimetry to determine the enthalpies of neutralization of acidic functional groups in humic substances and have combined that data with pH titration data to obtain ΔG, ΔH, and ΔS values for ionization of the acidic functional groups. The usual difficulties in locating potentiometric endpoints were encountered, leading these researchers to simply assume that their humic acid contained equal concentrations of carboxyl and phenolic groups. None of these studies has taken advantage of titration calorimetry as a technique for the determination of concentrations of acidic functional groups and for the simultaneous determination of both ΔH_a and pK_a values for ionization of the acidic functional groups of humic substances. Although pK_a values are generally regarded as being quite characteristic of a particular functional group, it is not widely recognized that ΔH_a values are also quite useful in determining the nature of an acidic functional group. In Figure 1, the relationship between ΔH_a and pK_a for a large number of carboxylic acids, phenols, and ammonium ions is plotted. The location and size of the domain of each functional group was determined by the mean value and standard deviation of ΔH_a and pK_a values computed from available data (26) utilizing data for all tabulated compounds containing only carbon, hydrogen, oxygen, and reduced nitrogen. It is apparent from Figure 1 that an acidic functional group is identified at least as readily by its ΔH_a value as by its pK_a value. Of course, if both pK_a and ΔH_a are known for the ionization of an acidic functional group, the task of identification of the functional group is further simplified. Furthermore, once pK_a and ΔH_a values are known at one temperature (e.g., 25°C) it is possible to calculate pK_a values at other temperatures using standard thermodynamic relationships.

As pointed out by Christensen et al. (27), the shape of a thermometric titration curve (heat evolved vs. moles of titrant) is determined by both the enthalpy change (ΔH) and equilibrium constant (K) for the reaction taking place in the calorimeter. In reactions for which $\log K > 4$ (e.g., the reaction of a carboxylic acid with hydroxide ion for which $\log K \simeq 10$), the equilibrium favors the formation of products so strongly that the added increment of titrant is essentially completely reacted. In this case the amount of heat evolved is determined by the ΔH for the reaction and the amount of added titrant. The resulting thermometric titration curve consists of two straight line segments which intersect sharply at the equivalence point. In such reactions, the K-value cannot be determined from the thermometric titration curve. However, the endpoint of the titration and the ΔH of the reaction are readily obtained. In reactions for which $\log K < 4$ (e.g., the reaction of phenols with hydroxide ion for which $\log K \simeq 4$), a small fraction of an added increment of titrant will remain unreacted at equilibrium. In this case, the amount of heat evolved is determined by the ΔH for the reaction, the amount of added titrant, and the K-value for the reaction (which determines what fraction of added titrant will be reacted at equilibrium). The resulting thermometric titration curve is nonlinear, showing pronounced curvature particularly near the equivalence point. Using a curve-fitting procedure, such nonlinear titration

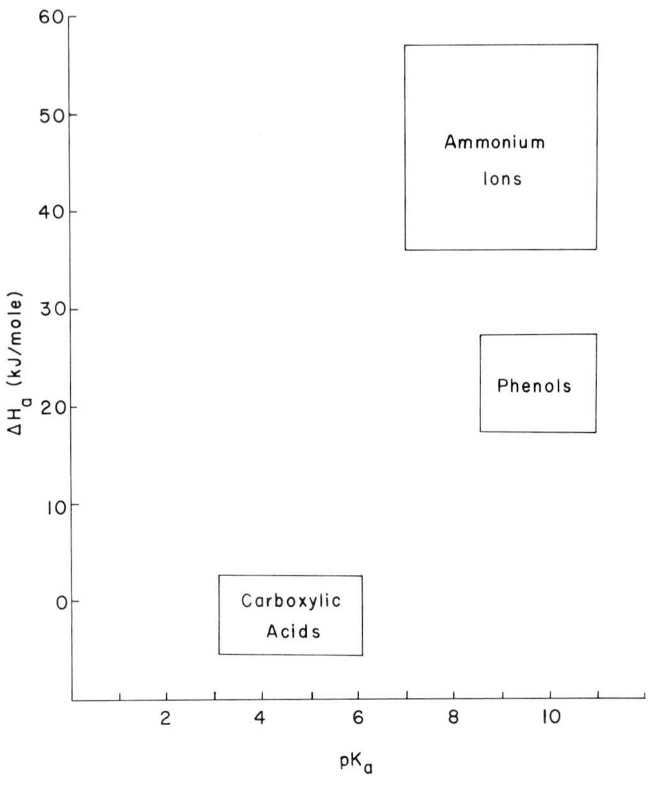

Figure 1. Relationship between the enthalpy of ionization (ΔH_a) and pK_a for major classes of acidic functional groups

curves can be analyzed to obtain the "best" ΔH and K values for the reaction taking place in the calorimeter and an approximate value for the endpoint of the titration.

As part of an overall evaluation of techniques for characterization of the acid-base and metal-ligand reactions of humic substances, the method of titration calorimetry was recently used in this laboratory to characterize the acidic functional groups of humic acid (28). It was possible to determine the concentrations of carboxyl and titratable phenolic hydroxyl groups, their respective average enthalpies of ionization (ΔH_a), and the average pK_a of the titratable phenolic hydroxyl groups. By comparison with simple organic acids, it was shown that the titratable phenolic hydroxyl groups (which constitute about one-third of the phenolic hydroxyl groups in soil humic acid) were probably not ortho to carboxyl groups and thus could not participate in metal complexation via salicylate-like ligands.

Because of the considerable potential of titration calorimetry as an analytical technique for characterization of the acidic functional groups of humic substances, our studies have been extended to river water humic substances. In this paper, results are presented for the thermochemical characterization of the acidic functional groups of river water humic substances from two quite different river systems: 1) the Satilla River in southeastern Georgia, and 2) the Williamson River in southern Oregon.

Experimental Procedure

Isolation and Characterization of River Water Humic Substances. River water humic substances were isolated from the Williamson River in southern Oregon and from the Satilla River in southeastern Georgia using a modification of the procedure described by Mantoura and Riley (29). The Williamson River is a small stream which flows through a semi-arid environment dominated by basalt, volcanic glass, and pumice. The river is a quite dilute (Ca, Mg, Na)-HCO_3 system, deriving most of its dissolved load from rock weathering. Prior to entering Klamath Marsh, the river contains only 1-3 mg/l of total organic carbon (TOC) and is not visibly colored. The highly colored river water flowing out of the marsh contains 20-40 mg/l of TOC and is slightly acidic (pH 6-7). Filtration studies have shown that at least 80 percent of the organic matter will pass through a 0.025 μm membrane filter. Approximately 250 liters of river water was collected, deaerated with prepurified N_2, and transported to the laboratory where the river water was centrifuged in a continuous flow system (rotor speed and flow rate were adjusted to achieve particle size fractionation similar to that obtained by filtration through 0.45 μm membrane filters). The centrifugate was acidified to pH 1.8 with concentrated HCl and passed through a column of pre-washed Rohm and Haas XAD-7 macrorecticular resin.

Spectrometric analysis of the river water before and after passage through the resin (absorbance at 420 nm, sample adjusted to pH 10 with NaOH) indicated that 97-99 percent of the colored organic matter was adsorbed on the resin. The adsorbed humic substances were then eluted from the resin with aqueous triethylamine (0.15 M). The use of an organic base minimizes the likelihood of modification of humic substances in an aerobic alkaline solution. Furthermore, a tertiary amine was used to minimize condensation reactions with carbonyl groups which would increase the nitrogen content of humic substances. Excess organic base is readily removed in a rotary evaporator and any triethylamine hydrochloride can be removed by chloroform extraction. These procedures yield a solution of approximately neutral pH which contains the triethylammonium carboxylate salts of river water humic substances. The triethylammonium ion is removed with a cation exchange resin (H^+-form) using either batch or column procedures. The major loss of humic substances occurs in the desalting step due to the precipitation of the less soluble "humic acid" component of river water humic substances in the acidic solution produced during this step. The amount of precipitated material depends on the amount of "humic acid" originally in the river water and on the concentration of river water humic substances in the solution to be desalted. The overall procedure resulted in a 70 percent recovery of river water humic substances with the following elemental composition: 47.1% carbon, 4.2% hydrogen, 1.8% nitrogen, and 4.2% ash.

The Satilla River in southeastern Georgia has previously been described (9, 12). Briefly, the Satilla River flows through highly weathered, poorly drained, often swampy terrain. The inorganic chemical composition of this very dilute Na-Cl type river indicates that atmospheric precipitation provides most of the dissolved salts in the river. The color and content of humic substances generally increases downstream and averages about 50 mg/l of TOC at the sample site used in this study. The river is quite acidic, with pH ranging from 4.0 to 4.5. The procedures for collection and isolation of river water humic substances from the Satilla River are essentially the same as those used in the Williamson River. The final isolated product had the following elemental composition: 50.1% carbon, 3.5% hydrogen, 0.8% nitrogen, and 0.0% ash.

Total acidity (meq/g) and carboxyl groups (meq/g) were determined by the barium hydroxide method and the calcium acetate exchange method, respectively (14). Phenolic hydroxyl groups (meq/g) were calculated as the difference between total acidity and carboxyl groups. These determinations were carried out under a nitrogen atmosphere to minimize absorption of carbon dioxide.

Titration Calorimetry of River Water Humic Substances. The concentration of river water humic substances used in all experiments was 750 mg/l. While Satilla River humic substances (SR-HS) were readily dissolved in distilled water, Williamson River humic substances (WR-HS), which contain a larger percentage of "humic acid", were not completely soluble in distilled water. As in the previous study on soil humic acid (28), however, the sample could be completely dissolved by shaking with 2 mmol NaOH in 800 ml H_2O followed by addition of 2 mmol HCl and adjustment of the final volume to 1000 ml with H_2O. The WR-HS solutions thus contained 2 mmol of NaCl, while the SR-HS solutions were salt-free.

In order to minimize the heat of dilution of titrant, 4.93 N NaOH was used in all experiments. All calorimetric measurements were made at 25.0 $\pm$ 0.1^{o}C.

The titration calorimetry system used in this research has previously been described (28, 30). Using a discontinuous enthalpimetric titration procedure, the amount of heat evolved is determined for a series of titrant additions. By summation of the individual data points, a composite curve of total heat evolved vs. mmol of titrant can be constructed. Through a previously described curve-fitting procedure (28), the endpoint of the titration, the enthalpy change (ΔH), and the equilibrium constant (K) for the neutralization reaction taking place in the calorimeter can be determined. The ΔH_a and K_a values for ionization of an acidic functional group may be readily calculated by combining the experimental ΔH and K for the neutralization reaction with ΔH_w(+55.8kJ/mole) and K_w(1.00×10^{-14}) for the ionization of water ($\Delta H_a = \Delta H + \Delta H_w$; $K_a = KK_w$). All calculations were carried out on the Harris System 200 Model 220 computer at Portland State University.

Results and Discussion

Concentrations of Acidic Functional Groups in River Water Humic Substances. The results of the total acidity, carboxyl group, and phenolic hydroxyl group determinations for WR-HS and SR-HS are given in Table I. For comparison, the previously reported values for soil humic acid (28) are included. These values are within the range of values usually reported for these functional groups (14, 31).

Thermometric Titration of River Water Humic Substances. The thermometric titration curves of WR-HS and SR-HS are given in Figure 2. In both instances, the initial slope of approximately -56 kJ/mol clearly indicates that carboxyl groups are being titrated (see Figure 1, with $\Delta H_a \simeq -0.2$ kJ/mol) while the non-linear portion of the titration curve indicates that a much weaker acidic functional group (presumably phenolic hydroxyl groups) is being titrated in the latter part of the titration. Through graphical extrapolation of initial and final slopes and through the previously described curve-fitting procedure (28),

Table I. Concentrations of acidic functional groups in river water humic substances and soil humic acid.

Sample[a]	Method	Concentration (meq/g) Total Acidity	Carboxyl Groups	Phenolic Groups
WR-HS	pH Titration[b,c,d]	9.5	5.1	4.4
WR-HS	Calorimetry	n.d.	3.3	1.2
SR-HS	pH Titration[b,c,d]	11.3	6.9	4.4
SR-HS	Calorimetry	n.d.	4.4	0.8
Soil HA	pH Titration[b,c,d]	6.6	4.4	2.2
Soil HA	Calorimetry	n.d.	4.1	0.7

n.d. = not determined

[a]Williamson River humic substances (WR-HS); Satilla River humic substances (SR-HS); soil humic acid (Soil HA).

[b]Total acidity determined by the barium hydroxide method (14).

[c]Carboxyl groups determined by the calcium acetate exchange reaction (14).

[d]Phenolic groups calculated as total acidity less carboxyl groups (14).

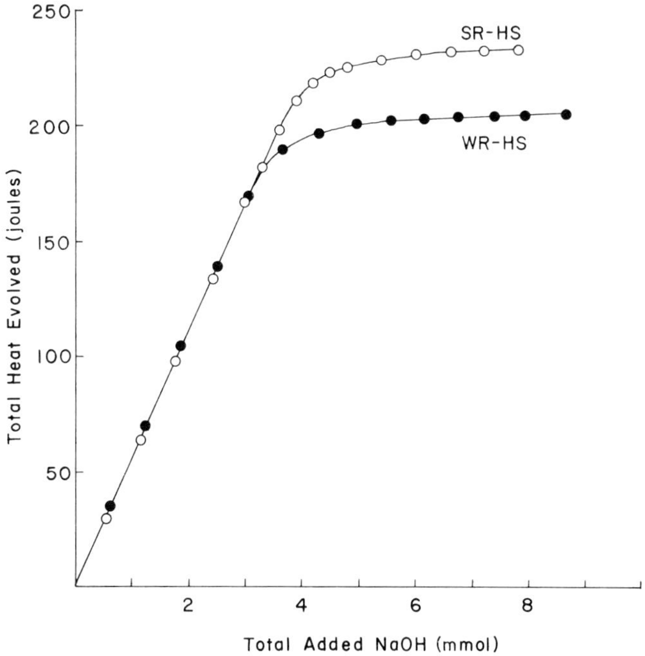

Figure 2. Thermometric titration curves for neutralization of (●) Williamson River humic substances and (○) Satilla River humic substances with NaOH

the concentrations of carboxyl groups and titratable phenolic hydroxyl groups in WR-HS and SR-HS were determined. These results are given in Table I along with the corresponding values which were previously reported for soil humic acid (28). In all three samples, the carboxyl content as determined by titration calorimetry is less than the value obtained by the calcium acetate exchange reaction. The discrepancy is greatest for SR-HS, less for WR-HS, and almost negligible for soil humic acid, clearly decreasing with increasing humic acid content of the sample. These results suggest that the calcium acetate exchange reaction seriously overestimates the carboxyl content of river water humic substances (and possibly soil fulvic acids as well). While most humic acid can be precipitated from $Ca(OAc)_2$ solution as a calcium salt and subsequently removed by filtration, very little river water humic material is precipitated under these conditions. In back-titrating the filtered $Ca(OAc)_2$ solution to pH 9.8, it is quite likely that some phenolic hydroxyl groups in the unprecipitated humic substances would be titrated, resulting in an over-estimation of the carboxyl content of the sample. Since phenolic groups are calculated as the difference between total acidity and carboxyl groups, most estimates of the phenolic content of humic substances are probably too low. Although titration calorimetry data have been obtained on only three samples of humic substances, it appears that humic acids and fulvic acids have similar concentrations of carboxyl groups and that the higher concentrations of carboxyl groups in fulvic acids which are frequently encountered in the literature (14) are possibly artifacts of the calcium acetate exchange reaction.

Dubach et al. (20) recognized the fact that any pK-dependent method of functional group analysis cannot be specific for a particular functional group because of the overlap of pK_a ranges for major classes of acidic functional groups in humic substances. They thus sought methods of analysis were were pK-independent. Using diborane to determine total active hydrogen, these researchers observed good agreement between total active hydrogen (carboxyl, phenolic, and alcoholic groups) and total acidity (probably carboxyl and phenolic groups). Thus, they concluded that humic substances contained essentially no alcoholic hydroxyl groups. It then followed that total hydroxyl groups (phenolic and alcoholic groups) determined by acetylation was equal to phenolic groups. Carboxyl groups were then calculated as the difference between total active hydrogen and total hydroxyl groups. The calculated values were always less than the values obtained by the barium acetate exchange reaction (this reaction is the same as the calcium acetate exchange reaction). Dubach et al. (20) concluded that some phenolic groups were titrated in the barium acetate exchange reaction, resulting in an overestimation of carboxyl groups.

More recently, Stevenson and Goh (21) conducted a careful study of the infrared spectra of humic substances and their methylated derivatives. They reported that "gross inadequacies" existed in conventional methods for determining functional groups in humic substances (the calcium acetate exchange reaction for carboxyl group determinations was specifically mentioned).

In a study of the applicability of the calcium acetate exchange reaction to humic substances, Schnitzer and Gupta (19) found that the calcium acetate reaction gave similar results to the quinoline-copper carbonate decarboxylation reaction. Thus, they recommended that the calcium acetate reaction be used for the determination of carboxyl groups in humic substances. In that same study, several model compounds were also examined. In almost every instance where the model compound contained both carboxyl and phenolic groups, the calcium acetate exchange reaction seriously overestimated (30-150 percent) the carboxyl content of the compound. This error was attributed to the reaction of calcium acetate with phenolic groups. Since hydroxybenzoic acids have been regularly isolated in structural studies of humic substances by numerous researchers (for review, see Schnitzer and Khan (14)), it seems quite reasonable that these structural moieties would introduce errors in the determination of carboxyl groups by the calcium acetate exchange reaction.

The phenolic hydroxyl content obtained by titration calorimetry is not expected to agree with the value obtained potentiometrically, since some of the phenolic groups which react with barium hydroxide in the total acidity determination ($pH > 13$) are too weakly acidic to be titrated with sodium hydroxide under our experimental conditions (max pH 11.5). As was previously demonstrated, phenolic hydroxyl groups which are ortho to carboxyl groups (e.g., o-hydroxybenzoic acid) cannot be titrated under these experimental conditions while more remote phenolic hydroxyl groups (e.g., p-hydroxybenzoic acid) are readily titrated (28). While the concentrations and relative abundances of carboxyl and titratable phenolic groups are similar for all three samples, the total acidities (as determined by the barium hydroxide method) are quite different. Since the total titratable concentration of acidic functional groups obtained by titration calorimetry is less than the total acidity of the sample, river water humic substances appear to contain a significant concentration of weakly acidic phenolic hydroxyl groups (those functional groups which were not titrated in the calorimeter). If these weakly acidic phenolic groups are, in fact, ortho to carboxyl groups, it might be anticipated that river water humic substances, having more chelating sites, would have a larger metal complexation capacity than does soil humic acid.

The average ΔH_a values for ionization of carboxyl groups and titratable phenolic groups and the average pK_a of the titratable phenolic groups of river water humic substances are given in Table II. The previously reported ΔH_a and pK_a values for soil humic acid (28) are included for comparison. As previously mentioned, the equilibrium constants for neutralization of car-

Table II. Thermodynamic parameters for ionization of the acidic functional groups of river water humic substances and soil humic acid.

Sample[a]	Functional Group	ΔH_a (kJ/mole)[b,c]	pK_a[c]
WR-HS	Carboxyl	+0.6 $\pm$ 1.2	n.d.
WR-HS	Phenolic	-11.8 $\pm$ 7.0	10.1 $\pm$ 0.2
SR-HS	Carboxyl	- 0.6 $\pm$ 0.5	n.d.
SR-HS	Phenolic	-32.0 $\pm$ 3.4	10.5 $\pm$ 0.1
Soil HA	Carboxyl	3.8 $\pm$ 3.3	n.d.
Soil HA	Phenolic	- 1.7 $\pm$ 9.2	10.5 $\pm$ 0.3

n.d. = not determined

[a]Williamson River humic substances (WR-HS); Satilla River humic substances (SR-HS); soil humic acid (Soil HA).

[b]The tabulated enthalpy of ionization (ΔH_a) of the functional group is calculated from the experimentally determined enthalpy of neutralization (ΔH) using the equation: $\Delta H_a = \Delta H + \Delta H_w$, where ΔH_w = + 55.8 kJ/mole is the enthalpy of ionization of water.

[c]Average of three determinations for WR-HS and SR-HS; average of seven determinations for Soil HA.

boxyl groups with sodium hydroxide are too large to permit determination of the average pK_a of carboxyl groups in humic substances. In all three samples, the average ΔH_a value for ionization of carboxyl groups is clearly within the normal range of values observed for carboxyl groups (see Figure 1).

While the average pK_a values of the titratable phenolic hydroxyl groups of the three samples are typical of phenols in general, the corresponding ΔH_a values are very unusual, being much too exothermic (see Figure 1). At present, there is no good explanation for the observed results, but it is possible that some other reaction is occurring simultaneously with the ionization of the phenolic hydroxyl groups (e.g., solvation effects, a change in the tertiary structure of the humic polymer as a consequence of the accumulation of negative charge on the polymer, or the binding of sodium ions to the negatively charged polymer). Further attempts to characterize these phenolic groups and to determine average pK_a values for ionization of carboxyl groups are currently underway in our laboratory.

Summary and Conclusions

The calorimetrically determined concentrations of carboxyl and titratable phenolic groups in river water humic substances are very similar to those values previously reported for soil humic acid (28). A comparison of these values with corresponding values determined by pH titration methods indicates that the traditional calcium acetate exchange reaction seriously overestimates the carboxyl content of river water humic substances. Since phenolic groups are calculated as the difference between total acidity and carboxyl groups, most estimates of the phenolic content of humic substances are probably too low. In attempting to construct models for metal complexation in natural waters, other researchers should exercise caution in using acidic functional group values determined by these methods.

The calculated ΔH_a and pK_a values confirm the generally accepted opinion that river water humic substances contain carboxyl groups and a small amount of moderately acidic phenolic groups. River water humic substances also appear to contain a significant concentration of weakly acidic phenolic hydroxyl groups. These phenolic groups are possibly in close proximity to carboxyl groups and could therefore participate in metal chelation reactions via salicylate-like ligands. The concentration of these functional groups is greater in river water humic substances than in soil humic acid, suggesting that river water humic substances may have a greater metal complexation capacity than soil humic acids.

Acknowledgements

The author wishes to thank Dr. J.H. Reuter and Dr. M. Ghosal of the Georgia Institute of Technology for providing the Satilla River humic substances and much of the total acidity and carboxyl group data which were obtained by the barium hydroxide and calcium acetate methods, respectively. The author also wishes to express his gratitude to E.L. Lenox, who conducted several of the calorimetry experiments at Portland State University.

Abstract

Calorimetric titrations have confirmed the generally accepted opinion that river water humic substances contain carboxyl groups and a small amount of more weakly acidic (presumably phenolic-hydroxyl) groups. The carboxyl content obtained by titration calorimetry is significantly lower than the value determined by the calcium acetate exchange reaction. These results suggest that some phenolic hydroxyl groups are sufficiently acidic to react with calcium acetate, causing that method to over estimate carboxyl group in river water humic substances.

Literature Cited

1. Gächter, R., Lum-Shue,-Chan, K., and Chau, Y.K. Complexing capacity of the nutrient medium and its relation to inhibition of algal photosynthesis by copper. Schweiz. Z. Hydrol. 35, 252-261 (1973).
2. Davey, E.W., Morgan, M.J., and Erickson, S.J. A biological measurement of the copper complexation capacity of seawater. Limnol. Oceanogr. 18, 993-997 (1973).
3. Andrew, R.W., Hodson, P.V., and Konasewich, D.E. "Toxicity to Biota of Metal Forms in Natural Water." Workshop Standing Committee on the Scientific Basis for Water Quality Criteria of the International Joint Commission's Research Advisory Board, Duluth, Minnesota, 329 p., 1976.
4. Westall, J.C., Zachary, J.L., and Morel, F.M.M. MINEQL - A computer program for the calculation of chemical equilibrium composition of aqueous systems, Water Qual. Lab. Tech. Note 18, R.M. Parsons Laboratory, Mass. Inst. Tech., Cambridge, 1976.
5. Sillen, L.G., and Martell, A.E. "Stability Constants of Metal-Ion Complexes," Spec. Publ. No. 17, 754 p., The Chemical Society, London, 1964.
6. Wetzel, R.G. Organic carbon cycle and detritus. p. 538-621, in "Limnology," Chapter 17. Saunders. Philadelphia, 1975.
7. Lamar, W.L. Evaluation of organic color and iron in natural surface waters. U.S. Geol. Survey Prof. Paper 600-D, D24-D29 (1968).
8. Midwood, R.B., and Felbeck, G.T., Jr. Analysis of yellow organic matter from fresh water. J. Am. Water Works Assoc. 60, 357-366 (1968).

9. Reuter, J.H., and Perdue, E.M. Chemical characterization of dissolved organic matter and its influence on the chemistry of river water. Completion Report OWRR Project No. A-026-GA, Georgia Inst. of Tech., Atlanta, 33 p., 1972.
10. Reuter, J.H., and Perdue, E.M. Importance of heavy metal-organic matter interactions in natural waters. Geochim. Cosmochim. Acta 41, 325-334 (1977).
11. Leenheer, J.A., and Malcolm, R.L. Fractionation and characterization of natural organic matter from certain rivers and soils by free-flow electrophoresis. U.S. Geol. Survey Water Supply Paper 1817-E. 14 p. (1973).
12. Beck, K.C., Reuter, J.H., and Perdue, E.M. Organic and inorganic geochemistry of some coastal plain rivers of the southeastern United States. Geochim. Cosmochim. Acta 38, 341-364 (1974).
13. Weber, J.H., and Wilson, S.A. The isolation and characterization of fulvic acid and humic acid from river water. Water Res. 9, 1079-1084 (1975).
14. Schnitzer, M., and Khan, S.U. "Humic Substances in the Environment." 327 p. Marcel Dekker, New York, 1972.
15. Flaig, W., Beutelspacher, H., and Rietz, E. Chemical composition and physical properties of humic substances, p. 1-211, in Gieseking, J.E., editor, "Soil Components, Vol. 1: Organic Components," Springer, New York, 1975.
16. Gamble, D.S. Titration curves of fulvic acid: the analytical chemistry of a weak acid polyelectrolyte. Can. J. Chem. 48, 2662-2669 (1970).
17. Gamble, D.S. Potentiometric titration of fulvic acid: equivalence point calculations and acidic functional groups. Can. J. Chem. 50, 2680-2690 (1972).
18. Wilson, D.E., and Kinney, P. Effects of polymeric charge variations on the proton-metal ion equilibria of humic materials. Limnol. Oceanogr. 22, 281-289 (1977).
19. Schnitzer, M., and Gupta, U.C. Determination of acidity in soil organic matter. Soil Sci. Soc. Amer. Proc. 29, 274-277 (1965).
20. Dubach, P., Mehta, N.C., Jakab, T., Martin, F., and Roulet, N. Chemical investigations on soil humic substances. Geochim. Cosmochim. Acta 28, 1567-1578 (1964).
21. Stevenson, F.J., and Goh, K.M. Infrared spectra of humic and fulvic acids and their methylated derivatives: Evidence for nonspecificity of analytical methods for oxygen-containing functional groups. Soil Sci. 113, 334-345 (1972).
22. Jespersen, N.D., and Jordan, J. Thermometric enthalpy titration of proteins. Anal Lett. 3, 323-334 (1970).
23. Ragland, J.L. The use of thermometric titrations in soil chemistry studies. Soil Sci. Soc. Am. Proc. 26, 133-137 (1962).

24. Khalaf, K.Y., MacCarthy, P., and Gilbert, T.W. Application of thermometric titrations to the study of soil organic matter-II. Humic acids. Geoderma 14, 331-340 (1975).
25. Choppin, G.R., and Kullberg, L. Protonation thermodynamics of humic acid. J. Inorg. Nucl. Chem. 40, 651-654 (1978).
26. Larson, J.W., and Hepler, L.G. Heats and entropies of ionization, p. 1-44, in Coetzee, J.F., and Ritchie, C.D., editors, "Solute-Solvent Interactions", Marcel Dekker, New York, 1969.
27. Christensen, J.J., Wrathall, D.P., Oscarson, J.O., and Izatt, R.M. Theoretical evaluation of entropy titration method for colorimetric determination of equilibrium constants in aqueous solution. Anal. Chem. 40, 1713-1717 (1968).
28. Perdue, E.M. Solution thermochemistry of humic substances-I. Acid-base equilibria of humic acid. Geochim. Cosmochim. Acta 42, 1351-1358 (1978).
29. Mantoura, R.F.C., and Riley, J.P. The analytical concentration of humic substances from natural waters. Anal. Chim. Acta 76, 97-106 (1975).
30. Perdue, E.M. "Thermodynamics of Acid-Base Equilibria. Substituted Anilinium Ions, Pyridinium Ions and Thiophenol." Ph.D. Thesis, Georgia Inst. Tech., Atlanta, 1973.
31. Stevenson, F.J., and Butler, J.H.A. Chemistry of humic acid and related pigments, p. 534-557, in Eglinton, G., and Murphy, M.T.J., editors, "Organic Geochemistry: Methods and Results." Springer, New York, 1969.

RECEIVED November 16, 1978.

Conditional Stability Constants for Copper Ions with Ligands in Natural Waters

CONSTANT M. G. VAN DEN BERG and JAMES R. KRAMER
Department of Geology, McMaster University, Hamilton, Ontario, Canada L8S 4M1

Heavy metals generally occur in natural waters in the forms of inorganic (1, p. 238-299) and organic complexes (2, p. 297-313) and adsorbed, or surface complexed, on charged colloids. This results in a lowering of the free metal ion concentration. Frequently copper is used for experiments with aquatic organisms since it produces effects at very low concentrations, possibly at levels as found in natural waters. Kamp Nielsen (3) and Steeman Nielsen and Wium-Anderson (4) determined a depression and delay of algal growth at copper concentrations as low as 2×10^{-8} M. Millimolar concentrations of sodium and potassium reduce the effect of copper (3). Also, the presence of colloidal $Fe(OH)_3$ and excretion of organic material appears to affect the toxicity (4). The concentration of free copper in seawater upwelling from subsurface waters, may be high enough to suppress plankton growth (4). Hutchinson (5, p. 817) mentioned this possibility sometime earlier. Davey et al. (6) used the sensitivity of a diatom to cupric ion as a tool to determine the copper complexing capacity of seawater.

Apparently the toxicity of copper to plankton depends upon the free metal concentration, as is shown in experiments with varying chelator concentrations (7, 8). Calculations using the REDEQL computer model for metal speciation (8, 9) related data from toxicity experiments to free metal concentrations. Partial growth inhibition is found in the activity range 4×10^{-11} to 2×10^{-9} M copper (7, 8), and effects on diatoms are calculated to be linearly dependent upon the free copper concentration, when p$[Cu^{+2}]$ is between 8 and 12 (9).

0-8412-0479-9/79/47-093-115$05.00/0
© 1979 American Chemical Society

Methods to Measure the Free Metal Ion Concentration

Evidently it is more important to determine the cupric ion concentration than the total copper concentration in natural waters. Polarographic methods have been used to measure the ligand concentration (complexing capacity) (10) and to determine stability constants for some strong chelators (11). The plating step, however, strips the metal out of its complex and introduces a systematic error which becomes larger when the complex is weaker. Additionally, adsorption effects on the electrode may obscure the measurement (12).

The concept of the method used here goes back at least to 1922, to the work of Gunther-Schulze (13, 14). She used a natural zeolite as an ion exchanger to determine complexation of copper by inorganic anions, such as Cl^-, SO_4^{2-} and Br^-, and even in her largest dilution (5×10^{-2} M) she still found $CuCl_2$ complexes. The basic change that was made after 56 years was to measure ligands at six orders of magnitude lower concentration. With the development of synthetic, organic ion exchange resins, the ion exchange method has been applied frequently to the determination of stability constants for complexes of organic anions and metal cations, e.g. Schubert (15, 16), Schnitzer and Hansen (17), Gamble et al. (18), Ardakani and Stevenson (19). These ion exchange resins have a strong affinity for the metal ions, however, so one needs either a very strong complexing ligand or a high concentration of a weak complexing ligand to be able to readily measure the affinity of the ligand for the metal ion. In addition, polarographic (a.s.v.) measurements of the filtrate of a Chelex-100 suspension showed that fragments or molecules are released which pass the 0.45 μm filter and are able to complex copper or in some other way obscure the polarographical measurement of copper at pH 6.

As an alternative to the strong, organic, ion exchange resins, inorganic oxides with intermediate ion exchange properties were considered. Manganese dioxide (δ-MnO_2) prepared as described (20) was chosen because of its stability over a large pH range and because of its rather straightforward ion exchange capability. It is negatively charged at the pH-range of importance to natural water, pH$>$3, in which vicinity the pH of zero point of charge is (21, 22, 23).

The authors' procedure has been described in detail (20), however a short description is given here also. To 450 ml of a filtered (0.45 μm) sample a small quantity (3 ml) of aged MnO_2 dispersion is added, to a concentration of 42 μm. Constant ionic strength (0.01 M KNO_3), constant temperature (25°C) and a fixed pH are maintained. Nitrogen is bubbled continuously to remove all carbonates from solution.

Copper is added in nine steps from 0.5 to 16 μM; one hour equilibrium is allowed after each addition before a subsample (30 ml) is filtered and acidified. Total dissolved copper is measured in the filtrate by d.p.a.s.v. Calibration of MnO_2 with Cu is carried out at the same concentration, temperature, ionic strength and pH conditions. The cupric ion concentration can then be calculated using the Langmuir equation. Mass balance from the measured total Cu and the cupric ion gives the complexed copper concentration, CuL; the total ligand concentration and the conditional stability constant, K_L, for the formation of CuL, $Cu^{+2} + L = CuL$, are then calculated by plotting (Cu^{+2}) vs. $\frac{(Cu^{+2})}{(CuL)}$ and using $\frac{(Cu^{+2})}{(CuL)} = \frac{1}{K_L\,(L_{total})} + \frac{(Cu^{+2})}{(L_{total})}$. The ligand concentration is obtained from the slope and the conditional stability constant is obtained from the slope divided by the Y-axis intercept. The correlation coefficient is calculated from a least squares analysis of all the data. The method has been tested on some reasonably well characterized ligands (20) and appears to work well at ligand concentrations found in natural waters (0.2 μM). The conditional stability constants obtained for naturally occurring ligands fall within the range of stability constants for known and tested ligands.

The results of titrations with copper of a number of lakes and rivers using this method, are given here. An attempt has been made to correlate the data to other factors such as UV-spectrophotometric measurements for organic content and pH.

Sample Description

Samples, usually two liters, were taken, filtered as soon as possible (0.45 μm Millipore) and stored in the dark under refrigeration. Samples were collected from the following fresh water environments:

- dystrophic waters: low pH (4.6) brown coloured waters, containing much organic matter, flushed out of soils.

The Rivers Dickie No. 5,6 and 10, Red Chalk No. 3 and 4 and Lake Dickie belong to this group.

- medium alkalinity (~0.5 meq/L), low productivity, non-polluted: Lake Windy
- high alkalinity (~2 meq/L), non-polluted: Lake Huron
- high alkalinity (~2 meq/L), medium high production (no bloom): Gloucester Pool, Lake Ontario, Onaping River (flows out of Lake Onaping).
- rather heavily polluted with metals, low alkalinity (~0.1 meq/L) and having a medium high production: Whitewater lake.
- FA: supplied to us by Schnitzer in dried form (15). It has been extracted from the soil.

Results and Discussion

Conditional Stability Constants and Complexing Capacity. The results of copper titrations in presence of MnO_2 of a number of natural waters are given in Table I. Initial analyses done in contact with air atmosphere are less accurate (low correlation coefficients) due to the large amount of copper carbonate complexation. Correction involves the subtraction of the calculated $CuCO_3^o$ concentration from the measured dissolved copper concentration to obtain Cu^{+2} and CuL. In all cases the correlation coefficients are calculated for nine data points and are thus comparable among themselves.

The log conditional stability constants of small, monoprotic acids can be assumed to be linearly and on a one-to-one basis dependent upon the pH, due to competition between H^+ and Cu^{+2}, until the pH reaches the value of the acidity constant. Since the complexation really is composed of two competitive reactions:

1. $Cu^{+2} + L^- = CuL,\ K_L$

and 2. $HL = H^+ + L^-,\ K_a$

When pH > K_a, equation 1 is sufficient and K_L is constant ($= K_{L_o}$).

Table I. Complexing capacities and conditonal stability constants of natural waters. In most cases only one type of complexing site per ligand molecule was found of relevance to the natural water system. Original sample pH shown in brackets, r = correl. coeff., original, undiluted, ligand concentration given in brackets.

	pH	log K_L	Complexing Capacity	r
*Gloucester Pool	8.4	9.3	0.51 μM	0.90
*Lake Huron	8.3	9.2	0.20	0.70
*Whitewater Lake	8.0	8.6	0.68	0.98
*Onaping River	7.8	8.6	0.38	0.95
*Windy Lake	6.6	7.2	0.20	0.57
Lake Ontario	8.4	9.5	0.33	0.97
Lake Ontario	7.4(8.4)	8.6	0.34	0.986
Dickie No. 5	8.4(4.6)	8.5	5.35(20)	0.98
Dickie No. 5	7.6(4.6)	7.8	2.47(20)	0.987
Dickie No. 6	7.6(4.6)	7.8	5.75	0.98
Dickie No. 10	7.6(4.9)	7.8	4.95(10.9)	0.991
Lake Dickie	7.6(4.6)	7.8	2.19	0.996
Red Chalk #3	7.6(6.3)	7.7	3.35	0.991
Red Chalk #4	7.6(4.7)	7.9(K_{L_1})	2.43	0.992
		7.2(K_{L_2})	5.93	0.998
Fulvic Acid	7.6(4.6)+	7.8	2.24	0.986

*has been determined in presence of carbonate, in air atmosphere.
+pH of FA dissolved in distilled water.

When pH < K_a, K_L, as used here, becomes:

$$K_L = \frac{[CuL]}{[Cu^{+2}][HL]}, \text{ or } K_L = \frac{K_{L_o} \cdot K_a}{[H^+]}.$$

Now log K_L is linearly pH dependent. But since the acidity constant, K_a, is unknown, the authors were unable to calculate K_{L_o}.

Large polyelectrolytic molecules, however, approach colloidal particles in their behaviour. Then K_L determines the equilibrium situation for the reaction between a metal ion and charges on the colloidal particle. So K_L should vary with the degree of neutralization and the ionic strength to the same extent as the ionization constant, but in opposite direction (24). Accordingly it has been found in several cases that log K_L varied with the pH but not with a slope equal to one. Takamatsu and Yoshida (25) found that the overall constants for the formation of the ML_2 complex (L = HA),

$$\beta_2 = \frac{[ML_2]}{[M][L]^2}$$, vary as follows for different metal ions:

$$Cu^{2+} : \log \beta_2 = 8.65 + 0.65 \text{ (pH-5)}$$

$$Pb^{2+} : \log \beta_2 = 8.35 + 0.30 \text{ (pH-5)}$$

$$Cd^{2+} : \log \beta_2 = 6.25 + 0.63 \text{ (pH-5)}$$

For better understanding of such correlations it is necessary to know the acidity constants involved. Generally, these have been determined at high ligand concentrations (around 10^{-3} M) and rather low pH. The acidity constants also have been found to be very pH dependent and ionic strength dependent, again fitting the comparison with colloidal particles. Thus Gambe (26) found log constants around 3.6 for K_1 and 4.3 for K_2 at pH 3 and pH 4 respectively, for FA's. Coleman *et al.* (27) found log K_1 = 5.5 for peat.

All these considerations limit the data with which the results of this study can be compared, since they were determined at a higher pH. Results of similar studies are summarized in Table II. The determinations by Branica (28) and Shuman and Woodward (11) were actually performed for conditions of seawater and lakewater, respectively. The other authors preconcentrated their ligands. Polarographical methods have the disadvantage that the complex may split up during measurement, unless the measurement is performed much faster than the kinetics of the complex dissociation. The apparent stability constants thus determined may well be too small. Mantoura and Riley (30), who worked at a pH similar to the present experiments and who also used a system where ligand and metal are in equilibrium with each other, found constants quite similar to those of the present study. The value determined by van Dijk (31) is mentioned to show how a comparatively simple method can produce the order of magnitude of the stability constant. He compared the strength of some known complexing agents with that of HA in the presence of an ion exchange resin.

Effect of pH on the Conditional Stability Constant. A number of waters have been analysed at their original pH, some have been adjusted to pH 7.6 for intercomparison, and some have been analysed twice at different pH's. The results are plotted in Figure 1.

The dystrophic waters all give very similar values at pH 7.6, and the log constant increases with a slope close to one with the pH: for Dickie No. 5, log K_L is 8.5 at pH 8.4, and 7.8 at pH 7.6

The other samples provided generally higher constants. The K_L determined for Lake Ontario, is an order of magnitude higher than FA. Again, a shift with the pH with a slope close to one was observed: Lake Ontario, log K_L is 9.5 at pH 8.4, and 8.6 at pH 7.4.

The authors observed that the conditional stability constants still increase at the highest pH's measured (pH 8.6). Apparently the ligand is at least still partially ionized so there has to be a log dissociation constant larger than 8.6. By specifically blocking active groups on organic matter, Schnitzer and Skinner (32) showed that metals are bound by simultaneous action of acidic carboxyl groups and phenolic hydroxyl groups. One could speculate that the hydroxyl groups have a rather high dissociation constant since they are less acidic.

Table II. Comparison of conditional stability constants for complexes of copper with naturally occurring organics

Ligand	Log Constant	pH	Method	Ionic Strength	Reference
Seawater	7.46	8.2	d.p.p.	0.7	(28)
HA	5.1-6.2	6.8(?)	dialysis	0.01	(29)
FA(1)	8.80 (K_1)	8.0	gelpermeation	0.01	(30)
	8.05 (K_2)				
FA(2)	8.51 (K_1)				
	7.16 (K_2)				
Lakewater	4.5-5.7	6.5	a.s.v.	?	(11)
HA	7	6	competition	0.1	(31)
FA	7.8	7.6	ion exchange	0.01	this paper

(1) FA extracted from lake water

(2) FA extracted from peat.

The ligands that were measured in the first group of (dystrophic) waters may well be derived from soils. Fulvic acids, being the smallest and most soluble of the humic compounds, are flushed out most easily and form a major part of dissolved organics in natural fresh waters (33). The similarity of the measured stability constants of FA and dystrophic waters suggests that similar (carboxyl) groups may be responsible for binding and that the remaining part of the molecules is not very much different from each other.

Adsorption of the Metal Complex on MnO_2. Adsorption of the metal-organic complex, or part of it, on the ion exchange medium would affect the determination of the stability constant in that the dissolved copper concentration will be decreased. Previous studies using ion exchange resins observed no adsorption effects (15) or do not mention it. The present results with known ligands (20) show that adsorption, if any, of the complex does not affect the measurement and the results. It was observed, however, that the organic matter has some sort of surface active effect on the MnO_2 dispersion: subsamples, taken during the titration with copper, of waters containing a high ligand concentration (10 μM) take more time, up to about twice as long to be filtered than samples containing a very low ligand concentration (0.3 μM). Also, the type of sample seems to affect the filter speed. A sample of Lake Ontario, containing 0.5 μM ligands, has roughly the same effect as a 5 μm FA solution. Apparently the size and complexity of the molecules determines in what way the MnO_2 dispersion will be stabilized.

From Coulombic-forces point of view one can speculate that there is some adsorption at very low copper concentration $[Cu_T] < [Lig_T]$, when most of the ligands are complexed and have a positive charge, while the MnO_2 still has its negative charge. As soon as $[Cu_T] > [Lig_T]$, both the complex and the surface of the MnO_2 will have a positive charge and will repel each other electrostatically. This will be discussed further in a later publication.

Molecular Weight of FA. FA was supplied to us by Schnitzer in dried form (32). If one assumes the complexation of one copper ion by one FA-molecule, one can calculate the molecular weight of FA to be 841, which compares well to the value of 950 as found by Schnitzer (34).

Determination of Conditional Stability Constants for Mixed Ligands and for Complexes other than 1:1. Schubert's (15) study of complex ions by an ion exchange technique was commented upon by I. Feldman, who mentioned that the method works only for 1:1 complexes. This is inherent to any method, however, which does not vary the ligand concentration. In case of a mixture of ligands, an average stability constant is determined (35).

The determination of two stability constants for 8-Hydroxyl-Quinoline (20) in the present study shows that the ion exchange method is capable of accurately determining two sites or two ligands if these two sites are present in equal concentration, and if the sites result in different enough stability constants relative to discerning slope changes. In other cases the graphical representation of the titration will be slightly curved concavely. Except in one case, the results produced almost straight lines with high correlation coefficients showing that one site or ligand is at least in a very dominant position. Contradictory results have been published earlier. Ardakani and Stevenson (19) found only 1:1 complexes of metals with HA and observed only a single complexing site. Schnitzer and Hansen (17) observed the same for FA (from soil). Mantoura and Riley (30) found two sites on FA extracted from water.

Light Absorption at 260 nm and Complexing Capacity. The concentration of humic matter in water has been related to measurements of light absorption at different wavelengths at 365 nm and at 250 nm. Scanning spectrophotometric measurements were made from 220 to 380 nm. The peak height generally decreases going to higher wavelengths and the peak was rather small to make accurate measurements at 365 nm. In most samples, however, a plateau that lasted from 255 to 265 nm, and sometimes from 250 to 270 nm, was found. It was then decided to measure the absorption at the height of this plateau, at 260 nm. The low pH samples were brought to pH 7 by the addition of sodium bicarbonate buffer until 10^{-3} M bicarbonate was present. The results are given in Figure 2, as absorption vs. complexing capacity. The data scatter widely which suggests that it is not possible to accurately predict the complexing capacity from an absorption

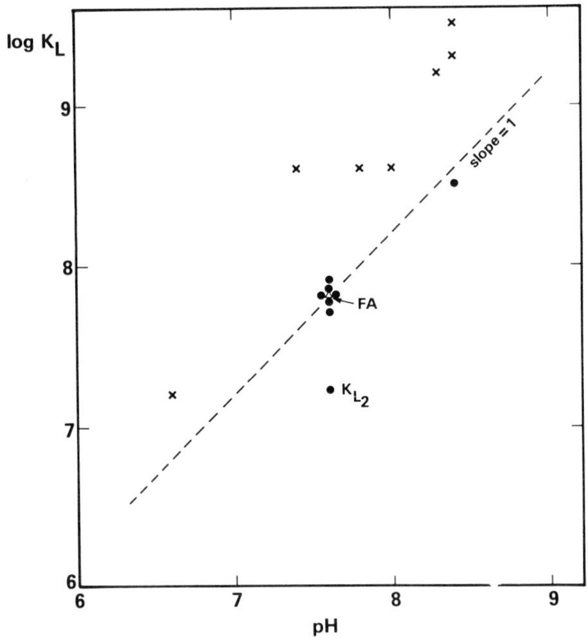

Figure 1. Conditional stability constants vs. pH at which they have been determined. (●) = Dystrophic waters and FA; (×) = other waters.

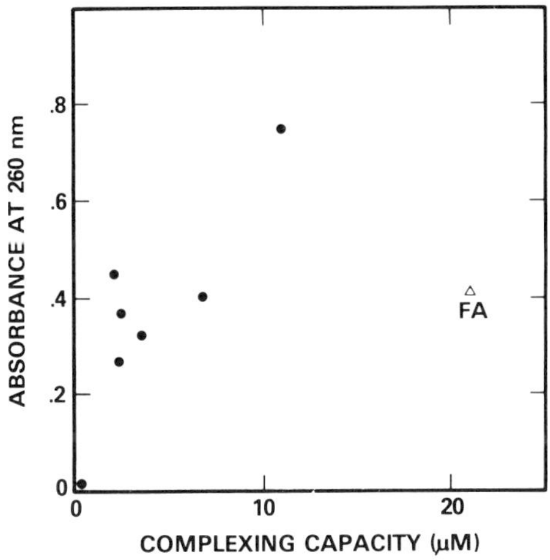

Figure 2. Light absorption at 260 nm vs. complexing capacity

measurement. One cannot say more than that high absorption might indicate a high complexing capacity. Possibly only part of the absorption is caused by complexing material. The rest is caused by other organic matter.

Complexation of Cu by some Known Ligands and Naturally Occurring Ligands at pH 8.3. The effectiveness in lowering the free metal ion concentration by some known complexing ligands and by FA and ligands as in Lake Ontario water is shown in Figure 3. The relationships have been calculated at pH 8.3, as in seawater, and for 25°C and 0.01 M ionic strength. The stability constants of the known ligands have been corrected for the pH using their acidity constants as found in Sillen and Martell (36). For FA and Lake Ontario the constants have been taken from Table I and adjusted for pH 8.3. In the calculations, 10^{-7} M complexing ligands and 10^{-8} M total dissolved copper were assumed. These concentrations are similar to values found in open ocean water which has about 1 mg. L^{-1}. C (37); if the organic carbon occurred as a compound with the Mw. of FA, it would be 10^{-6} M ligands, but because of the very long residence time the molecules probably are more complex and heavier than FA. Histidine, which has been reported to occur at levels of 10^{-8} M (38) to 10^{-7} M (39), would bring the free copper ion down to a very low level of $10^{-10.8}$ M; NTA which might enter the sea as a result of pollution, would bring it down to 10^{-12} M, but it is biodegradable and is not produced in situ. Organic matter occurring in Lake Ontario would produce a $p[Cu^{+2}]$ of 10.5. Anderson and Morel (8) calculated with the REDEQL computer model a $p[Cu^{+2}]$ of 9.6 by inorganic complexation. The presence of only 10^{-7} M organic complexing ligands may well be sufficient to diminish the free ion concentration by an order of magnitude. An inorganic particulate such as MnO_2 would not be a successful competitor for cupric ions under these circumstances.

Complexation of Cu at Varying pH. In order to compare the effects of complexation by lake organics at different pH's, the fraction (α) of total dissolved copper (Cu_T) was calculated which is in the form of the free metal ion

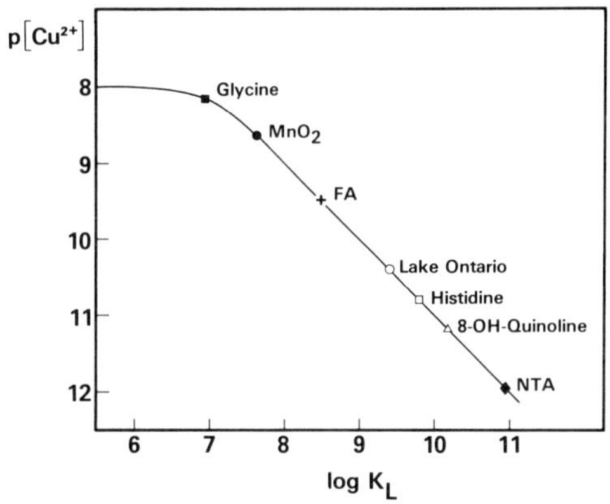

Figure 3. The effect of complexation at pH 8.3 by naturally occurring ligands and some artificial ligands (on Cu^{+2}) shown against the conditional stability constant (K_L) at this pH. Conditions for the plot are total ligand concentration L_T = 10^{-7}M, total copper concentration $Cu_T = 10^{-8}$M, 25°C, $\mu = 0.01$.

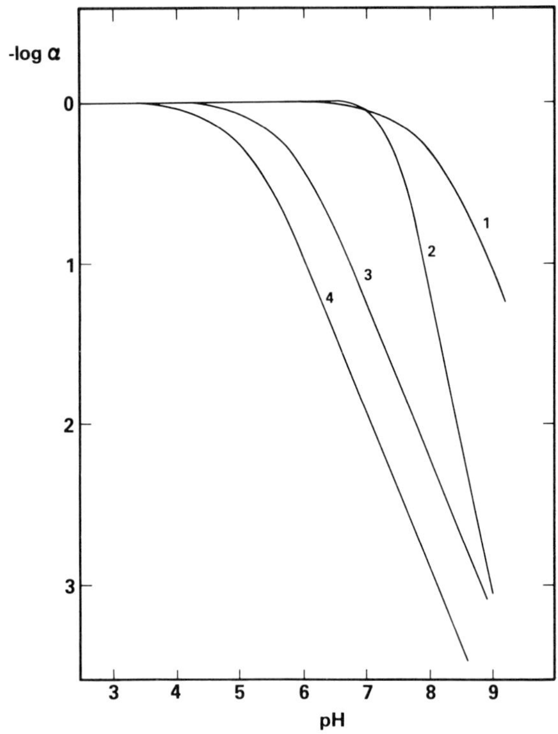

Figure 4. Model for complexation of copper by lake organics. $Ligand_{total}$ = 10^{-6}M, log K_{FA} = 7.8, and log $K_{Ontario}$ = 8.8 at pH 7.6, log K_{CuOH^+} = 6.0, log $K_{CuCO_3°}$ = 6.8, pCO_2 = $10^{-3.5}$, μ = 0.01, 25°C. $(Cu^{+2}) = \alpha \cdot (Cu_T)$, $\alpha = 1/(K_L \cdot [L] + 1)$, $Cu_T = 2 \times 10^{-7}$M. 1:L = OH^-, 2:L = $CO_3^=$, 3:L = FA, 4:L = ligand in Lake Ontario.

(Cu^{+2}): $\left[Cu^{+2}\right] = \alpha \cdot \left[Cu_T\right]$ and $\alpha = \dfrac{1}{K_L \cdot \left[L\right] + 1}$.

From the results, in Figure 4, one can see that until about pH 9, organics are much more important in complexing copper than are carbonate and hydroxyl ion. At a lower pH of 6, and with organics with a stability constant as found in Lake Ontario, 90% of the copper present is complexed by 10^{-6} M organics, whereas at a pH of 8, 99.9% of the copper is complexed.

Conclusions

Conditional stability constants for complexation of natural organics with copper have values, at pH 7.6, between $10^{7.8}$, for FA, and $10^{8.8}$, for ligands in Lake Ontario.

Generally, only one complexing site per molecule has been found with a stability constant of importance at the concentrations found in natural waters, for FA and other naturally occurring ligands.

Organic matter occurring in natural waters is a significant complexing ligand for free metal ions.

Acknowledgements

This work was supported by grants from the National Research Council of Canada and Inland Waters Subvention Program, Environment Canada.

Abstract

Toxicity to copper in natural waters may be related to the cupric ion concentration. Therefore, a method is developed using MnO_2 as weak ion exchanger, to assess the complexing capacities and conditional stability constants for compounds in natural waters. Constants found are in the range of $10^{7.8}$ for fulvic acid to $10^{8.8}$ for a ligand in Lake Ontario, at pH 7.6, 25°C, and 0.01 M ionic strength. Calculation of the complexation of copper by 10^{-6} M naturally occurring ligands, at different pH's, shows that the free metal ion concentration is lowered considerably between pH 5 and 9.

Literature Cited

1. Stumm, W. and Morgan, J.J. "Aquatic Chemistry," 583 p., Wiley-Interscience, New York, 1970.
2. Schnitzer, M. Metal-organic matter interactions in soils and waters, p. 297-313, in Faust, S.D. and Hunter, J.V. (Ed.), "Organic Compounds in Aquatic Environments," Marcel Dekker, New York, 1971.
3. Kamp Nielsen, L. The influence of copper on the photosynthesis and growth of Chlorella pyrenoidosa, Dan. Tidsskr. Farm. 43, 249-254 (1969).
4. Steeman Nielsen, E. and Wium-Andersen, S. Copper ions as poison in the sea and in freshwater, Mar. Biol. 6, 93-97 (1970).
5. Hutchinson, G.E. "A Treatise on Limnology," v. 1, 1015 p., Wiley & Sons, New York, 1957.
6. Davey, E.W., Morgan, M.J. and Erickson, S.J. A biological measurement of the copper complexation capacity of seawater, Limnol. Oceanogr. 18, 993-997 (1973).
7. Sunda, W.G., and Guillard, R.R. Relationship between cupric ion activity and the toxicity of copper to phytoplankton, J. Mar. Res. 34, 511-529 (1976).
8. Anderson, D.M., and Morel, F.M.M. Copper sensitivity of Gonyaulax tamarensis, Limnol. Oceanogr. 23, 283-295 (1978).
9. Jackson, G.A., and Morgan, J.J. Trace metal-chelator interactions and phytoplankton growth in seawater media: theoretical analysis and comparison with reported observations, Limnol. Oceanogr. 23, 268-282 (1978).
10. Chau, Y.K., Gachter, R., and Lum-Shue-Chan, K. Determination of the apparent complexing capacity of lake waters, J. Fish. Res. Board Can. 31, 1515-1519 (1974).
11. Shuman, M.S., and Woodward, G.P., Jr. Stability constants of copper-organic chelates in aquatic samples, Environ. Sci. Technol. 11(8), 809-813 (1977).
12. Brezonik, P.L., Brauner, P.A., and Stumm, W. Trace metal analysis by anodic stripping voltammetry: effect of sorption by natural and model organic compounds, Water Res. 10, 605-612 (1976).
13. Günther-Schulze, A. Die ermittlung der Selbstkomplex bildung in Wässerigen Lösungen von Kupfersalzen mit Hilfe des Permutits, Z. Electrochem. 28, 89-99 (1922).
14. Günther-Schulze, A. Die dissoziation der chloride zweiwertiger metalle in Wässerigen Lösung, Z. Electrochem 28, 387-389 (1922).

15. Schubert, J. Ion exchange studies of complex ions as a function of temperature, ionic strength, and presence of formaldehyde, J. Phys. Chem. 56, 113-118 (1952).
16. Schubert, J. The use of ion exchangers for the determination of physical-chemial properties of substances, particularly radiotracers in solution, I. Theoretical, J. Phys. Coll. Chem. 52, 340-350 (1948).
17. Schnitzer, M., and Hansen, E.H. Organo-metallic interactions in soils: 8. An evaluation of methods for the determination of stability constants of metal-fulvic acid complexes, Soil Sci. 109, 333-340 (1970).
18. Gamble, D.S., Schnitzer, M. and Hoffman, I., Cu^{2+}-fulvic acid chelation equilibrium in 0.1 m KCl at 25.0°C, Can. J. Chem. 48, 3197-3204 (1970).
19. Ardakani, M.S., and Stevenson, F.J. A modified ion exchange technique for the determination of metal-soil organic matter complexes, Soil Sci. Soc. Amer. Proc. 36, 884-890 (1972).
20. van den Berg, C.M.G., and Kramer, J.R. Determination of complexing capacities and conditional stability constants for copper in natural waters using MnO_2, Anal. Chim. Acta (in press).
21. Morgan, J.J., and Stumm, W. Colloid chemical properties of manganese dioxide, J. Coll. Sci. 19, 347-359 (1964).
22. Murray, J.W. The interaction of metal ions at the manganese dioxide-solution interface, Geochim. Cosmochim. Acta 39, 505-519 (1975).
23. Gray, M.J., Malati, M.A., and Raphael, M.W., The point of zero charge of manganese dioxides, J. Electronanal. Chem., 135-140 (1978).
24. Gregor, H.P., Luttinger, L.B., and Loebl., E.M. Metal-polyelectrolyte complexes. I. The polyacrylic acid-copper complex, J. Phys. Chem. 59, 34-39 (1955).
25. Takamatsu, T., and Yoshida, I. Determination of stability constants of metal-humic acid complexes by potentiometric titration and ion-selective electrodes, Soil Sci. 125, 377-386 (1978).
26. Gamble, D.S. Titration curves of fulvic acid: the analytical chemistry of a weak acid polyelectrolyte, Can. J. Chem. 48, 2662-2669 (1970).
27. Coleman, N.T., McClung, A.C., and Moore, D.P. Formation constants for Cu(II)-peat complexes, Science 123, 330-331 (1956).
28. Branica, M. personal communication (1978).

29. Guy, R.D., and Chakrabarti, C.L. Studies of metal-organic interactions in model systems pertaining to natural waters, Can. J. Chem. 54, 2600-2611 (1976).
30. Mantoura, R.F.C., and Riley, J.P. The use of gel-filtration in the study of metal binding by humic acids and related compounds, Anal. Chim. Acta 78, 193-200 (1975).
31. van Dijk, H. Cation binding of humic acids, Geoderma 5, 53-67 (1971).
32. Schnitzer, M., and Skinner, S.I.M., Organo-metallic interactions in soils. 4. Carboxyl and hydroxyl groups in organic matter and metal retention, Soil Sci. 99, 278-284 (1965).
33. DeHaan, H. Limnologische aspecten van humus verbindingen in het Tjeukemeer, Ph.D. Thesis, University of Gröningen, Netherlands, 1975.
34. Hansen, E.H. and Schnitzer, M. Molecular weight measurements of polycarboxylic acids in water by vapor pressure osmometry, Anal Chim. Acta 46, 247-254 (1969).
35. Allen, H.E., Crosser, M.L., and Brisbin, T.D. Metal speciation in aquatic environment, p. 33-57, in Andrew, R.W., Hodson, P.V., and Konasewich, D.E. (Ed.), "Toxicity to Biota of Metal Forms in Natural Water," Proc. of a Workshop held in Duluth, Minnesota, 1975 , 1976.
36. Sillen, L.G., and Martell, A.E. "Stability Constants of Metal-Ion Complexes," Special Publication No. 17, 754 p., Chem. Soc., London, 1964.
37. Le Williams, P.J. Biological and chemical aspects of dissolved organic material in water, p. 301-364, in Riley, J.P., and Skirrow, G. (Ed.), "Chemical Oceanography," v. 2, Academic Press, London, 1975.
38. Parks, G.A. Adsorption in the marine environment, p. 241-308, in Riley, J.P., and Skirrow, G. (Ed.), "Chemical Oceanography," v. 1, Academic Press, London, 1975.
39. Clark, M.E., Jackson, G.A., and North, W.J. Dissolved free amino acids in Southern California coastal waters, Limnol. Oceanogr. 17, 749-758 (1972).

RECEIVED November 16, 1978.

7

Ion Exchange on Humic Materials—A Regular Solution Approach

JOSEPH P. FRIZADO[1]

Department of Geological Sciences, Northwestern University, Evanston, IL 60201

Humic and fulvic acids are a large portion of the organic matter found in natural environments. They are polymeric molecules whose molecular weights range from several hundred to several million (1). The structure of these organic molecules is thought to include a large number of hydrogen bonds, an aromatic core, and large numbers of functional groups (2, p.137-198). This very open structure is capable of having cation exchange sites due to the geometry of the distribution of functional groups. The ability of humic materials to bind cations has been studied in detail (2, p.203-249; 3, 4, 5, 6, 7). They have been found to be excellent ion exchangers, with exchange capacities similar to that of smectites (4). The concentrations of cations associated with marine humic materials have been found to vary from tens of parts per million to several percent for any given cation (8).

Sea water contains a much lower concentration of dissolved organic matter than river water. More than half of this dissolved organic load is of a humic nature (9). These dissolved organic acids tend to flocculate as the salinity increases (10). Hair and Bassett (11) have observed an increase in the particulate humic acid load of an estuary as one approaches the sea. Although no studies of the distribution of humic materials throughout an estuarine system have been performed, it would appear that estuaries and their sediments in particular, act as a major sink for the dissolved and particulate humic materials. Nissenbaum and Kaplan (12) have observed that terrestrial humic materials are not deposited at great distances from shore in the marine system. A study of the flux of particulate carbon through the Chesapeake Bay comes to a similar conclusion (13). Sholkovitz (14) has shown that a large portion of the flocculated material is of a humic nature with large concentrations of

[1] Current address: Department of Geography and Earth Sciences, University of North Carolina at Charlotte, UNCC Station, Charlotte, NC 28223.

0-8412-0479-9/79/47-093-133$05.00/0
© 1979 American Chemical Society

associated cations. He believes that only a small fraction of the dissolved organics are flocculated within the estuary. His experiments used filtered water samples and neglected any effects of clay-organic interactions.

The present study attempts to define the ion exchange properties of humic materials as a function of ionic strength. Due to the flocculating effects of a salinity gradient, the distribution of humic materials between the dissolved and particulate loads within an estuary are not well known. The humics' ion exchange parameters have been documented at low ionic strengths and in marine environments. At the lower ionic strengths, some variation in the exchange parameters has been observed (3). In order to explain the effects of humic materials on the distribution of trace metals within an estuary, one must know the effects of transport through a salinity gradient upon the cations associated with the organic molecules.

Experimental Methods

The northern portion of Chesapeake Bay was studied. This upper section of the bay can be classified as a classic salt-wedge type estuary (15). The humic materials used in this study were extracted from several cores of sediments. Two one-meter cores were taken at each of two locations (39°14' N, 76°14' W; 38°56' N, 76°25' W) with a gravity corer. The cores of sediments were then divided into sections and squeezed under nitrogen to remove most of the interstitial water. The squeezed sediments were then dried, weighed and washed with distilled water. The sediments were then washed with 0.5N HCl to remove iron hydroxides, carbonates and exchangeable cations from the clays. Some of the fulvic acids are lost in this acid wash procedure. The humic materials were then extracted from the remaining solids by a 0.1N NaOH solution.

A large volume of concentrated humic material solution was needed to perform the ion exchange experiments. The UV spectra of the different humic material solutions were found to be similar. The extinction coefficients, as measured at 270 mμ, were also found to be identical. Since the samples seemed to be of a similar nature, a mixture of the samples from the four cores was made. The mixture was then dialyzed against distilled deionized water for a period of three weeks. The outer solution was replaced periodically. The molecular weight cutoff value for the dialysis membrane ($Spectrapor_R$ 6) was approximately two thousand. The concentration of the interior humic material solution was monitored by UV absorption. Less than five percent of the original humic material solution passed through the membrane and was lost to the outer solution.

The ion exchange properties of the humic materials were studied by potentiometric titrations. The humic materials were converted to the acid form by acidification of the solution to a pH of less than 2 and dialyzed against 0.01N HCl solution for another week. At this low pH, the functional groups involved in the ion exchange sites should be in H-form. It has been hypothesized that carboxylic acid and phenolic hydroxide groups are involved in the sites (2, p.203-249). Both of these groups should be in H-form at this pH. The ionic strength, I, of the reaction solution was maintained by the concentration of the chloride salt of the exchangeable cation being studied. The titrant was a solution of the same metal's hydroxide. In the case of the divalent cations, the hydroxide solutions were dilute enough to affect the ionic strength of the reaction solution upon addition. The values given for Ca^{+2} and Mg^{+2} exchanges cover a range of ionic strengths rather than a single value as with the monovalent cations. The titrations were performed under CO_2-free conditions. Additions of titrant were made until the pH was greater than 10. For monovalent cations, a constant pH was attained approximately fifteen minutes after an addition of titrant. Divalent cations took approximately three times longer to come to equilibrium.

Results

Some of the titration curves for the Na^+:H^+ and K^+:H^+ systems are shown in Figure 1. At least two distinct sites are visible throughout the series of titrations. At lower ionic strengths, other deflections were observed. These deviations may be due to ion exchange sites with slightly different characteristics than the two major sites. They could also be due to sample inhomogeneities, but this is unlikely in that 1) the deviations only appear at lower ionic strengths and 2) the number of deviations seems to decrease as the ionic strength increases. At ionic strengths greater than 0.3 only two sites were observed in the monovalent series of titrations. Evidence will be discussed elsewhere to suggest that the humic molecules undergo configurational changes as the ionic strength changes. This effect could alter the distribution of exchange site energies and thereby affect the titration curves and explain the low ionic strength deviations.

Back-titrations of the reaction solutions were also performed. The back titrant was 0.01N HCl solution. The shapes of the titration curves were similar for both the forward and back titrations. The number of milliequivalents added to reach the equivalence points was different for each titration pair. In the back titrations, the exchange capacities of the molecules were slightly smaller. The smaller exchange capacities can be explained by irreversible colloid formation. These colloids would remove some of the exchange sites from contact with the solution. A titration curve for K^+ shows the two sites prevalent in the Na^+ series. The first site's pK differs from its Na^+ analog by 0.7 units. The second site's pK is essentially the same as the Na^+ counterpart. This indicates that the lower pK site can differentiate between K^+ and Na^+, while the second site cannot.

Divalent cation titration curves are shown in Figure 2. All of these titrations exhibited only one deflection. None of the lower ionic strength titration curves had any deviations or additional sites similar to that observed in the monovalent series.

Derivation of Exchange Constants

The titration data were used to determine the thermodynamic equilibrium exchange constants for the reactions involved. To simplify matters, we will only discuss the $Na^+:H^+$ system. The titration curves were modeled by including all of the activities of the components involved. The exchange reaction can be written in the following forms:

$$\text{Hum-H} + Na^+ \rightleftharpoons \text{Hum-Na} + H^+ \qquad (1)$$

$$K = \frac{a_{H^+}\lambda_{\text{Hum-Na}}\, N_{\text{Hum-Na}}}{a_{Na^+}\lambda_{\text{Hum-H}}\, N_{\text{Hum-H}}} \qquad (2)$$

where λ is the rational activity coefficient for the exchange site occupied by the subscripted cation, a is the activity of the aqueous species, N is the mole fraction of the humic material exchange sites that is occupied by the noted cation, and K is the equilibrium constant for the reaction. ($N_{\text{Hum-Na}}$, a_{Na^+} and a_{H^+}) were evaluated from the titration data. The activity of the Na^+ was held constant by the NaCl solution used to maintain the ionic strength. The activity of the H^+ was measured by electrode. The endpoint for the conversion of a site to the Na-form was found by obtaining the point of maximum slope for the plot of pH versus base added. The equivalence point for the same site was found at the minimum slope. The difference in base added between the two points is equal to one half of the ion exchange capacity of the site in question. $N_{\text{Hum-Na}}$ is equal to the amount of base added minus the amount of base needed to initiate the exchange reaction, divided by the total capacity of the site.

The rational activity coefficients cannot be evaluated in any simple manner. Following the model of Truesdell and Christ (16), a regular solution approach to the problem can lead to expressions for the rational activity coefficients. If the exchange sites have the same charge and approximately the same size, then a symmetrical solid solution will be formed where the rational activity coefficients for the two components are given by:

$$\log \lambda_{\text{Hum-Na}} = \frac{\omega}{2.303RT}\left(N_{\text{Hum-H}}\right)^2 \qquad (3)$$

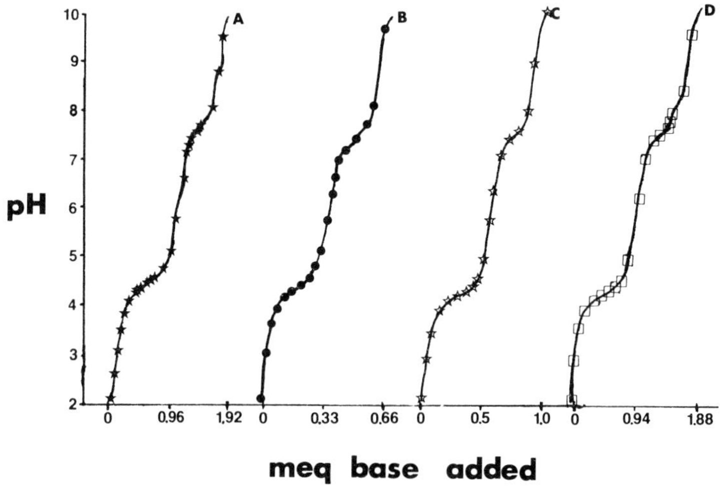

Figure 1. Titration curves for monovalent cation–hydrogen exchange. (A) Na^+ at 0.1 I; (B) Na^+ at 0.3 I; (C) Na^+ at 0.5 I; (D) K^+ at 0.1 I. Every third data point is shown. Interpolation is by a cubic spline method.

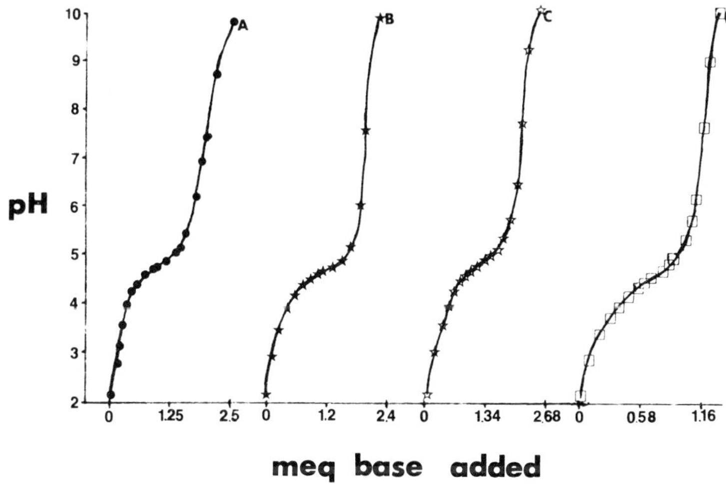

Figure 2. Titration curves for divalent cation–hydrogen exchange. (A) Ca^{+2} at 0.1 I; (B) Ca^{+2} at 0.3 I; (C) Ca^{+2} at 0.5I; (D) Mg^{+2} at 0.5 I. Every second data point is shown. Interpolation is by a cubic spline method.

$$\text{and} \qquad \log \lambda_{Hum-H} = \frac{\omega}{2.303RT} (N_{Hum-Na})^2 \qquad (4)$$

where ω is the interaction parameter for the solution, R is the gas constant and T is the temperature in Kelvin. The interaction parameter is an expression of the potential energy due to the mixture of the two different sites. For any pair of sites (Hum-Na and Hum-H in this case), ω should be a constant over an ionic strength change. In these experiments, ω was found to vary somewhat over the entire range of ionic strengths. By using these expressions for the rational activity coefficients, Equation 2 can be transformed into the following:

$$pK' = pH + \log a_{Na^+} + \log \frac{N_{Hum-Na}}{1-N_{Hum-Na}}$$

$$= \left(pK + \frac{\omega}{2.303RT} \right) - \frac{2\omega}{2.303RT} N_{Hum-Na} \qquad (5)$$

in which the unknowns (pK, ω) are on the right-hand side of the equation. If the regular solution model holds true, then a plot of pK' versus N_{Hum-Na} will be linear. Plots for the Na^+: H^+ (Figure 3) and Ca^{+2}:$2H^+$ (Figure 4) systems are given. Least-squares lines have been drawn in. The number of points used to generate these lines and their associated correlation coefficients are given in Table I. The slope and intercept of these lines can be used to evaluate pK and ω. Restrictions on the applicability of the above equations allow the model to be used only in the central region of N_{Hum-Na} values. The deviations of those points with value of N_{Hum-Na} near 0 and 1 are due to these restrictions.

It is interesting to note that the equilibrium exchange values for the reactions are variable, as shown in Figure 5. In the case of Na^+:H^+, as the ionic strength increased, the humic molecules preferred H^+ over Na^+ to a greater extent. Changes in the distribution of exchange site energies could be altered by polymerization reactions. The exchange capacities of the humic materials do not change drastically with ionic strength as one would expect if polymerization reactions were removing exchange sites. Another possible explanation is that the configurations of the organic molecules have undergone slight changes. As the molecular configurations are altered, the cation exchange involves a new solid phase. The interaction parameter would probably not be altered greatly by such a change in configuration. The thermodynamic constant for the reaction could easily be altered by such a process.

Effects of Configuration on Exchange Energies

Configurational changes in humic materials have been documented only recently. Viscosity measurements (17), SEM photomicrographs (18) and Sephadex gel filtration (19) have all been used

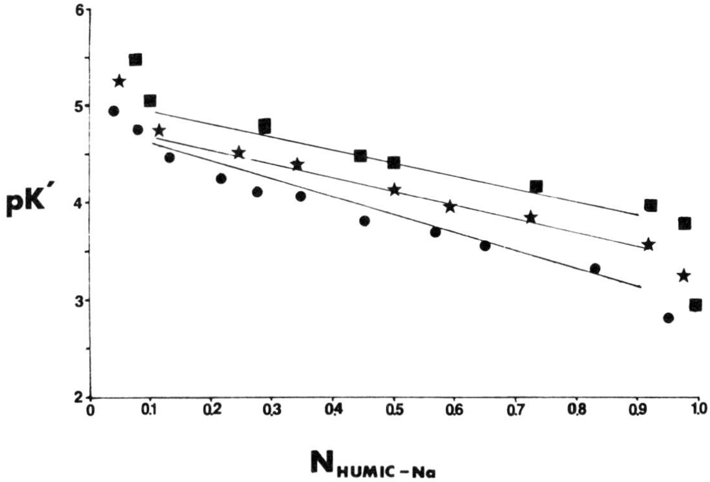

Figure 3. Plot of pK′ vs. $N_{Hum\text{-}Na}$ for the first sodium exchange site. pK′ = pH + log a_{Na}-log ($N_{Hum\text{-}Na}/N_{Hum\text{-}H}$). (●) is for the 0.3 I experiment; (★) is for 0.5 I; (■) is for 0.6 I. The least-squares lines have been drawn in.

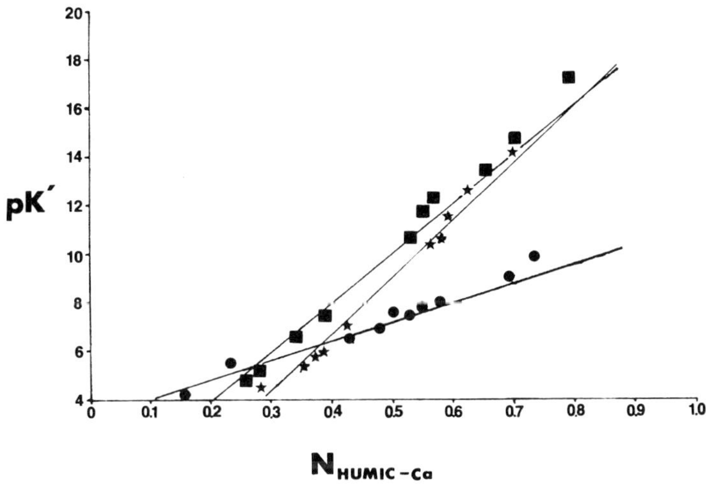

Figure 4. Plot of pK′ vs. $N_{Hum\text{-}Ca}$ for the calcium exchange site. pK′ = 2pH + log a_{Ca}-log ($N_{Hum\text{-}Ca}/N_{Hum\text{-}H_2}$). (●) 0.3 I experiment; (★) 0.5 I; (■) 0.6 I. The least-squares lines have been drawn in.

TABLE I - Exchange Reactions Parameters

Cation	Ionic Strength	Exchange Capacity meq/ g	No.Pts. used	Correl. Coeff.	ω* kJ/mole	pK**	ΔG_r kJ/mole
Na	0.6	0.35	10	0.93	-5.0	4.3	24.5
		0.69	7	0.91	-2.4	6.8	38.5
	0.5	0.46	9	0.95	-4.2	4.1	23.6
		0.71	11	0.94	-2.0	6.8	38.5
	0.4	0.32	12	0.94	-4.0	4.0	22.8
		0.61	9	0.94	-3.5	6.6	37.6
	0.3	0.25	8	0.96	-3.7	3.9	22.1
		0.54	11	0.92	-3.2	6.4	36.6
	0.2	0.27	15	0.95	-5.6	3.8	21.5
		0.80	8	0.96	-2.8	6.3	36.0
	0.1	1.19	19	0.92	-3.0	3.4	19.3
		0.91	7	0.94	-2.8	6.1	34.8
K	0.1	0.74	13	0.93	-5.4	2.7	15.4
		0.95	11	0.94	-2.3	6.2	35.4
Ca***	0.6	2.58	26	0.96	63.8	10.0	57.0
	0.5	1.69	28	0.95	71.7	9.1	51.9
	0.3	2.26	21	0.96	25.2	7.3	41.6
	0.1	2.10	23	0.93	82.2	8.8	50.2
Mg***	1.0	1.52	19	0.83	-30.9	9.5	54.2
	0.5	1.28	22	0.98	4.3	9.6	54.8
	0.1	1.40	18	0.94	5.8	9.4	53.7

* ω is found by taking the linear least-squares slope of the pK' versus N type plot, multiplying by 2.303RT and dividing by -2. All of the experiments were performed at 298 Kelvin.

** The monovalent cations have two possible sites. The parameters of the sites are listed sequentially.

*** The normality of the divalent cations hydroxide solutions were low, so the ionic strength listed is the initial value. During the course of the titration, the ionic strength was diminished. The entire range for each titration is depicted in Figure 5.

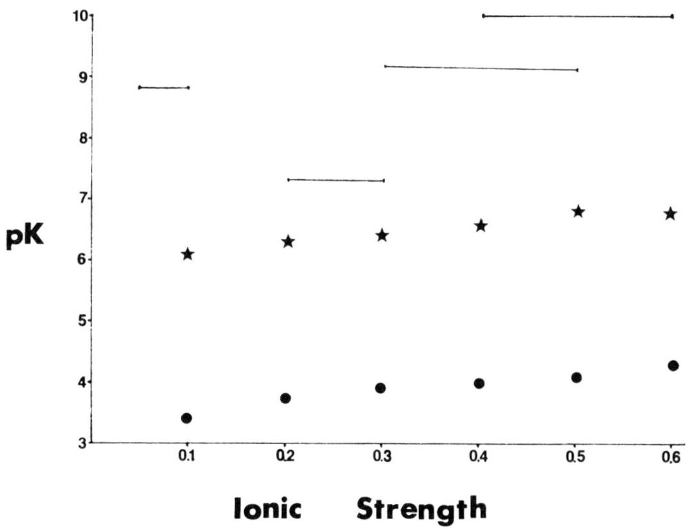

Figure 5. Plot of pK vs. ionic strength. (●) First Na exchange site; (★) second Na exchange site; (——) Ca exchange site. The nature of the divalent cation experiments was such that the ionic strength could only be held within the noted range of values.

to document changes in the configuration of humic acid molecules as a function of pH and ionic strength. As the ionic strength increases, the humic molecule is compressed nonuniformly. The effect of this compression on the exchange sites can be approximated by using the theories advanced by Eisenman (20) in his work on cation selectivity in ion exchangers.

Eisenman (20) derived a simple model which explained the observed selectivity sequences for monovalent cations in various ion exchanges. He only considered a coulombic attraction and the free energy difference between the glass and aqueous phases for each cation studied. For negative singly charged sites, the energy of attraction for monovalent cations is greatest when the sites are infinitely spaced. As the distance between exchange sites decreases, repulsive forces must be taken into consideration lessening the exchange energy. The other factor which must be taken into consideration is the anionic field strength. As the configuration of an actual exchange site is altered, the anionic field strength is changed. Either of these factors can cause the variation of pK with ionic strength observed for the Na^{+}:H^{+} system.

In the case of divalent:monovalent exchange, the site spacing becomes more important (21). Calculations using the Truesdell and Christ (21) equations for divalent-monovalent cation exchange have shown that the two processes, compression of the ion exchange site and the alteration of the distribution of sites on the molecule can act upon the free energy of exchange in different manners. It is a combination of these two opposite effects that can cause the minimum value for the pK of the Ca^{+2} exchange. By utilizing other multivalent cations in a similar number, it is possible to begin to discern qualitative changes in the configuration of the humic molecule.

Summary

Ion exchange properties of humic materials found within natural environments must be known to understand the speciation and distribution of trace metals. Although this paper deals with these properties with regards to major cations, the effects of the humic materials on trace metals is far greater (8, 22). The effects of major-minor cation exchanges as the humic materials pass through a salinity gradient and during diagenesis must also be studied. Due to the changes in ionic strength, both the amounts of associated metals and the division of humic materials between dissolved and particulate forms within an estuary are extremely variable.

It was found that the humic material ion exchange properties can be explained by a regular solution model similar to that of Truesdell and Christ (16) for clays. The thermodynamic constants for the exchange reactions studied were found to be different for each ionic strength. Changes in the configurations of the organic molecules could cause the observed variations. Other evidence (17,

18, 19) suggests that the humic molecules undergo some kind of compression during an increase of the ionic strength of the solution. Using Eisenman's work on cation selectivity of electrodes (20), it can be shown that the observed variations of the thermodynamic values can be due to a small compression of the organic molecules. In order to estimate the effects of humic materials upon the distribution and speciation of metals, the ion exchange parameters for the cations in question must be known, as well as their variations with ionic strength. The variations of these parameters for the major cations have been shown to be somewhat significant. These variations can also be used to determine the changes in the spacing and anionic field strength of the ion exchange sites on the molecule. With more information of this type, it will be possible to predict the variation of the exchange parameters for other metals, including trace metals.

Estuaries are the major pathway of materials from the rivers to the marine environment. In order to understand how dissolved and particulate organic matter within the estuary affect the speciation of cations within this environment, the ion exchange parameters as a function of ionic strength must be studied. In addition to the physical transfer of material between the dissolved and particulate forms, the salinity variations also affect the ion exchange abilities of these organic molecules. These two major processes can affect the organic material distribution and ability to bind metals, and hence the overall distribution of a given trace metal.

Acknowledgments

The author would like to thank Dr. Owen P. Bricker of the Maryland Geological Survey for his invaluable help in obtaining the samples used in this study. Support for this study was provided by NSF Grant EAR76-12279A01. I would also like to thank Dr. Robert M. Garrels for his continuing and vigorous contributions to this study.

Abstract

Humic and fulvic acids comprise a major portion of naturally occurring organic matter. They are high molecular weight polymers that can act as ion exchangers. Although humic materials have been found in both marine and fresh water environments, evidence suggests that significant amounts of terrigenous humic materials do not reach the marine environment. Humic materials are deposited by the mixing of fresh water and seawater in estuaries. The reactions of these organic molecules with cations were studied by potentiometric titrations. Humic substances were extracted from sediments of Chesapeake Bay. Ion exchange reactions between H and Ca, Mg, K and Na were studied in solutions of various fixed ionic strengths. The chloride salt of the H-displacing cation controlled the ionic strength. Analysis of the titration curves demonstrates that the organic molecule ion exchange sites fit a regular solution model. Thermodynamic constants for some of the exchange reactions are given here. The exchange parameters varied with ionic strength. These variations apparently are related to changes in configuration of the humic molecules.

Literature Cited

1. Ogura, N. Molecular weight fractionation of dissolved organic matter in coastal sea water by ultrafiltration, Mar. Bio. 24, 305-312 (1974).
2. Schnitzer, M. and Khan, S. "Humic Substances in the Environment," 327 p. Marcel Dekker, New York, 1972.
3. Broadbent, F. and Lewis, T. Soil organic matter-metal Complexes; 4. Nature and properties of exchange sites, Soil Sci. 91, 393-399 (1961).
4. Rashid, M. Role of humic acids of marine origin and their different molecular weight fractions in complexing di- and tri-valent metals, Soil Sci. 111, 298-306, (1971).
5. Gamble, D., Schnitzer, M., and Hoffman, I. Cu^{+2}-fulvic acid chelation equilibrium in 0.1m KCl at 25.0°C, Can. J. Chem. 48, 3197-3204 (1970).
6. Stevenson, F. Binding of metal ions by humic acids, p. 519-541, in Nriagu, J., ed., "Environmental Biogeochemistry," Ann Arbor Sci. Pub., Ann Arbor, Michigan, 1976.
7. Schatz, A., Schatz, V., Schalscha, E., and Martin, J. Soil organic matter as a natural chelating material, Compost Sci. 4, 25-28 (1964).
8. Nissenbaum, A., and Swaine, D. Organic matter-metal interactions in Recent sediments; The role of humic substances, Geochim. Cosmochim. Acta 40, 809-816 (1976).
9. Handa, N. Land sources of marine organic matter, Mar. Chem., 341-359 (1977).
10. Ong, H., and Bisque, R. Coagulation of humic colloids by metal ions, Soil Sci. 106, 220-224, (1968).
11. Hair, M. and Bassett, C. Dissolved and particulate humic acids in an east coast estuary, Estuarine Coastal Mar. Sci. 1, 107-111 (1973).
12. Nissenbaum, A., and Kaplan, I. Chemical and isotopic evidence for the in-situ origin of marine humic substances, Limnol. Oceanogr. 17(4), 570-582 (1972).
13. Biggs, R. and Flemer, D. The flux of particulate carbon in an estuary, Mar. Bio. 12, 11-17, (1972).
14. Sholkovitz, E. Flocculation of dissolved organic and inorganic matter during the mixing of river water and sea water, Geochim. Cosmochim. Acta 40, 831-845 (1976).

15. Prichard, D. Chemical and physical oceanography of the Chesapeake Bay, p. IVA 1-28, in Schubel, J., ed., "The Estuarine Environment," A.G.I. Short Course Lecture Notes, Washington, D.C., 1971.
16. Truesdell, A., and Christ, C. Cation exchange in clays interpreted by regular solution theory, Am. J. Sci. 266, 402-412 (1968).
17. Mkuherjee, P. and Lahiri, A. Studies on the rheological properties of humic acid from coal, J. Colloid Sci. 11, 240-243 (1956).
18. Chen, Y., and Schnitzer, M. Scanning electron microscopy of a humic acid and of a fulvic acid and its metal and clay complexes, Soil Soc. Am. Proc. 40, 682-686, (1976).
19. Reuter, J. Inferred configurational changes of aquatic humic substances, p. 1140, Geol. Soc. Am. Abstr. 9, Seattle National Mtg, 1977.
20. Eisenman, G. On the elementary origin of equilibrium ion specificity, p. 163-179, in Kleinzeller, A. and Kotyk, A., ed., "Membrane Transport and Metabolism," Academic Press, New York, 1961.
21. Truesdell, A. and Christ, C., Glass electrodes for calcium and other divalent cations, p. 293-322, in Eisenman, G., ed., "Glass Electrodes for Hydrogen and Other Cations," Marcel Dekker, Inc., New York, 1967.
22. Reuter, J. and Perdue, E. Importance of heavy metal-organic matter interactions in natural waters, Geochim. Cosmochim. Acta 41, 325-244 (1977).

RECEIVED November 16, 1978.

8

Chemical Speciation of Copper in River Water

Effect of Total Copper, pH, Carbonate, and Dissolved Organic Matter

WILLIAM G. SUNDA and PETER J. HANSON

National Marine Fisheries Service, Southeast Fisheries Center, Beaufort Laboratory, Beaufort, NC 28516

There is increasing evidence that the availability of aqueous trace metals to a number of organisms is determined by free metal ion activity rather than the total concentration of metal in solution (1-6). The chemical associations of trace metals with inorganic and organic ligands are major factors that control metal ion activities and thus bioavailability.

In this study, we investigate the complexation of copper by inorganic and organic ligands in the water of two rivers in coastal North Carolina. An ion-selective electrode was used to differentiate between free and bound cupric ion in titrations of river waters and model solutions. Stability constants were determined in chemically defined solutions for complexation of copper by hydroxide and carbonate ions, the two major inorganic copper complexing ligands in most natural waters (7). From these results, total copper in the river water was partitioned into organic and inorganic forms and stability constants for complexation of copper by natural organic ligands were calculated. Finally, models were calculated which predict the variations in chemical speciation of copper resulting from changes in the chemical parameters: pH, carbonate alkalinity, concentration of dissolved organic matter, and concentration of total dissolved copper.

The rivers sampled were the Newport and Neuse. The Newport River is a small coastal plain river (mean discharge approximately 0.4 to 11.2 m^3 sec^{-1}) with a watershed of approximately 340 km^2. The Neuse River, in contrast, is a larger river (mean discharge approximately 130 m^3 sec^{-1}), originating in the piedmont region with a drainage basin of approximately 1.1×10^4 km^2. The Newport River water used in this investigation is characterized by a high concentration of dissolved organic carbon (15 mgC ℓ^{-1}), low pH (6.0), low carbonate alkalinity (0.06 mM) and relatively low concentrations of alkaline earth cations (0.14 mM Ca and 0.03 mM Mg). The Neuse water had an appreciably lower concentration of dissolved organic carbon (3.0 mgC ℓ^{-1}), higher pH (6.8), higher

0-8412-0479-9/79/47-093-147$08.50/0
This chapter not subject to U.S. copyright
Published 1979 American Chemical Society

carbonate alkalinity (0.15 mM), and about the same concentrations of calcium and magnesium (0.13 mM Ca and 0.04 mM Mg).

Materials and Methods

Water samples were collected from the Newport River upstream from Newport, North Carolina and from the Neuse River several miles downstream from Kinston, North Carolina. Standard procedures for handling water samples were adopted to minimize changes in the natural state of the samples during sampling and storage. Portions of the samples were filtered through glass fiber filters (Gelman A-E) within 24 hr of collection. A portion of the filtered river water was exposed to high intensity ultraviolet radiation (LaJolla Scientific Model PO-14 Photooxidation Unit) to photooxidize organic matter (8). Calcium and magnesium concentrations were determined in filtered samples by flame atomic absorption spectrophotometry (Perkin Elmer 403). Dissolved organic carbon was determined using a CHN analyser (F and M 185) on the residue left after freeze drying samples of acidified, filtered river water. Carbonate alkalinity of filtered and UV-treated samples was calculated from pH and P_{CO_2} data for samples equilibrated with air.

The chemical speciation of copper in river water and model solutions was investigated by a titration technique in which cupric ion activities were measured at constant pH as the total copper concentration ($[Cu_{TOT}]$) was varied by incremental additions of $CuSO_4$. pCu(-log cupric ion activity) was measured with a cupric ion-selective electrode (Orion 94-29) and pH with a glass electrode (Beckman 39301) both coupled to a single junction Ag/AgCl reference electrode (Orion 90-01) in a temperature controlled (25 ± 0.5°C) water bath. Total copper concentrations in the titrated solutions were determined directly by atomic absorption spectrophotometry (Perkin Elmer 603) using a graphite furnace (Perkin Elmer 2200). Measurement of total copper concentrations is necessary because of adsorptive loss of copper from solution onto container and/or electrode surfaces.

The following procedure was followed for all copper titrations at 25°C and constant pH. The three electrodes were first preconditioned for 30 min in a solution at pH 8 containing 0.1 M Tris base, 0.05 M HCl and sufficient $CuSO_4$ to achieve a pCu of 13.0 to 13.5. The electrodes were then rinsed several times with distilled water and placed for 30 min in a portion of the solution to be titrated. The electrodes were then placed in a fresh 70 mℓ portion of the same solution contained in a 100 mℓ borosilicate glass beaker and titrated with $CuSO_4$. Sufficient time was allowed for the electrodes to reach steady state potentials after each copper addition. At no copper addition, 60 min was allowed. For copper concentrations $\leq 10^{-7}$ M and $> 10^{-7}$ M, measurements were made 30-60 min and 20-30 min, respectively, after each copper addition. After reaching steady state, pCu and pH values were recorded and the solution was subsampled (from 0.01 to 1.0 mℓ) for total copper determination. pH was held constant during the titration by controlling P_{CO_2}. At high concentrations of added copper in river water samples (> 2 x 10^{-5} M) it was necessary to add sufficient quantities of NaOH to neutralize hydrogen ions released by the complexation of weak acid functional groups.

Titrations were performed on untreated, filtered, and UV-treated filtered river water samples at in situ and adjusted pH values. The effect of pH on copper speciation was investigated by titration of filtered Newport River water at pH 7.0 and filtered Newport and Neuse waters at pH 8.0. Newport River water was adjusted to pH 7.0 by decreasing the partial pressure of CO_2 from the initial ambient value of about 10 times the atmospheric level. To adjust the pH to 8.0, sodium bicarbonate was added to bring the river water samples to a concentration of 0.5 mM with subsequent adjustment of P_{CO_2}. Titrations were also conducted at pH 7.0 in model solutions consisting of 0.01 M KNO_3 and 0.1 mM $NaHCO_3$ with and without the addition of 0.75 µM histidine to test electrode behavior in solutions of known chemistry.

Stability constants for the complexation of copper by hydroxide ion (Cu hydrolysis) were determined from measurements of pCu and p[Cu_{TOT}] as a function of pH in solutions containing 0.01 M KNO_3 and 1.0 and 2.5 µM $CuSO_4$. These solutions were first purged with nitrogen at pH < 6 to remove CO_2 and then closed to the atmosphere. Measurements were made under a nitrogen atmosphere as the solutions were titrated with small quantities (<25 µℓ per addition) of a concentrated solution of KOH. For the determination of stability constants for carbonate complexes, measurements of pCu and p[Cu_{TOT}] were made as a function of pH in a solution containing 0.01 M $NaHCO_3$ and 5 µM $CuSO_4$. The pH and thus the carbonate ion activity was varied by adjusting P_{CO_2}.

Cupric ion activities and cupric ion concentrations were determined using the Nernst equation from the differences in potential between the test solutions and a standard solution consisting of 10^{-5} M $CuSO_4$ and 0.01 M KNO_3 at pH 5.4 ± 0.3. Values of cupric ion activity in test solutions were based on a cupric ion activity coefficient of 0.68 in the standard solution as calculated from the extended Debye-Huckel equation. For measurements in defined solutions containing 0.01 M KNO_3 or 0.01 M $NaHCO_3$ cupric ion concentrations could be directly computed via the Nernst equation because activity coefficients were the same in both test and standard solutions.

Total copper concentrations in aliquot samples of the titration solutions were measured by electrothermal atomic absorption using direct injection of the acidified (HNO_3) sample into the graphite furnace. Standards were prepared at the same acid strength as the samples and copper values were obtained by the comparator method. Copper concentrations in filtered (Gelman A-E) Newport and Neuse waters were 0.011 µM and 0.025 µM, respectively, and were well within analytical capability. Thus, no sample preconcentration was required. The precision as estimated by replicate analyses was about ± 0.00062 μM, (reported as one standard deviation) which is equivalent to a 1% relative standard deviation at the midrange of the standards.

Computation of Stability Constants for Hydroxo and Carbonato Complexes. Stability constants for the formation of hydroxo complexes were computed by linear regression from the following equation:

$$\frac{[Cu_{TOT}] - [Cu^{2+}]}{a_{OH^-}\ [Cu^{2+}]} = \overset{*}{\beta}_{Cu(OH)_2}\ a_{OH^-} + \overset{*}{K}_{CuOH^+} \tag{1}$$

where $[Cu_{TOT}]$ and $[Cu^{2+}]$ are the measured concentrations of total dissolved copper and cupric ion respectively and a_{OH^-} is the activity of hydroxide ion as computed from the measured pH and the ion product of water. $\overset{*}{K}_{CuOH^+}$ and $\overset{*}{\beta}_{Cu(OH)_2}$ are stability constants for the formation of mono and dihydroxo complexes as defined by the equations:

$$\overset{*}{K}_{CuOH^+} = \frac{[CuOH^+]}{[Cu^{2+}]\ a_{OH^-}} \tag{2}$$

$$\overset{*}{\beta}_{Cu(OH)_2} = \frac{[Cu(OH)_2]}{[Cu^{2+}]\ a^2_{OH^-}} \ . \tag{3}$$

Mass balance for total soluble copper is expressed by the equation:

$$[Cu_{TOT}] = [Cu^{2+}] + [CuOH^+] + [Cu(OH)_2] \ . \tag{4}$$

Equation 1 is derived by algebraic combination of equations 2, 3 and 4.

The left term of equation 1 was computed for each data set ($[Cu^{2+}]$, $[Cu_{TOT}]$, pH) and then regressed as a linear function of a_{OH^-}. The y-intercept and the slope of the regression line are $\overset{*}{K}_{CuOH^+}$ and $\overset{*}{\beta}_{Cu(OH)_2}$, respectively.

A similar linear regression procedure was used for the determination of carbonato stability constants. Here the equation regressed was:

$$([Cu_{TOT}] - ([Cu^{2+}] + [Cu^{2+}]\sum_{1}^{2} a^{i}_{OH^-}\ \overset{*}{\beta}_{i}))/\ [Cu^{2+}]\ a_{CO_3{}^{2-}} = \overset{*}{\beta}_{Cu(CO_3)_2^{2-}}\ a_{CO_3^{2-}} + \overset{*}{K}_{CuCO_3} \ . \tag{5}$$

The term $[Cu^{2+}] \sum_{i}^{2} a^{i}_{OH^-} \beta^{*}_{i}$ represents the concentration of mono plus dihydroxo complexes as comptued from stability constants (Table I). The activity of carbonate ion was computed from the concentration of added $NaHCO_3$ (0.01 M) and carbonate acidity constants at infinite dilution and 25°C reported in Stumm and Morgan (9). The stability constants obtained from the slope and y-intercept of the regression of equation 6 are defined in terms of the concentrations of Cu^{2+}, $CuCO_3$, and $Cu(CO_3)_2^{2-}$ and the activity of carbonate ion. Stability constants for hydroxo and carbonato complexes were corrected to infinite dilution using activity coefficients calculated from the extended Debye-Hückel equation.

Scatchard plots. Stability constants for the binding of Cu by a model ligand (histidine) and by natural organic ligands in river water were computed using Scatchard plot diagrams as described previously by Mantoura and Riley (10). The general equation for this analysis was:

$$\frac{[CuL_i]}{[Cu^{2+}]} = K_c[L_{i\text{-}TOT}] - K_c[CuL_i] \qquad (6)$$

where $[CuL_i]$ and $[L_{i\text{-}TOT}]$ are the concentrations of copper ligand complex and total ligand. K_c is a conditional stability constant valid for a given set of chemical conditions of pH, ionic strength and concentration of competing metal ions (Ca, Mg, etc):

$$K_c = \frac{[CuL]}{[Cu^{2+}]\ ([L_i] + \Sigma\ [H_nL] + \Sigma\ [MeL])} \qquad (7)$$

$\Sigma\ [H_nL]$ is the concentration of all protonated forms of a weak acid ligand and $\Sigma\ [MeL]$ is the sum of all complexes of the ligand with metals other than copper, particularly Ca and Mg. The concentration of organically bound copper was computed to be equal to the total measured concentration of copper in solution minus the computed concentrations of inorganic species: Cu^{2+}, $CuOH^+$, $Cu(OH)_2$, $CuCO_3$ and $Cu(CO_3)_2^{2-}$. Concentrations of these species were calculated from hydroxo and carbonato stability constants determined in this study (Table I). Activity coefficients used in these calculations were computed from the extended Debye-Huckel equation using estimates of ionic strength based on the measured concentrations of Ca^{2+}, Mg^{2+}, carbonate alkalinity and, where applicable (titrations at pH 8.0) the added concentration of $NaHCO_3$.

The Scatchard plot analyses for the copper titration of histidine is straight forward since there is only a single one-to-one complex formed. Here a plot of $[CuL]/[Cu^{2+}]$ as a function of [CuL] should be linear with an x-intercept equal to the concentration of histidine and a slope equal to $-K_c$. However, Scatchard plots derived from titration data for river water showed markedly non-linear curves indicating the presence of several binding sites with different stability constants. Here an iterative stripping procedure was adopted to estimate total concentration of individual ligands (more likely groups of ligands or binding sites) and conditional stability constants.

Chemical Speciation Models. Using the stability constants derived by us for copper complexes with hydroxo and carbonato ligands (Table I) and for natural organic ligands (Table II), the Newport and Neuse Rivers were modeled for copper speciation as a function of pH, total copper, carbonate alkalinity and total dissolved organic matter. Speciation models were calculated from the equation:

$$[Cu_{TOT}] = [Cu^{2+}] + [CuOH^+] + [Cu(OH)_2] + [CuCO_3] + [Cu(CO_3)_2^{2-}] + [CuL_{TOT}] \qquad (8)$$

where

$$[Cu^{2+}] = a_{Cu^{2+}} / \gamma_{Cu^{2+}}$$

$$[CuOH^+] = a_{Cu^{2+}}\, a_{OH^-}\, K_{CuOH^+} / \gamma_{CuOH^+}$$

$$[Cu(OH)_2] = a_{Cu^{2+}}\, a^2_{OH^-}\, \beta_{Cu(OH)_2} / \gamma_{Cu(OH)_2}$$

$$[CuCO_3] = a_{Cu^{2+}}\, a_{CO_3^{2-}}\, K_{CuCO_3} / \gamma_{CuCO_3}$$

$$[Cu(CO_3)_2^{2-}] = a_{Cu^{2+}}\, a^2_{CO_3{}^{2-}}\, \beta_{Cu(CO_3)_2^{2-}} / \gamma_{Cu(CO_3)_2^{2-}}$$

$$[CuL_{TOT}] = a_{Cu^{2+}} \sum_1^N [L_{i-TOT}]\, K_{c-i} / (1 + K_{c-i}\, a_{Cu^{2+}})$$

K and β denote zero corrected stability constants, K_{c-i} are conditional stability constants valid under the same chemical conditions as for their determination, $[L_{i-TOT}]$ is the concentration of the ith organic ligand (ligand group or binding site) of N total ligands resolved by Scatchard analyses, and γ is the single ion activity coefficient. Polynuclear and mixed ligand complexes were not considered significant and, thus, not included in the model.

Results and Discussion

Hydroxo and carbonato constants. $p[Cu^{2+}]$ and $p[Cu_{TOT}]$ were measured as functions of pH for the KOH titrations of solutions containing 0.01 M KNO_3 and initial concentrations of 2.5 μM and 1.0 μM $CuSO_4$. $p[Cu^{2+}]$ increased with increasing pH. At least some of the decrease in the measured cupric ion concentration could be accounted for by a decrease in dissolved copper due to adsorption (Figure 1). Adsorption increased with increasing pH in the range approximately 7.0 to 9.0 but remained constant at pH values above this range. Up to 92% of the dissolved copper was lost from solution due to adsorption onto containers or electrode surfaces. Consistent values for adsorption were found as the pH was first increased and then decreased indicating that adsorption is reversible (Figure 1).

A plot of $p([Cu^{2+}]/[Cu_{TOT}])$ as a function of pH for three separate titrations fall on a single curve despite up to five-fold differences in measured dissolved copper concentration at a given pH (Figure 2). This behavior of the ratio $[Cu^{2+}]/[Cu_{TOT}]$ is indicative of the formation of mononuclear hydrolysis species and excludes the possibility that the observed reduction in free cupric ion may have been caused by precipitation of $Cu(OH)_2$ (solid) or the formation of polynuclear complexes. Analysis of data for $p[Cu^{2+}]$, $p[Cu_{TOT}]$ and pH in the pH range 7.7 to 10.8 indicated the presence of two hydrolysis species ($CuOH^+$ and $Cu(OH)_2$) whose stability constants are given in Table I. Our value of the stability constant for the monohydroxo complex ($10^{6.48}$) falls within the range of previously published values ($10^{6.0}$ to $10^{6.66}$) at 25°C and ionic strength approaching infinite dilution (11, 12, 13, 14).

Values of the stability constant for the dihydroxo complex reported in the literature are quite variable ranging from $10^{10.7}$ to $10^{14.3}$ (11, 12, 14, 15). This variability has resulted in considerable uncertainty as to the importance of copper dihydroxo species in natural waters. Paulson (14) discusses sources of error in previous experimental and computational procedures that can account for much of this variability. Our dihydroxo stability constant ($\log \beta_2 = 11.78$) at 25°C and infinite dilution is in close agreement with that reported by Paulson (14) ($\log \beta_2 = 11.80$) who also used a cupric ion-selective electrode technique and also corrected his data for adsorptive loss of copper.

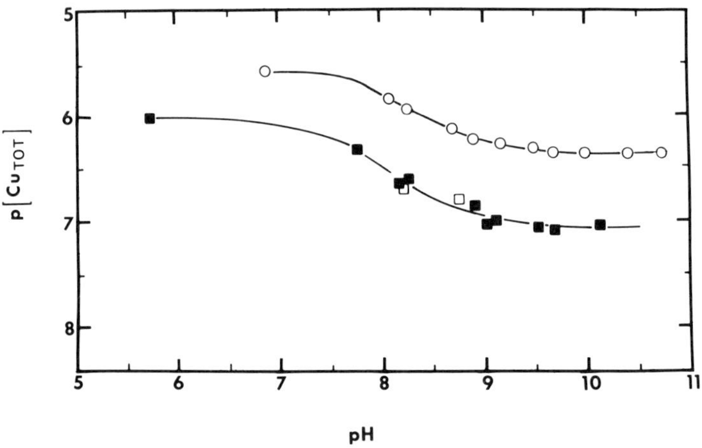

Figure 1. Loss of copper by surface adsorption as a function of pH for solutions containing 10mM KNO_3; (■) and (□) 1.0μM $CuSO_4$, and (○) 2.5μM $CuSO_4$; (■) increasing pH and (□) decreasing pH. pH varied by addition of concentrated KOH and HCl.

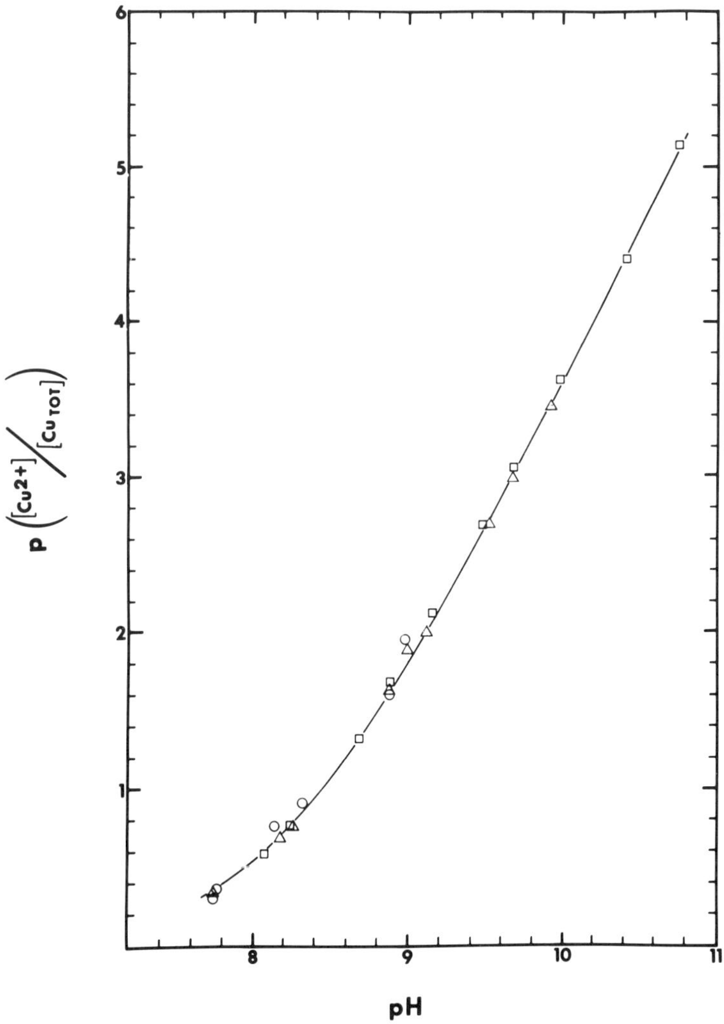

Figure 2. Titration data for complexation of copper by hydroxide. All solutions contained 10mM KNO_3; (△) 1.0μM $CuSO_4$, (○) 2.5μM $CuSO_4$ (first run), and (□) 2.5μM $CuSO_4$ (second run). Curve calculated according to hydroxo constants determined in this work (Table I).

Table I. Stability constants for reaction of copper with OH^-, CO_3^{2-}, and histidine at 25°C.

Ligand	Ionic strength	log K_1	log β_2	References
OH^-	0 corr.	6.48	11.78	This work
CO_3^{2-}	0 corr.	6.74	10.24	" "
Histidine	0.01	10.45*	-----	" "
	0.01	10.55	18.80	Sillen & Martell, (12)

* Calculated from a conditional stability constant at pH 7.00 of $10^{8.21}$ (Figure 5) and first and second hydrogen ion association constants of $10^{9.20}$ and $10^{6.00}$ (12).

Analysis of data for p[Cu^{2+}], p[Cu_{TOT}] and pH for the solution containing 0.01 M $NaHCO_3$ and 5 µM $CuSO_4$ are consistant with the presence of both mono and dicarbonato complexes (Figure 3). Constants for these complexes computed from titration data are given in Table I and are in good agreement with previously published values as reported in Sillen and Martel (11), Sunda (13) and Bilinski et al. (7). Adsorption of copper in the 0.01 M bicarbonate solutions was not as great (maximum adsorption 50%) as that which occured in the absence of carbonate ion, suggesting that carbonato complexes of copper are not as readily adsorbed as cupric ion and/or copper hydroxo species.

Model solutions: Titrations at constant pH and a variable p[Cu_{TOT}]. Data for the Cu titration of a model solution containing 0.75 µM histidine, 0.1 mM $NaHCO_3$ and 0.01 M KNO_3 at pH 7.0 (Figure 4) was analyzed by linear regression using a Scatchard plot diagram (Figure 5). This analysis yielded a stability constant in good agreement with previously published values (Table I). A theoretical curve calculated from this constant was in good agreement with the measured pCu and p[Cu_{TOT}] data throughout the titration range of p[Cu_{TOT}] (Figure 4) indicating good electrode behavior even at concentrations of Cu_{TOT} as low as $10^{-8.2}$ M. pCu and p[Cu_{TOT}] data did not agree as closely with the theoretically calculated curve for a titration of a similar solution containing no added chelator. The measured pCu values were in good agreement with values calculated on the basis of inorganic complexation for total copper concentrations $\geq 10^{-7}$ M, but deviated from predicted values below this concentration (Figure 4). A maximum difference between measured and calculated pCu values

Table II. Conditional Stability Constants for Complexation of Copper by Natural Organic Ligands

Natural Water	pH	Ligand	Ligand Concentration M	Moles of Cu Binding Sites / g organic carbon	log K_c	Reference
Newport River	5.95	L_1	5.0 E-7	3.3 E-5	9.0	This paper
		L_2	1.8 E-5	1.2 E-3	6.2	" "
		L_3	1.0 E-4	6.7 E-3	4.6	" "
	7.00	L_1	8.7 E-7	5.8 E-5	9.7	" "
		L_2	8.6 E-6	5.7 E-4	7.8	" "
		L_3	1.6 E-4	1.0 E-2	5.6	" "
	8.00	L_1	1.2 E-6	8.0 E-5	10.9	" "
		L_2	1.1 E-5	7.5 E-4	8.8	" "
		L_3	1.2 E-4	8.2 E-3	6.9	" "
Neuse River	6.78	L_1	2.1 E-7	7.0 E-5	9.5	" "
		L_2	1.9 E-6	6.4 E-4	7.3	" "
		L_3	3.0 E-5	9.9 E-3	4.9	" "
	8.00	L_1	2.7 E-7	8.9 E-5	11.2	" "
		L_2	2.9 E-6	9.6 E-4	8.3	" "
Lake Water	8.00	---	----------	4.8 E-4*	8.8	Mantoura & Riley,(10)
		---	----------	1.1 E-3*	8.0	" " " "
Soil Fulvic Acid	3.0	---	----------	2.4 E-3*	3.3	Schnitzer & Khan,(18)
	5.0	---	----------	2.4 E-3*	4.0	" " " "

* Data converted from number of binding sites per g of organic matter to binding sites per g of organic carbon assuming ratural fulvic acids contain 46% carbon (based on data in (18)).

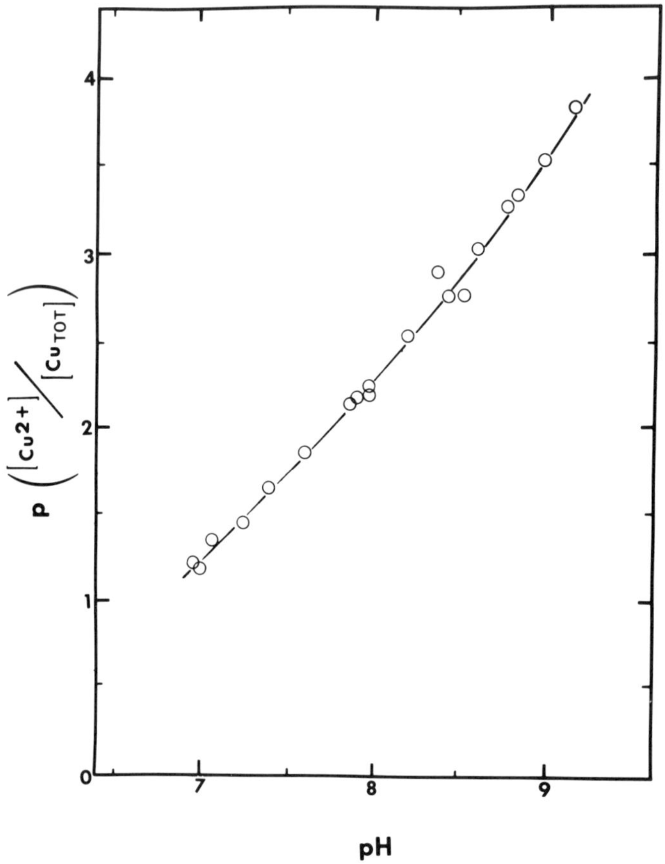

Figure 3. Titration data for complexation of copper by carbonate. Solution contained 10mM $NaHCO_3$ and 5μM $CuSO_4$. pH varied by adjusting P_{CO_2}. Curve calculated according to hydroxo and carbonato stability constants determined in this work (Table I).

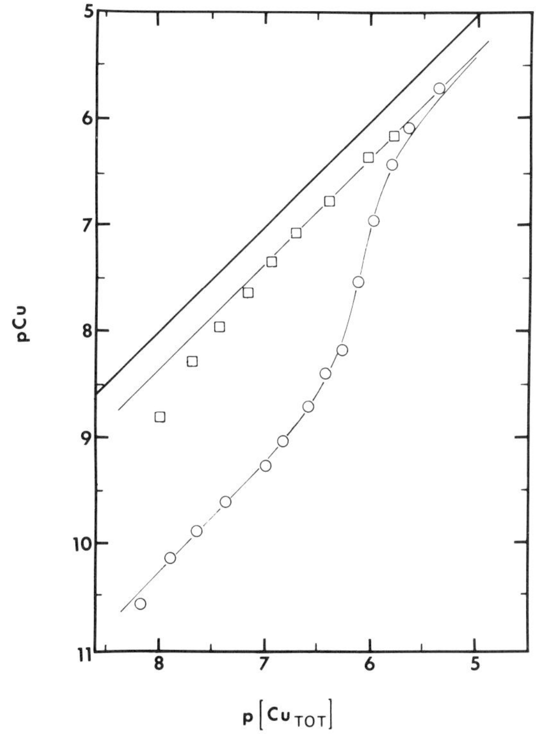

Figure 4. Copper titrations of model solutions at 25°C and pH 7.00. (□) Solution contained 0.01M KNO_3 and 0.1mM $NaHCO_3$ in distilled water. (○) Solution contained 0.75μM histidine, 0.01M KNO_3, and 0.1mM $NaHCO_3$. Solid lines through data points are theoretical curves calculated according to constants given in Table I. Dark solid line represents pCu = p[Cu_{TOT}].

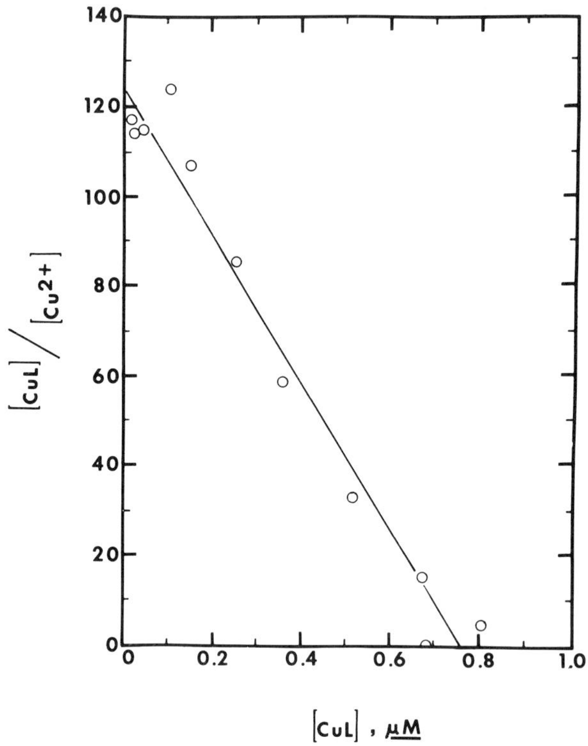

Figure 5. Scatchard plot for histidine model solution. The linear regression line for the data is shown.

(0.4 pCu units) occured at p[Cu_{TOT}] 8.0. This deviation from ideal behavior is apparently due to a disequilibrium between the solution titrated and the membrane surface of the cupric ion-selective electrode. Our experience has shown that 10^{-7} M total copper is often the lower limit for accurate determination of cupric ion activity. However, the exact limit of detection is determined by a variety of factors: complexing characteristics of the solution including kinetics of the complexation reactions, preconditioning of the electrode surface, and time allowed to establish equilibrium. The cupric ion electrode shows best behavior in well buffered solutions in which the kinetics of complexation reactions are fast, such as in the histidine solution.

Copper titrations of river water. Copper titration data indicates that copper is highly bound in both Newport and Neuse River water (Figures 6 and 7). A generally close agreement between titration curves of filtered and unfiltered samples is consistant with minor to negligible binding by particulate matter retained by glass fiber filters (the mean retention size of glass fiber filters is approximately 0.7 to 0.9 µm (16). Close agreement between measurements of background total dissolved copper in untreated Neuse River water (0.029 µM) and filtered Neuse River water (0.025 µM) also indicates that only a minor fraction of copper was associated with particulate matter.

Non-ideal behavior of the cupric ion electrode occurs at p[Cu_{TOT}] > 7 in the titrations of both filtered and unfiltered Newport River water at pH 5.95 and filtered water at pH 7.0 and 8.0. At low total copper concentrations, measured pCu values approach a constant value independent of the total copper in solution. Similar behavior was observed for filtered Neuse River water at pH 8.0, but not at pH 6.78.

As indicated by titration data (Figures 6 and 7), binding of copper in both Neuse and Newport River water decreases with increasing total copper in a manner consistant with a stepwise titration of a number of different ligands and/or binding sites. Binding of copper increases with increasing pH consistant with reactions with protonated weak acids.

Comparisons of Cu titrations of natural filtered river water and UV-irradiated filtered river water show a large decrease in binding of copper after photooxidation of organic matter. UV-treatment resulted in > 97% destruction of dissolved organic matter in both Neuse and Newport River waters, based on light absorption measurements in the wavelength range 300-500 nm and measurements of dissolved organic carbon. Measured pCu values in natural filtered Newport River water at pH 5.95 are appreciably higher than predicted for complexation to inorganic ligands (CO_3^{2-} and OH^-) alone by values ranging from 0.7 at p[Cu_{TOT}] 4.2 to 2.4 at p[Cu_{TOT}] 6.9. From calculations at p[Cu_{TOT}] 6.9 only 0.4% of the copper in solution is present as inorganic species, primarily Cu^{2+}, with the remaining 99% apparently bound to organic

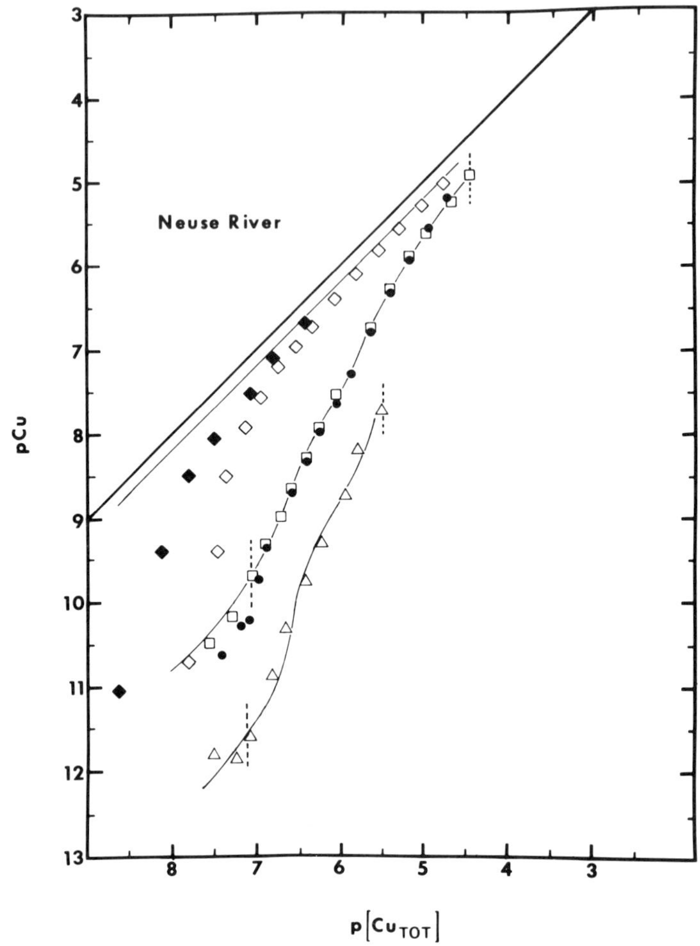

Figure 6. Copper titrations of Neuse River water at 25°C. (●) Untreated water at in situ pH 6.78; (□) glass–fiber filtered water at pH 6.78; (△) glass–fiber filtered water at pH 8.00; (◇) UV-treated glass–fiber filtered water at 6.78; (◆) twice filtered UV-treated water at pH 6.78, first filtration by glass–fiber prior to UV-irradiation, second filtration by membrane (0.2μm nuclepore) after irradiation. Model curves through data points were calculated according to stability constants determined in this work (Tables I and II). Dotted lines indicate limits on data used for calculation of conditional stability constants for organic binding.

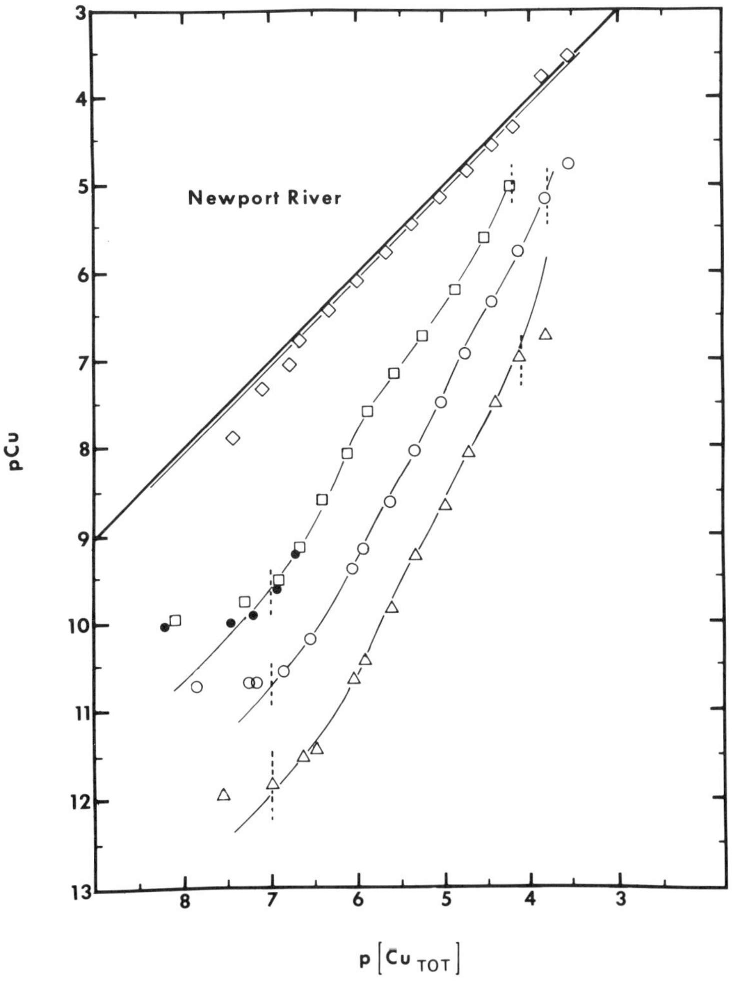

Figure 7. Copper titration of Newport River water at 25°C. (●) Untreated water at in situ pH 5.95; (□) glass–fiber filtered water at pH 5.95; (○) glass–fiber filtered water at pH 7.00; (△) glass–fiber filtered water at pH 8.00; (◇) UV-treated glass–fiber filtered water at pH 5.95. Model curves through data points were calculated according to stability constants determined in this work (Tables I and II). Dotted lines indicate limits on data used for calculation of conditional stability constants.

ligands. The predominance of binding by organic ligands is confirmed by an almost complete destruction of binding capability after UV-photooxidation (Figure 7). In the UV-treated Newport River water at pH 5.95, measured pCu values were in good agreement with those calculated from inorganic complexation in the range of p[Cu_{TOT}] 7 to 4. A slight increase in measured pCu of up to 0.4 units was observed relative to that predicted from inorganic complexation at p[Cu_{TOT}] > 7. This deviation from pCu values calculated from consideration of inorganic complexation at p[Cu_{TOT}] > 7 is similar to that observed in the model solution containing 0.01 M KNO_3 and 0.1 mM $NaHCO_3$ at pH 7 (Figure 4) and thus, may also be due to non-ideal electrode behavior at low total copper concentrations.

UV-treatment of filtered Neuse River water also caused a large decrease in the binding of copper at pH 6.78 and p[Cu_{TOT}] $\leq$ 7 consistant with predominance of binding of copper by organic ligands in the natural filtered river water. However, at values of p[Cu_{TOT}] > 7 in the photooxidized filtered Neuse River water (pH 6.78), the measured pCu values behave as if copper was highly bound by a site present at a low concentration ($\sim 10^{-7.5}$ M) with a relatively high stability constant, although we recognize that the cupric ion electrode may be exhibiting some degree of non-ideal behavior. Filtration of the UV-treated river water through a 0.2 µm filter (Nuclepore) reduced the background concentration of copper in the water from 0.025 µM to below the detection limit of total copper analysis (0.001 µM) indicating that the background copper was either adsorbed to or incorporated into filtrable particles. Titration of this refiltered water also gave a curve for measured pCu vs p[Cu_{TOT}] in agreement with a reduced level of copper binding. The nature of the particles that copper was apparently associated with is unknown as is the nature of the chemical association. The particles may have been hydrous metal oxides or a small fraction of organic matter resistent to photooxidation. We do not know whether the particles were initially present in the river water or were formed during photooxidation, or whether copper was adsorbed to the surface of the particles or incorporated into the particle matrix. All of these possibilities are completely consistant with the observed data. However, since the apparent degree of association of copper with the particles present in UV-treated water is small at p[Cu_{TOT}] $\geq$ 7, our conclusion that copper in the natural river water is primarily bound by organic matter at p[Cu_{TOT}] $\geq$ 7 is not affected significantly.

Analysis of Organic Binding. Scatchard plots for the binding of copper in filtered river water showed non-linearity indicative of the presence of binding sites having different stability constants (Figure 8). Similar non-linear behavior has also been observed in Scatchard plots for copper binding by natural organic

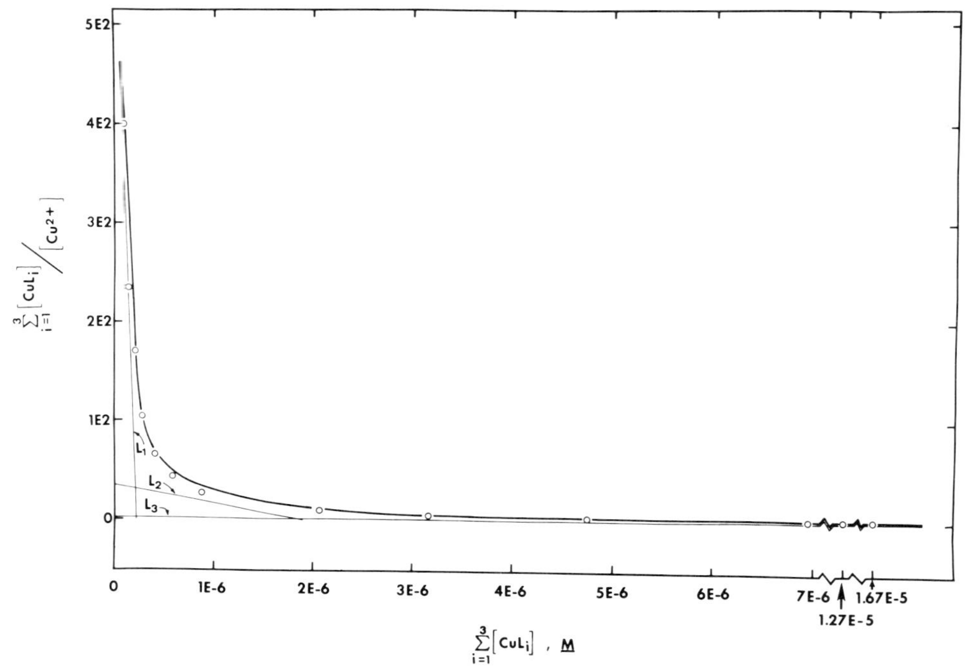

Figure 8. Scatchard plot for glass–fiber filtered Neuse River water at in situ pH 6.78 and 25°C. Linear segments shown for three resolved binding sites, L_1, L_2, and L_3. Model curve through data points was calculated from resolved stability constants and total concentrations for the three binding sites.

ligands from lake water and soil as measured by gel chromatography (10).

Scatchard plot analysis of data for river water titrations was conducted for values of $[Cu_{TOT}] \geq 10^{-7}$ M. This was done because of apparent non-ideal electrode behavior at $[Cu_{TOT}] < 10^{-7}$ M. Data at the highest concentrations of total copper in Newport River water at pH 7.0 and pH 8.0 were also excluded because pCu values approach those predicted for the formation of solid copper hydroxide, i.e., pCu 6.7 at pH 8.0 and 4.7 at pH 7.0.

In our analysis, we modeled the data for binding of copper by organic matter in accordance with the fewest number of binding sites required to account for the observed copper titration data. The titration data for each river and pH were analyzed separately by resolving the Scatchard plots into postulated one, two, three and four binding site models. The four possible models for each river and pH were tested against the data using a runs test and F-test (17) to detect and test the significance of systematic deviations. From these tests we conclude that at least three separate binding sites are required to model the Newport titration data at pH 5.95, 7.0 and 8.0 and the Neuse data at pH 6.78 (Figures 6 and 7). Two binding sites are required for the Neuse at pH 8.0 which is consistent with the restricted range of the titration data. The actual number of binding sites may be greater than the number resolved, since sites having similar stability constants cannot be resolved by the present technique. Values for total concentration of binding sites and conditional stability constants are given in Table II.

In general, a consistant set of conditional stability constants was obtained for the two rivers in which the constants increase with pH and decrease with the ratio of moles of binding sites per gram organic carbon (Figure 9). The relationship between stability constants and pH is similar for all three proposed binding sites in each river with values of $\Delta \log K_c / \Delta$ pH in the range 1.0 to 1.3. Mean values for $[L_{i-TOT}]$ for each of the i=1,2 or 3 types of binding sites are higher in the Newport River water relative to those for the Neuse by factors ranging from 4 to 5. Total ligand concentration $\Sigma [L_{i-TOT}]$, i.e., total binding capacity, is higher in the Newport by a factor of 5 which is in agreement with the five fold difference in dissolved organic carbon (DOC): 15 mgC ℓ^{-1} for the Newport and 3.0 mgC ℓ^{-1} for the Neuse. Thus, the general picture that emerges is that the binding characteristics of the organic matter in the Newport River water is similar to that in the Neuse with the major difference being the quantity of binding sites present as indicated by the difference in the quantity of dissolved organic carbon.

It is likely that most if not all of the observed binding of copper to organic matter results from complexation to fulvic or humic acids. Newport River water has a pronounced yellowish-brown color and shows continuously increasing light absorption with decreasing wave length (13) consistent with the presence of these substances. From unpublished data we know that nitrogen to carbon ratios (0.02 gN gC^{-1}), hydrogen to carbon ratios (0.11 gH gC^{-1}) and carboxyl group to carbon ratios (13 meq gC^{-1}) for the dissolved organic matter in Newport River water are similar to those of humic and fulvic acids extracted from soils (18). Similarly, Beck et al. (19) have reported that the dissolved organic matter in another coastal plain river of the southeastern United States, the Satilla, has chemical properties indistinguishable from those of fulvic acids. Postulated binding sites identified as L_1, L_2, and L_3 may represent individual groups of molecules within complex mixtures of humic and fulvic compounds or different sites on polyelectrolytic molecules (20) and/or colloidal particles.

The increase in log K_c with pH is typical of copper binding to weak acid functional groups. For the reaction of copper with a protonated binding site H_nL,

$$Cu + H_nL = CuL + nH \qquad (9)$$

observed slopes of Δ log K_c/ Δ pH (Figure 9) indicate values of n in the range of 1.0 to 1.3. A second possibility is that the observed increase in binding with pH results from an increase in negative charge on polyelectrolytic molecules or the surface of colloids (21). Both effects may be important.

Binding of metals to humic compounds is thought to occur primarily by chelation with sites consisting of a carboxyl group and an adjacent phenolic group (salicylate type sites) and/or with sites composed of two adjacent carboxyl groups on aromatic rings (phthalate groups) (18). Copper binding capacities for Newport River organic matter (10 $\pm$ 1 mmol gC^{-1}) at pH 7.0 and 8.0 and for the Neuse (11 $\pm$ 1 mmol gC^{-1}) at pH 6.8 are only slightly less than the content of carboxyl groups for Newport River organic matter (13 mmol gC^{-1}; Sunda, unpublished data). This is consistent with the binding of copper by carboxyl groups and indicates that there are sufficient carboxyl groups to account for the observed binding of copper. In addition, slopes (Δ log K_c/ Δ pH) of 1.0 to 1.3 observed for all three binding sites in the pH range 6 to 8 may result primarily from reactions in which a copper ion displaces a hydrogen ion from a phenolic group (equation 9), since these groups will be primarily protonated at neutral pH values whereas carboxyl groups will be mostly deprotonated. Therefore, of the proposed mechanisms for copper binding to humic compounds, our data are most consistent with copper binding to salicylate type chelation sites.

Our results show that conditional stability constants for binding of copper by natural organic ligands are highly dependent on the pH at which the measurement was conducted and the portion of the titration curve which was examined. Published stability constants for fulvic acids often vary considerably and Cheam (22) has reported that these apparent differences can be reasonably

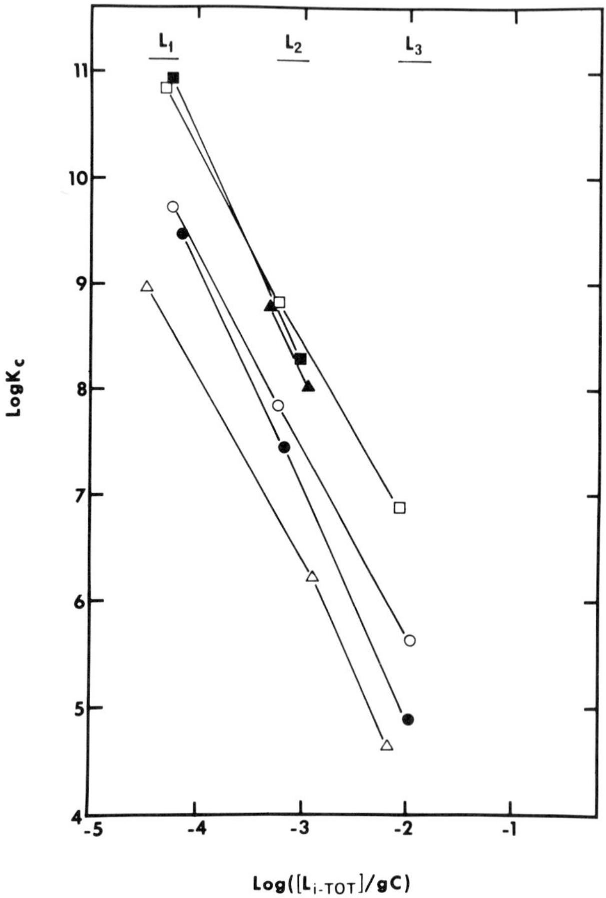

Figure 9. Relationship between the log of the conditional stability constant and the log of the binding site concentration per gram dissolved organic carbon for individual binding sites. (□) Newport River pH 8.00; (■) Neuse River pH 8.00; (○) Newport River pH 7.00; (●) Neuse River pH 6.78; (△) Newport River pH 5.95; (▲) Lake water (10).

accounted for by differences in pH and the ratio of $[Cu_{TOT}]$/gram organic matter. Our conditional constants at pH 8.0 for that portion of the binding sites designed as L_2 are in good agreement with stability constants measured by Mantoura and Riley (10) for binding of copper by organic matter from lake water at the same pH and similar ratio of moles of binding sites per gram organic carbon (Figure 9). Mantoura and Riley, who used a gel filtration technique to measure stability constants, found their data consistent with the presence of two separate binding sites. To model our titration data we usually had to postulate the existence of no fewer than three separate organic binding sites. The apparent difference between results of the two studies can be resolved if we note that our titrations usually covered 2 to 3 orders of magnitude change in bound copper concentration whereas, the experiments of Mantoura and Riley (10) covered only slightly greater than one order of magnitude. Thus, it was impossible for these investigators to determine the characteristics of copper binding outside of their experimental range.

Chemical Speciation Models for the Neuse and Newport River Waters. Copper speciation models for filtered Newport and Neuse River waters as a function of total copper concentration at *in situ* pH values (Figures 10a and 11a) indicate that soluble copper is greater than 98% bound to organic ligands in the range of total copper concentration normally encountered in river water, i.e., $p[Cu_{TOT}] \geq 7$. At these concentrations, copper is primarily complexed by organic binding site L_1, a site representing ligands present at low concentrations with high stability constants. Site L_1 accounts for about 1% of the total binding capacity but >90% of the bound copper at $p[Cu_{TOT}] \geq 7$. Only at extremely high concentrations of total copper, i.e., $p[Cu_{TOT}] \leq 3.7$ in the Newport and ≤ 4.5 in the Neuse, would inorganic species of copper become dominant. Increasing the pH of the Newport from 5.95 to 7.0 (Figures 10a and 10b) without changing carbonate alkalinity does not alter the order of dominance of copper species. However, in the natural range of total copper concentrations (i.e., $p[Cu_{TOT}] \geq 7$) increasing pH to 7.0 does increase the extent of total organic binding relative to inorganic species. This results from the strong pH dependence of conditional stability constants for copper complexes with natural organic ligands as discussed previously.

At *in situ* pH and *in situ* concentrations of dissolved copper in the Neuse (0.025 μM) and the Newport (0.011 μM), the models predict similar pCu values for the two rivers (10.4 and 10.6 for the Neuse and Newport). Thus in consideration of the major factors controlling pCu values in the two rivers, the higher concentration of organic binding sites and lower total dissolved copper in the Newport are almost exactly compensated for by the higher pH of the Neuse.

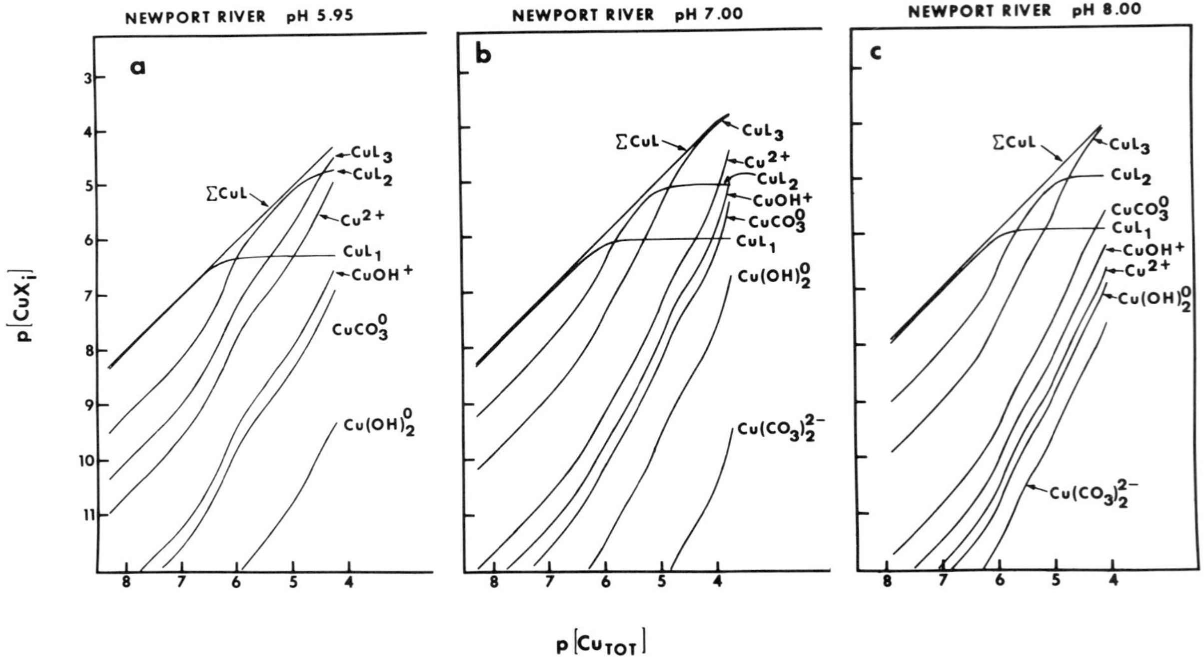

Figure 10. Chemical speciation model for dissolved copper in the Newport River at 25°C as a function of total copper concentration. (a) In situ pH 5.95, [Alk] = 0.05mM and I = 0.0005M; (b) pH 7.00, [Alk] = 0.05mM and I = 0.005M; and (c) pH 8.00, [Alk] = 0.55mM and I = 0.001M. DOC = 15 mgCL^{-1}.

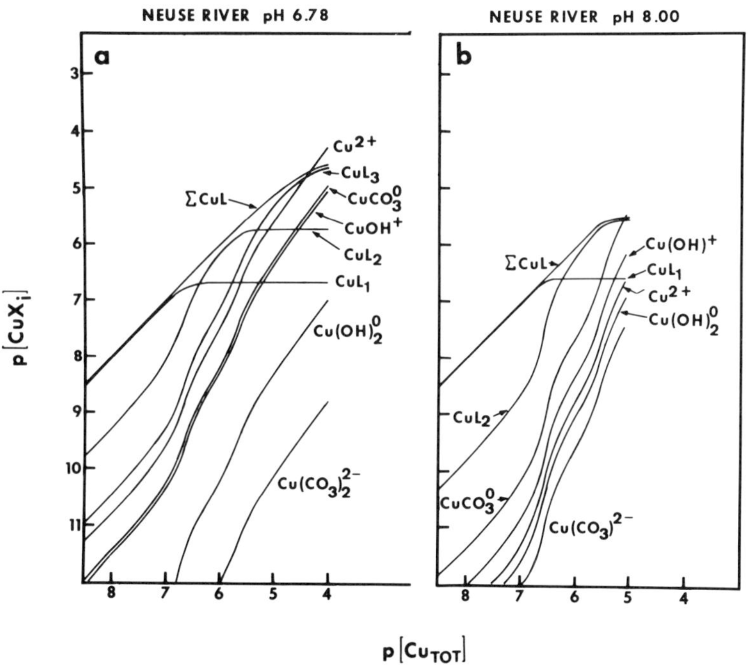

Figure 11. Chemical speciation model for dissolved copper in the Neuse River at 25°C as a function of total copper concentration. (a) In situ pH 6.78, [Alk] = 0.15mM and I = 0.0005M; and (b) pH 8.00, [Alk] = 0.65mM and I = 0.001M. DOC = 3 mgCL^{-1}.

Models for copper speciation as a function of total concentration of organic sites $p[L_{TOT}]$ (Figures 12 and 13) were calculated at $p[Cu_{TOT}]$ 7 with the assumption that total organic matter was varied without alteration of the relative concentrations of individual binding sites with respect to the total. Copper species are also shown versus dissolved organic carbon (DOC) using measured values of 15 and 3.0 mgC ℓ^{-1} for the Newport and Neuse, respectively. At *in situ* pH (Figures 12a and 13a), the dominance of organic bound species is again demonstrated by the observation that DOC would have to drop below approximately 0.4 and 0.2 mgC ℓ^{-1} for the Newport and Neuse respectively, before inorganic species of copper would dominate. The effect of pH on organic complexation is also evident as DOC in the Newport would have to be reduced to approximately 0.1 mgC ℓ^{-1} at pH 7.0 and 0.08 mgC ℓ^{-1} at pH 8.0 before inorganic species would dominate over organic species. Similarly, DOC for the Neuse would have to be reduced to 0.1 mgC ℓ^{-1} at pH 8.0.

In a recent review, Duce and Duursma (23) conclude that our knowledge of dissolved organic carbon in the world's rivers is limited. From available data for such rivers as the Amazon, Hudson, Mississippi, MacKenzie and others, dissolved organic carbon was found to range from 2 to 5 mgC ℓ^{-1}. These values are comparable to that found in the Neuse (3.0 mgC ℓ^{-1}), but lower than in the Newport (15 mgC ℓ^{-1}). Comparing this range of DOC concentration with the results of our work and assuming that the complexing characteristics of the organic matter remains approximately the same among watersheds, we would expect copper chemistry in the world's rivers to be dominated by copper-organic interactions. This conclusion agrees with that of Mantoura *et al.* (24) as based on their copper speciation model for river water.

Copper speciation in the Newport and Neuse Rivers was calculated as a function of carbonate alkalinity at several pH values for $p[Cu_{TOT}]$ 7 and the natural suite of organic ligands determined previously (Figures 14 and 15). The alkalinity range was selected to reflect that expected in the world's rivers. From the data of Livingstone (25), the alkalinity of the world's average river was calculated to be approximately 0.96 m*M*. Stumm and Morgan (9) present data that suggest carbonate alkalinities in the world's freshwater would fall in the approximate range 0.1 to 8 m*M*.

In the Newport River with a DOC of 15 mgC ℓ^{-1}, organic copper species clearly dominate over the full range of alkalinities (p[Alk] 5 to 2) at pH 5.95, 7.0 and 8.0. At $p[Cu_{TOT}]$ 7 and $p[L_{TOT}]$ 3.7 to 3.9, the organic binding site concentrations and stability constants are sufficiently large for organic matter to successfully outcompete inorganic ligands. Thus, the copper complexes with individual fractions of organic binding sites are essentially invariant with alkalinity (Figures 14a-c). As pH is increased the relative amounts of inorganic species of copper decline especially at low alkalinities. As alkalinity increases

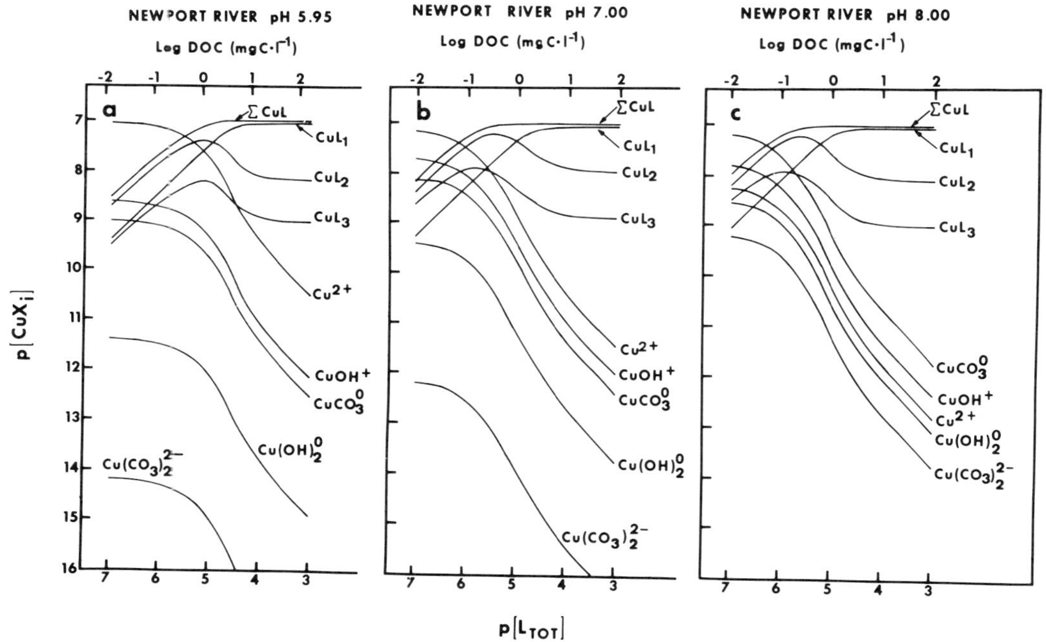

Figure 12. Chemical speciation model for dissolved copper in the Newport River at 25°C as a function of total organic binding site concentration. (a) In situ pH 5.95, [Alk] = 0.05mM and I *= 0.0005M; (b) pH 7.00, [Alk] = 0.05mM and* I *= 0.0005M; and (c) pH 8.00, [Alk] = 0.55mM and* I *= 0.001M.*

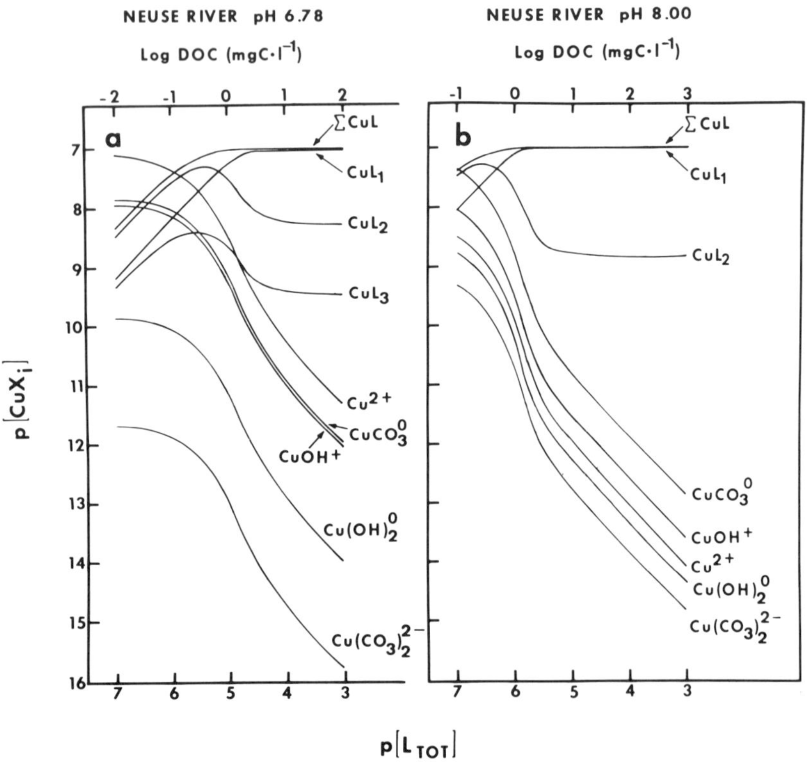

Figure 13. Chemical speciation model for dissolved copper in the Neuse River at 25°C as a function of total organic binding site concentration. (a) In situ pH 6.78, [Alk] = 0.15mM and I = 0.0005M; and (b) pH 8.00, [Alk] = 0.65mM and I = 0.001M.

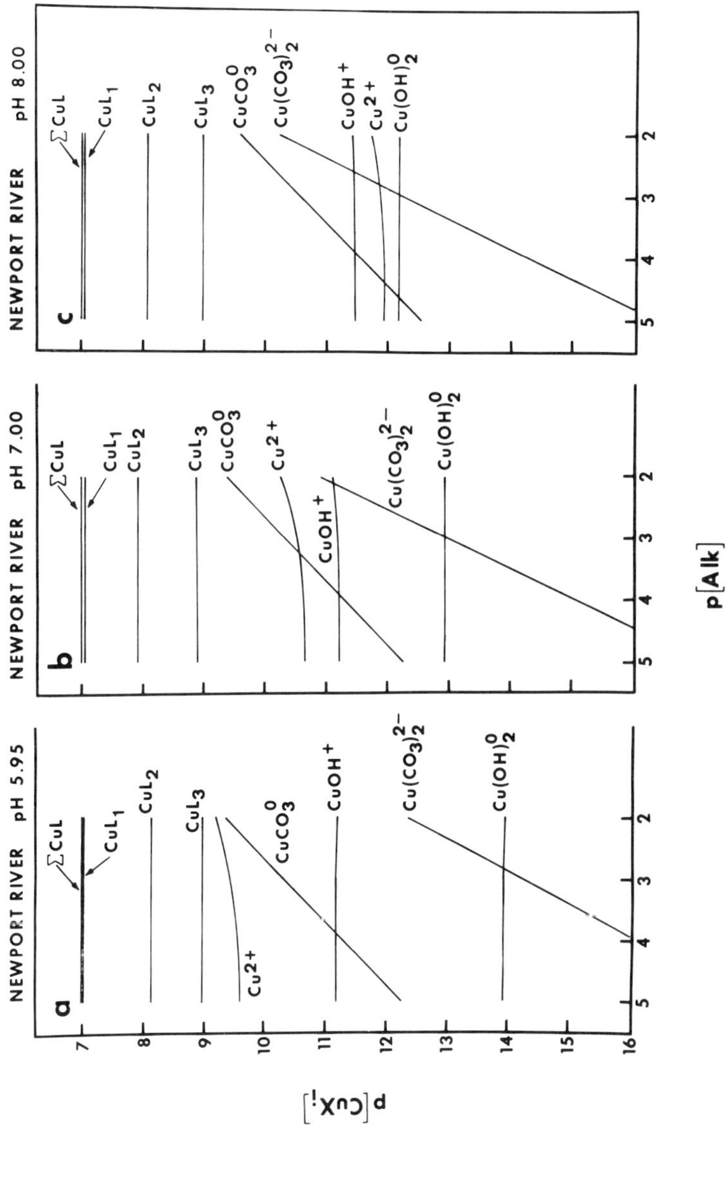

Figure 14. Chemical speciation model for dissolved copper in the Newport River at 25°C as a function of carbonate alkalinity at p[Cu_{TOT}] 7.

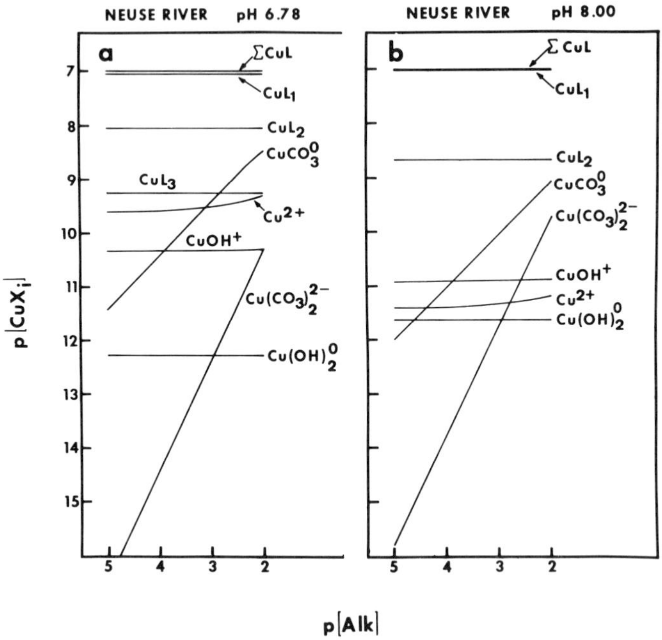

Figure 15. Chemical speciation model for dissolved copper in the Neuse River at 25°C as a function of carbonate alkalinity at p[Cu_{TOT}] *7*

the relative importance of the monocarbonato complex increases to dominate the inorganic species for [Alk] > $10^{-3.2}$ at pH 7.0 (Figure 14b) and [AlK] > $10^{-3.9}$ at pH 8.0 (Figure 14c). At *in situ* pH 5.95 (Figure 14a) free cupric ion is the dominant inorganic species at all alkalinities.

For the Neuse River, the copper speciation follows the same general pattern (Figures 15a and b) with clear dominance of organic species for all alkalinities (p[Alk] 5 to 2) at pH 6.78 and 8.0. The lower concentration of organic matter in the Neuse allows stronger competition by inorganic ligands for copper. At the *in situ* pH 6.78 and [AlK] > $10^{-2.8}$ the $CuCO_3$ complex outcompetes the most concentrated but least strongly complexing organic binding site L_3 (Figure 15a).

In our models, we have assumed that conditional stability constants are independent of alkalinity. This assumption is not strictly valid because increased alkalinity would usually be associated with increased concentrations of calcium and magnesium and these two cations may compete with copper for binding to natural organic ligands (24). Increased alkalinity will also result in increased ionic strength which may affect copper binding to organic matter. Previous measurements of copper binding in Newport River water using an ion-selective electrode have shown, however, that ionic strength has only a slight effect on the binding of copper by organic matter. Increasing the ionic strength of a sample of filtered Newport River water by KNO_3 addition from an initial value of ∿ 10^{-3} *M* to a value of 10^{-1} *M* at p[Cu_{TOT}] 6.0 and pH 7.7 caused the value of log ([Cu^{2+}]/[Cu-organic]) to increase by only 0.2 to 0.3 units (Sunda, unpublished data). In the same sample of Newport River water containing an ionic strength buffer (0.1 *M* KNO_3) at the same p[Cu_{TOT}] (6.0) and the same pH (7.7), increasing the calcium concentration by addition of $Ca(NO_3)_2$ from a backgound concentration of 0.26 *mM* to a value of 2.7 *mM* decreased values of log ([Cu^{2+}]/[Cu organic]) by less than 0.2 units (from an initial value of -3.56 to a value of -3.42). Likewise under the same experimental conditions, increasing the magnesium concentration by $Mg(NO_3)_2$ addition from background value of 0.04 *mM* to a concentration of 0.94 *mM* caused values of log ([Cu^{2+}]/[Cu-organic]) to increase by only 0.2 units. Since the majority of terrestrial waters have concentrations of calcium and magnesium within the concentration ranges investigated (9), the omission of the influence of changing concentrations of calcium and magnesium ions should have had only a minor influence on our binding models.

The general conclusion from the copper speciation models is that dissolved copper is predicted to occur in rivers principally as organic complexes over the range of total copper, pH, dissolved organic matter and alkalinity values encountered in world river waters. The principal variables controlling organic binding will be the concentration and composition of organic matter, pH and total copper concentration. In cases where the toxicity of dissolved copper to organisms is determined by free cupric ion activity, we would predict toxicity of dissolved copper to increase with decreasing dissolved organic matter and decreasing pH.

Acknowledgements

The authors wish to thank Lillian A. Flanagan and Jo Ann M. Lewis for their able assistance in the laboratory. We thank David R. Colby for helpful discussions during the preparation of this paper. The research on which this manuscript is based was supported by an interagency agreement between the Department of Energy and the National Marine Fisheries Service. This manuscript is contribution number 78-53B, Southeast Fisheries Center, National Marine Fisheries Service, NOAA, Beaufort, N.C., USA. Reference to trade names does not imply endorsement by the National Marine Fisheries Service.

Abstract

The complexation of copper by organic and inorganic ligands was investigated in the Neuse and Newport Rivers, two North Carolina rivers that have widely different concentrations of dissolved organic carbon (3 and 15 mgC ℓ^{-1}). Potentiometric titrations with a cupric ion electrode were used to measure complexation of copper by organic and inorganic ligands in river waters and chemically defined solutions. Stability constants for complexation of copper by OH^- and CO_3^{2-}, the dominant inorganic ligands for copper in natural waters, were determined. Constants at infinite dilution for $CuOH^+$, $CuCO_3$, and $Cu(CO_3)_2^{2-}$ ($10^{6.48}$, $10^{6.74}$, and $10^{10.24}$, respectively) agree favorably with previously published values. However, our constant for $Cu(OH)_2$ ($10^{11.78}$) is $\sim$ 2 orders of magnitude lower than values that have been widely used in equilibrium calculations of copper speciation in natural waters. As a result many published equilibrium models for copper appear to have overestimated the importance of the copper dihydroxo species resulting in some cases in an appreciable overestimation of the total level of copper complexation.

Copper was highly complexed by dissolved materials in Neuse and Newport river waters. A marked reduction in complexation following UV-photooxidation of organic matter indicated that copper was bound predominantly to organic ligands. Binding characteristics of the organic matter in the two rivers was similar and a pronounced increase in binding with pH suggested complexation of copper by protonated weak acids. Scatchard plot analysis of the Cu binding data indicated the presence of at least three organic binding sites in each river whose conditional stability constants increased with increasing pH and decreasing ratio of binding site concentration per gram organic carbon. Copper speciation models were computed for the range of pH, organic matter concentration, alkalinity and total copper typically found in rivers. These models predicted that the speciation of copper in the world's rivers will be dominated by complexes with natural organic ligands. Binding of copper by organic ligands should have a marked influence on biological and geochemical reactivity of copper in rivers, affecting important processes and phenomena such as toxicity and nutritional availability to organisms, adsorption onto surfaces, precipitation, and solid solution. Computational models for the chemical speciation of copper in natural waters that ignore organic complexation may inaccurately describe the chemistry of copper in terrestrial waters and perhaps marine waters as well.

Literature Cited

1. Sunda, W.G., and Guillard, R.R.L. The relationship between cupric ion activity and the toxicity of copper to phytoplankton. J. Mar. Res. 34, 511-529 (1976).
2. Anderson, D.M. and Morel, F.M.M. Copper sensitivity of Gonyaulax tamarensis. Limnol. Oceanog. 23, 283-295 (1978).
3. Sunda, W.G., Engel, D.W. and Thuotte, R.M. Effect of chemical speciation on toxicity of cadmium to grass shrimp, Palaemonetes pugio: importance of free cadmium ion. Environ. Sci. Tech. 12, 409-413 (1978).
4. Andrew, R.W., Biesinger, K.E. and Glass, G.E. Effects of inorganic complexing on the toxicity of copper to Daphnia magna. Water Res. 11, 309-315 (1977).
5. Jackson, G.A. and Morgan, J.J. Trace metal-chelator interactions and phytoplankton growth in seawater media: Theoretical analysis and comparison with reported observations. Limnol. Oceanog. 23, 268-282 (1978).
6. Sunda, W.G., and Lewis, J.M. Effect of complexation of natural organic ligands on the toxicity of copper to a unicellular alga, Monochrysis lutheri. Limnol. Oceanogr. (in press).
7. Bilinski, H., Huston, R. and Stumm, W. Determination of the stability constants of some hydroxo and carbonato complexes of Pb(II), Cu(II), Cd(II) and Zn(II) in dilute solutions by anodic stripping voltammetry and differential pulse polarography. Anal. Chim. Acta 84, 157-164 (1976).
8. Armstrong, F.A.J., Williams, P.M. and Strickland, J.D.H. Photo-oxidation of organic matter in seawater by ultraviolet radiation, analytical and other applications. Nature 211, 481-483 (1966).
9. Stumm, W. and Morgan, J.J. "Aquatic Chemistry," 583 p. Wiley Interscience, New York. 1970.
10. Mantoura, R.F.C. and Riley, J.P. The use of gel filtration in the study of metal binding by humic acids and related compounds. Anal. Chim. Acta 78, 193-200 (1975).

11. Sillen, L.G. and Martell, A.E. "Stability Constants," 754 p. Special Publication No. 17, The Chemical Society of London, 1964.
12. Sillen, L.G. and Martell, A.E. "Stability Constants," 865 p. Special Publication No. 25, The Chemical Society of London. 1971.
13. Sunda, W.G. "Relationship Between Cupric Ion Activity and the Toxicity of Copper to Phytoplankton." Ph.D. thesis, 168 p. Mass. Inst. Tech., Cambridge. 1975.
14. Paulson, A.J. "Potentiometric Studies of Cupric Hydroxide "Complexation." M.S. thesis, 102 p. Univ. Rh.I., Kingston, 1978.
15. Vuceta, J. and Morgan, J.J. Hydrolysis of Cu(II). Limnol. Oceanog. 22, 742-745 (1977).
16. Sheldon, R.W. Size separation of marine seston by membrane and glass-fiber filters. Limnol. Oceanogr. 17, 494-498 (1972).
17. Draper, N.R. and Smith, H. "Applied Regression Analysis," 407 p. John Wiley, New York. 1966.
18. Schnitzer, M. and Khan, S.U. "Humic Substances in the Environment." 327 p. Marcel Dekker, Inc. New York. 1972.
19. Beck, K.C., Reuter, J.H. and Perdue, E.M. Organic and inorganic geochemistry of some coastal plain rivers of the southeastern United States. Geochim. Cosmochim. Acta 38, 341-364 (1974).
20. Gamble, D.S. Titration curves of fulvic acid: the analytical chemistry of a weak acid polyelectrolyte. Can. J. Chem. 48, 2662-2669 (1970).
21. Wilson, D.E. and Kinney, P.J. Effects of polymeric charge variations on the proton-metal equilibria of humic materials. Limnol Oceanog. 22, 281-289 (1977).
22. Cheam, V. Chelation study of copper (II): fulvic acid system. Can. J. Soil Sci. 53, 377-382 (1973).
23. Duce, R.A. and Duursma, E.K. Inputs of organic matter to the oceans. Mar. Chem. 5, 319-339 (1977).
24. Mantoura, R.F.C., Dickson, A. and Riley, J.P. The complexation of metals with humic materials in natural waters. Est. Coast. Mar. Sci. 6, 387-408 (1978).
25. Livingstone, D.A. Chemical composition of rivers and lakes. U.S. Geological Survey Paper 440G, 64 p.(1963).

Disclaimer: The reviews expressed and/or the products mentioned in this article represent the opinions of the author(s) only and do not necessarily represent the opinions of the National Oceanic and Atmospheric Administration.

RECEIVED November 16, 1978.

9

Nickel Complexes with Soil Microbial Metabolites—Mobility and Speciation in Soils

R. E. WILDUNG, T. R. GARLAND, and H. DRUCKER

Battelle, Pacific Northwest Laboratories, Richland, WA 99352

It is well-established that inorganic physicochemical mechanisms play a predominant role in controlling trace element solubility in soils and sediments. However, soluble species of trace elements which hydrolyze in the neutral pH range, or tend to form cationic inorganic species with intermediate to high ionic potentials are often present in natural waters as organic complexes. Less is known of the form of trace elements in soil and sediment solutions, but on the basis of detailed reviews of inorganic and organic processes influencing trace element cycling, it recently has been concluded that dissolved organic matter at the sediment water interface serves to increase the concentration of trace elements in waters by decreasing sorption rate on the solid phase (1, 2).

Trace element organic complexes in soils may be generally categorized on the basis of their solubility (3), although considerable overlap likely occurs. The major classes of complexes are 1) relatively high molecular weight humic substances that have a high affinity for metals but are largely insoluble in soils, and 2) relatively low molecular weight nonhumic substances derived largely from microbial cells and metabolism that exhibit a range in solubilities in association with metals. The origin and properties of organic materials in both categories in relation to metal complexation in soils and sediments were recently overviewed (1, 4). Of the humic substances, the humates (alkali soluble, acid insoluble) and fulvates (alkali and acid soluble) constitute up to 90% of the soil organic matter (5). Both fractions exhibit high charge density due principally to acidic functional groups that lead to a strong pH dependent affinity for cations in solution and strong association with soil minerals and other organic constituents in soils (6). Although largely insoluble in soils, the fulvates, thought to be of lower molecular weight than the humates, may, in particular, have potential for formation of soluble complexes with metals (7). Nonhumate materials are also likely to be of importance in metal solubilization in soil. These consist of components of living cells, their

0-8412-0479-9/79/47-093-181$05.00/0
© 1979 American Chemical Society

exudates and the entire spectrum of degradation products which ultimately serve as building units for the soil humic fraction. Because of the generally high turnover rate of microorganisms and readily decomposable organic matter in soils, nonhumate materials, in contrast to humic substances, are inherently transitory and the relative quantities and composition in soil may be expected to vary with carbon sources and environmental conditions (8, 9).

Limited investigations (7) have attributed most of the titratable acidity in soil solution to the aliphatic acids (10) and amino acids (11). However, a wide range of organic acids and bases of microbial origin including simple aliphatic acids, carboxylic acids derived from monosaccharides, products of the citric acid cycle, and aromatic acids are likely present in soil solution (10, 12). Recent evidence indicates that soil microorganisms produce water soluble ligands with a high affinity for a range of metals (13, 14) and microbial activity influences Pu solubility in soil (15). Unfortunately, although the presence of organic complexes of Cu, Zn, and Mn in soil solution has been reported (16, 17) few investigations, summarized by Mortensen (18) and Stevenson and Ardakani (12), have identified specific water-soluble ligands capable of metal complexation in soil, and intact organometal complexes have not been isolated and identified. Thus, evidence in support of the presence of soluble metal complexes in soil is largely circumstantial and there is a paucity of knowledge regarding the nature, behavior, and role of organic ligands in geochemical cycling. The development of an understanding of the role of microorganisms in metal complexation has been limited by the complexity of soil, sediment and microbial systems and difficulties in the experimental separation of the effects of microbial processes from physicochemical processes in soils and sediments.

To aid in understanding the mechanisms of trace metal complexation by soil microorganisms, an experimental approach was developed which entailed 1) isolation of organisms from soil on the basis of trace metal tolerance and C (carbon) requirements, 2) examination of metal transformations and form on growth of isolated organisms *in vitro*, 3) determination of the mobility and form of stable metal complexes in soil, and 4) identification and detailed study of metal complexes exhibiting high metal affinities and soil solubility as determined from steps 1) through 3). This protocol was applied to the isolation and examination of metal complexes formed on growth of soil bacteria and fungi exposed to metals (Cd, Cr, Ni, Pu, Tl) with a range of properties and the chemistry of important Ni complexes was examined in detail.

Materials and Methods

Bacteria and fungi were isolated from soil on the basis of metal tolerance and C source, grouped on the basis of morphological, growth and physiological parameters and examined for ability to alter the solubility and form of metals in exocellular solutions and alter metal solubility in soil. Soluble Ni complexes were subsequently characterized in detail.

Isolation of Bacteria and Fungi from Soil Using Enrichment Techniques. Two enrichment procedures were employed to isolate microorganisms from soil: a tertiary enrichment, in which organisms were isolated from soil after incubation to log phase in the presence of added metal, and a two-phase enrichment, in which organisms were isolated from the unincubated soil and grown on a number of different C sources in the presence of metals.

In the tertiary enrichment procedure, soil (Ritzville silt loam) was amended with starch (1.0%), NH_4NO_3 (0.5%), and Cd, Cr, Ni, Tl (1, 10, and 100 μg/g) or Pu (0.05, 5 and 10 μCi/g). The soil, with amendments, was brought to 22% moisture and incubated (28°C) under aerobic conditions (continuous flow of CO_2 free air). At mid-log growth phase, as determined by CO_2 evolution rate, an aliquot (1 ml) of a 1:10 incubated soil:water slurry was inoculated into an enriched soil extract medium containing glucose (1%), NH_4NO_3 (0.5%), K_2HPO_4 (0.05%), soil extract (30%) and Cd, Cr, Ni, Pu, or Tl added in soluble form, to achieve Cd, Cr, Ni and Tl concentrations of 1, 10, and 100 μg/ml and Pu concentrations of 0.05, 5 and 10 μCi/ml. Controls were identical except metals were not added. The pH was adjusted to 7.0 for aerobic bacteria and 6.0 for fungi. Streptomycin sulfate (0.4%) was added to the fungal enrichment to prevent growth of bacteria. The inoculated cultures were incubated (28°C) with shaking (150 rpm) until maximum cell density was obtained and secondary enrichments were initiated by transferring an inoculum of each of the primary enrichments to fresh, enriched soil extract medium containing the metals at the same concentrations. The tertiary enrichment was then conducted in a manner similar to the secondary enrichment. Pure cultures of organisms were isolated at the end of each enrichment using successive pour plate and streaking until cultures differing in colony morphology were resolved. The pure cultures were maintained in stock on agar slants in the presence and absence of the metals at the concentrations used in isolation. The stock cultures were passed to fresh slants monthly in order to assure viability.

A two-phase enrichment procedure was utilized to select soil organisms on the basis of ability to metabolize classes of organic compounds in the first phase; and on ability to grow in the presence of metals in the second phase. In the first phase, aliquots of a standard mineral base medium (Table I, glucose used only in control) were separately amended with 1) aromatic acids,

Table I. Composition of standard mineral base medium employed in microbial studies.

Solution Designation	Volume per liter of Medium (ml)	Composition
A	100	685 ml of KH_2PO_4(0.2M), 315 ml of Na_2HPO_4(0.2M) diluted to 2000 ml
B	10	20.0 g $MgSO_4 \cdot 7H_2O$, 1.325 g $CaCl_2 \cdot 2H_2O$ dissolved in 1000 ml
C	1	50 mg $ZnSO_4 \cdot 7H_2O$, 50 mg $MnSO_4 \cdot H_2O$, 12 mg $CuSO_4 \cdot 5H_2O$, 15.9 mg $Co(NO_3)_2 \cdot 6H_2O$, 19.0 mg $Na_2B_4O_7 \cdot 10H_2O$, 235 mg $Na_2MoO_4 \cdot 2H_2O$, 10 ml Fe·EDTA solution (17.9 g $Na_2EDTA \cdot 2H_2O$ and 3.23 g KOH in 186 ml added to 13.7 g $FeSO_4 \cdot 7H_2O$ in 364 ml H_2O. Bubbled filtered air through solution overnight and adjusted pH to 5.0 with 8M HNO_3) dissolved in 100 ml
D	10	10 g NH_4Cl dissolved in 100 ml
E [1]	100	15 g glucose dissolved in 1000 ml
F	100	2.5 g yeast dissolved in 1000 ml

[1] Glucose was not used in studies of the effect of C sources.

including benzoate (0.5%), p hydroxy benzoate (0.5%), m hydroxy benzoate (0.5%), and tryptophan (0.11%); 2) organic acids, including succinate (0.5%), malate (0.5%), lactate (0.5%), and acetate (0.5%); 3) sugars, including glucose (0.5%), sucrose (0.5%), and fructose (0.5%); and 4) an organically rich medium containing yeast extract and peptone. Aliquots of these media were inoculated with a soil (1 g) and incubated with shaking (200 rpm) for 120 hours. From each of these primary enrichments, a secondary enrichment was performed by inoculating aliquots of sterile medium containing the same C sources and Cd, Cr,· Ni, Tl (1, 10, and 100 μg/ml) or Pu (0.05, 5 and 10 μCi/ml). The inoculated metal-containing medium was incubated (28°C) with shaking

(200 rpm) for 48 hr. Pure cultures were isolated from the media exhibiting microbial growth (turbidity) at the highest metal concentration using pour plate and successive streaking techniques. Morphologically different bacterial and fungal cultures were placed into stock and maintained as previously described. Bacterial and fungal enrichments differed only in that bacterial enrichments were conducted with media adjusted to pH 7.0 whereas the fungal enrichments were conducted with media adjusted to pH 6 and contained streptomycin sulfate (0.4%).

Morphological and Physiological Characteristics of Microorganisms Isolated from Soil. The 239 bacterial isolates obtained in pure culture from the two enrichment procedures were classified on the basis of morphological, growth and physiological parameters. These were measured using standard methods (19). Aliquots (2 ml) of stock cultures were transferred to a standard mineral base medium (Table I), incubated (28°C) on a rotary shaker (150 rpm) for 24 hr. Motility (wet mount) and gram reaction were measured. The culture was then used to inoculate an agar plate of standard mineral base medium which was cultured for an additional 24 hr and used for the catalase and oxidase tests. For the gram positive and spore-forming bacilli, additional tests were conducted on 24-48 hr cultures (standard mineral base agar). These included acetoin production by the Voges-Proskauer test; acid production from the mixed sugars, arabinose, mannose, and xylose; acid production from glucose; and starch hydrolysis. Spore location and morphology were determined in older cultures. Pigment production on starch agar was also used to subdivide certain of the spore-forming bacilli into major biotypes of *Bacillus mycoides*. Twenty strains considered representative of the cultures in the collection on the basis of these tests and chemical characteristics of excellular products (described below) were examined to determine if exocellular complexation of Ni altered Ni mobility in soil relative to inorganic forms.

A total of 250 fungal isolates were obtained in pure culture. Colony morphologies of many of the isolates were similar. On the basis of colony morphology on standard mineral base medium, organisms were separated into 59 types. These types were retained in stock to examine metal resistance characteristics and their ability to exocellularly modify Ni form and solubility in soil. Further classification to the species level is underway.

Modification of Nickel Form by Soil Microbial Isolates. To determine the ability of the microbial isolates to modify the chemical form of Ni, standard mineral base medium (15 liters) was prepared (Table I) and separated into aliquots (500 ml). The aliquots were frozen rapidly on dry ice and stored (-26°C) until used. Immediately prior to use, the aliquots were rapidly thawed at 37°C in a water bath, filter sterilized (0.2 μ) and separately amended with unlabeled Ni (as $NiCl_2$) and its radioisotope (^{63}Ni, 0.3 μCi/ml). In order to develop a bacterial inoculum for this medium, stock cultures were grown to log phase in standard mineral base medium (10 ml) without metal. An aliquot (0.25 ml) of this culture was used as an inoculum for the growth medium (10 ml) containing metal, and the cultures were incubated (28°C) with shaking (150 rpm) for 48 hr. The fungal inoculum was developed similarly, except fungi from stock were streaked on standard mineral base agar from the stock solution and incubated (28°C) until sporulation was evident. An aliquot (10 ml) of sterile buffer (Table I, Solution A) was added to the surface and the spore suspension pipetted aseptically into tubes. An inoculum (0.5 ml) of this spore suspension was added to the growth medium containing Ni (10 μg/ml). The organisms were incubated (28°C), with shaking (150 rpm) for 72 hr.

At the end of growth in the Ni-containing medium, the cells were separated by filtration (0.4 μ Nuclepore) and dry weights were determined. An aliquot of the filtrate (<0.4 μ) was refiltered (<0.01) and the Ni in this fraction was taken as soluble. Measurements on each culture also included initial and final pH, cell density, and the initial and final Ni concentration in the filtered (<0.4 and <0.01 μ) exocellular medium.

The extent of modification of the chemical form of Ni on microbial growth was initially assessed using thin-layer chromatography and electrophoresis. An aliquot of the filtered (<0.01 μ) growth medium (all Ni concentrations) was adjusted to pH 5 $\pm$ 0.1 and spotted (10 μl) on 0.1 mm cellulose (Mn 300) plates. Ascending thin-layer chromatography (TLC) was performed using a solvent containing ethanol, NH_4OH and water (6.6:1.5:1.0). Thin-layer electrophoresis (TLE) was performed (400 V) using 0.1 M HEPES buffer on a separate plate. After drying, the location of Ni on the plates was determined by autoradiography. Modification of Ni was ascertained by comparison of the autoradiographic pattern obtained before and after microbial growth.

Solubility of Exocellular Nickel Complexes in the Soil. To evaluate the effect of microbial metabolites on Ni solubility in soil, the soluble exocellular solution, separated after growth of the 20 selected bacterial isolates and all fungal isolates, was eluted through a column of soil. The concentrations and forms of Ni in the eluates were compared to the concentrations and forms of Ni in water and in the growth medium before and after microbial growth. A column (0.8 x 4 cm) was packed (1.5 g/cc) with sieved (<2 mm), air-dry Ritzville silt loam (1.0 g) and an aliquot of the filtered (<0.4 μ) exocellular medium (2 ml) was eluted through the soil. After 1 ml of eluant was collected, the Ni concentration was determined from the concentration of radiotracer. Metal form was characterized in soil eluates by TLC and TLE as previously described.

Chemical Form of Nickel Complexes Soluble in Soil. On the basis of Ni affinity in exocellular media, enhanced solubility in exocellular media, and solubility on elution through soil, several Ni complexes formed on fungal growth were selected for further chemical characterization. The growth medium before and after growth of organism 458 in the absence of Ni (Ni added after growth) and presence of Ni (0.1, 1.0, 10, 50 μg/ml) was subject to gel permeation chromatography (GPC) using a 2.6 x 72 cm column containing Sephadex, G-25 (superfine) and eluted with 0.02M NH_4OAc. The fractions collected after elution were freeze-dried to remove NH_4OAc and reconstituted in buffer to the original Ni concentration. The fractions were analyzed for ^{63}Ni, total organic C, and for the presence of selected complexes using TLC and TLE as previously described.

Results and Discussion

There are several general mechanisms whereby microorganisms may alter trace metal form in soil, including 1) direct transformation, such as alteration of valence state or alkylation, 2) indirect alteration, such as through interactions with microbial metabolites or alteration of the physicochemical environment such as pH and Eh, and 3) transport, such as uptake during cell growth and release on cell decomposition. The present investigations have been principally concerned with one mechanism, the interaction of metals with microbial metabolites, and to a lesser extent, transport, as this may have occurred in the systems under study.

Organisms isolated from soil on the basis of C source and metal presence were categorized on the basis of morphological (bacteria, fungi), physiological (bacteria), and chemical characteristics of exocellular complexes with metals (bacteria, fungi). Organisms representing major categories were further examined to determine the influence of organism growth on metal solubility and form and soil sorption of exocellular metal complexes. Growth of several fungi markedly increased Ni mobility in soil relative to inorganic Ni controls and the nature of these complexes was investigated in detail to provide insight into the factors influencing mobility and the importance of complexation on Ni solubility in soil.

Morphological and Physiological Characteristics of Microorganisms Isolated from Soil. Major taxonomic subdivisions of the soil bacterial isolates (20) included four groups: gram-positive, spore-forming bacilli (54%); gram-positive, nonsporeformers (17%); gram-negative, oxidase-negative rods (22%); gram-negative, oxidase-positive rods (7%). The spore-forming bacilli were further subdivided into three types (I, II, III) using motility, spore shape and location in the sporocarp, anaerobic growth, and pigmentation on starch agar as criteria. Type I organisms were distinguishable on the basis of pigmentation, spherical spores and swollen sporocarps but were not readily classified. Types II and III were provisionally classified as biotypes of *B. cereus* var. *mycoides*. Groups other than the spore-forming bacilli contained genera provisionally identified as *Acinetobacter*, *Arthrobacter*, *Kurthia*, *Nocardia*, and *Pseudomonas*.

Taxonomic classification of the fungal isolates has not been completed but the isolates include genera provisionally identified as *Aspergillus*, *Cephalosporum*, *Penicillium*, *Mucor*, and *Fusarium*.

Modification of Nickel Form by Soil Microbial Isolates. Following isolation from soil and in conjunction with taxonomic characterization, pure cultures of bacteria and fungi were grown to stationary phase in the presence of Ni at several concentration levels. The exocellular solution from each culture was separated by filtration (<0.4 and <0.01 μ) and the Ni-associated components characterized using TLC and TLE.

Of 239 strains of metal-tolerant bacteria, 165 produced metabolites which complexed Ni. These fell into 13 categories based on the chemical properties of the complexes, but 136 strains were in a single category with growth resulting in identical Ni complexes (Table II). The chromatographic and electrophoretic properties of the Ni complexes resulting from bacterial growth did not appear to be related to the C source or the metal employed in the initial enrichment. The chemical nature of the complexes differed with organism type. However, growth of the gram-positive, spore-forming bacilli (Types I-III), approximately one-half of the collection, resulted in similar exocellular Ni complexes. Growth of gram-positive, nonspore-formers also resulted in similar complexes but TLC and TLE properties of the complexes differed from the spore-formers. The gram-negative, oxidase-negative rods all produced complexes unique to the organisms and gram-negative oxidase-positive rods did not alter

Table II. Soil microorganisms modifying nickel form in solution culture.

Organism	Number of Isolates	Isolates Modifying Nickel Form	Categories of Exocellular Complexes (TLC, TLE)
Bacteria	239	165	14 (136 isolates in single category)
Fungi	59	59	59

Ni form relative to inorganic Ni present in the medium. As in the case of the bacteria, the Ni complexes formed on fungal growth were not related to C source or metal presence; but in contrast to the bacteria, the fungi all modified the form of Ni differently (Table II). The relationship between exocellular Ni complexation and the type and physiology of these organisms is currently under study.

Solubility of Exocellular Nickel Complexes in Soil. To identify those complexes most important in mobilizing Ni in soil, the filtrates (<0.01) of the exocellular media from 20 representative bacterial cultures and all fungal cultures were passed through a column of the soil from which the organisms were isolated and the solubility, and chromatographic and electrophoretic properties of Ni in sterile water, sterile medium and exocellular medium were compared before and after elution through soil.

In the sterile water and growth medium controls, Ni was essentially 100% soluble after incubation, but after soil elution less than 0.1% and approximately 10.4% of the Ni^{2+}, respectively, remained in solution, i.e., most was sorbed by soil (Table III). The difference in Ni sorption between the sterile water and growth medium was due to the formation of soluble Ni complexes in the growth medium. After bacterial growth, the major fraction of Ni remained soluble and the quantity removed on elution of the exocellular solution through soil was equivalent to the growth medium control. After fungal growth, the solubility of Ni in the medium ranged from 32.6% to 100%. This was due to the association of Ni with fungal cells and metabolites which were initially not soluble in the growth medium. There was a general reduction in Ni solubility after soil elution due to sorption of soluble components, with the extent of the reduction dependent upon the organism. Of major significance, however, is the comparison of the solubility of Ni in sterile controls with fungal exocellular medium after soil elution. Growth of six of the fungi resulted in marked increases in the concentration of Ni in soil eluates, i.e., growth reduced soil sorption, increasing Ni mobility in soil (Table III).

Influence of Chemical Form on Solubility of Exocellular Complexes in Soil. Application of TLC and TLE to characterization of Ni complexes with fungal exocellular metabolites (bacterial studies will be reported elsewhere) before and after soil elution provided unique insight into the phenomena responsible for altering Ni solubility relative to inorganic Ni controls. The ^{63}Ni in the sterile water control was present as the inorganic ion which behaved as a single component on TLC (Figure 1). As indicated by solubility measurements (Table III), the Ni ion was essentially removed on elution through soil. The Ni in the sterile growth medium was present in a number of complexes, several

Table III. Influence of bacterial and fungal growth on the solubility and soil sorption of nickel.

Solution Culture 1/,2/	Fraction of Soluble (<0.01μ) Nickel 3/,4/ After Incubation	After Soil Elution
	--------%--------	
Sterile Controls		
Water	100	0.1
Growth medium	100	10.4
Bacteria	89-100	5-26
Fungi		
458	100	92.1
319	95.7	39.2
534	67.9	23.7
380	32.6	20.9
527	71.1	15.0
453	89.6	14.0
484	83.6	8.1
Other 52 isolates	44-100	4.5-11.0

1/ Water and growth medium incubated under sterile conditions.
2/ Values represent a range for 20 bacterial and 52 fungal cultures.
3/ Nickel initially present as Ni^{2+}.
4/ Percentage after soil elution based on soluble Ni after incubation.

of which were removed on soil elution (Figure 1), accounting for the major reduction in Ni concentration in soil eluates (Table III). Organisms 484 (growth did not alter Ni solubility in soil) and 458 (growth increased Ni solubility in soil) serve to illustrate the range of effects of fungal growth on Ni form and soil solubility. The reduction in solubility of Ni in the exocellular medium of organism 484 was due to the sorption of at least two major Ni-containing components (Figure 1). The component remaining accounted for residual solubility equivalent to that observed in the growth medium and also exhibited TLC characteristics similar to the component of the growth medium which was eluted through soil. In contrast, the major Ni-containing component in the exocellular medium of organism 458 was not sorbed by soil. Also, a portion of the Ni was associated with a component that was not detected before soil elution. This may have resulted from redistribution of Ni or separation from another soluble medium or soil component on soil elution.

Further insight into the mechanisms responsible for altering Ni solubility in soil was obtained on application of TLE to characterization of Ni-containing exocellular metabolites (Figure 2). As would be expected, the Ni^{2+} ion in sterile water migrated to the negative pole. The major Ni component in the sterile growth medium (Figure 1) migrated to the negative pole (Figure 2) indicating a positively charged component and accounting for its almost complete removal from solution (Table III) by the predominantly negatively-charged soil (21). Conversely, there were two negatively-charged components in the sterile growth medium which likely arose from the yeast extract (also containing

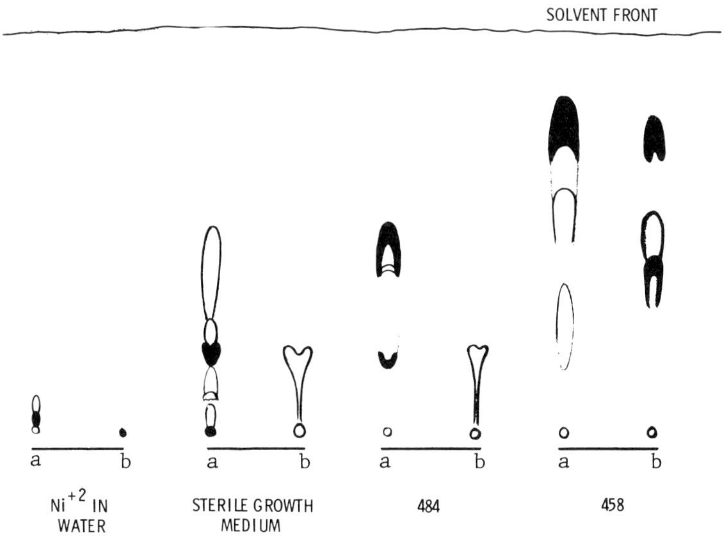

Figure 1. Thin-layer chromatographic behavior of nickel complexes with exocellular fungal metabolites (a) before and (b) after elution through soil. Nickel was visualized by autoradiography; intensity is directly related to nickel concentration.

metabolic products) used in the medium. These eluted through soil explaining the residual solubility of a portion of the Ni in the growth medium (Table III). In the case of isolate 484, the major portion of the Ni was present in a cationic complex near the origin which was sorbed by soil (Figure 2). The fraction soluble in soil consisted of a single, negatively-charged component which exhibited electrophoretic mobility and chromatographic characteristics (Figure 1) similar to the most electrophoretically-mobile, negative component of the growth medium. The reason for the lack of soil sorption of Ni in the exocellular medium of organism 458 was also apparent. The Ni in the exocellular medium (Table III and Figure 1) was associated primarily with neutral or negatively charged species which were not sorbed on elution through soil. The negatively charged component with high electrophoretic mobility observed in the growth medium and after growth of organism 484 was not observed in the exocellular medium of organism 458.

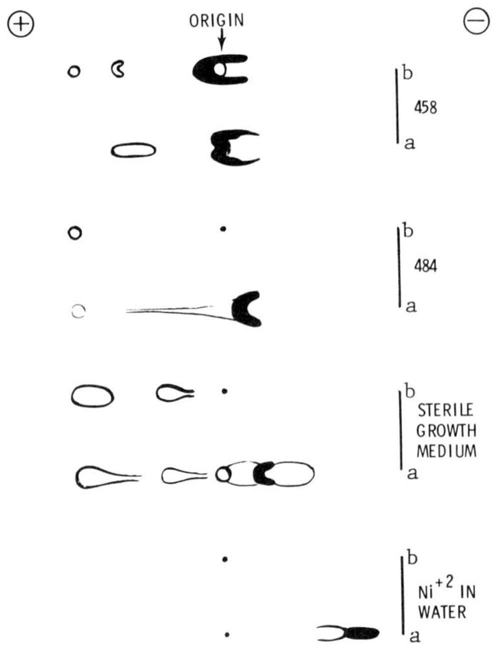

Figure 2. Thin-layer electrophoretic behavior of nickel complexes with exocellular, fungal metabolites (a) before and (b) after elution through soil. Nickel was visualized by autoradiography; intensity is directly related to nickel concentration.

However, a component of similar electrophoretic mobility was detectable after soil elution, suggesting that analogous to the TLC analysis, it may have arisen from redistribution of Ni in sufficient quantities to assure its detectability by autoradiography or from breakdown of a complex containing the medium-derived complex and a product of fungal metabolic origin. This component was also present in most of the other 57 cultures accounting for the 4.5 to 15% Ni solubility which remained on elution of the exocellular media through soil.

The improved detectability of the anionic component present in the medium on elution through soil emphasized the need to consider concentration and other kinetic factors in studies of this nature. The analytical studies were conducted over a Ni concentration range of 0.1 to 50 μg/ml and in systems in which Ni was added and equilibrated before and after microbial growth. This allowed some assessment of the effects of Ni on organism growth and the effects of time and Ni affinity on complex formation. Several components were not detectable in the sterile medium at concentration levels below 10 μg/ml due to competition for Ni in solution by ligands with greater Ni affinity (Figure 3). A similar situation occurred in the case of organism 458 in that the most electrophoretically-mobile anionic component predominated when organisms were grown at Ni concentrations of 0.1 and 1 μg/ml (Figure 3). At a Ni level of 10 μg/ml, the near neutral species which exhibited high soil mobility predominated; and at a Ni level of 50 μg/ml, the Ni was associated with a cationic complex which was readily sorbed by soil. This phenomena resulted from the presence of exocellular ligands in limited concentrations and exhibited different Ni affinities. Differences in Ni affinities were accentuated because the concentration of ^{63}Ni was held constant to avoid radiation effects leading to a greater association of ^{63}Ni, at lower total Ni concentrations, with ligands of limited concentration but highest Ni affinity. Regardless of Ni concentration, the complexes formed when Ni was present during growth were not different from those formed on addition of Ni after growth in the absence of Ni. Thus, the complexes formed during growth in the presence of Ni were not a unique response by the organism to Ni. Furthermore, their formation did not limit the quantity of Ni in solution, precluding formation of secondary complexes during growth.

<u>Chemical Form of Nickel Complexes Soluble in Soil</u>. Of the microorganisms investigated, which encompassed a broad range of bacteria and fungi differing in metal tolerance and metabolism, the fungi were most consistent in producing complex exocellular ligands with affinity for Ni. Furthermore, although the soil efficiently sorbed cationic exocellular Ni complexes, the growth of several fungi increased Ni solubility in soil relative to the growth medium (Table III) through production of neutral and anionic complexes (Figures 1 and 2). The most pronounced

increases (a factor of 9) occurred in growth of the fungal isolate 458 which produced a single, predominantly anionic complex with high affinity for Ni that maintained 92% of the Ni in solution on soil elution compared to 0.1% and 10.4% for Ni in sterile water and sterile growth medium, respectively. This complex was therefore subject to more intensive investigation.

In order to further define the chemical nature of the fungal complexes with high Ni affinity, the exocellular solution from organism 458 was subjected to GPC. Organic C was eluted over the entire separation range of the column (approximately 300 to 3000

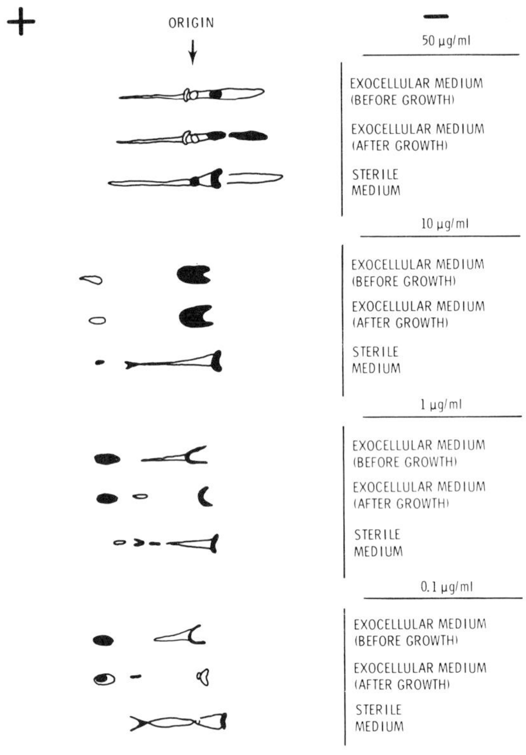

Figure 3. Influence of nickel concentration and time of addition (before and after growth) on the thin-layer electrophoretic behavior of nickel complexes with exocellular fungal (isolate 458) metabolites. Nickel was visualized by autoradiography; intensity is directly related to nickel concentration.

equivalent molecular weight) and included at least six major peaks (Figure 4). The Ni-containing fraction was eluted in two peaks which were sufficiently broad to be composites of several individual components. The average equivalent molecular weight of the minor (elution volume 230 to 260 ml) and major (elution volume 260 to 320 ml) Ni peaks were 900 and 350, respectively. The major Ni peak corresponded to an elution volume of relatively high organic C concentration.

Characterization of the eluates from GPC by TLC and TLE (Figure 5) indicated that the minor peak was a single, negatively-charged complex equivalent in electrophoretic mobility to the electrophoretically-mobile complex present in the sterile growth medium and the exocellular medium of most fungal isolates at 10 μg Ni/ml (Figure 2) and organism 458 at 0.1 and 1 μg Ni/ml (Figure 3). The major peak consisted of at least three components resolvable by TLC (Figure 5). The chromatographic mobility of one of the components was similar to the major Ni complex resolved prior to separation by GPC (Figures 1 and 5). It is,

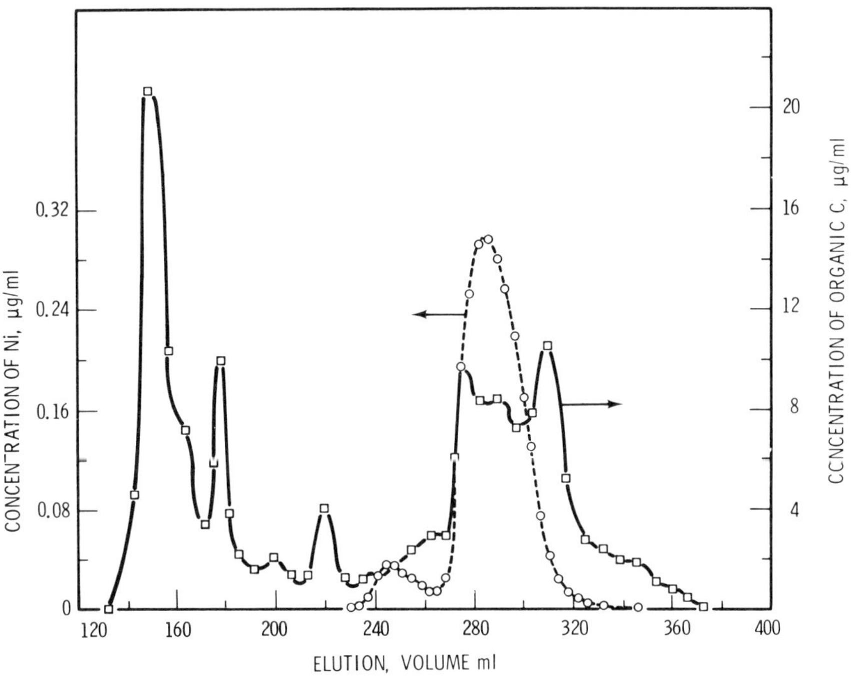

Figure 4. Separation by gel permeation chromatography (Sephadex G 25) of exocellular nickel and organic carbon after fungal growth (isolate 458, 10 μg Ni/mL)

therefore, likely that this was the negatively-charged component present in the GPC eluate and resolved by TLE (Fractions F-J, Figure 5). However, GPC resulted in the separation of a positively-charged component (Fractions F-P) with high affinity for Ni which was previously unidentified as an exocellular product of isolate 458. The GPC fractions containing the minor and major peaks were concentrated by lyophylization, eluted through soil and Ni complexes determined by TLC and TLE as previously described. The Ni associated with the minor peak was not sorbed (95% was eluted) as previously observed. The major peak, in contrast to direct application of the exocellular solution to soil (Table III; Figures 1 and 2) was sorbed by soil (20% was eluted) after separation by GPC likely due to the separation of a positive Ni-containing component from the original complex.

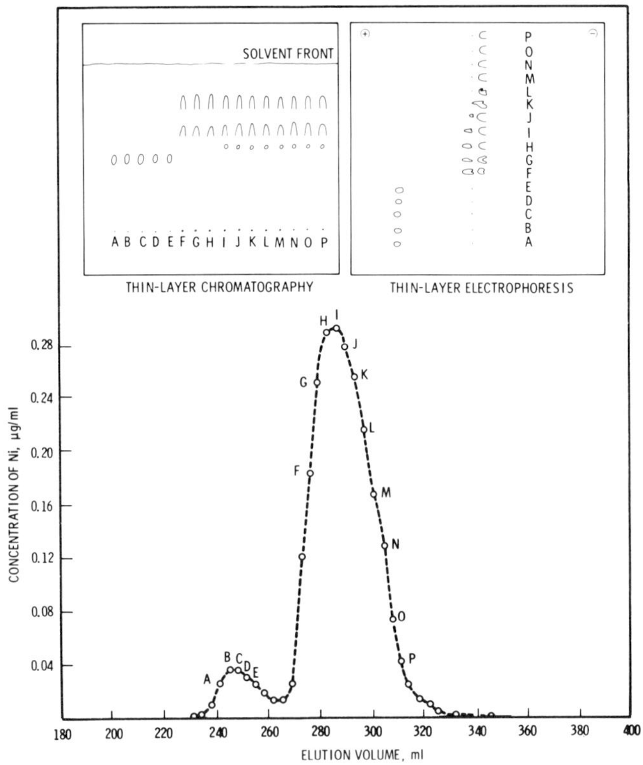

Figure 5. Thin-layer chromatographic and electrophoretic behavior of nickel complexes with exocellular fungal metabolites (isolate 458, 10 µg Ni/mL) after separation by gel permeation chromatography

The presence of a new, previously unidentified, positive component after separation may have resulted from redistribution of Ni to other complexes not observable by TLC and TLE prior to separation. However, redistribution is unlikely considering the relatively low concentration and/or Ni affinity of most ligands in the sterile growth medium compared to the demonstrated high concentration and Ni affinity of the major component, i.e., most of the Ni in the exocellular solution was associated with the complex and it was stable on TLC, TLE, and soil elution. Thus the exocellular complex responsible for mobilizing Ni in soil appeared to be a complex ligand comprised of several ligands capable of complexing Ni independently and exhibiting both positive and negative charge after separation from the original entity. These characteristics might be explained by a complex ligand comprised of several amino acids with molecular weights similar to glycine and bound by Ni. However, none of the components gave a positive ninhydrin reaction.

Further study will entail concentration and purification of the important intact complexes and subunits using recycling GPC and high pressure liquid chromatography, followed by detailed chemical characterization of the ligands using gas chromatography-mass spectrometry and estimation of Ni stability constants.

Importance of Nickel-Complexation by Microbes in Soil. There are a number of qualifications required in extrapolation of the results of these studies to intact soil. These arise primarily from the need to separate microbes from soil to investigate metal transformations in detail. Microorganisms exist in a dynamic equilibrium in soil with types and numbers varying markedly as a function of soil type and changes in environmental conditions such as carbon source, temperature, moisture, aeration and the presence of toxic substances (22). It is not known if the organisms isolated from soil exhibit the ability to effectively compete with other soil organisms over a range of soil conditions and, therefore, can produce metabolites in sufficient concentration to complex, in significant quantities, the Ni associated with different soil solutions or solid phases. Furthermore, the microbial systems studied undoubtedly contain cellular decomposition products as well as exocellular metabolites. The form and stability of these products may differ in a mixed population such as occurs in soil.

The present research was designed to optimize the opportunity to examine organisms that would 1) normally be present in soil under aerobic conditions, 2) reflect the presence of a broad range of C sources and, 3) be particularly competitive at elevated (pollution) levels of the metals. Emphasis was on identification of kinetically stable, soluble complexes with high affinity for Ni which exhibited a degree of commonality between organisms, optimizing the potential for identification of ligands important in controlling Ni chemistry in soil solutions. The studies demonstrated the formation of highly stable fungal metabolites with strong affinities for Ni and neutral and anionic charge characteristics. The metabolites were common to a number of organisms and controlled Ni solubility in the soil system under study. In order to evaluate the influence of microbial processes on Ni solubility in surface soils, further research is required to chemically characterize important ligands in detail sufficient to allow identification in soil and examination of the conditions of formation under a range of soil conditions.

Acknowledgments

This work was performed under Contract 211B00844 with the National Institute of Environmental Health Sciences.

Brand names are used to assist the reader in replicating the experiments and their use does not constitute endorsement by Battelle Memorial Institute.

Abstract

Soil bacteria and fungi, isolated in pure cultures from soil on the basis of metal tolerance and carbon requirements, produced soluble, exocellular metabolites which complexed inorganic Ni and altered Ni mobility on elution through soil columns. The chemical nature of the complexes, as determined by thin-layer chromatography, thin-layer electrophoresis and gel permeation chromatography, differed with organism type. Cationic, neutral, and anionic complexes exhibiting a range of chromatographic properties were resolved. Of 239 strains of metal-tolerant bacteria, 165 produced metabolites which complexed Ni. These organisms fell into 13 categories based on the chemical properties of the complexes, but 136 strains were in a single category with growth resulting in identical Ni complexes. In contrast, 59 strains of metal-tolerant fungi all modified the form of Ni differently. The mobility of complexed Ni on elution of exocellular solutions through soil was dependent upon the charge and molecular weight of the complex, the affinity of the ligand for Ni, and the concentration of Ni in solution. Neutral and anionic Ni complexes with apparent molecular weights less than 1200 arising from fungal metabolism were most mobile in soil increasing Ni mobility by up to 1000 times relative to inorganic Ni. A single Ni complex responsible for the most pronounced increases in Ni mobility in soil consisted of several ligands capable of complexing Ni independently and exhibiting both positive and negative charge after separation from the original complex ligand. The investigations indicate that microbial processes resulting in the formation of metabolites with high-affinity for Ni and neutral or anionic charge characteristics may be of major significance in controlling Ni behavior in surface soils.

Literature Cited

1. Jenne, E. A. Trace element sorption by sediments and soils -- sites and processes. Chp. 5. In: "Proceedings of Symposium on Molybdenum in the Environment". Vol. 2. M. Dekker, New York. 1977.
2. Jenne, E. A., and Luoma, S. N. Forms of trace elements in soils, sediments, and associated waters: an overview of their determination and biological availability. In: Drucker, H. and Wildung, R. E. (ed.). "Biological Implications of Metals in the Environment." ERDA CONF-750929. 1977.
3. Hodgson, J. F. Chemistry of the Micronutrient elements in soils. Adv. Agron. 15: 119-159 (1963).
4. Keeney, D. R., and Wildung, R. E. Chemical properties of soils. In: Elliott, L., and Stevenson, F. J. (ed.). "Soils for Management and Utilization of Organic Wastes and Waste Waters." Am. Soc. Agron., Madison, Wisconsin. 1977.
5. Kononova, M. M. "Soil Organic Matter". Permagon Press, New York. 1966.
6. Greenland, D. J. Interaction between clays and organic compounds in soils. Part 1. Mechanisms of interaction between clays and defined organic compounds. Soils Fert. 28: 415-425 (1965).
7. Geering, H. R., and Hodgson, J. F. Micronutrient cation complexes in soil solution: III. Characterization of soil solution ligands and their complexes with Zn^{2+} and Cu^{2+}. Soil Sci. Soc. Am. Proc. 33: 54-59 (1969).
8. Wildung, R. E., Garland, T. R. and Buschbom, R. L. The interdependent effects of soil temperature and water content on soil respiration rate and plant root decomposition in arid grassland soils. Soil Biol. and Biochem. 7: 373-378 (1975).
9. Wildung, R. E., Drucker, H., and Au, F. H. F. The relationship of microbial processes to the fate of transuranic elements in soil. In: White, M G. and Dunaway, P. B. (ed.). "Transuranics in Natural Environments" NVO-178. Nevada Appl. Ecol. Group, Las Vegas, Nevada. 1977.
10. Stevenson, F. J. Organic acids in soil. In: "Soil Biochemistry." M. Dekker, New York. 1967.
11. Bremner, J. M. Nitrogenous compounds. In: "Soil Biochemistry." M. Dekker, New York. 1967.
12. Stevenson, F. J., and Ardakani, M. S. Organic matter reactions involving micronutrients in soils. In: Mortvedt, J. J., Giordano, P. M., and Lindsay, W. L. (ed.). "Micronutrients in Agriculture." Am. Soc. Agron., Madison, Wisconsin. 1972.
13. Wildung, R. E., Drucker, H., Garland, T. R., and Pelroy, R. A. Transformation of trace metals by soil microorganisms, p. 141. "Agronomy Abstracts, ASA, SSSA, and CSSA Annual Meetings." Houston, Texas. 1976.

14. Wildung, R. E., Garland, T. R., and Drucker, H. Complexation of nickel by metal-resistant soil bacteria and fungi, p. 153. "Agronomy Abstracts, ASA, SSSA, and CSSA Annual Meetings." Los Angeles, California. 1977.
15. Wildung, R. E., and Garland, T. R. "The Relationship of Microbial Processes to the Fate and Behavior of Transuranic Elements in Soils, Plants, and Animals." DOE Report, PNL-2416. NTIS, Springfield, Virginia. 1977.
16. Hodgson, J. F., Lindsay, W. L., and Trierweiler, J. F. Micronutrient cation complexing in soil solution: II. Complexing of zinc and copper in displaced solution from calcareous soils. Soil Sci. Soc. Am. Proc. 30: 723-726 (1966).
17. Geering, H. R., Hodgson, J. F., and Sdano, C. Micronutrient cation complexes in soil solution: IV. The chemical state of manganese in soil solution. Soil Sci. Soc. Am. Proc. 33: 81-85 (1969).
18. Mortensen, J. L. Complexing of metals by soil organic matter. Soil Sci. Soc. Am. Proc. 27: 179-186 (1963).
19. Harrigan, W. F., and McCance, M. E. "Laboratory Methods in Microbiology" Academic Press, New York. 1966.
20. Pelroy, R. A., and Cresto, J. T. Taxonomy of soil bacteria isolated in the presence of heavy metals, p. 162. "Abstracts of the American Society for Microbiology Annual Meeting." Las Vegas, Nevada. 1978.
21. Wildung, R. E. "Soils of the Pacific Northwest Shrub-Steppe. Occurrence and Properties of Soils on the Arid Land Ecology Reserve, Hanford Reservation." ERDS Report BNWL-2272. NTIS, Springfield, Virginia. 1977.
22. Alexander, M. "Microbial Ecology." Wiley and Sons, New York. 1971.

RECEIVED November 16, 1978.

10

Stability Surface Concept

A Quantitative Model for Complexation in Multiligand Mixtures

PATRICK MAC CARTHY and GARON C. SMITH

Colorado School of Mines, Golden, CO 80401

The behavior of metal ions in the natural environment is determined to a large extent by their aqueous complexation chemistry. In a simple solution system, containing a single metal and a single ligand, and where only a 1:1 complex is formed, the equilibrium concentrations of all species (free metal, free ligand and complex) can be readily calculated from a knowledge of the stability constant, and the stoichiometric concentration of metal and ligand in solution (1, pp 112-114). Calculations become more cumbersome when a mixture of complexes of different stoichiometries are produced in the single metal, single ligand system (1, pp 114-119; 2). When many metal and many ligand species are simultaneously present in solution, the situation becomes vastly more complicated. Nevertheless, computational methods are available for calculating the equilibrium concentrations of all species present in such systems. These techniques involve successive refinement by iterative methods using digital computers. At present, many computer programs are available for calculating equilibrium speciation in multimetal-multiligand systems (for example references 3,4,5,6). Basically, these programs calculate the concentrations of all species existing at equilibrium. The required input data for proper use of these programs are the stoichiometric concentration of each metal and of each ligand species, in addition to the stability constants for all complexes formed in the mixture. The pH and ionic strength of the solution, the effects of solid phases, etc., can also be incorporated. The validity of the numbers resulting from such iterative calculations depends on the completeness and accuracy of the thermodynamic data and stoichiometric concentrations employed. Extensive compilations of thermodynamic data for complexes have been published (7-14), but the vast majority of these data relate to homogeneous, i.e. non-mixed ligand, complexes. However, there are many opportunities for mixed-ligand complexes to form in multiligand systems, and statistically, one may expect mixed ligand complexes to dominate the system. There are relatively little thermodynamic data pertaining to mixed-ligand complexes in the literature, and this

0-8412-0479-9/79/47-093-201$05.50/0
© 1979 American Chemical Society

seriously affects the computation of speciation in multiligand mixtures.

These problems would be encountered even in relatively simple systems containing a mixture of characterized, synthetic ligands. The situation is considerably more complicated in natural systems containing a multicomponent ligand mixture. Frequently, when dealing with such cases, the total number of ligand species, or their individual nature and concentration may not be known. This is particularly true for mixtures of natural organic matter which may contain phenolic, carboxylic, amino acid, etc. components, in addition to the non-specific humic substances. Humic substances are highly complex mixtures of organic materials formed in soil, sediments and natural waters (15,16,17). They result from the decay of plant debris, and play an active role in metal complexation in the environment. While much is known about their gross chemical properties, such as functional group content, molecular weight ranges, etc., very little is known about the specific nature of the individual ligands. Consequently, models for determining solute speciation cannot be satisfactorily applied to such natural systems in the conventional manner due to the lack of sufficient thermodynamic and stoichiometric data.

Does the above imply that all attempts to compute speciation in such natural systems are fruitless? While calculation of the *complete* aqueous speciation in these natural systems is not presently feasible, there *is* an alternative approach where one is primarily interested in distinguishing between the free and total complexed metal in solution. As a compromise measure and pragmatic approach to studying metal complexation in complicated data-deficient systems, we have developed an alternative model.

The purpose of this paper is to present a model for single metal-multiligand solution equilibrium which quantitatively describes the net binding of metal ions to the multiligand mixture *as a whole*. Equally important aspects of this model are that it yields a unified conceptual visualization of complexation in complicated multiligand mixtures, and provides a framework for interpreting the results of experiments on such systems. The model, as presented here, will be confined to mononuclear complexes, and it is assumed that no solid phases are present. The influence of pH on the complexation reactions is described. The model is illustrated by simulation of various metal-multiligand systems, and the properties of these systems are represented by means of three-dimensional plots, called stability surfaces.

Stability Function Concept

Consider the continuous addition of a single metal species to a ligand mixture represented by ΣLi where the total volume of the solution is maintained constant i.e. a metal-into-ligand *batch* titration. The following scheme with mononuclear complexes, including mixed ligand species, illustrates the reactions which can occur:

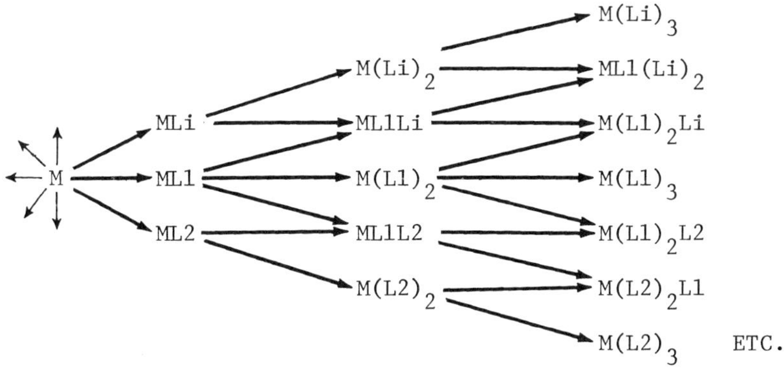

(1)

where L1, L2, . . . Li are the various ligand species. In order to describe the binding of the metal ion in the ligand mixture *as a whole*, an average stability quotient, or stability function, is defined for each stoichiometry represented in the system. This is analogous to the representation of each stoichiometry by a separate stability *constant* in a one-ligand system. The stability function for the 1:n complexes, $\mathcal{S}n$, is defined as follows:

$$\mathcal{S}n = \frac{\sum^{Tn} \left(\prod_{i=1}^{N} \omega_i^{bi} \right) \beta n(1_{b1}, 2_{b2}, \ldots, N_{bN})}{\left(\sum_{i=1}^{N} \omega_i \right)^n} \qquad (2)$$

where Tn = total number of complexes of stoichiometry, 1:n,
N = number of ligand species in system,
bi = number of times ligand Li appears in a particular complex, and

for any complex, $\sum_{i=1}^{N} bi = n$, the stoichiometry of the complex.

The ω_i terms are weighting factors representing the relative concentration of each free ligand species in the mixture and they are defined as follows:

$$\omega_i = \frac{[Li]}{[L1]} \qquad (3)$$

where [Li] represents the equilibrium concentration of the ith ligand species, and [L1] is the equilibrium concentration of any arbitrarily-designated reference ligand. $\beta n\ (1_{b1}, 2_{b2}, \ldots, N_{bN})$ represents the various stability products for complexes of 1:n stoichiometry in the mixture, and is defined as:

$$\beta n(1_{b1}, 2_{b2}, \ldots, N_{bN}) = \frac{[M(L1)_{b1}(L2)_{b2} \ldots (LN)_{bN}]}{[M]\left(\prod_{i=1}^{N} [Li]^{bi}\right)} \qquad (4)$$

Tn is given by the combinatorial formula (18, p. 13):

$$Tn = \frac{(N + n - 1)!}{(N - 1)!n!} \qquad (5)$$

and represents the total number of statistically possible combinations of ligands surrounding the metal ion for each stoichiometry. A complete description of any system requires that thermodynamic data be available for *all* complex species computed from eqn. 5. It is clear from this equation that in multiligand systems, with stoichiometries greater than 1:1, mixed-ligand species greatly outnumber the homogeneous complexes. As already pointed out above, there is a severe dearth of thermodynamic data for mixed ligand complexes. In most models, those species for which data are available are taken into consideration, and all other species are simply ignored! As a result of ignoring the *majority* of species in solving the system equations, one must seriously question the validity and value of the computed concentrations for the limited number of species with known thermodynamic data. While this model cannot include complexes for which constants are unavailable, it reminds the user that the final values which are calculated may deviate significantly from the real numerical answers. The model also suggests that, when there is a reasonable method to approximate constants for mixed ligand species, a more accurate picture may be obtained through their use. The logic behind the definition of Sn (eqn. 2) is further discussed in an earlier publication (19).

The stability function is basically a weighted-average stability product (20). It is clear that the stability function defined by equation 2 "weights" each ligand according to its free concentration, and the number of times that particular ligand occurs in the complex, as well as the stability products of the complexes which it forms with a given metal. One reason for choosing this particular definition of Sn can be seen by substituting the equilibrium *constant* expression (eqn. 4) into the stability function definition (eqn. 2) to yield:

$$\mathcal{S}n = \frac{\sum^{Tn} [M(L1)_{b1}(L2)_{b2} \cdots (LN)_{bN}]}{[M]\left(\sum_{i=1}^{N} [Li]\right)^{n}} \quad (6)$$

The similarity of the stability function to the conventional stability constant is evident from a comparison of equations 6 and 4. Equation 6 does not require knowledge of the concentration of the individual ligand or complex species in the equilibrium mixture. All of the complexes of a given stoichiometry are grouped together into one quantity, $\sum^{Tn} [M(L1)_{b1}(L2)_{b2} \cdots (LN)_{bN}]$ and all of the ligand species are also grouped together, ($\sum_{i=1}^{N} [Li]$). Other definitions of an average stability function, which are also logically sound, require that the individual [Li] and $[M(L1)_{b1}(L2)_{b2} \cdots (LN)_{bN}]$ concentrations be known in order to evaluate the defined function. If such detailed information were available for a multiligand mixture, the conventional approach to metal-multiligand equilibrium, discussed in the Introduction, would be more appropriate. The stability function is specifically defined to eliminate the need for such detailed knowledge of the ligand mixture and thereby to render it applicable to "unknown" ligand mixtures. Equation 6 can be expressed in words in the following manner: $\mathcal{S}n$ represents the total complexed metal concentration of each stoichiometry, MLi_n, divided by the product of the free metal ion concentration and the sum of the free ligand concentrations, raised to the power of the stoichiometry, n.

Generally the ligands, Li, are the conjugate bases of acids, and thus the overall complexation equilibrium is affected by the pH of the solution. When the hydrolytic reactions of the complexes, ligands and free metal are to be taken into consideration, the definition of $\mathcal{S}n$ becomes somewhat more complicated. In that case, the sum of all complex species in the numerator of eqn. 6 must also include all hydrolyzed forms of the various complexes; the free metal concentration term in the denominator must include all hydrolyzed forms of the metal not complexed to any of the ligands, Li; and the term in the denominator involving the sum of all free ligand species must also incorporate the various protonated forms of the ligands. These factors *are* incorporated into our model but are not shown specifically in eqn. 6 so as not to obscure the basic concepts involved in the definition of the stability function. Thus, pH effects are directly incorporated into our model.

Graphical Representation of Stability Function

Stability Profiles. A fundamental understanding of equations 2 and 6 provides a theoretical basis for a more comprehensive appreciation of metal-multiligand equilibrium. The following discussion is best presented in terms of a metal-into-ligand batch titration (defined under Stability Function Concept). As the fraction of a particular ligand which is complexed becomes small,

$$\omega_i \rightarrow W_i \tag{7}$$

where W_i is a constant and is defined as the ratio of the *stoichiometric* ligand concentrations, C_L:

$$W_i = \frac{C_{Li}}{C_{L1}} \tag{8}$$

When equation 8 is satisfied for *all* ligand species, the stability function adopts constant, limiting values, given by:

$$\lim_{\omega_i \rightarrow W_i} S_n^* = \frac{\sum^{Tn} \left(\prod_{i=1}^{N} W_i^{\,bi} \right) \beta n(1_{b1}, 2_{b2}, \ldots, N_{bN})}{\left(\sum_{i=1}^{N} W_i \right)^n} = \text{CLASP-}u \tag{9}$$

These constant, limiting values of the stability function have been previously discussed by MacCarthy (20), and have been given the name CLASP values (conditional, limiting average stability products). For reasons which will become evident later, we will refer to them here as upper CLASP values, or CLASP-*u*. In practice, eqns. 7, 8 and 9 may be applicable when *at least one* of the following conditions is satisfied:

(a) There is trace metal ion in the presence of a very large excess of each ligand species,
(b) All complexes are sufficiently weak that they are largely dissociated, and
(c) The system is sufficiently dilute that all complexes are largely dissociated.

On the other hand, when each of the ligand species is largely in the complexed form (i.e. $[MLi] \simeq C_{Li}$), ω_i approaches a different, limiting value, given by:

$$\omega_i = W_i \left(\frac{\beta 1(1)}{\beta 1(i)} \right) \tag{10}$$

where β1(i) is the stability product for the 1:1 complex with the ith ligand, Li, and β1(1) is the stability product of the 1:1 complex with reference ligand, L1. When equation 10 is satisfied by *all* ligand species, the stability function adopts the constant, limiting value:

$$\lim_{\omega_i \to W_i \left(\frac{\beta 1(1)}{\beta 1(i)} \right)} S_n = \frac{\sum^{T_n} \left\{ \prod_{i=1}^{N} \left(W_i \frac{\beta 1(1)}{\beta 1(i)} \right)^{b_i} \right\} \beta n(1_{b1}, 2_{b2}, \ldots, N_{bN})}{\left(\sum_{i=1}^{N} W_i \frac{\beta 1(1)}{\beta 1(i)} \right)^n} = \text{CLASP-}l \tag{11}$$

We will refer to these limiting values of the stability function as lower CLASP values, or CLASP-*l*. In practice, equations 10 and 11 apply when *both* of the following experimental conditions are satisfied:

(a) The metal is present in excess over the total ligand concentration, and
(b) All of the conditional stability constants (21,22) are sufficiently large to maintain all of the ligand species largely in the complexed form at the particular solution concentration and pH.

The stability function varies continuously between the two limiting values of CLASP-*u* and CLASP-*l* as the solution composition is varied. This is illustrated in Figure 1(a) which shows the stability profile (19) for the 1:1 complexes, in the metal-into-ligand batch titration of a two-ligand mixture. This profile is one slice of the three-dimensional surface depicted in Figure 3(a), which will be discussed later. The simple type of profile illustrated in Figure 1(a) is frequently obtained. However, in other cases, more complicated profiles, such as that shown in Figure 1(b) are found. Figure 1(b) is a slice of the three-dimensional surface of Figure 4(a). In this case, an intermediate plateau is observed, in addition to the two limiting plateau regions. Such an intermediate plateau is called a pseudo-CLASP, and occurs when some of the ligand species in the mixture are essentially totally complexed while the other ligand species are largely uncomplexed. In other words, some ligands are in the

CLASP-*u* condition while others are in the CLASP-*l* condition. It should be pointed out that conditional stability constants (21, 22) rather than absolute stability constants, are actually employed in equations 9 and 11 in order to take the effects of solution pH into account. Equations 2, 4, 6, 9 and 11 may appear complicated to the reader. Simple examples of their application to specific ligand mixtures are provided in the Appendix at the end of this paper.

Stability Surfaces. In order to provide an overall picture of multiligand complexation behavior, the stability function should be plotted for all values of metal-multiligand solution composition. This is easily achieved by representing the solution composition as a point on a two-dimensional grid, one axis of which is C_M, the stoichiometric metal concentration, and the other axis, C_{LT}, the stoichiometric total ligand concentration. The composition of any individual solution, prepared by mixing the metal ion and the multi-ligand mixture, can be depicted as a point on this two-dimensional coordinate system. The changes in the composition of this mixture resulting from various experimental operations, such as dilution, batch titration, regular titration, Job's continuous variations procedure etc., are described by specific paths along this grid. The stability function for each stoichiometry, or any other solution property such as absorbance, free metal ion concentration, etc., can then be plotted on a third axis orthogonal to the other two. Thus, the stability function can be represented as a three-dimensional surface, $\mathcal{S}n(C_M, C_{LT})$. These are called stability surfaces. The variation of the different solution properties during the course of an experimental run can then be depicted as the upward projection of the experimental path from the two-dimensional grid onto the surface. This conveniently allows one to simulate and pictorially represent the results of various types of experiments on metal-multiligand mixtures.

Figure 2 shows the two-dimensional grid with the paths of various experimental procedures indicated. It is important to recognize that this grid has logarithmic axes, in order to accommodate a wide range of experimental conditions. Stability surfaces, and other types of surfaces, are illustrated in the next section. The three-dimensional surfaces are conveniently generated by first producing a set of profiles for metal-into-ligand batch titrations at various concentrations of the total ligand mixture. These profiles are then stacked up one behind the other to generate the three-dimensional surface. A new computer program, SYSTAB, to generate system stability profiles and surfaces, as well as other types of surfaces, will be published elsewhere.

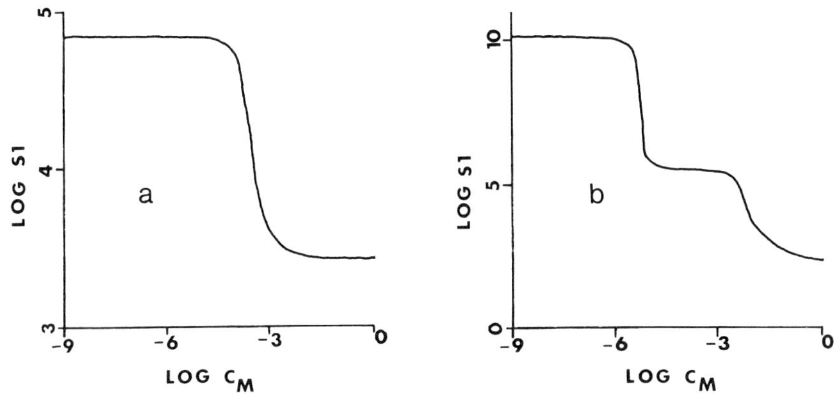

Figure 1. Stability profiles for metal-into-ligand batch titrations: (a) 1:1 profile for system described in Table I, with $C_{LT} = 1.0 \times 10^{-3}$M and pH = 10.0; (b) system described in Table III, with $C_{LT} = 1.0 \times 10^{-2}$M and pH = 7.0

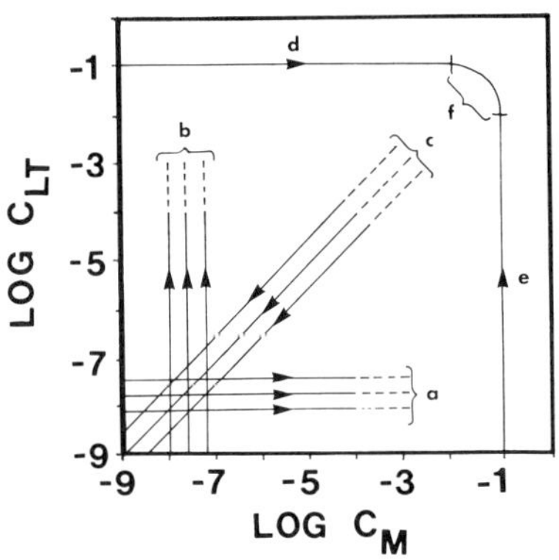

Figure 2. Two-dimensional grid representing solution composition and various experimental pathways: (a) metal-into-ligand batch titration, (b) ligand-into-metal batch titration, (c) dilution, (d) symmetrical (i.e. equimolar metal and ligand solutions) metal-into-ligand regular titration, (e) symmetrical ligand-into-metal regular titration. (d) and (e) also correspond to symmetrical continuous variations plots, with most of the experimental points generally lying in the region (f).

Discussion, and Illustration of Stability Surface Concept

The usefulness and applications of system stability surfaces will now be illustrated by computer-simulation of various metal-multiligand equilibrium systems.

Case 1. A simple two-ligand system is chosen first, consisting of zinc(II) in the presence of a mixture of ethylenediamine and oxalate ligands at pH values of 10.00, 7.72 and 3.00. This system is chosen because it is one of the relatively few for which quite complete thermodynamic data are available, inclusive of mixed ligand species (23). Table I describes the system where L1 is oxalate and L2 is ethylenediamine. The stability surfaces corresponding to this two-ligand mixture, over the specified solution composition and pH values, are shown by the various plots in Figure 3. There is a separate stability surface for each stoichiometry and each pH value. Each point on these surfaces represents the contribution of the particular stoichiometry to the strength of binding of the metal ion to the ligand mixture as a whole, in a solution whose composition is defined by the C_M and C_{LT} coordinates of the point.

Curves (a), (b) and (c) of Figure 3 show the stability surfaces for the three stoichiometries at pH 10.00. At this pH, both of the ligands are essentially totally deprotonated, and hydrogen ions do not compete effectively with the metal ions for ligand sites. The variation of the stability functions, with respect to solution composition, is evident from these surfaces, indicating that the strength of binding of metal ions to the multiligand mixture (as defined by eqn. 2) does depend on solution composition. The function varies from an upper, limiting plateau (CLASP-*u*) (20) to a lower, limiting value (CLASP-*l*). The locations of the CLASP-*u* and CLASP-*l* regions on the C_M-C_{LT} grid are consistent with the earlier discussion of stability profiles.

It is interesting to note that in a series of measurements of the distribution of metal between free and complexed forms, where all measurements are confined to one of the plateau regions, the stability function is behaving as a stability constant. Consequently, the two-ligand mixture is acting as a single ligand solution under these experimental conditions. The *apparent* "stability constant" incorporates a contribution from all constituent ligands in the mixture, and can be readily calculated from equations 9 or 11 for a known ligand system. If measurements are carried out on an unknown ligand mixture in one of the CLASP regions, a "stability constant" will be obtained, and equations 9 and 11 provide a theoretical interpretation of such results. This conclusion is of particular relevance to certain measurements on soil and environmental samples. Schubert's ion-exchange method is a technique for determining the stability constants of complexes, and involves measurements in solutions having a trace concentration of metal ions in the presence of a large excess of

Table I

System Description for Case I

Largest stoichiometric metal concentration	1 M
Smallest stoichiometric metal concentration	1×10^{-9} M
Number of C_M points	36
Largest stoichiometric total ligand concentration	1 M
Smallest stoichiometric total ligand concentration	1×10^{-9} M
Number of C_{LT} points	36
Total number of points on surface	1296
Number of ligands in system	2
Stoichiometric ligand weighting factors	
W_1	1
W_2	1
Stoichiometries of complexes in system	1:1, 1:2, 1:3
Number of complexes of 1:1 stoichiometry	2
Number of complexes of 1:2 stoichiometry	3
Number of complexes of 1:3 stoichiometry	4
Stability products for 1:1 complexes	
β1(1)	2.75423×10^{3}
β1(2)	5.12861×10^{5}
Stability products for 1:2 complexes	
β2(1,1)	3.01995×10^{6}
β2(1,2)	1.62181×10^{9}
β2(2,2)	7.07946×10^{10}
Stability products for 1:3 complexes	
β3(1,1,1)	1.73780×10^{7}
β3(1,1,2)	5.75440×10^{10}
β3(1,2,2)	2.04174×10^{12}
β3(2,2,2)	4.78630×10^{12}
Proton association products for ligands	
K1(1)	1.84502×10^{4}
K2(1)	3.44220×10^{5}
K1(2)	8.50000×10^{9}
K2(2)	6.03500×10^{16}
Formation constant for 1:1 metal hydroxide:	
H1	1.00000×10^{4}
pH values used	10.00, 7.72, 3.00

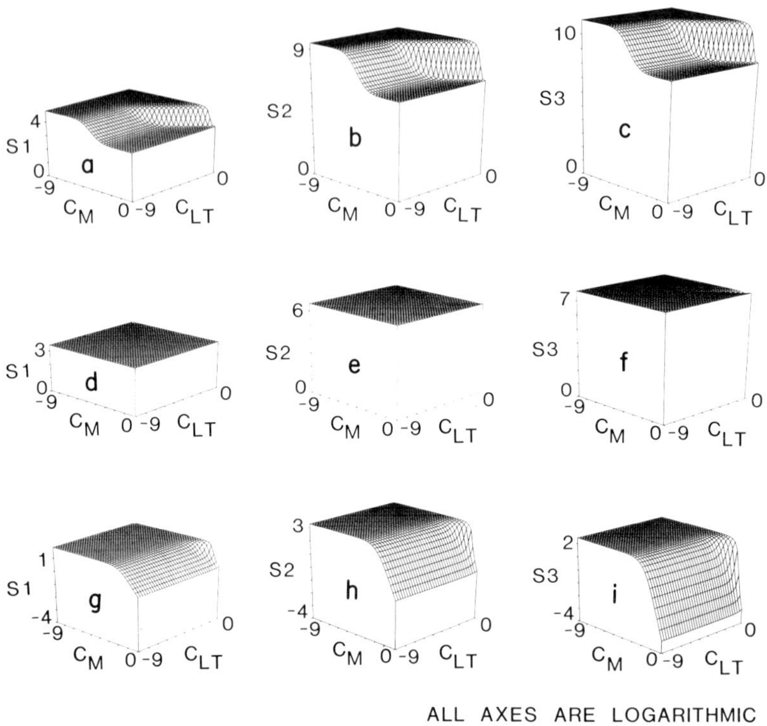

Figure 3. Stability surfaces for system described in Table I: (a), (b), (c) at pH 10.00; (d), (e), (f) at pH 7.72; (g), (h), (i) at pH 3.00; (a), (d), (g) are 1:1 surfaces; (b), (e), (h) are 1:2 surfaces; (c), (f), (i) are 1:3 surfaces

ligands (24). Whereas Schubert's method was developed for single ligand systems, the method has frequently been applied to environmental, multiligand mixtures, without a theoretical basis for doing so, or for providing a valid interpretation of the data. This gap has been filled in by the analysis presented here. Measurements carried out on multiligand mixtures by Schubert's ion-exchange method may yield constant values for stability products. The CLASP-*u* region unambiguously defines the valid range of "Schubert's conditions" for a multiligand mixture. Measurements which extend outside the range of Schubert's condition will lead to a variability in the results, as evident from the stability surface. The above analysis is of relevance in that the occurrence of trace metal ion in the presence of a large excess of ligand mixture is a commonly-encountered environmental situation. These arguments also apply to measurements carried out under CLASP-*l* conditions. However, this is not as important a region from an environmental point of view as the CLASP-*u* region. In addition, precipitation of complexes frequently occurs in this region, a factor which is not yet taken into consideration in the present model.

Papers dealing with the application of Schubert's method to environmental systems have been reviewed by MacCarthy and Mark (25). Other experimental methods which have been applied to multiligand mixtures in the past are Job's method of continuous variations and Bjerrum's potentiometric titration method. Unlike Schubert's technique, these two latter methods do not limit their measurements to one of the CLASP regions but, extend them over a considerable range of solution compositions and thus span a range of stability function values. This factor has generally been overlooked in the study of such systems. The misleading results which Job's method of continuous variations can give when applied to such systems have already been discussed (19,22).

The stability surfaces show, that in general, researchers dealing with the same system, but operating under experimental conditions corresponding to different points on the composition grid, may report different values for the stability "constant." In fact, simply diluting a system will generally lead to a variability in the resulting data, as is also evident from these surfaces.

Using the conventional approach, it is difficult to develop a conceptual feeling for the reactions occurring in a metal-multiligand system. The conventional approach yields the equilibrium concentration of all species, and in a multicomponent system it is virtually impossible to simultaneously comprehend all of the equilibrium reactions which are occurring. The approach developed here combines all of the reactions of a given stoichiometry into a single stability function which can be simply described in terms of a three-dimensional surface. This provides a simplified and unified visualization of the net equilibrium in a multicomponent mixture. The constant values which these functions adopt under limiting (and some intermediate) conditions, thus allowing multi-ligand mixtures to be mathematically analyzed as a single-ligand system, is a further conceptual simplification which emerges from this model. As this model provides an overall view of the complexation ability of the *system*, it could be termed a system approach. The conventional method focuses on the equilibrium concentrations of *all* species, and thus, could be called a species or component approach.

Curves (d), (e) and (f) of Figure 3 depict the three profiles at pH 7.72, where the 1:1 conditional stability constants of zinc with oxalate and ethylenediamine are identical (Table II). At this pH value, the two-ligand mixture is behaving as if only a single ligand were present in solution regardless of the composition of the mixture (as far as the 1:1 stability surface is concerned). Thus, the stability surface for the 1:1 complexes (curve (d)) is a plane parallel to the two-dimensional grid. Interestingly, the stability surfaces for the 1:2 and 1:3 complexes are also *almost* completely planar at pH 7.72 (curves (e) and (f)). The data for curve (e) do not form an exact plane, but this is not evident from the figure. Some degree of non-planarity is evident in surface (f). Thus, at pH 7.72, the two-ligand mixture as a whole, can be treated essentially as a single ligand system at *all* solution compositions, insofar as the distribution of metal between free and total complexed forms in solution is concerned.

Curves (g), (h) and (i) of Figure 3 show the stability surfaces at pH 3.00. Note that the CLASP-*u* region for $S3$ is lower than that for $S2$. This is the only example of such a trend, in this figure. This is due to the fact that the conditional stability products for the 1:3 complexes are smaller than those for the 1:2 complexes (Table II) because of the enhanced influence of alpha-coefficients (21,22) on reactions of higher stoichiometry. The hydrogen ion concentration at pH 3.00 prevents CLASP-*l* conditions from being attained over the C_M-C_{LT} range covered in Figure 3 (cf. condition (b) following eqn. 11).

Case 2. In this case a four-ligand mixture (Table III) is considered, where only 1:1 complexes form. In addition, absorptivities are assigned to each of the complex species in order to obtain an absorbance surface. Unlike Case 1, none of the numerical values used in Case 2 (stability products, absorptivities, etc.), while realistic, correspond to any specific, real system.

The stability surfaces depicted in Figure 3 are quite simple in shape. As mentioned previously (during discussion of Stability Profiles), a situation may be reached where some of the ligand species are almost totally complexed while the others are largely uncomplexed. This can lead to a temporary constancy of the stability function over an intermediate range of C_M/C_{LT} ratios, giving plateau values called pseudo-CLASP's. Surface (a)

Table II

Conditional Stability Products for the Zinc Complexes of Oxalate (L1) and Ethylenediamine (L2) at pH's 10.00, 7.72 and 3.00

Complex	Conditional Stability Product		
	pH 10.0	pH 7.72	pH 3.00
β1(1) cond.	$1.37711x10^{3}$	$2.73898x10^{3}$	$1.39142x10^{2}$
β1(2) cond.	$1.38566x10^{5}$	$2.73898x10^{3}$	$8.49691x10^{-6}$
β2(1,1) cond.	$1.50997x10^{6}$	$3.00216x10^{6}$	$7.70751x10^{3}$
β2(1,2) cond.	$4.38183x10^{8}$	$8.65835x10^{6}$	$1.35743x10^{-3}$
β2(2,2) cond.	$1.03358x10^{10}$	$2.02972x10^{6}$	$1.94323x10^{-11}$
β3(1,1,1) cond.	$8.68895x10^{6}$	$1.72695x10^{7}$	$2.24064x10^{3}$
β3(1,1,2) cond.	$1.55473x10^{10}$	$3.07101x10^{8}$	$2.43319x10^{-3}$
β3(1,2,2) cond.	$2.98087x10^{11}$	$5.85171x10^{7}$	$2.83127x10^{-11}$
β3(2,2,2) cond.	$3.77598x10^{11}$	$7.36687x10^{5}$	$2.17663x10^{-20}$

Table III

System Description for Case II

Largest stoichiometric metal concentration	1 M
Smallest stoichiometric metal concentration	$1x10^{-9}$ M
Number of C_M points	36
Largest stoichiometric total ligand concentration	1 M
Smallest stoichiometric total ligand concentration	$1x10^{-9}$ M
Number of C_{LT} points	36
Total number of points on surface	1296
Number of ligands in system	4
Stoichiometric ligand weighting factors	
W_1	1
W_2	0.1
W_3	1
W_4	0.001
Stoichiometries of complexes in system	1:1
Number of complexes of 1:1 stoichiometry	4
Stability products and molar absorptivities for 1:1 complexes	
β1(1)	$2.75x10^{3}$
β1(2)	$9.81x10^{3}$
β1(3)	$3.18x10^{7}$
β1(4)	$3.72x10^{3}$
ε1	$3.14x10^{3}$
ε2	$8.48x10^{3}$
ε3	$2.41x10^{5}$
ε4	$6.45x10^{2}$
Proton association products for ligands	
K1(1)	$5.71429x10^{4}$
K1(2)	$1.00000x10^{10}$
K1(3)	$4.76190x10^{8}$
K1(4)	$2.24719x10^{6}$
pH value used	7.00

of Figure 4 illustrates this effect. Part (d) of Figure 4 is a contour map representation of the same surface. The contour lines connect points of equal height above the grid surface, and they are positioned at equal increments along the vertical axis (which is logarithmic). Thus, the density of lines on the surface is a measure of the slope at each point. Contour maps provide an alternative method of representing these surfaces, and when used in conjunction with the 3-D plots, help to clarify certain surficial features. The two areas in plot (d) which are relatively free of lines correspond to the CLASP-*u* and pseudo-CLASP regions, which are parallel to the two-dimensional grid. A very gentle slope is still evident at the lower right hand corner of the contour map indicating that CLASP-*l* has not yet been fully reached.

Plots (b) and (e) show the free metal surface of the system (plotted as pM). These show pM increasing (i.e. free metal concentration decreasing) very rapidly as one moves more into the CLASP-*u* region (i.e. trace metal in presence of large ligand excess). The metal-buffering capability of the system is clearly depicted in the contour map (e). For example, the lines, or portions of lines, which are parallel to the C_{LT} axis, describe those solution compositions for which the metal is buffered during ligand-into-metal titrations (cf. Figure 2). Similarly, the lines or segments which lie at 45° angles to the C_M axis represent metal buffering with respect to dilution of the system (cf. Figure 2). The metal buffering against dilution occurs in the region corresponding to excess ligand over metal. This is consistent with the theory of metal buffering by 1:1 complexes.

Plots (c) and (f) show the absorbance surface for this system. As evident from the axis designation, plot (c) is being viewed from a different direction to (a) and (b), in order to optimize the presentation of its three-dimensional aspects.

A comparison of the three contour maps of Figure 4 shows certain features in common. There is a strong correlation between curves (d) and (e); this is generally true for systems in which only 1:1 complexes are present.

The three-dimensional surfaces presented in Figures 3 and 4 relate to relatively simple multiligand mixtures (Tables I and III, respectively). Even such simple systems can become quite complicated if all of the possible equilibrium reactions are to be taken into consideration. These simple systems were chosen for illustration in this paper so that the reader could acquire a more complete appreciation of the total complexation system. However, this model is not limited to such simple cases, and we have applied it to multiligand mixtures and systems which are vastly more complicated than those described in Tables I and III.

Conclusions

The stability function is a useful concept for describing

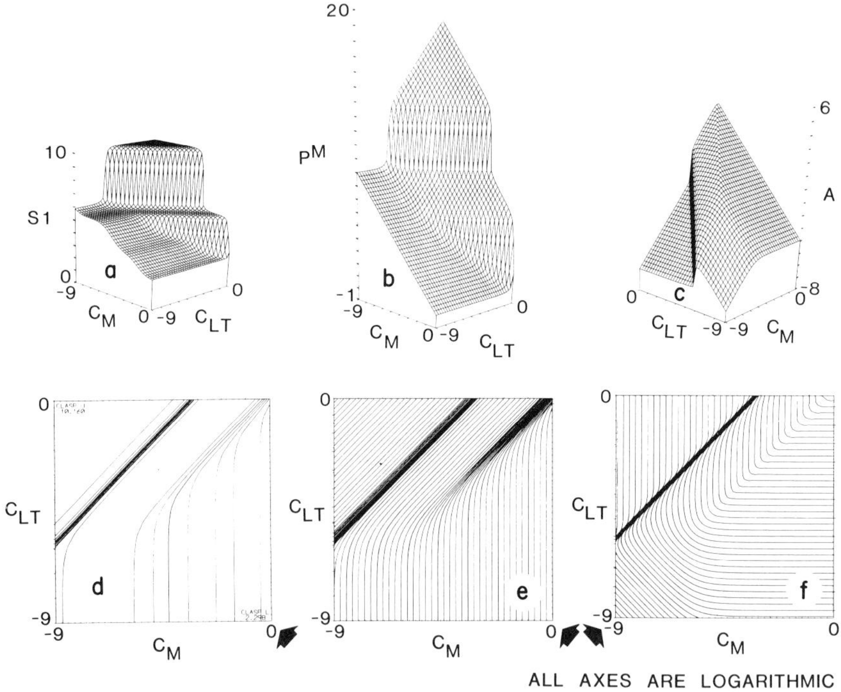

Figure 4. Various surfaces relating to the four-ligand mixture described in Table III: (a), (b), and (c) are the stability (§ 1), free metal (pM), and absorbance surfaces, respectively at pH 7.0. Plots (d), (e), and (f) are the contour diagrams corresponding to (a), (b), and (c), respectively. Note that surface (c) is being viewed from a different direction compared with (a) and (b). The arrows on the contour maps (d), (e), and (f) indicate the corners from which the corresponding perspective representations (a), (b), and (c) are being viewed.

metal-multiligand equilibrium. It provides a conceptually simple picture of the metal-binding properties of complicated ligand mixtures. The stability function adopts constant, limiting values under certain experimental conditions. This function, and other solution properties, can be readily depicted by means of three-dimensional surfaces which clearly illustrate the influence of solution composition and pH on the overall system behavior.

Acknowledgement

Acknowledgement is made to the donors of the Petroleum Research Fund, administered by the American Chemical Society, for partial support of this research. We thank Kim Mestecky for typing this manuscript.

Abstract

Methods are currently available for computing the equilibrium concentrations of all species in multimetal-multiligand mixtures. However, this is possible only when the stoichiometric concentrations of all metals and ligands are known, in addition to the stability constants for all complex species. Frequently, such detailed information is not available for the complicated ligand mixtures occurring in natural environments. The objective of this paper is to develop a unified, conceptual view of metal-multiligand equilibrium. For such systems, it is desirable to have some quantitative measure of the strength of binding of the metal ions to the ligand mixture *as a whole*. This can be achieved through introduction of the system stability function, which is a weighted-average of the stability constants for all complexes of a given stoichiometry in the system. In addition, it quantitatively describes the binding of metal ions to a multiligand mixture. The stability function has a unique value for each solution composition and stoichiometry, and can thus be represented as a three-dimensional surface, $\mathcal{S}(C_M, C_{LT})_{pH}$ where $\mathcal{S}$ is the stability function, and C_M and C_{LT} are the stoichiometric metal and total ligand concentrations, respectively. These surfaces clearly represent the metal-binding ability of the mixture as a whole and provide an insight into the complexation behavior of natural multiligand systems. They also illustrate how the stability functions adopt constant, limiting values under certain experimental conditions. Other solution properties, such as absorbance, free metal concentration, etc., can be similarly represented by means of three-dimensional surfaces. In addition to providing a unified picture of the net complexation in multi-

Appendix 1

Simplified presentation of Equations 2, 4, 6, 9 and 11. Equations 2, 4, 6, 9 and 11 of this paper may appear somewhat abstract and complicated. This is simply because they are written in a general mathematical format which incorporates mixed ligand in addition to the homogeneous complexes. The meaning of these equations can be clarified by means of a specific example: consider the $\mathcal{S}2$ function for a simple two-ligand mixture (from eqn. 2). The only 1:2 complexes in this system are $M(L1)_2$, $M(L1)(L2)$ and $M(L2)_2$. Thus,

$$\mathcal{S}2 = \frac{\omega_1^{\,2}\omega_2^{\,0}\beta2(1_2,2_0)+\omega_1\omega_2\beta2(1_1,2_1)+\omega_1^{\,0}\omega_2^{\,2}\beta2(1_0,2_2)}{(\omega_1+\omega_2)^2}$$

$$= \frac{\omega_1^{\,2}\beta2(1_2)+\omega_1\omega_2\beta2(1_1,2_1)+\omega_2^{\,2}\beta2(2_2)}{(\omega_1+\omega_2)^2} \qquad (2)'$$

which is consistent with equation 2. Equations 9 and 11 are essentially of the same form as equation 2. The stability product (eqn. 4) of the homogeneous complex, $M(L1)_2$, is given by:

$$\beta2(1,1) = \frac{[M(L1)_2(L2)_0]}{[M][L1]^2[L2]^0} \equiv \frac{[M(L1)_2]}{[M][L1]^2} \qquad (4)'$$

and the stability product (eqn. 4) of the mixed ligand complex M(L1)(L2) is given by:

$$\beta2(1,2) = \frac{[M(L1)(L2)]}{[M][L1][L2]} \qquad (4)''$$

For the two-ligand system, equation 6 takes the form:

$$\mathcal{S}2 = \frac{[M(L1)_2(L2)_0]+[M(L1)(L2)]+[M(L1)_0(L2)_2]}{[M]([L1]+[L2])^2}$$

$$= \frac{[M(L1)_2]+[M(L1)(L2)]+[M(L2)_2]}{[M]([L1]+[L2])^2} \qquad (6)'$$

ligand mixtures, these surfaces permit one to evaluate the results of various experimental methods which are applied to the study of complexation in ligand mixtures. This paper describes the basic mathematical theory and chemical principles underlying the stability function, and its use is illustrated through computer-simulation of complexation in various metal-multiligand systems.

Literature Cited

1. Kolthoff, I.M., Sandell, E.B., Meehan, E.J. and Bruckenstein, S., "Quantitative Chemical Analysis", 1199 p. MacMillan, London, 1969.
2. Daul, C. and Goel, J.J., Numerical estimation of equilibrium concentrations. J. Chem. Soc., Faraday Trans., I: 73, 985-990 (1977).
3. Perrin, D.D., Multiple equilibria in assemblages of metal ions and complexing species: A model for biological systems. Nature 206, 170-171 (1965).
4. Perrin, D.D. and Sayce, I.G., Computer calculation of equilibrium concentrations in mixtures of metal ions and complexing species. Talanta 14, 833-842 (1967).
5. Ingri, N., Kakolowicz, W., Sillen, L.G. and Warnquist, B., High speed computers as a supplement to graphical methods-V. HALTAFALL, a general program for calculating the composition of equilibrium mixtures. Talanta 14, 1261-1286 (1967).
6. I, Ting-Po and Nancollas, G.H., EQUIL--A general computational method for the calculation of solution equilibria. Anal. Chem. 44, 1940-1950 (1972).
7. Sillen, L.G. and Martell, A.E., "Stability Constants of Metal-Ion Complexes", 754 p. Spec. Publ. No. 17, Chemical Society, London, 1964.
8. Sillen, L.G. and Martell, A.E., "Stability Constants of Metal-Ion Complexes", 865 p. Suppl. No. 1, Spec. Publ. No. 25, Chemical Society, London, 1971.
9. Yatsimirskii, K.B. and Vasil'ev, V.P., "Instability Constants of Complex Compounds", 214 p. Pergamon, New York, 1961.
10. Smith, R.M. and Martell, A.E., "Critical Stability Constants, Vol. 1: Amino Acids", 469 p. Plenum, New York, 1974.
11. Smith, R.M. and Martell, A.E., "Critical Stability Constants, Vol. 2: Amines", 415 p. Plenum, New York, 1974.
12. Smith, R.M. and Martell, A.E., "Critical Stability Constants, Vol. 3: Other Organic Ligands", 495 p. Plenum, New York, 1977.
13. Smith, R.M. and Martell, A.E., "Critical Stability Constants, Vol. 4: Inorganic Complexes", 257 p. Plenum, New York, 1976.
14. Kragten, J., "Atlas of Metal-Ligand Equilibria in Aqueous Solution", 781 p. Halsted Press, New Jersey, 1978.
15. Kononova, M.M., "Soil Organic Matter" (T.Z. Nowakowski and A.C.D. Newman, Translators), 2nd Eng. Ed., Pergamon, New York, 1966.

16. Schnitzer, M. and Khan, S.U., "Humic Substances in the Environment", 327 p., Dekker, New York, 1972.
17. Gjessing, E.T., "Physical and Chemical Characteristics of Aquatic Humus", 120 p., Ann Arbor Science, Ann Arbor, 1976.
18. Liu, C.L., "Introduction to Combinatorial Mathematics", 393 p., McGraw-Hill Book Co., New York, 1968.
19. MacCarthy, P. and Smith, G.C., Metal-binding by ligand mixtures: a quantitative model, in Hemphill, D.D., ed., Proc. 12th Conf. "Trace Substances in Environmental Health," Missouri, June 1978 (in press).
20. MacCarthy, P., An interpretation of stability constants for soil organic matter-metal ion complexes under Schubert conditions. J. Environ. Sci. Health A12, 43-59 (1977).
21. Ringbom, A., "Complexation in Analytical Chemistry", 395 p., Interscience, New York, 1963.
22. MacCarthy, P. and Mark, Jr., H.B., An evaluation of Job's method of continuous variations as applied to soil organic matter-metal ion interactions. Soil Sci. Soc. Amer. J. 40, 267-276 (1976).
23. Kanemura, Y., and Watters, J.I., Acidimetric studies of cadmium and zinc complexes with ethylenediamine, oxalate, and their mixtures using the glass electrode. J. Inorg. Nucl. Chem. 29, 1701-1709 (1967).
24. Schubert, J., The use of ion-exchangers for the determination of physical-chemical properties of substances, particularly radiotracers in solution: I. J. Phys. and Colloid Chem. 52, 340-350 (1948).
25. MacCarthy, P. and Mark, Jr., H.B., A further examination of the Schubert ion-exchange method as applied to soil organic matter-metal ion interactions, pp 197-212 in Drucker, H. and Wildung, R.E., ed., "Biological Implications of Metals in the Environment," ERDA Conf.-750929, Nat'l. Technical Info. Service, Springfield, VA, (1977).

RECEIVED November 16, 1978.

Adsorption

11

Applicability of the Local Equilibrium Assumption to Transport through Soils of Solutes Affected by Ion Exchange

RONALD V. JAMES and JACOB RUBIN

Water Resources Division, U.S. Geological Survey, Menlo Park, CA 94025

In an attempt to deal with such unwanted substances as radioactive and chemical wastes, disposal sites are often used that are hydraulically connected with usable water supplies via subsurface transport routes. To manage these wastes effectively, it is desirable to have the capability of predicting the course of solute transport along these connecting routes.

Subsurface solute transport is affected by hydrodynamic dispersion and by chemical reactions with soil and rocks. The effects of hydrodynamic dispersion have been extensively studied (1, 2, 3, 4). Chemical reactions involving the solid phase affect subsurface solute transport in a way that depends on the reaction rates relative to the water flux. If the reaction rate is fast and the flow rate slow, then the local equilibrium assumption may be applicable. If the reaction rate is slow and the flux relatively high, then reaction kinetics controls the chemistry and one cannot assume local equilibrium. Theoretical treatments for transport of many kinds of reactive solutes are available for both situations (5-10).

It is often desirable, where applicable, to use the local equilibrium assumption when predicting the fate of subsurface solutes. Advantages of this approach may include 1) data such as equilibrium constants are readily available, as opposed to the lack of kinetic data, and 2) for transport involving ion exchange and adsorption, the mathematics for equilibrium systems are generally simpler than for those controlled by kinetics. To utilize fully these advantages, it is helpful to know the flow rate below which the local equilibrium assumption is applicable for a given chemical system. Few indicators are available which allow determination of that critical water flux.

The work presented here addresses this question for a laboratory ion-exchange system that is relatively simple. Effluent concentration histories were obtained for calcium and chloride ions during miscible displacement of calcium chloride solutions through vertical columns containing homogeneous, repacked sandy soil that was water-saturated. Calcium self-exchange was the

0-8412-0479-9/79/47-093-225$05.00/0
This chapter not subject to U.S. copyright
Published 1979 American Chemical Society

only reaction considered. The flow rate was kept constant during each experiment but it had a large range of variation among the experiments.

Experimental

Calcium chloride solutions (pH = 6.2) labeled with ^{45}Ca or ^{36}Cl were displaced vertically downward through columns of homogeneous, repacked, water-saturated sandy soil by a chemically identical solution labeled with ^{36}Cl or ^{45}Ca, respectively. Constant water fluxes, and solution activities of 1 to 15 $\mu Ci/dm^3$, were used. Calcium solutions were analyzed by titration with disodium dihydrogen ethylenediamine tetraacetate to a murexide end point (11). The activity of radioactively labeled solutions was obtained by liquid scintillation techniques. Concentrations of adsorbed calcium were calculated from isotope dilution. The concentration of calcium chloride in the influent solution was 0.08 N. Because exchange of calcium for itself was the only chemical process affecting transport, the calcium chloride concentration remained constant at 0.08 N throughout each experiment, both within the column and in the effluent.

The soils employed in this study were Delhi (11) and Oakley (12) sands, in which most of the clay appears to be present as coatings on the sand particles. Characteristics of each soil are shown in Table I. Oakley sand is quite acidic. This can be explained (13, p. 282-289) by the substantial aluminum component of the exchangeable cations.

Table I

Characteristics of Experimental Soils

Soil	Porosity (cm^3/cm^3)	Bulk density (g/cm^3)	pH** (in 0.01 M $CaCl_2$)	Cation exchange capacity* (meq/g)
Delhi sand	0.40	1.6	6.1	0.04
Oakley sand	0.34	1.8	4.2	0.02

*Exchange capacity values are averages. Specific values were obtained for each column and used as explained in the text.
**Determination of soil pH as described by Black (15).

Each Oakley column was pre-leached, removing about half the adsorbed aluminum. As subsequent leaching continued during repetitive experiments, exchangeable aluminum was removed and the effluent pH increased. In all of the individual experiments described here, the concentration of aluminum in the effluent was so low (less than 3 x 10^{-5} M) that the increases in adsorbed calcium resulting from removal of aluminum were negligible and the effluent pH increased less than 0.2 units. Therefore, in each given experiment the concentration of adsorbed calcium was taken to be constant. The aluminum concentration was determined colorimetrically (14). Delhi sand is a nearly neutral soil and no pH change occurred during the experiments using it. The concentration of adsorbed calcium was determined for each column by graphical integration of the area between plotted curves showing the calcium and chloride effluent histories. The exchangeable calcium thus determined was used in subsequent theoretical predictions that were compared with the empirical results obtained from the corresponding columns.

The columns used in these studies and the general experimental techniques employed are described by James and Rubin (16). The soil columns were 10 cm long and 5 cm in diameter. Special care was taken in packing the columns to avoid radial and longitudinal particle-size segregation (17). Fluid volumes in the columns averaged 73 cm^3 for Oakley sand and 81 cm^3 for Delhi sand.

Results are presented as effluent histories in which the relative effluent concentration (REC) of labeled calcium is plotted against time and is given by:

$$\mathrm{REC} = \frac{c_e - c_o}{c_\infty - c_o} \qquad (1)$$

where c_o, c_∞, and c_e are the concentrations of labeled calcium in the soil at zero time, in the displaced solution, and in the effluent at time t, respectively.

Theoretical

One-dimensional transport through soils of calcium affected by equilibrium-controlled self-exchange is described by:

$$\theta \frac{\partial c}{\partial t} + \rho \frac{\partial s}{\partial t} = D \frac{\partial^2 c}{\partial z^2} - q \frac{\partial c}{\partial z} \qquad (2)$$

where θ = porosity of the medium, ρ = bulk density of the medium (g/cm^3), z = distance from input end of column (cm), t = time (sec), c = concentration of labeled solute in the soil solution at z and t (meq/cm^3), s = concentration of adsorbed solute at z and t (meq/g), D = hydrodynamic dispersion coefficient (cm^2/sec), and q = water flux (cm/sec). The equilibrium relation is

$$\frac{sc^*}{cs^*} = 1 \qquad (3)$$

where the asterisk denotes unlabeled calcium. Noting that $c+c^*$

is the total concentration of calcium in solution, C_t, and that, with the concentration of adsorbed calcium negligibly affected by the presence of aluminum, s+s* can be taken to be the exchange capacity of the soil, λ, it follows from Equation 3 that

$$s = \frac{\lambda}{C_t} c \tag{4}$$

For the conditions used in this study C_t and λ are constant, therefore, partial differentiation of Equation 4 yields

$$\frac{\partial s}{\partial t} = \frac{\lambda}{C_t} \frac{\partial c}{\partial t} \tag{5}$$

Combining Equation 5 with Equation 2 gives the equation

$$(\theta + \rho \frac{\lambda}{C_t}) \frac{\partial c}{\partial t} = D \frac{\partial^2 c}{\partial z^2} - q \frac{\partial c}{\partial z} \tag{6}$$

which in this study was solved numerically for effluent concentrations as a function of time using the method and boundary conditions of Rubin and James (7). The numerical method is needed because experimental columns of this type are best represented as two consecutive layers: the soil column and the void space of the apparatus passages, respectively. The two-layer description is necessary in order to account for dispersion induced by the apparatus (16). The dispersion coefficients used at each flow rate are given in Table II.

Porosity, bulk density, and water flow rate were determined independently of the elution histories, as described previously (16). The dispersion coefficients for the soil were obtained by trial and error determination of the D values giving the best agreement between the predicted and the empirical chloride-effluent histories. The predicted histories were obtained from the solution to Equation 2 with $\partial s/\partial t = 0$. To find the dispersion coefficients for the end caps (16), a length of 0.2 cm and a porosity of 1.0 were used. The water flux in the end cap is the same as that in the soil.

Results and Discussion

For both soils studied, comparison of calcium-effluent histories predicted by the solution to Equation 6 with those obtained from experimental columns gave good agreement only for the lowest flow rates. For the three higher water fluxes, more apparent dispersion was observed than could be explained by predictions that assume local equilibrium. Examples of these comparisons are shown in Figures 1 and 2.

Table II

Variation with Water Flux of Soil and Apparatus Dispersion Coefficients

A. Delhi Sand

q (cm/sec)	D (cm^2/sec)	D_A* (cm^2/sec)
1.7×10^{-5}	5.0×10^{-6}	4×10^{-6}
1.8×10^{-4}	2.7×10^{-5}	3×10^{-5}
1.7×10^{-3}	3.5×10^{-4}	7×10^{-5}
1.8×10^{-2}	3.5×10^{-3}	6×10^{-4}

B. Oakley Sand

q (cm/sec)	D (cm^2/sec)	D_A* (cm^2/sec)
1.7×10^{-5}	3.0×10^{-6}	4×10^{-6}
1.7×10^{-4}	6.6×10^{-5}	3×10^{-5}
1.7×10^{-3}	1.0×10^{-4}	7×10^{-5}
1.7×10^{-2}	2.0×10^{-3}	6×10^{-4}

*The subscript A indicates the dispersion coefficient used for the layer which describes the dispersion induced by the apparatus.

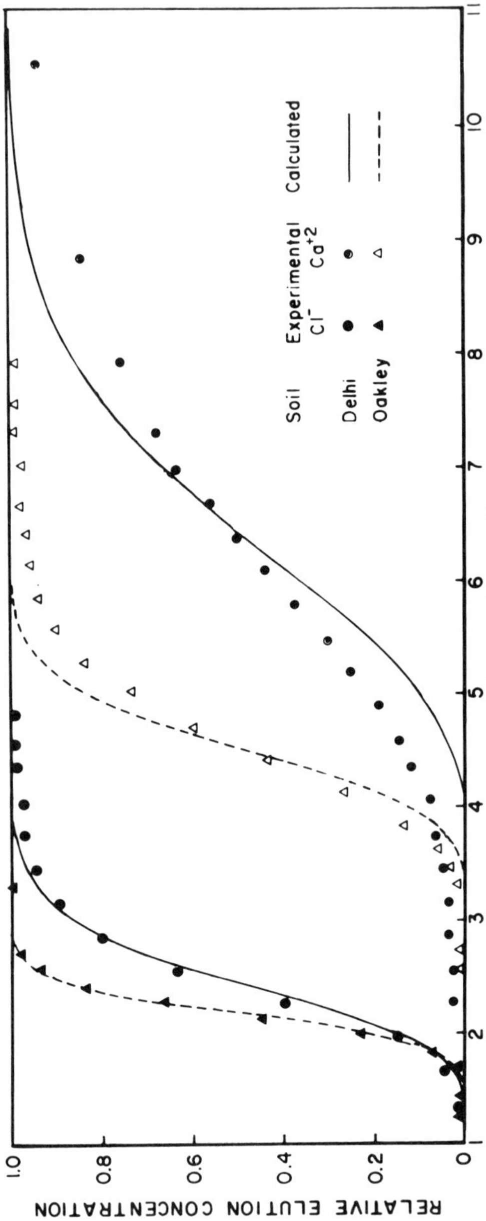

Figure 1. Elution histories for Delhi and Oakley sands with water flux of 1.7 × 10^{-3} cm/sec

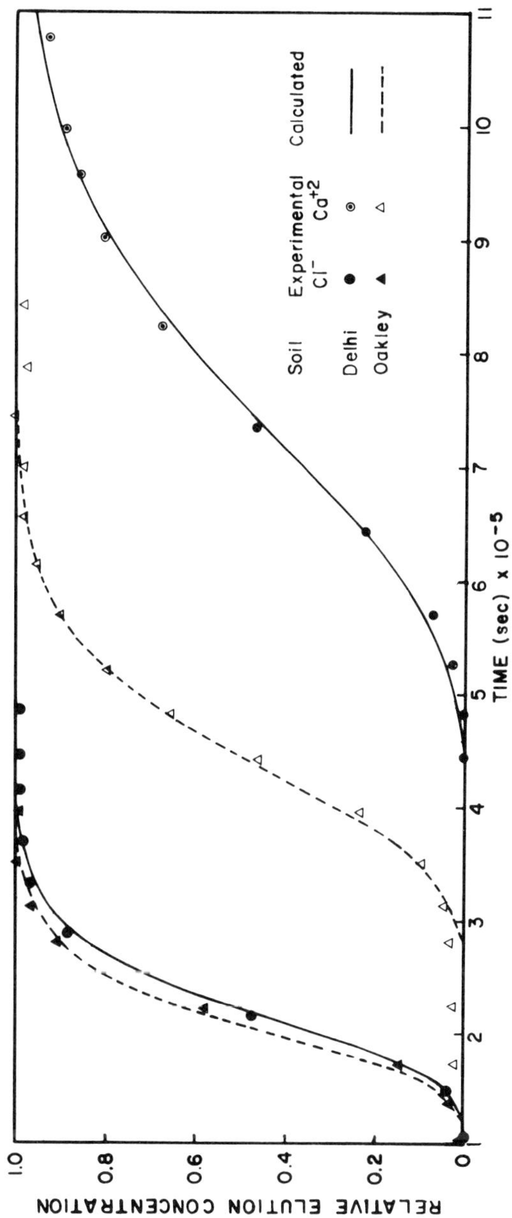

Figure 2. Elution histories for Delhi and Oakley sands with water flux of 1.7 × 10^{-5} cm/sec

The molecular-diffusion coefficient for calcium in dilute aqueous solutions of calcium chloride, $D_{m\ell}$, is about 1 x 10^{-5} cm^2/sec (18,p.700). The molecular-diffusion coefficient in the soil, D_{ms}, can be estimated by the relation $D_{ms} \sim D_{m\ell}\theta/T$, where T is the tortuosity, assumed here to be about 1.4. The estimated molecular-diffusion coefficient of calcium for the soils employed in this study is, therefore, about 3 x 10^{-6} cm^2/sec. Comparing this value with the dispersion coefficients in Table II, one finds that at the lowest flow rate studied for each soil the coefficients are similar, but at higher fluxes they differ by orders of magnitude. Thus, when the hydrodynamic dispersion coefficient is nearly the same as the estimated molecular-diffusion coefficient in the soil, the equilibrium assumption applies. When the dispersion coefficient is significantly larger, the local equilibrium assumption does not apply and a kinetics-based model presumably is indicated.

In order to further substantiate this conclusion, it is of interest to compare it with the prediction obtained from a simple theoretical model. Glueckauf's well-known transport model (19, p. 449-453), supplemented by the more modern concept of hydrodynamic dispersion, is well suited for this purpose. The model simulates dispersion-affected solute transport with ion exchange for which diffusion processes are rate limiting. In his development, Glueckauf assumes: 1) exchange takes place in porous sperical particles with radius r and porosity θ_p; 2) within these particles, water and solute movement occur only by molecular diffusion; 3) the reaction rate is determined by the rates of film and intraparticle diffusion, which are proportional to the differences between the theoretical equilibrium and actual concentrations of solution and adsorbed solutes, respectively; and 4) the relevant equilibrium relations are described by Equation 4.

Using the modified Glueckauf model described above, and employing his simplifying mathematical assumptions (the correctness of which has been confirmed by the authors with the aid of numerical methods), one obtains

$$(\theta + \rho \frac{\lambda}{C_t}) \frac{\partial c}{\partial t} = \bar{D} \frac{\partial^2 c}{\partial z^2} - q \frac{\partial c}{\partial z} \tag{7}$$

which is similar to Equation 6 with D replaced by the effective dispersion coefficient $\bar{D}$, which is given by:

$$\bar{D} = D + F_p + F_f \tag{8}$$

when F_p and F_f describe the contributions of intraparticle and film diffusion, respectively, and are expressed by

$$F_p = \frac{0.10\rho\lambda C_t q^2 r^2}{D_{m\ell}\theta_p(\theta C_t+\rho\lambda)^2}$$

$$F_f = \frac{0.133\rho^2\lambda^2 q^2 r^2}{D_{m\ell}(\theta C_t+\rho\lambda)^2(1+70qr)} \quad (9)$$

As q decreases, the terms F_p and F_f become small compared to D, and Equation 7 becomes more like Equation 6. This indicates that the local equilibrium assumption is applicable when q is sufficiently small.

Using the relationship

$$D = D_{ms} + \alpha q \quad (10)$$

where α is an empirical constant (1), one finds that when the dispersion coefficient is nearly equal to the molecular-diffusion coefficient in the soil, αq is much smaller than D_{ms}, giving

$$q \ll \frac{D_{ms}}{\alpha} = \frac{D_{m\ell}\theta}{\alpha T} \quad (11)$$

Inequality 11 was substituted into Equation 8, together with reasonable values of other parameters and 0.1 cm < α < 0.2 cm as α was found to be in this study. This leads to the conclusion that, for systems in which r is less than 0.1 cm, the local equilibrium assumption is applicable (i.e., q is sufficiently small) when D is nearly equal to D_{ms} as observed in the experiments. In soils in which the exchanging particles are not spherical, r would represent approximately the mean diffusion path within clay aggregates or within clay coatings on coarse particles.

For soils without appreciable clay aggregation, the experimental results and theoretical analysis described here indicate that when diffusion is rate-limiting, the ratio of the hydrodynamic dispersion coefficient to the estimated soil self-diffusion coefficient may be a useful indicator of the applicability of the local equilibrium assumption. For reacting solutes in laboratory columns such as those used in this study, systems with ratios near unity can be modeled using equilibrium chemistry. The experimental results indicate that when the ratio is considerably larger than one, another relationship, presumably one involving kinetics, must be used.

Abstract

Miscible displacement of calcium-chloride solutions through water-saturated laboratory soil columns was studied for a wide range of constant water-flow rates. Calcium- and chloride-effluent histories were obtained. Calcium self-exchange was the only reaction considered. Calcium-effluent histories were compared with predictions from a one-dimensional solute-transport model, assuming local chemical equilibrium. Good agreement between the predictions and the data was obtained for the slowest flow rate studied, but not for the higher fluxes. Thus, the local equilibrium assumption applies when the ratio of the hydrodynamic dispersion coefficient to the estimated molecular-diffusion coefficient is near unity. This conclusion is further substantiated by comparison with the results of a theoretical analysis using the relatively simple transport model for solutes affected by ion exchange that has been developed by Glueckauf (19). It is suggested that the data obtained for the higher water fluxes that yield a coefficient ratio much greater than unity cannot be described by assuming local equilibrium, but must be modeled using another relationship, presumably one involving kinetics.

Literature Cited

1. Perkins, T. K., and Johnston, O. C., A review of diffusion and dispersion in porous media, Soc. Petrol. Eng. J., 228 (March), 70-84 (1963).
2. Pfannkuch, H. O., Contribution A l'Etude des Deplacement de Fluides Miscibles dans un Milieu Poreux, Paris Instit. Franc. due Petr. Rev., 18 (2), 215-270 (1963).
3. Nunge, R. J., and Gill, W. N., Mechanisms affecting dispersion and miscible displacement, Indust. Eng. Chem., 61 (9), 33-49 (1969).
4. Bear, J., "Dynamics of Fluids in Porous Media", 764 p., American Elsevier Publishing Company, Inc., New York, 1972.
5. Cameron, D. R., and Klute, A., Convective-dispersive solute transport with a combined equilibrium and kinetic adsorption model, Water Resour. Res., 13 (1), 183-188 (1977).
6. Skopp, J., and Warrick, A. W., A two-phase model for the miscible displacement of reactive solutes in soils, Soil Sci. Soc. Am. Proc., 38 (4), 545-550 (1974).
7. Rubin, J., and James, R. V., Dispersion-affected transport of reacting solutes in saturated porous media: Galerkin method applied to equilibrium-controlled exchange in unidirectional steady water flow, Water Resour. Res., 9 (5), 1332-1336 (1973).
8. Reiniger, P., "Movement and Exchange of Sodium and Calcium in Calcareous and Gypseous Soils", Ph.D. Thesis, Hebrew Univ., Jerusalem, 1970.
9. Lai, S.-H., and Jurinak, J. J., Numerical approximation of cation exchange in miscible displacement through soil columns, Soil Sci. Soc. Am. Proc., 35 (6), 894-899 (1971).
10. Peterson, E. E., "Chemical Reaction Analysis", 276 p., Prentice-Hall, Inc., New Jersey, 1965.
11. Cole, R. C., Gardner, R. A., Koehler, L. F., Anderson, A. C., Bartholomew, O. F., and Retzer, J. L., Soil Survey of the

Bakersfield area, California, USDA, Bureau of Plant Industry, Soils, and Agricultural Engineering, Series 1937, Rept. no. 12, 1945.

12. Carpenter, E. J., and Cosby, S. W., Soil survey of Contra Costa County, California, USDA, Bureau of Chemistry and Soils, Series 1933, Rept. no. 26, 1939.
13. Black, C. A., "Soil-Plant Relationships", 2nd ed., 792 p., John Wiley and Sons, Inc., New York, 1968.
14. Brown, E., Skougstad, M. W., and Fishman, M. J., Methods for collection and analysis of water samples for dissolved minerals and gases in "Techniques of Water-Resources Investigations of the United States Geological Survey", Book 5, Chapter A1, U.S. GPO, Washington, D.C., 1970.
15. Black, C. A., ed., "Methods of Soil Analysis, Part 2, Chemical and Microbiological Properties", Chapter 60, American Society of Agronomy, Inc., Madison, Wisconsin, 1965.
16. James, R. V., and Rubin, J., Accounting for apparatus-induced dispersion in analyses of miscible displacement experiments, Water Resour. Res., 8 (3), 717-721 (1972).
17. Ripple, C. D., James, R. V., and Rubin, J., Radial particle-size segregation during packing of particulates into cylindrical containers, Powder Tech., 8, 165-175 (1973).
18. Harned, H. S., and Owen, B. B., "The Physical Chemistry of Electrolytic Solutions", 803 p., Reinhold Book Corporation, New York, 1958.
19. Helfferich, F., "Ion Exchange", 624 p., McGraw-Hill, New York, 1962.

Disclaimer: The reviews expressed and/ or the products mentioned in this article represent the opinions of the author(s) only and do not necessarily represent the opinions of the U.S. Geological Survey.

RECEIVED November 16, 1978.

12

Adsorption of Lead from Solution on the Quartz- and Feldspar-Containing Silt Fraction of a Natural Streambed Sediment

DAVID W. BROWN

U.S. Geological Survey, Menlo Park, CA 94025

The extensive use of lead antiknock additives in gasoline has made lead perhaps the most widely distributed toxic heavy metal in the urban environment and has greatly increased its availability for solution in natural waters. It is important for this reason to know whether its introduction into surface and ground waters by rainfall and runoff will make it available for solution or whether chemical processes will place a safe upper limit on its solubility.

A model for predicting the concentration of lead in surface or ground water must accurately reflect the physical and chemical processes that affect the distribution of lead between the solid and dissolved phases. In a natural water system the chemical processes that can greatly vary the concentration of lead are those of complex formation, solution-precipitation, and adsorption.

This paper presents a comprehensive method for evaluating the surface characteristics needed to predict the adsorption behavior of lead on the silt fraction of sediments from a natural streambed. Also presented is a quantitative description of lead adsorption on such material primarily as a function of pH between 4 and 8, the range most commonly encountered in natural waters.

The predicted pH-dependence is then compared with experimental results at two different total lead concentrations. Special emphasis is placed here on the effect of the presence of adsorbed lead and other cations on the pH-dependency of this adsorption, particularly that which results from the effect of the adsorbed cations on electrostatic potential and charge density at the solution-surface interface.

Chemical Processes Affecting Aqueous Lead Concentration

Complex Ion Formation. Formation of complexes of lead with the various anions such as chloride, fluoride, carbonate, bicarbonate, and hydroxide increases the concentration of lead in natural waters by preventing lead from taking part in other chemical reactions, primarily adsorption, that would lower its

0-8412-0479-9/79/47-093-237$06.00/0
This chapter not subject to U.S. copyright
Published 1979 American Chemical Society

concentration. The important complexes in most surface and ground-water systems are those of hydroxide and carbonate. The known monomeric hydroxy complexes are $PbOH^+$, $Pb(OH)_2{}^0$, $Pb(OH)_3{}^-$, and $Pb(OH)_4{}^{2-}$. The most important of these is $PbOH^+$, as the latter three become significant only above the pH of most natural waters, and because $PbOH^+$, a cation, is much more readily adsorbed to negatively charged surfaces than are uncharged $Pb(OH)_2{}^0$ and the anions $Pb(OH)_3{}^-$ and $Pb(OH)_4{}^{2-}$. Thermodynamic data indicate that the Pb^{2+} species predominates to about pH 7, followed by $PbOH^+$ to about pH 10, $Pb(OH)_2{}^0$ between pH 10 and 11, then by $Pb(OH)_3{}^-$ and $Pb(OH)_4{}^{2-}$ at higher pH. Since our interest is in the pH range of commonly encountered natural waters, we use only the thermodynamic data applicable to the Pb^{2+}-$PbOH^+$ reaction. Lind (1) gives the formation constant for $PbOH^+$ (for which the equilibrium expression is $[PbOH^+][H^+][Pb^{2+}]^{-1}$) as $10^{-7.23}$. Hem and Durum (2) give the lead hydroxide solubility product $^*K_{s0}$ (written $[Pb^{2+}][H^+]^{-2}$) as $10^{8.16}$.

Adsorption and Cation Exchange. The role of adsorption of lead cations onto streambed or aquifer material must be considered in any model used to predict the aqueous lead concentration in natural waters, particularly in ground-water systems. Hem (3) and Hem and Durum (2) note that the concentrations of lead in surface and ground waters in the United States are far below those which would be predicted from equilibrium with solid mineral species. It therefore appears that precipitation-dissolution equilibria are not the chemical processes controlling lead concentrations in natural waters. The large surface areas and cation exchange capacities of clays and silts provide a sufficiently large sink for lead ions such that any accurate equilibrium model used for predicting aqueous lead in natural systems would have to consider this effect.

Most natural waters come into contact with a large surface area of clay, silt, and sand. At moderate values of pH, the silicate surfaces contain a number of negatively charged adsorption sites to which cations may be tightly bound; the upper limit to the number of cations that may be held to the surface at a given time is called the cation exchange capacity (CEC). Despite the use of this term, we should realize that on one hand cation exchange and adsorption of cations could be considered to be two different processes, while on the other hand they may well be two different attempts at explanation of the same phenomena. What is commonly referred to as cation exchange is actually indistinguishable from simultaneous adsorption of one ion and desorption of the same number of equivalents of another. (This often happens when all adsorption sites are initially filled, and further adsorption of other cations can occur only if some of the currently adsorbed ions are desorbed.) The main difference between the cation exchange and electrical double-layer adsorption models lies mainly in that the cation exchange model is a mass-action approach to the competition among different cations for a fixed number of negatively charged sites; this results in the imposition of the requirement of constant charge of surface cations. The electrical double-layer approach does not impose this requirement because the magnitude of the negative surface charge is variable and influenced by specific adsorption of potential-determining H^+ and OH^- ions. It is for this reason that a version of the double-layer model of cation adsorption shall be used here in attempting to reconcile pH-dependent adsorption of lead on a fraction of a natural streambed sediment.

The major feature of the James and Healy (4) model is the description of adsorption by a Langmuir isotherm whose equilibrium constant is based upon a free energy of adsorption made up of what are referred to as coulombic (ΔG_i^{coul}), solvation (ΔG_i^{solv}), and chemical (ΔG_i^{chem}) terms.

Of these three contributions to the free energy, the most important in terms of its effect on the pH-dependence of adsorption is the coulombic term ΔG_i^{coul}. This is a measure of the interaction of the charged ion with the Gouy potential $\Delta\psi_{x_i}$ at the plane of adsorption. This surface potential is assumed by James and Healy (4) to vary in a Nernstian manner with pH, assuming a value which is negative for a pH above the pH_{pzc} (or PZC) and positive for a pH below the pH_{pzc}.

The solvation term ΔG_i^{solv} represents the change in free energy which results from the replacement, on an ion as it adsorbs, of water molecules from the bulk of the solution with water molecules at the interfacial region having a lower dielectric constant. James and Healy's (4) expression for ΔG_i^{solv} was further refined (prior to publication) by Levine (5). This, according to Wiese et al. (6), constitutes a more accurate and rigorous theoretical analysis of the changes in solvation energy that accompany adsorption. Accordingly, the Levine (5) expression was used here in calculations which were used to attempt to predict lead solubility.

The chemical term ΔG_i^{chem} is assumed to be a constant which applies for all species of a given heavy metal (e.g., Pb^{2+}, $PbOH^+$, $PbCl^+$, etc.), and which takes into account effects such as those of Van der Waals and London dispersion forces which do not easily lend themselves to theoretical interpretation.

The model recently proposed by Bowden et al. (7) avoids the use of the assumption that the surface potential varies in a Nernstian fashion with pH. In the VSC-VSP (variable surface charge-variable surface potential) model, the surface charge density σ_s is produced as the result of a chemical interaction between surface sites specific for adsorption only of the potential-determining ions H^+ and OH^-. These ions are recognized as having finite sizes, resulting in a maximum adsorption density N_s of such ions which can be physically located on the surface.

The surface, interfacial, and diffuse charge densities (respectively, σ_s, σ_i, and σ_d) and the electrostatic (Gouy) potentials at the surface and adsorption planes (respectively ψ_s and ψ_d)

are related by the equations

$$\sigma_s = \frac{FN_s[K_H[H^+]e^{-F\psi_s/RT} - K_{OH}[OH^-]e^{+F\psi_s/RT}]}{1 + K_H[H^+]e^{-F\psi_s/RT} + K_{OH}[OH^-]e^{+F\psi_s/RT}} , \qquad (1)$$

$$\sigma_s = -\sqrt{8000\ \varepsilon_0\varepsilon_{H_2O}\ RT\mu}\ \sinh\left(\frac{|z|F\psi_d}{2RT}\right) , \qquad (2)$$

$$\sigma_s + \sigma_i + \sigma_d = 0 \text{ , and} \qquad (3)$$

$$\psi_s - \psi_d = \frac{\sigma_s \underline{d}}{\varepsilon_{int}\varepsilon_0} . \qquad (4)$$

The terms K_H and K_{OH} are the specific adsorption constants for the potential-determining ions H^+ and OH^-, respectively; ε_{H_2O} and ε_{int} are the dielectric constants of water (78.5 at 25°C) and the interfacial region, respectively; μ is ionic strength; $|z|$ is the mean valence of the supporting electrolyte; and $\underline{d}$ is the distance between the surface and the average plane of adsorption. (The terms F, R, T, and ε_0 are the Faraday, the universal gas constant, absolute temperature, and permittivity of free space, respectively.) These equations, in the order listed, describe the specific adsorption of the potential-determining ions that give rise to surface charge, the dependence of diffuse layer charge on ionic strength and Gouy potential, overall electroneutrality of the system, and a linear decrease in potential over the interfacial region proportional to the surface charge density. The value of the interfacial charge density σ_i due to cation adsorption is obtained from the total surface concentrations of all adsorbed ions. Each ion's surface concentration is calculated from equilibrium considerations as in the James and Healy (4) model, using ψ_d to calculate the coulombic free-energy term ΔG_i^{coul} Simultaneous solution of the appropriate equilibria along with the above equations is normally accomplished by a computerized algorithm. The manner in which the various surface parameters are related to one another in the VSC-VSP model is shown in Figure 1 along with a similar scheme for the James and Healy (4) model.

In the VSC-VSP model the charge density σ_s of adsorbed cations exerts an effect on the surface and Gouy potentials which can be quite profound where relatively large amounts of cations are adsorbed. This effect, not considered in the classical double-layer theory as presented by James and Healy (4), can help us to explain, for example, a rise in the increase in adsorption with increased pH that is different from that otherwise predicted.

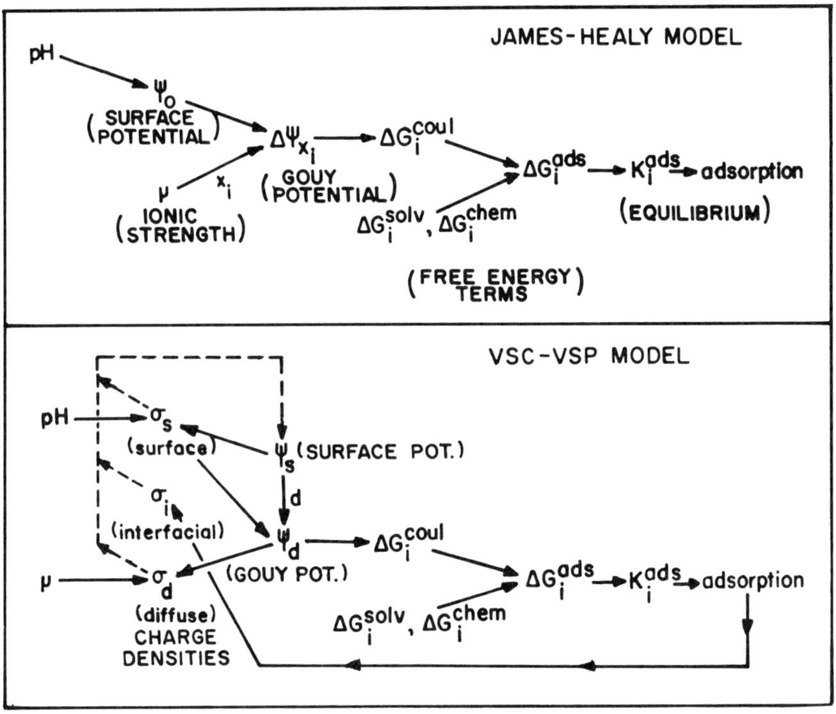

Figure 1. Relationships between charge densities, potentials, and free energies in James–Healy and VSC–VSP models

Preparation and Characterization of Sediment

Experimental. The sediment used in this study was obtained from Colma Creek, at Serramonte Boulevard between El Camino Real and Junipero Serra Boulevard, in the city of Colma, San Mateo County, California. Colma Creek was chosen because its entire course occurs within an area of urbanization, and the sediments are therefore of the type which normally come in contact with lead and other heavy metals. The main sources of lead, atmospheric fallout and rainfall runoff contain particulate matter from automobile exhaust emissions. Several shovels full of the bottom material were placed in a plastic container. In the laboratory, several kilograms of the material were wet-sieved, and the fraction passing through a 200-mesh sieve (particle diameter less than 74 μm) was placed in a 1-liter graduated cylinder containing a 1 M sodium phosphate solution. The silt fraction settles in this medium while the finer clay particles remain suspended. After several hours, this clay suspension was then decanted and a portion of the material saved for X-ray diffraction, as were portions of the sand and silt fractions.

Adsorption experiments were carried out using only the silt fraction because this fraction lends itself to batch experiments and is similar in composition to the sand fraction, consisting primarily of quartz and feldspar. Furthermore, it was felt that the application of double-layer theories of adsorption would be less appropriate on the clay fraction, where the effects of valence deficiencies resulting from isomorphic substitution in actual clays are probably more important.

After treatment with hydrogen peroxide in 0.3 M hydrochloric acid at 70°C for 30 minutes in order to remove adsorbed organic matter, the silt fraction (4 to 74 μm diameter) was washed repeatedly with distilled deionized water to constant conductivity in order to remove any adsorbed acid. The specific surface area of this material was measured using Lawrie's (8) 1,10-phenanthroline adsorption method. The cation exchange capacity per gram was determined by Chapman's (9) method, in which the surface is first saturated with sodium by three successive washings with 1.0 M sodium acetate solution, followed by three successive washings with 1 M ammonium acetate, which is saved and analyzed for the displaced sodium.

For the purpose of determining how the specific adsorption of potential-determining H^+ and OH^- ions is affected by pH and background electrolyte concentration, it is common practice to titrate portions of the adsorbent with acid and base at various ionic strengths. Amounts of adsorbent with cation exchange capacity equal to 1.00×10^{-4} eq were therefore placed in each of two 25.0 mL CO_2-free solutions of 0.001 M $NaClO_4$, two of 0.01 M $NaClO_4$, and two of 0.10 M $NaClO_4$. Each solution was separately titrated with 0.100 N sodium hydroxide or perchloric acid. Between each addition of acid or base in these six titrations,

3 minutes of magnetic stirring were allowed for the pH reading to equilibrate.

The dielectric constant of the adsorbent material was calculated using the Looyenga (10) equation given by Lal and Parshad (11). Using these equations, the dielectric constant of a component of a mixture can be calculated from the overall dielectric constant of a given mixture, which is normally determined from the capacitance of a calibrated cell containing the mixture. Such a cell was constructed from phenolic copper-clad boards 6 cm square, placed about 1 mm apart, and sealed at the edges with epoxy glue. Capacitances were measured with a 1000-Hz AC capacitance bridge, and the cell was calibrated for fluids of known dielectric constants. Finally 1.131 grams of adsorbent (0.469 cm^3) was placed in the 2.714 cm^3 cell which was then filled with cyclohexane (dielectric constant 2.015 at 25°C) and shaken vigorously during measurement.

Results and Discussion. The data below show the weight fractions and cation exchange capacity contributions of the gravel, sand, silt, and "clay" fractions of the sediment sample. (The clay fraction actually contains minerals other than clays and is in this context a size category only.)

Material	Diameter (μm)	CEC, meq/100 g	Percent by weight	Percent by CEC
Gravel	>2000	0	2.6	0
Sand	74-2000	3.6	94.7	72.9
Silt	4-74	32.8	1.9	13.3
"Clay"	<4	80.5	0.8	13.8

It can be seen that the sediment sample consisted almost entirely of sand, which contributed nearly three-fourths of the total cation exchange capacity. It must be remembered, however, that in a surface water environment the silt and "clay" fractions will be in closer and more direct contact with the flowing water (much of this material will in fact be suspended in it) than will be the sand underneath, a large quantity of which was unavoidably scooped up during sampling.

The X-ray diffraction patterns for the sand and silt fractions had a nearly identical predominance of quartz and feldspar peaks. The "clay" fraction, however, showed X-ray peaks from quartz and feldspars as well as for chlorite/montmorillonite. Our use of the silt fraction in the following adsorption experiments is supported by the similarity in composition (i.e., mostly quartz and feldspars and no clays) between the silt and sand fractions. The silt fraction, more easily suspended than sand, was an indicator of the adsorption characteristics that might logically be expected of the sand fraction as well.

The specific surface area of the peroxide-treated silt fraction was found by 1,10-phenanthroline adsorption to be 72.2 m^2 g^{-1}

and the cation exchange capacity by sodium saturation was measured as 3.42×10^{-4} eq g^{-1} or 34.2 meq/100 g.

The VSC-VSP model uses the maximum potential-determining ion adsorption density N_s, a kind of maximum adsorption capacity for potential-determining ions H^+ or OH^-, which is assumed to be the same for both ions. For our purposes, we are concerned with the maximum OH^- capacity since the pH range of most natural waters lies above our measured pH_{pzc} of pH 4.3. Bowden et al. (7) estimate N_s at 10^{-5} eq m^{-2} on the basis that any surface-charge densities greater than this are unrealistic because the potential-determining ions would be less than 5 Å apart, and lateral coulombic forces would become excessive.

The amount of hydroxide adsorbed during a sodium hydroxide titration of an adsorbent-material slurry was calculated by comparing the slurry conductivity with that of a blank. Conductivity was used here because it is a more accurate indicator of aqueous hydroxide concentration than is pH, particularly at high pH (11 to 14) where moderate electrode error is probable. For the purpose of determining aqueous OH^-, conductivity was used in preference to calculation (NaOH added, less the earlier calculated adsorbed OH^-) because any errors in calculated adsorbed hydroxide (OH^-_{ads}) would also have appeared in the calculated aqueous hydroxide activity $[OH^-]$. This would have resulted in twice the error in the $(OH^-_{ads})/[OH^-]$ term used to determine N_s in a Langmuir plot. (The existence, due to specific OH^- adsorption, of a ratio of Na^+ to OH^- slightly higher in the slurry bulk solution than in the conductivity titration blank's normal 1:1 ratio should have caused little error; the excess of Na^+ ions in the former was concentrated mostly in the first few angstroms of the diffuse double layer and therefore contributed little to the conductivity. Even if this were not true, the fact that OH^- has four times the specific conductance of Na^+ would still make this type of error rather small; conductivity is much more a measure of the activity of OH^- than of Na^+.) As a first approximation of the previously described adsorption of potential-determining ions, hydroxide ion adsorption at high pH can be described by a Langmuir isotherm, such that $(OH^-_{ads}) = N_s K[OH^-]/(1 + K[OH^-])$, where K is an equilibrium constant which includes the electrostatic repulsion term $e^{F\psi_s/RT}$. Rearrangement of this expression implies, for a constant value of K, that a plot of (OH^-_{ads}) versus $(OH^-_{ads})/[OH^-]$ should give a straight line whose intercept equals the maximum capacity of OH^- adsorption, N_s. Such a plot is shown as Figure 2.

Figure 2 shows that the data points toward the lower right of the graph, for the beginning of the titration, do not give a straight line while those at the upper left, for the completion of the titration, do. The reason for this is that the pH in the latter half of the titration varied more gradually than in the first half. Because the equilibrium constant K implicitly includes the electrostatic repulsion term which is a function of ψ_s and hence of pH, the more constant pH toward the end of the

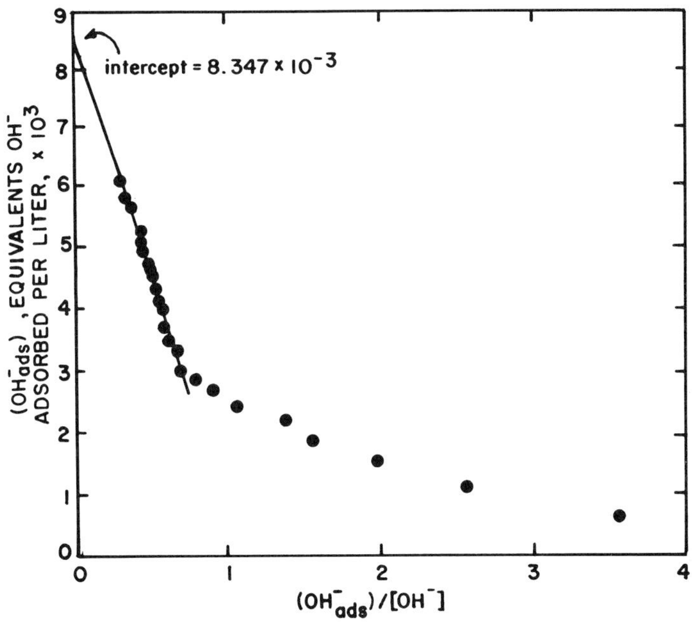

Figure 2. Langmuir plot of hydroxide adsorption

titration gave rise to a more constant K, giving a straight line plot. The least-squares intercept for the last 15 data points plotted in the upper left is 8.35×10^{-3} eq L^{-1} or 2.23×10^{-4} eq in the final slurry volume of 26.7 mL. For the 0.292 g adsorbent sample used, this becomes 7.63×10^{-4} eq g^{-1} in comparison to the Lawrie (8) cation exchange capacity of 3.42×10^{-4} eq g^{-1}, a ratio of 2.23 to 1.

A maximum hydroxide adsorption density of 7.63×10^{-4} eq g^{-1} on an adsorbent with 72.2 m^2 g^{-1} gives rise to a surface charge density of 1.02 C m^{-2} (1 charge per 16 Å^2 of surface), which is approximately equal to the Bowden et al. (7) theoretical limit of 0.965 C m^{-2}.

The calculated surface charge density for the adsorbent at 0.001 M ionic strength is plotted as a function of pH in Figure 3, where it is zero at pH 4.3, the pH_{pzc} at this ionic strength. For the similar titrations at 0.01 and 0.10 M ionic strengths, the values of the pH_{pzc} were found to be 4.0 and 3.7, respectively. This indicates that the pH at which adsorption of H^+ and OH^- are equal decreases with increasing $NaClO_4$ concentration, further indicating the possibility of a specific potential-determining interaction of the surface with electrolyte ions. Sodium ions might have been specifically adsorbed in order to produce this result. Since it is not clear exactly how the pH_{pzc} varies as a function of ionic strength, we will use 4.3 for our pH_{pzc}. This is the value obtained at the lowest (0.001 M) ionic strength. For the purpose of attempting to reconcile theory with experiment, this value is probably the one which should be used because our experimental work will be done at an ionic strength near 10^{-3} M, approximating the ionic strength of many natural waters.

Using the approach of Atkinson et al. (12), it can be shown that where the surface potential is small and where the OH^- adsorption density Γ_{OH^-} is much larger than the H^+ adsorption density Γ_{H^+}, a linear relationship exists such that

$$\log_{10}(\Gamma_{OH^-} - \Gamma_{H^+}) - pH + 2pH_{pzc} =$$

$$\log_{10}K_H^{ads}N_s - \frac{F(\Gamma_{OH^-} - \Gamma_{H^+})}{2.303}\sqrt{\frac{1}{2000\varepsilon_0\varepsilon_{H_2O}RT\mu}} \quad (5)$$

The pH at which the approximations necessary to derive this relation are valid lie within a small pH range slightly above the pH_{pzc}. At constant ionic strength a plot of the quantity $\log_{10}(\Gamma_{OH^-} - \Gamma_{H^+}) - pH + 2pH_{pzc}$ versus $\Gamma_{OH^-} - \Gamma_{H^+}$ for data within this pH range should produce a straight line of intercept $\log_{10}K_H^{ads}N_s$. Figure 4 shows such a plot for the data taken at 0.001 M ionic strength, yielding an intercept of -2.52. Titrations at 0.01 and 0.10 M ionic strengths gave similar plots yielding intercepts of -2.47 and -3.38, respectively. While the values of the intercept at the two lower ionic strengths are very nearly equal, the value of -3.38 at the high ionic strength of

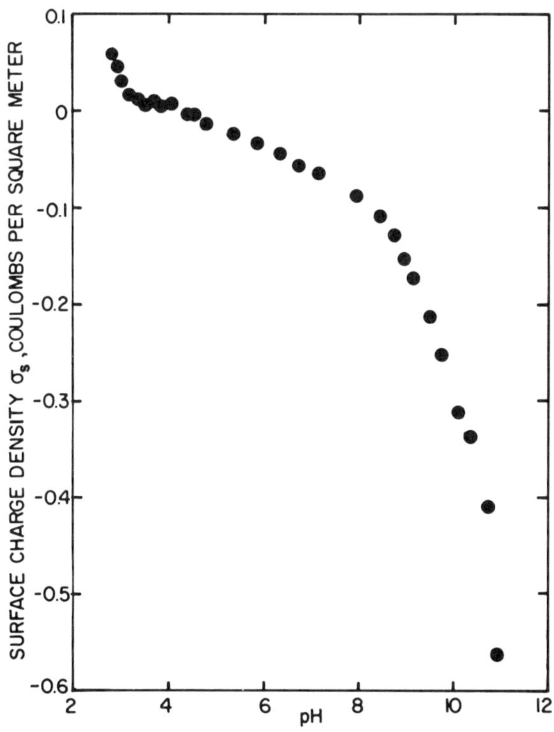

Figure 3. Surface charge density vs. pH at 0.001M ionic strength

0.10 departs drastically from those of the lower ionic strengths. This may also be an indication that the sodium perchlorate electrolyte may have interacted with the adsorbent in a manner for which we have not yet theoretically accounted. Adsorption of sodium ions, which becomes appreciable at high concentrations near 0.10 M is a possible explanation. In this paper we again use the value (-2.52) obtained at the 0.001 M ionic strength at which we can expect the least interference from the electrolyte. This intercept is equal to $\log_{10}K_H^{ads}N_s$; since N_s was earlier found to be 1.06 X 10^{-5} eq m^{-2}, K_H^{ads} is equal to 285, or $10^{2.45}$. Given the pH_{pzc} of 4.3, K_{OH}^{ads} by comparison calculates as $10^{7.85}$, more than five orders of magnitude higher than K_H^{ads}.

The value of the dielectric constant of the adsorbing solid is very important in determining the magnitude of the change in free energy of solvation undergone by the adsorbed ion. This term is used in both the James and Healy (4) and Levine (5) expressions for the solvation free-energy term. As a result, the dielectric constant of the solid is extremely important in determining the relative values of $\Delta G_{Pb^{2+}}^{solv}$ and $\Delta G_{PbOH^+}^{solv}$, and hence the relative proportions of Pb^{2+} and $PbOH^+$ adsorption.

The cyclohexane-adsorbent mixture gave a capacitance corresponding to an overall dielectric constant of 3.267. Based on the adsorbent volume fraction, the Looyenga (10) equation yielded a value of 16.4 for the dielectric constant of the adsorbent.

In the strict sense, the dielectric constant for the bulk particles as measured here may be incorrect as applied to a hydrated surface environment which has undergone substantial changes in electrical properties upon addition of water. Levine (5), for example, uses a value of 10 for the dielectric constant of a solvated quartz surface when the actual capacitor plate value for the solid is only 4.3. The effect of a variation in the value for the solid dielectric constant on the pH-dependence of adsorption, however, is minimal, requiring only a very small change in the value of the fitting parameter ΔG_i^{chem} in order to produce the same results. For this reason, our use of the measured value of 16.4 for the solid should suffice, producing little error as compared to estimating a value for the solvated solid.

Adsorption of Lead

Experimental. In order to study the effect of pH and total lead concentration on adsorption, slurries were prepared with pH between 2 and 6 and containing total lead concentrations (ΣPb) of 1.0 X 10^{-4} and 5.0 X 10^{-4} M. These concentrations, while large in comparison to those normally found in natural waters, were used mainly for two reasons.

First, adsorption sites on sediments in natural waters containing moderate to large concentrations of calcium, magnesium, and potassium will probably be moderately or near-saturated with these adsorbed species, and this large surface concentration of

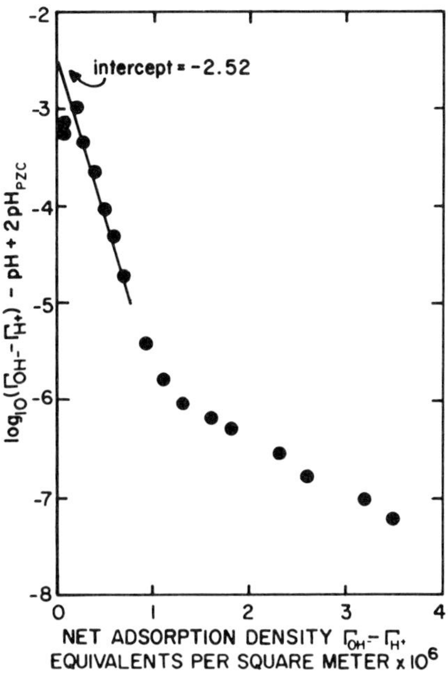

Figure 4. Extrapolation of linear portion of 0.001M ionic strength titration data for $\log_{10}K_H^{ads}N_s$ determination

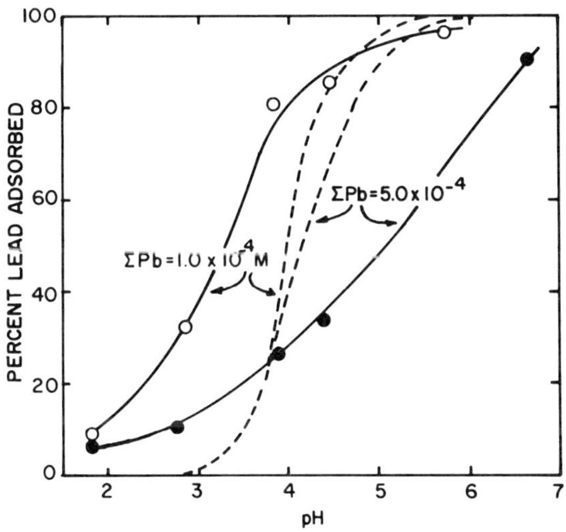

Figure 5. Comparison of lead adsorption data (points and solid connecting lines) with best-fitting James–Healy model predictions (broken lines)

cations near one plane may exert a perturbing effect on the nature of adsorption of other cations, lead included. As an alternative to a rigorous determination of this effect by adsorption studies for each of these ions, we instead use large total lead concentrations; the adsorption of this quantity of lead should give rise to an adsorption density of charged ions of sufficient magnitude to study such an effect.

Second, a total lead concentration, approaching the maximum that can be accommodated by the adsorbent, will cause the rate of increase in adsorption with pH to decrease. The adsorption will rise more gradually as pH is increased, rather than suddenly shifting from a very low to a very high adsorption; the latter is characteristic of many studies of adsorption of hydrolyzable heavy metals. This rise in adsorption with pH is called the "adsorption edge." Adsorption which increases only gradually with an increase in pH (i.e., a gradual adsorption edge) will give rise to a larger amount of useful data between the two extremes of near-zero and near-total adsorption, at which experimental uncertainties tend to more greatly affect interpretation of the data. For adsorption which increases extremely rapidly with pH (i.e., a steep adsorption edge), all that need be done in fitting the James and Healy (4) model to the data is to find the proper value of the chemical free-energy term $\Delta G_{Pb}{}^{chem}$ which will describe the pH at which the sudden increase in adsorption occurs. For a large total lead concentration, however, we will be able to determine whether the model will still properly describe the shape of the adsorption edge.

Ten separate portions of adsorbent of 0.146 g (5.0×10^{-5} eq) each were washed to constant conductivity with deionized water in order to remove any adsorbed ions. They then were each placed in 50 mL of solution to give a CEC per unit volume of 1.0×10^{-3} eq L^{-1} which contained either 1.0×10^{-4} or 5.0×10^{-4} M lead perchlorate in CO_2-free deionized distilled water. The pH was then adjusted, by the addition of 1.0 normal sodium hydroxide or perchloric acid as appropriate, to values of approximately 2, 3, 4, 5, and 6 for both the 5.0×10^{-4} and 1.0×10^{-4} M sets of solutions. These slurries were then placed in a CO_2-free atmosphere at 25°C for 48 hours, during which they were magnetically stirred for 5 minutes every half hour.

After 48 hours, the slurries were removed from the CO_2-free atmosphere and the pH measured. Aliquots of 5 mL were then removed and centrifuged at 30 000 G's for 5 minutes in order to remove any suspended particulate matter. The supernatants were then analyzed for lead and sodium by atomic absorption.

Results and Discussion. Table I gives the analyses and resultant lead adsorption for the 1.0×10^{-4} and 5.0×10^{-4} M slurries. The percent lead adsorbed is plotted as a function of pH in Figure 5, where the results are also compared with the best-fitting predictions of the James and Healy (4) model. It can be

Table I--Adsorption of lead from 50 mL solution onto 0.146 gram of adsorbent as function of pH and total lead content, and comparison of experimental and predicted adsorption of lead

pH at 48 hours	$\sum Pb$, total lead	C_{Pb}, dissolved lead	Ionic strength[1]	Percent adsorption of lead		
	Moles/liter			Experimental	James-Healy model[2]	VSC-VSP model[3]
1.83	1.0×10^{-4}	8.98×10^{-5}	0.015	9.3	0.087	12.6
2.86	1.0×10^{-4}	6.76×10^{-5}	0.0017	32.4	0.68	35.2
3.83	1.0×10^{-4}	1.90×10^{-5}	3.8×10^{-4}	80.7	28.0	72.4
4.47	1.0×10^{-4}	1.45×10^{-5}	6.6×10^{-4}	85.5	85.5	89.2
5.73	1.0×10^{-4}	3.4×10^{-6}	6.0×10^{-4}	96.6	99.9	99.7
1.85	5.0×10^{-4}	4.68×10^{-4}	0.016	6.4	0.098	7.42
2.78	5.0×10^{-4}	4.46×10^{-4}	0.0031	10.7	0.59	16.9
3.90	5.0×10^{-4}	3.69×10^{-4}	0.0014	26.2	28.4	26.5
4.38	5.0×10^{-4}	3.32×10^{-4}	0.0013	33.6	63.9	32.6
6.66	5.0×10^{-4}	4.68×10^{-5}	0.0024	90.6	99.9	95.3

[1] Includes analyses for sodium present as a result of NaOH added in order to adjust the pH.
[2] $\Delta G_{Pb}^{chem} = -7.94$ kcal/mol and $x_i = 3.96$ Å.
[3] $\Delta G_{Pb}^{chem} = -14.05$ kcal/mol and $\underline{d} = 2.17$ Å.

seen that the rate of increase of percent adsorption with increasing pH is gradual even in the 1.0 X 10^{-4} M slurries. The rise in percent adsorption with pH, for this data, is still much more gradual than that which would be predicted using the model of James and Healy (4). The curves shown in Figure 5 for this model were generated using a computer program which selects the best value of ΔG_{Pb}^{chem} in order to minimize the differences between predicted and experimental percent adsorption. The value for ΔG_{Pb}^{chem} obtained in this manner was -7.94 kcal mol^{-1}, a result leading to an average error between predicted and experimental data of 15.5 percent adsorption. Figure 5 shows that the optimum agreement obtainable between experiment and predictions of this model is poor.

The data at total lead concentration of 1.0 X 10^{-4} M can by itself be fitted to the James and Healy (4) model with a minimal average deviation of only 7.7 percent adsorption between theory and experiment. The value of ΔG_{Pb}^{chem} (-12.18 kcal mol^{-1}) used for fitting this data, however, produces very poor agreement between theory and experiment (an average deviation of 31.4 percent adsorption) for the data at 5.0 X 10^{-4} M total lead, as is shown by Figure 6. The James and Healy (4) model therefore cannot reconcile the data obtained under these two sets of experimental conditions.

The fitting of predicted adsorption and experimental data was performed using Levine's (5) expression for the solvation free-energy term and his suggested value of 30 for the interfacial dielectric constant. Other workers, however, have used values as low as 6.0 for this parameter. The effect of a variation in the interfacial dielectric constant on the fit of the predicted results is minimal, the only difference being in the value calculated for the other fitting parameter, the chemical free-energy term. Use of 6.0 for the interfacial dielectric constant gives almost identical predictions, with an average error between predicted and experimental percent adsorption of 16.1 percent. (The chemical free-energy term arrived at in this instance was -9.46 kcal mol^{-1}.)

Experimental lead adsorption data are compared in Figure 7 with the best-fitting theoretical adsorptions, using the optimized parameters ΔG_{Pb}^{chem} = 14.05 kcal mol^{-1} and d = 2.17 Å in the VSC-VSP model. It is clear that a much better fit is obtained using this model than by using that of James and Healy (4). The average deviation in percent adsorption between that found experimentally and that predicted was found to be a very low 3.4 percent, compared to the rather poor 15.5 percent average deviation obtained with the best-fitting James and Healy (4) model.

Like the James and Healy (4) model, the VSC-VSP model also requires an estimate of the magnitude of chemical interactions, which usually must be determined by comparison of experimental data with adsorption predicted at various values of ΔG_{Pb}^{chem}. This model, however, also requires a knowledge of the "average"

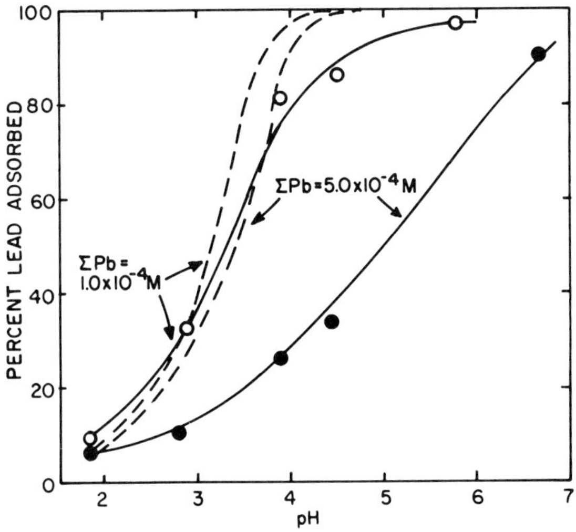

Figure 6. Comparison of lead adsorption data (points and solid connecting lines) with James–Healy model predictions (broken lines) fitted to 1.0 × 10^{-4}M data

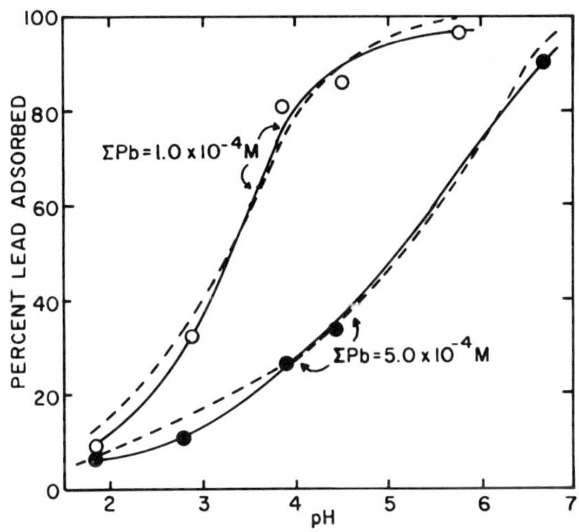

Figure 7. Comparison of lead adsorption data (points and solid connecting lines) with best-fitting VSC–VSP model predictions (broken lines)

distance of approach *d* of an adsorbed cation to the surface. In reality, any model involving only one or two planes of adsorption near the surface will be an oversimplification of the real situation, particularly on surfaces as complex and irregular as oxides and silicates. Such a concept is nevertheless incorporated in the VSC-VSP model because of its ability to approximately describe adsorption phenomena. The concept of an "average" distance from the surface of an adsorbed cation does not easily give rise to a method for calculating such distance, except for that by which comparison of experimental data with theoretical predictions of adsorption for different values of *d* will yield a value which gives the best agreement. What this means, then, is that the VSC-VSP model as used here has two fitting parameters, ΔG_{Pb}^{chem} and *d*, while the James and Healy (4) model has one, namely, ΔG_{Pb}^{chem}. (It should be noted that allowing the adsorbed ion's distance of approach x_i, as well as ΔG_{Pb}^{chem}, to vary in the curve-fitting procedure for the James and Healy (4) model does not result in any significantly better fit between experiment and theory for this model. The problems illustrated in figures 5 and 6 are still encountered; in contrast to experiment, no significant effect of total lead concentration on the adsorption edge position is predicted.)

Bowden *et al*. (7) use a curve-fitting approach of this sort to estimate the inner-layer capacitance, of which the value of *d* is a measure. They compare their experimental data, plotted as pH versus surface charge density, with a family of theoretical curves calculated using various capacitances, selecting the capacitance giving the best agreement with experiment. It seems reasonable to assume that the value of *d* should lie somewhere between the crystal ionic radius (1.20 Å) of the Pb^{2+} ion and the crystal ionic radius plus the diameter of the first layer of attached water molecules or primary hydration sheath (3.96 Å), as is normally used in the James and Healy (4) model for Pb^{2+} and $PbOH^+$ ions.

In incorporating Levine's (5) expression for the solvation free-energy term into the VSC-VSP model, we were faced with the dilemma of which value to use for the interfacial dielectric constant ε_{int}. Levine (5) uses a value of 30 in his solvation free-energy term, while Bowden *et al*. (7) base their use of 6.0 for this term on work by Hasted *et al*. (13). Instead of speaking of a dielectric constant which continuously varies with distance from the surface, we should rather speak of dielectric constants of each of the successive shells of adsorbed water molecules. The water shells very near the surface will have low dielectric constants, and the more distant shells will have higher ones. Since the dielectric constant of a dipole is really a measure of the freedom of the dipole to become oriented in response to changes in electrostatic field strength, it would seem reasonable that the first layer of water molecules adsorbed to the charged surface (as the result at least of Van der Waals and London dispersion forces) would be rather low as a result of their attachment to the surface and resulting immobility. Such water molecules would also be affected by the electrical field strength charge of the adsorbed cations, tending to further lower the dielectric constant. Bockris et al. (14) for this reason give 6.0 as the dielectric constant value for the first such layer.

It therefore seems reasonable to say that as a first approximation, the interfacial dielectric constant within the first layer of adsorbed water molecules may be taken as the lower value of 6.0. It is this lower value which we used in the Levine (5) expression for the solvation free-energy term, and in the surface potential-Gouy potential relation in the VSC-VSP model in order to obtain the predicted adsorption shown in Figure 7 and tabulated in Table I. Our value for d of 2.17 Å lies well within the diameter (2.76 Å) of the first adsorbed layer of water within which the approximation of a 6.0 dielectric constant should be valid. We should also note that this 2.17 Å value of d lies well within the crystal ionic radius of Pb^{2+} and the hydrated radius used by James and Healy (4).

As in the earlier James and Healy (4) model calculations here, the use of different values for the interfacial dielectric constant did not greatly affect the final agreement between theory and experiment, and resulted mainly in a change in the value of the fitting parameters. Use of 30, instead of 6.0, for the interfacial dielectric constant, gave an average error between predicted and experimental data of only 6.1 percent adsorption; this is still better agreement with experiment than that predicted by the James and Healy (4) model, but not quite as good as the result obtained using 6.0 for the interfacial dielectric constant. (Fitting parameters here were found to be -11.37 kcal mol^{-1} for ΔG_{Pb}^{chem}, and 3.00 Å for d.) This result appears to lend support to the preferred use here of 6.0 for the interfacial dielectric constant.

Conclusion

While previous work has often been conducted under conditions where only trace quantities of lead or other heavy metals have been placed in contact with an adsorbent, very few of these approaches have dealt with the problems faced as the adsorbent sites begin to be filled. The usefulness of the VSC-VSP model in taking this into account is illustrated here by demonstration of the effect of charged adsorbed species on the electrostatic potential which acts on the adsorbing ions. When a given number of equivalents of adsorbent are placed in contact with a comparatively large number of moles of cations, some of which will attach to the adsorbent, adsorption will be further opposed in two ways. First, of course, the process of adsorption will reduce the number of sites available for further adsorption. Second, the Gouy potential ψ_d is said by Bowden et al. (7) to decrease from the surface potential value in linear proportion to the surface charge density σ_s. Since increased cation adsorption works to make the interfacial charge density σ_i more positive and the opposing surface charge density σ_s more negative, the potential ψ_d at the plane of cation adsorption should become more positive. This effect is shown in Figure 8 for adsorbent slurries of the type studied here. The result is a potential which is very different from that predicted by classical double-layer theory, and which seems consistent with the charge-reversal phenomena discussed by James and Healy (15, Fig. 1, p. 55). It can be seen from Figure 8 that a total lead concentration of 5.0 X 10^{-4} M (equal to half the CEC of 1.0 X 10^{-3} eq L^{-1}) has a profound effect on ψ_d, rendering it positive and in some instances increasing with pH at values of pH above the pH_{pzc}. The VSC-VSP model, therefore, predicts a surface potential which is not only dependent on pH and ionic strength, but also on the total surface charge density of any adsorbed cations. (As the adsorbed cation concentrations approach zero, however, the dependence of the surface potential on pH approaches that predicted by James and Healy (4).) The result is a pH-dependent potential which actually tends to desorb rather than adsorb cations as pH increases.

Adsorption nevertheless increases with pH mainly as the result of the increase with pH in the aqueous activity of the more readily adsorbed $PbOH^+$ at the expense of Pb^{2+}, the latter ion having a higher solvation energy barrier. The VSC-VSP model, as applied to our experimental conditions and material, predicts that at least five times more $PbOH^+$ will be adsorbed beyond pH 4.

While natural waters will certainly not contain total lead or other heavy metal concentrations so large as the 5.0 X 10^{-4} M value used here to illustrate the effect of cation adsorption on adsorption potential, sediments in contact with natural waters will undoubtedly have a large fraction of their adsorption sites occupied with major cations such as Ca^{2+}, Mg^{2+}, K^+, or Na^+. The presence of cations such as these near the plane at which heavy metal cations adsorb should have somewhat the same effect in opposing further cation adsorption as did the large surface concentrations of adsorbed lead here. If we are to view studies such as the one here with an eye toward eventual prediction of heavy metal adsorption in the presence of the other major and minor solute species normally encountered in natural waters, the use of the VSC-VSP model will be necessary in order to assist us in more precisely accounting for the interactions between the various adsorbed species.

Figure 9 shows the pH-dependence of lead adsorption which would be predicted using the chemical free-energy term that gives the best fit in the James and Healy (4) model with the data in Table I. It can be seen that the adsorption edge positions are rather similar for total lead concentrations which differ by an order of magnitude or more. Only when the total lead concentration closely approaches the CEC is it predicted that it becomes

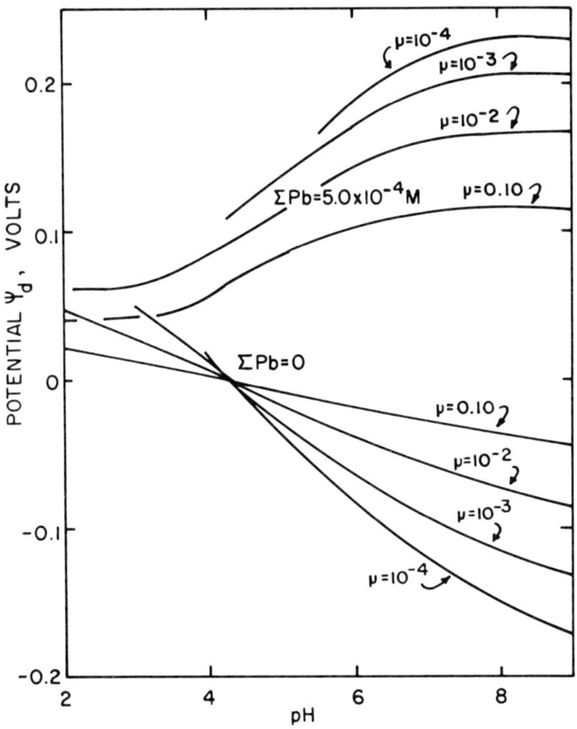

Figure 8. Effects of pH, ionic strength, and total lead on potential at the 2.17 Å Stern plane, as predicted by VSC–VSP model

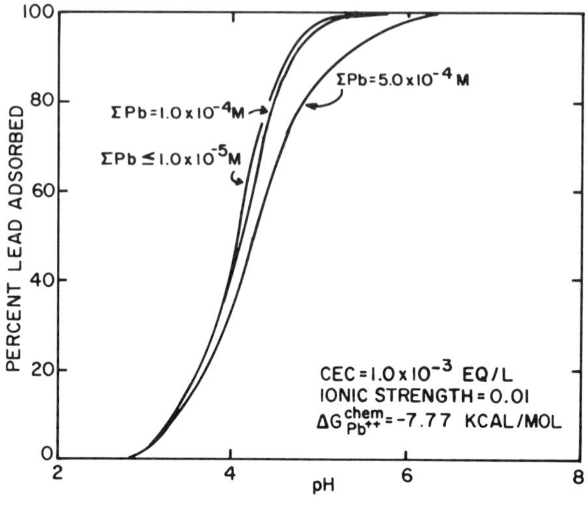

Figure 9. *Effect of total lead concentration on adsorption edge as predicted by the James–Healy model*

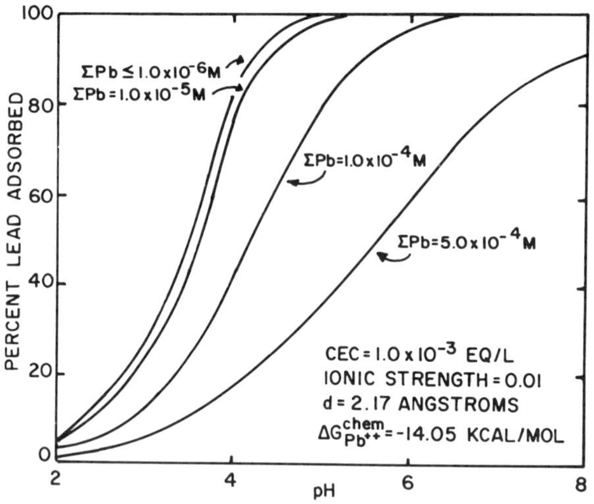

Figure 10. *Effect of total lead concentration on adsorption edge as predicted by the VSC–VSP model*

slightly more difficult to increase adsorption by raising the pH.

Benjamin (16), however, has recently shown that the position of the adsorption edge is significantly affected by the total heavy metal concentration to a much greater degree than that predictable by classical double-layer models of adsorption. Figure 10, showing the much larger effect of total lead concentration on the pH-dependence predicted by the VSC-VSP model, is consistent with these observations. As can be seen, the predicted shift in the adsorption edge is much larger, in some cases exceeding a pH unit. This illustrates the effect of the adsorbed lead cations in making more positive the adsorption potential ψ_d and thereby opposing further adsorption at higher surface concentrations.

In summary, the fact that the VSC-VSP model predicts adsorption edge shift and the effect of charge reversal on the pH-dependence of adsorption, neither of which can be accounted for on the basis of the model of James and Healy (4), lends support to its use in accurately describing heavy metal adsorption. The agreement between theory and experiment shown in Figure 7 additionally supports the use of the VSC-VSP model.

Abstract

The James and Healy model for adsorption of metals at the oxide/solution interface predicts a rise in increased pH which is much more abrupt than that observed for higher concentrations of lead adsorbing on the quartz-feldspar silt fraction of a streambed sediment from Colma Creek in San Mateo County, Calif. Experimental observations are found to be more consistent with the variable surface charge-variable surface potential (VSC-VSP) model which considers the non-Nernstian dependence of surface potential on pH by accounting for the effect of adsorbed metal cations on the Gouy potential at the plane of adsorption. The VSC-VSP model predicts an adsorption pH-dependence which agrees with experiment here. It also predicts phenomena observed by other workers, namely, the dependence of adsorption edge position on total heavy metal concentration and the effect on further adsorption of adsorption-caused charge reversal.

Literature Cited

1. Lind, C. J. Polarographic determination of lead hydroxide formation constants at low ionic strength, Environ. Sci. Technol. (in press).
2. Hem, J. D., and Durum, W. H. Solubility and occurrence of lead in surface water, J. Amer. Water Works Assoc. 65, 562-568 (1973).
3. Hem, J. D. Geochemical controls on lead concentrations in stream water and sediments, Geochim. Cosmochim. Acta 40, 599-609 (1976).

4. James, R. O., and Healy, T. W. Adsorption of hydrolyzable metal ions at the oxide-water interface III. A thermodynamic model of adsorption, J. Colloid Interface Sci. 40, 65-81(1972).
5. Levine, Samuel, Remarks *on* Discreteness of charge and solvation effects in cation adsorption at the oxide/water interface, by Wiese, G. R., James, R. O., and Healy, T. W., Discussions of the Faraday Soc. 52, 41A-42A (1971).
6. Wiese, G. R., James, R. O., and Healy, T. W. Remarks *on* Discreteness of charge and solvation effects in cation adsorption in the oxide/water interface, by Wiese, G. R., James, R. O., and Healy, T. W., Discussions of the Faraday Soc. 52, 42A-42B (1971).
7. Bowden, J. W., Posner, A. M., and Quirk, J. P. Ionic adsorption on variable charge mineral surfaces. Theoretical-charge development and titration curves, Aust. J. Soil Sci. 15, 121-136 (1977).
8. Lawrie, D. C. A rapid method for the determination of approximate surface areas of clays, Soil Sci. 92, 188-191 (1961).
9. Chapman, H. D., Cation exchange capacity, p. 891-901, *in* Black, C. A., ed., Agronomy Ser. 9, part 2, Amer. Soc. Agron., 1965.
10. Looyenga, H. Dielectric constants of heterogeneous mixtures, Physica 31, 401-406 (1965).
11. Lal, K., and Parshad, R. Test and utilization of the Fricke and Pearce equations for dielectric correlation between powder and bulk, J. Physics D: Applied Physics 7, 455-461 (1974).
12. Atkinson, R. J., Posner, A. M., and Quirk, J. P. Adsorption of potential-determining ions at the ferric oxide-aqueous electrolyte interface, J. Phys. Chem. 71, 550-558 (1967).
13. Hasted, J. B., Ritson, D. M., and Collie, C. H. Dielectric properties of aqueous ionic solutions, Parts I and II, J. Chem. Phys. 16, 1-21 (1948).
14. Bockris, J. O'M., Devanthan, M. A. V., and Muller, K. On the structure of charged interfaces, Proc. Royal Soc. Ser. A 274, 55-79 (1963).
15. James, R. O., and Healy, T. W. Adsorption of hydrolyzable metal ions at the oxide/water interface II. Charge reversal of SiO_2 and TiO_2 colloids by adsorbed Co(II), La(III), and Th(IV) as model systems, J. Colloid Interface Sci. 40, 53-64 (1972b).
16. Benjamin, Mark, Adsorption of cadmium, copper, zinc, and lead on oxide surfaces in model natural systems, Ph.D. Thesis, Stanford Univ. (in press).

Disclaimer: The reviews expressed and/or the products mentioned in this article represent the opinions of the author(s) only and do not necessarily represent the opinions of the U.S. Geological Survey.

RECEIVED November 16, 1978.

13

Oceanic Elemental Scavenging

PETER G. BREWER and WEI MIN HAO

Woods Hole Oceanographic Institution, Woods Hole, MA 02543

The phenomenon of scavenging, or adsorption onto solid surfaces, is frequently invoked by marine chemists as an important control on the distribution of the chemical elements in seawater. Kranskopf (1) in an early paper commented on the importance of adsorption as a control on the abundance of minor elements in seawater, and other papers, too numerous to mention, have since ascribed geochemical importance to this process. Unfortunately, attempts to use surface chemical theory, such as that given in Stumm and Morgan (2) or in the elegant review by Parks (3), have had mixed success. One approach has been to study the surface chemistry of a well-defined solid phase, such as δ-MnO_2 (4) or illite and beachsand (5), in the laboratory and to apply these data to a set of field observations. However, the particles present in seawater and assumed to be responsible for scavenging are not well characterized and are of complex composition (6, 7, 8), and the validity of applying laboratory results from pure phases is questionable. Here we attempt a different approach; firstly in collating the various scavenging rate constants which have been derived, putting these on a common basis and examining their chemical correlations, and secondly in asking what surface chemical properties must be attributed to deep ocean particulate matter in order to explain the observed effects. A knowledge of these properties would not only have considerable academic merit, but would be of great practical use in predicting the fate of other chemical species of radionuclides in the deep ocean.

Most of the observed oceanic elemental removal rates obtained from field studies result from a complex mix of biological, physical and chemical processes and extracting the component due to any one idealized process, such as adsorption, is difficult. A possible exception to this generality lies in the various papers which describe scavenging as an *in situ* process operating in the deep ocean by means of a one-dimensional advection-diffusion-scavenging model (9, 10). We now realize that the data set (11) used by Craig (10) was partially deficient; however, his papers were exemplary and the concepts have since found wide application

0-8412-0479-9/79/47-093-261$05.00/0
© 1979 American Chemical Society

(12, 13). In such a model, the distribution of a chemical element is controlled by mixing between an upper and lower boundary and some production or removal process. The assignment of a mechanism to explain this process is to some extent intuitive (for instance, a deficit over atmospheric equilibrium is attributed to respiration); however, for many metal ions, the explanation of the observed removal rate by an adsorptive mechanism in the deep aphotic ocean appears to be most likely.

Methods and Data

Thorium. The short residence time of thorium in seawater and its affinity for solid phases have long been recognized. Broecker *et al.* (14) have examined the residence time of thorium in surface seawater and its implications regarding the fate of reactive pollutants. Krishnaswami *et al.* (12) have examined the isotopic composition of deep Pacific particulate matter samples obtained by large volume *in situ* pumping. From their observations of the ^{238}U-^{234}Th and ^{234}U-^{230}Th pairs, the activity of the ^{230}Th on filters and the observation that on ~20% of the ^{234}Th in seawater exists in a particulate state, they derive a rate constant for adsorption onto a solid phase of 8×10^{-8} sec^{-1} or $10^{0.4}$ yr^{-1}.

Lead. The collection and analysis of seawater samples for stable lead presents severe problems (see the report of C. Patterson's group (15)), and all reliable estimates of the rate constant for the removal of lead derive from observations of the ^{210}Pb-^{226}Ra disequilibrium (9) in the deep sea. Craig *et al.* (9) derived a scavenging rate constant for ^{210}Pb of ~1.8×10^{-2} yr^{-1}, based upon the half-life of ^{210}Pb (22 yrs) and its observed approximately 50% depletion from its parent ^{226}Ra. This depletion has been amply confirmed (16, 17, 18); however, Bacon *et al.* (13) have shown that the ^{234}Pb activity of marine particulate matter is inadequate to account for the observed deficiency and postulated adsorption at the sediment-water interface as a dominant mechanism. Their removal rate constant for adsorption by sinking particles was estimated to be in the range $10^{-2.3}$ to $10^{-3.15}$ yr^{-1}. We will take $10^{-2.7}$ yr^{-1} as an average rate constant.

Copper. Craig (10) and Brewer (19) have postulated that copper is scavenged in deep ocean water. More recent data given by Boyle *et al.* (20) reveal significantly lower concentrations and convincing evidence for the scavenging process. Boyle *et al.* (20) have used an advection-diffusion-scavenging model and calculate a removal rate constant of 10^{-3} yr^{-1}. As pointed out by Craig (10) and Brewer and Murray (21), such models yield net rates giving the difference between production from decomposing planktonic debris and consumption due to adsorption onto more stable particles. The rate constant of 10^{-3} must then be a lower limit; the upper limit is unknown but may be estimated by postulating that, were no adsorption to occur, copper might be a linear function of phosphate and summing the postulated production and observed consumption rates. The removal rate constant for copper obtained in this way is 20% higher than the value given by Boyle et al. (20) and in the range $10^{-3.05}$ yr^{-1}.

Nickel. The marine geochemistry of nickel has been discussed by Sclater et al. (22) and in Boyle (23). The general trend of the vertical profiles is to show an increase with depth and a partial co-variance with the nutrients phosphate and silicon. Scavenging of nickel was not discussed in these papers; however, the chemical similarities of Pb, Cu and Ni would suggest that some adsorption must occur, the question being to what degree. We have used the data given by Boyle (23) for GEOSECS Station 219 in the Bering Sea and treated this in the same way as for copper, examining the deviations from the most strongly biophilic element, phosphate. In this way, we estimate a maximum removal rate for nickel of 0.65 x 10^{-3} nmol kg^{-1} yr^{-1} and a maximum removal rate constant of $10^{-4.2}$ yr^{-1}.

Cadmium. The data available for cadmium show marked increases with depth, and Martin et al. (24) have observed a strong correlation of cadmium with phosphate in Californian coastal waters. Boyle (23) gives cadmium data for GEOSECS Station 219. The observed correlation with phosphate indicates that scavenging of Cd^{2+} must be very small and our estimate of the rate constant is an upper limit of $\sim 10^{-5.25}$ yr^{-1}.

Discussion

The rate constants for removal, presumably by adsorption, given in the preceding section were all obtained from deep Pacific data and were derived from the application of a one-dimensional advection-diffusion-scavenging model. There are geographical differences; however, the differences in properties such as pH and the amount of suspended matter (P. G. Brewer, unpublished data, 1978) will not be large. If the derived rate constants have any validity, they should exhibit some correlation with important chemical properties relating to adsorption. We know little of the specific details of the solid-solution interface in natural waters. Neihof and Loeb (25, 26) have examined the surface charge of particulate matter in sea water and the role of adsorbed organic matter in determining this charge. In their experiments, all solid surfaces assumed a moderately electronegative charge in natural sea water; in U.V. irradiated, organic-free water, the solids exhibited their own characteristic charges. Loeb and Niehof (27) have further obtained optical data on this adsorbed film which are consistent with a polymeric macromolecular character. It is generally accepted that pH is the master variable governing the extent of adsorption of metal ions at oxide-water interfaces (2), and that adsorption increases with increasing pH.

Within these broad constraints there is considerable latitude in choosing a model which might realistically describe the *in situ* adsorptive process. James and Healy (28) have developed an important model in which the free energy of adsorption (ΔG ads) results from the difference in electrostatic contributions between coulombic attraction (ΔG coul) and repulsion due to change in solvation energy (ΔG solv). The complete process is governed by

$$\Delta G \text{ ads} = \Delta G \text{ coul} + \Delta G \text{ solv} + \Delta G \text{ chem} \tag{1}$$

where ΔG chem is an adjustable parameter which appears as a small difference between the two large terms. A point of interest is that in this model $M_{aq}Z+$ ions show minimal tendency to adsorption due to unfavorable ΔG solv terms. O'Connor and Kester (5) discussed the James and Healy model, but opted for a model in which $M_{aq}Z+$ was exchanged for surface bound hydrogen ion:

$$M_{aq}^{Z+} + H\underline{X} = M\underline{X}^{(Z-1)+} + H^{+} . \tag{2}$$

The reaction was determined by an equilibrium constant (K_A) given as:

$$K_A = \frac{[M\underline{X}^{(Z-1)+}]\ a_{H^+}}{a_{M_{aq}^{Z+}}\ [H\underline{X}]} . \tag{3}$$

Schindler (29, 30) has proposed a similar model in that M_{aq}^{Z+} ions are adsorbed, yet adsorption is understood in terms of surface complex formation with deprotonated surface OH-groups as ligands. His schematic example using Si as a typical oxide surface is:

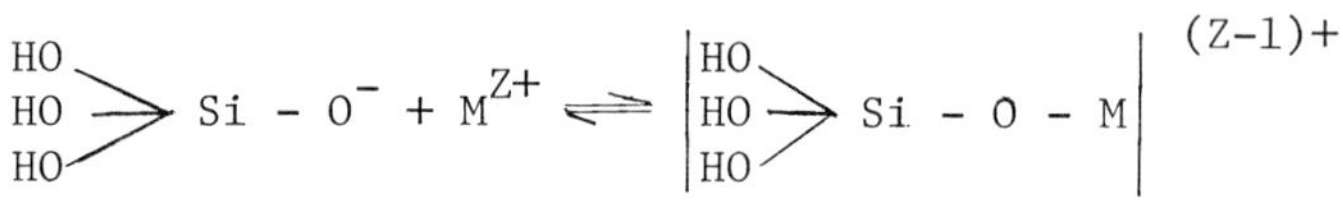
(4)

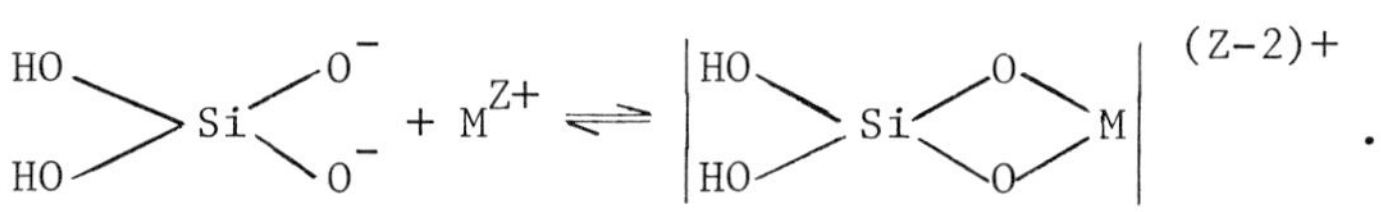
(5)

The free OH^- ion is a powerful ligand. Dugger *et al*. (31) have pointed out that the ligand properties of the surface OH groups are not greatly modified by the attached silicon.

If the Schindler model is correct, then the derived rate constants should bear a significant relationship to the strength of the interaction of the metal ion with OH^-. In Table I and Figures 1 and 2, we show the rate constants derived from field

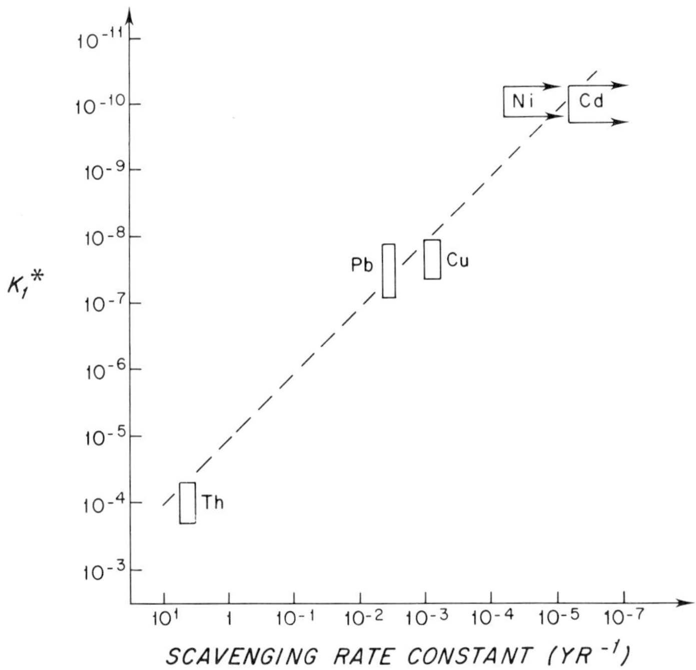

Figure 1. Plot of the scavenging rate constants derived from advection–diffusion–scavenging models against the stability constant for simple hydroxo complexes, $^*K_1$

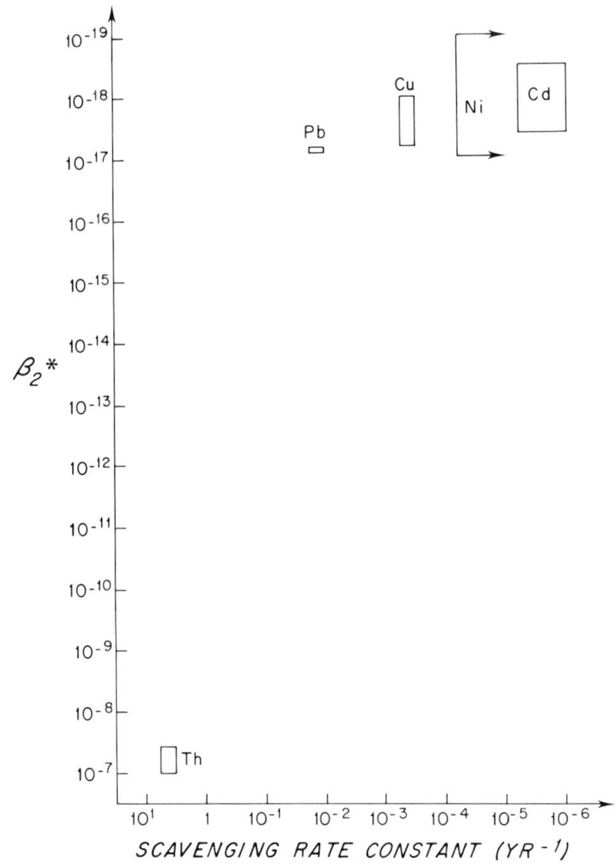

Figure 2. Plot of the scavenging rate constants derived from advection–diffusion–scavenging models against the stability constant $^*\beta_2$ for reaction with two hydroxo groups

measurements compared to the stability constants for hydroxo-complexes $^*K_1$ and $^*\beta_2$. The correlation lends considerable support to the argument that the interaction with surface OH-groups is a dominant control on the in situ adsorptive process. The rate constants derived here may be criticized in that they contain several assumptions; however, the geochemist is afforded some relief in an equally wide choice of constants. The constants used by Schindler (29) are also given in Table I. The value for $^*\beta_2$ Th in Figure 2 falls significantly away from the other ions reflecting the comparison of tetravalent and divalent ions and the intensity of interaction with a second OH-group.

TABLE I

Comparison of the adsorption rate constants derived from field studies with values for $^*K_1$ and $^*\beta_2$ selected (a) in this paper and (b) by Schindler (29).

Element	Rate Constant (yr^{-1})	(a) Log $^*K_1$	(a) Log $^*\beta_2$	(b) Log $^*K_1$	(b) Log $^*\beta_2$
Th	$10^{0.4}$	-3.7 → 4.3	-7.7	-----	-----
Pb	$10^{-2.7}$	-7.1 → 7.7	-17.2	-7.76	-17.2
Cu	$10^{-3.06}$	-7.35 → -8	-18	-8	-18
Ni	$<10^{-4.2}$	-9.85 → -10.25	-19	-10	-19
Cd	$<10^{-5.25}$	-9.7 → -10.25	-19	-----	-----

The constants were taken from the compilations of Sillen and Martell (34, 35).

The data shown in Figure 1 permit simple predictions to be made. For instance, Co^{2+} with $^*K_1 = 10^{-9.85}$ should have a scavenging rate constant of approximately 10^{-5} yr^{-1}, while Sc^{3+} ($^*K_1 = 10^{-5.1}$) should have a rate constant of $\sim 10^{-0.5}$ yr^{-1}, if the correlations generally hold true. These simple predictions should not be pressed too far, and it should be pointed out that the scavenging rates for Sb, Sc and Ni derived by Craig (10) do not follow the correlations in Figures 1 and 2.

Application of a Scavenging Model

The good agreement observed between adsorption and the stability of interaction with OH-groups leads us to pursue the Schindler (29, 30) model further. The reader should consult the original papers for details. Briefly, the model gives the percentage A_i of adsorbed metal i as

$$A_i = \frac{100\, a \sum^{n} {}^{*}\beta^{s}_{n}\, h^{-n}\, C^{n}}{1 + \sum^{n}\sum^{j} \beta_{nj}\, [L_j]^{n} + a \sum^{n} {}^{*}\beta^{s}_{n}\, h^{-n}\, C^{n}} \qquad (6)$$

where a is the amount of suspended matter (kg l^{-1}), h is the concentration of H^+-ions (mol l^{-1}), L_j is the concentration of free ligand and the constant β_{nj} is given by

$$\beta_{nj} = \frac{[M_i\,(L_j)_n]}{[M_i^{Z+}]\,[L_j]^{n}} \qquad (7)$$

and the constant ${}^{*}\beta^{s}_{n}$ represents the appropriate equilibrium constant for the interaction of the $[M_i^{Z+}]$ ion with surface OH-groups as in Equations 4 and 5. C is the concentration of free surface OH-groups (mol kg^{-1}) on particulate matter. The total concentration of surface OH-groups (C_o) is related to C by

$$C_o = C + \sum^{i}\sum^{n} \{(\equiv Si\text{-}O)_n\, M_i^{(z-n)+}\} \ . \qquad (8)$$

In practice under seawater conditions, the free concentrations of the ligands SO_4^{2-} and CO_3^{2-}, and the free surface OH-groups are all governed by the interaction with Mg^{2+}.

Knowing something of the flux of particulate matter, it is then easy to derive an expression for the residence time of an element with respect to adsorption ($\tau_{r(i)}$)

$$\tau_{r(i)} = \frac{\tau_p\,(1 + \sum^{n}\sum^{j} \beta_{nj}\, [L_j]^{n} + a \sum^{n} {}^{*}\beta^{s}_{n}\, h^{-n}\, C^{n})}{a \sum^{n} {}^{*}\beta^{s}_{n}\, h^{-n}\, C^{n}} \qquad (9)$$

where τ_p is the residence time of particles.

Schindler (29, 30) took whole-ocean residence times for various elements, as given by Goldberg (32), and found reasonable agreement between these and the residence times with respect to adsorption calculated from his model. His choice of properties took h equal to $10^{-8.1}$, τ_p for particulate silica to be 2 x 10^2 years and a to be 2 x 10^{-6} kg of suspended silica l^{-1} (2 mg silica l^{-1}). The term C_o was taken to be 1.0 mol kg^{-1} of silica, which led to a value of C of 0.509 mol kg^{-1}, indicating that approximately 50% of the surface silanol groups are occupied by adsorbed Mg^{2+}. The choice of silica as a model oxide surface is not realistic; however, in this application it has the distinct advantage that its surface properties have been well studied, enabling estimates of $^*\beta_n^s$ to be made (31).

In applying this model to the elements for which scavenging residence times were derived earlier, we know $\tau_{r(i)}$, and a from various papers (1). τ_p can be estimated from papers such as Krishnaswami et al. (12), Tsunogai et al. (33) and Bishop et al. (8), and the terms relating to the dissolved phase, h and β_{nj} and L_j are well known (see Table II). The unknowns are C, the number of free OH-groups on marine particulate matter, and $^*\beta_n^s$, the series of constants representing the stability of interaction of metal ions with the naturally occurring surfaces. We shall assume that, as a first approximation, marine particulate matter is not dissimilar to silica and that $^*\beta_n^s$ is a linear function of $^*\beta_n$ (31), thus

$$\log {}^*\beta_n^s = X \log {}^*\beta_n \quad . \tag{10}$$

TABLE II

Formation constants, K_1 and β_2, of metal ions with different ligands used in this paper

Element	CO_3^{2-}		Cl^-		SO_4^{2-}
	Log K_1	Log β_2	Log K_1	Log β_2	Log K_1
Mg	2.18	----	----	----	1.22
Th	----	----	0.25	-1.08	3.32
Pb	7.06	9.09	0.9	1.36	2.7
Cu	6.73	9.83	0.74	0.78	0.95

The constants were taken from the compilations of Sillen and Martell (34, 35)

Results

In solving Equation 9, we have taken the pH of deep Pacific water to be 7.8, the amount of particulate matter to be 15 µg l^{-1} (P. G. Brewer, unpublished data, 1978) and the residence time of particulate matter to be 3.65 years (consistent with the mean settling velocity of 2 x 10^{-3} cm sec^{-1} given by Krishnaswami et al. (12) and the boundary conditions of the various advection-diffusion scavenging models). By taking any two of the three Th-Pb-Cu data sets and substituting into Equation 9, we can derive X and C. The value of X thus obtained is 0.87, which is somewhat higher than the value reported by Schindler (29) for silica as

$$\log {}^{*}\beta_n^s = 0.077 + 0.624 \log {}^{*}\beta_n .$$

The value of 0.87 would indicate that the surface hydroxyl groups of marine particulate matter are closer in their chemical properties to free OH-groups than are silanol groups, and are less hydrolyzed. However, the value of C obtained is 0.76×10^4 mol kg^{-1}! Clearly, this result is unreasonable since the free OH-groups on the particulate matter surfaces would weigh 1.3×10^2 more than the particulate matter itself; the value of C_o is even larger. The result is not greatly sensitive to changes in pH and settling velocity; changing the pH by ± 0.3 pH units changes C by about a factor 2, while decreasing the settling velocity to 10^{-3} cm sec^{-1} only increases C to 1.07×10^4 mol kg^{-1}. The value of X obtained is the same even when changing the above parameters. Various other manipulations of the data set were tried with the most reasonable result being obtained as follows: let the surface OH-groups be less than the amount of particulate matter ($\leq$ 59 mol kg^{-1} of particulate matter) and let the principal ion adsorbed from seawater, magnesium, also be adsorbed to an extent which is less than the amount of particulate matter ($\leq$ 41 mol kg^{-1} of particulate matter) and solving Equation 9 so as to find the minimum value of surface $[OH^- + Mg^{2+}]$. The result obtained is X equal to 0.51 and C equal to 4.68 mol kg^{-1}, the greatly increased stability of interaction with the surface OH-groups leading to a lesser molar concentration being required. However, the value for C is still very high and a fit to the observed residence times of Pb and Cu only can be obtained; a very poor fit for Th, Ni and Cd is found.

Conclusions

The generally good correlation, seen in Figure 1, between ${}^{*}K_1$ and the scavenging rate constants would seem to indicate that adsorption is indeed important in the abyssal marine environment. However, the unrealistically high concentration of surface groups required on marine particulates suggests that unless the concepts of surface chemistry outlined here and our estimates of the flux of marine particulate matter are seriously in error, sinking particles cannot be responsible for the observed effects. If, for instance, the estimate given by Schindler (29) for C of 0.509 mol kg^{-1} is physically reasonable, then the discrepancy is approximately 2×10^4, and sinking particles could produce less than 0.1% of the observed scavenging. A similar conclusion was reached by Bacon *et al.* (13) who suggested that the deep ocean ^{210}Pb removal was controlled by a dual process of adsorption onto sinking particles and removal at the sediment-water interface, possibly onto highly active surfaces such as Mn and Fe oxides. This latter process was suggested to account for > 90% of the observed removal.

The conclusion that sinking particulate material produces little scavenging in deep water is not surprising. We typically think of adsorptive equilibria being reached on a time scale of minutes to hours; the particles falling through the water column should reach equilibrium long before arriving at the upper boundary of the advection-diffusion models from which our estimates of removal rates are obtained. The particles in the adsorption model represented by Equation 9 appear de novo in the deep ocean and thus would tend to over-estimate the importance of deep scavenging. The concept of horizontal diffusion towards a boundary and adsorption at the sediment-water interface may have much merit and may account for the observed removal processes described here. These calculations should not be taken as definitive, but rather as a first attempt at a systematic treatment of deep-sea scavenging data, constrained by the concepts of surface chemistry.

Acknowledgements

We wish to thank O. C. Zafiriou and J. W. Murray for helpful comments. This work was supported by the National Science Foundation under Grant OCE76-2177; and by the U. S. Department of Energy under Contract No. E(11-1)-3566.

Abstract

The use of one-dimensional advection-diffusion models to describe trace metal data from the deep ocean reveals that for many elements in situ consumption occurs. This consumption is usually regarded as being due to scavenging, or adsorption onto sinking particles, and "scavenging residence times", τ_ψ, have been given for Pb (9), Th (12) and Cu (20). We have attempted to reconcile these calculated removal rates with the known abundance and flux of marine particulate matter to derive the surface chemical properties of marine particulate matter required to produce the observed effects. The calculations are based upon a model proposed by Schindler (29, 30) in which interaction with the free M_{aq}^{Z+} ion with surface OH-groups is the adsorptive mechanism. Given a particulate matter concentration of 15 μg kg^{-1} (1) and settling velocities of 10^{-3} to 10^{-4} cm sec^{-1}, we find that surface Oh-group concentrations of the order of 10^4 mol kg^{-1} would be required. Since this would require that the surface groups weigh 10^2 to 10^3 times more than the particles themselves, we conclude that, unless estimates of the particulate flux are seriously in error, sinking particulate matter lacks the capacity to produce the inferred scavenging effect. This is not inconsistent with the conclusions of Bacon et al. (13) who suggested that removal of Pb at the sediment-water interface greatly exceeded that fraction scavenged by particles.

Literature Cited

1. Krauskopf, K. B. Factors controlling the concentration of thirteen rare metals in sea water. Geochim. Cosmochim. Acta 9, 1-32 (1956).
2. Stumm, W. and Morgan, J. J. "Aquatic Chemistry". 583 p. Wiley-Interscience, New York, 1970.
3. Parks, G. A., Adsorption in the marine environment, p. 241-308, in Riley, J. P. and Skirrow, G. ed., "Chemical Oceanography." Vol. 1, 2nd Ed., Academic Press, London, 1975.
4. Murray, J. W. The surface chemistry of hydrous manganese dioxide. J. Colloid. Interface Sci. 46, 357-371 (1974).
5. O'Connor, T. P. and Kester, D. R. Adsorption of copper and cobalt from fresh and marine systems. Geochim. Cosmochim. Acta 39, 1531-1543 (1975).
6. Spencer, D. W. Brewer, P. G. and Sachs, P. L. Aspects of the distribution and trace element composition of suspended matter in the Black Sea. Geochim. Cosmochim. Acta 36, 71-86 (1972).
7. Brewer, P. G., Spencer, D. W., Biscaye, P. E., Hanley, A., Sachs, P. L., Smith, C. L., Kadar, S., and Fredericks, J. The distribution of particulate matter in the Atlantic Ocean. Earth Planet. Sci. Lett. 32, 393-402 (1976).
8. Bishop, J.K.B., Edmond, J. M., Ketten, D. R., Bacon, M. P. and Silker, W. B. The chemistry, biology and vertical flux of particulate matter from the upper 400 m of the equatorial Atlantic Ocean. Deep-Sea Res. 24, 511-548 (1977).
9. Craig, H., Krishnaswami, S. and Somayajulu, B.L.K. ^{210}Pb-^{226}Ra: radioactive disequilibrium in the deep sea. Earth Planet. Sci. Lett. 17, 295-305 (1973).
10. Craig, H. A scavenging model for trace elements in the deep sea. Earth Planet. Sci. Lett. 23, 149-159 (1974).
11. Spencer, D. W., Robertson, D. E., Turekian, K. K., and Folsom, T. R. Trace element calibrations and profiles at the GEOSECS test station in the northeast Pacific Ocean. J. Geophys. Res. 75, 7688-7696 (1970).
12. Krishnaswami, S., Lal, D., Somayajulu, B.L.K., Weiss, R. F., and Craig, H. Large-volume in situ filtration of deep Pacific waters: mineralogical and radioisotope studies. Earth Planet. Sci. Lett. 32, 420-429. (1976).
13. Bacon, M. P., Spencer, D. W., and Brewer, P. G. $^{210}Pb/^{226}Ra$ and $^{210}Po/^{210}Pb$ disequilibrium in seawater and suspended particulate matter. Earth Planet. Sci. Lett. 32, 277-296 (1976).
14. Broecker, W. S., Kaufman, A. and Trier, R. The residence time of thorium in surface seawater and its implications regarding the rate of reactive pollutants. Earth Planet. Sci. Lett. 20, 35-44 (1973).

15. Meeting Report. "Interlaboratory lead analyses of standardized samples of seawater," Mar. Chem. 2, 69-84 (1974).
16. Somayajulu, B.L.K. and Craig, H. Particulate and soluble ^{210}Pb activities in the deep sea. Earth Planet. Sci. Lett. 32, 268-276 (1976).
17. Thomson, J. and Turekian, K. K. ^{210}Po and ^{210}Pb distributions in ocean water profiles from the eastern South Pacific. Earth Planet. Sci. Lett. 32, 297-303 (1976).
18. Nozaki, Y. and Tsunogai, S. ^{226}Ra, ^{210}Pb, ^{210}Po disequilibria in the western North Pacific. Earth Planet. Sci. Lett. 32, 313-321 (1976).
19. Brewer, P. G. Minor elements in seawater, in Riley, J. P. and Skirrow, G., ed. "Chemical Oceanography", Vol. 1, 2nd Ed., Academic Press, London, 1975.
20. Boyle, E. A., Sclater, F. R. and Edmond, J. M. The distribution of dissolved copper in the Pacific. Earth Planet. Sci. Lett. (in press).
21. Brewer, P. G. and Murray, J. W. Carbon, nitrogen and phosphorous in the Black Sea. Deep-Sea Res. 20, 803-818 (1973).
22. Sclater, F. R., Boyle, E. A. and Edmond, J. M. On the marine geochemistry of nickel. Earth Planet. Sci. Lett. 31, 119-128 (1976).
23. Boyle, E. A."The Marine Geochemistry of Trace Elements." Ph.D. Thesis,Mass. Inst. Tech.-Woods Hole Ocean.Inst., 156 p. 1976.
24. Martin, J. H., Bruland, K. W. and Broenkow, W. W. Cadmium transport in the California current 159-184, in Windom, H. L. and Duce, R. A., ed. "Marine Pollutant Transfer", Lexington Books, Lexington, Mass., 1976.
25. Neihof, R. and Loeb, G. The surface charge of particulate matter in seawater. Limnol. Oceanogr. 17, 7-16 (1972).
26. Neihof, R. and Loeb, G. Dissolved organic matter in seawater and the electric charge of immersed surfaces. J. Mar. Res. 32, 5-12 (1974).
27. Loeb, G. I. and Neihof, R. A. Adsorption of an organic film at the platinum-seawater interface. J. Mar. Res. 35, 283-291 (1977).
28. James, R. O., and Healy, T. W. Adsorption of hydrolyzable metal ions at the oxide-water interface. II. Charge reversal of SiO_2 and TiO_2 colloids by adsorbed Co(II), La(III) and Th(IV) as model systems. J. Colloid. Interface Sci. 40, 53-64 (1972).
29. Schindler, P. W. Removal of trace metals from the oceans: a zero order model. Thalassia Yugoslavica 11 (1/2), 101-111 (1975).
30. Schindler, P. W. The regulation of trace metal concentrations in natural water systems: a chemical approach, in "Atmospheric Contribution to the Chemistry of Lake Waters." Proc. First Speciality Symposium, Internat'l Assoc. Great Lakes Res. 1975.

31. Dugger, D. L., Stanton, J. H., Irby, B. N., McConnel, B. L., Cummings, W. W. and Waatman, R. W. The exchange of twenty metal ions with the weakly acidic silanol group of silica gel. J. Phys. Chem. 68, 757-760 (1964).
32. Goldberg, E. D. Minor elements in sea water, p. 163-196, in Riley, J. P. and Skirrow, G., ed. "Chemical Oceanography" Vol. 1, 1st Ed., Academic Press, London, 1965.
33. Tsunogai, S., Nozaki, Y. and Minagawa, M. Behavior of heavy metals and particulate matters in sea water expected from that of the radioactive nuclides. J. Oceanogr. Soc. Japan 30, 251-259 (1974).
34. Sillen, L. G. and Martell, A. E. "Stability Constants of Metal-Ion Complexes." 754 p. Special Publication No. 17. The Chemical Society, London, 1964.
35. Sillen, L. G. and Martell, A. E. "Stability Constants of Metal-Ion Complexes." 865 p. Special Publication No. 17. The Chemical Society, London, 1964.
36. Schindler, P. W., Furst, B., Dick, R. and Wolf, P. Ligand properties with Fe^{3+}, Cu^{2+}, Cd^{2+} and Pb^{2+}. J. Colloid. Interface Sci. 55, 469-475 (1976).

RECEIVED November 16, 1978.

14

Surface of Goethite (αFeOOH) in Seawater

LAURIE BALISTRIERI and JAMES W. MURRAY

Department of Oceanography, University of Washington, Seattle, WA 98195

The mechanism of adsorption has been extensively studied in order to evaluate its importance in the regulation of the concentration of certain species in natural waters. In particular, adsorption on iron and manganese oxides has been proposed as the chemical mechanism which controls the concentration of some trace metals in the world's oceans (1, 2) and the enrichment of certain trace metals in ferromanganese nodules (3).

The most commonly reported solid forms of iron and manganese oxides in marine sediments and ferromanganese nodules are goethite (αFeOOH) and hydrous manganese dioxide (birnessite, todorokite and δMnO_2)(4, 5, 6). The surface chemistry of hydrous manganese dioxide has been previously reported (7, 8, 9). As an extension of that work we have investigated the surface properties of goethite (αFeOOH). The primary concerns of this work are to evaluate the effect of the major ions of seawater on the titratable charge of αFeOOH and to quantitatively evaluate the capacity of the solid for these ions.

Various theories have been proposed to describe and interpret the adsorption of metal ions at hydrous oxide interfaces (10). Most models have stressed either the double layer structure and ion-solvent interactions (11, 12, 13) or surface coordination reactions with amphoteric functional groups (14, 15, 16). Recently Davis et al. (17) and Davis and Leckie (18, 19) have proposed a comprehensive model that combines both of these approaches. No attempt has been made, however, to model surface interactions or to describe the distribution of surface species in a complex natural water such as seawater. We present here experimentally determined acidity constants and binding constants of Na^+, K^+, Ca^{+2}, Mg^{+2}, Cl^- and $SO_4^=$ with goethite. With these data we can now calculate the distribution of the major species on the surface of goethite in seawater. This approach will form the basis for modeling trace metal adsorption in seawater and determining the competitive effects of the major ions with each other and with trace metals.

0-8412-0479-9/79/47-093-275$06.00/0
© 1979 American Chemical Society

Methods

Goethite was synthesized according to the method of Atkinson *et al.* (20). Goethite forms after the hydrolysis and aging of ferric nitrate or perchlorate solutions and consists of double chains of linked $[Fe(O,OH)_6]$ octahedra. The double chains are further cross linked to adjacent double chains through corner sharing of oxygen atoms to give orthorhombic symmetry (6). The oxide was stored in distilled, deionized water around pH 7.5. The oxide was identified by the occurrence of major peaks at 4.18, 2.69, and 2.44 Å (Cu radiation with curved crystal monochrometer) in the X-ray diffraction pattern (6). The surface area was 48.5 ± 0.2 $m^2\ g^{-1}$ as determined by N_2 adsorption by the B.E.T. method (21). SEM pictures of the oxide revealed needle-shaped crystals approximately 1 micron in length and 0.2 microns wide. The value for the total surface sites (FeO_T) was taken from Yates' (22) work on αFeOOH. Yates (22) determined FeO_T to be equal to 27.8 μmol m^{-2} by tritum exchange. Further details of the solid's preparation, identification, and basic surface characteristics can be found in Balistrieri (23).

Potentiometric titrations were done in various concentrations of NaCl, KCl, $MgCl_2$, $CaCl_2$, and Na_2SO_4. In addition, titrations were done in mixed electrolyte solutions of different ionic strength containing the major ions of seawater in their appropriate seawater proportions (24). Potentiometric titrations involve measuring the amount of acid which is consumed or released by the solid as a function of pH. For a given pH, this is experimentally accomplished by determining the equivalents of acid consumed or released by the solid in a supporting electrolyte and subtracting from that the equivalents of acid consumed or released only by the supporting electrolyte solution. Care must be taken to keep the system free of CO_2. Further description of the experimental procedures for potentiometric titrations can be found elsewhere (25, 26, 27, 28).

The amount of acid consumed or released by the solid varies with the ionic strength of the supporting electrolyte and the pH of the solution. The pH at which the adsorption density of the solid is independent of the ionic strength is termed the pH(PZC) or the pH point of zero charge (10). The charge as a function of pH is calculated relative to the pH(PZC).

Interpretation of Potentiometric Titrations for αFeOOH in a Single Electrolyte Solution

The titratable surface charge measurements can be interpreted to give a quantitative assessment of the interactions between the solid and the supporting electrolyte ions. Potentiometric titrations measure the adsorption or release of protons and the model developed by Yates *et al.* (29) and Davis *et al.* (17) proposes reactions between oxide surface groups and supporting electrolyte ions which account for this measured proton change. According to their model the measured surface charge is a result of the ionization of the surface functional groups (Equation 1 and 2)

$$Fe\text{-}OH_2^+ \overset{K_1^{INT}}{\rightleftharpoons} Fe\text{-}OH + H_s^+ \quad (1)$$

$$Fe\text{-}OH \overset{K_2^{INT}}{\rightleftharpoons} Fe\text{-}O^- + H_s^+ \quad (2)$$

and the interaction of supporting electrolyte ions with the oxide surface (Equation 3 and 4)

$$Fe\text{-}OH + cation_s \overset{*K_{cation}^{INT}}{\rightleftharpoons} Fe\text{-}O^-\text{-}cation + H_s^+ \quad (3)$$

$$Fe\text{-}OH + anion_s + H_s^+ \overset{*K_{anion}^{INT}}{\rightleftharpoons} Fe\text{-}OH_2^+\text{-}anion\ . \quad (4)$$

The subscript s denotes isolated ions on the surface. The intrinsic equilibrium constants (K^{INT}) are determined at zero charge and potential conditions, thereby eliminating the electrostatic field effects.

The interaction of a cation with a neutral oxide group results in the release of a proton, while the association of an anion results in the adsorption of a proton. Accordingly, the formation of a negative site from a neutral site involves the release of a proton and the formation of a positive site involves the adsorption of a proton. Therefore, the titratable surface charge determined by potentiometric titration is a measure of both the formation of surface-ion complexes and the ionization of surface functional groups, and

$$\sigma_o = F\ [\{FeOH_2^+\} + \sum_n X_n\ \{FeOH_2^+ - anion_n\} - \{FeO^-\} - \sum_m Y_m\ \{FeO^- - cation_m\}] \quad (5)$$

where σ_o = titratable surface charge in μcoul cm^{-2}
F = Faraday's constant
{ } = surface species concentration in mol cm^{-2}
X = number of protons consumed by the formation of an individual anion complex
Y = number of protons released by the formation of an individual cation complex

$\sum_n$ = summation of all anion complexes

$\sum_m$ = summation of all cation complexes.

The intrinsic constants are thermodynamic constants written for reactions occurring at a hypothetical isolated site on the surface. Actual activities on the surface cannot be directly determined but Q or apparent stability quotients can be calculated based on measurable bulk concentrations. The intrinsic constants and apparent stability quotients are related by considering the electrostatic correction for an ion in solution near the surface compared to an isolated ion on the surface. In an idealized planar model, Ψ_o is the mean potential at the plane of surface charge created by the ionization of the surface functional groups and the formation of surface complexes and Ψ_β is the mean potential at the plane of adsorbed counter ions at a distance β from the surface (17). The electrostatic interaction energies at the surface and at a distance β are expressed as exponentials. Therefore:

$$(H^+)_s = (H^+) \exp(-e\Psi_o/kT) \tag{6}$$

$$[\text{cation}]_s = [\text{cation}] \exp(-e\Psi_\beta/kT) \tag{7}$$

$$[\text{anion}]_s = [\text{anion}] \exp(e\Psi_\beta/kT) \tag{8}$$

e = electronic charge

Ψ_o, Ψ_β = mean potentials

k = Boltzmann constant

T = temperature

The equilibrium constants which define Equations 3 and 4 may therefore be written as:

$$*K^{INT}_{cation} = \frac{\{\text{Fe-O}^-\text{cation}\}\,(H^+)_s}{\{\text{Fe-OH}\}\,[\text{cation}]_s} = \frac{\{\text{Fe-O}^-\text{cation}\}\,(H^+)}{\{\text{Fe-OH}\}\,[\text{cation}]} \exp([e\Psi_\beta - e\Psi_o]/kT) = *Q_{cation} \exp([e\Psi_\beta - e\Psi_o]/kT) \tag{9}$$

$$*K^{INT}_{anion} = \frac{\{\text{Fe-OH}_2^+\ \text{anion}\}}{\{\text{Fe-OH}\}\,[\text{anion}]_s(H^+)_s} = \frac{\{\text{Fe-OH}_2^+\ \text{anion}\}}{\{\text{Fe-OH}\}\,[\text{anion}](H^+)} \exp([e\Psi_o - e\Psi_\beta]/kT) = *Q_{anion} \exp([e\Psi_o - e\Psi_\beta]/kT) \tag{10}$$

{ } = concentration in mol cm^{-2}

[ ] = concentration in mol l^{-2}

( ) = activity in mol l^{-2}

$*K^{INT}$ = intrinsic equilibrium constant

*Q = apparent stability quotient

The apparent stability quotients are determined by utilizing the potentiometric titration data and the value for the total surface sites (FeO_T). The procedure will be illustrated by considering $\alpha FeOOH$ in NaCl. In this case,

$$\sigma_o = F[\{FeOH_2^+\} + \{FeOH_2^+Cl^-\} - \{FeO^-\} - \{FeO^-Na^+\}] \quad (11)$$

and

$$FeO_T = F[\{FeOH_2^+\} + \{FeOH_2^+Cl^-\} + \{FeOH\} + \{FeO^-\} + \{FeO^-Na^+\}]. \quad (12)$$

At higher electrolyte concentrations the dominant contribution to the surface charge is the formation of surface complexes. For a negative surface (pH > pH(PZC)) and for values of pH away from the pH(PZC), the surface charge is approximated by the formation of cation complexes, i.e.,

$$\sigma_o \approx - F\{FeO^-Na^+\}. \quad (13)$$

In terms of fractional ionization this is written as:

$$\alpha_- \approx - \frac{\sigma_o}{FeO_T} . \quad (14)$$

Under these conditions the neutral sites can be approximated as the total sites minus the cation surface complexes, i.e.,

$$\{FeOH\} \approx \frac{FeO_T}{F} - \{FeO^-Na^+\} \approx 1 - \alpha_- . \quad (15)$$

Using these approximations the apparent stability quotient for Na is:

$$*Q_{Na} = \frac{\alpha_-}{(1 - \alpha_-)} \frac{(H^+)}{[Na^+]} \quad (16)$$

and the intrinsic constant for Na is written as:

$$*K_{Na}^{INT} = \frac{\alpha_- \; (H^+)}{(1 - \alpha_-)[Na^+]} \exp \frac{[e\Psi_\beta - e\Psi_o]}{kT} \quad (17)$$

or in the logarithmic form:

$$p*K_{Na}^{INT} = pH - \log \frac{\alpha_-}{1 - \alpha_-} + \log[Na^+] + \frac{e(\Psi_o - \Psi_\beta)}{2.3 \; kT} \quad (18)$$

$$= p*Q_{Na} + \frac{e(\Psi_o - \Psi_\beta)}{2.3 \; kT} . \quad (19)$$

If $p*Q_{Na}$ is plotted as a function of charge, then $p*K_{Na}^{INT}$ can be obtained by extrapolation to zero charge.

The major assumption in this procedure is that the surface

charge is due solely to the formation of cation surface complexes. In reality, the surface charge of a negative surface is produced by the formation of both negative charged sites (FeO^-) and cation surface complexes (FeO^-Na^+). Thus, the apparent stability quotients and the intrinsic equilibrium constants are functions of both the surface charge and the electrolyte concentration. Extrapolation only to zero charge will not yield a precise intrinsic equilibrium constant. For this reason, James et al. (30) and Davis and Leckie (19) developed a double extrapolation procedure to obtain intrinsic equilibrium constants. The double extrapolation method involves extrapolations to zero charge and either zero concentration for the intrinsic acidity constants (K_1^{INT} and K_2^{INT}) or 1 M concentration for the intrinsic-ionization constants ($*K^{INT}_{cation}$ and $*K^{INT}_{anion}$).

Results

The pH(PZC) and the titratable surface charge of goethite as a function of pH were determined for the various electrolytes from the potentiometric titration data. The pH(PZC) was 7.5 in NaCl and KCl. It shifted to 7.1 in $CaCl_2$, 5.0 in $MgCl_2$, and 8.5 in Na_2SO_4. The titratable surface charge data as a function of pH are shown in Figures 1 through 4. The charge data for αFeOOH in NaCl and KCl were the same. Titration curves were reproducible to 2%. Titrations done over 6½ months indicate a decrease in charge with time although the pH(PZC) remained constant. The decrease in charge with time increased with distance from the pH(PZC) and appeared to have stabilized after 5 months. This change can be attributed to the aging of the surface. All the results reported here were completed on well-aged goethite.

In addition, we propose plausible reactions for the ion-goethite interactions. The intrinsic equilibrium constants describing these associations have been calculated by the double extrapolation method (19, 30).

The data for the determination of the intrinsic equilibrium constants for Na and Cl are shown in Figures 5 and 6. For Na (Figure 5), the acidity quotients, pH-log ($\alpha_-/1-\alpha_-$), are plotted as a function of the fractional ionization, α_-, and the log of the electrolyte concentration. The concentration term is multiplied by an arbitrary constant in order to separate the curves. The acidity quotients calculated from the potentiometric titration data as a function of $\alpha_- - 0.05 \log [Na^+]$ are represented by the filled circles. For each ionic strength the points are extrapolated to $\alpha_- = 0$. These extrapolated points are designated by open squares. These extrapolated points are then further extrapolated to 1 M electrolyte concentration. The open circle is the value for $p{*}K_{Na}^{INT}$ at zero charge and 1 M electrolyte concentration.

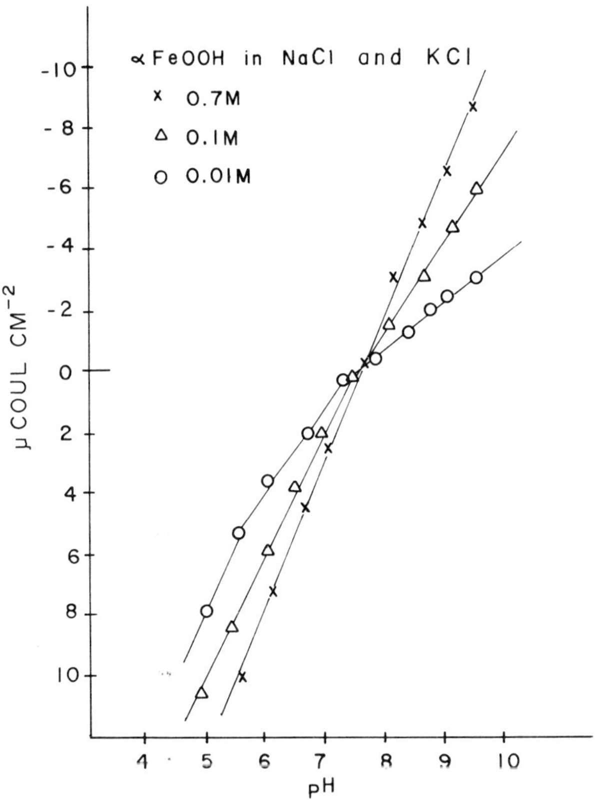

Figure 1. Charge vs. pH for goethite in NaCl and KCl

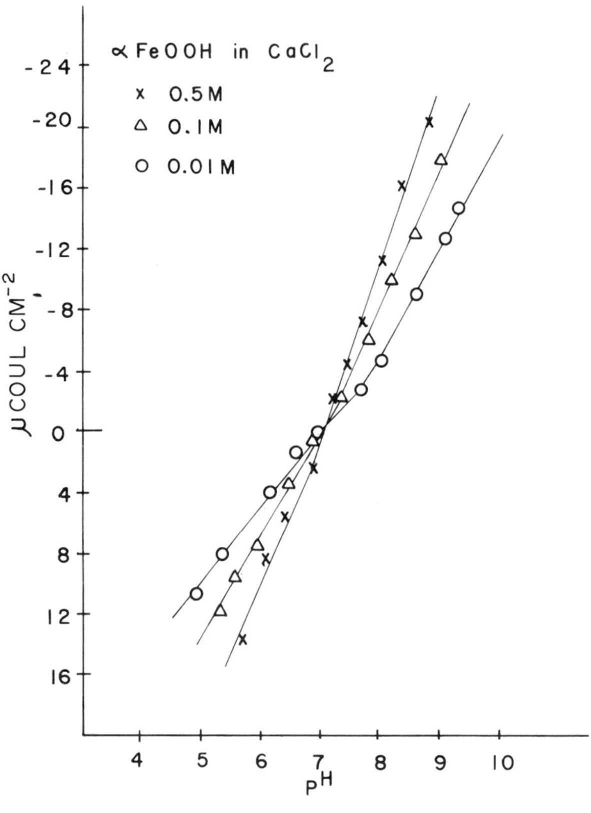

Figure 2. Charge vs. pH for goethite in $CaCl_2$

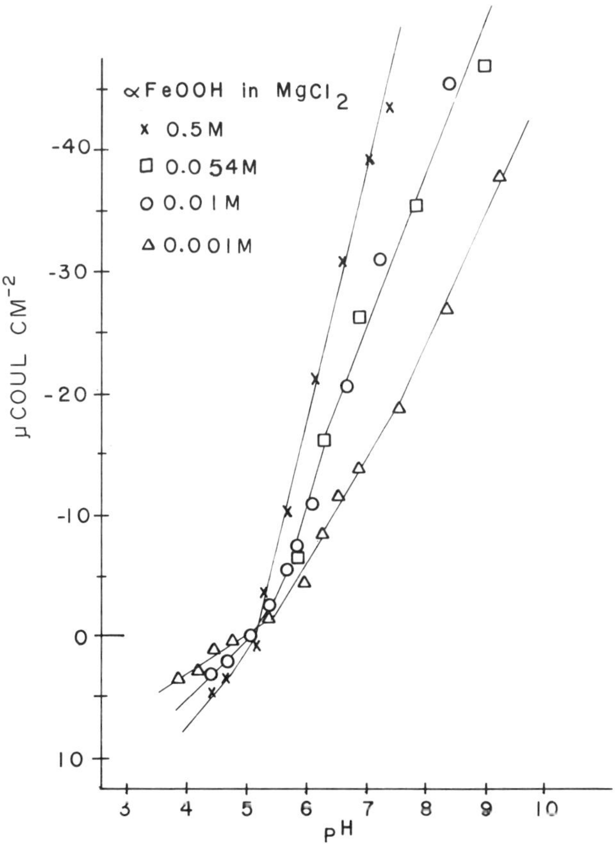

Figure 3. Charge vs. pH for goethite in $MgCl_2$

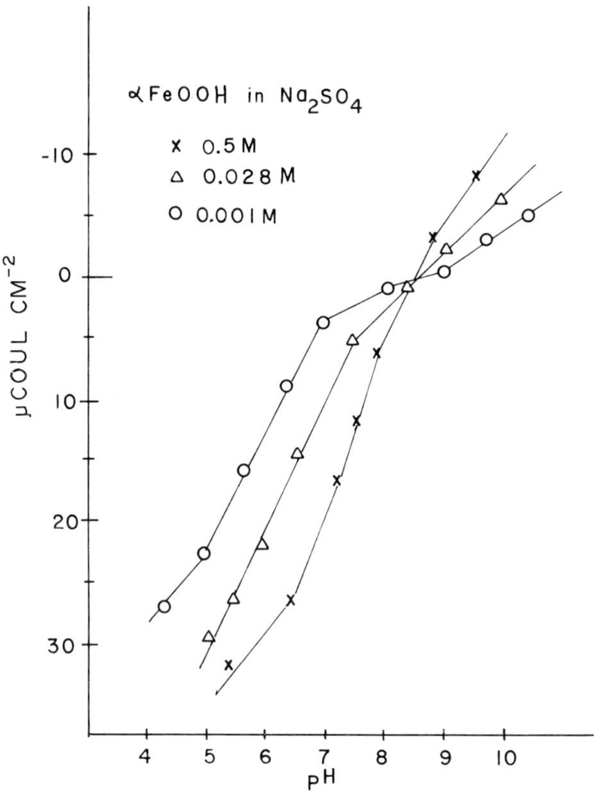

Figure 4. Charge vs. pH for goethite in Na_2SO_4

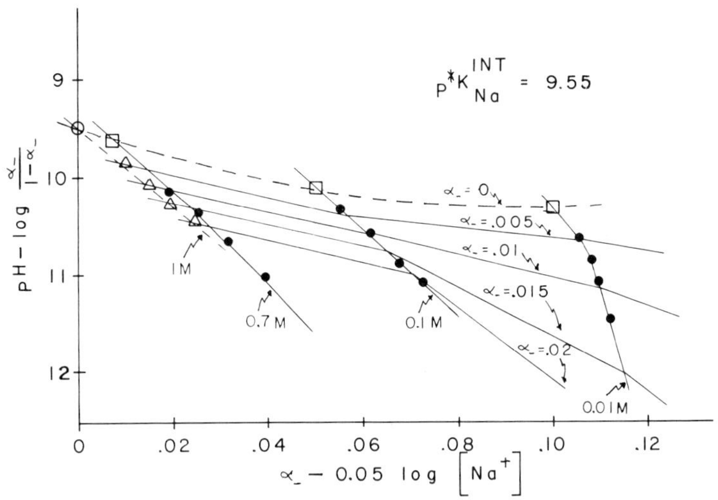

Figure 5. $p^*K_{Na}^{INT}$ *determination by the double extrapolation method. Note that the concentration is multiplied by an arbitrary constant in order to separate the curves.*

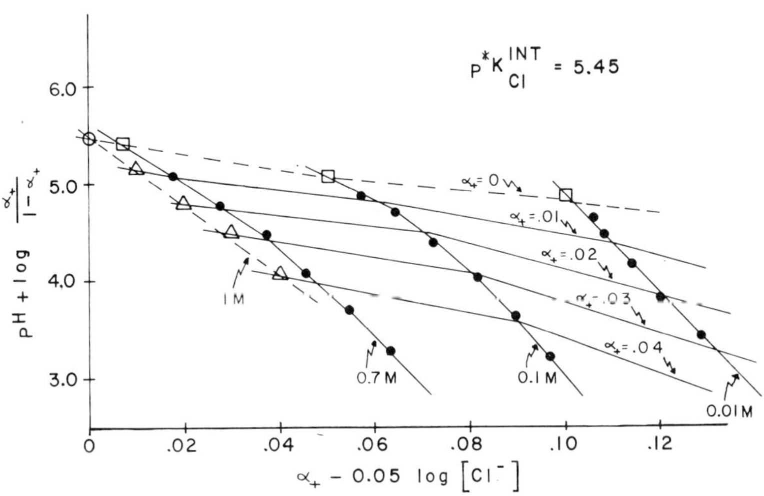

Figure 6. $p^*K_{Cl}^{INT}$ *determination by the double extrapolation method. Note that the concentration is multiplied by an arbitrary constant in order to separate the curves.*

The value of $p^*K^{INT}_{Na}$ can be checked by considering a second set of extrapolations. Curves of constant α_- are extrapolated to points at which the electrolyte concentration is 1 M. These points are then extrapolated to zero charge conditions. Once again, the open circle represents $p^*K^{INT}_{Na}$ at zero charge and 1 M electrolyte concentration.

The charge asymmetry in the $MgCl_2$, $CaCl_2$ and Na_2SO_4 titration data (Figures 2, 3, 4) suggests the release of two protons when a surface complex is formed with Mg^{+2}, Ca^{+2} and $SO_4^=$. For cations, the release of two protons can be described by the formation of either a bidentate complex (Equation 20) or the adsorption of a cation which either hydrolyzes on the surface or in solution (Equation 21).

$$2\ \text{Fe-OH} + \text{cation}^{-2}_s \rightleftharpoons \begin{matrix}\text{FeO}^- \\ \text{FeO}^-\end{matrix} > \text{cation}^{-2} + 2\ H^+_s \tag{20}$$

$$\text{FeOH} + \text{cation}^{+2}_s + H_2O \rightleftharpoons \text{FeO}^- - \text{cation OH}^+ + 2\ H^+_s \tag{21}$$

The $SO_4^=$ data can be described by either the formation of a bidentate complex (Equation 22)

$$2\ \text{FeOH} + SO_{4\,s}^{-2} + 2\ H^+_s \rightleftharpoons \begin{matrix}\text{FeOH}_2^+ \\ \text{FeOH}_2^+\end{matrix} > SO_4^{-2} \tag{22}$$

or a complex suggested by Davis and Leckie (19)

$$\text{FeOH} + SO_{4\,s}^{-2} + 2\ H^+_s \rightleftharpoons \text{FeOH}_2^+ - HSO_4^- \tag{23}$$

Using the data presented here we have no basis for distinguishing between these alternative reactions. The possible reactions for the interactions of the ions with the goethite surface as well as the corresponding estimate of the intrinsic equilibrium constants determined by the double extrapolation method are summarized in Table I. The intrinsic acidity constants, pK_1^{INT} and pK_2^{INT} (Equation 1 and 2), for αFeOOH were also determined by the double extrapolation method. A discussion of those results can be found in Davis and Leckie (19).

The values of K^{INT}_{ion} or the equilibrium constants describing the free energy of association of a hypothetical, isolated charged site with an ion are determined from a combination of the intrinsic constants for the ionization and surface complex reactions (Equations 24 and 25).

$$\text{FeO}^- + \text{cation}_s \overset{K^{INT}_{cation}}{\rightleftharpoons} \text{FeO}^- - \text{cation}\ ;\ K^{INT}_{cation} = \frac{^*K^{INT}_{cation}}{K_2^{INT}} \tag{24}$$

TABLE I

Summary of Complex-Ionization Reactions and Intrinsic Constants

	Reaction	p^*K^{INT}
1.	$FeOH + Na^+ \rightleftarrows FeO^-Na^+ + H^+$	9.6
2.	$FeOH + K^+ \rightleftarrows FeO^-K^+ + H^+$	9.6
3.	$2\ FeOH + Mg^{2+} \rightleftarrows (FeO^-)_2Mg^{2+} + 2\ H^+$	11.9
4.	$FeOH + Mg^{2+} + H_2O \rightleftarrows FeO^-MgOH^+ + 2\ H^+$	12.2
5.	$2\ FeOH + Ca^{2+} \rightleftarrows (FeO^-)_2Ca^{2+} + 2\ H^+$	15.8
6.	$FeOH + Ca^{2+} + H_2O \rightleftarrows FeO^-CaOH^+ + 2\ H^+$	16.5
7.	$2\ FeOH + SO_4^{2-} + 2\ H^+ \rightleftarrows (FeOH_2^+)_2SO_4^{-2}$	-14.1
8.	$FeOH + SO_4^{2-} + 2\ H^+ \rightleftarrows FeOH_2^+ - HSO_4^-$	-14.4
9.	$FeOH + Cl^- + H^+ \rightleftarrows FeOH_2^+Cl^-$	- 5.5

10.	$FeOH_2^+ \rightleftarrows FeOH + H^+$	$-\log K_1^{INT} = 4.9$
11.	$FeOH \rightleftarrows FeO^- + H^+$	$-\log K_2^{INT} = 10.4$

		$\log K^{INT}$
12.	$FeO^- + Na^+ \rightleftarrows FeO^-Na^+$	0.8
13.	$FeO^- + K^+ \rightleftarrows FeO^-K^+$	0.8
14.	$2\ FeO^- + Mg^{2+} \rightleftarrows (FeO^-)_2Mg^{2+}$	8.9
15.	$2\ FeO^- + Ca^{2+} \rightleftarrows (FeO^-)_2Ca^{2+}$	5.0
16.	$2\ FeOH_2^+ + SO_4{}^{2+} \rightleftarrows (FeOH_2^+)_2\ SO_4{}^{-2}$	4.4
17.	$FeOH_2^+ + Cl^- \rightleftarrows FeOH_2^+\ Cl^-$	0.6

$$FeOH_2^+ + anion_s \rightleftharpoons FeOH^+ - anion \; ; \; K_{anion}^{INT} = K_1^{INT} * K_{anion}^{INT} \quad (25)$$

The equilibrium constants, K_{Na}^{INT}, K_K^{INT}, K_{Cl}^{INT} have values less than 10. Solution ion-pair equilibrium constants which describe electrostatic interactions range from 1 to 10 (31). This suggests that the interactions of Na^+, K^+ and Cl^- with the goethite surface are basically electrostatic. Although not directly comparable with K_{Na}^{INT}, K_K^{INT} and K_{Cl}^{INT}, the equilibrium constants K_{Mg}^{INT}, K_{Ca}^{INT} and $K_{SO_4}^{INT}$ are much larger and indicate stronger, more specific interactions.

For Ca^{+2}, Mg^{+2}, and $SO_4^=$ there is a shift in the pH(PZC) which is expected in situations of nonsymmetrical specific adsorption (17). A change in the pH(PZC) due to specific adsorption is accompanied by a shift in the pH_{IEP} in the opposite direction (10). Under these conditions, extrapolation to zero charge to obtain $*K^{INT}$ does not necessarily correspond with extrapolation to zero potential. Thus there may be some error involved in the intrinsic constants for Ca^{+2}, Mg^{+2} and $SO_4^=$. This problem is currently being investigated using a version of the solution equilibrium computer program MINEQL (32) modified to include charge and mass balances for the surface (17, 18, 19, 30). These errors are probably not very large because of the good agreement between our measured and calculated charge in seawater which will be discussed in the next section.

Predictions of Titratable Charge and Surface Species Distributions of Goethite in Seawater

The complexation constants of the individual major seawater ions with αFeOOH determined in single salt solutions can be used to predict the titratable charge and surface species distribution of goethite in seawater. This prediction can then be compared with the experimentally determined charge of goethite in a mixed seawater type electrolyte.

There are two alternatives available for calculating the surface species distribution in a sample or a mixed electrolyte solution. One approach is the solution equilibrium computer program MINEQL (32) as modified to include surface species by Davis *et al*. (17). The surface species distribution is calculated by simultaneously solving the equations for charge, potential, total surface sites and individual surface species. Values for FeO_T, K_1^{INT}, K_2^{INT}, $*K_{cation}^{INT}$, $*K_{anion}^{INT}$, and integral capacitances (C_1 and C_2) of the inner regions of the double layer are necessary. Unfortunately at this time there is no clear way to choose values for the capacitances in mixed electrolyte solutions. This is the most powerful approach, however, and ultimately will have the most widespread application. The second approach, which we have used here, utilizes apparent stability quotients rather than intrinsic constants and does not require knowledge of the integral capacitances. This method, however, requires values for the apparent stability quotients for each specific set of concentrations and as a result has less flexibility than the MINEQL program.

A total of nine possible types of surface sites on goethite can occur when it is suspended in a solution containing the major ions of seawater. These are the negative, neutral, and positive surface sites caused by the ionization of the surface functional groups and the six surface complexes caused by the interactions of the six (Na, Mg, Ca, K, Cl, SO_4) major ions of seawater with goethite. The distribution of these surface sites will vary with the pH and concentrations of the individual ions in the solution.

The concentrations of the various surface sites ($Fe{-}OH_2^+$, Fe-OH, $Fe{-}O^-$, $Fe{-}O^-Na^+$, $Fe{-}O^-K^+$, $Fe{-}O^-MgOH^+$, $Fe{-}O^-CaOH^+$, $Fe{-}OH_2^+Cl^-$, $Fe{-}OH_2^+\ HSO_4^-$) were determined by simultaneously solving the apparent stability quotients describing the formation of these sites and the equation describing the total surface sites (Equation 18).

$$FeO_T = \{Fe{-}O^-\} + \{Fe{-}OH\} + \{Fe{-}OH_2^+\} + \{Fe{-}O^-Na^+\} + \{Fe{-}O^-MgOH^+\} + \{Fe{-}O^-CaOH^+\} + \{Fe{-}O^-K^+\} + \{Fe{-}OH^+Cl_2^-\} + \{Fe{-}OH_2^+\ HSO_4^-\} \qquad (26)$$

The equations for the formation of the Mg^{2+} and Ca^{2+} hydrolysis complexes were chosen rather than the bidentate reactions because Davis et al. (17) suggest a similar reaction for Mg adsorption on rutile. Justification for this species can be presented by considering the effect of the surface charge on the first hydrolysis constant of Mg^{+2} and Ca^{+2} (12, 19). The effect of a charged surface with a lowered dielectric constant is to increase the hydrolysis constant for a cation and decrease the acidity constant for an anion.

The values for the apparent stability quotients ($^*Q_{Na}$, $^*Q_{Mg}$, $^*Q_{Ca}$, $^*Q_K$, $^*Q_{Cl}$, $^*Q_{SO_4}$) were determined as described earlier using the titration data of goethite in the individual electrolytes at a given pH and at the concentration of the specific ion in seawater, rather than at the ionic strength of seawater. Thus in these calculations we neglect the effect of ionic strength on the constant. We also assume that the activity coefficients for the surface sites are equal to one. These appear to us to be serious deficiencies in the present calculations and in a final model they will have to be corrected. As will be shown, however, the agreement between measured and calculated charge is quite good in spite of these deficiencies. The concentrations of the free ions in the mixed electrolyte were calculated by considering the formation of solution complexes ($MgSO_4^{\circ}$, $CaSO_4^{\circ}$, $NaSO_4^-$, KSO_4^-) (33).

The titratable charge is mathematically related to the surface species distributions (Equation 27).

$$\sigma_o = F\ [2\{FeOH_2^+ - HSO_4^-\} + \{FeOH_2^+Cl^-\} + \{Fe\text{-}OH_2^+\} - \{Fe\text{-}O^-Na^+\} - \{FeO^-K^+\} - 2\ \{FeO^-MgOH^+\} - 2\ \{FeO^-CaOH^+\} - \{FeO^-\}] \quad (27)$$

Table II is a summary of the surface species distributions with pH. These were used to calculate a titratable charge. The effect of the ionized surface species (FeO^- and $FeOH_2^+$) on the titratable charge and surface species distributions is less than the effect of the potassium complexes. Also included in Table II are the % contributions of the individual complexes to the total calculated charge. In Figure 7 the calculated charge is compared with the titratable charge determined by the potentiometric titration of αFeOOH in a major seawater ion electrolyte. Also included in Figure 7 is a compilation of the titration data used in determining the calculated charge.

Discussion

An examination of the potentiometric titration data of goethite in the individual electrolyte solutions permits a qualitative assessment of the solids capacity for the major ions of seawater. In more concentrated electrolyte solutions the charge primarily represents the formation of surface complexes and the magnitude of the charge is indicative of the amount or strength of complexation. There are two observations to be made of the potentiometric titration data of goethite. First, the absolute magnitude of the charged sites in the pH range of 4 to 9.5 does not exceed 5.7 mol cm^{-2} of solid. The total surface sites are 27.8 mol cm^{-2} of solid and, therefore, the neutral Fe-OH sites are the dominant surface sites rather than the goethite - major seawater ion complex sites. Second, the absolute magnitude of the charges of goethite in the individual electrolytes suggests that Mg and SO_4 bind more strongly than Na, K, Ca or Cl. Together these observations suggest the following order for the strength of ion-binding with goethite:

$$H \gg Mg \simeq SO_4 > Ca > Cl \simeq Na \simeq K$$

The sequence of $Mg^{+2} > Ca^{+2}$ is the reverse of the normal affinity sequence (Hofmeister series) which is observed on most clays and on MnO_x. The observed sequence on αFeOOH is that expected when interactions between the adsorbed ions and the surface sites are greater than hydration effects (31).

In an idealized planar model, σ_o is the charge at the oxide's surface caused by the ionization of the surface functional groups and the formation of the surface complexes. This is

TABLE II

Distribution of Surface Species on Goethite in Seawater as a Function of pH

% Sites	pH 5	6	7	8	9
FeOH	97.2	96.7	94.9	92.5	87.7
FeO^- $MgOH^+$	--	1.8	4.4	6.7	8.1
FeO^- $CaOH^+$	--	--	--	0.7	1.8
FeO^- Na^+	--	--	--	--	1.5
FeO^- K^+	--	--	--	--	0.8
$FeOH_2^+$ HSO_4^-	2.7	1.6	0.7	0.1	--
Calculated[1] Charge (μcoul cm^{-2})	14.5	-0.1	-19.9	-38.8	-59.5

% Charge[2]	pH 5	6	7	8	9
FeO^- $MgOH^+$	--	52.9	86.1	88.7	73.4
FeO^- $CaOH^+$	--	--	--	9.3	16.0
FeO^- Na^+	--	--	--	--	6.8
FeO^- K^+	--	--	--	--	3.7
$FeOH_2^+$ HSO_4^-	(100)	(47.1)	(13.9)	(2.0)	--

[1] Calculated (or net) charge is sum of negative and positive charge

[2] ( ) indicate the % contribution of positive charge to the total absolute charge

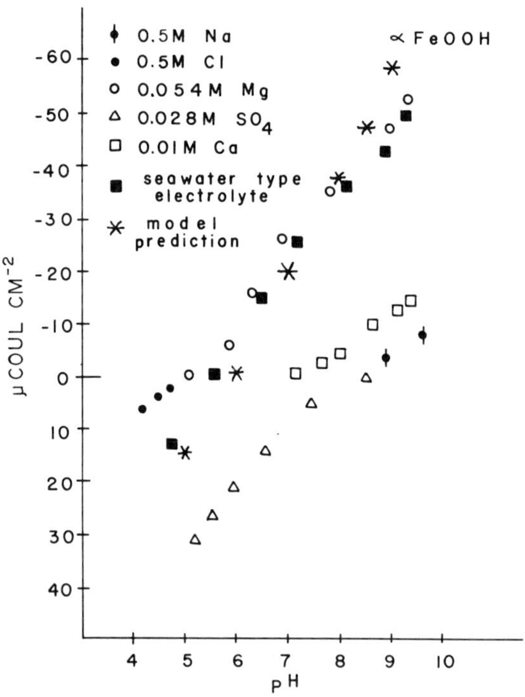

Figure 7. Calculated and measured charge for goethite in a seawater-type electrolyte and compilation of titration data used to obtain the calculated charge

TABLE III

Comparison of Various Forms of Iron Oxide

References	Solid Phase	Salt	pH(PZC)	Condition	σ_o (coul cm^{-2})
Atkinson et al. (20)	αFeOOH	KCl	7.5	pH 9, 0.1 M KCl	- 3.99
	αFe_2O_3	KCl	9.2	pH 9, 0.1 M KCl	+10.0
Hingston et al. (35)	αFeOOH	NaCl	7.7	pH 9, 0.1 M NaCl	- 8.75
Yates et al. (34)	αFeOOH	KNO_3	7.5	pH 9, 0.1 M KNO_3	- 5.0
	αFeOOH	K_2SO_4	8.1	pH 6, 10^{-3}M K_2SO_4	+ 8.0
Present work	αFeOOH	NaCl, KCl	7.5	pH 9, 0.1 M NaCl/KCl	- 4.2
		$MgCl_2$	5.0	pH 9, 0.05 M $MgCl_2$	-48.5
		$CaCl_2$	7.1	pH 9, 0.05 M $CaCl_2$	-19.0
		Na_2SO_4	8.5	pH 6, 10^{-3}M Na_2SO_4	+10.54
Breeuswma et al. (37)	αFe_2O_3	KCl	8.5	pH 9, 0.1 M KCl	- 1.0
	αFe_2O_3	$Mg(NO_3)_2$	6.5	pH 9, 0.05 M $Mg(NO_3)_2$	-66.0
	αFe_2O_3	$Ca(NO_3)_2$	6.5	pH 9, 0.05 M $Ca(NO_3)_2$	-28.0
Davis et al. (18)	am. $Fe(OH)_3$	$NaNO_3$	7.9	pH 9, 0.1 M $NaNO_3$	- 5.0
Murray (39)	βFeOOH	NaCl	8.1	pH 9, 0.1 M NaCl	- 7.25

the charge measured by potentiometric titration. In addition, there is the plane of adsorbed counter-ions and σ_β is the charge caused by the presence of these ions. The ions in the diffuse part of the double layer "see" the combined effect of σ_o and σ_β. Therefore the shift in the pH(PZC) to 5.5 for goethite in seawater and the corresponding high negative charge (-36.2 μcoul cm^{-2}) at pH 8 determined by potentiometric titration cannot be used to evaluate goethite's electrostatic influence on ions in seawater.

A comparison of our results with other investigators' work on goethite and other forms of iron oxides is shown in Table III. Atkinson et al. (20), Yates et al. (34) and our work on αFeOOH indicate excellent agreement for pH(PZC) and charge in the appropriate electrolyte solutions. Hingston et al. (35) charge data for goethite tend to be higher in magnitude. Davis et al. (17) calculations for pK_1^{INT}, pK_2^{INT}, $p{*}K_K^{INT}$ for goethite agree very well with our results for goethite. A comparison of the αFe_2O_3, αFeOOH, βFeOOH and am. $Fe(OH)_3$ data indicates similarities in the values of the pH(PZC). This contrasts with the results of Healy et al. (36) for manganese oxides where the values for the pH(IEP)(pH of the isoelectric point) as determined by electrophoresis ranged from pH 1.5 for σ-MnO_2, pH 1.8 for Mn(II) manganite, pH 4.5 for αMnO_2, pH 5.5 for γ-MnO_2, to pH 7.3 for β-MnO_2. Breeuwsma and workers (27, 37, 38) found the reverse Hofmeister series for αFe_2O_3 and in addition Mg adsorbed much stronger than Ca. Our data on goethite indicate the same conclusions. However, the values of the pH(PZC) for αFeOOH in Mg and Ca solutions were 5.0 and 7.1, respectively, while for αFe_2O_3 pH(PZC) was 6.5 in both Mg and Ca solutions.

Conclusions

A quantitative assessment of ion-binding with goethite was obtained from an application of the Davis et al. (17), James et al. (30) and Davis and Leckie (19) model to potentiometric titration data. By comparison with solution complex equilibrium constants, the intrinsic constants for the association of Na, K, and Cl with charged αFeOOH sites (Equations 12, 13 and 17 of Table I) indicate primarily electrostatic interactions. The associations of Mg, SO_4, and Ca with charged goethite sites suggest stronger or more specific bonds.

The simplified mass and proton balance model determined what the surface species distribution of goethite would be in a mixed, seawater type electrolyte. This surface species distribution was used to calculate a surface charge for goethite. This calculated surface charge successfully predicted the titratable surface charge, as determined by potentiometric titration, of goethite in a seawater major-ion electrolyte (Figure 7). These surface species distributions indicate that Fe-OH sites are the primary sites and that the formation of Mg and SO_4

complexes with goethite account for the formation of the majority of the charge (Table II).

Abstract

Potentiometric titrations of goethite (αFeOOH) have been performed in various concentrations of NaCl, KCl, $MgCl_2$, $CaCl_2$, Na_2SO_4 and a mixed electrolyte containing the major ions of seawater in their seawater proportions. From this data we have calculated apparent stability quotients. A method of double extrapolation has been used to calculate intrinsic acidity and surface complex equilibrium constants. Using these constants we calculate the titratable charge and surface species distribution of goethite in seawater. At pH 8 the calculated charge is -38.8 μcoul cm^{-2} and 90% of the titratable charge is due to FeO^-MgOH^+ surface complexes. The calculated charge over the pH range 5-9 is in excellent agreement with the measured charge in the seawater type electrolyte.

Literature Cited

1. Bacon, M. P., Spencer, D. W., and Brewer, P. G., $^{210}Pb/^{226}Ra$ and $^{210}Po/^{210}Pb$ disequilibria in seawater and suspended particulate matter. Earth Planet. Sci. Lett. 32, 277-296 (1976).
2. Boyle, E. A., Edmond, J. M., and Sholkovitz, E. R., The mechanism of iron removal in estuaries. Geochim. Cosmochim. Acta 41, 1313-1324 (1977).
3. Murray, J. W. and Brewer, P. G., The mechanisms of removal of iron, manganese, and other trace metals from sea water. p. 291-326 in Glasby, G. P., ed., "Marine Manganese Deposits," Elsevier, Amsterdam, 1977.
4. Dymond, J., Corliss, J. B., Heath, G. R., Field, C. W., Dasch, E. J., and Veeh, H., Origin of metalliferous sediments from the Pacific Ocean. Bull. Geol. Soc. Am. 84, 3355-3372 (1973).
5. Heath, G. R. and Dymond, J., Genesis and transformation of metalliferous sediments from the East Pacific Rise, Bauer Deep, and Central Basin, Northwest Nazca Plate. Bull. Geol. Soc. Am. 88, 723-733 (1977).
6. Burns, R. G. and Burns, V. M., Mineralogy of ferromanganese nodules. p.184-248 in Glasby, G. P., ed., "Marine Manganese Deposits," Elsevier, Amsterdam, 1977.
7. Murray, J. W., "The Interaction of Metal Ions at the Hydrous Manganese Dioxide-Solution Interface." Ph.D. Thesis, Mass. Inst. of Tech.-Woods Hole Ocean. Inst., 1973.
8. Murray, J. W., The interaction of ions at the manganese dioxide-solution interface. Geochim. Cosmochim. Acta 39, 505-519 (1975).

9. Murray, J. W., The interaction of cobalt with hydrous manganese dioxide. Geochim. Cosmochim. Acta 39, 635-647 (1975).
10. Parks, G. A., Adsorption in the marine environment. p. 241-308 in Riley, J. P. and Skirrow, G., ed., "Chemical Oceanography," Vol. 1, Academic Press, New York, 1975.
11. James, R. O. and Healy,T. W., Adsorption of hydrolyzable metal ions at the oxide-water interface. I. Co(II) adsorption on SiO_2 and TiO_2 as model systems. J. Coll. Interface Sci. 40(1), 42-52 (1972).
12. James, R. O. and Healy, T. W., Adsorption of hydrolyzable metal ions at the oxide-water interface. II. Charge reversal of SiO_2 and TiO_2 colloids by adsorbed Co(II), La(III), and Th(IV) as model systems. J. Coll. Interface Sci. 40(1), 53-64 (1972).
13. James, R. O. and Healy, T. W., Adsorption of hydrolyzable metal ions at the oxide-water interface. III. A thermodynamic model of adsorption. J. Coll. Interface Sci. 40(1), 65-81 (1972).
14. Stumm, W., Huang, C. P. and Jenkins, S. R., Specific chemical interaction affecting the stability of dispersed systems. Croatica Chem. Acta 42, 223-245 (1970).
15. Schindler, P. W., Furst, B., Dick, R., and Wolf, P. U., Ligand properties of surface silanol groups. I. Surface complex formation with Fe^{3+}, Cu^{2+}, Cd^{2+}, and Pb^{2+}. J. Coll. Interface Sci. 55(2), 469-475 (1976).
16. Stumm, W., Hohl, H., and Dalang, F., Interaction of metal ions with hydrous oxide surfaces. Croatica Chem. Acta 48(4), 491-504 (1976).
17. Davis, J. A., James, R. O., and Leckie, J. O., Surface ionization and complexation at the oxide-water interface. 1. Computation of electrical double layer properties in simple electrolytes. J. Coll. Interface Sci. 63(3), 480-499 (1978).
18. Davis, J. and Leckie, J. O., Surface ionization and complexation at the oxide-water interface. 2. Surface properties of amorphous iron oxyhydroxide and adsorption of metal ions. J. Coll. Interface Sci. (in press)
19. Davis, J. A. and Leckie, J. O., Speciation of adsorbed ions at the oxide/water interface. Adv. Chem. Series, (in press).
20. Atkinson, R. J., Posner, A. M., and Quirk, J. P., Adsorption of potential-determining ions at the ferric oxide-aqueous electrolyte interface. J. Phys. Chem. 71(3), 550-558 (1967).
21. Brunauer, S., Emmett, P. H., and Teller, E., Adsorption of gases in multimolecular layers. J. Amer. Chem. Soc. 60, 309-319 (1938).
22. Yates, D. E., "The Structure of the Oxide/Aqueous Electrolyte Interface." Ph.D. Thesis, Univ. of Melbourne, 1975.

23. Balistrieri, L. S., "The Basic Surface Characteristics of Goethite (αFeOOH)." M.S. Thesis, Univ. of Washington, Seattle, 1977.
24. Wilson, T. R. S., Salinity and the major elements of seawater. p. 365-413 in Riley, J. P. and Skirrow, G., ed., "Chemical Oceanography," Vol. 1 (2nd Ed.), Academic Press, New York, 1975.
25. Parks, G. A. and deBruyn, P. L., The zero point of charge of oxides. J. Phys. Chem. 66, 967-972 (1962).
26. Blok, L. and de Bruyn, P. L., The ionic double layer at the ZnO/solution interface. I. The experimental point of zero charge. J. Coll. Interface Sci. 32(3), 518-526 (1970).
27. Breeuwsma, A. and Lyklema, J., Physical and chemical adsorption of ions in the electrical double layer on hematite (αFe_2O_3). J. Coll. Interface Sci. 43(2), 437-448 (1973).
28. Huang, C. P. and Stumm, W., Specific adsorption of cations on hydrous γ-Al_2O_3. J. Coll. Interface Sci. 43(2), 409-420 (1973).
29. Yates, D. E., Levine, S., and Healy, T. W., Site binding model of the electrical double layer at the oxide/water interface. J. Chem. Soc. London Faraday Trans. 1, 70, 1807-1818 (1974).
30. James, J. O., Davis, J. A., Leckie, J. O., Computer simulation of the conductometric and potentiometric titrations of the surface groups on ionizable latexes. J. Coll. Interface Sci. 65(2), 331-344 (1978).
31. Stumm, W. and Morgan, J. J., "Aquatic Chemistry," 583 pp., Wiley-Interscience, New York, 1970.
32. Westall, J. C., Zachary, J. L., and Morel, F. M. M., MINEQL, a computer program for the calculation of chemical equilibrium composition of aqueous systems. Water Qual. Lab., Tech. Note No. 18, Dept. of Civil Eng., Mass. Inst. of Tech., Cambridge, 1976.
33. Garrels, R. M. and Thompson, M. E., A chemical model for sea water. Amer. J. Sci. 260, 57-66 (1962).
34. Yates, D. E. and Healy, T. W., Mechanism of anion adsorption at the ferric and chromic oxide/water interfaces. J. Coll. Interface Sci. 52(2), 222-228 (1975).
35. Hingston, F. J., Posner, A. M., and Quirk, J. P., Adsorption of selenite by goethite. Adv. Chem. Series 79, 82-90 (1968).
36. Healy, T. W., Herring, A. P., and Fuerstenau, D. W., The effect of crystal structure on the surface properties of a series of manganese oxides. J. Coll. Interface Sci. 21, 435-444 (1966).

37. Breeuwsma, A. and Lyklema, J., Interfacial electrochemistry of hematite (α-Fe_2O_3). Disc. Farad. Soc. 52, 324-333 (1971).
38. Breeuwsma, A., "Adsorption of ions on hematite (α-Fe_2O_3)." Ph.D. Thesis, Agricultural Univ., Wageningen, Netherlands, 1973.
39. Murray, J. W., βFeOOH in marine sediments. E⊕S 59, 411-412 [Abstract] (1978).

RECEIVED November 16, 1978.

15

Speciation of Adsorbed Ions at the Oxide/Water Interface

JAMES A. DAVIS[1] and JAMES O. LECKIE

Environmental Engineering and Science, Department of Civil Engineering, Stanford University, Stanford, CA 94305

Ionization of the oxide/water interface and the resultant electrical double layer have been studied intensively by a variety of techniques within the last decade. Although many electrical double layer and adsorption models have been proposed, few are sufficiently general to consider surface equilibria in complex electrolyte solutions. Recently we proposed a comprehensive adsorption model for the oxide/water interface which can simultaneously estimate adsorption density, surface charge, and electrokinetic potential in a self-consistent manner (1, 2, 3). One advantage of the model was that it could be incorporated within the computer program, MINEQL (4), by adding charge and mass-balance equations for the surface.

Our approach combined concepts from models which emphasize specific chemical interactions of solutes with oxide surfaces (5, 6) and those with well-defined electrical double layer structure (7, 8, 9, 10). Since the model and computational procedure are general, we were able to describe surface and solution equilibrium in simple electrolyte solutions (1) and in more complex solutions containing dilute solutes and swamping electrolyte (2, 3).

Although the stoichiometry of surface reactions may sometimes be deduced by potentiometric titration, the physical and chemical nature of bonding at the oxide/water interface is not well understood. Consequently, it is often necessary to make assumptions regarding the surface speciation and charge distribution of adsorbed ions. In our previous work (1, 2, 3) we assumed that the charge of all specifically adsorbed ions was located in a single plane, σ_β (Figure 1). This assumption greatly simplified the mathematics required to determine the equilibirum distribution of ions and resultant charge and potential within the compact layer.

[1] Current address: Swiss Federal Institute of Technology EAWAG, CH-8600 Dübendorf, Switzerland.

0-8412-0479-9/79/47-093-299$05.00/0
© 1979 American Chemical Society

In this paper we examine the assumptions of our previous modeling approach and present new model calculations which consider alternative assumptions. In addition, we discuss the physicochemical factors which affect the formation of surface complexes at the oxide/water interface, in particular the effect of decreasing dielectric strength of the solvent. Finally, to demonstrate the general applicability of the model we present modeling results for a complex electrolyte system, where adsorption of a metal-ligand complex must be considered.

Generalized Adsorption Model

Electrical Double Layer. In order to model the structure of the electrical double layer (EDL) of oxide colloids, it is necessary to formulate 1) the reactions which result in the formation of surface charge (σ_o), and 2) the potential and charge relationships in the interfacial region. It has been generally assumed that surface charge (σ_o), defined experimentally by the net uptake of protons by the surface, results from simple ionization of oxide surface sites (5, 6, 11, 12, 13), i.e.,

$$SOH_2^+ \underset{\longrightarrow}{\overset{K_{a_1}^{int}}{\longleftarrow}} SOH + H_s^+ \qquad (1)$$

$$SOH \underset{\longrightarrow}{\overset{K_{a_2}^{int}}{\longleftarrow}} SO^- + H_s^+ \qquad (2)$$

where the subscript s refers to a surface concentration (1). However, Davis et al. (1) proposed that the principal mechanism of surface charge development in simple electrolyte solutions is the reaction of electrolyte ions with ionizable surface sites, i.e.,

$$SOH + Na_s^+ \underset{\longrightarrow}{\overset{{}^*K_{Na^+}^{int}}{\longleftarrow}} SO^- \!-\! Na^+ + H_s^+ \qquad (3)$$

$$SOH + H_s^+ + Cl_s^- \underset{\longrightarrow}{\overset{1/{}^*K_{Cl^-}^{int}}{\longleftarrow}} SOH_2^+ \!-\! Cl^- \ . \qquad (4)$$

Formation of these surface complexes occurs in addition to simple ionization of surface groups via Equations 1 and 2. Smit et al. (14, 15) have recently shown experimentally that Na^+ is specifically adsorbed within the compact layer of silica and alumina surfaces. The experimental evidence supported a site-binding model for the EDL of non-porous oxides (1, 7), rather than gel layer models proposed by other workers (16, 17, 18).

Our modeling approach was first used to describe the EDL properties of well-characterized, crystalline oxides (1). It was shown that the model accounts for many of the experimentally observed phenomena reported in the literature, e.g. the effect of supporting electrolyte on the development of surface charge, estimates of differential capacity for oxide surfaces, and measurements of diffuse layer potential. It is important to note that a Nernstian dependence of surface potential (ψ_o) as a function of pH was not assumed. The interfacial potentials (ψ_o, ψ_β, ψ_d in Figure 1) are independent variables in the model and the equilibrium values result from the simultaneous numerical solution of all surface equilibria. In subsequent work, the model was applied to a complex oxide precipitate, i.e., amorphous iron oxyhydroxide, and polymer colloids (2, 19). Balistrieri and Murray (20) have used the model to describe the surface charge of goethite in a synthetic seawater solution.

An important development of the model presented by Davis et al. (1) was a method for estimating intrinsic surface ionization and complexation constants. It was suggested that intrinsic surface complexation constants, e.g. $^*K^{int}_{Na^+}$ and $^*K^{int}_{Cl^-}$, could be estimated from potentiometric titrations of concentrated electrolyte solutions (0.1-0.5 M). This followed from the conclusion that the dominant reactions which account for surface charge (σ_o) in moderately concentrated solutions are the formation of surface complexes by electrolyte ions, Equations 3 and 4. Intrinsic acidity constants, K^{int}_{a1} and K^{int}_{a2}, were estimated from potentiometric titrations of dilute electrolyte solutions (10^{-2} or 10^{-3}M), in which it was assumed surface ionization reactions would be more important in the measured surface charge (σ_o).

Recently James et al. (19) proposed an improved method for determining intrinsic surface stability constants. In the former approach (1), apparent stability quotients were extrapolated to a hypothetical condition of zero charge and potential to estimate intrinsic constants. In the improved method (19), a double extrapolation is made. For example, to determine intrinsic acidity constants, apparent stability quotients are extrapolated to zero charge and infinitely dilute electrolyte concentration. Thus, any contribution to the surface charge (σ_o) from complex formation with electrolyte ions is avoided. Intrinsic surface complexation constants are estimated by extrapolations to zero charge and 1 M electrolyte concentration.

To date the double extrapolation technique has only been applied to ionizable latexes (19). Since this technique avoids some of the assumptions of our earlier methods (1), it would also be useful for a determination of the intrinsic surface stability constants of oxides. Figures 2 and 3 show sample calculations of pK^{int}_{a1} and pK^{int}_{a2} for α-FeOOH in NaCl, using the experimental

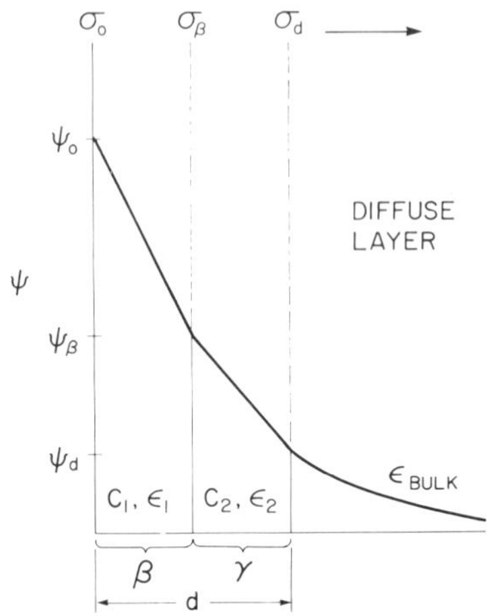

Figure 1. Schematic of the charge distribution at an idealized planar surface and the potential decay away from the surface

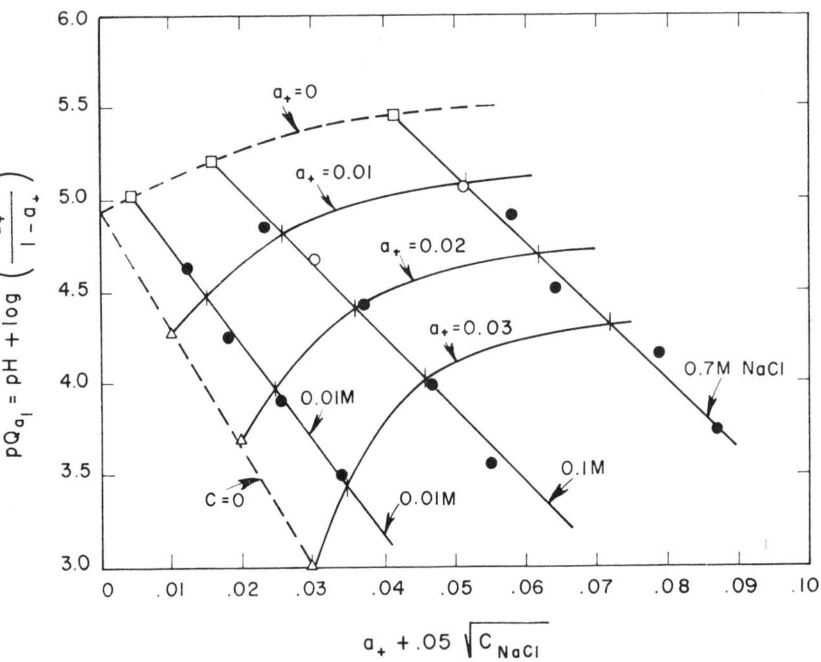

Figure 2. Variation of the apparent acidity quotient (pQ_{a_1}) of α-FeOOH as a function of surface charge and concentration of supporting electrolyte, NaCl. Lines are contours for constant electrolyte or constant surface charge. The condition $\alpha = C = 0$ yields $pK_{a_1}{}^{int}$. a—FeOOH/NaCl. Balistrieri and Murray (1978).

data of Balistrieri and Murray (20). Apparent acidity quotients (pQ_a, filled circles) are calculated from the experimental data and plotted as a function of fractional ionization, α_+ or α_-, plus a concentration-dependent term, to separate the curves at each electrolyte concentration. For each electrolyte concentration, a curve is drawn through the filled circles to the condition of zero charge, $\alpha = 0$ (open squares). A second extrapolation is made through these points (open squares) to the vertical axis, where $\alpha = 0$ and $C = 0$ (infinitely dilute electrolyte). An important advantage of this method is that one may check the result by reversing the order of extrapolation. Additional curves are plotted in Figures 2 and 3 for various values of constant charge, e.g. $\alpha = 0.01$. These curves are extrapolated to the condition $C = 0$ (open triangles). A curve is drawn through these points to the vertical axis where $C = 0$ and $\alpha = 0$. The two extrapolations (the dashed curves in Figures 2 and 3) should meet at the same point on the vertical axis, which is pK_a^{int}. It is demonstrated here that the technique works well for α-FeOOH, $pK_{a1}^{int} = 4.9$ and $pK_{a2}^{int} = 10.4$. Using the method suggested by Davis *et al.* (1) for the data at 0.01 *M* NaCl, the estimates would have differed by ~0.3 log units. Presumably the error would have been less for titrations in more dilute solutions of NaCl.

Adsorption of Metal Ions. Significant advances have been made in recent years in the development of phenomenological models to describe trace metal adsorption at the oxide/water interface. In particular, the concepts of surface ionization and complexation introduced by Stumm, Schindler, and co-workers (5, 6, 11, 12, 13) have aided in the understanding of complex adsorption phenomena. As a further development of the surface complexation approach, Davis *et al.* (1) introduced exponential terms to the mass-law expressions for surface equilibria to account for the effect of the electrostatic field at the interface. By including the surface equilibria of metal ions with those of the site-binding EDL model, one may consider adsorption of major electrolyte ions and dilute metal ions simultaneously. Thus, calculations reflect the net interactions of all ions at the surface, and the effect of a change in electrolyte composition or concentration can be assessed.

The modeling approaches used to describe the surface reactions of metal ions differ in their definition of surface structure and the charge/potential relationships within the compact layer of the EDL (2, 5, 6). In our previous calculations (2) we have assumed that the average location of the center of adsorbed metal ions is the ψ_β plane (Figure 1), and that ions are attracted or repelled by the equilibrium potential in that plane, ψ_β. With this assumption, we found that metal ion uptake was best described by a combination of surface reactions involving aquo metal ions and their hydrolytic complexes, *i.e.*,

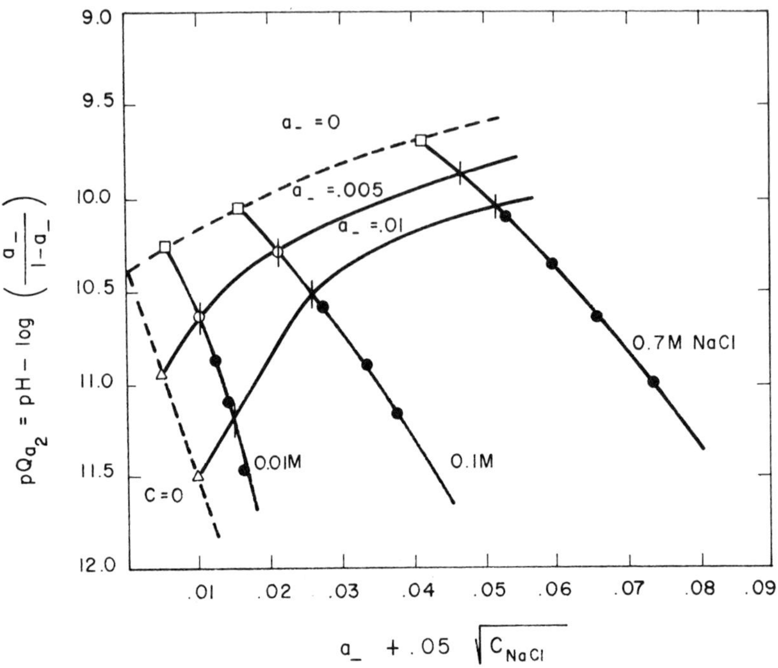

Figure 3. Variation of the apparent acidity quotient (pQ_{a_2}) of α-FeOOH as a function of surface charge and concentration of supporting electrolyte, NaCl. Lines are contours for constant electrolyte or constant surface charge. The conditions $\alpha = C = 0$ yields $pK_{a_2}{}^{int}$. a—FeOOH/NaCl. Balistieri and Murray (1978).

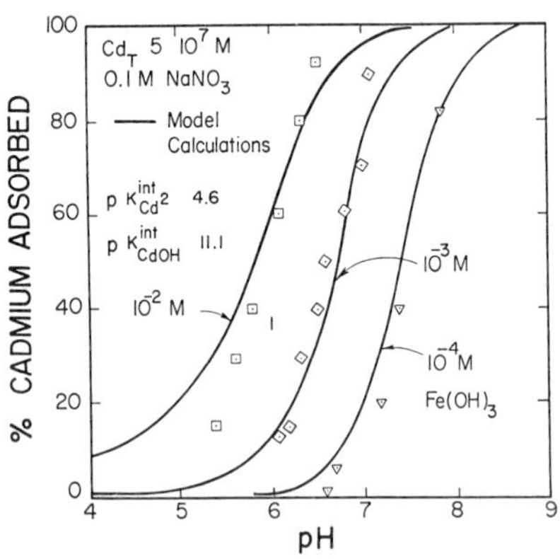

Figure 4. Predictive model calculation of Cd(II) adsorption on amorphous iron oxyhydroxide as a function of pH and amount of solid substrate present. Cd_T 5 10^7M, 0.1M $NaNO_3$. (——) Model calculations. $pK_{Cd^{2}}{}^{int}$—4.6; $pK_{CdOH}{}^{int}$—11.1.

$$SOH + M_s^{z+} \underset{}{\overset{*K^{int}_{M^{z+}}}{\rightleftharpoons}} SO^- \text{—} M^{z+} + H_s^+ \tag{5}$$

$$SOH + M_s^{z+} + H_2O \underset{}{\overset{*K^{int}_{M(OH)}}{\rightleftharpoons}} SO^- \text{—} MOH^{z-1} + 2H_s^+ \ . \tag{6}$$

This approach successfully described the experimental results of several adsorption studies with various metal ions and oxide substrates (2). In addition, one can make predictive calculations of metal ion uptake, if the surface parameters of an oxide/electrolyte can be estimated. For example, Figure 4 shows predicted and experimental adsorption behavior of Cd(II) on amorphous iron oxyhydroxide. Surface stability constants for Cd(II) were estimated (2) from an experimental study of Cd(II) uptake by α-FeOOH (21, 36).

Previously we showed that model calculations that consider complexation of metal ions by bidentate surface sites were in poor agreement with experimental adsorption data (2). These complexes have been proposed in other modeling studies (5, 6). Our calculations assumed that the charges in the surface complex

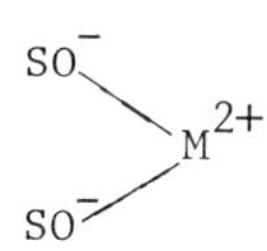

could be represented by a charge of -2 in the σ_o plane and +2 in the σ_β plane. However, the occurrence of nephelauxetic effects in coordination compounds indicate that the actual charges of coordinated metal ions can be less than their formal oxidation number (23). Thus, the charge of a metal ion complexes by oxide surface sites may be better represented by some fractional charge (2δ+) rather than an integer charge (2+). Consequently, the mass-law expression for surface complex formation would have different coefficients for ψ_o and ψ_β , *i.e.*,

$$\left[\begin{matrix} SO^{\delta-} \\ \quad\searrow M^{2\delta+} \\ SO^{\delta-}\nearrow \end{matrix}\right] = \frac{\left[\begin{matrix} SO^- \\ \quad > \\ SO^- \end{matrix}\right][M^{2+}]}{[H^+]^2} \exp[(2\delta e\psi_o - 2\delta e\psi_\beta)/kT] \; {}^*\beta^{int}_{M^{2+}} . \qquad (7)$$

Figure 5 shows experimental data (21) and model calculations of Cd(II) adsorption on TiO_2. The dashed curve in Figure 5 represents model calculations with non-integer charges, using the modified mass-law expression, Equation 7. Calculations with $\delta = 0.5$ in Equation 7 can also simulate the experimental data, giving a somewhat stronger pH dependency (steeper slope) as compared to our calculations using integer charge (solid line in Figure 5). The success of the partial charge approach may be fortuitous, since the coefficients of ψ_o and ψ_β with $\delta = 0.5$ in Equation 7 are the same as written for adsorption of a monohydroxometal(II) complex. There is presently no way of knowing the degree of the nephelauxetic effects of surface complexes, if any; however it appears to be conceptually possible. In view of the difficulty in making appropriate corrections for partial charge, the earlier modeling approach (2) is simpler and less tenuous conceptually. In those calculations, one need only consider integer charges located in the σ_o and σ_β planes.

Adsorption of Anions. The general nature of the adsorption model and computational method allow one to describe the uptake of anions also (3). Similar to the approach for metal ions, we included a term in the mass-law expression to correct for the effect of potential on surface equilibria. Although adsorption of some anions (e.g. chloride, nitrate, syringic acid, thiosulfate) can be simulated by one surface reaction (24), formation of two surface complexes is probable for other anions, e.g. chromate, selenate (3). Model calculations were more consistent with experimental adsorption data when the following surface reactions were considered, i.e.

$$SOH + H^+_s + A^{2-}_s \underset{}{\overset{1/{}^*K^{int}_{A^{2-}}}{\rightleftarrows}} SOH_2^+{-}A^{2-} \qquad (8)$$

$$SOH + 2H^+_s + A^{2-}_s \underset{}{\overset{1/{}^*K^{int}_{HA^-}}{\rightleftarrows}} SOH_2^+{-}HA^- \qquad (9)$$

For example, Figure 6 shows experimental data and model calculations for sulfate adsorption on amorphous iron oxyhydroxide. The dashed curve represents a model calculation using Equation 8 only, with $p^*K^{int}_{SO_4^{2-}} = 10.1$. At low pH the calculated adsorption is less than observed experimentally. At pH < 5, protolysis of adsorbed sulfate ions becomes significant, and Equation 9 must be included in model calculations. The solid curve in Figure 6 considers both surface equilibria, with $p^*K^{int}_{SO_4^{2-}} = 10.1$ and $p^*K^{int}_{HSO_4^-} = 15.6$. Analysis of the data of Balistrieri and Murray (20) for potentiometric titrations of α-FeOOH in Na_2SO_4 solutions yields a similar value for $p^*K^{int}_{HSO_4^-} = 15.0$. Previous model calculations which considered complexation of anions by bidentae surface sites were in poor agreement with the experimental data (3).

A significant problem in surface complexation models is the definition of adsorption sites. The total number of proton-exchangeable sites can be determined by rapid tritium exchange with the oxide surface (25). Although surface equilibria are usually written in terms of one surface site, e.g. Equations 5, 6, 8, 9, adsorption isotherms for many ions show that the number of molecules adsorbed at maximum surface coverage (Γ_{max}) is less than the total number of surface sites. For example, uptake of Se(VI) and Cr(VI) ions on $Fe(OH)_3$(am) at Γ_{max} equals 1/3 and 1/4 the total number of surface sites (J. A. Davis and J. O. Leckie, unpublished data, 1978). These results may be interpreted in two ways: 1) the adsorbed ions cover a larger area than a proton-exchangeable site, or 2) not all surface sites determined by ^3H exchange are available for surface complexation with large anions.

The number of surface sites covered by an adsorbate is important in model calculations when the amount adsorbed approaches the maximum surface coverage, Γ_{max}, and available surface sites become a limiting factor in complex formation. Although one could write surface equilibria with a larger coefficient for surface sites, e.g.

$$4SOH + CrO^{2-}_{4_s} + H^+_s \rightleftarrows \left.\begin{matrix} SOH \\ SOH_2^+ \\ SOH \\ SOH \end{matrix}\right] CrO_4^{2-} \qquad (10)$$

it would be inappropriate to use an exponent of four for the concentration of available surface sites, [SOH], in the mass-law expression. We have solved this problem by using different coefficients for surface sites in the mass-law and mass-balance

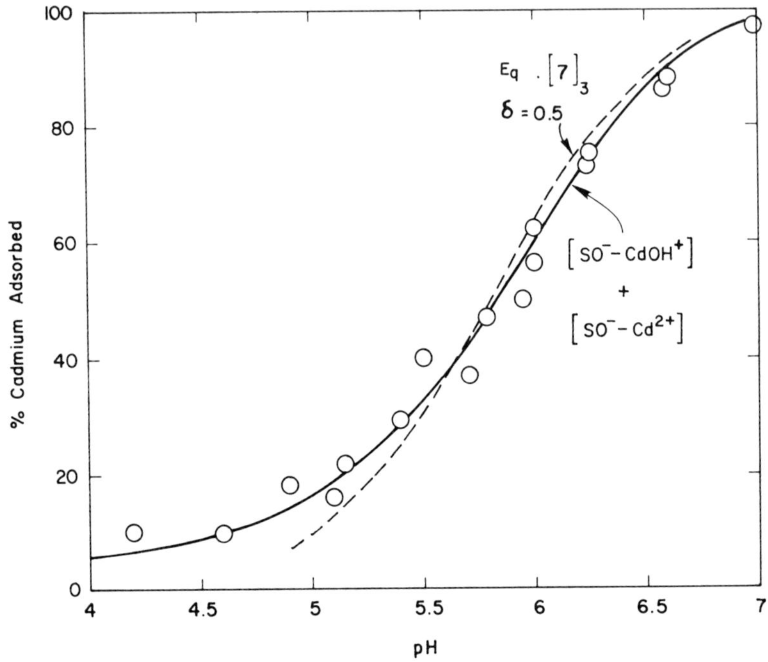

Figure 5. Experimental data (21, 22) and model calculations of Cd(II) adsorption on TiO_2 in 0.01M KNO_3. (○) $Cd(II)_T$—2×10^{-4}M. TiO_2—200m²/L. 0.01M KNO_3, 25°C. Stiglich (1976). (- - -) Model; (——) calculations.

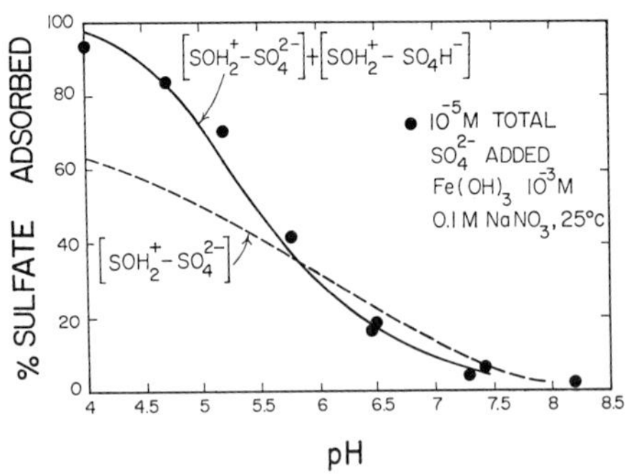

Figure 6. Experimental data and model calculations of sulfate adsorption on amorphous iron oxyhydroxide as a function of pH. (●) 10^{-5}M total SO_4^{2-} added, $Fe(OH)_3$ 10^{-3}M, 0.1M $NaNO_3$, 25°C.

equations. For example, for each Cr(VI) ion adsorbed, four surface sites are removed from the total available for other surface reactions. However, the mass-law equation is dependent on [SOH], rather than $[SOH]^4$. This is equivalent of assuming a tetradentate site. Since all equations are solved simultaneously by numerical techniques, this formulation does not present any difficult mathematical problems. This approach is very useful for modeling competition between adsorbing anions and adsorption of large organic compounds (24). Figure 7 shows experimental data and model calculations for adsorption of glutamic acid on amorphous iron oxyhydroxide. Model calculations are in reasonable agreement with the experimental data using a coefficient of eighteen for the surface site coverage of one adsorbed glutamate molecule. Calculations with a coefficient of one for surface site coverage yielded almost no dependence of percent adsorbed on total glutamate concentration. Of course, experimental estimates of surface site coverage are preferable for model calculations. The empirically derived number for surface site coverage in the example above should be viewed with caution. In particular, the microporous nature of $Fe(OH)_3$(am) causes a problem in interpretation since there may be interior surface sites unavailable for complexing large ions such as glutamate.

Adsorption of Metal-Ligand Complexes. In multicomponent electrolyte solutions containing several cations and anions, it has been shown that complex formation in solution can have a significant effect on adsorption of solutes. For example, MacNaughton and James (26) demonstrated that Hg(II) adsorption on α-quartz was greatly decreased by the formation of chloromercury(II) complexes in solution. However, Davis and Leckie (27) have shown that other metal-ligand complexes may be adsorbed, e.g. $AgS_2O_3^-$, Ag-ethylenediamine, Cu-glutamate. Thus, in a mixed electrolyte system, one may be required to consider surface equilibria of metal ions, anions, and metal-ligand complexes simultaneously.

Such complex systems can be modeled if the surface equilibria of each solute are well understood from separate experiments. A good example for illustration is the $Ag(I)/S_2O_3^{2-}/Fe(OH)_3$ system. First, the adsorption behavior of each solute alone was studied experimentally (27). Modeling results for Ag(I) and thiosulfate have been previously reported (2, 23). Thus, in the complex system, the only unknown surface equilibria are those of metal-ligand complexes, e.g.

$$SOH + H_s^+ + Ag_s^+ + S_2O_3^{2-} \underset{\longrightarrow}{\overset{1/{}^*K^{int}_{AgS_2O_3^-}}{\longleftarrow}} SOH_2^+{-}S_2O_3Ag^- \quad (11)$$

In Equation 11 we have assumed that the ligand portion of the complex, $AgS_2O_3^-$, binds to a positively charged surface site, since

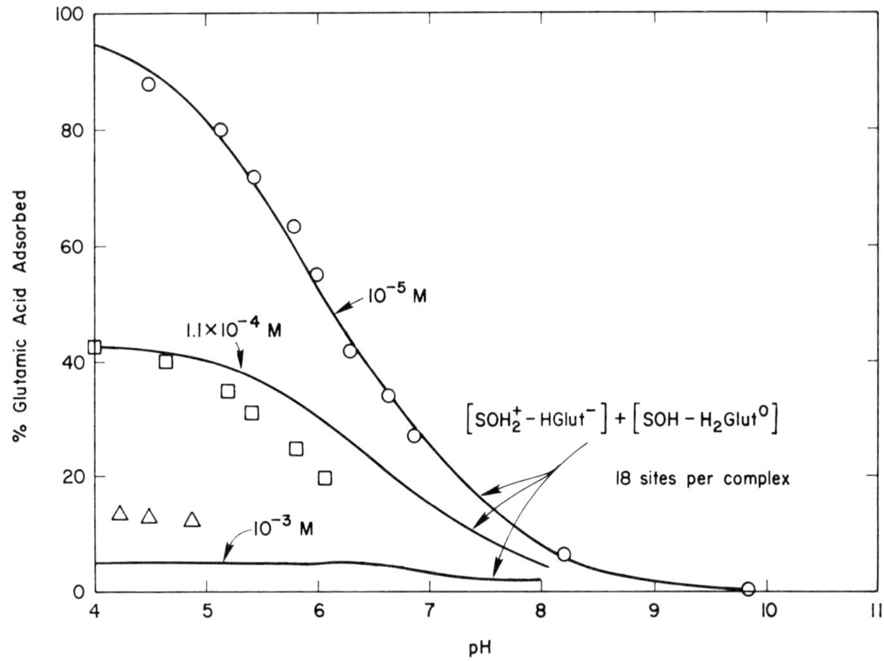

Figure 7. Experimental data and model calculations of glutamic acid adsorption on amorphous iron oxyhydroxide as a function of pH and total glutamate added. For model calculations a surface site coverage of 18 sites/adsorbed glutamate molecule was assumed. $Fe(OH)_3$(am), 10^{-3}M; 0.1M $NaNO_3$, 25°C. Glutamic (○) 10^{-5}M; acid (□) 1.1 × 10^{-4}M; added (△) 10^{-3}M. (——) Model calculation.

the pH dependence of $AgS_2O_3^-$ adsorption is very similar to $S_2O_3^{2-}$ adsorption (27). Other metal-ligand complexes, e.g. Cu-ethylenediamine, may attach to oxide surfaces via the metal atom (28).

By adding Equation 11 to the surface equilibria previously determined for Ag(I) and $S_2O_3^{2-}$ alone, one can obtain a reasonable description of Ag(I) adsorption in the complex system (Figure 8). Although the agreement between calculated and experimental adsorption density is not perfect, the degree of success is encouraging since so little is known about the formation of ternary surface complexes. The difference between $p^*K^{int}_{AgS_2O_3^-}$ (19.5) and $p^*K^{int}_{S_2O_3^{2-}}$ (10.0) is $10^{9.5}$, indicating that Ag(I) complexation by adsorbed thiosulfate ions is energetically similar to complexation of Ag^+ by thiosulfate in solution ($\sim 10^{8.9}$).

Speciation of Adsorbed Ions

The modeling results suggest that the physico-chemical state of the oxide/water interface has a significant effect on the speciation of ions at the surface. For example, adsorbed metal ions may be more easily hydrolyzed (i.e. at lower pH) than aquo metal ions in bulk solution. Consider the following surface equilibria of Cu(II) (2), i.e.,

$$SO^-{-}Cu^{2+} + H_s^+ \rightleftharpoons SOH + Cu_s^{2+}, \qquad -\log {}^*K^{int} = 4.1 \qquad (12)$$

$$SOH + Cu_s^{2+} + H_2O \rightleftharpoons SO^-{-}CuOH^+ + 2H_s^+, \qquad \log {}^*K^{int} = -9.0 \qquad (13)$$

$$SO^-{-}Cu^{2+} + H_2O \rightleftharpoons SO^-{-}CuOH^+ + H_s^+, \qquad \log K^{int} = -4.9 \qquad (14)$$

Hydrolysis of an adsorbed Cu(II) ion can be compared with the following reactions in homogeneous solution (29), i.e.,

$$Cu^{2+} + H_2O \rightleftharpoons CuOH^+ + H^+, \qquad \log K \leq -8 \qquad (15)$$

$$CuOH^+ + H_2O \rightleftharpoons Cu(OH)_2^o(aq) + H^+, \quad \log K \leq -5.7 \ . \qquad (16)$$

We cannot be certain whether Equation 15 or 16 should be compared with the acidity of the adsorbed Cu(II) ion, Equation 14. The surface bond will undoubtedly affect hydrolysis of an adsorbed Cu(II) ion, and perhaps release of the second proton from the hydration sheath of $Cu^{2+}(aq)$ may be more appropriate for comparison. Such comparisons are of limited value, however, since uncertainty in the available thermodynamic data for hydrolysis of metal ions is rather large (30). Nonetheless, with the present data it would appear that metal ions are more easily hydrolyzed within the compact layer of the EDL.

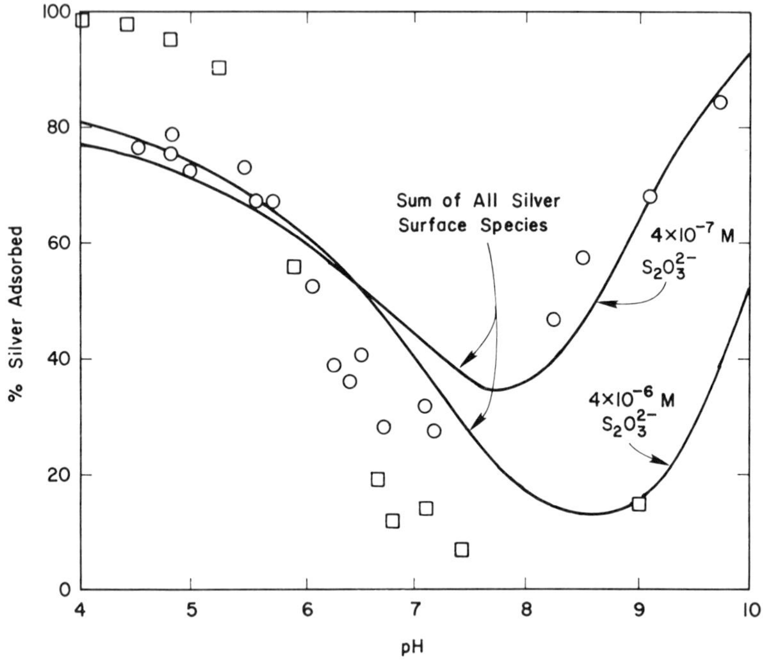

Figure 8. Experimental data and model calculations of Ag(I) adsorption on amorphous iron oxyhydroxide as a function of pH and thiosulfate added. Ag^{+}, $AgOH^{\circ}$, *and* $AgS_2O_3^{-}$ *are silver(I) adsorbing species.* (○) 4×10^{-7}M *and* (□) 4×10^{-6}M—$S_2O_3^{2-}$ *added.* Ag_T, 4×10^{-7}M; $Fe(OH)_3$, 10^{-3}M; 0.1M $NaNO_3$, 25°C. (——) *Model calculation.*

Differences in the speciation of other ions at the surface can be noted. Using an analysis similar to that above for metal ions, one finds that adsorbed anions are less acidic than in bulk solution. For example, it was shown in Figure 6 that protolysis of adsorbed sulfate ions becomes significant in the pH range 4-5, whereas in solution bisulfate is formed at much more acidic conditions (pH ~ 2). Complexes formed by supporting electrolyte, e.g. Na^+, with oxide surface sites have greater stability constants ($\log K^{int}$ ~ 0.5-1.7) than observed for complex formation with oxyanions in solution (log K ~ 0.0) (1).

It is clear from the foregoing that there are some differences in complex formation within the compact layer of the interface as compared to bulk water. An explanation of these phenomena requires a thorough understanding of the effect of solvent molecules on complex formation. When two oppositely charged ions associate to form a single species, a moderately large and unfavorable negative entropy change (ΔS) would be expected, reflecting the disappearance of a particle in the system. However, in most cases, formation of complexes in water results in a positive entropy change (31). The positive entropy change is caused by a compensation of charges, which increases the mobility of solvent molecules exposed to the electrostatic field surrounding the ions. The entropy change is accompanied by a change in enthalpy (ΔH), which may be positive or negative, and which is generally correlated with the A- or B-character of the interaction (32).

Fortunately, the association reactions of interest to our model studies have been thoroughly studied in aqueous solution. Perhaps the most important generalization that can be made about these associations is that the stability of the complex is mostly or entirely due to a large and positive entropy change. For example, protolysis of sulfate ion is a very endothermic process (ΔH = 5.6 Kcal/mol), but the reaction occurs ue to the large positive entropy change ($\Delta S = 26.7\ cal \cdot deg^{-1} mol^{-1}$) (33). Hydrolysis of transition metal ions, e.g. Fe^{3+}, Cr^{3+}, is slightly exothermic, but a positive entropy change contributes about 70 to 80 percent of the free energy of bond formation ($-\Delta G$) at 25°C (32). Similarly, the association reactions of alkali and halide ions are endothermic and are driven only by positive entropy changes. Thus, the most important factor governing the stability of these complexes is the release of water molecules from the ions concerned.

An oxide surface in water is covered with a layer of highly structured, chemisorbed water molecules (34, 35). When exposed to an electric field caused by charging of the surface, these water molecules approach dielectric saturation. As a consequence, the dielectric strength (ε) of the solvent within the compact layer of the interface is lowered. The dependency of log K (stability constant) at constant temperature varies linearly with the reciprocal of the dielectric constant (ε) of the media (36, 37), resulting in a displacement of the equilibrium towards the adduct in association reactions (32). As ε decreases, the stability of complexes formed primarily by positive entropy changes will increase. Thus, the stabilities of such complexes formed at the oxide/water interface should be increased, since a neutralization of charge will increase the mobility of solvent molecules exposed to the electrostatic field of the EDL.

From this perspective the conclusions of our modeling studies can be better understood. In all cases, the surface equilibria proposed favor neutralization of surface charge (σ_o). Surface complexes formed by supporting electrolyte ions decrease the "effective surface charge" ($\sigma_o - \sigma_\beta$) and thus increase the mobility of solvent molecules. The decreased dielectric strength of the solvent within the compact layer is consistent with the observed behavior for changes in hydrolysis and protolysis of adsorbed ions. Adsorbed metal ions are more easily hydrolyzed, because the entropy gained from the reaction is greater than in bulk solution. While we have not attempted to define this energy contribution explicitly as in other models (10), our conclusions regarding surface speciation are similar using the surface complexation approach (2). Formation of charged or relatively polar complexes is accompanied by a smaller entropy increase than if complete charge neutralization occurs (31). Thus, adsorption of hydrolyzed or protolyzed forms of ions should be favored over complexation by bidentate surface sites.

For metal ions there may be an additional entropy factor involved in surface complexation reactions. A metal ion, such as Pb^{2+} or Cu^{2+}, which is hexacoordinated in the aquo state but which normally forms four-coordinate complexes, may lose an additional two water molecules upon association with a ligand (31). Opposing this favorable entropy change is a simultaneous loss of configuration entropy of the ligand. However, for oxide surface sites this loss will be negligible. Thus, the entropy gained by liberation of these water molecules may be greater in surface reactions than in bulk solution.

Summary

A comprehensive adsorption model has been proposed which can describe the uptake of cations, anions, and metal-ligand adducts in complex electrolyte systems. An important advantage of the model is that surface charge and diffuse layer potential are also calculated and may be compared with experimental quantities, if available. Calculations reflect the net interactions of all ions at the surface, and the effect of a change in composition or concentration of the supporting electrolyte can be assessed. Intrinsic ionization and complexation constants for supporting electrolyte ions may be determined from potentiometric titrations at several electrolyte concentrations. Surface sites coverage can be empirically derived from experimental adsorption data in systems approaching maximum surface coverage, Γ_{max}.

Refinement of the model awaits further experimental work on the physico-chemical nature of surface bonding of ions within the compact layer of the EDL. At present our conclusions concerning the speciation of adsorbed ions are supported by 1) enthalpy/entropy arguments for analogous reactions in solution, and 2) a limited knowledge of the solvent medium of the compact layer of the EDL.

Abstract

The results of a newly proposed model for adsorption at the oxide/water interface are discussed. The modeling approach is similar to other surface complexation schemes, but mass-law equations are corrected for the effect of the electrostatic field. In this respect, this model bridges the gap between those models that emphasize physical interactions. The general applicability of the model is demonstrated with comparisons of calculations and experimental data for adsorption of metal ions, anions, and metal-ligand complexes. Intrinsic ionization and surface complexation constants can be determined with an improved double extrapolation technique.

By comparison with analogous reactions in solution, it is shown that the stabilities of complexes formed at oxide surfaces are governed primarily by large positive entropy changes. The most important factor in the change of entropy is the increased mobility of solvent molecules after complex formation. The entropy change of complexes within the compact layer of the EDL will be larger than analogous reactions in solution, due to the decreased dielectric strength of water. Thus, reactions which lead to a neutralization of charge, e.g. hydrolysis of metal ions, protolysis of anions, have greater stability constants at the surface than in bulk water.

Literature Cited

1. Davis, J. A., James, R. O., and Leckie, J. O., Surface ionization and complexation at the oxide/water interface: 1. Computation of electrical double layer properties in simple electrolytes. J. Coll. Interface Sci. 63, 480-499, (1978).
2. Davis, J. A., and Leckie, J. O., Surface ionization and complexation at the oxide/water interface: 2. Surface properties of amorphous iron oxyhydroxide and adsorption of metal ions. J. Coll. Interface Sci., in press (1978).
3. Davis, J. A., and Leckie, J. O., Surface ionization and complexation at the oxide/water interface: 3. Adsorption of anions. J. Coll. Interface Sci., accepted for publication (1978).

4. Westall, J. C., Zachary, J. L., and Morell, F. M. M., MINEQL, a computer program for the calculation of chemical equilibrium composition of aqueous systems. Tech. Note 18, Water Qual. Lab., Dept. of Civil Eng., MIT, Cambridge, MA, 1976.
5. Schindler, P. W., Fürst, B., Dick, R., and Wolf, P. U., Ligand properties of surface silanol groups. I. Surface complex formation with Fe^{3+}, Cu^{2+}, Cd^{2+}, and Pb^{2+}. J. Coll. Interface Sci. 55, 469-475 (1976).
6. Hohl, H., and Stumm, W., Interaction of Pb^{2+} with hydrous γ-Al_2O_3. J. Coll. Interface Sci. 55, 281-288 (1976).
7. Yates, D. E., Levine, S., and Healy, T. W., Site binding model of the electrical double layer at the oxide/water interface. J. Chem. Soc. Faraday Trans. I, 70, 1807-1818 (1974).
8. Bowden, J. W., Bolland, M. D. A., Posner, A. M., and Quirk, J. P., Generalized model for anion and cation adsorption at oxide sufaces. Nature, Phys. Sci. 245, 81-83 (1973).
9. Bowden, J. W., Posner, A. M., and Quirk, J. P., Ionic adsorption on variable charge mineral surfaces. Theoretical-charge development and titration curves. Austral. J. Soil Res. 15, 121-132 (1977).
10. James, R. O., and Healy, T. W., Adsorption of hydrolyzable metal ions at the oxide-water interface. III. A thermodynamic model of adsorption. J. Coll. Interface Sci. 40, 65-81 (1972)
11. Stumm, W., Huang, C. P., and Jenkins, S. R., Specific chemical interaction affecting the stability of dispersed systems. Croatia Chemica Acta 42, 223-245 (1970).
12. Schindler, P. W., and Kamber, H. R., Die Acidität von Silanolgruppen. Helv. Chim. Acta 51, 1781-1786 (1968).
13. Schindler, P. W., and Gamsjäger, H., Acid-base reactions of TiO_2 (anatase)-water interface and point of zero charge of TiO_2 suspensions. Kolloid Z. Z. Polym. 250, 759-763 (1972).
14. Smit, W., Holten, C. L. M., Stein, H. N., de Goeij, J. J. M., and Theelen, H. M. J., A radiotracer determination of the adsorption of sodium ion in the compact part of the double layer of vitreous silica. J. Coll. Interface Sci. 63, 120-128 (1978).
15. Smit, W., and Stein, H. N. J. Electroanal. Chem., (in press).
16. Lyklema, J., The structure of the electrical double layer on porous surfaces. J. Electroanal. Chem. 18, 341-348 (1968).
17. Wright, H. J. L., and Hunter, R. J., Adsorption at solid-liquid interfaces. I. Thermodynamics and the adsorption potential. Austral. J. Chem. 26, 1183-1189 (1973).
18. Perram, J. W., Hunter, R. J., and Wright, H. J. L., The oxide/solution interface. Austral. J. Chem. 27, 461-475 (1974).
19. James, R. O., Davis, J. A., and Leckie, J. O., Computer simulation of the conductometric and potentiometric titrations of the surface groups on ionizable latexes. J. Coll. Interface Sci. 65, 331-344 (1978).

20. Balistrieri, L., and Murray, J. W., The surface of goethite (α-FeOOH) in seawater, in Jenne, E. A., ed., "Chemical Modeling--Speciation, Sorption, Solubility, and Kinetics in Aqueous Systems," Amer. Chem. Soc., 1978 (this volume).
21. James, R. O., Stiglich, R. J., and Healy, T. W., Analysis of models of adsorption of metal ions at oxide/water interfaces. Farad. Disc. Chem. Soc. 59, 142-156 (1975).
22. Stiglich, P. J., "Adsorption of Cadmium(II) Complexes at the Oxide/Water Interface," 196 pp., M.S. Thesis, Univ. Melbourne, Australia, 1976.
23. Jørgensen, C. K., "Absorption spectra and chemical bonding in complexes," 352 pp., Addison-Wesley, Reading, MA, 1962.
24. Davis, J. A., "Adsorption of trace metals and complexing ligands at the oxide/water interface," 286 pp., Ph.D. Thesis, Stanford Univ., Stanford, CA, 1977.
25. Yates, D. E., and Healy, T. W., The structure of the silica/electrolyte interface. J. Coll. Interface Sci. 55, 9-19 (1976).
26. MacNaughton, M. G., and James, R. O., Adsorption of aqueous Hg(II) complexes at the oxide/water interface. J. Coll. Interface Sci. 47, 431-440 (1974).
27. Davis, J. A., and Leckie, J. O., Effect of adsorbed complexing ligands on trace metal uptake by hydrous oxides. Environ. Sci. Tech., in press (1978).
28. Bourg, A. C. M., and Schindler, P. W., Ternary surface complexes: 1. Complex formation in the system silica-Cu(II)-ethylenediamine. Chimia 32, 166-168 (1978).
29. Vuceta, J., and Morgan, J. J., Hydrolysis of Cu(II). Limnol. Oceanogr. 22, 742-746 (1977).
30. Baes, C. F., Jr., and Mesmer, R. E., "The Hydrolysis of Cations," 489 pp., Wiley-Interscience Publications, N.Y., 1976.
31. Nancollas, G. H., "Interactions in Electrolyte Solutions," 214 pp., Elsevier Publishing Company, N.Y., 1966.
32. Schwarzenbach, G., Electrostatic and non-electrostatic contributions to ion association in solution. Pure Appl. Chem. 24, 307-334 (1970).
33. Nair, V. S. K., and Nancollas, G. H., Thermodynamics of ion association. Part V. Dissociation of the bisulfate ion. J. Chem. Soc., 4144-4147 (1958).
34. Bérubé, Y. G., and de Bruyn, P. L., Adsorption at the rutile-solution interface. II. Model of the electrochemical double layer. J. Coll. Interface Sci. 28, 92-105 (1968).
35. McCafferty, E., and Zettlemoyer, A. C., Adsorption of water vapor on α-Fe_2O_3. Disc. Faraday Soc. 52, 239-254 (1971).
36. Denison, J. T., and Ramsey, J. B., The free energy, enthalpy, and entropy of dissociation of some perchlorates in ethylene chloride and ethylidene chloride. J. Am. Chem. Soc. 77, 2615-2621 (1955).
37. Fuoss, R. M., Ionic association. I. Derivation of constants from conductance data. J. Am. Chem. Soc. 79, 3301-3303 (1957).

RECEIVED November 16, 1978.

Evaluation and Estimation of Thermodynamic and Kinetic Data

Critical Review of Plutonium Equilibria of Environmental Concern

JESS M. CLEVELAND
U.S. Geological Survey, Water Resources Division, Lakewood, CO 80225

The complex interrelationships of three types of chemical equilibria, namely oxidation-reduction, hydrolysis, and complexation, as well as polymerization, a nonequilibrium process, determine the nature and speciation of plutonium in aqueous environmental systems. This paper presents a selective, critical review of the literature describing these processes. Although most research has been conducted under non-environmental conditions--that is, macro concentrations of plutonium and high acidities--the results in some cases are applicable to environmental conditions. In other cases the behavior is different, however, and care should always be exercised in extrapolating macro data to environmental conditions.

Oxidation-Reduction Potentials

The oxidation-reduction behavior of plutonium is described by the redox potentials shown in Table I. (For the purposes of this paper, the unstable and environmentally unimportant heptavalent oxidation state will be ignored.) These values are of a high degree of accuracy, but are valid only for the media in which they are measured. In more strongly complexing media, the potentials will change. In weakly complexing media such as 1 M $HClO_4$, all of the couples have potentials very nearly the same; as a result, ionic plutonium in such solutions tends to disproportionate. Plutonium is unique in its ability to exist in all four oxidation states simultaneously in the same solution. Its behavior is in contrast to that of uranium, which is commonly present in aqueous media as the uranyl(VI) ion, and the transplutonium actinide elements, which normally occur in solution as trivalent ions.

Hydrolysis

It is important to emphasize the often-overlooked fact that reactions between a metal ion and water molecules (hydration) or

0-8412-0479-9/79/47-093-321$05.00/0
This chapter not subject to U.S. copyright
Published 1979 American Chemical Society

Table I Plutonium Formal Potentials (in Volts) at 25° (1)

In 1 M HClO:

$$Pu^{0} \xrightarrow{2.03} Pu^{3+} \xrightarrow{-0.9819} Pu^{4+} \xrightarrow{-1.1702} PuO_2^{+} \xrightarrow{-0.9164} PuO_2^{2+}$$

$$Pu^{4+} \xrightarrow{-1.0433} PuO_2^{2+}$$

$$Pu^{3+} \xrightarrow{-1.0228} PuO_2^{2+}$$

In neutral solution (pH = 7):

$$Pu^{3+} \xrightarrow{0.63} Pu(OH)_4 \cdot yH_2O(s) \xrightarrow{-1.11} PuO_2^{+} \xrightarrow{-0.77} PuO_2(OH)_2(aq)$$

$$Pu(OH)_4 \cdot yH_2O(s) \xrightarrow{-0.94} PuO_2(OH)_2(aq)$$

In 1 M OH^-:

$$Pu(OH)_3 \cdot xH_2O \xrightarrow{0.95} Pu(OH)_4 \cdot yH_2O \xrightarrow{-0.76} PuO_2(OH)(aq) \xrightarrow{-0.26} PuO_2(OH)_3^-(aq)$$

$$Pu(OH)_4 \cdot yH_2O \xrightarrow{\sim -0.4} PuO_2(OH)_3^-(aq)$$

hydroxide ions (hydrolysis) are actually complexation reactions that do not differ in kind from those with acid anions and organic ligands. Hydrolysis will be considered as a separate process, however, because it can lead to precipitation and/or formation of polymeric species. Each plutonium oxidation state hydrolyzes by the successive addition of hydroxide ions as the pH is increased, the final product in each case being an insoluble hydroxide precipitate. The various oxidation states decrease in their tendency to hydrolyze in the order

$$Pu^{4+} >> PuO_2^{2+} > Pu^{3+} > PuO_2^{+} .$$

The first hydrolysis constant for plutonium(III), that is, the equilibrium constant for the reaction

$$Pu^{3+} + H_2O \rightleftarrows Pu(OH)^{2+} + H^{+},$$

has been determined from acid-base titration curves at an ionic strength (I) of 0.069 M to be 7.5×10^{-8} (2). The solubility product of the neutral hydroxide precipitate, $Pu(OH)_3$, is reported to be 2×10^{-20} (3, p. 299).

Hydrolysis of plutonyl(VI) has been investigated extensively, but not all the data are relevant to environmental concentrations. Potentiometric data at I = 1 M and 25° have been interpreted to indicate the formation of three species, $PuO_2(OH)^{+}$, $(PuO_2)_2(OH)_2^{2+}$, and $(PuO_2)_3(OH)_5$, with respective overall hydrolysis constants of 1.1×10^{-6}, 3.1×10^{-9}, and 6.9×10^{-23} (4). Another potentiometric study (5), at I = 3 M and 25°, yielded the following overall hydrolysis constants:

$$PuO_2(OH)^{+},\ K = 5.0 \times 10^{-7};\ (PuO_2)_2(OH)_2^{2+},\ K = 5.74 \times 10^{-9};$$

$$(PuO_2)_3(OH)_5^{+},\ K = 2.0 \times 10^{-22};\ (PuO_2)_4(OH)_7^{+},\ K = 7.68 \times 10^{-30}.$$

The polynuclear species would not be expected at environmental plutonium concentrations, at which the most common hydrolysis product is $PuO_2(OH)^{+}$. The cited hydrolysis constants for this species are in satisfactory agreement.

Least hydrolyzed of the four oxidation states is plutonyl(V), which has a reported first hydrolysis constant of 2×10^{-10} at I = 3×10^{-3} M and 25° (2, pp. 478-499).

Plutonium(IV) is the most readily hydrolyzed of the four oxidation states, but only the first hydrolysis constant is known with any confidence. For the reaction

$$Pu^{4+} + H_2O \rightleftarrows Pu(OH)^{3+} + H^{+}$$

the best value for the first hydrolysis constant, K_1, is 0.031, as determined potentiometrically at I = 1 M and 25° (6). Stepwise values for K_1 to K_4 have been calculated from TTA extraction data at I = 1 M to be 0.35, 0.18, 5.0 X 10^{-4}, and 5.0 X 10^{-7}, respectively (7), but results are questionable. Successive hydrolysis constants suffer from such an accumulation of errors that they become virtually meaningless. In particular, the value for K_4, if it refers to the equilibrium between $Pu(OH)_3^+$ and solid $Pu(OH)_4$, yields values for the equilibrium concentration of plutonium(IV) in near-neutral solutions that are too high by many orders of magnitude. On the assumption of a regular progression of the stabilities of successive hydrolysis products, overall values for K_1 to K_4 (that is, for the formation of $Pu(OH)_n$ from $Pu(OH)_{n-1}$) were estimated to be 0.32, 5.0 X 10^{-3}, 5.0 X 10^{-6}, and 3.2 X 10^{-10} (8), but these results cannot be taken seriously in view of the unverified assumptions made in their calculation.

The reported solubility product of $Pu(OH)_4$, 7 X 10^{-56} as measured by pH titration (3, p. 300), is an exceedingly small number. If it were a true representation of the concentration of plutonium in solution, at pH 7 there would be only 7 X 10^{-28} mole of plutonium per liter; thus the equilibrium concentration of plutonium in neutral water would be about one atom per 2400 liter and there would be no problem with plutonium-contaminated ground water. The solubility product does not accurately define the concentration of plutonium in aqueous solutions because it merely states the concentration of the Pu^{4+} ion. At pH values of environmental interest, plutonium will be present not primarily as Pu^{4+}, but as species such as $Pu(OH)_2^{2+}$, $Pu(OH)_3^+$, unionized $Pu(OH)_4$, colloidal polymeric forms to be discussed later, as well as other oxidation states formed by disproportionation at low acidities. Thus, the total plutonium concentration will be much higher than that described by the solubility product of $Pu(OH)_4$.

Complexes

A brief review of methodology is in order before embarking on a discussion of plutonium complexes. Determination of stability constants of metal complexes depends on the measurement of the effect produced by the particular ligand on an equilibrium involving the uncomplexed metal ion. The most common methods are ion exchange, solvent extraction, spectrophotometry, solubility, potentiometry, and polarography, of which only the last two give values based on activities rather than concentrations. For this and other reasons, these two methods generally yield more thermodynamically significant constants. Unfortunately, they have been rather infrequently used in studying plutonium complexes. The least reliable method, solubility, suffers from the difficulty of achieving equilibrium solubilities and from problems in properly interpreting the data.

In cases where the stability constants of plutonium complexes are not known, it is possible to obtain a rough estimate by comparison with analogous metal ions (that is, La^{3+} for Pu^{3+}, Th^{4+} for Pu^{4+}, NpO_2^+ for PuO_2^+, UO_2^{2+} for PuO_2^{2+}). With one exception, this approach is not applied in this paper because of space limitations. Moreover, attempting to extrapolate stability constants from one element to another is a shortcut that can easily be abused. The analogous ions differ significantly from the respective plutonium ions in such properties as ionic potential and oxidation-reduction behavior, which can result in differences in complexation.

Two types of stability constants should be defined. For the reaction of one or more ligands, L, with a metal ion, M, the <u>stepwise</u> stability constant is expressed by the relation

$$K_n = \frac{[ML_n]}{[ML_{n-1}][L]},$$

whereas the <u>overall</u> stability constant is defined as

$$\beta_n = \frac{[ML_n]}{[M][L]^n}.$$

It follows that $K_n = \frac{\beta_n}{\beta_{n-1}}$. Moreover, stability constant equations involving protonated species, such as undissociated or partially dissociated acid, may be converted to the above expressions by application of the appropriate acid dissociation constant.

Standard deviations are not given, since they merely establish the reproducibility of a given method and do not necessarily reflect the accuracy of the data.

<u>Carbonate Complexes</u>. Of the many ligands which are known to complex plutonium, only those of primary environmental concern, that is, carbonate, sulfate, fluoride, chloride, nitrate, phosphate, citrate, tributyl phosphate (TBP), and ethylenediaminetetraacetic acid (EDTA), will be discussed. Of these, none is more important in natural systems than carbonate, but data on its reactions with plutonium are meager, primarily because of competitive hydrolysis at the low acidities that must be used. No stability constants have been published on the carbonate complexes of plutonium(III) and plutonyl(V), and the data for the plutonium(IV) species are not credible. Results from studies on the solubility of plutonium(IV) oxalate in K_2CO_3 solutions of various concentrations have been interpreted (<u>9</u>) to indicate the existence of complexes as high as $Pu(CO_3)_8^{12-}$, a species that is most unlikely from both electrostatic and steric considerations. From the influence of K_2CO_3 concentration on the solubility of $Pu(OH)_4$ at an ionic strength of 10 <u>M</u>, the stability constant of the complex $Pu(CO_3)^{2+}$ was calculated (<u>10</u>) to be 9.1 X 10^{46} at 20°. This value

is based on poor experimental methodology and incorrect interpretation of the data, and is best ignored. From comparisons with other plutonium complexes and with carbonate complexes of analogous metal ions, it appears to be high by about 30 orders of magnitude.

In the case of plutonyl(VI), somewhat more plausible results have been obtained. Data on the solubility of plutonyl(VI) hydroxide in $(NH_4)_2CO_3$ solutions of varying concentrations have been interpreted (11) to reveal the presence of three complexes, $PuO_2(CO_3)_2^{2-}$, $PuO_2(CO_3)(OH)^-$, and $PuO_2(CO_3)(OH)_2^{2-}$, with stability constants of 6.7 X 10^{13}, 4.5 X 10^{22}, and 2.3 X 10^{22}, at I = 1 M and 20°. The value for the dicarbonato species is reasonably consistent with published constants for the analogous uranyl(VI) species, but the constants for the hydroxy species need further verification.

Sulfate Complexes. Knowledge of the sulfate complexes of plutonium(III) and (IV) has advanced somewhat in recent years. Ion exchange data indicated the presence of two complexes of the trivalent ion, $Pu(SO_4)^+$ and $Pu(SO_4)_2^-$, with respective stepwise stability constants of 44.7 and 43.7 at I = 2 M and 25° (12). It is unlikely K_1 and K_2 would have such similar values, and hence K_2 must be regarded with suspicion. Because these values were the same in both 1 M and 2 M $HClO_4$, it was concluded that no protonated species were formed. This value of K_1 takes precedence over earlier data (13) indicating the existence of two complexes, $Pu(SO_4)^+$, with K_1 = 18.1 and $Pu(HSO_4)_2^+$, with K_1 = 9.9, at I = 1 M and 28°. In the latter case all determinations were at a constant acidity of 1 M, and therefore the presence of the protonated complex $Pu(HSO_4)_2^+$ was not demonstrated by studies at varying acidity.

Numerous determinations have been made of the stability constants of sulfate complexes of plutonium(IV), and the results vary by three orders of magnitude. The most plausible values (14) have come from careful ion exchange studies at I = 2 M and 25°, which indicate the presence of two species, $Pu(SO_4)^{2+}$ and $Pu(SO_4)_2$, with stepwise stability constants of 6.6 X 10^3 and 5.8 X 10^2, respectively. This K_1 is in satisfactory agreement with the value 4.6 X 10^3 obtained potentiometrically at I = 1 M and 25° (6). Spectrophotometric and electrophoretic studies indicate that plutonyl(VI) forms complexes containing as many as four sulfate groups, with anionic species predominating at sulfate concentrations above 1 M (15). The monosulfato complex has a reported stability constant of 14.4 as determined by extraction studies at I = 2 M and 25° (16).

Fluoride Complexes. Fluoride is known to complex plutonium strongly, but quantitative data on these environmentally important complexes are limited. Cation exchange studies (17) yielded values of 4.5 X 10^7 at I = 1 M and 7.9 X 10^7 at I = 2 M for the stability constant of the monofluoro complex of plutonium(IV), which are in satisfactory agreement with the value 1.2 X 10^8 obtained from extraction data at I = 2 M and 25° (18). The latter study also gave 3.4 X 10^{14} as the overall stability constant, β_2, of PuF_2^{2+} under the same conditions. (The values cited in references 19 and 20 have been corrected for the ionization constant of HF to convert them to stability constants as defined in this discussion.) Four fluoro complexes of plutonyl(VI) have been identified from cation exchange studies (19) at I = 1 M, with overall stability constants as follows: PuO_2F^+, β_1 = 130; PuO_2F_2, β_2 = 1.4 X 10^4; $PuO_2F_3^-$, β_3 = 1.2 X 10^6; $PuO_2F_4^{2-}$, β_4 = 2.0 X 10^6. These results are somewhat clouded by the authors' failure to demonstrate the absence of plutonium(IV), a potential source of error in this system. A lower value, β_1 = 12, has been reported from extraction data at I = 2 M and 25° (16).

Chloride Complexes. Because of marine disposal or storage of actinide wastes in salt formations, the relatively weak chloro complexes of plutonium could become important over extended time periods. Two plutonium(III) complexes, $PuCl^{2+}$ and $PuCl_2^+$, have been reported (20), the former occurring at chloride concentrations above 2 M and the latter in the presence of 8 M or greater chloride ion concentration. No stability data were given. Plutonium(IV) forms complexes containing from one to six chloride ligands. Quantitative values exist for the three lower complexes in 4 M $HClO_4$ at 20° as determined by cation exchange: $PuCl^{3+}$, K_1 = 1.4; $PuCl_2^{2+}$, K_2 = 1.2; $PuCl_3^+$, K_3 = 0.1 (21). Other values for K_1 agree satisfactorily with the above: 1.38 by potentiometry (22), and 1.42 by an extraction technique at I = 2 M and 25° (23); thus the value for K_1 seems well established. Values for K_2 show more variation: 0.49 by potentiometry (22) and 0.16 by extraction (23). Agreement between the ion exchange and potentiometric values is sufficiently good to suggest that K_2 is somewhere in the range 0.5 to 1.2.

Potentiometric studies (3, pp. 312-13) at unspecified temperature and ionic strength have indicated K_1 for the plutonyl(V) complex PuO_2Cl to be 0.67, a value that should be accepted with caution in view of the experimental vagueness. Plutonyl(VI) forms complexes containing up to three or four chloride ions, but quantitative data have been reported for only the mono and dichloro species. The following values have been found, all by spectrophotometry, for the stability constant of PuO_2Cl^+: 1.25 at I = 2 M and 25° (24); 0.73 at I = 1 M and 23° (3, pp. 312-13); 0.56 at I = 1 M and 20° (25). Again, agreement is reasonably good, although it is regrettable that all values were obtained by the same technique. The single reported stability constant, K_2, for PuO_2Cl_2, also by spectrophotometry, is 0.35 at I = 2 M and 25° (24).

Nitrate Complexes. Although there is spectrophotometric evidence for the existence of nitrate complexes of plutonium(III), they are unstable because of oxidation to plutonium(IV). The latter forms complexes containing as many as six nitrate groups, and they are important in the chemical processing of plutonium by ion exchange and solvent extraction. Values for K_1 are in good agreement: 5.5 by cation exchange at I = 4 M and 20° (21); 5.3 by extraction at I = 6 M (26); 4.4 by extraction at I = 2 M and 25° (23). Strong disagreement exists among the values for K_2: 23.5 by ion exchange at I = 4 M and 20° (21); 3.0 by extraction at I = 2 M and 25° (23); 0.96 at I = 6 M, also by extraction (26). The ion exchange value is probably the least reliable because of compounding of errors; on this basis the most plausible K_2 is in the range 1 to 3. Values for K_3 become even less reliable; of the two reported, 15 by ion exchange at I = 4 M and 20° (21) and 0.33 by extraction at I = 6 M (26), neither can be accepted with assurance. Mono-, di-, and trinitrato complexes of plutonyl(VI) have been identified, but they are weak. Of the two plausible values reported for K_1, that is, 0.93 by extraction at I = 4.1 M (27) and 0.25, also by extraction, at I = 4.6 M and 25° (28), the latter is more reliable because of more careful experimental controls and the more valid assumptions made in its calculation.

Phosphate Complexes. Of all the systems described in this paper, the most difficult to evaluate is that of the phosphate complexes of plutonium. Many of the stability constants were calculated from solubility measurements, which often yield data of dubious accuracy. The values that have been obtained are reported in Table II, and many appear to be too high, a common error in solubility determinations. The ion exchange values for the acid phosphate complexes of plutonium(III) are the most plausible. The data for plutonium(IV) and plutonyl(VI) should be used with caution.

Citrate Complexes. Because of their ability to form sterically favored chelate structures, many organic ligands complex plutonium more strongly than inorganic anions. One of the strongest naturally occurring chelating agents is citrate, and it forms a number of very stable complexes with plutonium. Three plutonium(III) complexes have been reported (32), $Pu(C_6H_5O_7)$, $Pu(H_2C_6H_5O_7)_2^+$, and $Pu(H_2C_6H_5O_7)_3$, with respective stability constants of 7.3×10^8, 4.0×10^6, and 1.0×10^{10}, but these values are suspect because they differ by several orders of magnitude from constants for citrate complexes of other actinides and lanthanides. The apparent error is primarily a result of failure to allow for plutonium(III) hydrolysis and complexing by sulfite formed from the sodium formaldehyde sulfoxylate used as a reducing agent, and by improper interpretation of the data. More reliable data are available for the plutonium(IV) complexes, which are present at citrate concentrations as low as 10^{-15} M (33). Stability constants of the unprotonated species $Pu(C_6H_5O_7)^+$

Table II Phosphate Complexes of Plutonium

Complex	Stability Constant	Method	Reference
Pu (III)			
$Pu(PO_4)$	$K_1 = 1.5 \times 10^{19}$ at I = 0.5 M and 20°	Solubility	(29)
$Pu(H_2PO_4)^{2+}$	$K_1 = 30.2$ at I = 1 M and 20°	Cation exchange	(29)
$Pu(H_2PO_4)_2^+$	$K_2 = 5.25$ at I = 1 M and 20°	Cation exchange	(29)
$Pu(H_2PO_4)_3$	$K_3 = 5.01$ at I = 1 M and 20°	Cation exchange	(29)
$Pu(H_2PO_4)_4^-$	$K_4 = 3.98$ at I = 1 M and 20°	Cation exchange	(29)
Pu(IV)			
$Pu(HPO_4)^{2+}$	$K_1 = 8.3 \times 10^{12}$	Solubility	(30)
$Pu(HPO_4)_2$	$K_2 = 6.7 \times 10^{10}$	Solubility	(30)
$Pu(HPO_4)_3^{2-}$	$K_3 = 4.8 \times 10^9$	Solubility	(30)
$Pu(HPO_4)_4^{4-}$	$K_4 = 6.3 \times 10^9$	Solubility	(30)
$Pu(HPO_4)_5^{6-}$	$K_5 = 6.3 \times 10^8$	Solubility	(30)
Pu(VI)			
$PuO_2(HPO_4)$	$K_1 = 1.5 \times 10$ at I = 0	Unspecified; presumably solubility	(31)
$PuO_2(H_2PO_3)^+$	$K_1 = 200$ at I = 0	Solubility	(31)

and $Pu(C_6H_5O_7)_2^{2-}$ have been determined by three separate methods with generally good agreement, as shown in Table III.

Table III Stability Constants of Citrate Complexes of Plutonium(IV) at I = 0.5 M and 25°

Method	β_1	β_2	Reference
Spectrophotometric	5.3×10^{14}	1.6×10^{30}	(33)
Potentiometric	3.5×10^{15}	1.0×10^{30}	(34)
pH titration	6.9×10^{15}	1.0×10^{29}	(34)

TBP Complexes. Tributyl phosphate (TBP) does not normally occur in the environment, but its use in nuclear fuel reprocessing makes it a probable constituent of radioactive wastes and hence a possible ground water contaminant. TBP reacts with nitrate complexes of plutonium(IV) and (VI) to form the species $Pu(NO_3)_4 \cdot 2TBP$ and $PuO_2(NO_3)_2 \cdot 2TBP$, which although not exceptionally stable, are useful in separations chemistry because of their solubility in organic solvents and insolubility in aqueous media. By contrast, the plutonium(III) complex, $Pu(NO_3)_3 \cdot 3TBP$, is poorly extracted by organic solvents. The stability constant of the neptunium(IV) complex, defined as

$$K = \frac{[Np(NO_3)_4 \cdot 2TBP]}{[Np][NO_3]^4[TBP]^2},$$

is reported to be 130 at I = 2 M (35), and the value for the plutonium(IV) complex should be similar. Although these complexes and their solvent are water-insoluble, they could exist in ground water in an emulsified form.

EDTA Complexes. Ethylenediaminetetraacetic acid (EDTA) and its homologues form the most stable known complexes of plutonium. This discussion will be limited to EDTA, which is most likely to be found in the environment as a result of its use as a medium for the addition of soluble iron to soils. The equilibrium constant for formation of the 1:1 chelate of plutonium(III), as given by the expression

$$K = \frac{[PuY^-][H^+]^2}{[Pu^{3+}][H_2Y^{2-}]}$$

(where H_4Y would represent the unionized EDTA molecule), was found in 0.1 N KCl solution to be 3.9×10^{18} at pH 1.5 by spectrophotometry (36) and 1.32×10^{18} at pH 3.3 and 20° by ion exchange (37). Another ion exchange study, at I = 1 M and pH values from 1.2 to 3.4, yielded a stability constant for PuY^- of 2.28 X 10^{17} and indicated the presence at pH 1.5 to 2.0 of another species, PuHY, with a stability constant of 1.61 X 10^9 (38). A stability constant of 1.6 X 10^{12} for PuY^- formation from Pu^{3+} and Y^{4-} at I = 0.1 M and 20° has been calculated from spectrophotometric and polarographic data (39).

Determination of the stability constants for EDTA chelates of plutonium(IV) is rendered difficult by hydrolysis of the metal ion at fairly low pH values and by protonation of the EDTA molecule in more strongly acid solution. Thus attempts to determine the equilibrium constant in the expression

$$K = \frac{[PuY][H^+]^2}{[Pu^{4+}][H_2Y^{2-}]}$$

at pH 3.3 in 0.1 N KCl solution gave values of 4.57 X 10^{17} by ion exchange (37) and 1.26 X 10^{17} by spectrophotometry (36), but these numbers describe the complexing of a partially hydrolyzed plutonium species, rather than the Pu^{4+} ion. In 1 M HNO_3 a spectrophotometrically determined value of 1.59 X 10^{24} was reported (36), but this, too, is suspect because it fails to allow for the apparent protonation of EDTA(H_4Y) to form species such as H_5Y^+ and H_6Y^{2+} (40). The dissociation constants of these species were determined and combined with other ionization constants of EDTA to calculate a more accurate value for the concentration of Y^{4-} ions in 1 M HNO_3 and from this the stability constant of PuY at I = 1 M and 25° was calculated to be 5.7 X 10^{25} (41). A more recent value of 4.0 X 10^{25} at I = 0.1 M determined by spectrophotometry and polarography (39) is in excellent agreement.

Even plutonyl(V) is strongly complexed by EDTA. The stability constant of PuO_2Y^{3-} has been determined by three methods: 7.7 X 10^{12} by spectrophotometry at 20° (42); 8 X 10^{12} by potentiometry in 0.1 M KCl at room temperature (43); 1.5 X 10^{10} by ion exchange at I = 0.05 M (44). The first two values, by virtue of their close agreement, take precedence. The stability constant of the plutonyl(VI) chelate, PuO_2Y^{2-}, is somewhat higher than that for plutonyl(V). Two different spectrophotometric studies yielded values of 1.07 X 10^{16} at pH 4.0 (36) and 4.0 X 10^{14} (42), both at 20°, whereas an ion exchange-derived value of 2.46 X 10^{16} at pH 3.3 and 20° has been reported (37). Although these values indicate a high degree of stability for the complex, the metal ion is unstable toward reduction (36, 42). EDTA reduces plutonyl(VI) to plutonyl(V), or plutonium(IV) if an excess of ligand is present (45). However, because all oxidation states are strongly complexed by EDTA, the reduction does not result in the release of uncomplexed plutonium ions.

Polymerization

The equilibria discussed above are useful in characterizing plutonium and predicting its behavior in aqueous systems, but only if it is in true solution. This is frequently the case for plutonyl(V) and (VI), but much less common with plutonium(III) and (IV). In particular, plutonium(IV) is very susceptible to hydrolysis and polymerization at pH values above 1 (46). In 0.009 M plutonium(IV) solutions at 25°, polymerization increased from essentially zero in 0.1 M HNO_3 (pH 1) to 98 percent in 0.04 M HNO_3 (pH 1.4). Although polymerization would not occur at such low pH values with the concentrations of plutonium found in natural waters, it would be expected at the higher pH values most common in the environment. Whether the resulting polymer remains dispersed in the aqueous medium or precipitates depends on a number of factors, including its molecular weight, pH, temperature, and the type and concentration of anions in solution; in general, the presence of polymer will result in a higher concentration of plutonium in the aqueous phase than would otherwise be the case. It is not in true solution, however, but is present in colloidal form. Values for its molecular weight range from 4000 to 10^{10} (47), depending on conditions of formation. Formation of the polymer is relatively rapid, generally occurring within a matter of minutes (47); depolymerization, on the other hand, requires hours to days, depending on acidity and temperature (48), and is even slower for aged polymers (47). Because of this, polymer formation is often considered to be irreversible. The presence of a large molar excess of citrate retards polymerization and enhances depolymerization (49), doubtless because of complex formation. The sizes of the colloidal particles vary, but generally increase with aging, increased ionic strength, and the presence of bicarbonate at environmental concentrations (50).

The nature of the polymer has not been firmly established. It has generally been considered an intermediate hydrolysis product in which partially hydrolyzed plutonium species are linked by hydroxide or oxide bridges into long chains. In this view, the effect of aging is to increase the size of the polymer units. A more recent conclusion (51), supported by spectrophotometric, x-ray diffraction, electron diffraction, and electron microscopy data, is that the polymer consists of small, discrete primary particles (size 0.5 to 2.0 nm), which may be amorphous or crystalline, and secondary particles which are aggregates of the primary particles. Colloidal sols would thus consist of dispersions of the primary particles in aqueous media. In this concept the slower depolymerization of aged polymer is the result of the conversion of amorphous particles to crystalline particles on standing. X-ray diffraction patterns of the crystalline particles correspond to that for PuO_2. A similar pattern is found for precipitated $Pu(OH)_4$ (47), leading to the suggestion that the primary particles, which are presumably hydrated PuO_2, are also the first particles to form during precipitation (51).

Dispersed polymer can be adsorbed from aqueous media onto surfaces with which it comes in contact. Among solids whose surfaces can adsorb plutonium are numerous minerals, glass, and stainless steel. Glass reportedly can adsorb 1.6 μg colloid/cm^2 of surface, and the adsorption by steel is much higher (47). Adsorption is highly variable, depending on colloid concentration, pH, age of polymer, nature of presence of anions and complexing ligands, and nature of the surface. It is important to bear in mind that this adsorption is strictly a surface phenomenon and is generally irreversible; the rate of desorption is dependent on depolymerization and solubility considerations. Because of the nature of the process and its irreversibility, it cannot be described by equilibrium ion exchange parameters, such as distribution coefficients [K_d]. The two processes are completely unrelated; interestingly enough, the polymer is reportedly not adsorbed by ion exchange resins (46). Accordingly, it is important for researchers to establish the absence of polymeric species in a given solution before using it to determine equilibrium distribution values.

Areas Needing Further Research

With the exception of a few areas such as polymerization, the aqueous chemistry of plutonium has received much study. However, most of the current knowledge applies to macro concentrations of the element in strong acid solutions, and its application to the picogram levels and neutral pH values of environmental systems requires utmost caution. Since hydrolysis is relatively more important, it is possible that the complex species formed are not simple metal-ligand moieties, but rather involve hydroxylated metal ions in complex species not previously identified. The present state of knowledge of plutonium chemistry under environmental conditions does not appear adequate to permit chemical modeling with any degree of confidence. To reach this desired objective, additional data, determined under simulated environmental conditions--not extrapolated from grossly different conditions--are needed.

Several areas of research seem to merit top priority:

1. attempt to verify published stability constants of environmental interest at lower metal concentrations and higher pH;
2. determine stability constants that are not currently known, the prime example being the plutonium-carbonate system;
3. assess the interplay of complexation, hydrolysis, and polymerization at environmental pH values, as these factors are important but not well understood under neutral conditions;
4. study the complex chemistry of plutonyl(V), which some workers believe to be an important species in ground waters;
5. attempt to elucidate the nature and behavior of polymeric species with the ultimate objective of developing quantitative, reproducible expressions for dispersion, precipitation,

adsorption, and desorption of the polymer. This is a big order, and may appear unrealistic to some. In at least one man's opinion, however, progress in these five areas is desirable, not only for meaningful modeling, but more importantly, for a true understanding of the behavior of plutonium in the environment.

Abstract

The behavior of plutonium in aqueous environmental systems is governed by the complex interrelationships of three types of chemical equilibria - oxidation state, hydrolysis, and complexation - and by polymerization, which is a non-equilibrium process; and our ability to understand its environmental behavior will depend on our knowledge of these fundamental chemical reactions. Oxidation potentials are known with satisfactory precision. Hydrolysis constants and hydroxide solubility products are of varying but generally dubious reliability. The situation with regard to complex stability constants is somewhat variable. Complexes with strong chelating legands such as citric acid, EDTA, and DTPA have been quantitatively described with reasonable accuracy, but data on plutonium complexes with such environmentally-important ions as carbonate and phosphate are meager and questionable. Least understood of all are the irreversible polymerization reactions of plutonium which play so large a part in its aqueous chemistry. Data for all of these reactions will be reviewed and evaluated, accompanied when necessary by a critique of the methodology. The environmental consequences of this information will be assessed; for example the fallacies of attempting to use solubility products to calculate plutonium concentrations, or of employing equilibrium ion exchange concepts to describe non-equilibrium surface adsorption of polymer will be emphasized. Moreover, since most of the literature refers to macro concentrations of plutonium and frequently to relatively strong acid solution, great caution is necessary in extrapolating to environmental conditions. In conclusion, suggestions will be made of areas of plutonium chemistry that most urgently require further study.

Literature Cited

1. Cleveland, J. M. "The Chemistry of Plutonium," 653 p. Gordon and Breach Science Publishers, Inc., New York, 1970.
2. Kraus, K. A., and Dam, J. R., p. 466-477, *in* Seaborg, G. T., Katz, J. J., and Manning, W. M., eds., "The Transuranium Elements," McGraw-Hill Book Co., New York, 1949.
3. Katz, J. J., and Seaborg, G. T. "The Chemistry of the Actinide Elements," John Wiley and Sons, Inc., New York, 1957.
4. Cassol, A., Magon, L., Portanova, R., and Tondello, E. Hydrolysis of plutonium(VI): acidity measurements in perchlorate solutions, Radiochim. Acta, 17, 28 (1972).

5. Schedin, U. Hydrolysis of metal ions. 62. Plutonyl ion in sodium perchlorate medium, Acta Chem. Scand., A29, 333 (1975).
6. Rabideau, S. W., and Lemons, J. F. The potential of the Pu(III)-Pu(IV) couple and the equilibrium constants for some complex ions of Pu(IV), J. Am. Chem. Soc., 73, 2895 (1951).
7. Metivier, H., and Guillaumont, R. Hydrolysis of plutonium(IV), Radiochem. Radioanal. Lett., 10, 27 (1972).
8. Baes, C. F., Jr., and Mesmer, R. E. "The Hydrolysis of Cations," 458 p., John Wiley and Sons, Inc., New York, 1976.
9. Gel'man, A. D., and Zaitsev, L. M. Carbonate and oxalato-carbonate complex compounds of plutonium(IV), Russ. J. Inorg. Chem., 3, 47 (1958).
10. Moskvin, A. I., and Gel'man, A. D. Determination of the composition and instability constants of oxalate and carbonate complexes of plutonium(IV), Russ. J. Inorg. Chem., 3, 198 (1958).
11. Gel'man, A. D., Moskvin, A. I., and Zaitseva, V. P. Carbonate compounds of plutonyl, Soviet Radiochem., 4, 138 (1962).
12. Fardy, J. J., and Buchanan, J. M. An ion exchange study of the sulfate complexes of plutonium(III), J. Inorg. Nucl. Chem., 38, 579 (1976).
13. Nair, G. M., Rao, C. L., and Welch, G. A. Plutonium(III)-sulfate complexes, Radiochim. Acta, 7, 77 (1967).
14. Fardy, J. J., and Pearson, J. M. Ion exchange study of the sulfate complexes of plutonium(IV), J. Inorg. Nucl. Chem., 36, 671 (1974).
15. Pozharskii, V. G. Behavior of plutonyl in hydrochloric and sulfuric acid solutions, Soviet Radiochem., 8, 259 (1966).
16. Patil, S. K., and Ramakrishna, V. V. Sulfate and fluoride complexing of uranium(VI), neptunium(VI), and plutonium(VI), J. Inorg. Nucl. Chem., 38, 1075 (1976).
17. Krylov, V. N., and Komarov, E. Investigation of the complex formation of Pu(IV) with the fluoride ion in solutions of $HClO_4$ by the ion-exchange method, Soviet Radiochem., 11, 94 (1969).
18. Bagawde, S. V., Ramakrishna, V. V., and Patil, S. K. Aqueous TTA complexing of neptunium(IV) and plutonium(IV), J. Inorg. Nucl. Chem., 38, 2085 (1976).
19. Krylov, V. N., Komarov, E. V., and Puchlenkov, M. F. Complex formation of Pu(VI) with the fluoride ion in solutions of $HClO_4$, Soviet Radiochem., 10, 705 (1968).
20. Marcus, Y. Anion exchange of metal complexes. XV. Anion exchange and amine extraction of lanthanides and trivalent actinides from chloride solutions, J. Inorg. Nucl. Chem., 28, 209 (1966).
21. Grenthe, I., and Noren, B. On the stability of nitrate and chloride complexes of plutonium(IV), Acta Chem. Scand., 14, 2216 (1960).

22. Rabideau, S. W., Asprey, L. B., Keenan, T. K., and Newton, T. W., p. 361-73, "Proceedings of the Second United Nations International Conference on the Peaceful Uses of Atomic Energy, Geneva, 1958," Vol. 28, United Nations, Geneva, 1958.
23. Bagawde, S. V., Ramakrishna, V. V., and Patil, S. K. Complexing of tetravalent plutonium in aqueous solutions, J. Inorg. Nucl. Chem., 38, 1339 (1976).
24. Newton, T. W., and Baker, F. B. Chloride complex ions of Pu(VI), J. Phys. Chem., 61, 934 (1957).
25. Rabideau, S. W., and Masters, B. J. Kinetics of the reaction between Pu(VI) and Sn(II) in chloride-perchlorate solution, J. Phys. Chem., 65, 1256 (1961).
26. Zebroski, E. L., and Neumann, F. K. data quoted in C. G. Suits and K. H. Kingdon, "Progress Report No. 33, April 1-30, 1949," USAEC report KAPL-184, May 20, 1949.
27. Ghosh Mazumdar, A. S., and Swaramakrishnan, C. K. A study of the nitrate and the chloride complexes of plutonium(VI) by solvent extraction technique using TTA as the chelating agent, J. Inorg. Nucl. Chem., 27, 2423 (1965).
28. Heisig, D. L., and Hicks, T. E. The distribution of Pu(VI) and Pu(III) in thenoyltrifluoroacetone-benzene-nitric acid mixtures, USAEC Report UCRL-1664, Feb. 2, 1952.
29. Moskvin, A. I. Investigation of the complex formation of trivalent plutonium, americium, and curium in phosphate solutions, Soviet Radiochem., 13, 688 (1971).
30. Denotkina, R. G., Moskvin, A. I., and Shevchenko, V. B The composition and dissociation constants of phosphate complexes of plutonium(IV) determined by the solubility method, Russ. J. Inorg. Chem., 5, 731 (1960).
31. Moskvin, A. I. Complex formation of the actinides with anions of acids in aqueous solutions, Soviet Radiochem., 11, 447 (1969).
32. Moskvin, A. I. Zaitseva, V. P., and Gel'man, A. D., Investigation of the complex formation of trivalent plutonium with anions of acetic, citric, and tartaric acids by the ion exchange method, Soviet Radiochem., 6, 206 (1964).
33. Nebel, D. Spectrophotometric studies of equilibrium of plutonium(IV)-citrate in aqueous solution, Z. Physik. Chem. Leipzig, 232, 161 (1966).
34. Nebel, D. Potentiometric studies of the equilibrium of plutonium citrate in an aqueous solution, Z. Physik. Chem. (Leipzig), 232, 368 (1966).
35. Moskvin, A. I. Complex formation by neptunium(IV) and plutonium(IV) in nitrate solutions, Russ. J. Inorg. Chem., 16, 405 (1971).
36. Foreman, J. K., and Smith, T. D. The nature and stability of the complex ions formed by ter-, quadri-, and sexa-valent plutonium ions with ethylenediaminetetraacetic acid (EDTA). Part II. Spectrophotometric studies, J. Chem. Soc., (Pt. 2), 1758 (1957).

37. Foreman, J. K., and Smith, T. D. The nature and stability of the complex ions formed by ter-, quadri-. and sexa-valent plutonium ions with ethylenediaminetetraacetic acid. Part I. pH titrations and ion exchange studies, J. Chem. Soc., (Pt. 2), 1752 (1957).

38. Gel'man, A. D., Moskvin, A. I., and Artuykhin, P. I. Composition and dissociation constants of Pu(V) and Pu(III) complexes with ethylenediaminetetraacetic acid, Soviet J. At. Energy, 7, 667 (1961).

39. Cauchetier, P., and Guichard, C. Electrochemical and spectrophotometric study of the complexes of plutonium ions with EDTA. I. Plutonium(III) and (IV), Radiochim. Acta, 19, 137 (1973).

40. Klygin, A. E., Smirnova, I. D., and Nikol'skaya, N. A. The solubility of ethylenediaminetetraacetic acid in ammonia and hydrochloric acid and its reaction with uranium(IV) and plutonium(IV), Russ. J. Inorg. Chem., 4, 1279 (1959).

41. Krot, N. N., Ermolaev, N. P., and Gel'man, A. D. The behavior of ethylenediaminetetraacetic acid in acid solutions and its reaction with uranium(IV), Russ. J. Inorg. Chem., 7, 1062 (1962).

42. Cauchetier, P., and Guichard, C. Electrochemical and spectrophotometric study of plutonium complexes with EDTA. Plutonium(V) and (VI), J. Inorg. Nucl. Chem., 37, 1771 (1975).

43. Kabanova, O. L. Plutonium(V) complexes with ethylenediaminetetraacetic acid, Russ. J. Inorg. Chem., 6, 401 (1961).

44. Gel'man, A. D., Artyukhin, P. I., and Moskvin, A. I. A study of complex formation by pentavalent plutonium in ethylenediaminetetraacetate solutions by means of ion exchange, Russ. J. Inorg. Chem., 4, 599 (1959).

45. Kabanova, O. L., Danuschenkova, M. A., and Paley, P. N. The reactions of plutonium ions with ethylenediaminetetraacetic acid, Anal. Chim. Acta, 22, 66 (1960).

46. Costanzo, D. A., Biggers, R. E., and Bell, J. T. Plutonium polymerization. I. Spectrophotometric study of the polymerization of plutonium(IV), J. Inorg. Nucl. Chem., 35, 609 (1973).

47. Ockenden, D. W., and Welch, G. A. The preparation and properties of some plutonium compounds. Part V. Colloidal quadrivalent plutonium, J. Chem. Soc., 3358 (1956).

48. Brunstad, A. Polymerization and precipitation of plutonium(IV) in nitric acid, Ind. Eng. Chem., 51, 38 (1959).

49. Lindenbaum, A., and Westfall, W., Colloidal properties of plutonium in dilute aqueous solution, Int. J. Appl. Radiat. Isotop., 16, 545 (1965).

50. Andelman, J. B., and Rozzell, T. C. Plutonium in the water environment. I. Characteristics of aqueous plutonium, p. 118-137 *in* "Radionuclides in the Environment," Adv. Chem. Ser. 93, 1970.

51. Lloyd, M. H., and Haire, R. G. Studies on the chemical and colloidal nature of Pu(IV) polymer, *in* "Proceedings of the XXIVth IUPAC Congress, Hamburg, September 1973;" CONF 730927.2.

Disclaimer: The reviews expressed and/ or the products mentioned in this article represent the opinions of the author(s) only and do not necessarily represent the opinions of the U.S. Geological Survey.

RECEIVED November 16, 1978.

Estimated Free Energies of Formation, Water Solubilities, and Stability Fields for Schuetteite ($Hg_3O_2SO_4$) and Corderoite ($Hg_3S_2Cl_2$) at 298 K

GEORGE A. PARKS and DARRELL KIRK NORDSTROM[1]

Department of Applied Earth Sciences, Stanford University, Stanford, CA 94305

Mercury forms many double salts of which a few occur naturally; these include eglestonite, Hg_4OCl_2 and terlinguaite, Hg_2OCl (1), schuetteite, $Hg_3O_2SO_4$ (2), and corderoite, $Hg_3S_2Cl_2$ (3). Solubility and thermodynamic data useful in assessing the importance of these compounds in controlling the mercury content of natural waters are lacking. Only an enthalpy of formation for schuetteite is reported. We have estimated the standard free energies of formation of schuetteite and corderoite by first estimating their absolute entropies and the missing enthalpy of formation of corderoite. Independent estimates were derived from solubility data for schuetteite and from vapor-phase synthesis data for corderoite. The two sets of estimates are compared and the results were used to determine the stabilities of these minerals relative to the more common mercury minerals. The stability field diagrams and solubilities are used to comment briefly on conditions required for field occurrence.

Estimation of Entropies

The Debye theory of the heat capacities of solid elements (4) yields an expression for their entropies,

$$S = 3R(\ell nT - K_2 + K_1) \qquad (1)$$

in which K_1 is a universal constant. K_2 is a constant for each element and is determined by the interatomic force constant and atomic mass. Equation 1 assumes that the heat capacity and entropy are dominated by vibrational contributions. Electronic, disorder, rotational, structural, and mixing contributions are small for most monatomic solids, and are neglected. Postulates by Latimer (5) and Kopp (see Pitzer and Brewer (4)) make it possible to predict entropies of polyatomic solids or compounds. Latimer assumed that the contribution of interatomic force

[1] Current address: Department of Environmental Sciences, University of Virginia, Charlottesville, VA 22903.

0-8412-0479-9/79/47-093-339$05.00/0
© 1979 American Chemical Society

constants to entropy was small, and postulated that the entropy of each element in a solid compound could be approximately accounted for by its mass alone,

$$S = R\ell nM + S'_o \tag{2}$$

where S'_o has the same value for all elements. Kopp proposed that the heat capacity of a solid compound could be approximated by the stoichiometric sum of the heat capacities of its constituent elements. On the basis of Kopp's postulated additivity of heat capacities, Latimer suggested that the entropy of a solid compound, M_iA_j , could be approximated by the stoichiometric sum of the contributions of its constituent elements,

$$S_{M_iA_j} = iS_M + jS_A \ . \tag{3}$$

Latimer used Equation 2 to obtain entropy contributions for cations and Equation 3, together with experimental entropies of compounds and the cation contributions, to estimate anionic contributions. Naumov, Ryzhenko and Khodakovsky (6) extended Latimer's tables of constituent contributions. For estimating the entropy of a particular compound with Equation 3, Drozin (7) used anionic contributions derived from compounds of metals in the same periodic subgroup rather than overall averages as Latimer (5) had used. Equations 2 and 3 can be combined to express the entropy of one compound in terms of another, thus

$$S_{M_iA_j} = S_{N_iA_j} + \frac{3}{2} R \left(\ell n \frac{W_M}{W_N} \right) , \tag{4}$$

Pitzer and Brewer (4) suggest this approach for estimation if N_iA_j is similar, in terms of structure and properties, to M_iA_j. This approach should minimize error arising in neglect of non-mass-related entropy contributions.

If the heat capacities and entropies of solids are simply the weighted sums of those of their elemental constituents, then the entropy change should be zero for symmetrical reactions such as,

$$2HgX_2 + Pb_2OSO_4 = 2PbX_2 + Hg_2OSO_4 \tag{5}$$

in which the number of molecules produced is the same as the number consumed and the number of atoms in each product molecule is the same as the number in a corresponding reactant molecule (7). X is a monovalent anion in Equation 5. Using this principle, the entropy of Hg_2OSO_4 is given by

$$S_{Hg_2OSO_4} = S_{Pb_2OSO_4} - 2S_{PbX_2} + 2S_{HgX_2} \tag{6}$$

and can be estimated if the other entropies are known. This method should yield results similar to those obtained with Equation 4 but could introduce greater error since close similarity in structure and properties among four compounds of two different types is unlikely. For this reason Drozin (7) suggests using compounds of metals in the same periodic subgroup in comparisons utilizing Equation 6.

Tests of Estimation Methods

Equation 3. We estimated the error to be expected in the use of Equation 3 by comparing empirical entropies with those calculated using contributions tabulated by Latimer (5) without modification. The empirical data were taken from Hepler and Olofsson (8), Robie, Hemingway and Fisher (9) and the National Bureau of Standards Technical Note 270 series (11). Data for mercury are listed in Table I. Empirical and calculated entropies are compared

Table I. Thermodynamic Properties of Mercury Minerals and Compounds at 298.15K,* Including Estimates.*

		S^o, $JK^{-1}mol^{-1}$	ΔH^o, $kJ\ mol^{-1}$	ΔG^o, $kJ\ mol^{-1}$
	$Hg(\ell)$	+ 76.02	0.0	0.0
	$Hg^{2+}(aq)$	- 36.23	+170.16	+164.703
montroydite	HgO(c,red,orth.)	+ 70.29	- 90.83	- 58.555
calomel	$Hg_2Cl_2(c)$	+191.42	-265.579	-210.773
	$HgCl_2(c)$	+146.0	-225.9	-180.3
cinnabar	HgS(c,red)	+ 82.4	- 54.0	- 46.4
corderoite	$Hg_3S_2Cl_2(c)$	[301±17]	[(<-396.)]	[(<-332.)]
	$Hg_2SO_4(c)$	+200.66	-743.58	-626.34
	$HgSO_4(c)$	[142]	-707.5	[-594.]
schuetteite	$Hg_3O_2SO_4(c)$	+208 to [269±21]	-932.4	-737 to [-752]

*Data from Hepler and Olofsson (8), including estimated entropy and ΔG_f^o for $HgSO_4(c)$, unless otherwise specified. Brackets denote estimated entropies; parentheses denote estimated enthalpies of formation.

visually in Figure 1, using calcium, mercury and lead as test cases. The slope and intercept in each case are statistically indistinguishable from 1.0 and 0.0 respectively, as required by Equation 3; there is no justification in these data for use of an empirical correction to the calculated entropy. The standard error of the estimated entropy is about 17 JK^{-1} mol^{-1}(at a 95% confidence level) for mercury compounds.

If anionic entropy contributions are calculated from Pb(II) compounds alone instead of using Latimer's average anionic contributions, the standard error of the estimate for mercury compounds remains about the same. No significant improvement is obtained by using the constituent contribution suggested by Naumov et al. (6).

Equation 4. Among the heavier elements in the B subgroup of the periodic table, thallium, lead, and bismuth are notable because their compounds resemble those of the elements with atomic numbers two lower; thus the properties of Pb(II) compounds are similar to those of Hg(II) compounds, etc. (12). For this reason, we have chosen Pb(II) compounds as the principle bases for estimation of entropies of Hg(II) compounds with Equation 4.

To estimate the error expected in predictive use of Equation 4, we attempted to fit a straight line to the empirical entropies of Hg(II) compounds as though they were functions of the entropies of the corresponding Pb(II) compounds, as required by Equation 4. The slope should be 1.0 and the intercept, $1.5R\ell n(W_M/W_N)$, if the k' term is small for pairs of compounds of similar properties (4).

Results of these tests are shown in Figure 2. We found no statistical justification in any case for slopes different from 1.0. Using the Student-t test, there is no statistically significant difference, at the 90% confidence level, between the observed intercept and the atomic weight in Equation 4 for the Hg/Pb or Hg/Zn data sets. The standard error of an estimated entropy, using Pb(II) compounds as references, is about 21 JK^{-1} mol^{-1} (at 95% confidence level).

Equation 6. The entropy of schuetteite was estimated using the entropy of the corresponding lead compound and the entropies of the oxides, chlorides, bromides, iodides, and sulfides of Hg(II) and Pb(II) as the binary salts in Equation 6. There are insufficient data to estimate the error expected in estimated entropies for mercury compounds by this method. Instead, we will arbitrarily assign Drozin's estimate of the probable error, or about 17 JK^{-1} mol^{-1} (at 95% confidence level).

Estimates. Entropies estimated for schuetteite and corderoite with Equations 3, 4, and 6 are summarized in Table II. A simple average was chosen, rather than an average weighted to reflect expected errors, because the three estimates contain different sources of error. These averages are listed in Table I.

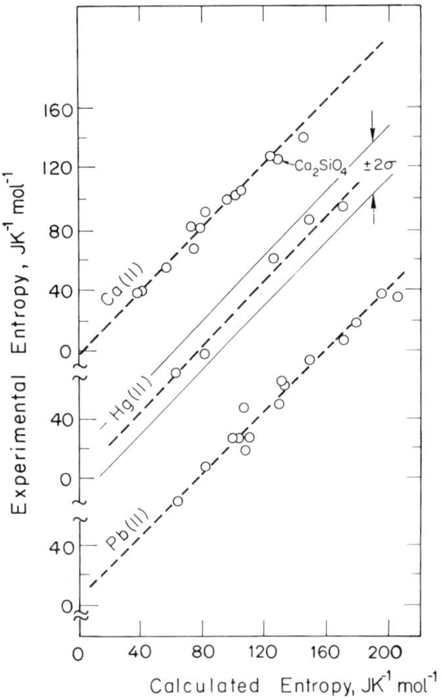

Figure 1. Comparison of experimental and calculated entropies of Ca(II), Pb(II), and Hg(II) compounds; a test of Equation 3

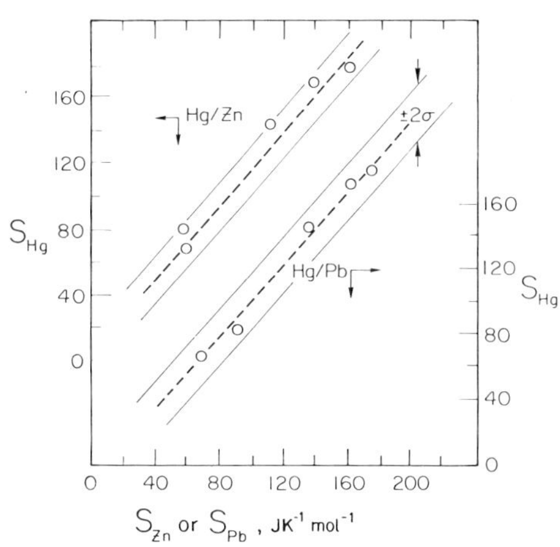

Figure 2. Correlation of entropies of Hg(II) compounds with entropies of corresponding compounds of Zn(II) and Pb(II); a test of Equation 4

TABLE II. Entropies of Hg(II) Compounds Estimated With Equations 3, 4, and 6.

	S^{o}_{298} J deg^{-1} mol^{-1}			
	Eqn. 3	Eqn. 4	Eqn. 6	mean
Schuetteite $HgSO_4 \cdot 2HgO$	269.4±17	274.5±21	263.6±46	269.0±21
Corderoite $HgCl_2 \cdot 2HgS$	302.9±17	....	....	302.9±17

Enthalpy of Formation

An experimental value for the standard enthalpy of formation of schuetteite is available; it is given in Table I. No enthalpy of formation is available for corderoite.

Many methods of estinating enthalpies of formation of binary compounds have been proposed (see, e.g., 13). Wilcox and Bromley (14), treating complex compounds analogous to corderoite as double salts, e.g.,

$$Hg_3S_2Cl_2 \simeq 2HgS \cdot HgCl_2 ,$$

suggest accepting the stoichiometric sum of the enthalpies of formation of the parent salts, for brevity designated "ΣΔH", as an estimate of the enthalpy of formation of the compound, thus

$$\Delta H^{o}_{f}(Hg_3S_2Cl_2) \overset{est}{=} 2\Delta H^{o}_{f}(HgS) + \Delta H^{o}_{f}(HgCl_2) \cong \Sigma\Delta H. \tag{7}$$

We have compared ΣΔH with the corresponding experimental ΔH^{o}_{f} for Pb(II) and Hg(II) oxide salts. In all cases, ΣΔH is less negative than the experimental ΔH^{o}_{f} ; we designate the difference "ΔΔH". ΔΔH is plotted as a function of the mole fraction of MO in the salt in Figure 3.

We will base our estimate of ΔH^{o}_{f} on the sum of ΣΔH and an estimated ΔΔH . It is difficult to estimate ΔΔH for $2HgS \cdot HgCl_2$ because we have no experimental data for sulfide-containing double salts. As a first approximation, we have chosen to set ΔΔH = -41.8 kJ mol^{-1}; the most negative value observed for the analogous oxide salts. The resultant ΔH^{o}_{f} is -381 kJ mol^{-1}; this will be revised after consideration of other data.

Free Energies of Formation

For the two compounds of interest we estimated $\Delta G^o_{f,298}$, using the equation

$$\Delta G^o = \Delta H^o + T\Delta S^o \quad (8)$$

with tabulated or estimated entropies and enthalpies.

Schuetteite. $\Delta G^o_{f,298}$ for schuetteite, estimated from Equation 8 is -751.6 ± 8. Another estimate can be derived from published solubility data. Cameron (15) synthesized $Hg_3O_2SO_4$ by reacting solid $HgSO_4$ with water. He determined the solubility of the material as prepared and after drying at 100°C. Vosburgh and Lackey (16) determined the solubility of $Hg_3O_2SO_4$ in sulfuric acid solutions. In both investigations, the solid was characterized only by its chemical analysis, and very little information was provided about solution compositions. The data of Vosburgh and Lackey are summarized in Table III. We have used these data to calculate estimates of the solubility product of schuetteite written for the reaction

$$Hg_3O_2SO_4 + 2H_2O = 3Hg^{++} + 4OH^- + SO_4^= \quad (9)$$

for which $K_{so} = (Hg^{++})^3(OH^-)^4(SO_4^=)$. Of course the equilibrium solution contains hydroxo sulfato complexes of mercury and HSO_4^- as well as H^+, OH^- and $SO_4^=$. The individual activities of the free ions, Hg^{++}, $SO_4^=$, and OH^- are needed to derive a value for K_{so}. These were calculated from the analytical solution compositions given in Table III using the computer program MINEQL (Westall, Zachary and Morel (17)) with stability constants selected from Martell and Smith (18). MINEQL corrects for all hydrolysis and complexation and for computed ionic strength. The resulting values of K_{so} are listed in Table III. The solubility data reported by Cameron (15) were rejected because our calculations showed his solutions to be much oversaturated with respect to HgO.

TABLE III. Solubility (Vosburgh and Lackey (16)) and Calculated Solubility Products for Schuetteite

Experimental concentrations, mol dL^{-1}, for four experiments:				
Total dissolved Hg,	0.007	0.0203	0.0131	0.001
Total sulfate,	0.0129	0.1518	0.0998	0.0123
Initial H_2SO_4 added,	0.0127	0.1450	0.0954	0.0119
Calculated:				
Ionic strength, I,	0.02_6	0.22	0.	0.02_5
$\log_{10}(\gamma_1)$	- 0.07	- 0.14	- 0.13	- 0.07
$\log_{10}K_{so}(I \to 0)$	-58.9_3	-57.8_9	-57.8_6	-58.3_4

The ΔG_f^o corresponding to our average K_{so} is $-737._8 \pm 4._2$, which is considerably less negative than the value estimated with ΔH_f^o and our estimated entropy. The entropy that schuetteite would have to have if ΔG_f^o were $-737._8$ and the NBS value for ΔH_f^o were correct is between 208 and 237 J K^{-1} mol^{-1}, the range reflecting probable error in the solubility data alone. This value differs from our estimated value by more than our expected error, leading us to suspect possible error in either the solubility data or the NBS ΔH_f^o in addition to error in the estimated entropies. Ordinarily we would prefer the experimentally based ΔG_f^o derived from K_{so} over an estimate, but in this case there is sufficient question about the experimental methods (e.g., no test for equilibrium is described) that we prefer to select a probable range for ΔG_f^o, -737 to -752 kJ mol^{-1}.

Corderoite. The free energy calculated from estimated entropy and enthalpy of formation for corderoite is -317.6 kJ/mole.

Carlson (19) synthesized corderoite using the reaction,

$$3HgCl_2(g) + 2H_2S(g) = Hg_3S_2Cl_2(s) + 4HCl(g) \qquad (10)$$

and used the conditions required for the reaction to proceed to calculate an approximate value for the free energy change associated with the process, -159.8 kJ mol^{-1} at 575°K. With this value and high temperature free energy data for the other reactants (9,20) we calculate $\Delta G^o_{f,575} = -274.9$ kJ mol^{-1} for corderoite. Finally, using this free energy of formation, our estimated S^o_{298}, heat capacity data (9,21), and the assumption that the heat capacity of corderoite is the stoichiometric sum of the heat capacities of HgS and $HgCl_2$, we calculated an estimate of $\Delta G^o_{f,298} = -332.2$ kJ mol^{-1}. Errors arising in estimates of entropies and heat capacities are likely to be smaller than those arising in estimates of ΔH_f^o, thus we believe that this value of ΔG_f^o contains less error from this source than the preceding estimate, -317.6. However, because Carlson provided no evidence that equilibrium was achieved in his experiment, -332.2 should be considered a limit to ΔG_f^o; the true value is probably more negative. If -332.2 were the correct ΔG_f^o, and our S^o correct, ΔH_f^o would have the value $-395._9$ kJ mol^{-1}; we consider this a more reliable estimate and include it in Table I.

As for schuetteite, it should be possible to derive another estimate of ΔG_f^o for corderoite from published solubility data. Karoglanov and Šagortschev (22) describe a series of experiments in which an alleged $Hg_3S_2Cl_2$ was synthesized by mixing solutions of mercuric chloride with solutions of sodium sulfide. The solid was not analyzed; its composition was inferred from the amounts of reagents added, the composition of the final solution and the assumption that all sulfur was precipitated. No details of method, such as measures taken to prevent loss or oxidation of H_2S, are provided. Providing no details whatever, these authors report that the solubility of "pure" $Hg_3S_2Cl_2$ is 0.0097 g per litre of

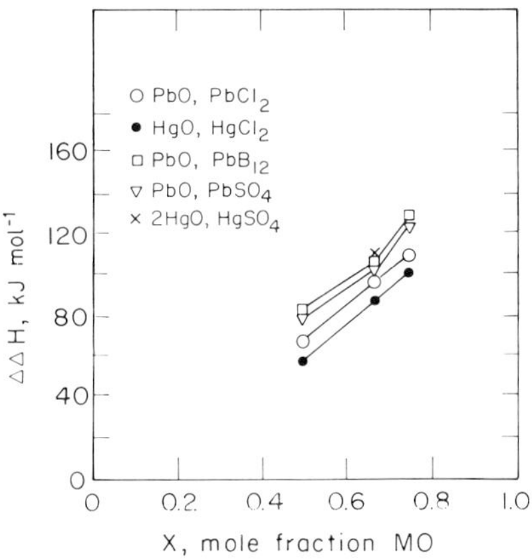

Figure 3. Excess enthalpies of formation of oxide salts of Pb(II) and Hg(II)

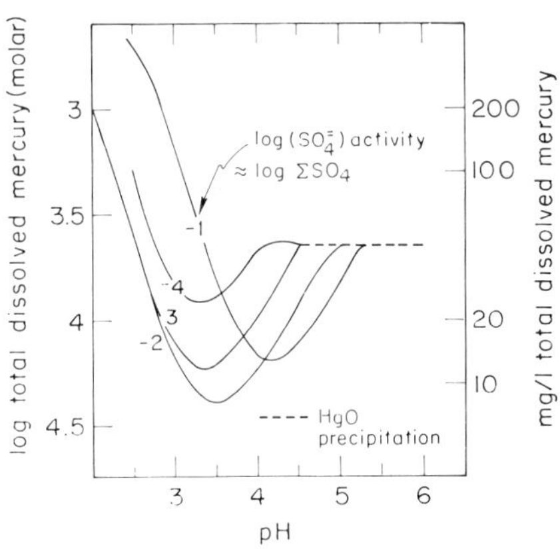

Figure 4. Calculated water solubility of schuetteite, $Hg_3O_2SO_4$, assuming $G_f°$ *= −752 kJ mol*$^{-1}$

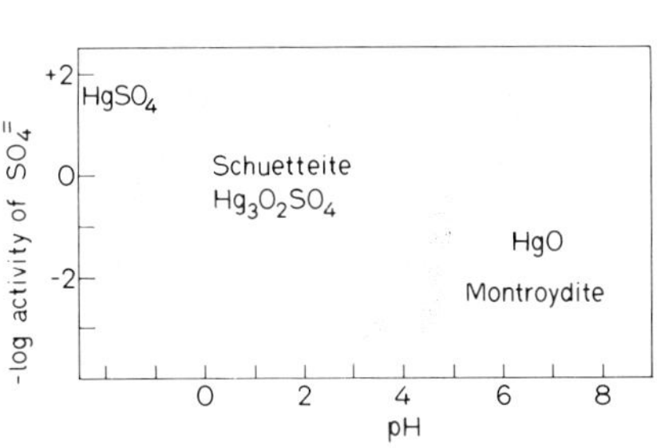

Figure 5. Relative stabilities of montroydite, HgO, schuetteite, $Hg_3O_2SO_4$, and $HgSO_4$ in a chloride-free system. Stippled zones represent estimated uncertainty in locations of the phase dominance boundaries.

water at 22°C. Owing to the lack of analyses for dissolved sulfur and the high probability of error due to oxidation or volatilization of H_2S, we chose not to calculate serious estimates of K_{so} and ΔG_f^o for corderoite from these data.

We are left with the conclusion that $\Delta G_{f,298}^o$ for corderoite is probably more negative than -332 kJ/mole.

Discussion

Because mercury forms relatively stable hydroxo, sulfato, and chloro complexes in solution, the solubilities and relative stabilities of mercury minerals depend heavily on ambient solution compositions. Figure 4 illustrates the calculated effect of pH and total sulfate on the solubility of schuetteite. The curves take into account all complexes for which data are available in Martell and Smith (18) and precipitation of HgO; they cross when changes in solution composition result in changes in dominance among complexes. Schuetteite is quite soluble relative to HgS and elemental mercury under common conditions.

Tentative stability field diagrams illustrating conditions under which $HgSO_4$, schuetteite, and corderoite are likely to form are given in Figures 5 and 6. These diagrams were constructed using our estimated free energies of formation together with other required data from Hepler and Olofsson (8) and Robie, Hemingway, and Fisher (9) and methods described by Garrels and Christ (23) and Stumm and Morgan (24).

$HgSO_4$ is unlikely to be found in nature, first, because it is unstable with respect to the oxide sulfates except under unrealistically high $SO_4^=$ and H^+ activities and, second, because it is so soluble ($\log K_{so} = -2.8$).

Formation of schuetteite requires acid, oxidizing conditions, as suggested by Bailey, *et al*. (2). However, these conditions are not unduly restrictive and might be expected in or near oxidizing sulfide ore deposits, mine dumps, or smelter waste (calcine) dumps (23,24). The solubility of schuetteite is high relative to that of the common rock forming minerals and heavy metal sulfides, i.e., a minimum of about $10^{-4.5}$ molar or roughly 10 mg Hg per litre. This solubility is not high with respect to that of some other common oxidized minerals however. The minimum solubility of gypsum, for example, is about $10^{-3.6}$ molar (as total dissolved Ca). Schuetteite may be more common than has been observed in oxidized zones of mercury deposits high in iron sulfides; it could easily be missed by visual inspection of samples containing jarosites or hydrous Fe(III) oxides.

The conditions needed for formation of corderoite required by our estimated ΔG_f^o (Figure 7) are incompatible with any realistic natural environment. A combination of low pH and mildly reducing conditions with a high ratio of chloride to sulfide or sulfate is conceivable in the anaerobic sediments of closed-basin lakes, but the lowest pH would be closer to 4 than to 2.5. The ΔG_f^o of

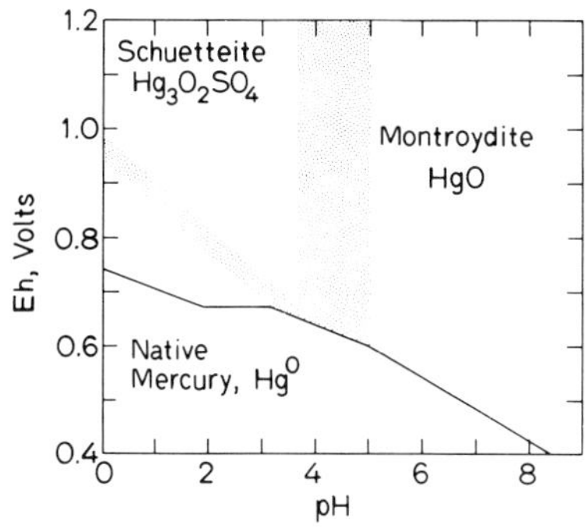

Figure 6. Relative stabilities of native mercury, Hg°, Hg_2SO_4, schuetteite, $Hg_3O_2SO_4$, and montroydite, HgO in a chloride-free system in which the activity of the predominant dissolved sulfur species is 0.01M. Stippled zones represent estimated uncertainty in locations of phase dominance boundaries.

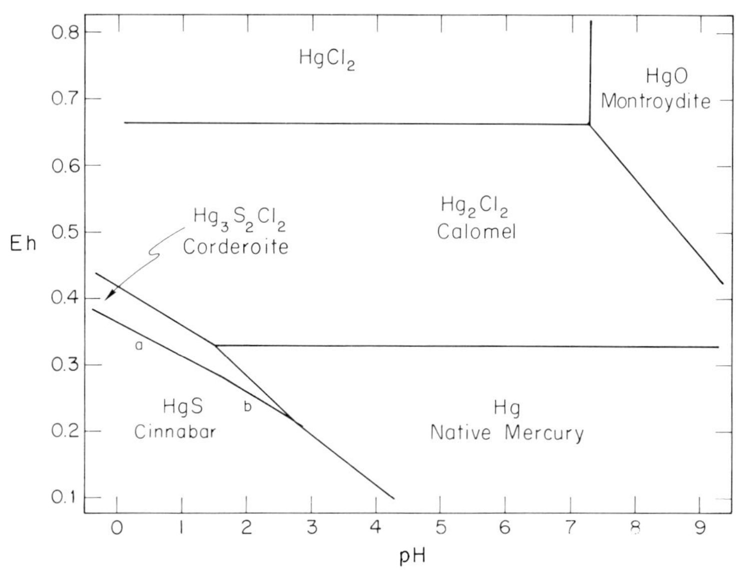

Figure 7. Tentative eH-pH stability field diagram for the system Hg–S–Cl–H_2O at 298 K. Activities of Cl^- and the dominant SO_4^{2-} species are 0.1M.

corderoite must be 40 kJ mol^{-1} more negative than we have estimated if the stability field for corderoite is to extend to pH values as high as four under the conditions of Figure 7.

Acknowledgements

We thank the Swiss Federal Institute for Water Supply, Water Treatment and Pollution Control for the opportunity for one of us (GAP) to spend uninterrupted time preparing the final form of this paper. We also thank Gordon E. Brown for criticizing our methods and John Burkstaller, Jim Davis, Al Umaña, and Vijay Tripathi for help with calculations.

Symbols

I	Ionic strength mol dL^{-1}
ΔG_f^o	Standard free energy of formation
ΔH_f^o	Standard enthalpy of formation
K_{so}	Solubility product
R	Gas constant, 8.31 JK^{-1} mol^{-1}
S^o	Standard entropy, JK^{-1} mol^{-1}
T	Temperature, K
W	Atomic weight
γ_1	Activity coefficient of monovalent ion

Abstract

With published and estimated entropies and enthalpies of formation, we have estimated free energies of formation of schuetteite ($Hg_3O_2SO_4$) and corderoite ($Hg_3S_2Cl_2$) at 298°K. These estimates were compared with others based on the solubility of schuetteite and conditions of vapor phase synthesis of corderoite, to yield tentative selection as follows; ΔG_f^o = -737 to -752 kJ mol^{-1} for schuetteite and ΔG_f^o < -332 kJ mol^{-1} for corderoite. Tentative water solubility curves and stability field diagrams derived from these data indicate that schuetteite should form only in acid, oxidizing, high sulfate, low chloride conditions such as are found in acid mine waters, while corderoite requires extremely acid, mildly reducing, low sulfur, high chloride conditions unlikely to be found in nature.

Literature Cited

1. Tunell, G., The geometry of mercury, in Wedepohl, K.H., ed., "Handbook of Geochemistry", Vol. II, pt. 2, Springer Verlag, Berlin, 1968.

2. Bailey, E.H., Hildebrand, F.A., Christ, C.L., and Fahey, J.J., Schuetteite, a new supergene mercury mineral. Am. Mineral., 44, 1026-1038 (1959).
3. Foord, E.E., Berendsen, P., and Storey, L.O., Corderoite, first natural occurrence of α-$Hg_3S_2Cl_2$ from the Cordero Mercury Deposit, Humboldt County, Nevada, Am. Mineral., 59, 652-655 (1974).
4. Pitzer, K. and Brewer L., "Thermodynamics", 2nd. ed., 723p. McGraw-Hill Publishing Co., New York, 1961.
5. Latimer, W.M., "Oxidation Potentials", 2nd. ed., 392p. Prentice Hall, Inc., New York, 1952.
6. Naumov, G.B., Ryzhenko, B.N., and Khodakovsky, I.L., "Handbook of Thermodynamic Data", (English transl. by Solesmani, G.J., ed. by Barnes, I., and Speltz, V.), 328p. N.T.I.S., Washington, D.C., 1974.
7. Drozin, N.N., Application of the Berthelot principle in the calculation of the standard entropies of solid inorganic compounds. Russ. J. Phys. Chem., 35, 879-881 (1961).
8. Hepler, L.G., and Olofsson, G., Mercury: thermodynamic properties, chemical equilibria, and standard potentials, Chem. Rev., 75, 585-602 (1975).
9. Robie, R.A., Hemingway, B.S., and Fisher, J.R., Thermodynamic properties of minerals and related substances at 298.15K and 1 bar (10^5 Pascals) pressure and at higher temperatures, U.S. Geol. Survey Bull., 1452, 456p. (1978).
10. Wagman, D.D., Evans, W.H., Parker, V.B., Halow, I., Bailey, S.M., and Schumm, R.H., Selected values of chemical thermodynamic properties, tables for the first thirty-four elements in the standard order of arrangement, U.S. Nat. Bur. Std., Tech. Note, 270-3, 264p. (1968).
11. Wagman, D.D., Evans, W.H., Parker, V.B., Halow, I., Bailey, S.M., and Schumm, R.H., Selected values of chemical thermodynamic properties, tables for elements 35 through 53, U.S. Nat. Bur. Stds., Tech. Note, 270-4, 141p. (1969).
12. Wells, A.F., "Structural Inorganic Chemistry", 3rd. ed., 1055p. Oxford University Press, London, 1962.
13. Kubachewski, O., Evans, E.L., and Alcock, C.B., "Metallurgical Thermochemistry", 4th. ed., 495p. Oxford, Pergamon Press, New York, 1967.
14. Wilcox, D.E., and Bromley, R.A., Computer estimation of heat and free energy of formation for simple inorganic compounds, Ind. Eng. Chem., 55, 32-39 (1963).
15. Cameron, C.A., On the action of water on mercuric sulfate, Analyst, 5, 144-146 (1880).
16. Vosburgh, W.C., and Lackey, O.N., A mercury-basic sulfate voltaic cell, J. Am. Chem. Soc., 52, 1407-1410 (1930).
17. Westall, J.C., Zachary, J.L., and Morel, F.M.M., MINEQL, a computer program for the calculation of chemical equilibrium compositions of aqueous solutions, Dept. Civil Engin., Mass. Inst. Tech., Note No. 18, 91 p. (1976).

18. Smith, Robert M., and Martell, A.E., "Critical Stability Constants", 257 p. Plenum Press, New York, 1976.
19. Carlson, E.E., The growth of HgS and $Hg_3S_2Cl_2$ single crystals by a vapor phase method, J. Cryst. Growth, 1, 271-277 (1967).
20. Stull, D.R., and Prophet, H., JANAF thermochemical tables, 2nd. ed., U.S. Nat. Bur. Stds. Ref. Data Ser., 37 , 1141p. (1971).
21. Kelley, K.K., Contributions to the data on theoretical metallurgy. XIII. High temperature heat content, heat capacity, and entropy data for the elements and inorganic compounds, U.S. Bur. Mines Bull., 584, 232p. (1960).
22. Karaglanov, Z., and Sagortschev, B., Über den Mechanismus von Fällungsvorgängen. XIII. Vorgänge, welche bei der Beteiligung von $HgCl_2$ verlaufen, Z. Anorg. u. Allgem. Chemie., 213, 155-160 (1933).
23. Garrels, R.M., and Christ, C.L., "Solutions, Minerals, and Equilibria", 450p. Harper and Row, New York, 1965.
24. Stumm, W., and Morgan, J.J., "Aquatic Chemistry," 583 p., Wiley-Interscience, New York, 1970.

RECEIVED November 16, 1978.

18

Techniques of Estimating Thermodynamic Properties for Some Aqueous Complexes of Geochemical Interest

DONALD LANGMUIR[1]

Department of Geosciences, Pennsylvania State University, University Park, PA 16802

A host of variables can influence the stability of an aqueous complex relative to its uncomplexed addends. Among these are: (a) the valences of the addends; (b) their distance of approach in the complex; (c) the degree of covalency of their bonding; (d) the number of ligands coordinating the cation or cations in the complex; (e) the packing arrangement of cations and ligands in the complex; and more subtle considerations, including (f) the polarizability or deformability of the electron cloud surrounding the addends; and (g) ligand field effects for transition metal complexes in particular (5, 6). Of these, the most important variables are the valences of the addends and their separation distance in the complex. As a rule, the stability of complexes formed with a given ligand increases with cation valence and decreases with cation radius. This is expected, based strictly on Coulombic arguments. Equations to predict the thermodynamic properties of complexes as functions of valence of addends and the reciprocal of their radii or separation distance abound in the literature. The best known of these are the simple Coulombic expression of Denison and Ramsey (7) and the equations of Bjerrum and of Gilkerson and Fuoss (8, 9). As the covalency and polarizability (etc.) of associating addends increases, however, the simple Coulombic models tend to fail. They do, nevertheless still suggest limiting values for the thermodynamic properties of complexation. Because of their convenience and general utility, the simple charge-distance equations are discussed at some length in this paper in connection with free energy and entropy data for sulfate, fluoride, and phosphate complexes.

Predicting the stability of complexes which owe their existance to important non-Coulombic contributions can be difficult. Where increased covalent bonding between cations and a given ligand is involved, plots of stability versus ionization potential or electronegativity (EN) of the cation are sometimes useful (5, 10). In any case, the best general approach is to compare the

[1] Current address: Department of Chemistry and Geochemistry, Colorado School of Mines, Golden, CO 80401

0-8412-0479-9/79/47-093-353$08.25/0
© 1979 American Chemical Society

stabilities of complexes formed with a common ligand where the cations considered exhibit similar or systematically trending behavior in complexation. Here, classifications distinguishing cations and ligands are useful. These include the class A, B, and C cation groupings of Schwarzenbach (2), and the hard and soft acid and base designations of Pearson (3) and Ahrland (4).

A variety of estimation and correlation techniques are presented in this paper. They include plots involving the thermodynamic properties of metal complexes with the ligands HCO_3^-, CO_3^{2-}, SO_4^{2-}, OH^-, SH^-, the halogens (especially F^-), and $H_nPO_4^{3-n}$ (n = 0-3), plotted against, for example: (a) the value of n in $H_nPO_4^{3-n}$; (b) the EN of the cation or EN order of ligands; (c) the ligand number (for complexes of U and Th); and (d) z_+z_-/d, where z_+ and z_- are the valences of the cation and ligand, respectively, and d their distance of separation in the complex.

The thermodynamic data plotted in the figures or discussed in the text have been obtained from references including Yatsimirskii and Vasil'ev (11), Ringbom (12), Sillén and Martell (13, 14), Christensen *et al*. (15), and Smith and Martell (16). Other sources have included Wagman *et al*. (17) for thorium complexes, and Langmuir (18) for uranium complexes. Whenever possible, the thermodynamic data are for 25°C and zero ionic strength (I = 0). When stability constants were reported for higher ionic strengths but could be corrected to I = 0, this has been done, using the extended Debye-Hückel equation or mean salt calculations (10, 19).

Complexes Formed by Hydrolysis of Cations

An appreciation of the nature and stability of complexes formed by hydrolysis of cations is important because: 1) such complexes are the predominant forms of occurrence and transport of many cations in natural waters; and 2) the occurrence of important amounts of other types of complexes involving the same cation requires that their ligands compete effectively with the ligands formed by hydrolysis of the cation. In general, hydrolysis of monovalent and divalent cations produces H_2O-cation complexes or aquo-cations. In contrast, smaller cations of valence 3+ to 6+ usually exist as OH^-- or O^{2-}- cation complexes in water.

The larger monovalent alkali metals Rb^+ and Cs^+ (6-fold crystallographic radii of 1.47 and 1.67 A, respectively) are thought to be essentially unassociated with water molecules (1). However, the smaller monovalent and most divalent cations form strong aquo-complexes. Thus, the reaction

$$Mg^{2+} + nH_2O = Mg(H_2O)_n^{2+}$$

has a Gibbs free energy (ΔG^o) of about -460 kcal/mol. (Here, n corresponds to the number of water molecules in two layers of coordinating water (1)). For most cations, the immediate hydration envelope around the cation contains six water molecules (20). In this first envelope, the water dipoles are oriented with their oxygens towards the cation and protons away (21). Second and third layers of water may also be significantly associated with the cation, although these are more loosely held. The strength of bonding between the inner sphere of water dipoles and a positively charged core ion increases qualitatively as the charge density of the core ion increases. The direct dependence of charge density on the valence (z_+) and its inverse dependence on ionic radius (r, in Angstroms) has led to the concept of ionic potential (Ip) which is defined as z_+/r. The valence of most geochemically important cations or cations in their oxy- and hydroxy-complexes is plotted against the cation radii in Figure 1. Most of the radii are from Shannon and Prewitt (22) and, when the data exist, are for the cations at their appropriate coordination numbers with oxygen or hydroxyl. The plot shows, in general, that at pH values between 2 and 12 and at 25°C, the larger monovalent and divalent cations occur as simple cations or aquo-ions. If we define a hydroxycation as a cation whose first OH^- complex predominates over the unhydrolyzed cation at some pH below 7, then Be, Cu, Sn, Hg, and Pb are all hydroxycations (23).

With increasing ionic potential of the cation, OH^- and then O^{2-}-bonding of the core cation becomes important. The OH^- and O^{2-}-bonded cations become predominant forms even in very acid waters. For the oxycations (e.g., UO_2^{2+}, VO^{2+}) and oxyanions (e.g., SO_4^{2-}, PO_4^{3-}, CO_3^{2-}), the cation-oxygen bond is a strong covalent one, so that these complexes persist as such in most reactions that form larger complexes. However, for di- and trivalent cations, the hydroxy complexes may be less stable than complexes formed with other ligands. The dashed lines in Figure 1 are drawn to roughly separate groups of cations by their tendency towards formation of OH^- and O^{2-} complexes in water. These lines clearly do not yield groupings based simply on ionic potential, which would plot as straight lines passing through zero for both variables in Figure 1.

The stoichiometries of known or presumed complexes formed by hydrolysis of cations have been listed in Table I. The best discussion of the stabilities and behavior of most of these species is given by Baes and Mesmer (23).

Inner and Outer Sphere Complexes

As noted above, most metal cations in pure water are complexed with H_2O, OH^-, or O^{2-}. Formation of a complex with another ligand then may involve displacing one or more water molecules from the coordination sheath of the cation by the ligand. Thus,

$$M(H_2O)_n + L = M(H_2O)_{n-1}L + H_2O$$

When one or more hydration spheres remain after complexation, then

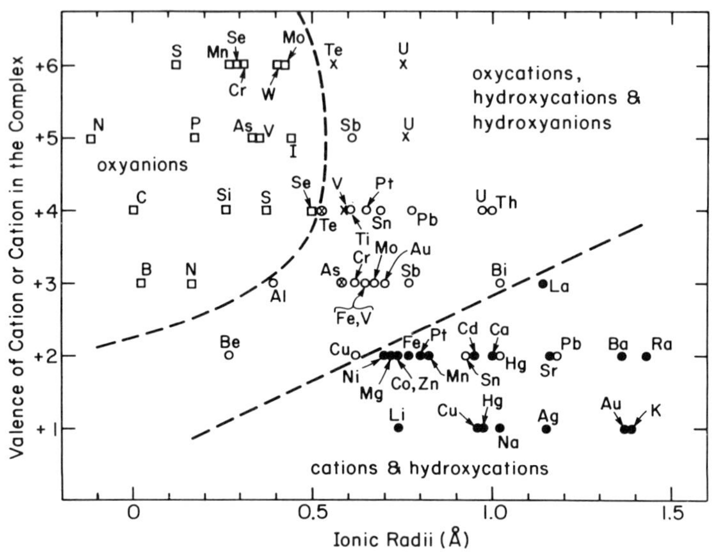

Figure 1. Valence of core cations in their aquo-complexes plotted against crystal radii of the cations. The radii are mostly from Shannon and Prewitt (22). (●) Cations; (○) hydroxycations and hydroxyanions; (×) oxycations; (□) oxyanions. See *caption to Table I.*

Table I. Species formed by hydrolysis of cations in water at 25°C, and pH 2-12. Only species which may exceed 10% of the total core cation concentration are listed. Cations are written as free species if they occur chiefly as such in the pH range 2-12. When a cation is less stable than its hydroxy or oxy-complex at a pH below 7, the hydroxy or oxy-complex is listed instead of the free cation. Most of the free cations occur as H_2O-complexes. Elements such as Cu(II), Fe(III) and Th(IV) which form polymers, are assumed to occur at $\leq 10^{-5}$ molal total concentrations. Brackets enclose possible values of n and/or m. Most of the tabulated species are discussed by Baes and Mesmer (23).

Valence of core cation	Cations and complexes
1:	Ag, Au, Cu, Hg, K, Li, Na
2+:	Ba, Ca, Cd, Co, Fe, Mg, Mn, Ni, Pt, Ra, Sr, Zn; $Be(OH)_n^{2-n}$[0-4], $Cu(OH)_n^{2-n}$[0-4], $Cu_2(OH)_2^{2+}$, $Hg(OH)_n^{2-n}$ [0-3], $Pb(OH)_n^{2-n}$[0-3], $Pb_3(OH)_4^{2+}$, $Sn(OH)_n^{2-n}$[0-3].
3+:	$Al(OH)_n^{3-n}$[0-4], AsO+, $As(OH)_n^{3-n}$[3,4], $Au(OH)_n^{3-n}$[2-4], $B(OH)_n^{3-n}$[3,4], $Bi(OH)_n^{3-n}$[0-4], $Bi_6(OH)_{12}^{6+}$, $Bi_9(OH)_n^{27-n}$[20-22], $Cr(OH)_n^{3-n}$[0-4], $Cr_3(OH)_4^{5+}$, $Fe(OH)_n^{3-n}$[0-4], $H_nNO_2^{2-n}$ [0,1], La^{3+}, Mo^{3+}, $Sb(OH)_n^{3-n}$[2-4], $V(OH)_n^{3-n}$[0-3].
4+:	$H_nCO_3^{2-n}$[0-2], $H_nSO_3^{2-n}$[0,1], $H_nSeO_3^{2-n}$[0-2], $H_nSiO_4^{4-n}$ [2-4] $Pb(OH)_n^{4-n}$[3-6?], $Pt(OH)_n^{4-n}$[1-4?] $Sn(OH)_4^{\circ}$? $Te(OH)_n^{4-n}$[3,4], $TeO(OH)_n^{n-4}$[2,3], $VO(OH)_n^{2-n}$[0,1], $Th(OH)_n^{4-n}$[0-4], $Th_2(OH)_2^{6+}$, $Ti(OH)_n^{4-n}$ [2-4], $U(OH)_n^{4-n}$ [1-5].
5+:	$H_nAsO_4^{n-3}$[0-3], $H_nPO_4^{n-3}$[0-3], $H_nVO_4^{n-3}$[1-4], IO_3^-, NO_3^- $Sb(OH)_n^{5-n}$[5,6], UO_2^+.
6+:	$H_nCrO_4^{2-n}$[0,1], $H_nMoO_4^{2-n}$[0-2], $H_nSO_4^{2-n}$[0,1], $H_nSeO_4^{2-n}$ [0,1], $H_nWO_4^{2-n}$[0-2], MnO_4^{2-}, $TeO_n(OH)_m^{6-2n-m}$ [nm = 06, 15, 24], $(UO_2)_n(OH)_m^{2n-m}$ [nm = 10, 11, 22, 35], $W_{12}O_{39}^{6-}$.

the complex is called outer sphere or an ion pair. Bonding is chiefly long range and electrostatic. The ion pair may persist through a number of collisions with water molecules. Examples of ion pairs are most monovalent and divalent oxyanion complexes formed with alkali metals and alkali earths (for example, $NaHCO_3^o$, $SrCO_3^o$, $CaNO_3^+$, and $MgHPO_4^o$). The list also includes most monovalent and divalent metal sulfate and chloride complexes. Available data suggest the following log K_{assoc} values for ion pairs: ML pairs ∿0 - 1; ML_2 or M_2L pairs ∿.7 - 1.3, and M_2L_2 pairs 2.3 - 3.2. For ion pairs, K_{assoc} values tend to be roughly constant for a given ligand with metal cations of identical valence and roughly the same size. For example, the divalent metal sulfate ion pairs formed with Ca, Mg, Ni, Zn, Cu, Co, Cd, Mn, Fe, and Cu have log K_{assoc} values from 2.28 to 2.35 (see Figure 2). This behavior reflects the largely electrostatic attraction between the ions, nearly independent of the detailed electron configuration of the cation (1).

When complexation involves displacement by the ligand of water molecules immediately adjacent to the cation so that the ligand contacts the cation, the complex is called inner sphere. Such complexes are generally more stable than outer sphere complexes or ion pairs. Actually, there are no pure "inner" or "outer" sphere complexes. For example, even in SO_4 complexes of Be, Mg, Zn, Ni, Co, and Mn, about 10% of the bonding is inner and 90% outer sphere. For Cr^{3+}-sulfate, the proportions become more than 70% inner and 30% outer sphere (1).

Electronegativity

Pauling (24) has defined electronegativity (EN) as the power of an atom to attract electrons. The concept is useful in comparing stabilities of inner sphere complexes when their bonding is significantly covalent. The degree of covalency (as opposed to ionicity) of bonding in the complex increases as the difference in EN (ΔEN) of the cation and ligand approaches zero. Based on bonds in crystals, Pauling computes that roughly when $\Delta EN < 1.7$, covalency predominates over ionicity. Electronegativity data for monatomic cations and ligands are available in Pauling (24), Gordy and Thomas (25), Allred (26), Wells (27), and Rosler and Lange (28). Unfortunately, such data are mostly lacking for polyatomic cations and ligands, although EN may be estimated or computed for such species from mineral solubility products (29). In the absence of such data, plots of stability constants for isostructural metal complexes formed with a common ligand against EN values for the metal ions are still useful. As shown in Figure 2, the stability of divalent metal complexes with HCO_3^- and SO_4^{2-} is practically independent of cation electronegativity, whereas the metal complexes formed with HPO_4^{2-} and CO_3^{2-} are increasingly dependent on EN, particularly for $EN > 1.5$. Based on the relative solubilities of their salts, EN values for CO_3^{2-},

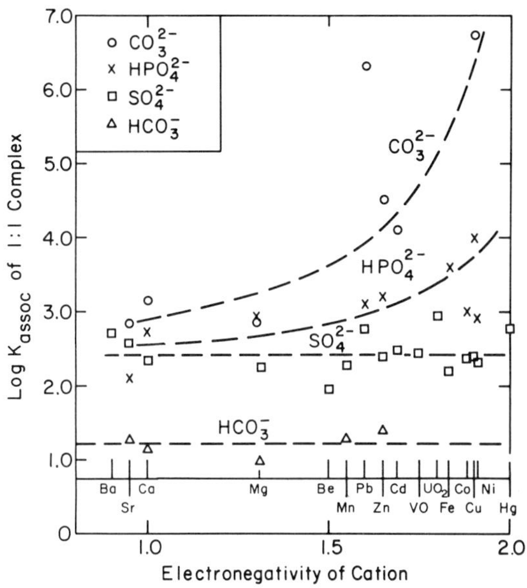

Figure 2. Stabilities of some 1:1 oxyanion complexes plotted against the electronegativity of the cation. Literature data have been corrected to I = 0 when necessary. Lines through the data for HCO_3^- and SO_4^{2-} complexes are for mean values. Curves drawn through the HPO_4^{2-} and CO_3^{2-} data have no statistical significance.

HPO_4^{2-}, and SO_4^{2-} should decrease from near 4 to about 3 in the order $SO_4^{2-} > HPO_4^{2-} > CO_3^{2-}$. Thus, covalency of bonding should be least for the sulfate complexes and greatest for the carbonate complexes, which are apparently for the most part of inner sphere character for cation EN values above 1.5.

For the alkaline earths, bonding with all four ligands shown in Figure 2 is largely ionic (i.e., independent of EN) because of their predominantly outer sphere character. The slight decrease in stability of these complexes generally, from Ba through Be, may reflect the ligands' closer approach to the relatively unhydrated Ba ion (Ip = 1.5), a proximity not possible in the case of Mg (Ip = 3.0) or Be (Ip = 5.7). These smaller ions strongly retain their waters of hydration in complex formation.

Schwarzenbach's and Pearson's Classifications

There are no simple rules involving addend size, valence, Ip, or EN that can explain all the observed stabilities of complexes. However, further insights are provided through approaches suggested by Schwarzenbach (2), and Pearson (3) and Ahrland (4). Pearson and Ahrland classify cations and ligands as 'hard' or 'soft' acids or bases. Soft infers the species' electron cloud is deformable or polarizable and may enter into electronically unique states in complexation. Hard addends are comparatively rigid and non-deformable and show an absence of electronic interactions in complexation.

Schwarzenbach (2) (see also Phillips and Williams (30); Nancollas, (1)) defines three classes of complexes. Class A includes metal cations which have noble gas configurations. These are listed below.

Cation (increasing covalency →)	↑ increasing covalency	Atomic Configuration
Li^+, Be^{2+}		He
Na^+, Mg^{2+}, Al^{3+}		Ne
K^+, Ca^{2+}, Sc^{3+}, Ti^{4+}		Ar
Rb^+, Sr^{2+}, Y^{3+}, Zr^{4+}		Kr
Cs^+, Ba^{2+}, La^{3+}		Xe

Class A cations have spherical symmetry and low polarizability and thus are 'hard spheres' (31). Pearson's hard acids include the above, but also Mn^{2+}, UO_2^{2+}, VO^{2+}, Cr^{3+}, Fe^{3+}, Co^{3+}, Ga^{3+}, Si^{4+}, U^{4+}, and Th^{4+}. These cations tend to form largely electrostatic bonds with ligands, especially when the ligands are hard (have low polarizability) and the cations monovalent or divalent. Important complexes are formed with hard ligands F^-, H_2O, and OH^-. Carbonate and bicarbonate complexes are also important but only for the monovalent and especially divalent cations. The stability of complexes is generally in the order $OH^- > F^-$, $CO_3^{2-} > HPO_4^{2-} >> NO_3^-$ and $PO_4^{3-} >> HPO_4^{2-} > SO_4^{2-}$. Complexes are usually not formed with S, N, C, Cl^-, Br^-, or I^-, because these species cannot compete with H_2O or OH^-. Complexes probably increase generally in stability in the order $I^- < Br^- < Cl^- < F^-$, although data are mostly lacking for the extremely weak I^- and Br^- complexes.

Because bonding is in large part electrostatic, when a single ligand is considered, K_{assoc} values are usually proportional to z_+/r_+ or to $z_+z_-/(r_+ + r_-)$ where the + and - subscripts denote cation and ligand charge and radii respectively (30). As z_+/r_+ (Ip) values increase, however, covalent bonding becomes important. For example, important covalency and cation deformation occurs when complexes are formed with Be^{2+} or species such as Fe^{3+}, Al^{3+}, and U^{4+} (6).

Schwarzenbach (2) defines class B metal cations as those with electron configurations of Ni^o, Pd^o, or Pt^o (31). The group includes Cu^+, Ag^+, Au^+, Zn^{2+}, Cd^{2+}, Hg^{2+}, Ga^{3+}, In^{3+}, Tl^{3+} and Sn^{4+}. Here and below the underlined species are the most important geochemically. Class B cations are considered soft spheres and are highly polarizable. Sometimes included in the list are Tl^+, Sn^{2+}, Pb^{2+}, and Bi^{3+}. Pearson (32) prefers to treat In^{3+} as a hard acid species, and Zn^{2+}, Pb^{2+}, Sn^{2+}, and Bi^{3+} as "borderline" rather than truly soft or hard cations. Class B cations form largely covalent bonds in complexation, so that the stability of their inner sphere complexes tends to increase with decreasing ΔEN values between the cation and ligand. Thus, stability constants for the monovalent cation complexes and Pb^{2+}, Cd^{2+}, and Hg^{2+} complexes increase in the orders $F^- < Cl^- < Br^- < I^-$ and $F^- < OH^- < SH^- < S^{2-}$ (see Figures 3 and 4). For Sn^{2+}, Zn^{2+}, and In^{3+}, complex stabilities are reversed for the halogens and increase in the order $I^- < Br^- < Cl^- < F^- < OH^-$. The latter three cations thus exhibit behavior more akin to cations of the class A group. As shown in Figure 2, the divalent class B cations form SO_4^{2-} complexes and probably also HPO_4^{2-} complexes.

For the reaction $2Cu^+ = Cu^o + Cu^{2+}$, $K = 10^6 = [Cu^{2+}]/[Cu^+]^2$, so that Cu^+ is relatively unimportant in solution. It occurs, however, in sulfides such as Cu_2S ($K_{sp} = 10^{-48.5}$, (16)). In solution, Cu^+ and Ag^+ usually form four-coordinated complexes. Au^+ is unimportant in natural waters.

Class B metals generally yield their strongest complexes when they associate with deformable or polarizable ligands such as I^-, SH^-, S^{2-}, Se^{2-}, and Te^{2-}. These ligands are termed soft acids by Pearson (32).

The transition element cations, which comprise class C (2), have 0-10 d subshell electrons in the M shell (first series) or N

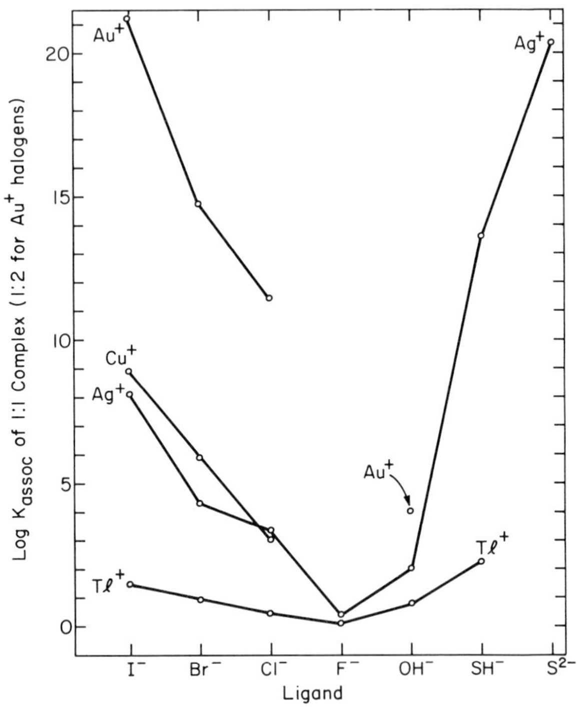

Figure 3. Stabilities of some class B monovalent cation complexes at I = *0 based on published data. EN values for Tl^+, Ag^+, Cu^+, and Au^+ are 1.5, 1.8, 1.8, and 2.3;* of *I^-, Br^-, Cl^-, F^-, OH^-, SH^-, and S^{2-} are 2.55, 2.8, 3.0, 3.95, 2.15, 2.3 (estimate), and 2.5, respectively (29).*

shell (second series), etc. These cations have partially filled 3d subshells either in the ground state or when ionized (Table II). Moving across the periodic chart from Sc to Cu, protons are added to the nucleus and electrons to the unfilled inner 3d subshell. Attraction of these inner electrons to the nucleus leads to an overall decrease in cation radii (Table III). The divalent ions are generally 6-fold coordinated in complexes. Cu^{2+} is an exception and, because of its small size and unique electronic configuration, tends to occur in distorted 4 or 6-fold coordination. Table III also lists Ip, EN values, and thermodynamic data for divalent cations of the first transition series. The near constancy of S^o for the transition metal cations indicates that their aquo-ion stability differences reflect chiefly ΔH_f^o differences. Typically, the stability order for inner sphere M^{2+} complexes (K_{assoc} values) follows the ΔG_f^o, ΔH_f^o, and EN sequences for the aquo-cations (11). Thus, for the M^{2+} cations of geochemical interest, complex stabilities increase in the order $Ca<Mn<Fe<Co<Ni<Cu>Zn$. This is the same as the order of increasing Ip, except for Cu^{2+}, which is preferentially stabilized in complexes because of Jahn-Teller effects (6). The above sequence has been called the Irving-Williams order. It is followed by stability constants with the majority of ligands, including oxalates, citrates, NTA, and EDTA (1, 11, 12), soil fulvic acid complexes (38), and also by the stability constants of the sulfide, telluride, and hydroxide minerals (16, 39). Although data are incomplete, available stability constants for M^{2+} - CO_3^{2-} and HPO_4^{2-} complexes (Figure 2) appear to follow the Irving-Williams order.

The Irving-Williams order results from an enhancement of stabilities for complexes formed with Fe^{2+}, Co^{2+}, Ni^{2+}, and Cu^{2+} because of electronic interaction between the cation and ligand, i.e., ligand field stabilization energies, LFSE's (6). Such effects do not occur with Mn^{2+} or Zn^{2+}, so that their complexes are generally less stable than those formed with Fe^{2+} and Cu^{2+}, respectively. Accordingly, a plot of ΔG^o, ΔH^o, or log K_{assoc} data for a particular ligand against atomic number will usually show minima at Ca^{2+}, Mn^{2+}, and Zn^{2+}. LFSE effects will raise the stabilities of intermediate complexes above the line connecting Ca^{2+}, Mn^{2+}, and Zn^{2+}. The Mn^{2+} to Zn^{2+} portion of such a plot is shown in Figure 5 for 1:1 M^{2+} - OH^- complexes. Stability data for these complexes reported by Yatsimirskii and Vasil'ev (11) and Hancock and Mariscano (40, not shown) follow the Irving-Williams order. These authors suggest log K_{assoc} values of 3.9 and 3.7 for $FeOH^+$, repectively, whereas data in Wagman et al. (36) yield 7.2, with Baes and Mesmer (23) and Smith and Martell (16) both suggesting 4.5. These latter three data sets violate the Irving-Williams order, and, in addition, the sets of Baes and Mesmer and Smith and Martell suggest no LFSE for $NiOH^+$, which seems unlikely. Whichever of these data are correct, it is clear that K_{assoc} for $FeOH^+$ from Wagman et al. (36) is too large.

Based on LFSE arguments, the stability constants of the tri-

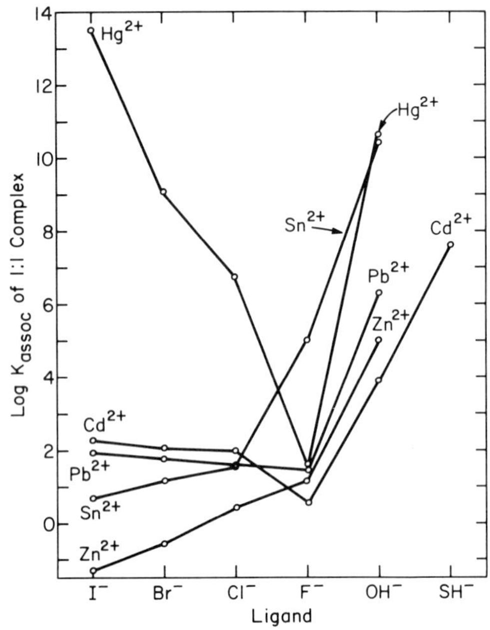

Figure 4. Stabilities of some class B divalent cation complexes at I = 0 based on published data. EN values for Hg^{2+}, Cd^{2+}, Pb^{2+}, Sn^{2+}, and Zn^{2+} are 1.8, 1.5, 1.6, 1.7, and 1.5, respectively (29).

Table II: Number of 3d electrons and common valence states in natural waters of elements of atomic numbers twenty to thirty. Parentheses indicate ion is rare or does not exist in natural waters as a complex former.

Element	No. of 3d electrons	(+) valence states	Comments
Ca	0	2	ion is class A
Sc	1	3	ion is class A
Ti	2	(2), (3), 4	4+ is class A
V	3	2, 3, 4, 5	----
Cr	5	(2), 3, 6	----
Mn	5	2, (3), 4	3+ in minerals only
Fe	6	2, 3	----
Co	7	2, (3)	3+ in minerals only
Ni	8	2	----
Cu	10	1, 2	1+ is class B
Zn	10	2	1+ is class B

Table III. Some properties of the divalent cations of elements with atomic numbers twenty to thirty, at 25°C and 1 atm pressure, Radii are in 6-fold coordination (33). EN values are from Allred (26). Thermodynamic data for Ca, Cu and Zn are from Parker *et al*. (34); Ti, V, and Cr from Latimer (35); Mn, Co, and Ni from Wagman *et al*. (36), and Fe from Tremaine *et al*. (37). Ti^{2+}, V^{2+}, and Cr^{2+} are unstable in water.

Cation	Ca	Ti	V	Cr	Mn	Fe	Co	Ni	Cu	Zn
Radius (A)	.99	.90	.88	.84	.80	.74	.72	.69	.72	.74
Ip	2.02	2.22	2.27	2.38	2.5	2.70	2.78	2.90	2.78	2.70
EN	1.00	1.54	1.63	1.66	1.55	1.83	1.88	1.91	1.90	1.65
ΔG_f^o (kcal/mol)	-132.24	(-75.1)	-54.7	-42.1	-54.5	-21.8	-13.0	-10.9	+15.68	-35.19
ΔH_f^o (kcal/mol)	-129.71	----	----	-33.2	-52.76	-22.1	-13.9	-12.9	+15.70	-36.66
S^o (cal/mol deg.)	- 12.79	----	----	----	-17.6	-25.6	-27.	-30.8	-23.21	-26.2

valent transition metal complexes often follow the Irving-Williams order Sc<Ti<V<Cr<Mn>Fe<Co<Ni<Cu>Ga (11). For the trivalent ions stable in nature waters, ΔG_f^o, ΔH_f^o, and EN data suggest the order of increasing complex stabilities should be V<Cr<<Fe<Co>>Ga. Among the trivalents, Sc^{3+}, Fe^{3+}, and Ga^{3+} form the sequence of minimal complex stabilities, reflecting the absence of LFSE effects, with intermediate cation complexes having greater stabilities because of such effects.

Exceptions to the two Irving-Williams orders for inner sphere complexes occur particularly when complexes with more than four ligands or large, assymetric ligands are compared (12). Stability constant sequences for weak outer sphere complexes, such as are formed with chloride (41) and fluoride (16, 40), do not obey the Irving-Williams orders.

As noted earlier, Pearson (3) lists cations Mn^{2+}, Ga^{3+}, Cr^{3+}, Co^{3+}, and Fe^{3+} in the hard acid class. The divalent cations Fe, Co, Ni, and Cu are described as "borderline" between the hard and soft acids.

Some Thermodynamic Considerations

The stability of a complex is usually described by its K_{assoc} value, where K_{assoc} = [ML]/[M][L] for a 1:1 complex, $\Delta G^o = -RT\ln K_{assoc}$, and $\Delta G^o = \Delta H^o - T\Delta S^o$. The stability of the complex depends on the relative and absolute contributions of ΔH^o and ΔS^o. Ahrland (4) observes that the strongest complexes are formed by the association of hard acid cations and hard base addends, such as F^-, OH^-, and O^{2-}, or by soft acid cations and soft base addends, such as S^{2-}, Se^{2-}, and I^-. Bonding in hard acid-hard base complexes is largely electrostatic and reflects a predominant $+\Delta S^o$ and usually minor $+\Delta H^o$ (endothermic) contribution to complexation. On the other hand, bonding in soft acid-soft base complexes is largely covalent, and the stability of the complex results from a large $-\Delta H^o$ (exothermic) contribution. The ΔS^o term is usually small in this case and may be of either sign. Complexes formed between hard-soft or soft-hard addends are generally weak.

Because most common species of geochemical interest are hard, the stability of complexes usually reflects a $+\Delta S^o$ contribution to their formation (42). This entropy effect can be qualitatively described in terms of structural changes consequent to complexation. Thus:

$$\underset{(+)}{\Delta S^o} = \underset{(-)}{\Delta S^o_{net\ chg}} + \underset{(-)}{\Delta S^o_{tr}} + \underset{(+)}{\Delta S^o_{rot}} + \underset{(+)}{\Delta S^o_{vibr}} + \underset{(+)}{\Delta S^o_{dehydr}}$$

where the signs in parentheses indicate the usual contribution of each term. There is an entropy reduction ($\Delta S^o_{net\ chg}$) due to the decrease in number of charged particles. Some of the translational entropy (ΔS^o_{tr}) in the addends is converted to vibrational (ΔS^o_{vibr}) and rotational entropy (ΔS^o_{rot}) in the complex. However, the chief effect is usually the + ΔS^o_{dehydr} term which reflects partial dehydration of the cation and/or ligand. Complexation leads to a breakdown of the structured water of hydration around the separate addends, with a resultant decrease in the order of the solution. Accordingly, the $+\Delta S^o_{dehydr}$ term is proportional to the number of water molecules displaced by the ligand. Thus, ΔS^o_{dehydr} (and $+\Delta S^o$ total) is greatest for multivalent ion complexation, since these species are initially the most hydrated. If little or no coordinating water is eliminated in complex formation, a net $-\Delta S^o$ may result because of the contributions of $-\Delta S^o_{net\ chg}$ and $-\Delta S^o_{tr}$. This is the case with cation-neutral ligand complexes, which consequently are weak. For the same reason, ΔS^o is less positive in ion pair formation than in inner sphere complexation.

The $+\Delta S^o$ term is usually greatest for the addition of the first ligand to the cation, decreasing for each additional ligand. Chelates (such as EDTA) are ligands that can bond the cation via more than one electron-donating atom in the ligand. This usually frees several cation-coordinating waters and so yields a large $+\Delta S^o$ contribution to complex formation (6, 43).

Correlation or Comparison Methods and Plots

Nieboer and McBryde (44) have developed what they term a correlating numerical index Q, which is a function of the ionic charge and electronegativity of a particular metal cation, and of the type of ligand complexing with it. Linear relations are obtained when Q is plotted against log K_{assoc} values for complexes formed by more than thirty-six metal cations, and ligands that include the halogens, OH^-, and a number of organic anions. Nieboer and McBryde show that Q is directly related to the hardness or softness (3, 4) of the cation, and to the thermodynamic properties of the complex.

Other researchers have derived equations to predict the thermodynamic properties of complexes based on the degree of softness or hardness of their addends. The earlier work in this area is reported on in papers collected by Pearson (3, 45). More recently, such research has been continued by Hancock and Marsicano (40), among others. These authors (see also Drago *et al.* in Pearson (3)) have suggested equations of the form log K_{assoc} (or $-\Delta H^o$) = $C_A C_B + E_A E_B$, where C and E are the contributions of covalent and ionic bonding, and subscripts A and B denote the acid and base, respectively. The right-hand terms are established empirically. Hancock *et al.* (46), Hancock and Marsicano (40), and Marsicano and Hancock (47) have used graphic methods to correlate the thermodynamic properties of metal complexes in terms of differences in the hardness or softness of the metals and associated ligands. Stability constants have been so correlated for Ag^+, Au^+, Cu^+, Cu^{2+}, Co^{2+}, Hg^{2+}, Fe^{3+}, and Bi^{3+}.

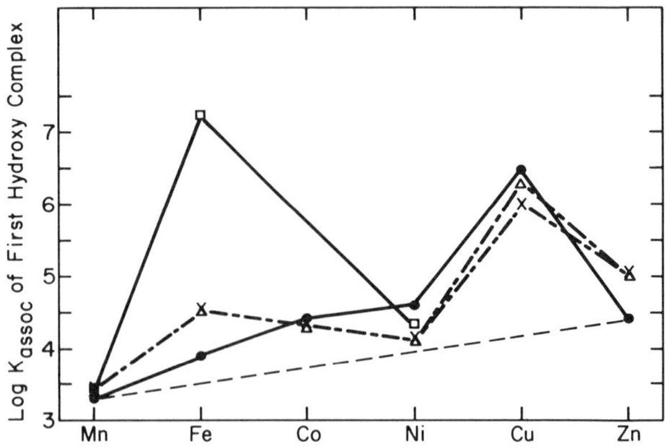

Figure 5. Stability constants of 1:1 hydroxy complexes of divalent metal cations of the first transition series. (□) Wagman et al. (36); (×) Baes and Mesmer (23); (●) Yatsimirskii and Vasil'ev (11); (△) Smith and Martell (16).

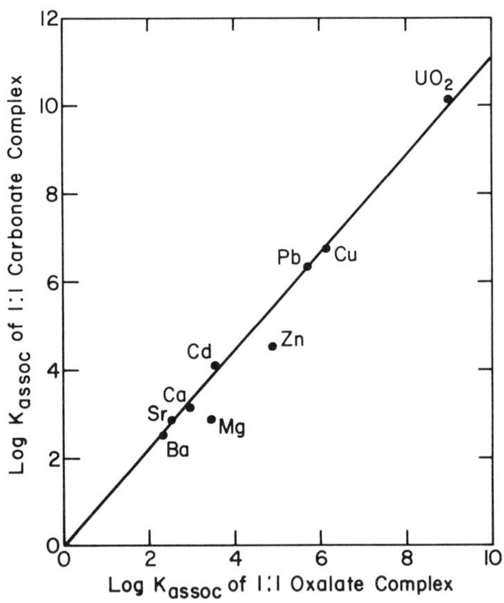

Figure 6. Stability constants of some 1:1 divalent metal carbonate complexes plotted against stability constants for the corresponding oxalate complexes. Data are from literature and are for I = 0. The equation of the line is log K_{assoc} ($MCO_3°$) = 1.11 × log K_{assoc} ($MC_2O_4°$).

Comparison plots such as these are a valuable predictive tool. Such plots are most useful when the cations or ligands chosen for comparison have similar properties, such as equal valence, similar size and geometry, and similar electron configurations and bonding properties. In this way, trends or differences in their plotted behavior can be more rationally understood. For example, Figure 6 compares the stability constants of oxalate and carbonate complexes with the same divalent cations. The equation of the line is $\log K_{MCO_3^{\circ}} = 1.11 \log K_{MC_2O_4^{\circ}}$. The plot indicates a strong correlation ($r^2 = 0.95$). Stability constant data for carbonate complexes is less available than that for the oxalate complexes. Yatsimirskii and Vasil'ev (11) have developed the equation $\log K_{assoc} = 2.5 + 0.47 \times B$ for the $MC_2O_4^{\circ}$ complexes, where B is 2.0, 3.0, 4.0, 4.8, 5.0, 5.2, 6.0, 6.9, and 8.5 for the divalent ions Mg, Mn, Fe, Co, Pb, Zn, Ni, Cu, and Co. The success of Figure 6 suggests that the stability of unknown carbonate complexes might be predicted using it. Thus, $\log K_{assoc} = 3.89$ for $MnC_2O_4^{\circ}$ (11) leads to $\log K_{assoc} = 4.32$ for $MnCO_3^{\circ}$.

The cations Fe^{3+} and Al^{3+} are both considered hard acids by Pearson (3). Thus, it is not surprising that a plot comparing stabilities of the isostructural complexes and minerals gives a strong correlation. The oxygen-donor species, in particular, plot nearly on a straight line. The almost exact linearity of the OH complexes is particularly interesting. The relations in Figure 7 suggest a means of estimating the stability constants of $AlH_2PO_4^{2+}$ and $AlHPO_4^{+}$, which have not been measured. The $\log K_{assoc}$ values for $FeH_2PO_4^{2+}$ and $FeHPO_4^{+}$ are 4.17 and 9.92 at zero ionic strength, respectively, based on Galal-Gorchev and Stumm (48). The ratio of the logs of the plotted Fe^{3+} versus Al^{3+} complexes with one OH^- and one SO_4^{2-} ligand are 1.32 and 1.37, respectively. Assuming an average ratio of 1.35 (slope of solid line in Figure 7), leads to 3.1 and 7.4 for the $\log K_{assoc}$ values of $AlH_2PO_4^{2+}$ and $AlHPO_4^{+}$, respectively. Only data for the MSO_4^{+} and MOH^{2+} complexes were used in this case. This is because they are the only 1:1 complexes formed with oxygen-donor ligands in the plot. Their bonding is thus more likely to be comparable to that in the 1:1 phosphate complexes.

A variety of other plots have been found useful for predicting unknown data and appraising published data. Figure 8 shows stability constants for metal orthophosphate complexes as a function of decreasing valence of the $H_nPO_4^{n-3}$ ligand (n = 0-3). The increasing strength of the complexes with increasing ligand valance is expected. Also expected is the grouping of cations by their valence for each ligand, and the observation that H^+ and

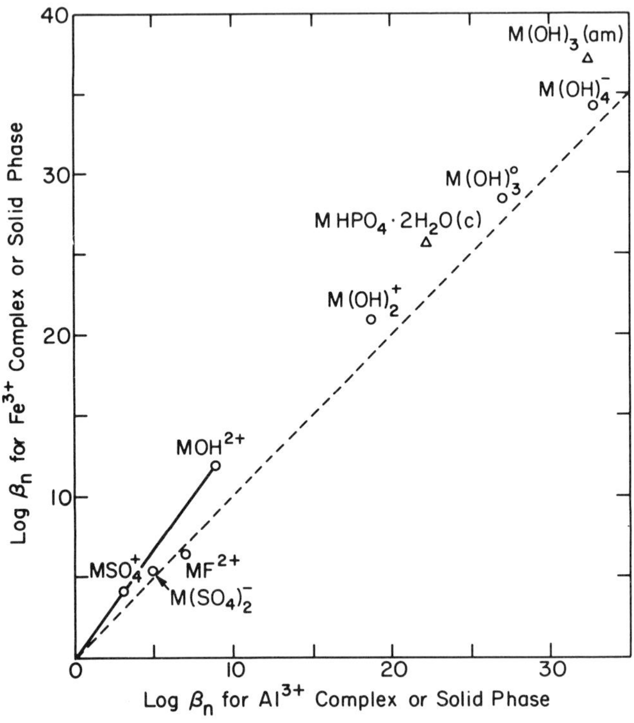

Figure 7. Cumulative stability constants for some Fe(III) complexes and solids at I = 0, plotted against the constants for corresponding Al(III) complexes and solids. The dashed line is drawn for reference through equal log β_n values for both metals.

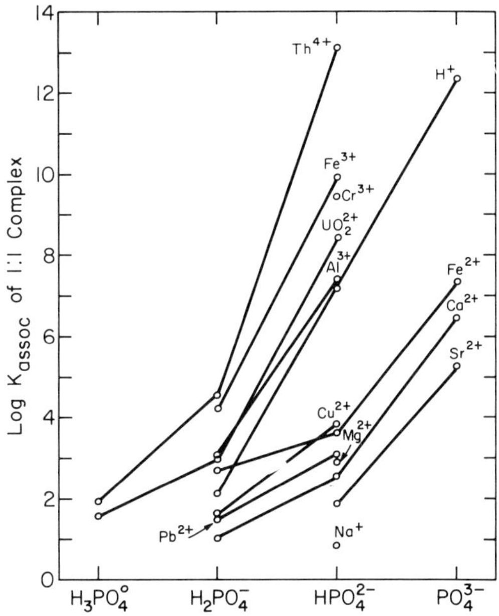

Figure 8. Stability constants of 1:1 complexes with $H_nPO_4^{n-3}$ ligands (n = 0–3). All stability constants are for I = 0, except values for $ThH_3PO_4^{4+}$ and $UO_2H_3PO_4^{2+}$, which were obtained at I = 2M. Stability constants for $CoHPO_4^{\circ}$, $NiHPO_4^{\circ}$, and $ZnHPO_4^{\circ}$ (log K_{assoc} values of 3.0, 2.9, and 3.2, respectively) are not plotted to avoid cluttering the figure. The constants for Al^{3+} species are estimates.

UO_2^{2+} behave like trivalent cations. The relative constancy of differences in log K_{assoc} values for cations of equal valence with the same ligands suggests that if a stability constant with one of these ligands is known, the other constants may be roughly predicted. The plot suggests, for example, that the published constant for $FeH_2PO_4^+$ may be too large.

Another useful approach is to plot the number of ligands (n) in monomeric complexes with a given ligand against the cumulative formation constant (β_n) of each complex. ($\beta_n = [ML_n]/[M][L]^n$, where valences are ignored for simplicity.) Such plots are given in Figures 9 and 10 for thorium and uranium complexes, respectively. The stepwise constants (as log K_n values) for the complexes where $K_n = [ML_n]/[ML_{n-1}][L]$ correspond to the differences in successive log β_n values (note $\beta_n = K_1K_2 \ . \ . \ . \ K_n$ (49). The steepness of the slope between successive cumulative constants is a measure of the size of the stepwise formation constant for the complex with the higher n value. Figure 9 shows that for weak complexes, such as form with chloride, complexes beyond ML are relatively weak and thus unimportant. In contrast, the more stable the ML species, the more likely that higher ligand-numbered complexes will also be important. The general parallelism of the lines through successive β_n values suggests that, when such lines cross, the constants responsible should be questioned. This casts doubt, for example, on β_2 for $UO_2(HPO_4)_2^{2-}$ in Figure 10.

The maximum value of n in a stepwise complex is constrained by the valences of the addends and maximum possible coordination number of the ligand around the cation (usually ≤ 6). Monovalent and divalent cations rarely associate with more than 2 ligands, whereas trivalent and more electropositive cations have maximum coordination numbers up to 6 with spherical ligands such as fluoride. Maximum coordination numbers of 3 or less are usual for oxyanions. Complexes above ML_n are rarely formed when ML_n has a negative charge in excess of 2. Complexes beyond 1:1 (and polymers) are most important at high ligand concentrations.

Beck (49) discusses equations for predicting K_n values in a stepwise series where two or more such values are known. For electrostatically bonded complexes $K_n/K_{n+1} = 10^{2\lambda}$ where λ is constant for a given M and L. This expression may be rewritten log $K_n = \log K_1 - 2\lambda(n-1)$. Such an approach has been used by Baes and Mesmer (23) and Langmuir (50) to predict the stabilities of hydroxy complexes.

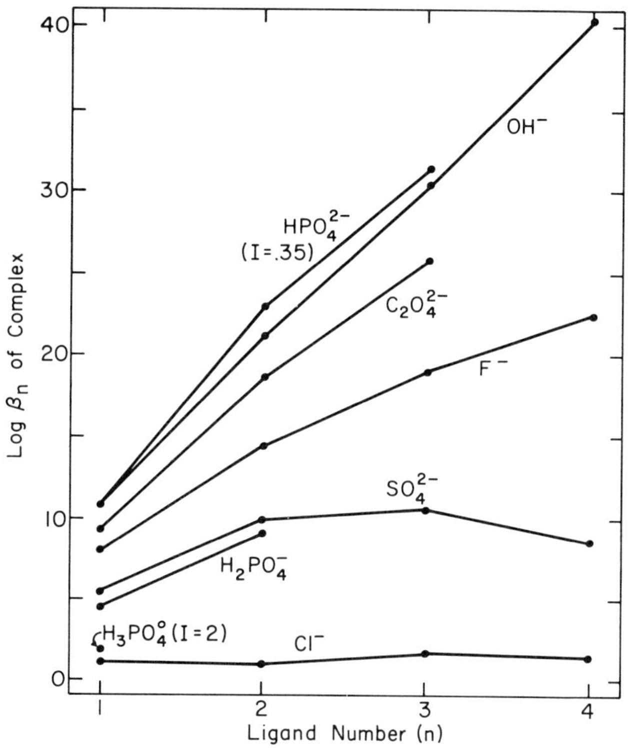

Figure 9. Cumulative formation constants of monomeric Th^{4+} complexes plotted against their ligand numbers. All the data are for I = 0 unless otherwise indicated.

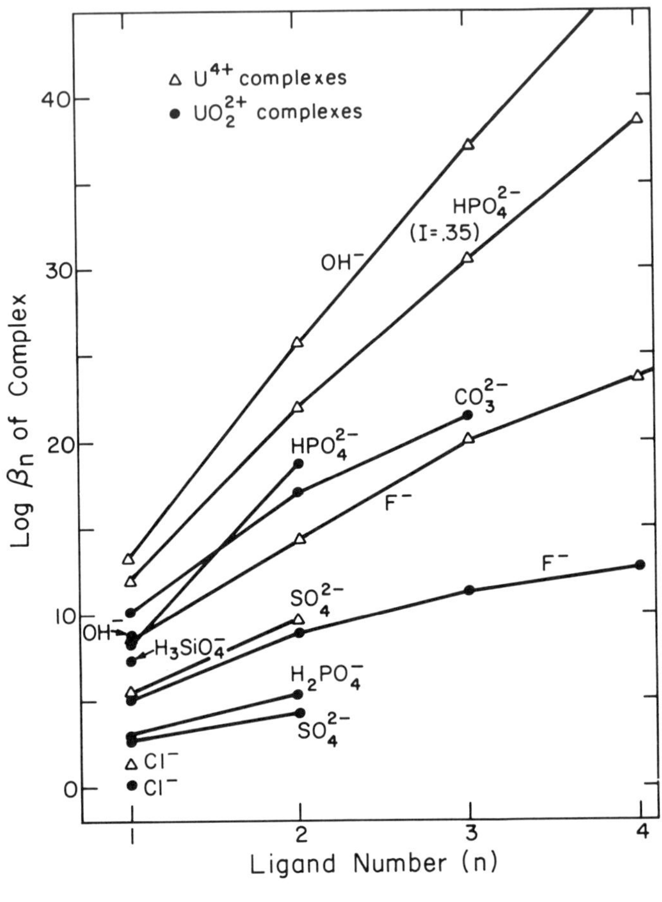

Geochim Cosmochim Acta

Figure 10. Cumulative formation constants of monomeric U^{4+} and UO_2^{2+} complexes plotted against their ligand numbers. All the data are for I = 0 unless otherwise indicated (50).

Thermodynamic Models Based on the Valences and Radii of the Addends

It has been observed that straight line plots often result when log K, $\Delta S°$, or $\Delta H°$ of association values for complexes formed within a given ligand are plotted against functions of cation and ligand charge, divided by a measure of the distance between cation and ligand in the complex. Such straight line relations have been taken by some as proof that the bonding results from strictly Coulombic forces (51). However, this is often not the case. It will be shown that straight line plots are obtained even when bonding is largely covalent.

The Simple Electrostatic Model. The simplest theoretical approach to predict the stability of a complex is to assume that the bonding is purely electrostatic, and that the separation distance (d) of the addends is the sum of their crystallographic (unhydrated) radii. These assumptions lead to

$$\Delta G° = z_+ z_- e_o^2 N_A / \varepsilon d \tag{1}$$

(7, 9), where z_+ and z_- are the valences of the cation and anion, respectively, e_o the electron charge, N_A Avogadro's number, and ε the dielectric constant of water. At 25°C, this expression reduces to

$$\Delta G° (\text{kcal/mol}) = -4.24 \times 10^{-8} \ (z_+ z_- / d) \tag{2}$$

or

$$\log K_{assoc} = 3.11 \times 10^{-8} \ (z_+ z_- / d) \tag{3}$$

where d is in centimeters. Because $\Delta S° = -\partial \Delta G° / \partial T$, one can show

$$\Delta S° = \Delta G° \ (\partial \ln \varepsilon / \partial T). \tag{4}$$

The term $(\partial \ln \varepsilon / \partial T)$ is almost constant near room temperature and equals 0.00454 K^{-1} at 25°C based on data in Akerlof and Oshry (52). (Note also $\partial \varepsilon / \partial T = -0.3556 \ K^{-1}$ near 25°C.)

The Bjerrum Model. Bjerrum (see Robinson and Stokes (19)) defined an "ion pair" as existing when two ions of opposite charge approached such that the mutual potential energy between them equalled 2kT (k is the Boltzman constant). At 25°C, this means that an "ion pair" exists if the ion separation distance is equal to or less than $3.57 |z_+ z_-|$A. For large b values, the model leads to

$$K_{assoc} = 4 \pi N_A d^3 e^b / 1000b \tag{5}$$

(53), where $b = z_+z_-e_o^2/\varepsilon dkT$. Expression 5 may be used to compute $\Delta G°$ of complexation via $\Delta G° = -RT\ln K_{assoc}$. In general, expression 5 is equivalent to

$$\Delta G°(cal/mol) = -112.8\ T + RT\ \ln\ [z_+z_-/d^4\varepsilon T] - 3.321 \times 10^{-3}\ [z_+z_-/d\varepsilon] \quad (6)$$

which at 25°C becomes

$$\Delta G°(cal/mol) = -39600 - 4.241 \times 10^{-5}\ [z_+z_-/d] + 1364\ \log\ [z_+z_-/d^4]. \quad (7)$$

Because $\Delta S° = -\partial\Delta G°/\partial T$, it may be shown

$$\Delta S°\ (cal/mol\ deg) = 114.8 - R\ \ln\ [z_+z_-/d^4\varepsilon T] - 0.3556\ RT/\varepsilon \quad (8)$$

and at 25°C

$$\Delta S°\ (cal/mol\ deg) = 132.1 - R\ \ln\ [z_+z_-/d^4] + 19.26 \times 10^{-8}\ [z_+z_-/d]. \quad (9)$$

The Fuoss Model. Fuoss (9) observed that because the Bjerrum d value would generally exceed the sum of the radii of the addends, Bjerrum's so-called "ion pairs" might not be in contact. Accordingly, he developed an equation based on the assumption that an "ion pair" existed only when oppositely charged ions were in contact. Based on this and other arguments (19, 54), Fuoss derived the expression

$$K_{assoc} = 4\pi N_A d^3 e^b/3000 \quad (10)$$

with terms defined as before. Robinson and Stokes (19) note that this expression may be linearized in the form

$$-\ln K_{assoc} = A - B/\varepsilon T \quad (11)$$

where $A = \ln\ [3000/4\pi N_A d^3]$ and $B = z_+z_-\ e_o^2/kd$ (see also Siebert and Hostetler (55)). Expression 10 is equivalent to

$$\Delta G°(cal/mol) = -RT\ \ln\ [2.524 \times 10^{21} d^3] - 3.321 \times 10^{-3}[z_+z_-/\varepsilon d] \quad (12)$$

and at 25°C

$$\Delta G°(cal/mol) = -29200 - 1364\ \log\ d^3 -4.241 \times 10^{-5}[z_+z_-/d]. \quad (13)$$

The corresponding entropy at 25°C is

$$\Delta S^\circ(\text{cal/mol deg}) = 97.94 + 4.576 \log d^3 + 19.26 \times 10^{-8}[z_+z_-/d]. \quad (14)$$

Application of the Models. Plotted in Figures 11 and 12 are measured and model-predicted ΔG° and ΔS° values for 1:1 sulfate complexes against $z_+z_-/(r_M + r_{SO_4})$. Clearly, the Fuoss equation better predicts the ΔG° data than does the Bjerrum equation or simple electrostatic equation. Not surprisingly, soft (class B) cations Tl^+, Ag^+, and Hg^{2+} have stabilities discordant from the borderline and hard cations. The Fuoss equation has been assumed by some to be a measure of outer sphere coulombic contributions to complexation (40). Its success at all valences is, in fact, a bit surprising, particularly since the Fuoss values are computed assuming constant, crystallographic ionic radii and the dielectric constant for bulk water (ε = 78.3). The cation-sulfate separation distance decreases with increasing valence of the cation, along with increasing covalency of bonding for 3+ and 4+ cations which form predominantly inner sphere complexes. Thus, true cation-SO_4 separation values must exceed the sum of the crystallographic radii for MSO_4^- ion pairs but approach this sum in the MSO_4^{2+} complexes.

Further, the dielectric constant of water associated with a complex is known to decrease as cation and ligand more closely approach each other (19, 56). Thus, Choppin and Unrein (57) suggest "effective" ε values of 57.0 for MF^{+2} and 40.8 for MF^{+3} complexes. The drop in both d and ε should increase the stability of multivalent cation complexes over monovalent ones. That the Fuoss equation roughly predicts ΔG° for 3+ and 4+ cation complexes, although ignoring real changes in d and ε, must therefore be considered fortuitous.

The Fuoss model yields more accurate ΔS° values for the sulfate complexes than the Bjerrum model. Most remarkable, however, is the even better overall accuracy of ΔS° calculated by the simple electrostatic model.

An extensive body of literature exists on the thermodynamic properties of fluoride complexes. As noted by Hefter (51), the electronic configuration of fluoride is such that bonds in its complexes should be governed by electrostatic effects "more than for any other ligand." Hefter, and Mesmer and Baes (58) present a variety of plots of $\log K_{assoc}$ and ΔS° against such functions as $z_+z_-/(r_M + r_F)$. Mesmer and Baes develop simple equations to predict $\log K_{assoc}$ and $\log \beta_n$ values for the fluorides. Unfortunately, these equations and those of Hefter are defined for 1 molal ionic strength.

Figure 13 shows ΔG° data for fluoride complexes at zero ionic

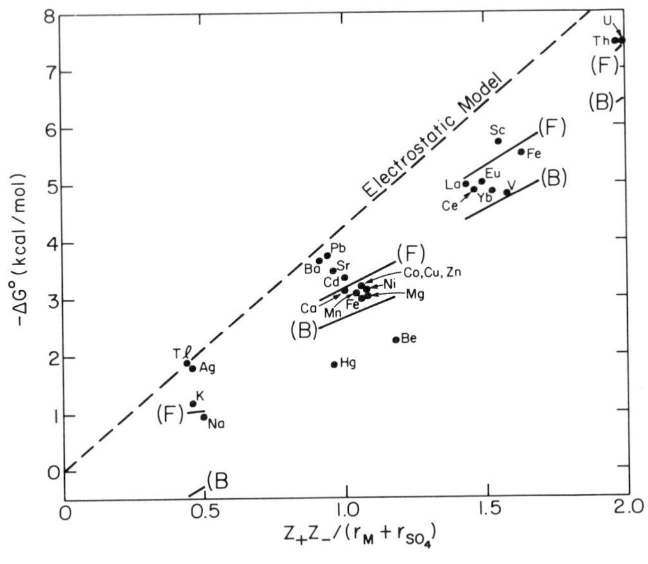

Figure 11. Empirical $-\Delta G°$ data for 1:1 metal sulfate complexes (I = 0) plotted against $z_+z_-/(r_M + r_{SO_4})$, where z_+ and z_- are the valence of cation and sulfate ion, r_M is the crystallographic radius in Angstroms of the cation in sixfold coordination (33), and $r_{SO_4} = 3.05$ A (59). The locus of $-\Delta G°$ values computed for the complexes by the simple electrostatic model is shown as a dashed line, and computed by the Fuoss and Bjerrum equations as lines labeled (F) and (B), respectively.

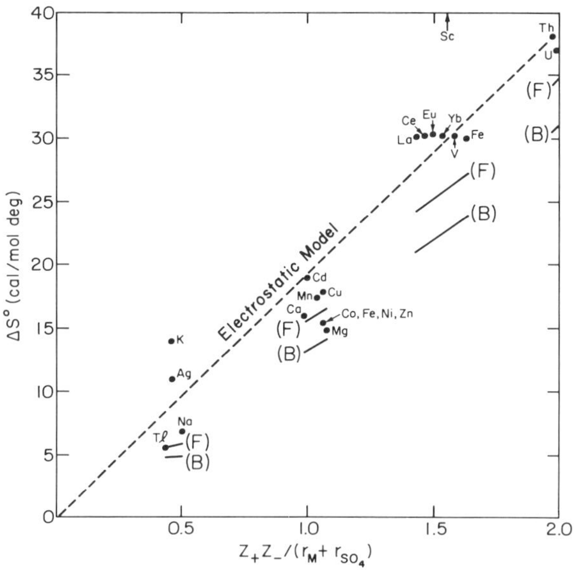

Figure 12. Empirical and model-predicted ΔS° values for 1:1 metal sulfate complexes (I = 0) plotted against $z_+z_-/(r_M + r_{SO_4})$. (F) Fuoss and (B) Bjerrum models.

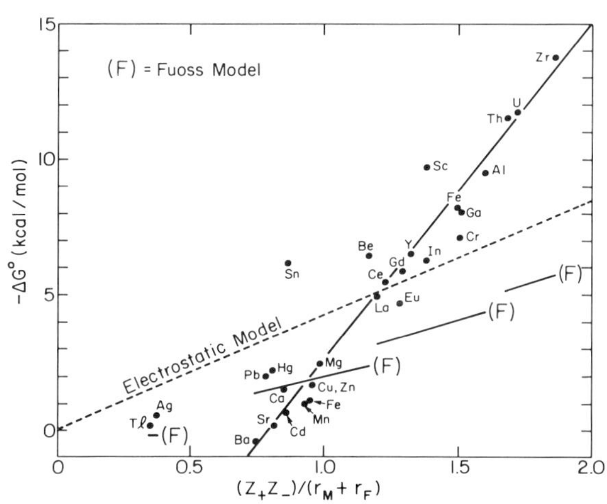

Figure 13. Empirical and model-predicted −ΔG° values for 1:1 metal fluoride complexes (I = 0) plotted against $z_+z_-/(r_M + r_F)(r_F) = 1.36$ Å from Ahrens (33). The solid straight line is regressed through the data for hard cations only. (F) Fuoss model.

strength versus $z_+z_-/(r_M + r_F)$. The line drawn through the data is an excellent fit, particularly if points for the soft cations (Tl^+, Ag^+, Cd^{2+}, Hg^{2+}, Ga^{3+}, In^{3+}) and the borderline cations (Pb^{2+}, Sn^{2+}, and Zn^{2+}) are ignored. Without these cations, the equation of the regression line through the data is $-\Delta G^o = -9.96 + 12.50\,[z_+z_-/(r_M + r_F)]$ with $r^2 = 0.95$. The hard 3+ and 4+ cation complexes are much more stable than predicted by the Fuoss or electrostatic models. These complexes evidently owe an important part of their stability to covalent bonding.

For the trivalent and quadrivalent actinide and lanthanide fluoride complexes, Choppin and Unrein (57) suggest an equation for ΔG^o which contains $z_+z_-/(r_M + r_F)\varepsilon$, with ε the adjustable parameter. They give ε = 79.8, 57.0, and 40.8 for MF^+, MF^{2+}, and MF^{3+} complexes. Introducing these values in the electrostatic or Fuoss model equations greatly improves the agreement between predicted and empirical data. However, it must be remembered that these ε values also include the effect of decreasing d values in the complex with increased valence of the cation.

Measured and predicted ΔS^o values are plotted against $z_+z_-/(r_M + r_F)$ in Figure 14. Again, surprisingly, the simple electrostatic model most closely reproduces the empirical data, except for Be^{2+} and Cu^{2+} and soft cations Ag^+, Cd^{2+}, and Hg^{2+}.

In Figure 15, ΔG^o data for $MHPO_4$ complexes are plotted against $z_+z_-/(r_M + r_{HPO_4})$. As with the fluorides, for M^{3+} comp-plexes, $-\Delta G^o$ rises rapidly above values predicted by the Fuoss (or electrostatic) models. This presumably reflects increased covalency of bonding with the multivalent cations. The curve is schematic, and has been drawn to suggest the trend of increasing covalency. No reliable ΔS^o or ΔH^o data are available for the HPO_4^{2-} complexes. However, emboldened by the success of the simple electrostatic equation for predicting ΔS^o in the case of sulfate and fluoride complexes, it is tempting to estimate ΔS^o for the phosphates in the same manner.

Conclusions

Some of the more important findings of this paper include the following:

1. The Fuoss equation is a good predictor of ΔG^o and ΔS^o values for complexes in which the bonding is chiefly electrostatic (most 1:1, 1:2, 2:1, and 2:2 complexes formed by hard acids and hard bases, including in this study F^-, SO_4^{2-}, and HPO_4^{2-} complexes).
2. The simple electrostatic model is the most accurate predictor of ΔS^o of complexation for trivalent and quadrivalent cation complexes with SO_4^{2-} and F^-.
3. Empirical thermodynamic data for complexes with a common ligand often plot along straight lines or smooth curves when the abscissa is z_+z_-/d. Such lines indicate important covalent bonding for 3+ and 4+ metal complexes with F^- and HPO_4^{2-}. The strong-

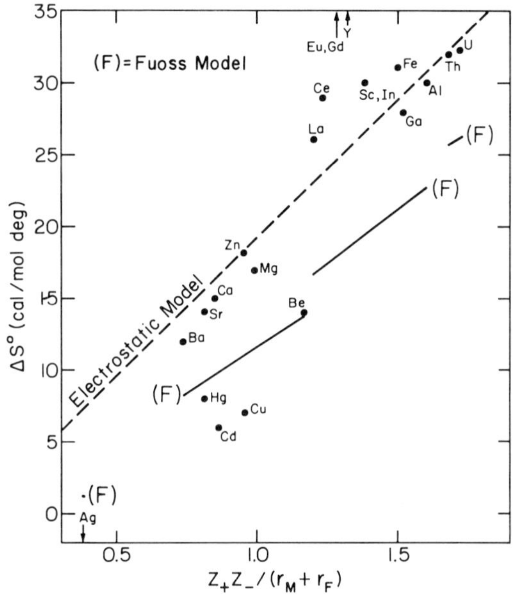

Figure 14. Empirical and model-predicted $\Delta S°$ *values for 1:1 metal fluoride complexes* ($I = 0$) *plotted against* $z_+z_-/(r_M + r_F)$. *(F) Fuoss model.*

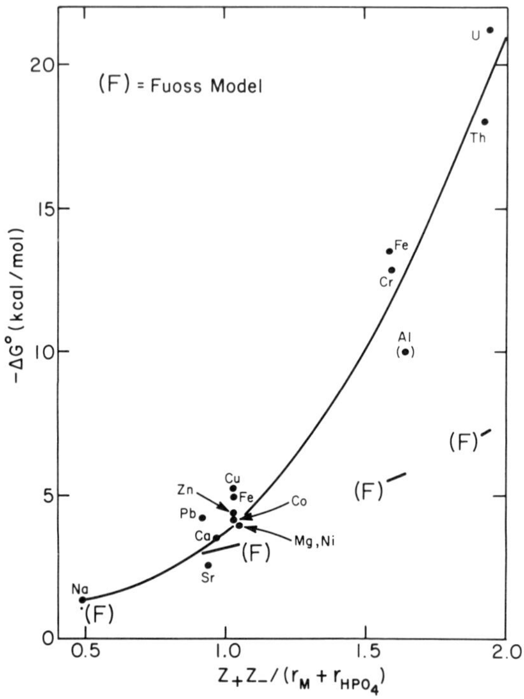

Figure 15. Empirical $-\Delta G°$ values for metal cation $-HPO_4$ complexes ($I = 0$) plotted against $z_+z_-/(r_M + r_{HPO_4})$, with $r_{HPO_4} = 3.15$ Å based on Izatt et al. (60) and Wells (27). $-\Delta G°$ for $AlHPO_4^+$ is an estimate. The smooth plotted curve has no statistical significance. (F) Fuoss model.

est linear correlations are obtained when such plots compare the stability of complexes with metal cations having the same general behavior (for example, hard or soft, or of class A or B).

4. Electronegativity is a useful plotting parameter, particularly for inner sphere complexes such as are formed with 3+ and 4+ metal cations or with soft cations or soft ligands.

5. Plots comparing the stabilities of complexes in which two similar cations or ligands occur are also valuable. For example, in this paper, plots of Fe(III) versus Al(III) complex stabilities and CO_3^{2-} versus $C_2O_4^{2-}$ complex stabilities led to estimated K_{assoc} values for $AlHPO_4^+$ and $MnCO_3^o$.

Thermodynamic data, and especially ΔS^o values, are generally unreliable or lacking for important phosphate complexes. Until such ΔS^o data is measured, it can be estimated with fair accuracy using the Fuoss equation for monovalent and divalent-bonded complexes and the electrostatic model when trivalent and quadrivalent addends are associated. Unfortunately, published ΔG^o and ΔS^o data on HS^-, S^{2-}, and Se and Te aquo-complexes are suspect or largely lacking (Barnes, H. L., Pennsylvania State University, personal communication, 1978). Both the stoichiometry and stability of such complexes remains in doubt. Once a few such data have been accurately measured, plots with EN (10) or Q (44) as a variable, or using hard and soft acid and base concepts (3, 4, 40) should permit the useful estimation of many as yet unknown values.

Acknowledgements

The aid of Graduate Research Assistant Joseph McNally in making the model calculations, and secretaries Margaret Biggers, Kim Mestecky, and Patricia Roberts for typing the manuscript is greatly appreciated. Dr. Antonio Lasaga suggested several improvements to an earlier version of this paper. Financial support for this research has been provided by the National Science Foundation through Grant No. AER77-06511.

Abstract

Geochemists and others concerned with the behavior of metals in natural waters are frequently confounded by a dearth of reliable thermodynamic data for aquo-metal complexes at zero ionic strength. Classifications of types of cations, ligands, and complexes which provide guidance in appraising or estimating such data for complexes include: the concept of inner versus outer sphere-type complexes (1); grouping metal cations into classes A, B, and C, based on their electron configurations (2); and the concept of hard and soft acids and bases (3, 4). Taking into account these classifications, and considering also the effects on complexation of valence, electronegativity, radii of the addends, and of ligand number, a host of graphic and other methods have been developed to relate and predict the free energy and entropy of formation of complexes. Such approaches, including the Fuoss and Bjerrum equations, have been used here to appraise published values and to estimate free energy and entropy of formation of metal complexes with hydroxyl, the halogens, bicarbonate, carbonate, sulfate, and orthophosphate among other ligands, and for a number of thorium and uranium complexes.

Literature Cited

1. Nancollas, G. H. "Interactions in Electrolyte Solutions," 214, Elsevier, New York, 1966.
2. Schwarzenback, G. The general, selective and specific formation of complexes by metallic cations, Adv. Inorg. Radiochem. 3, 257, 265-27 (1961).
3. Pearson, R. G., ed., "Hard and Soft Acids and Bases," 480 p. Dowden, Hutchinson and Ross, Stroudsburg, Pennsylvania, 1973.
4. Ahrland, S., Thermodynamics of the stepwise formation of metal-ion complexes in aqueous solution, Structure and Bonding 15, 167-188 (1973).
5. Phillips, C. S. H., and Williams, R. J. P. "Inorganic Chemistry," Vol. II, 693 p., Oxford Univ. Press, New York, 1966.
6. Cotton, F. A., and Wilkinson, G. "Advanced Inorganic Chemistry, A Comprehensive Text,"3rd Ed., 1145 p., John Wiley and Sons, Inc., New York, 1972.
7. Denison, J. T., and Ramsey, J. B., Free energy, enthalpy, and entropy of dissociation of some perchlorates in ethylene chloride and ethylidene chloride, J. Am. Chem. Soc. 77, 2615-2621 (1955).
8. Harned, H. S., and Owen, B. B. "The Physical Chemistry of Electrolyte Solutions," 3rd Ed., 803 p., Reinhold, New York, 1958.
9. Fuoss, R. M., Ionic association. III. The equilibrium between ion pairs and free ions, J. Am. Chem. Soc. 80, 5059-5061 (1958).
10. Garrels, R. M., Christ, C. L. "Solutions, Minerals, and Equilibria," 450 p., Freeman, Cooper and Co., San Francisco, 1965.
11. Yatsimirskii, K. B., Vasil'ev, V. P., "Instability Constants of Complex Compounds," 214 p., D. Van Nostrand Co. New York, 1966.
12. Ringbom, A., "Complexation in Analytical Chemistry," 395 p., John Wiley and Sons, New York, 1963.
13. Sillén, L. G., and Martell, A. E., "Stability Constants," 754 p., Chem. Soc., Spec. Publ. No. 17, Burlington House, London, (1964).
14. Sillén, L. G., and Martell, A. E., "Stability Constants of Metal-Ion Complexes, Supplement No. 1," 865 p., Chem. Soc. Spec. Publ, No. 25, Burlington House, London (1971).

15. Christensen, J. J., Eatough, D. J., and Izatt, R. M. "Handbook of Metal Ligand Heats," 2nd Ed., Rev., 495 p., Dekker, Inc., New York, 1975.
16. Smith, R. M., and Martell, A. E. "Critical Stability Constants, Inorganic Complexes," Vol. 4, 257 p., Plenum Press, New York, 1976.
17. Wagman, D. D., Schumm, R. H., and Parker, V. B. A Computer-Assisted Evaluation at the Thermochemical Data of the Compounds of Thorium, Natl. Bur. Stds. Info. Rept. 77-1300, 29 p. (1977).
18. Langmuir, D. Uranium solution-mineral equilibria at low temperatures with applications to sedimentary ore deposits, Geochim. Cosmochim. Acta 42, 547-569 (1978).
19. Robinson, R. A., and Stokes, R. H. "Electrolyte solutions," 2nd Ed., Rev., 571 p., Butterworths, London, 1959.
20. Ohtaki, H., Yamaguchi, T., and Maeda, M. The structures of hydrated divalent transition-metal ions in solution, p. 163-168, in Wanninen, E., ed., "Essays on Analytical Chemistry," Pergamon Press, New York, 1977.
21. Veillard, H., Hydration of the cations Al^{3+} and Cu^{2+}, a theoretical study, J. Am. Chem. Soc. 99(22), 7194-7199 (1977).
22. Shannon, R. D., and Prewitt, C. T. Effective ionic radii in oxides and fluorides, Acta Cryst. B25, 925-946, (1969).
23. Baes, C. F., Jr., and Mesmer, R. E. "The Hydrolysis of Cations," 489 p., John Wiley, New York, 1976.
24. Pauling, L. "The Nature of the Chemical Bond," 3rd Ed., 644 p., Cornell University Press, Ithaca, New York, 1960.
25. Gordy, W., and Thomas, W. J. O. Electronegativities of the elements, J. Chem. Phys. 24(2), 439-444 (1956).
26. Allred, A. L. Electronegativity values from thermochemical data, J. Inorg. Nucl. Chem. 17, 215-221 (1961).
27. Wells, A. F. "Structural Inorganic Chemistry," 3rd Ed., 1055 p., Clarendon Press, Oxford, 1962.
28. Rösler, H. J., and Lang, H. "Geochemical Tables," 468 p., Elsevier, New York, 1972.
29. Clifford, A. F. The electronegativity of groups, J. Phys. Chem. 63, 1227-1231 (1959).
30. Phillips, C. S. H., and Williams, R. J. P. "Inorganic Chemistry," Vol. I, 685 p., Oxford Univ. Press, New York, 1965.
31. Stumm, W., and Morgan, J. J. "Aquatic Chemistry," 583 p., John Wiley, New York, 1970.
32. Pearson, R. G. Hard and soft acids and bases, Part I: J. Chem. Educ. 45, 581-587 (1968).
33. Ahrens, L. H. The use of ionization potentials. Part I. Ionic radii of the elements, Geochim. Cosmochim. Acta 2, 155-169 (1952).
34. Parker, V. B., Wagman, D. D., and Garvin, D. Selected thermochemical data compatible with the CODATA recommendations NBSIR 75-968, Interim Report, Office of Std. Ref. Data,

Natl. Bur. Stds., Washington, D. C., 31 p. (1976).
35. Latimer, W. M. "Oxidation Potentials," 2nd Ed., 392 p., Prentice-Hall, Englewood Cliffs, New Jersey, 1952.
36. Wagman, D. D., Evans, W. H., Parker, V. B., Halow, I., Bailey, S. M., and Schumm, R. H. Selected values of chemical thermodynamic properties, Natl. Bur. Stds. Tech. Note 270-4, 141 p. (1969).
37. Tremaine, P. R., von Massow, R., and Shierman, G. R. A calculation of Gibbs free energies for ferrous ions and the solubility of magnetite in H_2O and D_2O to 300°C, Thermochim. Acta 19, 287-300 (1977).
38. Schnitzer, M. Metal-organic matter interactions in soils and waters, Ch. 13, p. 297-315, in Faust, S. D., and Hunter, J. V., ed., "Organic Compounds in Aquatic Environments," Marcel Dekker, New York, 1971.
39. Naumov, G. B., Ryzhenko, B. N., and Khodakovsky, I. L., "Handbook of Thermodynamic Data," 328 p., NTIS, Springfield, Virginia, 1974.
40. Hancock, R. D., and Marsicano, F. Parametric correlation of formation constants in aqueous solution. 1. Ligands with small donor atoms, Inorg. Chem. 17(3), 560-564 (1978).
41. Libûs, Z., and Tialowskia, H. Stability and nature of complexes of the type $M\ Cl^+$ in aqueous solution (M = Mn, Co, Ni, Zn), J. Soln. Chem. 4(12), 1011-1022 (1975).
42. Helgeson, H. C. "Complexing and Hydrothermal Ore Deposition," 128 p., Macmillan, New York, 1964.
43. Schwarzenback, G. Electrostatic and non-electrostatic contributions to ion association in solution, Pure and Applied Chem. 24, 307-334 (1970).
44. Nieboer, E., and McBryde, W. A. E. Free energy relationships in coordination chemistry. III. A comprehensive index to complex stability, Can. J. Chem. 51, 2512-2524 (1973).
45. Misono, M. and Saito, Y. Evaluation of softness from the stability constants of metal-ion complexes, Bull. Chem. Soc. Japan 43(12), 3680-3684 (1970).
46. Hancock, R. D., Finkelstein, N. P., and Evers, A. Linear free energy relationships in aqueous complex-formation reactions of the d^{10} metal ions, J. Inorg. Nucl. Chem. 36, 2539-2543 (1974).
47. Marsicano, F., and Hancock, R. D. The linear free energy relation in the thermodynamics of complex formation, Part 2, J. Chem. Soc. Dalton, 228-234 (1978).
48. Galal-Gorchev, H., and Stumm, W. The reaction of ferric iron with orthophosphate, J. Inorg. Nucl. Chem. 25, 567-74 (1963).
49. Beck, M. T. "Chemistry of Complex Equilibria," 285 p., Van Nostrand Reinhold, New York, 1970.
50. Langmuir, D. Uranium solution-mineral equilibria at low temperatures with applications to sedimentary ore deposits, Geochim. Cosmochim. Acta 42, 547-569 (1978).

51. Hefter, G. Simple electrostatic correlations of fluoride complexes in aqueous solutions, Coord. Chem. Rev. 12, 221-239 (1974).
52. Akerlof, G. C., and Oshry, H. I. The dielectric constant of water at high temperatures and in equilibrium with its vapor, J. Am. Chem. Soc. 72, 2844-2847 (1950).
53. Fuoss, R. M., and Kraus, C. A. Properties of electrolyte solutions. III. The dissociation constant, J. Am. Chem. Soc. 55, 1019-1028 (1933).
54. Bockris, J. O'M, and Reddy, A. K. N. "Modern Electrochemistry," 622 p., Plenum Press, New York, 1973.
55. Siebert, R. M., Hostetler, P. B. The stability of the magnesium bicarbonate ion pair from 10° to 90°C, Am. Jour. Sci. 277, 697-715 (1977).
56. Friedman, H. L., and Krishman, C. V. Thermodynamics of ionic hydration, p. 1-118, Ch. 1, in Franks, F., ed., "Water, a Comprehensive Treatise," Vol. 3, Plenum Press, New York, 1973.
57. Choppin, G. R., and Unrein, P. J. Thermodynamic study of actinide fluoride complexation, p. 97-107, in Muller, W. and Lindner, R., eds., "Transplutonium Elements," North-Holland Publ. Co., Amsterdam, 1976.
58. Mesmer, R. E., and Baes, C. F., Jr. Fluoride complexes of beryllium (II) in aqueous media, Inorg. Chem. 8(3), 618-626 (1969).
59. Izatt, R. M., Eatough, D., Christensen, J. J., and Bartholomew, C. H. Colorimetrically determined Log K, ΔH° and ΔS° values for the interaction of sulphate ion with several bi- and ter-valent metal ions, J. Chem. Soc. A, 47-53 (1969).

RECEIVED November 16, 1978.

Critical Review of the Equilibrium Constants for Kaolinite and Sepiolite

R. L. BASSETT—U.S. Geological Survey, Denver, CO 80225
Y. K. KHARAKA—U.S. Geological Survey, Menlo Park, CA 94025
D. LANGMUIR—Pennsylvania State University, University Park, PA 16802

Clay minerals generally form in three ways: 1) precipitation from solution; 2) alteration through weathering of primary minerals which have become unstable in a given environment; and 3) diagenetic or hydrothermal alteration of other minerals. In the last two cases, resultant clay-mineral chemistry usually reflects parent mineral composition. Chemical variability among clay minerals of a given group poses a formidable problem when one wishes to determine the thermodynamic properties of a given clay. Other factors must also be considered in the experimental determination of free energies of formation.

First, particle size may be a significant property, because most clays range from a few microns in effective diameter to collodial dimensions. In this size range, the free energy required to form the surface per mole of material can easily be two to three kilocalories (1).

Second, isomorphus substitution of aluminum for silicon in the tetrahedral layer, or iron, magnesium, titanium, lithium, and so forth, for aluminum in the octahedral layer, will affect the energetics of formation of the mineral. Such substitutions may vary from trace impurity to complete replacement.

Third, the degree of crystallinity or extent of disorder should be evaluated, whether due to (a) mode or conditions of formation, for example, low temperature as opposed to hydrothermal precipitation, or (b) techniques employed to prepare the sample for investigation, such as mechanical grinding. These factors tend to increase the solubility and make the material less stable than an ordered macroscopically crystalline sample.

Finally, clays such as the smectites almost invariably have a net negative structural charge because of isomorphous substitution of cations of lower charge than would be present in a balanced structure. In kaolinite, the amphoteric nature of the hydrated aluminum and silica surface contributes more to surface charge than does substitution. As a result of either substitution or surface dissociation, a region of counter ions (exchangeable and

0-8412-0479-9/79/47-093-389$05.00/0
This chapter not subject to U.S. copyright
Published 1979 American Chemical Society

adsorbed ions) surrounds the clay to balance the charge. Whether variations in the chemical composition of the layer of counter ions affects the free energy of formation, ΔG^o_f, of the clay is a question that has not yet been resolved.

Initially in this study, it was planned to critically evaluate ΔG^o_f data for complex clays, including chlorite, illite, and the smectites. However, there is much evidence that these clays dissolve incongruently so that the apparent equilibria in solution are determined by secondary phases, such as gibbsite, boehmite, amorphous silica, and ferric oxyhydroxides. The smectites are frequently the dominant clays in the colloidal size fraction in natural sediments. They have very large exchange capacities, and exhibit wide chemical variations. Usually, one or more of these factors have not been considered in the experimental solubility work. Even if appropriate corrections could be made, it is uncertain whether a ΔG^o_f value so obtained would have applicability to natural systems.

In part to avoid such problems, we have restricted this study to an appraisal of the stabilities of kaolinite and sepiolite; compared, for example, to chlorite and the smectites. These clays are well defined both chemically and structurally. Despite these characteristics, the published $\Delta G^o_{f,298.15}$ for kaolinite ranges from -900 kcal/mol to -908.07 kcal/mol, and for sepiolite, -1101.8 kcal/mol to -1105.6 kcal/mol.

Methods and Computational Scheme

Thermodynamic data, whether determined through calorimetry or solubility studies, are subject to refinement as more exact values for the components in the reaction scheme, or more complete description of the solution phases, become available. Many of the solubility studies on clays were done before digital-computer chemical equilibrium programs were available. One such program, SOLMNEQ, written by one of the authors (2) solves the mass-action and mass-balance equations for over 200 species simultaneously. SOLMNEQ was employed in this investigation to convert the chemical analytical data into the activities of appropriate ions, ion pairs, and complexes.

To derive free energy of formation data from solubility investigations, the solution phase must be in equilibrium with the solid phase and the activities of the ions must be known. The dissolution rate for clay minerals is extremely slow at 25°C; consequently, most studies have allowed equilibration times of several years. Because of the requirement for a long equilibration time, a number of researchers have fitted kinetic expressions to the rate of dissolution and extrapolated to the equilibrium value. In all cases in this study, the chemical analytical data measured at what appeared to be equilibrium, rather than extrapolated values derived from systems still changing in concentration with time, were used in the computation.

For the dissolution of kaolinite and sepiolite, the following reaction schemes were employed:

Kaolinite

$$Al_2Si_2O_5(OH)4_{(c)} + 6H^+ \rightleftarrows 2Al^{3+} + 2H_4SiO_4^o + H_2O \tag{1}$$

$$K_{eq} = \frac{a^2_{Al^{3+}} \cdot a^2_{H_4SiO_4^o} \cdot a_{H_2O}}{a^6_{H^+}} \tag{2}$$

Sepiolite

$$Mg_2Si_3O_{7.5}(OH).3H_2O_{(c)} + 4.5\ H_2O \rightleftarrows 2Mg^{2+} + 3H_4SiO_4^o + 4OH^- \tag{3}$$

$$K_{eq} = \frac{a^2_{Mg^{2+}} \cdot a^3_{H_4SiO_4^o} \cdot a^4_{OH^-}}{a^{4.5}_{H_2O}} \tag{4}$$

where a_i denotes activity of species i, and the activity of the solid phase is defined to be unity. Reaction schemes used by several authors (3, 4) were written to include the dominant aluminum complex present at the pH of the experiment. In this way the activity could be estimated, neglecting the other complexes. The approach taken in our study, utilizing the chemical equilibrium computer program, offers the advantage that regardless of the pH, the activities of all the identified ions are computed through the use of an internally consistant set of stability constants for the complexes and ion pairs. This being the case, all data were analyzed according to the equations shown below:

$$\Delta G^o_R = -RT \ln K_{eq}$$

where R = the gas constant,
T = temperature, in degrees kelvin, and
ΔG^o_R = standard Gibbs free energy of reaction.

$$\Delta G^o_R = \Sigma\Delta G^o_f(\text{Products}) - \Sigma\Delta G^o_f(\text{reactants}) \cdot$$

For kaolinite:

$$\Delta G^{o}_{f,298.15}(Al_2Si_2O_5(OH)_4)$$

$$= -\Delta G^{o}_{R} + 2\Delta G^{o}_{f}(Al^{3+}) + 2\Delta G^{o}_{f}(H_4SiO_4^{o}) + \Delta G^{o}_{f}(H_2O).$$

The second modification performed on the orignial data was the substitution of more recently determined values for the free energy of formation of the components. Hemingway and Robie (5, 6, 7) have recently revised the free energy of formation for the aqueous Al^{3+} ion to -116.97 $\pm$ 0.33. In most cases, the Al^{3+} value is -1.9 kcal/mol more negative than that previously used. The data employed in the computations are given in Table I, and in part represent a critical review of the existing thermodynamic data for aqueous species being conducted by one of the authors (D. Langmuir).

Table I.--A Partial List Of The Thermodynamic Data Employed In Computing The Revised Values For The Free Energy Of Formation Of Kaolinite And Sepiolite

Components	$\Delta G^{o}_{f,298.15}$ (kcal/mol)	$\Delta H^{o}_{f,298.15}$ (kcal/mol)	S^{o} (cal/deg mol)	Reference
Aluminum				
Al^{3+} (aq)	-116.97	-126.9	-73.3	(6)
$AlOH^{2+}$ (aq)	-166.8	-183.3	-40	1/
$Al(OH)_2^{+}$ (aq)	-217.7	--	--	(9)
$Al(OH)_3^{o}$ (aq)	-266.66	--	--	(9)
$Al(OH)_4^{-}$ (aq)	-312.0	-356.1	35	1/
Silicon				
$H_4SiO_4^{o}$ (aq)	-312.6	-348.30	45.1	1/
$H_3SiO_4^{-}$ (aq)	-299.18	-342.18	20.7	(10)
$H_2SiO_4^{2-}$ (aq)	-281.31	-330.7	-.7	1/
Others				
H_2O	-56.688	-68.315	16.71	(11)
OH^{-}	-37.60	-54.98	-2.56	(8)
H^{+}	0	0	0	(8)
Mg^{2+}	-108.70	-111.58	-32.98	(8)

1/ Langmuir, Donald, Pennsylvania State Univ. unpublished data, 1978).

Thermochemical Studies of Kaolinite. The earliest reported Gibbs free energy of formation ($\Delta G^o_{f,298.15}$) for kaolinite is -888.1 ± 0.7 kcal/mol, as computed by Barany and Kelley (12). Two kaolinites were employed in their study, one from a deposit near Murfreesboro, Ark., and the second from Alta Mesa, N. Mex. The free-energy value was obtained by combining the heat of solution measurements from hydrofluoric acid solution calorimetry of Barany and Kelley (12), with heat capacity determinations for kaolinite by King and Weller (13). The free-energy value is much too positive and can be made more realistic by incorporating revised and updated thermodynamic data into the reaction scheme. In the same study, Barany and Kelley also determined the heat of formation for gibbsite; Hemingway and others (5) redetermined the enthalpy and heat capacity of gibbsite (5, 7) and re-evaluated Barany and Kelley's results. Hemingway and others discuss the experimental work in detail and point out that the heat of solution measurements for both gibbsite and kaolinite appear to be correct; however, the heat of solution of $H_2O(l)$ + HF and $H_2O(l)$ + HCl in HF is in error. In addition, the reaction scheme originally employed required that the heat of solution for $AlCl_3 \cdot 6H_2O$ (aluminum chloride hexahydrate) be known; that value was incorrect because of the technique used in emplacing the sample in the ampules and loading them into the calorimeter. To avoid these sources of error, Hemingway and Robie (7) used only the heat-of-solution measurement of Barany and Kelley and wrote the reaction scheme to include gibbsite, which has well-known thermodynamic properties. Their recalculated value for the free energy of formation of kaolinite at 298.15°K is -908.07 kcal/mol (7), which is the most negative free energy reported to date and represents the only calorimetric value available. It should be noted that the mineralogical purity and crystallinity of the two kaolinites used by Barany and Kelley (12) is not known, as an X-ray examination was not conducted; however, the chemical analysis of the bulk material indicated that the SiO_2 and Al_2O_3 content was within 1½ percent of the theoretical composition of kaolinite.

Solubility Studies of Kaolinite. There have been numerous attempts to determine the free energy of formation of kaolinite from solubility studies (Table II). Polzer and Hem (3) reacted an API standard kaolinite from Lewistown, Mont., for 2 years with an acidic solution approaching equilibrium from undersaturation. The sample appears to have been very near to equilibrium, and these authors report a $\Delta G^o_{f,298.15}$ = -903 kcal/mol. Using SOLMNEQ to recompute the activities of the dissolved species from their experimental data, and employing the most recent thermodynamic data, the recalculated value is -907.76 kcal/mol. This is very close to -908.1 kcal/mol, which is the calorimetric value of Robie and others (8). Additional confidence may be placed in this value because the particles

Table II.--Original And Recomputed Data For The Free Energy Of Formation Of Kaolinite And Sepiolite

Material source	Size	Original $\Delta G^{o}_{f,298.15}$	Recomputed log K_{eq}	Recomputed $\Delta G^{o}_{f,298.15}$
Murfreesboro, Ark.----------	Bulk	[1]-888.1 (12)	-----	[2]-908.1
Alta Mesa, N. Mex.	Bulk			
Lewiston, Mont.	2.0-149.0 μm	-903.0 (3)	5.92	-907.8
England----------	Bulk	-903.8 (19)	6.70	-906.7
Georgia 1--------	Bulk	-903.6 (19)	6.88	-906.4
Georgia 2--------	Bulk	-903.4 (19)	7.08	-906.1
Idaho------------	Bulk	-902.9 (19)	7.36	-905.7
North Carolina---	Bulk	-902.9 (19)	7.33	-905.8
Georgia 3--------	Bulk	-902.7 (19)	7.37	-905.7
South Carolina---	Bulk	-902.5 (19)	7.61	-905.4
Montmorillonite[3]	.2-5.0 μm	-904.2 (20)	6.20	-907.4
Dry Branch, Ga.--	Bulk	[4]-905.8	7.38	-905.8
Keokuk, Iowa-----	Bulk	-903.6 (4)	7.54	-905.5
Varied[5]----------	Bulk	[5]-897.5 to -903.6 (4)	-----	-901.1 to -907.5
Selected value---	----------	---------	5.96	[3]-907.7
		Sepiolite		
Precipitated-----	----------	-1101.0 (24)	-37.4	-1101.0
Balmut, N.Y.-----	0.5-10 μm	-1105.6 (25)	-40.2	-1105.4
Amboseli, Kenya--	63-124 μm	-1105.6 (26)	-40.4	-1105.6
Selected value---	----------	----------	-40.4	[3]-1105.6

[1]Average value from five experimental runs on each clay.

[2]Accepted value of Hemingway (7) computed from the heat of solution data of Barany and Kelley (12).

[3]See text.

[4]Average value for 16 sample runs on same material (May, H. M., University of Wisconsin, personal communication, 1978).

[5]Range of values for kaolinite from 26 locations.

less than 2 µm diameter were removed to eliminate particle-size effects.

Reesman (4) investigated the solubility of numerous standard clay minerals, monitoring the approach to equilibrium from undersaturation for several months. The analytical data have been re-evaluated by us and yield free energies of formation ranging from -900.4 to -907.4 kcal/mol for kaolinite (Table II). Reesman employed a well crystallized kaolinite (Keokuk), which is generally associated with geodes found in Iowa. Even though the Keokuk kaolinite has been proposed as a reference mineral, due to its well-ordered and highly crystalline nature (14), the samples used by Reesman were bulk samples and probably contained very small particles which increased the solubility. An indication of this is the fact that he chose to centrifuge his samples for 8 hours to settle all the suspended colloidal material (4). The recomputed value obtained for Keokuk kaolinite is -905.5 kcal/mol (Table II). This value is slightly greater than 2 kcal/mol more positive than the calorimetric value, or the free energy of formation determined for Polzer and Hem's data, which had the <2.0 µm particles removed.

To determine the possible effect of small particles on the solubility of a similar mineral, gibbsite, Parks (1) used the equation of Enustun and Turkevich (15) and Schindler and others (16), shown below:

$$\Delta G = \frac{2}{3}\frac{M\alpha\bar{\gamma}}{\rho}\left(\frac{1}{d_2} - \frac{1}{d_1}\right)$$

where M = formula weight of solid,
ρ = density of solid,
$\bar{\gamma}$ = mean surface-free energy
d = particle dimension, $d_1 > d_2$, and
α = a shape factor (ratio of particle surface to particle volume multiplied by d, a value of 14 is used here).

There are no published values for the surface-free energy for clay minerals; however, the value for hydrated silica gel is approximately 120 ergs/cm^2 (17). Smith and Hem (18) report surface energies for the edge and face for gibbsite crystals as 483 and 140 ergs/cm^2, respectively. Parks determined that a surface energy of 270 ergs/cm^2 would be all that is required to explain the free-energy discrepancy in the gibbsite data that he evaluated which was due to 0.010 µm particles (1). Employing the same reasoning, assuming that 0.015 µm particles are present in the bulk samples used by Reesman in his study, then the 2.2 kcal/mol difference between the data of Reesman (4) and that of Polzer and Hem (3) would require a surface-free energy for kaolinite of 150 ergs/cm^2. This is certainly within the range of values one would expect for a surface composed of hydrated aluminum and silica.

Because the exact value for the surface-free energy and the minimum particle size in the bulk samples are not known, the position taken in this paper is that a free energy of formation for kaolinite of -905.5 kcal/mol derived from Reesman's data is the most positive value one should use for well-crystallized kaolinite. There are numerous uncertainties in the solubilities determined by Reesman (4) for the other kaolinites; many samples contained other clay mineral impurities; most were bulk samples; and some were mechanically crushed, which may have distorted the crystallinity and increased the solubility.

Kittrick (19, 20) performed two separate studies to determine the stability of kaolinite. In the first investigation, samples of bulk kaolinite from seven localities were equilibrated with a dilute acid solution for 2 years. Recomputed ΔG_f^o values for the seven kaolinites based on his final solution compositions range from -905.4 to -906.7 kcal/mol.

In the second study, Kittrick (20) reacted the 0.2 to 5 μm fractions of three montmorillonite clays with low pH (<3.47) solutions for 3 to 4 years. Under these conditions, montmorillonite is unstable with respect to kaolinite. It is uncertain whether the kaolinte formed through "precipitation," in which case the nucleation process may have produced numerous small particles, or it formed through the alteration of the pre-existing montmorillonite structure, which could have maintained the existing particle size or even increased it, with growth of the new phase.

The only other study for which the raw experimental data are available is that of May (see Table II footnote). In this investigation, 16 samples of Dry Branch kaolinite were equilibrated with solutions containing various concentrations of silica and aluminum. His data yield an average free energy of formation of -904.8 kcal/mol. The samples were bulk material with no reported minimum particle size (personal communication). This value is close to the free energies of formation obtained by Reesman and Kittrick for bulk samples (Table II) and seems to represent the upper stability limit.

In summary, the mean of solubility, precipitation, and calorimetric data for kaolinite free of particle size effects yields the value for the free energy of formation of -907.7 ± 0.4 kcal/mol.

Solubility Data for Sepiolite. Although sepiolite is not a major clay mineral in terms of worldwide abundance, it is commonly associated with deep-sea sediments, salt formations, evaporite and playa lakes, soils, and carbonate deposits (21, p. 140). In most instances, it appears that sepiolite is formed during a time of silica abundance and(or) high pH. Sepiolite was precipitated in the laboratory by Siffert and Wey (22) and Siffert (23) by raising the pH of a magnesium chloride solution saturated with amorphous silica. Insufficient data are available describing the equilibrium conditions. Wollast and others (24) reviewed the literature for data on the laboratory synthesis of sepiolite and found that the experimental data were qualitative only and too incomplete for determining thermodynamic values.

The work by Wollast and others (24) was intended to determine the equilibrium constant and free energy of formation for precipitated sepiolite. By adding silica to freshly collected seawater and maintaining the solution at pH 8.0, a hydrous magnesium silicate precipitated. The product was chemically similar to sepiolite, but the X-ray analysis indicated a relatively amorphous, fine-grained phase. These authors report $\Delta G^o_{f,298.15} = -1101.8$ kcal/mol with an equilibrium constant of $10^{-37.2}$, as defined by equation 4 . Assuming a general composition for seawater and recomputing the solution speciation with SOLMNEQ, K_{eq} becomes $10^{-37.4}$ and ΔG^o_f = -1101.5 kcal/mol for sepiolite. Because the precipitate was poorly defined, much uncertainty remains.

Christ and others (25) attempted to determine the stability of sepiolite at three temperatures (51°C, 70°C, and 90°C) approaching equilibrium from undersaturation. They used well-crystallized samples, obtained good reproducibility between samples runs, and apparently reached equilibrium. Our calculations with SOLMNEQ do not significantly alter the K_{eq} obtained by Christ and others ($10^{-40.2}$); however, use of a more recent value for the ΔG^o_f of magnesium ion of -108.7 kcal/mol (25) reduces the ΔG^o_f for sepiolite by approximately 200 calories and give -1105.4 kcal/mol (Table II). The three high-temperature values for the free energy of formation were fitted with a heat-capacity power function so that $\Delta F^o_{f,298}$ is an extrapolated value (25). Recent dissolution experiments at 25°C by Stoessell (26) using naturally occurring sepiolite from Amboseli, Kenya, suport the extrpolated value of Christ and others (23). Stoessell obtained a Keq equivalent to $10^{-40.4}$ for the reaction described by expression (4) yielding a ΔG^o_f (sepiolite) = -1105.6 kcal/mol.

Conclusions

The experimental values for the free energies of formation of kaolinite and sepiolite are given in Table II. The value of -907.7 ± 1.33 kcal/mol recommended for kaolinite, is the mean of three recomputed free energies of formation weighed equally in the computation, and was obtained from calorimetry, dissolution, and precipitation data. Several values in the -905 to -906.0 kcal/mol range probably reflect the more soluble nature of small particles typically present in bulk samples.

A free energy of formation for sepiolite of -1105.4 kcal/mol is based on extrapolation to 25°C of results from measurements made at 51, 70, and 90°C by Christ and others (25). In support of this, Stoessell (26) has determined a K_{eq} at 25°C which yields a ΔG^o_f at 25°C, only 200 calories more negative than that computed from the results of Christ and others. The value recommended here for the free energy of formation for sepiolite is -1105.6 ± 0.4 kcal/mol.

Abstract

Clay minerals are present in almost all surface-water and ground-water systems, and in many instances may be controlling the concentration of aluminum, silica, iron, magnesium, or other cations in solution. The thermodynamic data necessary to evaluate the state of reaction (saturation) are not available for some clay minerals, and for those minerals with published values, the data are in disagreement by as much as 10 kilocalories per mole for the same clay mineral. A critical review of the available data for kaolinite and sepiolite, incorporating both the most recent thermodynamic data for the components in the reaction schemes and a more complete computation for the solubility data, yields the values of -907.7 ± 1.3 and 1105.6 ± 0.4 kilocalories per mole for the free energy of formation of kaolinite and sepiolite, respectively.

Literature Cited

1. Parks, G. A. Free energies of formation and aqueous solubilities of aluminum hydroxides and oxide hydroxides at 25°C. Am. Min. 57, 1163-1189 (1972).
2. Kharaka, Y. K., and Barnes, I. SOLMNEQ: Solution-mineral equilibrium computations. U.S. Geol. Survey Computer Contr. Report PB-215 899, 81 p. (1973).
3. Polzer, W. L., and Hem J. D. The dissolution of kaolinite. J. Geophys. Research 70, 6233-6240 (1965).
4. Reesman, A. L. "A Study of Clay Mineral Dissolution." Ph. D. thesis, Univ. Missouri, 215 p., 1966.
5. Hemingway, B. S., Robie, R. A., Fisher, J. R., and Wilson, W. H. Heat capacities of gibbsite, $Al(OH)_3$, between 13 and 480 K and magnesite, $MgCO_3$, between 13 and 380 K and their standard entropies at 298.15 K, and the heat capacities of calorimetry conference benzoic acid between 12 and 316 K. U.S. Geol. Survey J. Research 5, 797-806 (1977).
6. Hemingway, B. S., and Robie, R. A. The entropy and Gibbs free energy of formation of the aluminum ion. Geochim. Cosmochim. Acta 41, 1402-1404 (1977).
7. Hemingway, B. S., and Robie, R. A. Enthalpies of formation of low albite ($NaAlSi_3O_8$), gibbsite ($Al(OH)_3$), and $NaAlO_2$; revised values for $\Delta H^\circ_{f,298}$ and $\Delta G^\circ_{f,298}$ of some aluminosilicate minerals. U.S. Geol. Survey J. Research 5, 413-429 (1977).
8. Robie, R. A., Hemingway, B. S., and Fisher, J. R. Thermodynamic properties of minerals and related substances at 298.15 K and 1 bar (10^5 pascals) pressure and at higher temperatures. U.S. Geol. Survey Bull. 1452, 456 p. (1978).
9. Baes, C. F., and Mesmer, R. E. "The Hydrolysis of Cations." 496 p. Wiley-Interscience, New York, 1976.

10. Busey, R. H., and Mesmer, R. E. Ionization equilibriums of silicic acid and polysilicate formation in aqueous sodium chloride solutions to 300°C. J. Inorg. Chem. 16, 2444-2446 (1977).
11. Robie, R. A., and Waldbaum, D. R. Thermodynamic properties of minerals and related substances at 298.15°K (25.0°C) and 1 atmosphere (1.013 bars) pressure and at higher temperatures. U.S. Geol. Survey Bull. 1259, 256 p. (1968).
12. Barany, R., and Kelley, K. K. Heats and free energies of formation of gibbsite, kaolinite, halloysite, and dickite. U.S. Bureau Mines Report Inv. 5825, 13 p. (1961).
13. King, E. G., and Weller, W. W. Low-temperature heat capacities and entropies at 298.15°K of diaspore, kaolinite, dickite, and halloysite. U.S. Bureau Mines Report Inv. 5810, 6 p. (1961).
14. Keller, W., Pickett, E. E., and Reesman, A. L. Elevated dehydroxylation temperature of the Keokuk geode kaolinite--a possible reference mineral, in Proceedings Internat. Clay Conf. 1, 75-85 (1966).
15. Enüstun, B. V., and Türkevich, J. Solubility of fine particles of strontium nitrate. J. Am. Chem. Soc. 82, 4502-4509 (1960).
16. Schindler, P. H., Hofer, F., and Minder, W. Loslichkeitsprodukte von metalloxiden und hydroxiden 10. Loslichkeitsprodukte von Zinkoxyd, Kupfer Hydroxid und Kupferoxid in Abhangigkeit von Teilchengrosse und Molarer Oberflache ein Beitrag zur Thermodynamik von Grenzflachenfest-flussing. Helv. Chem. Act. 48, 1201-1215 (1965).
17. Adamson, A. W. "Physical Chemistry of Surfaces." 698 p. Interscience, New York, 1960.
18. Smith, R. W., and Hem, J. D. Effect of aging on aluminum hydroxide complexes in dilute aqueous solutions. U.S. Geol. Survey Water-Supply Paper 1827-D, 51 p. (1972).
19. Kittrick, J. A. Free energy of formation of kaolinite from solubility measurements. Am. Min. 51, 1457-1466 (1966).
20. Kittrick, J. A. Precipitation of kaolinite at 25°C and 1 atm. Clays Clay Min. 18, 261-266 (1970).
21. Velda, B. "Developments in Sedimentology." 218 p. Elsevier, Amsterdam, 1977.
22. Siffert, B. Quelques réactions de la silice en solution: La formation des argiles. Mem. Ser. Carte Geol. Alsace-Lorraine 21, 86 p. (1962).
23. Siffert, B., and Wey, R. Synthese d'une sepiolite á temperature ordinaire. C. R. Acad. Sci. Paris 254, 1460-1462 (1962).
24. Wollast, R., Mackenzie, F. T., and Bricker, O. P. Experimental precipitation and genesis of sepiolite at earth surface conditions. Am. Min. 53, 1645-1662 (1968).
25. Christ, C. L., Hostetler, P. B., and Siebert, R. M. Studies in the system MgO-SiO_2-CO_2-H_2O (III): the activity-product constant of sepiolite. Am. J. Sci. 273, 65-83 (1973).

26. Stoessell, R. K. "Geochemical Studies of Two Magnesium Silicates, Sepiolite and Kerolite." Ph. D. thesis, Univ. Calif., Berkeley, 122 p., 1977.

Disclaimer: The reviews expressed and/ or the products mentioned in this article represent the opinions of the author(s) only and do not necessarily represent the opinions of the U.S. Geological Survey.

RECEIVED November 16, 1978.

20

Ion Exchange and Mineral Stability: Are the Reactions Linked or Separate?

J. A. KITTRICK

College of Agriculture Research Center, Washington State University, Pullman, WA 99164

From a general chemical point of view the most important class of minerals in agricultural soils are those having ion exchange capacity. These same minerals also possess most of the total inorganic surface area of soils and sediments. They thus have a major influence on water retention and on other physical properties of soils. They are also important in the retention of the many pollutants that find their way into soils and sediments, including Cd (e.g., (1)) and long-lived radiosotopes, including iodine-129, neptunium-237, and plutonium-239 (e.g., (2)).

Minerals that have ion exchange capacity are also important in the discovery, recovery and refining of petroleum. In petroleum refining, they are important chiefly as catalysts and adsorbents. In petroleum recovery they mainly affect reservoir permeability. In the search for petroleum they are chiefly important in drilling fluids and as marker horizons.

Much of the practical importance of minerals that have exchange capacity hinges on how they control the composition of waters they contact. Examples have been given for the soil solution (3), spring waters (4), lake waters (5,6), and the ocean.

A recent trend in the study of mineral stability has been toward the use of predictions based upon the principles of chemical thermodynamics. This has been prompted by the inability of previous empirical approaches to provide quantitative predictive models. Models based upon equilibrium thermodynamics require free energies of formation of minerals and the ions and molecules with which they are in equilibrium, plus equations and equilibrium constants that describe equilibrium conditions. With these, models based on equilibrium thermodynamics can provide insight into the relationship between various aqueous environments and associated minerals that are in equilibrium with them. Where the aqueous environment and associated minerals are not in equilibrium, equilibrium thermodynamics can furnish a frame of reference for understanding the kinetics of mineral alteration and formation. The approach was pioneered by R. M. Garrels and has been

0-8412-0479-9/79/47-093-401$05.00/0
© 1979 American Chemical Society

used increasingly for studies in sedimentary geochemistry (e.g., (8, 9)).

Minerals that have exchange capacity are usually rather complicated, so the determination of their stability by calorimetric or solubility methods has been difficult. An important problem, unique to these minerals, also limits progress in the understanding of their stability and the stability of other minerals that compete for the same elements. This problem concerns whether equilibria with structural ions and equilibria with exchangeable ions are linked or separate.

Almost all investigators of mineral equilibria tacitly assume that equilibria with structural ions and equilibria with exchangeable ions are part of the same reaction. Studies of ion exchange equilibria invariably assume that the two equilibria are separate. One group must be wrong, and the consequences are far from trivial.

The Assumption of Homogeneous Equilibrium for Montmorillonite. Montmorillonite will be used as an important example of a mineral that possesses ion exchange capacity. Consider the equilibrium of Na^+ saturated Belle Fourche montmorillonite with water as follows:

$$[(Si_{7.87}Al_{0.13})(Al_{3.03}Mg_{0.58}Fe^{3+}_{0.45})O_{20}(OH)_4]Na_{0.56} + 7.48\ H_2O + 12.55\ H^+ = 7.87\ H_4SiO_4 + 3.16\ Al^{3+} + 0.58\ Mg^{2+} + 0.45\ Fe^{3+} + 0.56\ Na^+. \quad [1]$$

One cannot tell by looking at an expression such as equation 1, however, whether the overall balanced equation applies to a homogeneous equilibrium, or to a heterogeneous equilibrium in which has been included a second equilibrium, perhaps associated with and dependent upon the first. If we assume homogeneous equilibrium for equation 1, and include H^+ with each cation to avoid the necessity of a separate H^+ variable, then

$$\log K_{eq} = 7.87 \log H_4SiO_4 + 3.16 (\log Al^{3+} - 3 \log H^+) + 0.58 (\log Mg^{2+} - 2 \log H^+) + 0.45 (\log Fe^{3+} - \log H^+) + 0.56 \log \frac{Na^+}{H^+} \quad [2]$$

where K_{eq} is the equilibrium constant and the activity of montmorillonite and water are assumed to be unity. The last term has been written as a ratio to simplify a later comparison. There is some question as to whether montmorillonites can come to equilibrium with acid aqueous solutions (10), although there is recent evidence that they can (11).

Equations 1 and 2 are written as if they apply to a

homogeneous reaction, where K_{eq} depends upon both structural and exchangeable ions. This K_{eq} implies that the stability of Belle Fourche montmorillonite (or of any mineral having exchangeable ions) depends upon the solution activity of whatever exchangeable ions it contains. Since natural exchangers usually contain several exchangeable ions, understanding the stability of these minerals in detail could become very complicated. To simplify the situation, a single exchangeable ion is usually assumed (e.g., (12)).

Investigators of ion exchange equilibria are usually interested in the details of exchangeable ion composition, so they seldom assume the presence of a single exchangeable ion. Furthermore, there do not seem to have been any ion exchange investigations that even suggest the possibility of linked equilibria, such as depicted in equation 2. If equation 2 were correct (as assumed by most investigators of mineral equilibria) then even a small change in the activity of the neutral molecule H_4SiO_4 under the right circumstances could make a large change in the activity of exchangeable Na^+. It is clear that, if equation 1 is correct, ion exchange equilibria cannot be understood without reference to equation 2.

The Assumption of Heterogeneous Equilibrium for Montmorillonite. If the dissolution of montmorillonite is heterogeneous as assumed by investigators of ion exchange equilibria, then it might be more suitable to write equation 1 as

$$[(Si_{7.87}Al_{0.13})(Al_{3.03}Mg_{0.58}Fe^{3+}_{0.45})O_{20}(OH)_4]^{0.56-} + 7.48\ H_2O +$$

$$12.55\ H^+ = 7.87\ H_4SiO_4 + 3.16\ Al^{3+} + 0.58\ Mg^{2+} + 0.45\ Fe^{3+}. \quad [3]$$

The corresponding expression for the equilibrium constant is

$$\log K_{eq} = 7.87 \log H_4SiO_4 + 3.16 (\log Al^{3+} - 3 \log H^+)$$
$$+ 0.58 (\log Mg^{2+} - 2 \log H^+) + 0.45 (\log Fe^{3+}$$
$$- 3 \log H^+) - 0.56 \log H^+. \quad [4]$$

In equation 3 the montmorillonite is considered to exist as a charged silicate, but the exchangeable ion is omitted. K_{eq} in equation 4 is then independent of exchangeable ion composition.

Equation 3 assumes that the exchange reaction coexists with the silicate equilibrium, but that the equilibrium constants of the two reactions are independent. If this is true, then something is amiss with respect to current methods for approximating the standard free energy of formation of minerals having exchangeable ions, because all such methods require inclusion of the exchangeable ion in the calculation (13, 14, 15). For a consequence that may be easier to visualize, consider a portion of a stability diagram from Helgeson, Brown and Leeper (16), as shown in Figure 1. Notice that the stability field of Na montmorillonite increases relative to kaolinite, as log Na^+/H^+ increases, in accord with equation 2. That is, montmorillonite becomes more stable as the activity of its exchangeable ion increases. If equation 4 were correct, the kaolinite-montmorillonite join would be vertical on such a diagram, because the stability of both minerals would be independent of Na^+ activity.

Homogeneous vs. Heterogeneous Equilibrium. Why Don't We Know Which is Right? Ion exchange equilibria are usually so much faster than mineral equilibria that it is easy to assume the two are not linked. However, mineral equilibria can be rapid if small amounts of equilibrating liquid are used (e.g., (17). Many investigators may not realize that their equations involve such an assumption. Perhaps the main reason the question of homogeneous vs. heterogeneous equilibrium is not settled, for minerals having ion exchange capacity, is that many investigators wish to apply mineral equilibria and ion exchange equilibria to immediate practical problems. They are not in a position to perform the necessary basic research, so merely making an assumption is an attractive way to proceed. If equation 1 represents a heterogeneous reaction, that fact can probably not be determined calorimetrically, since it does not appear to be possible to evaluate mineral equilibria and ion exchange equilibria separately with that method.

Considering minerals with large exchange capacities, it was not until 1968 that Reesman and Keller published the first experimental solubility work on montmorillonite stability (18). Since that time, solubility work on montmorillonite (19-26, 10) and on vermiculite (27, 28) has not specifically addressed the question of homogeneous vs. heterogeneous equilibrium. Such a determination was not even possible with the systems used by these investigators, though it should be possible using a proposed experimental system to be outlined later in this paper.

The System Montmorillonite-Solution

From equation 2 it is evident that in order to calculate K_{eq} for a reaction assumed to be homogeneous, it will be necessary to determine the equilibrium solution activity of H_4SiO_4, Al^{3+}, H^+, Mg^{2+}, Fe^{3+}, and Na^+. However, in order for the system to contain measureable amounts of Al^{3+} and Fe^{3+}, and for the Al ion species to be known with reasonable certainty, the pH of the system will probably have to be 4.0 or less (20). At this pH the dominant exchangeable ions will be Al^{3+} (and perhaps Fe^{3+}) not Na^+ (e.g., (29), page 284). Rewriting equation 1 for exchangeable Al^{3+} and homogeneous equilibrium we have:

$$[(Si_{7.87}Al_{0.13})(Al_{3.03}Mg_{0.58}Fe^{3+}_{0.45})O_{20}(OH)_4]Al_{0.19} + 7.48\ H_2O + 12.55\ H^+ = 7.87\ H_4SiO_4 + 3.35\ Al^{3+} + 0.58\ Mg^{2+} + 0.45\ Fe^{3+},\quad [5]$$

with the corresponding equilibrium constant expression

$$\log K_{eq} = 7.87 \log H_4SiO_4 + 3.35 (\log Al^{3+} - 3 \log H^+) + 0.58 (\log Mg^{2+} - 2 \log H^+) + 0.45 (\log Fe^{3+} - 3 \log H^+). \quad [6]$$

Comparing equation 6 with equation 2 we find that, since Na^+ is no longer the exchangeable ion, K_{eq} is no longer a function of $\log Na^+/H^+$. Furthermore, since Al^{3+} is now the exchangeable ion, the coefficient of ($\log Al^{3+} - 3 \log H^+$) in equation 6 is greater than in equation 2 because the amount of Al^{3+} in equation 5 is greater than in equation 1. By rearranging equation 6 and dividing through by 3.35, we find that

$$3pH - pAl^{3+} = 2.35pH_4SiO_4 - 0.17(2pH - pMg^{2+}) - 0.13 (3pH - pFe^{3+}) + 0.30\ pK_{eq} \quad [7]$$

where p indicates the negative logarithm of the respective terms. Since heterogeneous equilibrium, on the other hand, is independent of the exchangeable ion, we may start with equation 3 . Equation 4 can then be rearranged as was equation 6 to give

$$3pH-pAl^{3+} = 2.49pH_4SiO_4 + 0.18(3pH-pMg^{2+}) + 0.14(3pH-pFe^{3+}) -0.32pK_{eq} - 0.18pH. \quad [8]$$

Comparing equation 8 with equation 7 we see that the two largest differences in coefficients are 2.35 vs. 2.49 for pH_4SiO_4 and zero and 0.18 for pH. If equation 7 can be distinguished from equation 8 experimentally, it will have to be through use of these coefficients. In Figure 2, $3pH-pAl^{3+}$ vs. pH_4SiO_4 for equations 1 and 8 is graphed at constant Mg^{2+}, Fe^{3+} and pH. Considering estimated precision of analyses and appropriate ranges in solution composition, it would not be possible to distinguish a slope of 2.35 from one of 2.49.

Based upon previous results with Belle Fourche montmorillonite (20), it is not likely that the pH range will exceed one unit for equilibrium samples in a system where both Al^{3+} and Fe^{3+} activities can be determined and where Al^{3+} is the dominant exchangeable ion. This means that $\log K_{eq}$ for heterogeneous equilibrium in equation 4 would differ from $\log K_{eq}$ for homogeneous equilibrium in equation 6 by a maximum of about 0.2 due to the pH term. Since the average deviation of $\log K_{eq}$ for Belle Fourche montmorillonite was ± 0.2 for an equilibrium pH range of less than one unit (20), it is evident that equation 7 cannot be distinguished from equation 8 on the basis of the pH term. Thus,

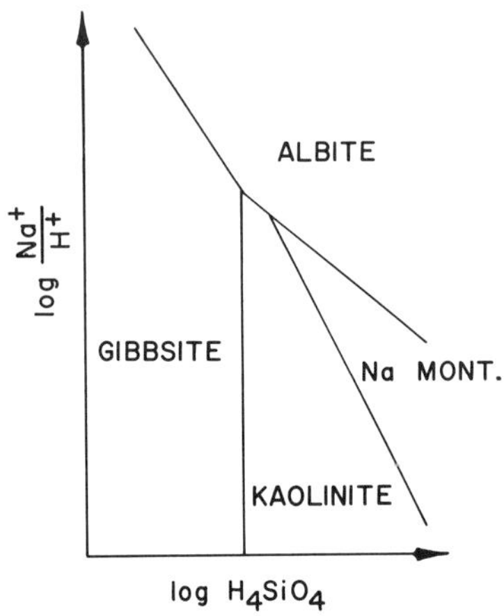

Figure 1. Some minerals in the system $HCl–H_2O–Al_2O_3–Na_2O–SiO_2$ at 25°C. After Helgeson et al. (16).

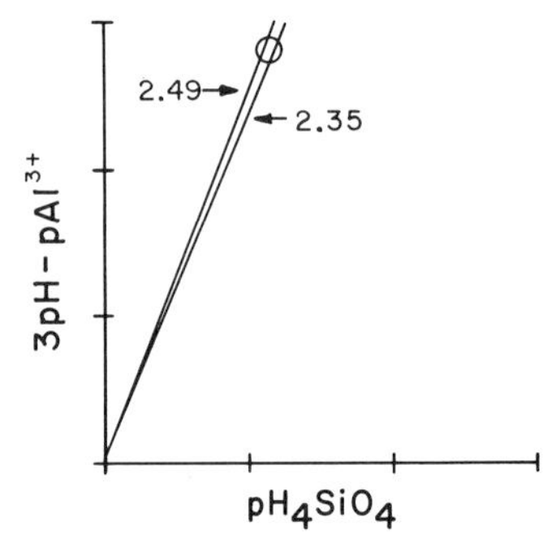

Figure 2. Comparison of Equations 7 and 8 for constant Mg^{2+}, Fe^{3+}, and pH. The slope of 2.35 on the basis of analyses having the precision indicated by the circle.

under the experimental conditions required to determine all necessary ions in the system montmorillonite-solution, it does not appear possible to experimentally distinguish homogeneous equilibrium from heterogeneous equilibrium.

The System Montmorillonite-Gibbsite-Hematite-Solution

The experimental difficulty preventing a choice between homogeneous or heterogeneous equilibrium arises from the fact that Al^{3+} and Fe^{3+} occupy mineral exchange sites when their activity in solution is high enough to measure. The key to surmounting this difficulty is the addition to the system of standard minerals of known stability (e.g., gibbsite, $Al(OH)_3$, and hematite, Fe_2O_3 which will control equilibrium Al^{3+} and Fe^{3+} at very low activities which can be calculated from the equilibrium pH even though they cannot be measured. Since Al^{3+} and Fe^{3+} cannot be measured, gibbsite and hematite of known stability cannot serve as internal indicators of sample equilibrium.

If the equilibrium pH is kept near 7, Al^{3+} and Fe^{3+} activities in solution will be very low and their occupation of exchange sites will be negligible. Also, Mg^{2+} levels will be relatively low, because only small amounts of the montmorillonite need dissolve to saturate the equilibrium solution. If Sr^{2+} is used as the exchangeable ion and is added in large enough amounts, it will compete effectively with Mg^{2+} for exchange sites. Thus, the exchange sites should be occupied almost exclusively by a single exotic ion, providing a maximum effect on montmorillonite stability with a minimum of complications due to exchangeable ions also present in the montmorillonite framework.

Homogeneous Equilibrium. For homogeneous equilibrium including gibbsite, hematite and exchangeable Sr^{2+}, the dissolution equation would be

$$[(Si_{7.87}Al_{0.13})(Al_{3.03}Mg_{0.58}Fe^{3+}_{0.45})O_{20}(OH)_4]Sr_{0.28} + 17.62\ H_2O + 1.72\ H^+ = 7.87\ H_4SiO_4 + 3.16\ Al(OH)_3 + 0.58\ Mg^{2+} + 0.22\ Fe_2O_3 + 0.28\ Sr^{2+}. \quad [9]$$

Assuming the activity of solid phases and water to be unity, then

$$pK_{eq} = 7.87\ pH_4SiO_4 + 0.58\ pMg^{2+} + 0.28\ pSr^{2+} - 1.72\ pH.$$

Grouping terms and rearranging gives

$$2pH\text{-}pSr^{2+} = 28.11pH_4SiO_4 - 2.07(2pH\text{-}pMg^{2+}) - 3.57pK_{eq}. \quad [10]$$

If samples are prepared so that the equilibrium pH ranges from 6 to 8 and pSr^{2+} is maintained at 3 for all samples, then the range in $2pH - pSr^{2+}$ will be 4. This range will correspond to a change in pH_4SiO_4 of 0.14, which is negligible. However, a range in $2pH^+ - pSr^{2+}$ of 4 corresponds to a range of approximately 2 in $2pH - pMg^{2+}$. This in turn corresponds to a change in pMg^{2+} of approximately 6, over the pH range 6 to 8. This change in pMg^{2+} can be measured easily.

Heterogeneous Equilibrium. For heterogeneous equilibrium and exchangeable Sr^{2+}, there should be no relationship between $pH-1/2pSr^{2+}$ and the other parameters in equation 10. The relationship that probably should be observed can be derived as follows:

$$[(Si_{7.87}Al_{0.13})(Al_{3.03}Mg_{0.58}Fe^{3+}_{0.45})O_{20}(OH)_4]^{0.56} + 17.62\ H_2O$$
$$+ 1.72\ H^+ = 7.87\ H_4SiO_4 + 3.16\ Al(OH)_3 + 0.58\ Mg^{2+}$$
$$+ 0.22\ Fe_2O_3. \qquad [11]$$
$$pK_{eq} = 7.87\ pH_4SiO_4 + 0.58\ pMg^{2+} - 1.72\ pH.$$

Grouping terms and rearranging gives

$$pH = 14.05\ pH_4SiO_4 - 1.03(2pH-pMg^{2+})-1.79pK_{eq} \qquad [12]$$

For a range in pH from 6 to 8 the anticipated change in pH_4SiO_4 is once again negligible (at 0.14). As with equation 10, however, a range in pH of 2 units corresponds to an easily-measured range in pMg^{2+} of 6 units.

Dissolution Equations. In equation 10, the effect of variable experimental values of pH_4SiO_4 can be eliminated by adjusting all pH_4SiO_4 values to a single arbitrary value through the coefficient 28.11. Equation 10 thus represents a straight line of slope -2.07 on $2pH-pSr^{2+}$ axes with an intercept of $28.11\ pH_4SiO_4-3.57pK_{eq}$, as indicated in Figure 3. Similarly, equation 12 can be represented by a straight line on pH vs $2pH - pMg^{2+}$ axes, with a slope of -1.03 and an intercept of $14.05\ pH_4SiO_4-1.79pK_{eq}$, as shown in Figure 4. If overall montmorillonite equilibrium corresponds either to equations 9 and 10 or to equations 11 and 12, it should be possible to distinguish between the two by noting which relationship is most closely approached when the data are plotted according to Figures 3 and 4.

It may be that overall montmorillonite equilibrium, whether it be homogeneous or heterogeneous, is not adequately represented by the formulas considered thus far (e.g., 30). However, if equilibrium is carefully approached from undersaturation and supersaturation over a sufficiently broad range of solution conditions, it whould be possible to accurately determine the slopes of lines similar to those shown in Figures 3 and 4. These slopes represent the coefficients in appropriate dissolution equations. Thus, it seems likely that proper equations can

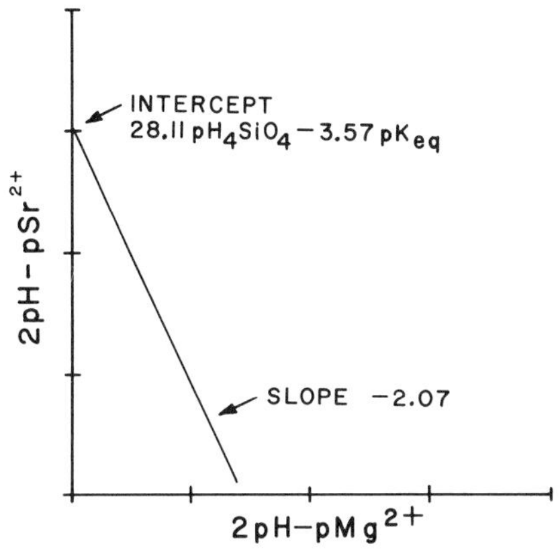

Figure 3. Experimental values conforming to Equation 10 should lie along the indicated line for homogeneous equilibrium.

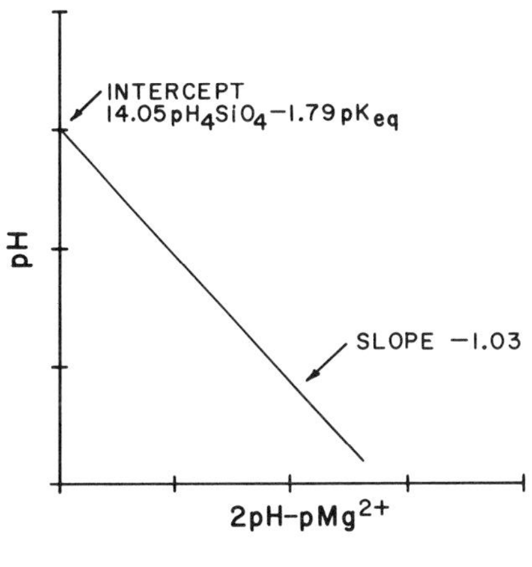

Figure 4. Experimental values conforming to Equation 12 should lie along the indicated line for heterogeneous equilibrium.

be derived from careful experimental results. Such experimentally-derived equations should show whether or not stability and exchange equilibria are linked. They should also show exactly how dissolution equations for minerals with exchangeable ions must be written in order that true equilibrium constants will be obtained.

An array of analyses along a line of appropriate slope may in fact be one indicator of sample equilibrium. More reliable indicators would be 1) approaching the solubility line from both undersaturation and supersaturation and 2) the return of samples to a solubility line after the equilibrium has been perturbed.

Abstract

Minerals with ion exchange capacity strongly influence the chemical and physical properties of soils and sediments, and the composition of natural waters. The stability and exchange characteristics of these minerals have important practical and theoretical implications to agriculture, pollution control, and petroleum production. Understanding the stability and exchange characteristics of these minerals is presently limited because of uncertainty as to whether equilibria involving mineral structural ions are 1) separate from, or 2) linked to, equilibria involving exchangeable ions. Practical and theoretical applications of mineral and exchange equilibria have proceeded upon one or the other of these opposite assumptions, guaranteeing a large body of questionable data.

Calorimetric methods cannot distinguish between separate or linked equilibria in the same sample. For the mineral montmorillonite (an aluminum silicate containing Mg, Fe and other elements), under the restricted conditions where all common ions are measurable, solubility methods also cannot determine whether the two equilibria are separate or linked. However, when gibbsite, $Al(OH)_3$, and hematite, Fe_2O_3, are allowed to control Al^{3+} and Fe^{3+} in equilibrium with montmorillonite (at low levels that cannot be measured but which can be accurately calculated), it should be possible to experimentally determine whether or not the two equilibria are linked, using solubility methods.

Literature Cited

1. Garcia-Miragaya, J. and Page, A.L. Influence of ionic strength and inorganic complex formation on the sorption of trace amounts of Cd by montmorillonite. Soil Sci. Soc. Am. Proc. 40, 658-663 (1976).
2. de Marsily. G., Ledoux, E., Barbreau, A., and Margat, J. Nuclear waste disposal: Can the geologist guarantee isolation? Science 197, 519-527 (1977).
3. Weaver, R.M., Jackson, M.L., and Syers, J.K. Magnesium and silicon activities in matrix solutions of montmorillonite-

containing soils in relation to clay mineral stability. Soil Sci. Soc. Am. Proc. 36, 85 (1972).
4. Garrels, R.M. and Mackenzie, F.T. Origin of the chemical compositions of some springs and lakes. p. 222-242, in Gould, R.F., ed., "Equilibrium Concepts in Natural Water Systems." Adv. Chem. Ser. 67, Washington, D.C., 1967.
5. Kramer, J.R. Equilibrium models and composition of the Great Lakes. p. 243-254, in Gould, R.F. ed., "Equilibrium Concepts in Natural Waters Systems." Adv. Chem. Ser. 67. Washington, D.C., 1967.
6. Sutherland, J.C. Silicate mineral stability and mineral equilibria in the Great Lakes. Environ. Sci. Tech. 4, 826-833 (1970).
7. Mackenzie, F.T. and Garrels, R.M. Chemical mass balance between rivers and oceans. Am. J. Sci. 264, 507-525 (1966).
8. Helgeson, H.C. and Mackenzie, F.T. Silicate-sea water equilibria in the ocean system. Deep-Sea Res. 17, 877-892 (1970).
9. Rai, D. and Serne, R.J. Plutonium activities in soil solutions and the stability and formation of selected plutonium minerals. J. Environ. Qual. 6, 89-95 (1977).
10. Churchman, G.J. and Jackson, M.L. Reaction of montmorillonite with acid aqueous solutions. Geochim. Cosmochim. Acta 40, 1251-1259 (1976).
11. Kittrick, J.A. Solubility product of Belle Fourche and Colony montmorillonites in acid aqueous solutions. Soil Sci. Soc. Am. J. 42, 524-528 (in press).
12. Garrels, R.M. and Christ, C.L. "Solutions, Minerals and Equilibria." 435 p. Harper and Row, New York, 1965.
13. Tardy, Y. and Garrels, R.M. A method of estimating the Gibbs energies of formation of layer silicates. Geochim. Cosmochim. Acta 38, 1101-1116 (1974).
14. Nriagu, J.O. Thermo-chemical approximations for clay minerals. Am. Min. 60, 834-839 (1975).
15. Mattigod, S.V. and Sposito, G. Modifications of Nriagu's method for the estimation of ΔG°_{298} of smectites and vermiculites. Agron. Abstr. p. 188. 1977.
16. Helgeson, H.C., Brown, T.H., and Leeper, R.H. "Handbook of Theoretical Activity Diagrams Depicting Geologic Systems Involving Aqueous Phases at One Atm and 0° to 300°C. 253 p. Freeman, Cooper and Co., San Francisco, 1969.
17. Adams, F. Soil Solution. p. 441-481, in Carson, E.W., ed., "The Plant Root and Its Environment." 691 p. Univ. Virginia Press, Charlottesville, VA., 1974.
18. Reesman, A.L. and Keller, W.D. Aqueous solubility studies of high-alumina and clay minerals. Am. Min. 53, 929-942. (1968).
19. Kittrick, J.A. Soil minerals in the Al_2O_3-SiO_2-H_2O system and a theory of their formation. Clays Clay Min. 17, 157-167 (1969).

20. Kittrick, J.A. Stability of montmorillonite: I. Belle Fourche and Clay Spur montmorillonites. Soil Sci. Soc. Am. Proc. 35, 140-145 (1971).
21. Kittrick, J.A. Stability of montmorillonite: II. Aberdeen montmorillonite. Soil Sci. Soc. Am. Proc. 35, 820-823. (1971).
22. Wildman, W.E., Whittig, L.D., and Jackson, M.L. Serpentine stability in relation to formation of iron rich montmorillonite in some California soils. Am. Min. 56, 587-602 (1971).
23. Weaver, R.M., Jackson, M.L., and Syers, J.K. Magnesium and silicon activities in matrix solutions of montmorillonite-containing soils in relation to clay mineral stability. Soil Sci. Soc. Am. Proc. 35, 823-830 (1971).
24. Huang, W.H. and Keller, W.D. Gibbs free energies of formation calculated from dissolution data using specific mineral analyses III: Clay minerals. Am. Min. 58, 1023-1028 (1973).
25. Carson, C.D., Kittrick, J.A., Dixon, J.B., and McKee, T.R. Stability of soil smectite from a Houston black clay. Clays Clay Min. 24, 151-155 (1976).
26. Misra, U.K. and Upchurch, W.J. Free energy of formation of beidellite from apparent solubility measurements. Clays Clay Min. 24, 327-331 (1976).
27. Kittrick, J.A. Mica-derived vermiculites as unstable intermediates. Clays Clay Min. 21, 479-488 (1973).
28. Henderson, J.H., Doner, H.E., Weaver, R.M., Syres, J.K. and Jackson, M.L. Cation and silica relationships of mica weathering to vermiculite in calcareous Harps soils. Clays Clay Min. 24, 93-100 (1976).
29. Black, C.A. "Soil-Plant Relations." 792 p. 2nd Ed. John Wiley and Sons, New York, 1968.
30. Lippmann, F. The solubility products of complex minerals, mixed crystals, and three-layer clay minerals. N. Jb. Miner. Abs. 130, 243-263 (1977).

RECEIVED November 16, 1978.

Silica Apparent Solubilities and Rates of Dissolution and Precipitation

For ca. 25 Common Minerals at 1°–2°C, pH 7.5–8.5 in Seawater

DAVID C. HURD, CHARLES FRALEY, and JAMES K. FUGATE

Hawaii Institute of Geophysics, University of Hawaii, Honolulu, HI 96822

This paper is basically an outgrowth of an earlier paper (1) wherein the author considered the possible effects of glacial weathering on the silica budget of Antarctic waters. In that study a number of finely ground rocks, considered to be typical of those now being glacially eroded from the Antarctic continent and deposited nearby, were allowed to react with 1-2°C seawater having dissolved silica concentrations typical of Antarctic surface and bottom waters. It was found that a number of rocks either adsorbed silica from solution or dissolved so slowly as to provide negligible input to the silica budget. This finding was somewhat different from those of earlier investigators (2, 3) who had postulated that these same rocks might well be a significant source of dissolved silica to the world oceans and comparable to the annual input by rivers. Constructive criticism of Hurd's manuscript by C. V. Clemency and P. E. Calkin State Univ. N.Y., Buffalo, personal communication 1977) suggested the need for studying individual minerals as well as the above rocks observed by Hurd. The following study is, in part, a response to that criticism.

We have chosen to look at the data from a particular point of view: how much reaction would occur during a 0-6 week period for a given mineral at various dissolved silica concentrations. We wanted to try to begin to understand the following questions by such an approach:

1. Would any of the 25 common silica-containing minerals we were studying release or precipitate dissolved silica over the range of dissolved silica values commonly encountered in open ocean waters or in sediment pore waters, and if so at what rates and magnitudes would these reactions occur?

0-8412-0479-9/79/47-093-413$08.25/0
© 1979 American Chemical Society

2. If any of the above minerals were to be the only mineral injected into the bulk of ocean waters, could it measurably affect the overall silica budget? If so, to what extent?

3. If we were to look at all of the minerals under their conditions of maximum dissolution rate (i.e. at near-zero dissolved silica concentrations and/or from freshly ground material) can we create what will be analogous to a chemical weathering sequence for these minerals in seawater at low temperatures? If we can, how does this arrangement relate to those currently in existence for fresh waters?

4. Is there any simple way to relate mineral structure to solubilities and release rates? Will this ranking allow us to suggest the order of importance of these minerals with respect to input or removal of silica to the silica budget?

The results and discussions of our experiments involving the above questions will be considered in two sections along with a Hawaii Institute of Geophysics data report. The first section, dealing with our proposed method for considering the problem of estimating alumino-silicate solubilities in seawater at 1-2°C, pH 7.6-8.3 and of a method for estimating particle dissolution rates, will be discussed and criticized in this paper. The second section, dealing with the application of these calculations to the silica cycle in the oceans will appear in the near future. The data report which will be a compilation of the dissolved silica and pH measurements, and X-ray diffraction and elemental analyses of each mineral sample, as well as a number of the first-order flux and dissolution rate calculations will be available by Spring 1979. A copy may be obtained by writing to: Publications Office, Hawaii Inst. of Geophysics, 2525 Correa Road, Honolulu, Hawaii 96822.

Methods

Mineral samples were initially broken up using a Diamond@ rock crusher. Using an agate mortar and pestle the coarse sand and smaller sized particles were further ground to a fine powder, which was then screened through a 160 mesh sieve. The resulting sample particles were all less than 100 microns in diameter.

Then, 2.00 ± 0.02 grams of a given powdered sample were weighed out and transferred to a clean, labeled, polyethylene bottle. Approximately 75 ± 3 cm^3 portions of 1-3°C, 33 ± 2 ‰ salinity, filtered seawater, whose dissolved silica concentration had been adjusted to the desired value, were added to each of the samples. The sample bottles were then placed on a shaker table, shaken at 100-150 shakes/min, and kept at 1-3°C for the duration of the experiment. After the appropriate sampling interval had passed, the samples were removed from the shaker table and centrifuged at 1000-2250 rpm (300-1000 g) for 4-15 minutes depending on the sample. The clear supernatant was poured off into clean plastic bottles, filtered through 0.45 micron Millipore® or Gelman® filters and analyzed for dissolved silica according to the method described by Strickland and Parsons (4). Fresh, cold 75 cm^3 portions of the corresponding seawater solutions were added to the samples and shaking was resumed.

Silica standard solutions were prepared by fusing a weighed amount of powdered quartz with Na_2CO_3 in a platinum crucible, following a procedure described in Maxwell (5). The resulting fused material was dissolved in distilled-deionized water and diluted to a concentration of either 10 or 20 x 10^{-3} molar SiO_2. A couple of pellets of NaOH were added to the solution to maintain basicity. The standard solutions used in the analyses were dilutions of these SiO_2 solutions, using 0.70 N NaCl as the diluting liquid. A set of standards was done with each set of analyses, and the precautions of Fanning and Pilson (6) adhered to, as well as those of Strickland and Parsons (4).

Those solutions which were buffered (pH range 7.6-7.9) had the pH adjusted by the addition of enough sodium bicarbonate to make the solution 10 mM. Those solutions which were unbuffered generally remained in the pH range 8.3-8.5. The pH measurements were made after centrifugation but before filtration of each sample.

Specific surface areas were generally measured at the beginning and end of each run. The one exception to this was the experimental series pH 8.3-8.5, ca. 100 μM $Si(OH)_4$ wherein the seawater solutions were divided roughly in half after six weeks. One of the portions was continued in the experiment using 37 to 38 cm^3 portions of the spiked seawater. The other half of the sample was centrifuged, and the solid washed several times with distilled water and dried. This latter sample was then used to determine the "half-way" point specific surface area. Difficulties attendant with this procedure are described in the results section.

Results

Release Rate Curves. Figures 1a, b and c show a portion of the raw data involving dissolution and precipitation of silica in pH 8.3-8.5, 1-3°C seawater solutions for the minerals olivine, albite and kaolinite respectively. These figures show the difference between the initially adjusted value of dissolved silica in the extracting solution and whatever the value was at the time of solution change. This particular approach of using a series of seawater extractions all having the same initial dissolved silica concentration for a given experiment was an attempt to imitate the situation of a given mineral suspended in open ocean seawater, which has a certain dissolved silica value. In that there is so little suspended silicate or alumino-silicate material per unit volume in the deep oceans (7, and references therein), the dissolution or precipitation of such material would not immediately have a significant effect on the water surrounding it. It is obvious to us that our experiments imperfectly mimic the desired set of conditions. A graph of instantaneous concentration of dissolved silica in the extracting solution versus time would show a series of ramps above the initial dissolved silica value for dissolution and below it for precipitation. The concentration of dissolved silica in solution is always changing but is periodically returned to the same reference point at the time of solution change to see if the process of dissolution or precipitation of dissolved silica will continue at that reference value. In Figure 1a it appears that the change from net dissolution to net precipitation occurs somewhere between 200 and 475 μM dissolved silica in the pH range 8.3-8.5. In Figure 1b, the changeover occurs near 50 μM for albite and in Figure 1c, between 2 and 50 μM for kaolinite. The type of information to be gained from the above figures is useful but limited unless we know the amount of available reactive surface for each mineral per unit volume of extracting solution.

Surface Area Data. Table I summarizes the surface area data for the rocks and minerals studied. There are three groups of surface area values. Those in the column labeled "initial" are those values obtained from dry-ground samples which had no contact with seawater or any other fluid. Those values in the 5th and 6th columns refer to numbers obtained from the pH 8.3-8.5, ca. 100 μM dissolved silica experiment

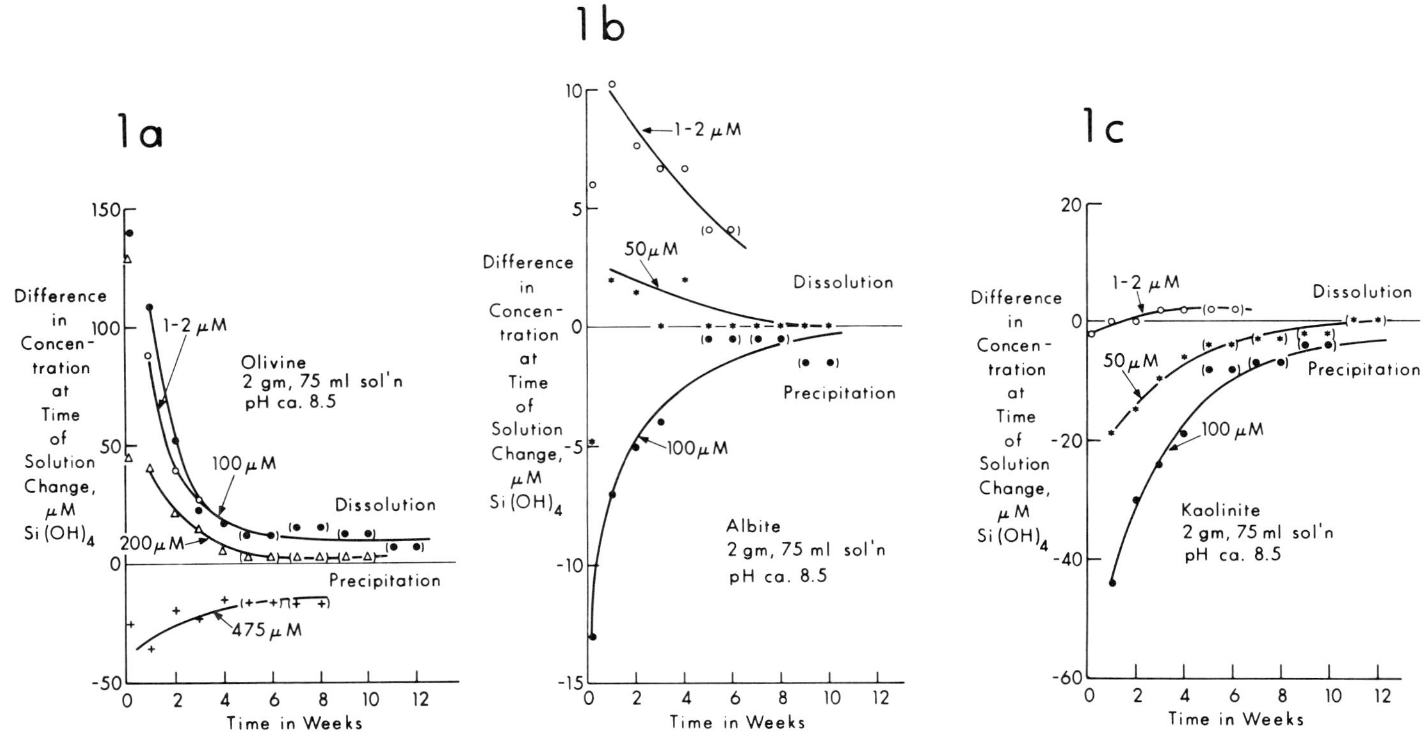

Figure 1(a, b, c). Difference in concentration at time of solution change, μM $Si(OH)_4$ vs. time in weeks. Note that each data point represents a solution change. Where two points are in parentheses, a single point had been taken at a two-week interval and the value halved for ready comparison with previous one-week sampling intervals. Note also the order of magnitude range in concentration differences between olivine and albite, even though the total areas available for reaction are comparable in both cases.

TABLE I

Specific surface areas of mineral samples in m^2/gm before (initial) and after (others) 6-12 weeks reaction with 1-2°C, pH 7.6-8.5 seawater. The dissolved silica concentrations below indicate the nominal value in the extracting solutions.

Mineral	Initial	Ca. 1-5 µM (unbuffered)	Ca. 1-5 µM (buffered)	Ca. 50 µM (unbuffered)	Ca. 100 µM 6 wks	Ca. 100 µM 12 wks	Ca. 100 µM (buffered)	Ca. 200 µM (unbuffered)	Ca. 400/800 -
Quartz (1 HF)	0.29	-	.29	-	-	-	.31	-	-
Quartz (2 HF)		-	.27	.29 (buff)	-	-	.28	-	-
Apua Pt. Lava	0.31	0.32	0.53	-	-	0.22	0.49	-	-
Olivine	0.43	0.38	~1.1	-	2.7	0.53	1.03	1.17	0.63(800)
Hyaloclastite	0.56	0.56	-	-	-		0.64	0.58	-
Nepheline	0.61	0.45	0.77	-	0.99	0.39	0.68	-	-
Albite	0.62	0.43	0.68	-	0.66	0.38	0.49	-	-
Reticulite	0.63	-	0.63	-	-		-	-	-
"Serpentine"	0.70	0.73	0.97	-	1.20	0.63	0.97	-	-
Hornblende	0.72	0.72	1.2	-	1.20	0.51	0.89	0.67*	-
Anorthite	0.72	0.60	0.84	-	-		0.81	-	-
Tremolite	0.73	0.84	1.1	-	0.93	0.59	1.16	-	1.25(400)
K-feldspar	0.75	0.61	0.61	-	-		0.81	-	-
Diopside	0.75	0.87	1.2	-	1.44	0.88	1.23	-	1.28(400)
Quartz	0.86	0.70	0.69	-	0.72	0.50	0.77	0.75	-
Epidote	0.88	0.68	0.53	-	0.90	0.52	0.61	-	-
Orthoclase	0.90	0.67	0.88	-	1.08	0.74	0.84	-	-
Kyanite	0.94	0.66	0.84	-	1.13	0.47	0.55	-	-
Obsidian	0.98	0.96	0.94	-	-		0.97	-	-
Hypersthene	1.1	1.06	1.3	-	2.3	0.82	1.1	1.54	3.4(400)
Chlorite	4.0	4.0	3.3	-	4.0	2.5	0.80	-	-
Biotite	5.5	5.9	-	6.4	9.5	12.7	-	-	-
Kaolinite	8.6	8.0	8.3	-	9.3	8.2	8.4	-	-
Muscovite	9.4	8.4	10.0	-	-		8.2	-	-
Bentonite (N.E.W.)	33	10.4	8.9	-	7.7	-	8.7	9.0	16.7(400)/10.7(800)
Illite	34	20	31	25	25	36	34	-	-
Montmorillonite	65.7	55	53	8.1	-		33	-	69.7(400)
Bentonite (C.R.C.)	68.4	33	43	-	27	46	54	-	44.3(400)/40.6(800)
Montmorillonite	75.6	-	-	-	-	43	-	-	-

wherein: the values in the 5th column refer to the half of the suspension which was poured off after six weeks, and the 6th column refers to the remaining solid whose area was determined after an additional 4-6 weeks. The remaining numbers refer to samples which had reacted with the seawater solutions for at least six weeks, and which were washed briefly with distilled water, dried and analyzed.

There are many troubling inconsistencies in the numbers in Table I.

The observant reader will note that in many cases, the average of the columns 5 and 6 data for a given mineral may be very close to the column 1 value, even though the numbers in columns 2 and 3 are quite different from each other. This prompted an experiment which produced the data in Table II. In this experiment, two grams of freshly ground mineral were placed in a small plastic bottle with 75 cm^3 of seawater; the bottle and contents vigorously shaken for ca. 60 seconds and approximately half of the suspension poured off as rapidly as possible. Then the specific surface area of each portion was determined. The data in Table II suggest that there is a great deal of inhomogeneity in freshly ground samples having specific surface areas of ca. < 1 m^2/gm. Larger surface areas (i.e. ca. > 20 m^2/gm) are seemingly much less affected probably because of the longer settling times of the smaller particles. These results suggest that unless a given sample has been previously size sorted, samples of the suspension should not be periodically removed by pouring off small volumes because the surface area of solid:volume of solution may change in an unpredictable manner.

Yet another difficulty arises when the data in columns 2-4, 7 and 8 are compared with column 1 and the averages of columns 5 and 6. Differences among the former columns of 100% are common although a number of the minerals show fairly small scatter. This suggests that the variability from one sample to the next for the same mineral is important, especially in view of the fact that no attempt to size separate the finest particles from our freshly ground samples was made. We also do not appear to have enough information to conclusively say whether there should be consistent increases or decreases in specific surface area depending on whether dissolution or precipitation of dissolved silica occurs. It is likely that intersample specific surface area variability swamps most of the effects of chemical reaction during our sampling periods.

TABLE II

Data from experiments showing effects of initial specific surface area and particle size distributions on decanted mineral slurry specific surface area values.

Mineral	Specific Surface Area, A* m^2/gm	Specific Surface Area, B** m^2/gm
Albite	.59	.35
Olivine	1.0	0.56
Diopside	1.2	0.59
Bentonite (N. E. Wyoming)	24.3	23.8
Bentonite (Castle Rock, Colo.)	43.8	44.2
Montmorillonite	54.4	53.6

*Refers to solid material poured off after ca. 30-60 sec. shaking time.

**Refers to solid material remaining behind after separation into two halves by decanting.

Effects of Grinding on Solubility and Dissolution Rate. As mentioned in the introduction, this research is an outgrowth of a previous work (1) wherein freshly-ground but otherwise untreated material was allowed to react with low-temperature seawater having various dissolved silica concentrations. The process of grinding surely produces very small particles which may have higher surface energy because of their size alone and may also disturb the actual surface of larger mineral grains especially where bonds have been broken. These two possible effects would then manifest themselves as both higher solubilities and high dissolution rates which would remain in effect until either the very small particles had dissolved or the stresses on the mineral surfaces had been relieved. Since we made no effort either to remove small particles or relieve surface stresses by cleaning procedures except for our quartz sample, it is entirely possible that both our estimates of solubility (termed "apparent" solubility for this reason) and dissolution rates may be higher than the same samples which had been treated to eliminate or reduce the above difficulties. We suggest as a working hypothesis that the ease or difficulty with which a given mineral cleaves under pressure may directly correspond to the extent to which a given mineral exhibits higher surface energy effects after grinding. This hypothesis is not tested in this paper but is under investigation at this time. In any case, the best we can say about the following discussions of solubility and dissolution rates is that they are greater than or equal to those of unstrained minerals and should be regarded as upper limit values.

Calculations for Flux to and from the Surface of the Solid. There are a number of valuable and interesting studies which have dealt with the detailed kinetics of dissolved silica and assorted cation release from alumino-silicate mineral surfaces (8, and references therein). We, however, have used the simplest possible equations to describe what are probably more complex reactions because these simple calculations are more suited to those problems mentioned in the introductory section. We would hope that other investigators who have such interests (detailed kinetics) would make use of the information in our forthcoming data report.

Although many minerals do not uniformly dissolve when subjected to chemical attack, it may be useful to consider dissolution or release rates as though such a process were happening. It may be further useful to attempt to estimate rates of removal in terms of some sort of layer, whose thickness may approximate the average dimension of a single molecular unit of this substance. In that we consider dissolution or precipitation of silica in our reactions in terms of mass/area time units, we converted these numbers to the above layer removal or production concept in the following manner.

We assumed that the volume of a single molecular or formula weight unit could be estimated by dividing the molar volume of the mineral by Avogadro's number (6.023×10^{23} molecules/mole). Although not strictly correct, we also assumed that this molecular unit (not to be confused with a unit cell) could also be approximated by a cube. In our calculations we then assumed that the length of one edge of this cube was equivalent to the thickness of a molecular layer of this substance. Interestingly enough the range of such thickness' was small: 3.4 Å for quartz through 7.7 Å for tremolite, with an average of 5.6 Å for the samples as a whole. We do not deny that differential solution will occur from different crystal faces, but in that we generally measured initial and final surface areas, we felt that a reasonable average effect should result. In any case, in that etching and pitting is known to occur, we have suggested the term "apparent number of layers removed or extracted of silica." To obtain these values, the volume of a layer 1 cm^2 x 1 layer thickness was calculated; this number divided by the molar volume of the mineral gave the number of moles of that mineral in that volume. This number of moles was then multiplied by the number of silicon atoms in the molecular formula to give moles of silica in this volume. In that flux to and from the mineral surface is calculated in terms of moles silica per cm^2 per second, when the number of moles of silica per layer is divided by the cumulative flux, the "apparent number of layers removed or extracted of silica" results.

Figure 2 shows the results of such calculations for a number of our mineral samples. In this figure it will be noted that there is a 3-order-of-magnitude range in the values at the six week time interval. This suggests that the factor of two range in estimated layer thickness' may in fact not be as important as other considerations which are referred to in the discussion sections.

There are several other things which should be noted about Figure 2. Because of the low dissolved

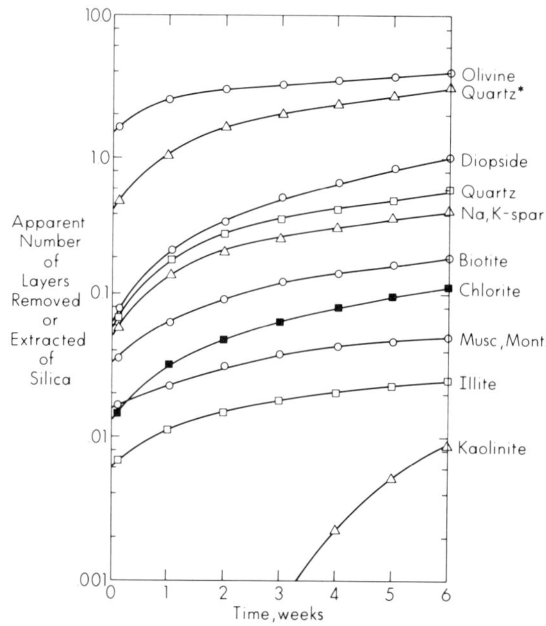

Figure 2. Apparent number of layers removed or extracted of silica vs. time in weeks. This semilog plot is basically a summary graph giving the range of removal rates for the various minerals involved.

a

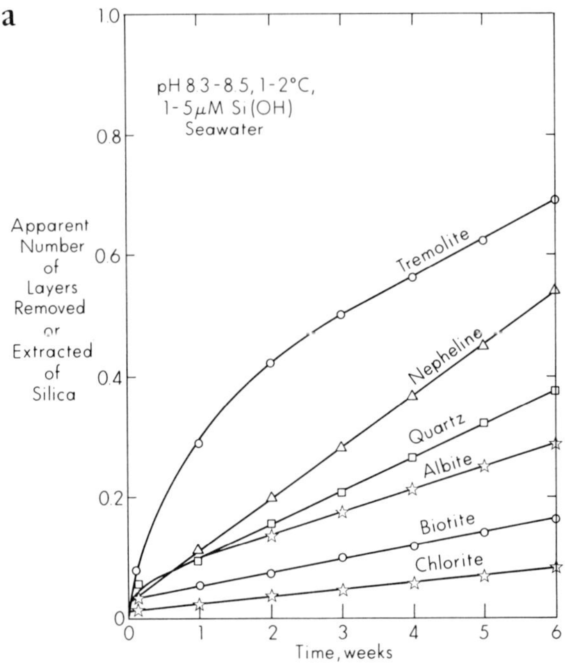

Figures 3 (a, above; *b,* right). *Apparent number of layers removed or extracted of silica vs. time in weeks. These two figures show expanded scale drawings of a portion of the data from Figure 2 on linear graph paper. The linear sections (i.e., after 7–10 days) all show about ±25% scatter which is more apparent at the upper portion of each curve than the lower. Quartz* stands for freshly ground untreated quartz, while Quartz (with no asterisk) represents quartz which had been treated with HF and NaOH as discussed in the text.*

b

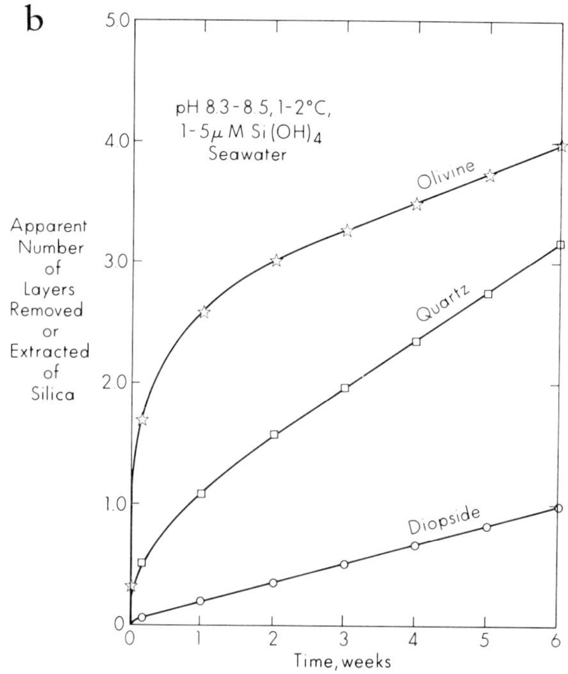
pH 8.3-8.5, 1-2°C,
1-5μM Si(OH)4
Seawater
Olivine
Quartz
Diopside
Apparent Number of Layers Removed or Extracted of Silica
Time, weeks
5.0
4.0
3.0
2.0
1.0
0
0
1
2
3
4
5
6

silica concentration, i.e. none of the minerals studied being near saturation with respect to dissolved silica, all of the minerals are dissolving at their maximum rates with respect to this system. These rates are also different for each mineral and usually vary with time, being relatively rapid initially and decreasing 1 to 2 orders of magnitude during the first week to 10 days. After this initial period, the release rates are uniformly linear for virtually every mineral studied, as also shown in Figure 3a and b. This linearity makes for several simplifying assumptions regarding the long-term, maximum dissolution rates that a given mineral would have, and also this allows predictions regarding minimum survival times for certain size particles of a given mineral.

If we assume for a moment that the linear dissolution rates (LDR) are constant as long as the particle survives and that the flux (F) from the surface may be described in terms of moles silica/cm^2 sec, then this surface flux times the molar volume (MV) of the mineral divided by the moles of Si per molecule (#Si) gives the rate at which the radius of a spherical particle or half the length of a cube will disappear by the dissolution process (9).

$$LDR = (F)(MV)(\#Si)^{-1} \qquad (1)$$

In general such rates range from 0.1 to several microns per thousand years, suggesting that even at the highest long-term dissolution rates, a few hundred to a few thousand years would be required for these minerals to dissolve completely, even at their maximum long-term rates under the above conditions.

Methods of Estimating Solubilities. Estimating the solubility of a given mineral in seawater appears to us to be a difficult task. A number of investigations have used varying approaches for a limited number of minerals (10, 11, 12, 13, 26). Problems arise when dealing with minerals which appear to dissolve incongruently in that it is difficult to know whether the solution is at equilibrium with the true mineral surface or some unknown, albeit thin, type of secondary product layer or somewhere balanced between the two. In addition, many investigators study the dissolution of a given mineral under constantly changing conditions of pH and other dissolved constituents. We felt that this situation was not representative of the true conditions in the open ocean wherein the suspended load of minerals is usually so low that immediate total dissolution of the silicate portion would hardly affect a noticeable change in concentration of nearly all of the dissolved species. As a result, we continually changed our solutions in order to bring the solution portion of the system back to the same value, even though the surface of the solid portion might be changing considerably.

Figures 4a-z show one way of treating our silica precipitation and dissolution data. The horizontal axis represents the number of moles of silica per cm^2 (x 10^{10}) which have been released from the surface (dissolution = negative values) or either taken up by the surface or precipitated by another unknown phase in solution (precipitation = positive numbers) during a 0-6 week period (open circles, triangles and squares) or a 2-6 week period (closed circles, triangles and squares). Circles also show the pH of the solution to be in the range 7.5-7.9, triangles show the range to be in the range 8.3-8.5, and a few hybrid symbols show the pH range 8-8.3. The vertical scale gives the nominal concentration of dissolved silica in the extracting seawater solutions in 10^{-6} moles/liter, remembering that excursions from these values occur between every solution change.

In an approach analogous to the Siever and Woodford (14) "cross-over" method for estimating solubilities of clay minerals, we have plotted cumulative mass flux per unit area versus nominal dissolved silica concentration in our extracting solutions as described above. We are calling the range of points at which lines connecting these values cross the vertical axis our "apparent or estimated solubility" values with respect to dissolved silica for a given mineral and pH range. In fact, we do not know whether these values represent the true solubility of the mineral or some combination of the mineral surface and secondary product layer, because we did not simultaneously determine cation gain or loss by the mineral as well. It would be premature at this stage to attempt to calculate solubility products in that we only know that at the "cross-over" point there is no net change in dissolved silica concentrations in solution. This type of "solubility" information should, however, be extremely useful from a pragmatic standpoint with respect to the silica cycle and the fate of suspended silica-containing minerals in seawater.

There are a number of possible sources of error in the above approach for estimating "apparent solubilities" as we have done.

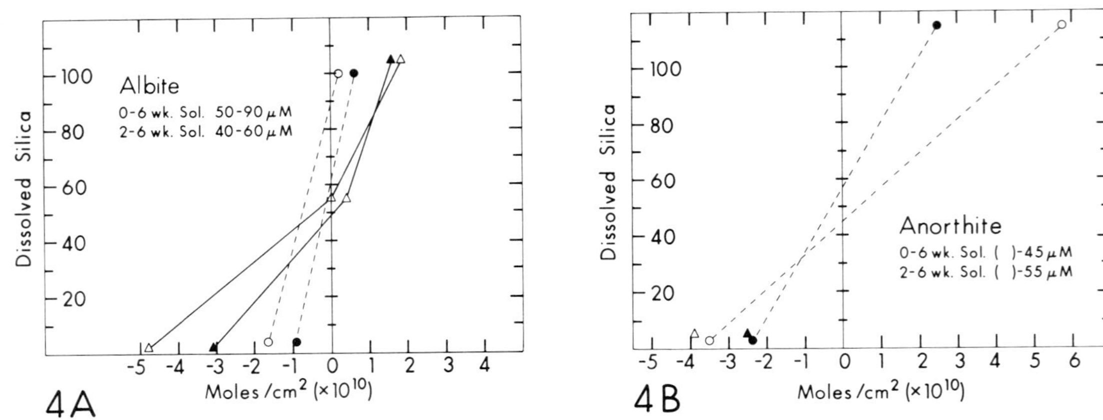

Figure 4(a–z). These figures show plots of initially adjusted concentration of dissolved silica in the extracting seawater solutions in μM $Si(OH)_4$ *vs. moles of silica/cm²(× 10¹⁰) either dissolved from (negative values) or precipitated out of solution (positive values) during a 0–6 week interval (open circles, triangles, squares) or 2–6 week interval (closed circles, triangles, squares) at pH 7.5–7.9 (circles and squares) and pH 8.3–8.5 (triangles).*

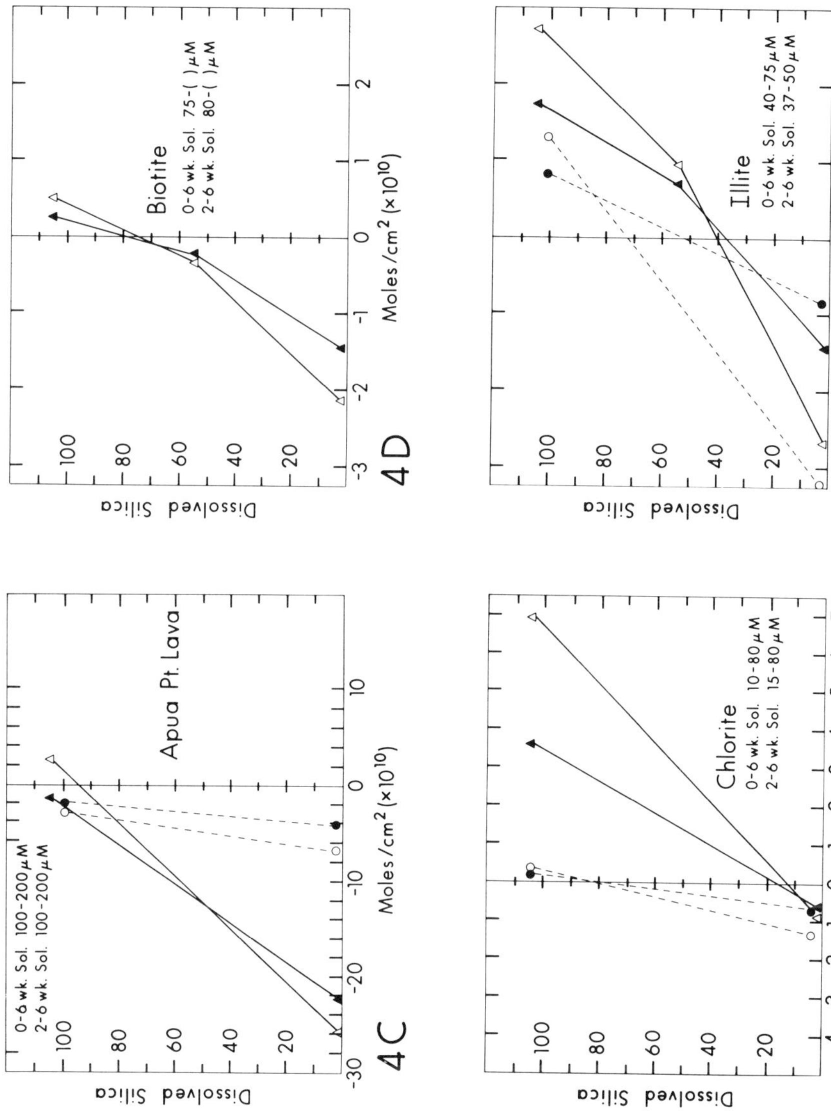
4C
Apua Pt. Lava
0-6 wk. Sol. 100-200 μM
2-6 wk. Sol. 100-200 μM
Dissolved Silica
Moles/cm² (×10¹⁰)
4D
Biotite
0-6 wk. Sol. 75-( ) μM
2-6 wk. Sol. 80-( ) μM
Dissolved Silica
Moles/cm² (×10¹⁰)
4E
Chlorite
0-6 wk. Sol. 10-80 μM
2-6 wk. Sol. 15-80 μM
Dissolved Silica
Moles/cm² (×10¹⁰)
4F
Illite
0-6 wk. Sol. 40-75 μM
2-6 wk. Sol. 37-50 μM
Dissolved Silica
Moles/cm² (×10¹⁰)

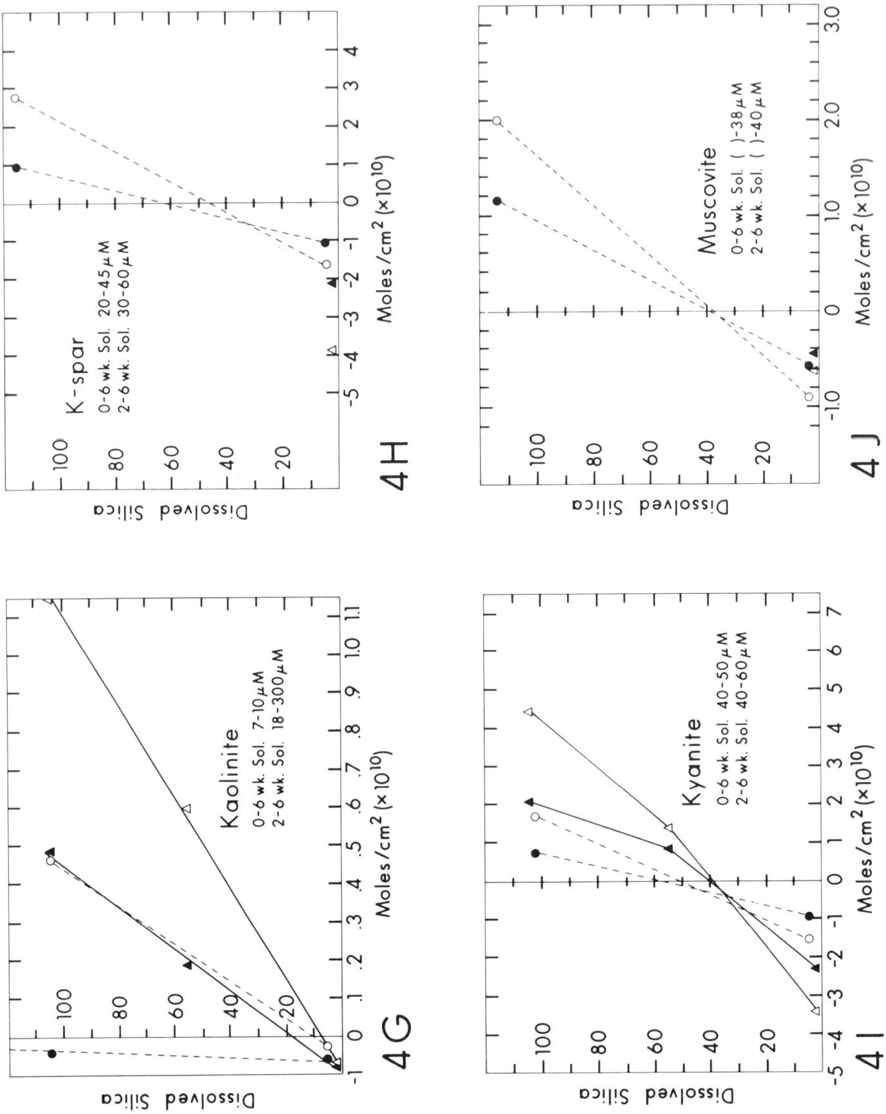
4G
Kaolinite
0-6 wk. Sol. 7-10 μM
2-6 wk. Sol. 18-300 μM
Dissolved Silica
Moles/cm² (×10¹⁰)
4H
K-spar
0-6 wk. Sol. 20-45 μM
2-6 wk. Sol. 30-60 μM
Dissolved Silica
Moles/cm² (×10¹⁰)
4I
Kyanite
0-6 wk. Sol. 40-50 μM
2-6 wk. Sol. 40-60 μM
Dissolved Silica
Moles/cm² (×10¹⁰)
4J
Muscovite
0-6 wk. Sol. ( )-38 μM
2-6 wk. Sol. ( )-40 μM
Dissolved Silica
Moles/cm² (×10¹⁰)

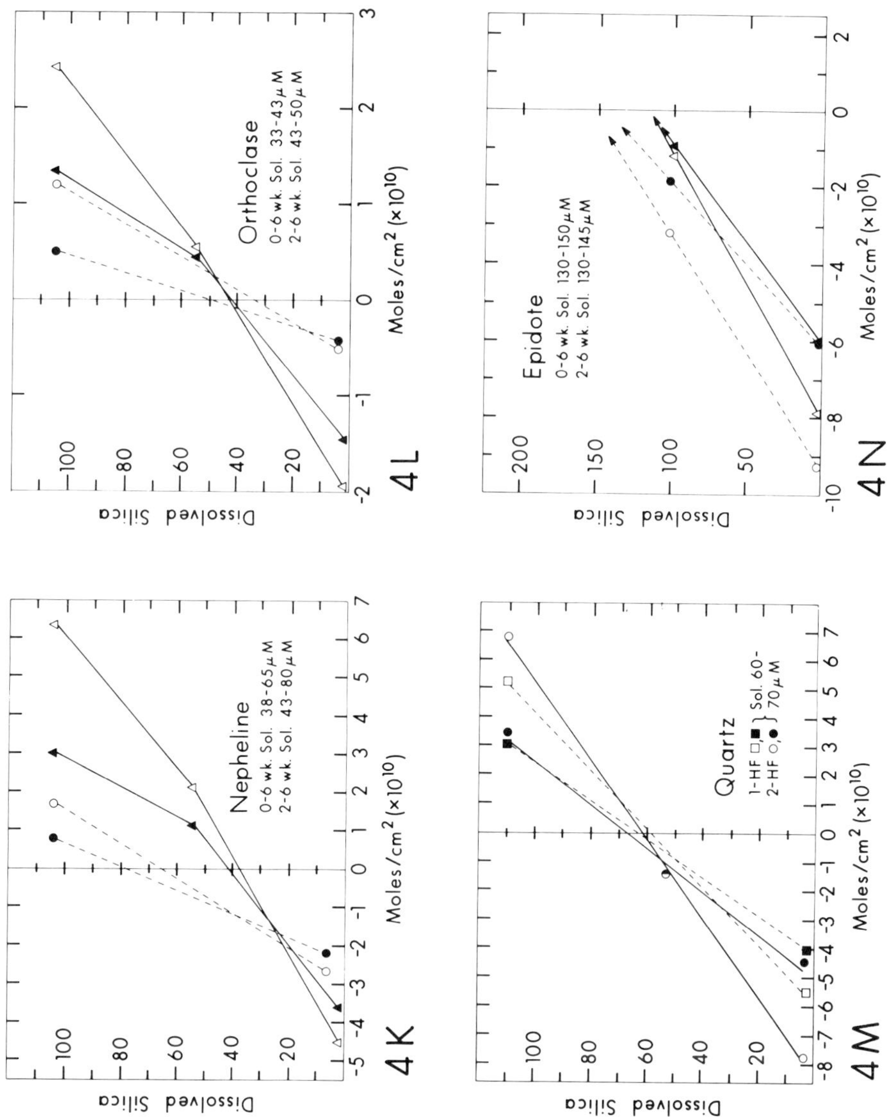

4K
Nepheline
0-6 wk. Sol. 38-65 μM
2-6 wk. Sol. 43-80 μM
4L
Orthoclase
0-6 wk. Sol. 33-43 μM
2-6 wk. Sol. 43-50 μM
4M
Quartz
1-HF □, ■ } Sol. 60-
2-HF ○, ● } 70 μM
4N
Epidote
0-6 wk. Sol. 130-150 μM
2-6 wk. Sol. 130-145 μM
Dissolved Silica
Moles/cm² (×10¹⁰)

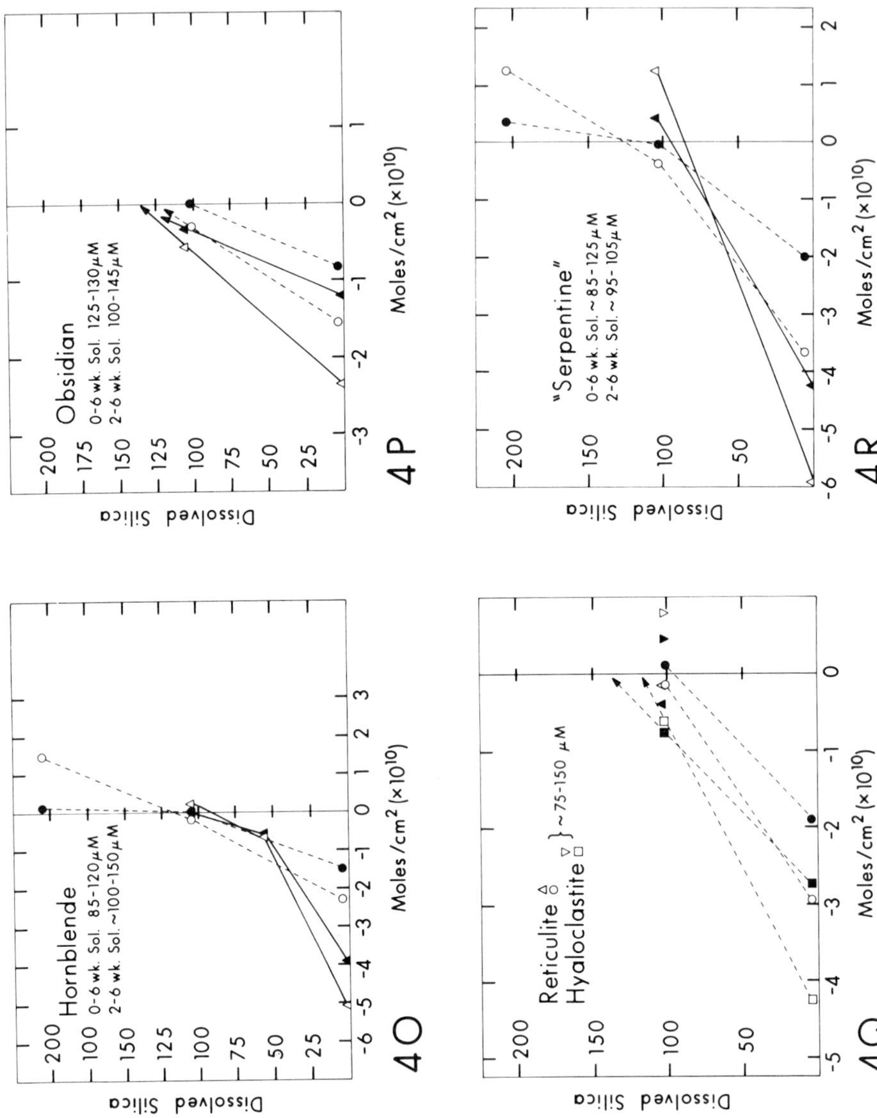
Hornblende
0-6 wk. Sol. 85-120μM
2-6 wk. Sol. ~100-150μM
4O
Obsidian
0-6 wk. Sol. 125-130μM
2-6 wk. Sol. 100-145μM
4P
Reticulite
Hyaloclastite
~75-150 μM
4Q
"Serpentine"
0-6 wk. Sol. ~85-125μM
2-6 wk. Sol. ~95-105μM
4R
Dissolved Silica
Moles/cm² (×10¹⁰)

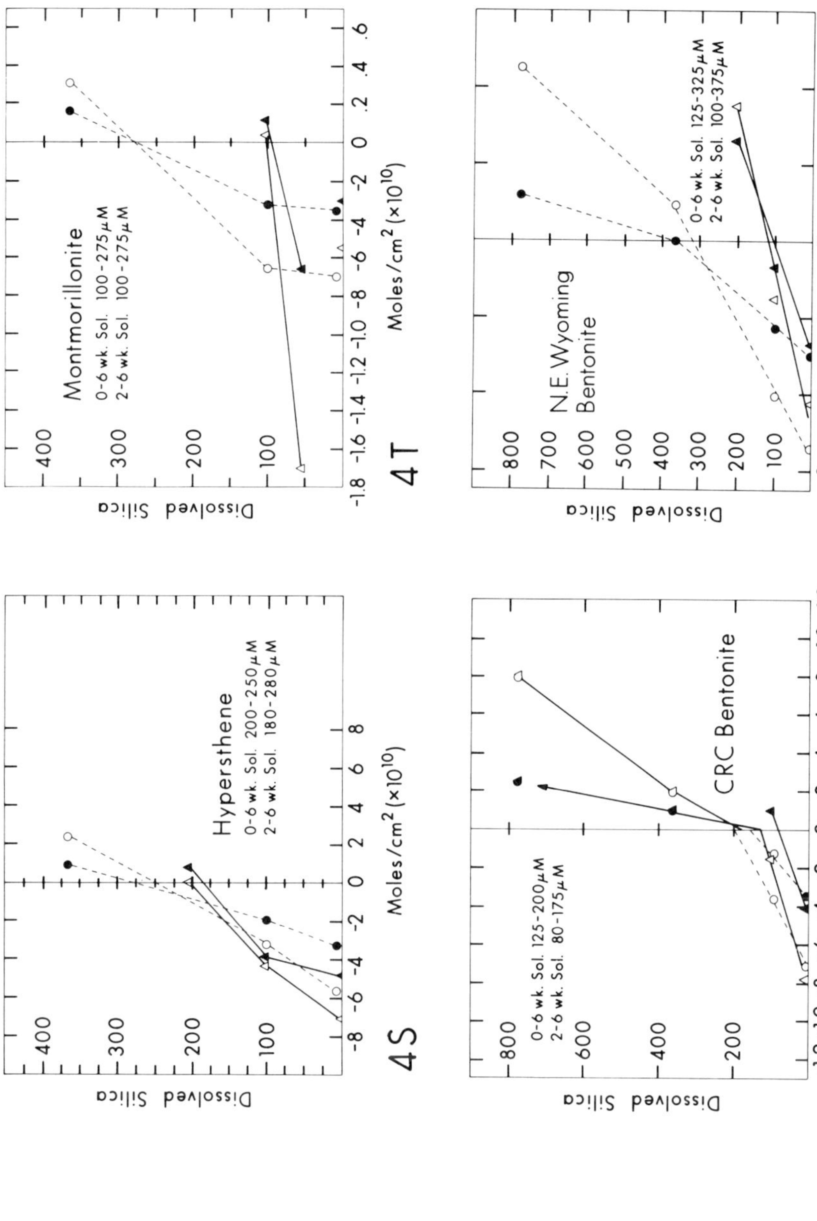
Montmorillonite
0-6 wk. Sol. 100-275 μM
2-6 wk. Sol. 100-275 μM
Dissolved Silica
Moles/cm² (×10¹⁰)
4T
N.E. Wyoming Bentonite
0-6 wk. Sol. 125-325 μM
2-6 wk. Sol. 100-375 μM
Dissolved Silica
Moles/cm² (×10¹⁰)
4V
Hypersthene
0-6 wk. Sol. 200-250 μM
2-6 wk. Sol. 180-280 μM
Dissolved Silica
Moles/cm² (×10¹⁰)
4S
CRC Bentonite
0-6 wk. Sol. 125-200 μM
2-6 wk. Sol. 80-175 μM
Dissolved Silica
Moles/cm² (×10¹⁰)
4U

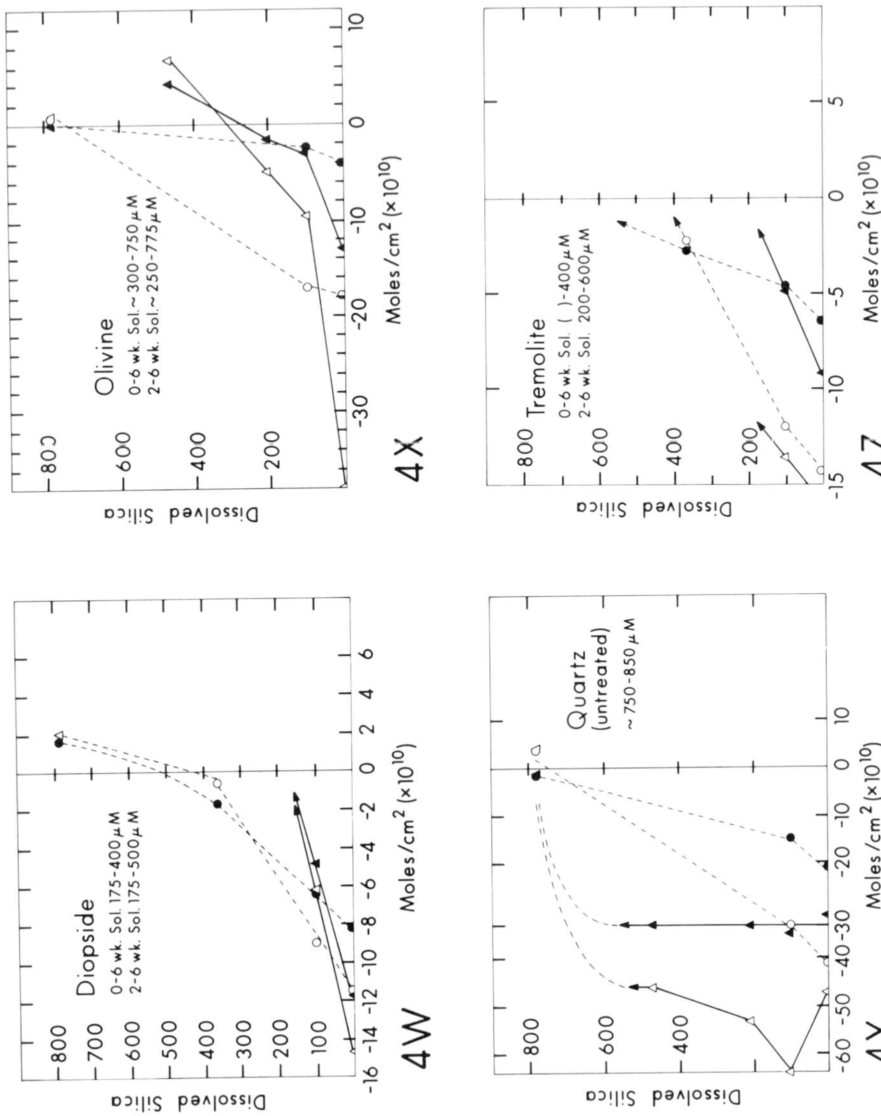
Olivine
0-6 wk. Sol. ~300-750 μM
2-6 wk. Sol. ~250-775 μM
Dissolved Silica
Moles/cm² (×10¹⁰)
4X
Tremolite
0-6 wk. Sol. ( )-400 μM
2-6 wk. Sol. 200-600 μM
Dissolved Silica
Moles/cm² (×10¹⁰)
4Z
Diopside
0-6 wk. Sol. 175-400 μM
2-6 wk. Sol. 175-500 μM
Dissolved Silica
Moles/cm² (×10¹⁰)
4W
Quartz
(untreated)
~750-850 μM
Dissolved Silica
Moles/cm² (×10¹⁰)
4Y

The most obvious error source has already been mentioned: the possible existence of very small particles having high surface energies and/or the possible existence of a disturbed surface layer on the larger mineral particles. Both of these effects would yield higher solubility values, depending on how readily very small particles (ca. less than 50 angstroms radius, below which surface energy effects become appreciable) and/or stressed surfaces are generated during the grinding process.

Another major source of error is the specific surface area data for each mineral and the necessity for determination of values for the same sample before and after reaction. Since total silica changes in solution are normalized on a per unit area basis, if the areas are incorrectly estimated the intercept with the vertical axis will be displaced to some extent.

Figures 2 and 3a and b also show that fluxes per unit area are much greater during the first 7 to 14 days than during subsequent time intervals. For this reason we have estimated solubilities both including and excluding this time period (Figures 4a-z), but in general the estimates are not significantly different from one another.

It might also be argued for a number of minerals that the paucity of data precludes <u>any</u> estimate of "apparent" solubilities. This argument is valid to the extent of the degree of accuracy required by the critic. We undertook to survey a large number of common minerals in a general way within a finite set of time constraints. We have no doubt that future workers will improve upon our data, but we also feel that reasonable estimates along with their uncertainties are better than no data at all.

<u>Layers Lost Vs. Apparent Solubility Estimates.</u> Figure 5 shows a summary data plot wherein the vertical axis gives the apparent number of layers removed or extracted of silica (0-6 week data set, dissolved silica concentration 1-5 μM, pH range 7.5-8.5) and the horizontal axis gives our estimated "apparent" solubility values in μM dissolved silica. Each mineral is represented by a box because of the effect of pH on both the actual dissolution rate or surface flux and on our "cross-over" or "apparent" solubilities. This can also be seen in Figures 4a-z, wherein higher pH values usually result in higher dissolution rates and lower solubilities.

Various minerals have been grouped according to their structural types (i.e. ino-, neso-, soro-,

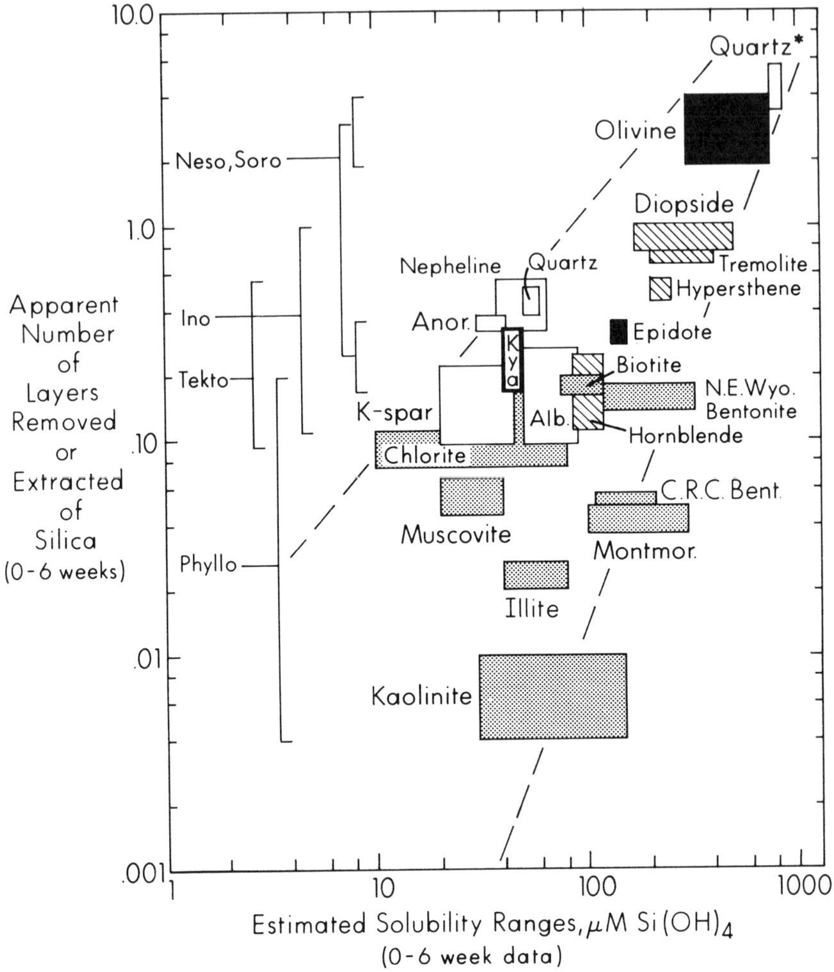

Figure 5. Summary plot of apparent number of layers removed or extracted of silica (0–6 weeks) vs. estimated solubilities with respect to dissolved silica, μM, using the 0–6 week data in Figure 4(a–z). The diagonal dashed lines suggest general trends in the data, while the narrow solid line at 100 μM is the mean concentration of dissolved silica in seawater. Brackets and various types of shading separate minerals into various groups: phyllosilicates, crosshatching; tectosilicates, open; inosilicates, diagonal lines; neso- and soro-silicates, solid.

tekto-and phyllo-silicates) and although there exists a range of values for each group, it appears that both the solubilities and dissolution rates are controlled in part by the structural arrangement of the silica tetrahedra. The following groups are listed in order of decreasing solubility and dissolution rates: neso, soro > ino > tekto ≥ phyllo silicates. It is interesting and perhaps not surprising to note that this parallels the same general order as Goldich weathering sequence for fresh-water environments (15) and other weather sequences (16, 17, 18, 25), in spite of the possible effects on solubility and dissolution rate produced by grinding. Although several of the above authors have considered the concept of specific surface area of various minerals as a factor in their discussions of weathering or have attempted to normalize their data by using the same size range material in their experiments, it appears that we are the first to use measured specific surface areas on such a wide range of minerals and to normalize our data in terms of these surface area values. It appears also that our approach is successful in that Figure 5 shows in a simple way that solubility and maximum dissolution rates are related in a simple way, and that structure and solubility are also closely tied together.

Discussion

The method and results described in this paper present a particular approach toward describing the magnitude and rate of silica precipitation and dissolution for a number of common silica-containing minerals under conditions frequently found in the oceans. We have suggested that once a given mineral is in suspension in seawater, that the conditions of temperature and pH do not rapidly change from one day to the next, nor does the dissolved silica in the water column surrounding the particle. For these reasons we have chosen an approach which attempts to maintain constant temperature, pH and dissolved silica concentrations by frequently changing the extracting solutions. Both pH and dissolved silica do vary in our solutions, by relatively large amounts at the beginning of each run and considerably less so after several weeks; however, the initial conditions are continually re-introduced. The response of a given mineral and whatever temporary or semi-permanent secondary product layer is formed at its surface are constantly responding to a relatively narrow range of solution conditions.

Dissolution rates at low dissolved silica concentrations are generally proportional to the distance of a given mineral from saturation with respect to dissolved silica (Figure 5). This would suggest that the energy of activation regarding silica release might not be too different (ca. ± 2-3 kcal/mole) from one mineral to the next. The logic behind this is as follows: for first-order dissolution of a substance, the maximum flux from the surface in moles/cm^2 sec, is obtained when the dissolved silica concentration in solution is much smaller than whatever the solubility of that substance is under the given set of conditions being studied. We suggest that the linear portion (2-6 weeks) of Figures 3a and b can be described in such a way. Further we suggest that this maximum long-term dissolution flux from the surface when divided by our "apparent" solubility estimates for the corresponding conditions will produce a value for the heterogeneous rate constant which we will call K_{LT} which has the units of cm/sec. The range of values of K_{LT} (see Table 3) when inserted in an equation which relates heterogeneous first-order rate constants to free energy of activation (19) shows only a 2-3 kcal/mole range.

Mineralogical structural types may have an effect on silica release rates as has been mentioned earlier. That is, isolated silica tetrahedra and small units (neso-, soro-silicates) release silica more quickly than single or double chains (ino-silicates) which in turn are more reactive than three dimensionally linked tetrahedra (tecto-silicates) or tetrahedra arranged in sheets (phyllo-silicates). But such structures are not the only cause for variations in silica release rates.

An important factor in controlling mineral dissolution rates are the steps which involve the hydrolysis of cations at the mineral surface and their subsequent removal into the solvent (20, and references therein).

One semi-quantitative approach toward estimating the ease of hydrolysis of the cations in a given mineral is the determination of its abrasion pH (21, 22). When a cation on the surface of a mineral is hydrolyzed, the net effect is the consumption of protons. A high value for the abrasion pH of a given mineral would then suggest that the cations in the structure of that mineral would be readily hydrolyzed.

Figure 6 shows a plot of apparent number of layers removed or extracted of silica versus the abrasion pH of various minerals, the latter using the data

TABLE III

Estimation of long-term dissolution rate constants (K_{LT}) from 2-6 week surface flux averages and "apparent" mineral solubilities.

Mineral	Surface Flux, (K_{LT} X Csatapp.) moles SiO_2/cm^2 sec		Apparent Solubility (Csatapp.) moles SiO_2/cm^3		Long-Term Dissolution Rate Constant, (K_{LT}) cm/sec	
	pH 8.3-8.5 ($X10^{16}$)	pH 7.6-7.9 ($X10^{16}$)	pH 8.3-8.5 ($X10^9$)	pH 7.6-7.9 ($X10^9$)	pH 8.3-8.5 ($X10^{10}$)	pH 7.6-7.9 ($X10^{10}$)
Albite	1	.34	50	75	20	4.5
Anorthite	.8	.8	50	50	16	16
N.E.W. Bentonite	.4	.4	100	350	0.4	0.1
Biotite	.47	-	75		6.3	
Chlorite	.2	.23	15	80	15	2.8
Diopside	4	2.3	175	450	23	5
Hornblende	1.4	.48	100	125	14	3.8
Hypersthene	1.5	1	190	260	7.8	3.8
Illite	.046	.025	40	65	1.2	.32
K-spar	.66	.35	50	50	13	7
Kaolinite	.033	.022	10-100	10-100	.3-3	.2-2
Kyanite	.82	.32	40	5.5	21	5.8
Montmorillonite	.094	.11	100	275	.94	.4
Muscovite	.115	.18	50	40	3-5	4.5
Nepheline	1.2	.67	40	70	30	9.6
Olivine	3.9	1.1	300	750	18	1.5
"Orthoclase"	.46	.14	~ 40	~ 40	12	3.5
Quartz (untr.)	9.7	6	~800	800	12	7.5
Quartz (1 HF)	1.3	-	60	60	22	-
Quartz (2 HF)	1.3	-	60	60	22	-
"Serpentine"	1.3	.6	90	125	14	4.8
Tremolite	2.6	1.6	200-400	400-600	7-13	2.5-4

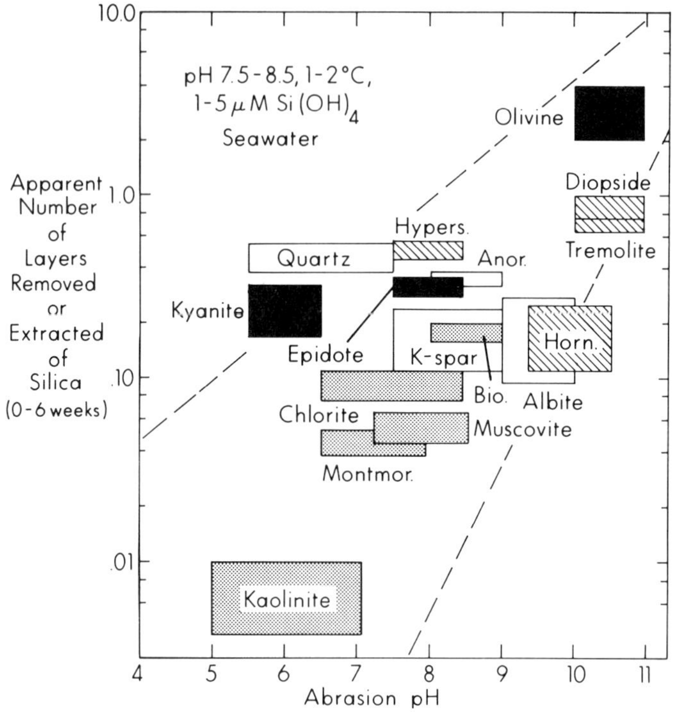

Figure 6. Apparent number of layers removed or extracted of silica (0–6 week data) vs. abrasion pH. Readers should note that abrasion pH determined on different mineral samples and different reaction conditions than data in vertical axis (21).

of Stevens and Carron (21). The abrasion pH data may not be strictly comparable in that they were determined in distilled water at room temperature, i.e. values determined in seawater might not be exactly the same as those in distilled water. The general trends should, however, remain the same, and the correlation between abrasion pH and removal rates is as striking as that between solubility and removal rates. It will be an interesting problem to determine if the various silica tetrahedra arrangements make the cations more or less available by their structure or if the processes of cation hydration and tetrahedral structure breakdown are separate from one another.

It is interesting to note in those samples where sufficient data points exist, that the rates of silica precipitation are generally less than those of dissolution when equal degrees of undersaturation and supersaturation with respect to silica are compared. This may be because 1) a different reaction order exists, 2) the activation energy of mineral formation may be higher than for mineral dissolution, or 3) insufficient amounts of cations (especially aluminum) may be available for the formation of additional mineral. In any case, the magnitudes and rates of silica precipitation are still on the same order as those for dissolution, at least during our six-week study periods.

In general, the apparent solubility values are reasonable when compared with extrapolated estimates of solubilities determined at room temperature. One of the basic difficulties lies in the fact that we have been unable to locate other studies which actually attempted to do low temperature (i.e. 1-2°C) solubility work using alumino-silicate minerals, thus we have no other analogous lab studies with which to compare our numbers. A few of our apparent solubilities are worth singling out for discussion.

a. Kaolinite - the 0-6 week estimated values are surely too low and the 2-6 week values are more correct. While the X-ray diffraction patterns of the sample generally match well with the powder file standards, elemental analyses of the sample show about a 10 wt. % excess of Al_2O_3. This aluminum oxide phase could well be responsible for the initial delay in silica release values during the first two weeks of sampling. An estimate of 30-150 μM is more reasonable.

b. Quartz - it has been known for some time that fine grinding of minerals forms either a disturbed surface layer or a large number of small, high surface energy particles which increase the solubility of a given mineral considerably (8, and references therein). Quartz is no exception as has been shown numerous times (23, 24, and references therein).

We have used our proposed method to study the rates of dissolution and precipitation of silica on both freshly ground quartz (Figure 4y) and quartz which had been treated once with HF and NaOH (24) (Figure 4m, 1-HF) and then treated again (Figure 4m, 2-HF). The freshly ground quartz gave an apparent solubility value which was essentially identical to acid-cleaned biogenic silica and silica gel (1, 19), 800-900 μM $Si(OH)_4$ at 1-2°C. The once and twice cleaned samples gave virtually identical solubilities, 60-70 μM at 1-2°C. This latter figure compares quite well with extrapolated data from higher temperatures and appears to be the first 1-2°C determination of the solubility of this mineral.

c. Although not minerals as such, the volcanic glasses also show some interesting effects. Both low-silica glasses (Figure 4c, q) and high silica glasses (Figure 4p) show "apparent" solubilities between 100 and 200 μM $Si(OH)_4$. In that a glass would not be expected to have a well defined solubility because it does not have a crystalline framework for its component elements, it was surprising to us that relatively narrow ranges of dissolved silica cross-over values were found for these substances. These low "apparent" dissolved silica solubilities may begin to explain the presence of unaltered glass in sediments (28) which surely have dissolved silica pore water concentrations higher than 200 μM $Si(OH)_4$ (27).

Acknowledgments

The following people have been quite helpful with respect to supplying samples and critical advice; they are in no way responsible for any of the errors contained herein, however: Drs. Michael Garcia, John Sinton, Kost Pankiwskyj, Erwin Suess, Rich Holdren, and Bob Berner. This research was supported by the Office of Naval Research. Hawaii Institute of Geophysics Contribution No. 917.

Abstract

A method for the determination of the solubilities of ca. 20 common silica-containing minerals is proposed, wherein the release rate of dissolved silica is monitored in terms of mol cm^{-2} sec^{-1} from the mineral surface, using a series of extractions with pH 7.5-8.5 seawater and having the dissolved silica concentration adjusted to 5, 50, 100 and 200 x 10^{-6}mol l^{-1} (the range of dissolved silica concentrations commonly found in the deep oceans).

These extractions reveal both long-term dissolution rates and precipitation magnitudes for a given mineral at a given dissolved silica concentration. Preliminary results suggest the following solubility ranges with respect to dissolved silica in seawater at 1-2°C for the following minerals: albite, 55-110 μM; kaolinite, 10-300 μM; hornblende, 50-120 μM; nepheline, 35-75 μM, chlorite, 40-80 μM; biotite, 60-120 μM; illite, 40-70 μM; montmorillonite, 80-250 μM; hyperstene, 200-275 μM; bentonite, 115-340 μM; epidote, 150-200 μM; tremolite, 200-275 μM; olivine, 300-700 μM; quartz, variable but between 50 and 800 M. The above approach may allow us to predict a stability sequence or weathering series for these minerals in seawater and deep-sea sediments having a variety of dissolved silica concentrations.

Literature Cited

1. Hurd, D. C., The effect of glacial weathering on the silica budget of Antarctic waters, Geochim. Cosmochim. Acta, 41, 1213-1222 (1977).
2. Schutz, D. G. and Turekian, K. K., The investigation of the geographical and vertical distribution of several trace elements in sea water using neutron activation analysis, Geochim. Cosmochim. Acta, 27, 259-313 (1965).
3. Burton, J. D. and Liss, P. S., Processes of supply and removal of dissolved silicon in the oceans, Geochim. Cosmochim. Acta, 37, 1761-1773 (1973).
4. Strickland, J. D. H. and Parsons, T. R., A Practical Handbook of Seawater Analysis, pp. 65-70. Bull. Fish. Res. Bd. Can. 167 (1968).
5. Maxwell, J. A., "Rock and Mineral Analysis," 584 pp., Wiley-Interscience, New York, 1968.
6. Fanning, K. A. and Pilson, M. E. Q., On the spectrophotometric determination of dissolved silica in natural waters, Anal. Chem., 45(1), 136-140 (1973).
7. Chester, R., Cross, D., Griffiths, A. and Stoner, J. H., The concentrations of "aluminosilicates" in particulates from some surface waters of the world ocean, Mar. Geol., 22, M59-M67 (1976).
8. Busenberg, E. and Clemency, C. V., The dissolution kinetics of feldspars at 25°C and 1 atm CO_2 partial pressure, Geochim. Cosmochim. Acta, 40, 41-49 (1976).

9. Lawson, D., Hurd, D. C. and Pankratz, H. S., Silica dissolution rates of decomposing phytoplankton assemblages at various temperatures, Am. J. Sci. (in press).
10. Mackenzie, F., Garrels, R. M., Bricker, O. and Bickley, F., Silica in seawater: control by silica minerals, Science, 155, 1404-1405 (1967).
11. Siever, R., Establishment of equilibrium between clays and seawater, Earth Planet. Sci. Lett., 5, 106-110 (1968).
12. Maynard, J. B., Kinetics of silica sorption by kaolinite with application to seawater chemistry, Am. J. Sci., 275, 1028-1048 (1975).
13. Dayal, R., Kinetics of silica sorption and clay dissolution reactions at 1 and 670 atm., Geochim. Cosmochim. Acta, 41, 135-141 (1977).
14. Siever, R. and Woodford, N., Sorption of silica by clay minerals, Geochim. Cosmochim. Acta, 37, 1851-1880 (1973).
15. Goldich, S. S., A study in rock weathering, J. Geol., 49, 17-58 (1938).
16. Pettijohn, F. J., Persistence of heavy minerals and geologic age, J. Geol., 49, 610-625 (1941).
17. Jackson, M. L., Tyler, S. A., Willis, A. L., Bourbeau, G. A. and Pennington, R. P., Weathering sequence of clay size minerals in soils and and sediments, J. Phys. Colloid Chem., 52, 1237-1260 (1948).
18. Fieldes, M. and Swindale, L. D., The chemical weathering of silicates in soil formation, J. Sci. Tech. New Zealand, 36, (Sect. B) 140-154 (1954).
19. Hurd, D. C. and Theyer, F. Changes in the physical and chemical properties of biogenic silica from the central equatorial Pacific--I. Solubility, specific surface area, and solution rate constants of acid-cleaned samples, 211-230, in (Gibb, Jr., T. R. P., editor) "Analytical Methods in Oceanography," Adv. Chem. Ser. 147, 1975.
20. Loughnan, F. C., "Chemical Weathering of the Silicate Minerals," 154 pp., Elsevier, New York, 1969.
21. Stevens, R. E. and Carron, M. K., Simple field test for distinguishing minerals by abrasion pH, Am. Mineralogist, 33, 31-49 (1948).
22. Grant, W. H., Abrasion pH, an index of chemical weathering, Clays Clay Min., 17, 151-155 (1969).

23. Sosman, R. B., "The Phases of Silica," 388 pp., Rutgers Univ. Press, New Brunswick, N. J., 1965.
24. Henderson, J. H., Syers, J. K. and Jackson, M. L., Quartz dissolution as influenced by pH and the presence of a disturbed surface layer, Israel J. Chem., 8, 357-372 (1970).
25. Keller, W. D., "The Principles of Chemical Weathering," 88 pp., Lucas Bros., Columbia, Mo., 1955.
26. Lerman, A., Mackenzie, F. and Bricker, O. P., Rates of dissolution of aluminosilicates in seawater, Earth Planet. Sci. Lett., 25, 82-88 (1975).
27. Hurd, D. C., Interactions of biogenic opal, sediment and seawater in the central equatorial Pacific, Geochim. Cosmochim. Acta, 37, 2257-2282 (1973).
28. Scheidegger, K. F., Volcanic ash layers in deep-sea sediments and their petrological significance, Earth Planet. Sci. Lett., 17, 397-407 (1973).

RECEIVED November 16, 1978.

Dissolution Kinetics of Silicate Rocks—Application to Solute Modeling

ART F. WHITE and HANS C. CLAASSEN

U.S. Geological Survey, Denver, CO 80225

Silicate minerals comprise more than 75 percent of the rocks with which ground water comes in contact (1). Quantitative knowledge of the weathering rates of these minerals could provide an important means of estimating the total residence time a water has been in contact with a given silicate rock, the effective surface area of the aquifer, and other information pertinent in defining a given hydrogeologic system (for a specific example, see Claassen and White's report in these proceedings).

Since the work of Daubreé in 1867 (2), workers have experimented to define the dissolution rates of silicate minerals. This paper will briefly review the types of reaction rates that have been observed, followed by more detailed discussion concerning the dependence of these rates on constituents in the aqueous media, and the implications for solute modeling.

The Rate Expression

Silicate dissolution can generally be described by one of two rate expressions (3,4):

$$Q = Q_1^o + k_1 t, \qquad (1)$$

or

$$Q = Q_p^o + k_p t^{\frac{1}{2}}, \qquad (2)$$

where Q (mol/cm^2) is the mass transfer of a species into the aqueous solution per unit surface area of the solid, Q_1^o and Q_p^o (mol/cm^2) are the linear and parabolic mass transfers extrapolated to time t, equal to zero, and k_1 ($mol/cm^2 sec$) and k_p ($mol/cm^2 sec^{\frac{1}{2}}$) are the respective linear and parabolic rate constants.

Figures 1 and 2, respectively, show the mass transfer of sodium, Q_{Na}, plotted against time (secs), and the square root of time ($secs^{\frac{1}{2}}$), for dissolution experiments involving laboradorite, a plagioclase feldspar, and obsidian, a volcanic glass. Both

0-8412-0479-9/79/47-093-447$06.75/0
This chapter not subject to U.S. copyright
Published 1979 American Chemical Society

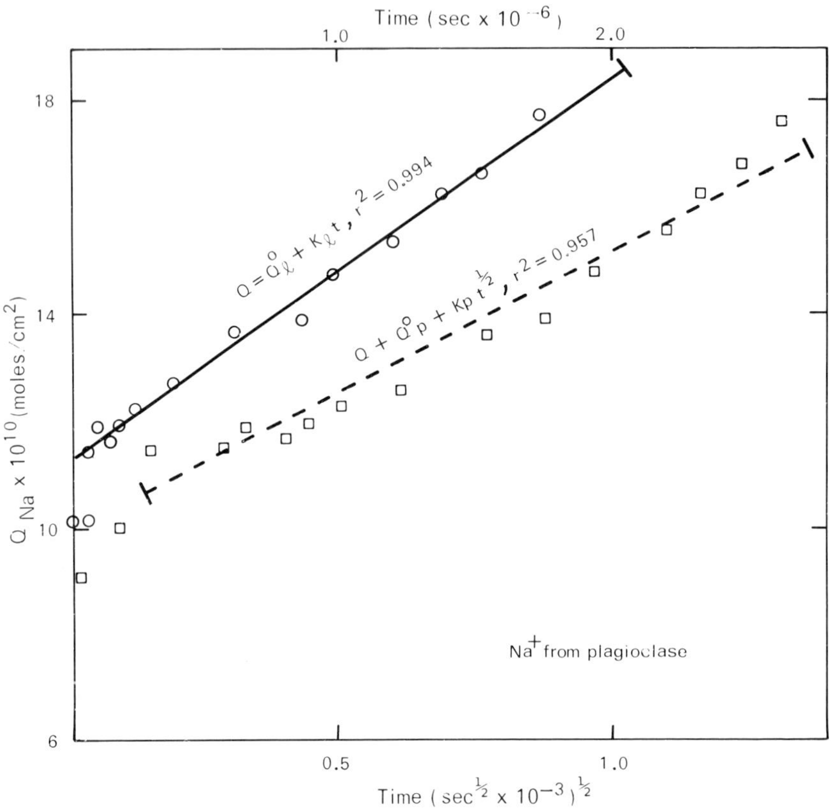

Figure 1. Mass transfer rates of sodium from a plagioclase feldspar into solution at pH 6.0 and 25°C

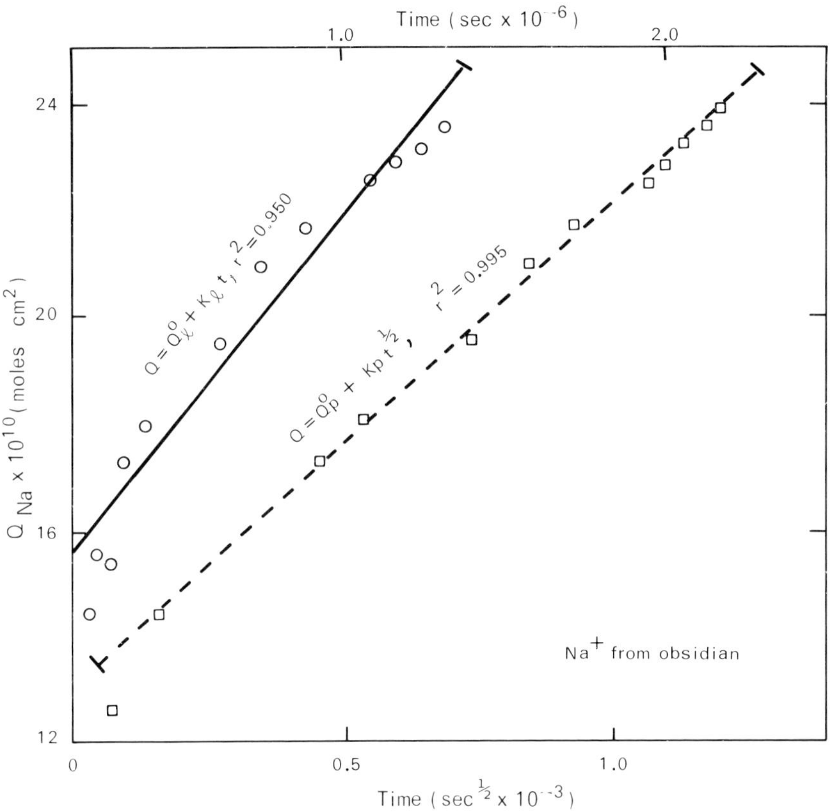

Figure 2. Mass transfer rates of sodium from obsidian into solution at pH 6.0 and 25°C

experiments were conducted under a nitrogen atmosphere at 25°C ± 0.1 using a Metrohm pH stat titrator and .01 N HCl. The solid lines through the data are linear regression fits for the linear rate expression (equation 1), and the dashed lines regression fits for the parabolic rate (equation 2). Data points at short reaction times which represent neither linear nor parabolic rates are excluded from the statistical fit. Based on the regression coefficient, r^2, the linear rate best describes plagioclase dissolution, and the parabolic rate best describes glass dissolution. The difference in rate expression under identical experimental conditions emphasizes the role of silicate structure and composition in determining the dissolution mechanism.

Figures 1 and 2 show that the intercepts Q_l^o and Q_p^o at t = 0 are not zero. Rapid initial mass transfer into solution at short times is attributed to exchange between aqueous hydrogen ions and cations situated at or near the fresh silicate surface. Tamm (5) and Garrels and Howard (6) have demonstrated the reversibility of such surface exchange between hydrogen and potassium ions in potassium feldspars. However, in the case of magnesium silicates, Luce and others (4) showed that such exchange was not stoichiometric, owing to establishment of surface charge and the electric double layer.

Three basic reaction mechanisms have been invoked by various workers to explain the long-term rate expressions (equations 1,2):

I. The dissolution rate is controlled by reaction of the unaltered silicate with aqueous hydrogen ions at the interface between the two phases (7, 8, 9).

II. The dissolution rate is controlled by interdiffusion of hydrogen or hydronium ions and species contained in lattice sites within the interior of the silicate phase. This process results in a leached layer consisting mainly of silica and alumina. Such a layer may retain the original silicate structure (10, 11) or may represent a collapsed or hydrated layer (12, 13).

III. The dissolution rate is controlled by diffusion through a continuously growing precipitate layer that forms on the silicate surface. Such a protective barrier has been postulated to consist of an amorphous alumino-silica phase (14) or a mono- or multi-phase crystalline alumino-silicate assemblage (15).

Assuming a constant surface area, dissolution at a solution-solid interface (Case I) results in linear kinetics in which the rate of mass transfer is constant with time (equation 1). Analytical solutions to the diffusion equation result in parabolic rates of mass transfer (4, 16) (equation 2). This result is obtained whether the boundary conditions are defined so diffusion occurs across a progressively thickening, leached layer within the silicate phase (Case II), or across a growing precipitate layer forming on the silicate surface (Case III). Another case of linear kinetics (equation 1) may occur when the rate of formation of a metastable product or leached layer at the fresh silicate surface becomes equal to the rate at which this layer is destroyed at the aqueous interface (4, 11, 17); that is, such steady state diffusion across a layer of constant thickness results in a linear mass transfer rate (equation 1).

The above hypotheses are based principally on rate expressions describing changes in experimental aqueous compositions such as those shown on Figures 1 and 2. As indicated, more than one reaction mechanism can be invoked to explain the observed rate expressions. Substantiating evidence, particularly regarding the condition of the solid state during dissolution, is meager. The existence of diffusion in leached layers of artificial silicate glass has been confirmed by step-wise dissolution of the glass with hydrofluoric acid (12, 13, 18). The extent of the structural collapse of this layer was dependent on the size ratio of the substituting ions and cations originally contained in the silicate structure. The occurrence of lithium diffusion through the leached layer of an artificial silicate glass was confirmed using an ion-sputtering technique (19).

The occurrence of a precipitate layer controlling the rate of diffusion is more tenuous. Various amorphous precipitates, in addition to crystalline phases, including boehmite, kaolinite, and halloysite, have been shown to form on silicate surfaces during dissolution (20, 21, 22, 23). However, based on a summary of surface characteristics, Petrović (24) concluded that such phases generally occur as individual particles on the silicate surface, and not as layers coating the surface. The resulting high porosity would tend to exclude such material as a diffusion-limiting medium.

Petrović and others (25), using X-ray photoelectron spectroscopy, failed to detect the presence of either a leached or precipitated layer on alkali feldspars reacted at 82°C.

The derivative of mass transfer with respect to time for equations 1 and 2 is, respectively

$$\partial Q/\partial t = k_1, \tag{3}$$

and

$$\partial Q/\partial t = \tfrac{1}{2}k_p t^{-\frac{1}{2}}. \tag{4}$$

The rate of mass transfer will be constant with time for the linear rate and will continually decrease for the parabolic rate. For long times, such as those encountered in hydrogeologic environments, the linear rates of mass transfer may be expected to dominate over the parabolic rates. Although the duration of experimental reactions is short compared to such times, evidence exists that a shift from parabolic to linear rates does occur. In an artificial silicate glass at elevated temperatures (60-100°C), Rana and Douglas (3) found that change from parabolic to linear kinetics occurred between 6 minutes and 66 hours, depending on the glass composition. Busenberg and Clemency (26) noted that changeover during the reaction of plagioclase feldspar at 25°C occurred after approximately 20 days (26). One of the problems in postulating linear reactions for all natural long-term processes lies in the problem of congruency. Almost all experimental reactions, whether parabolic or linear, are incongruent; that is, the ratios of mass transfers of chemical species into solution are not directly proportional to the molecular composition of the dissolving phase (14).

Incongruent mass transfer in the case of parabolic kinetics based on diffusion control can be readily explained by differences in the diffusion coefficients of chemical species. However, the proposed mechanisms invoked to explain linear kinetics involve the dissolution of the bulk silicate phase, which, over extended periods, should result in the chemical components being transferred to solution from the silicate in their stoichiometric ratios (25). However, many field studies have shown that silicates weather incongruently in nature (27). This is supported by our studies of ground waters associated with vitric rhyolite tuffs.

Effect of Specific Aqueous Ions on the Rate Expression

One parameter necessary for modeling silicate dissolution kinetics is the influence of aqueous chemistry on the rate expression. Lagache and others (7) found that for K-feldspar dissolution at 200°C, the rate of release of silica and alumina was inversely proportional to their respective concentrations in aqueous solution. On the other hand, Wollast (28) found that the release of silica and alumina from K-feldspar at room temperature was independent of aqueous concentrations. Wollast suggested that the reaction rate is influenced by aqueous concentrations only when those concentrations approach saturation with respect to a product phase. However, extensive data collected by the authors (29) on vitric tuffs and obsidian showed that the resulting parabolic rate expressions for cations, far from equilibrium with a product phase, are strongly dependent on the concentrations of certain cations in coexisting aqueous solution.

The aqueous-dissolution experiments were performed in the following manner. Individual 25-, 50-, 100-, and 200-g portions of glass were added to 2-L volumes of deionized water in polyethylene Erlenmeyer flasks. The flasks were placed in constant-temperature water baths. The mixtures were agitated by stirring paddles at a rate just sufficient to keep the glass in suspension. Surface area determinations before and after individual experiments indicated no surface area change due to abrasion during agitation.

Commercial grade CO_2 and compressed air, controlled by flow meters, were continually mixed prior to being bubbled through the solutions. The resulting partial pressures of CO_2, which equilibrated with the water, determined the pH ranges of the experiments.

Equilibration of the solutions was accomplished prior to addition of solid and verified by pH measurements. Immediately after addition of the glass phase, the pH of the system increased, indicating rapid hydrogen ion consumption. After approximately 24 hours, the rate of consumption decreased. This effect, coupled with the buffering effect of increasing bicarbonate concentrations, confined subsequent pH increases to an average of about 0.3 units for the remainder of a given experiment. Twenty 5-ml aliquots of the homogenized suspension were withdrawn with a plastic syringe at appropriate time intervals. Samples were filtered through 0.45-μm membrane filters and analyzed for major dissolved constituents.

Figure 3 shows a plot of the mass transfer data for sodium from vitric Rainier Mesa tuff plotted against the square root of time in seconds for four different surface-to-volume ratios at 25°C and an average pH of 4.75. Experimental data are presented in Table I. The solid lines represent the linear regression fits to the data. Because the parabolic mass transfer is normalized per unit surface area, the reaction rate, k, ideally should be independent of the total surface area used in a given experiment. However, Figure 3 clearly shows that the reaction rates are lower for experiments involving greater total surface areas. A greater surface area produces a larger total mass transfer, and, for a fixed solution volume, a greater aqueous concentration. This increased concentration apparently inhibits the release rate of sodium from the glass, as shown on Figure 3.

The inhibition of the release of sodium can be demonstrated more directly by addition of sodium chloride to the aqueous solutions, prior to introduction of the vitric tuff. In Figure 4, data labeled C, containing no sodium chloride, cover the same experiment shown on Figure 3. Data labeled E and F are, respectively, for experiments in which 1.5×10^{-4} and 3.0×10^{-4} mol/L of NaCl were added to solution. Experimental data are presented in Table II. The data show only the sodium mass transfer from the glass and not sodium introduced as NaCl. The solid lines represent linear regression fits to the data. As shown in Figure 4, increasing additions of NaCl progressively decrease the slope of the regression line, and, consequently, the parabolic rate constant.

Figure 5 shows calcium data for experiments C, G, and H, in which, respectively, no calcium chloride, 1.5×10^{-4}, mol/L $CaCl_2$, and 3.0×10^{-4} mol/L $CaCl_2$ were initially added to solution. Decreases in calcium rate constants with increases in $CaCl_2$ are smaller than the decreases in the sodium rate constants with comparable additions of NaCl (Figure 4). Note, however, the strong inhibition of added calcium on the initial-exchange part of the reaction. Figure 5 shows that, at short times, experiments G and H produced a net-negative mass transfer in which calcium ions were sorbed from solution onto the glass surface.

Experiments on the vitric tuff showed that the rate of mass transfer of an ion from the glass was inhibited specifically by

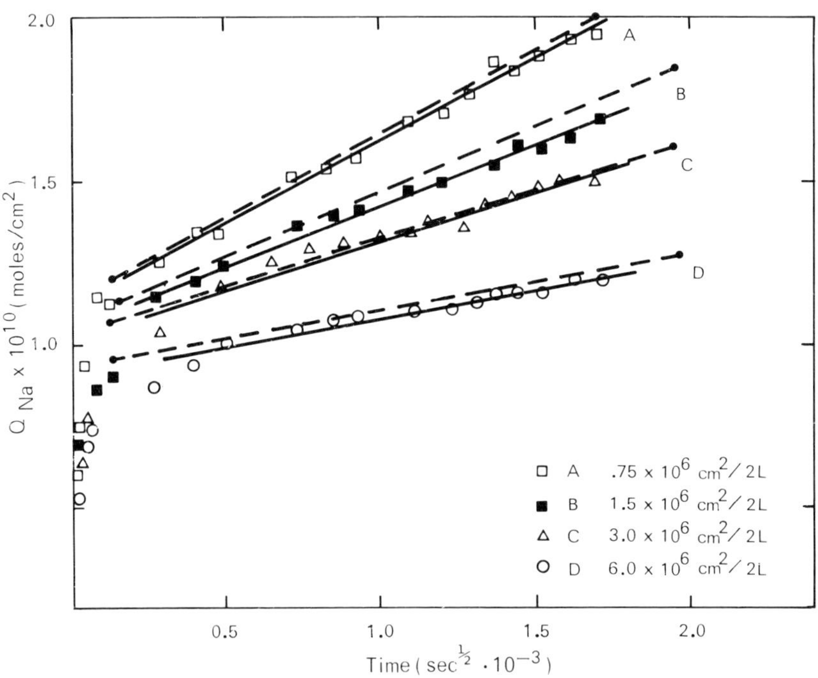

Figure 3. Effect of surface to volume ratios on inhibition of sodium mass transfer from a rhyolitic tuff at an average pH of 4.83 and 25°C. Solid lines are statistical fits to data points. Dashed lines predict inhibition based on diffusion model.

Table I.--Experimental Data Listed As Time, t ($s^{\frac{1}{2}}.10^{-3}$), Mass Transfer Of Sodium, Q_{Na} ($mol/cm^2 \times 10^{10}$), And pH, For Experiments Of Different Total Surface Areas

Experiment	A			B			C			D		
Solid surface area - solution volume ratio	(0.38×10^6 cm^2/L)			(0.75×10^6 cm^2/L)			(1.5×10^6 cm^2/L)			(3.0×10^6 cm^2/L)		
	t	Q_{Na}	pH	t	Q_{Na}	pH	t	Q_{Na}	pH	t	Q_{Na}	pH
Sample No.												
1	0.0078	0.61	4.23	0.0078	0.56	4.38	0.0078	0.61	4.51	0.0078	0.54	4.77
2	.024	.75	4.38	.025	.72	4.44	.025	.69	4.58	.024	.59	4.84
3	.055	.93	4.37	.052	.85	4.49	.052	.77	4.59	.051	.69	4.96
4	.094	1.15	4.39	.095	.88	4.53	.084	.88	4.64	*	*	*
5	.138	1.12	4.40	.127	.93	4.49	.118	.91	4.64	.103	.74	5.22
6	.294	1.25	4.46	.289	1.15	4.57	.285	1.01	4.73	.278	.87	4.87
7	.415	1.33	4.46	.412	1.20	4.61	.409	1.11	4.73	.404	.94	4.93
8	.495	1.33	4.43	.522	1.25	4.60	.517	1.14	4.74	.516	1.00	4.92
9	.702	1.52	4.53	.722	1.28	4.63	.722	1.19	4.79	.727	1.04	4.83
10	.825	1.57	4.47	.842	1.39	4.65	.842	1.19	4.78	.842	1.07	4.96
11	.921	1.57	4.51	.937	1.41	4.67	.937	1.25	4.82	.937	1.09	4.99
12	1.09	1.68	4.54	1.10	1.47	4.68	1.10	1.30	4.85	1.10	1.11	5.02
13	1.21	1.71	4.49	1.22	1.49	4.65	1.22	1.33	4.81	1.22	1.11	5.00
14	1.30	1.76	4.54	1.35	1.52	4.72	1.32	1.35	4.85	1.32	1.13	5.02
15	1.37	1.87	4.54	1.38	1.55	4.71	1.39	1.35	4.84	1.38	1.15	5.01
16	1.44	1.84	4.60	1.44	1.60	4.70	1.44	1.38	4.82	1.45	1.17	4.99
17	1.52	1.89	4.59	1.53	1.60	4.70	1.55	1.38	4.86	1.53	1.17	4.99
18	1.63	1.95	4.60	1.64	1.63	4.70	1.64	1.41	4.92	1.64	1.20	5.07
19	1.71	1.95	4.54	1.72	1.68	4.75	1.72	1.41	4.84	1.71	1.20	5.00

* Data not available.

Table II.--Experimental Data Listed As Time, t ($s^{1/2}.10^{-3}$), Mass Transfer Of Sodium, Q_{Na} ($mol/cm^3 X 10^{10}$) And Calcium, Q_{Ca} ($mol/cm^3 X 10^{10}$), And pH, For Experiments Containing Initial Additions Of NaCl And $CaCl_2$. All Experiments Have A Solid Surface Area To Solution Volume Ratio Of 0.75 X 10^6 cm^2/L

Experiment	E				F			
	(1.5 X 10^{-4}N NaCl added)				(3.0 X 10^{-4}N NaCl Added)			
Sample No.	t	Q_{Na}	Q_{Ca}	pH	t	Q_{Na}	Q_{Ca}	pH
1	0.0078	0.48	0.29	4.46	0.0078	0.31	0.30	4.53
2	.030	.53	.33	4.51	.030	.57	.33	4.69
3	.052	.73	.41	(*)	.052	.61	.39	4.61
4	.120	.80	.47	4.61	.106	.79	.48	4.59
5	.281	.94	.54	4.75	.276	.79	.61	4.65
6	.400	1.09	.63	4.73	.391	.88	.63	4.76
7	.509	1.19	.70	4.80	.509	.92	.70	4.71
8	.712	1.23	.77	4.76	.710	1.00	.74	4.82
9	.835	1.27	.76	4.81	.840	1.07	.77	4.79
10	.885	1.27	.76	4.81	.871	1.07	.77	4.80
11	1.05	1.27	.83	4.85	1.05	1.09	.82	4.80
12	1.17	1.34	.83	4.90	1.17	1.09	.86	4.89
13	1.31	1.34	.89	4.90	1.34	1.14	.90	4.88
14	1.37	1.38	.92	4.83	1.37	1.09	.94	4.84
15	1.44	1.34	.88	4.86	1.44	1.14	.92	4.84
16	1.55	1.38	.90	4.83	1.54	1.14	(*)	4.98
17	1.63	1.45	.94	4.93	1.63	1.14	.93	4.90
18	1.73	1.45	(*)	5.02	1.74	1.18	.97	5.02
19	1.80	1.49	(*)	5.07	1.80	1.22	1.00	5.07

Table II.--Experimental Data Listed As Time, t ($s^{\frac{1}{2}}.10^{-3}$), Mass Transfer Of Sodium, Q_{Na} ($mol/cm^3 X 10^{10}$) And Calcium, Q_{Ca} ($mol/cm^3 X 10^{10}$), And pH, For Experiments Containing Initial Additions Of NaCl And $CaCl_2$. All Experiments Have A Solid Surface Area To Solution Volume Ratio Of 0.75 X 10^6 cm^2/L--Continued

Experiment	G				H			
	(1.5 X 10^{-4}N $CaCl_2$ added)				(3.0 X 10^{-4}N $CaCl_2$ added)			
	t	Q_{Na}	Q_{Ca}	pH	t	Q_{Na}	Q_{Ca}	pH
Sample No.								
1	0.0078	0.77	-0.12	4.47	0.0078	0.72	-0.46	4.45
2	.030	.79	-.53	4.58	.030	.72	-.39	4.50
3	.052	.85	.01	4.54	.052	.85	-.36	4.54
4	.0900	.93	.07	4.58	.0730	1.04	-.36	4.62
5	.309	1.00	.15	4.66	.268	(*)	-.14	4.60
6	.402	1.20	.29	4.68	.388	1.22	.11	4.67
7	.483	1.25	.30	4.74	.502	1.23	.02	4.67
8	.694	(*)	.37	4.71	.707	(*)	.07	4.71
9	.835	(*)	.37	4.77	.883	(*)	.07	4.81
10	.885	1.40	.41	4.76	.885	1.41	.07	4.76
11	1.05	1.40	.43	4.86	1.05	1.44	.13	4.78
12	1.16	1.46	.47	4.80	1.17	1.49	.16	4.85
13	1.31	(*)	.51	4.79	1.31	(*)	.19	4.79
14	1.38	1.49	.53	4.82	1.37	(*)	.19	4.89
15	1.43	1.49	.53	4.86	1.44	(*)	.19	4.94
16	1.55	1.49	(*)	4.87	1.55	1.56	(*)	4.77
17	1.64	1.49	.59	4.84	1.63	1.57	.29	4.82
18	1.73	(*)	.63	4.82	1.73	1.57	.35	4.91
19	1.81	(*)	.65	4.85	1.87	1.60	.38	4.86

* Data not available.

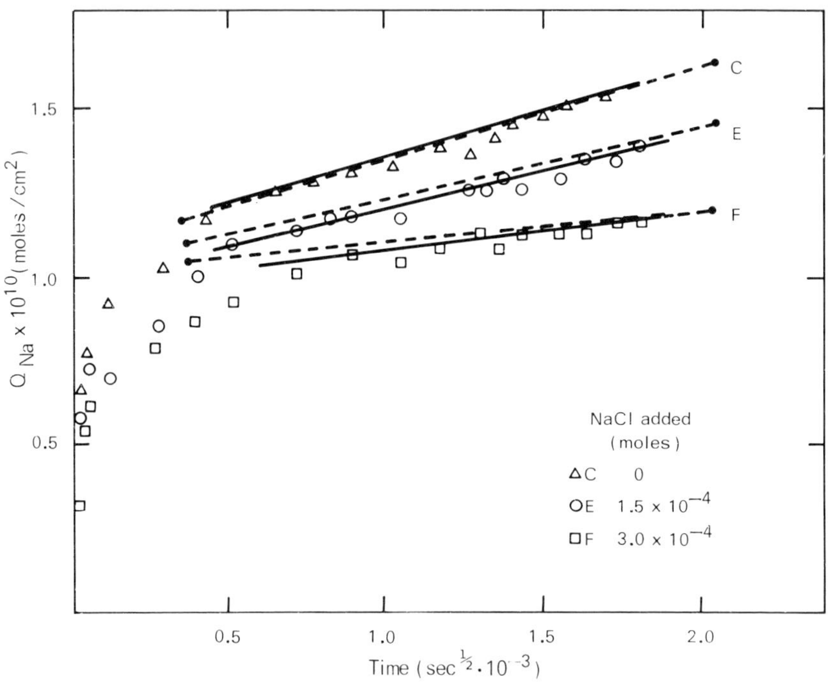

Figure 4. Effect of aqueous sodium chloride on inhibition of sodium mass transfer from a rhyolitic tuff at an average pH of 4.84, 25°C, and a surface area of 1.5×10^6 cm^2/L of solution. Solid lines are the statistical fits to data points. Dashed lines predict inhibition based on diffusion model.

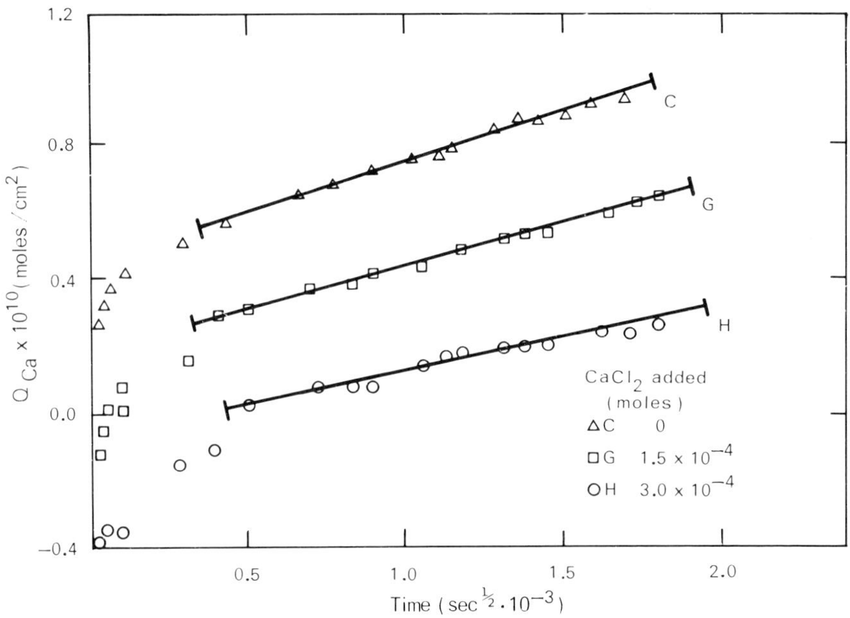

Figure 5. Effect of aqueous calcium chloride on the inhibition of calcium mass transfer from a rhyolitic tuff at an average pH of 4.84, 25°C, and a surface area of 1.5 × 10⁶ cm²/L of solution. Solid lines are statistical fits to the data points.

concentrations of that ion in solution. Figure 6 shows the mass transfer data for sodium in experiments C, G, and H, in which increasing concentrations of $CaCl_2$ were added. As is apparent, aqueous calcium concentrations do not affect the rate of sodium release. Analogously, Figure 7 shows that the calcium mass transfer for experiments C, E, and F are, at most, only slightly reduced by increasing additions of NaCl.

Intuitively, an increase in the aqueous concentration of a specific ion may be expected to increase the concentration of that species at the silicate surface and to decrease the diffusion gradient within the glass. Quantitative modeling of the diffusion process requires a solution to Fick's second law, which, in the case of sodium, is

$$\frac{\partial C_{Na^+}}{\partial t} = D_{Na^+} \frac{\partial^2 C_{Na^+}}{\partial x^2} . \tag{5}$$

The change in concentration of diffusable sodium, C_{Na^+}, with time, t, is related to the apparent sodium diffusion coefficient, D_{Na^+}, and the rate of change in concentration, C_{Na^+}, with distance, x, in the glass. This model assumes that diffusion occurs within a structurally-unaltered leached zone (Case IIa). The solution to the partial differential equation assumes a boundary condition:

$$C^{Surf}_{Na^+} = R[C^{Aq}_{Na^+}]^{1/P} \tag{6}$$

which relates the sodium concentration at the glass surface, $C^{Surf}_{Na^+}$, to the sodium concentration in the aqueous solution, $C^{Aq}_{Na^+}$, where R and P are empirical constants. This isotherm assumes no competitive effects from other aqueous ions, an assumption supported by the data in Figure 6, which represent ion concentration ranges appropriate for this discussion. When $P = 1$, equation 6 reduces to a Langmuir isotherm. When $P > 1$, equation 6 represents a Freundlich isotherm (30).

Crank (16) showed that equation 5 can be solved analytically for the boundary condition (equation 6) where: (1) $P = 1$ and the surface concentration is directly proportional to the aqueous concentration, and (2) $R = 0$ and the surface concentration is zero. The first solution results in diffusion which is dependent on the aqueous concentration, but produces mass transfer which is non-parabolic with time. The second solution results in diffusion which is independent of aqueous concentrations but is parabolic. This latter case has been used by Luce and others (4), Busenberg and Clemency (26), and others to describe diffusion and to calculate apparent diffusion coefficients for silicate minerals.

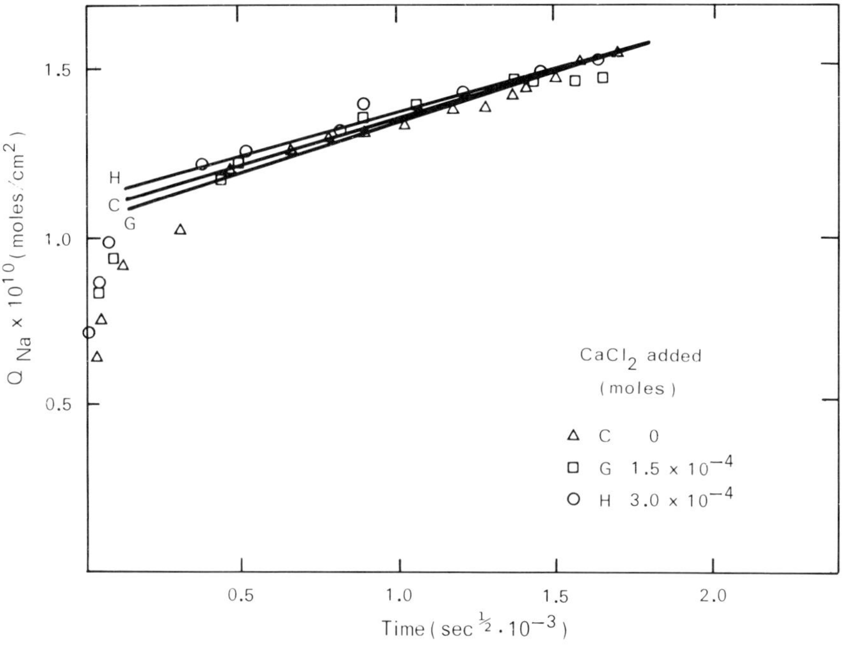

Figure 6. Effect of aqueous calcium chloride on the inhibition of sodium mass transfer from a rhyolitic tuff at an average pH of 4.84, 25°C, and a surface area of 1.5 × 10^6 cm^2/L of solution

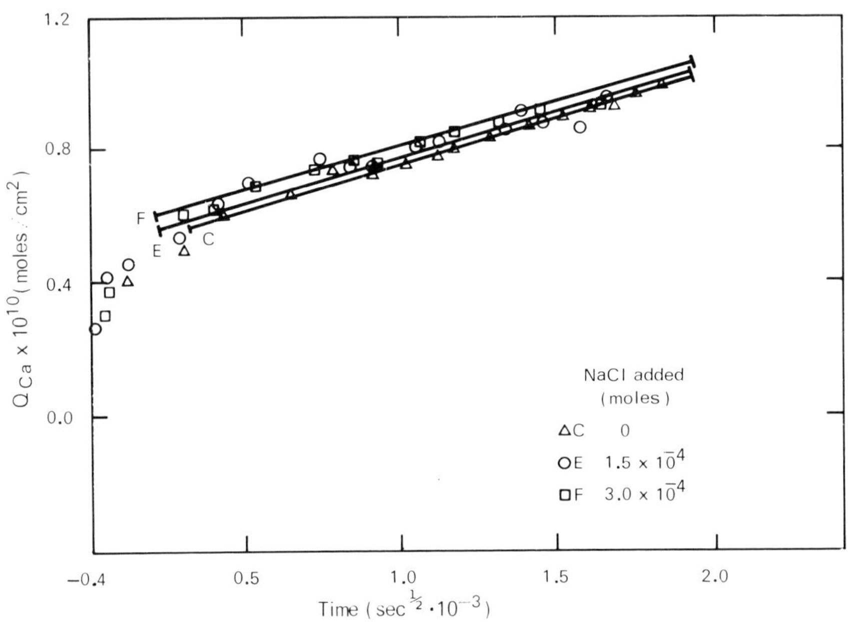

Figure 7. Effect of aqueous sodium chloride on the inhibition of calcium mass transfer from a rhyolitic tuff at an average pH of 4.84, 25°C, and a surface area of 1.5 × 10^6 cm^2/L of solution

Neither analytical solution [1 or 2] successfully describes the sodium data shown on Figures 3 and 4, which indicate that sodium diffusion is both parabolic and dependent on the aqueous sodium concentration. Therefore, a numerical solution was developed (29) with boundary conditions for the range $\infty > R > 0$ and $\infty > P > 1$ using the Crank-Nicolson implicit method (31). The initial conditions at time, $t = 0$, assume that the concentration of sodium in the glass is homogeneous and equal to the analytical concentration. The mass of sodium in the aqueous solution is equal to $Q^{o}_{Na^+}$ times the total surface area of glass. At $t > 0$, sodium is assumed to diffuse through a semi-infinite, one-dimensional glass slab, across a surface layer into a finite, homogeneous aqueous solution. The one-dimensional aqueous distance is determined by the ratio of glass surface area to aqueous solution volume. In the model, as sodium ions diffuse through the solid into the aqueous phase, the aqueous concentration (mol/cm^3) will increase more rapidly in the experiments with a shorter aqueous distance. This is analogous to the real experimental situation in which aqueous concentrations increase more rapidly in experiments possessing larger glass surface areas per unit volume of solution. From equation 6, the higher aqueous concentrations will result in higher sodium concentrations on the glass surface, which will retard the sodium flux from within the glass by decreasing the diffusion gradient (equation 5).

As sodium diffusion proceeds, the rate of mass transfer for experiments using different total surface areas of glass will continue to diverge. The dashed lines on Figure 3 show the best fit of the model to the experimental data, using an adsorption isotherm ,$C^{Surf}_{Na^+} = 0.090\ [C^{Aq}_{Na^+}]^{1/4.0}$, and an apparent sodium diffusion coefficient of 1.3×10^{-21} cm^2/sec. The agreement is good between rate constants based on the statistical fit to the data, and those predicted by the model (Figure 3).

In applying the diffusion model to experiments containing additions of NaCl (Figure 4), the aqueous distance is assumed constant because all experiments represented identical glass surface-to-aqueous volumes. The only variable in the experiments is the Q^{o} term, which defines the addition of NaCl to the aqueous solutions plus the amount released during the initial exchange phase of the experiments. Other experimental conditions, including pH and temperature, are identical to the variable surface-to-volume experiments (Figure 3) previously modeled: therefore, the same surface adsorption isotherm and apparent diffusion coefficient should be applicable. The dashed lines on Figure 4 show the fit of the model to the data. Again, the comparison is good between the rate constants based on the statistical fit of the data and those predicted by the model.

The effect of using the sodium adsorption isotherm (equation 6) to determine the apparent diffusion coefficient can be seen by comparison with the values determined assuming zero sodium concentrations at the surface. The latter case has been used by previous workers (4, 26) because it provides an analytical solution to equation 3. An apparent diffusion coefficient of 1.3 x 10^{-21} cm^2/s results from the surface adsorption model for the data shown on Figures 3 and 4. A zero sodium surface concentration representing a greater diffusion gradient results in an apparent diffusion coefficient of 2.2 x 10^{-22} cm^2/s for the experimental mass transfer data of experiment A (Figure 3), and 2.5 x 10^{-23} cm^2/s for the mass transfer data of experiment D. Neglect of surface adsorption effects in the case of dissolution of the vitric tuff would result in a significant underestimation of the apparent diffusion coefficient.

Effect of Aqueous pH on the Rate Expression

As previously discussed, the reaction of silicate rocks with water involves the consumption of aqueous hydrogen ions. Therefore, the concentration of available hydrogen ions, as indicated by the pH, should affect the reaction rates. One manifestation of pH control is the variation in the form of the rate expression. Luce and others (4) found that over a time span of approximately 100 hours, magnesium silicates reacted according to a parabolic rate expression within a pH range of 3.2 to 9.6. However, at low pH (1.65) the reaction rate appeared to be linear. Luce and others interpreted this as a shift from diffusion control at higher pH to surface reaction at low pH between abundant hydrogen ions and the silicate structure.

Perhaps an even more dramatic example of pH control on the form of the rate expression is shown in the case of labradorite. The experimental conditions have been previously discussed for the data shown in Figure 1. Figure 8 shows a plot of the regression coefficients (r^2) for the linear rate equation (solid line) and parabolic rate equation (dashed line) plotted against pH. Note that increases in the value of r^2 correspond to improvements in the statistical fit of the rate equations to experimental data. A value of 1.0 represents a perfect fit. Figure 8 shows that the parabolic fit is superior to the linear fit at low pH (3-4) and high pH (7-8). However, at pH 5 and 6, the regression coefficient for the parabolic rate decreases and there is a large increase in the coefficient for the linear rate. Figure 8 clearly indicates a shift between linear and parabolic rates as a function of pH.

Changes in pH also affect the magnitude of individual rate expressions. Wollast (28) found for K-feldspar a progressive decrease in the parabolic rate constant for silica with increases in pH over the range 4-10 at 25°C. In investigating magnesium silicate dissolution at 25°C, Luce and others (4) found a slight

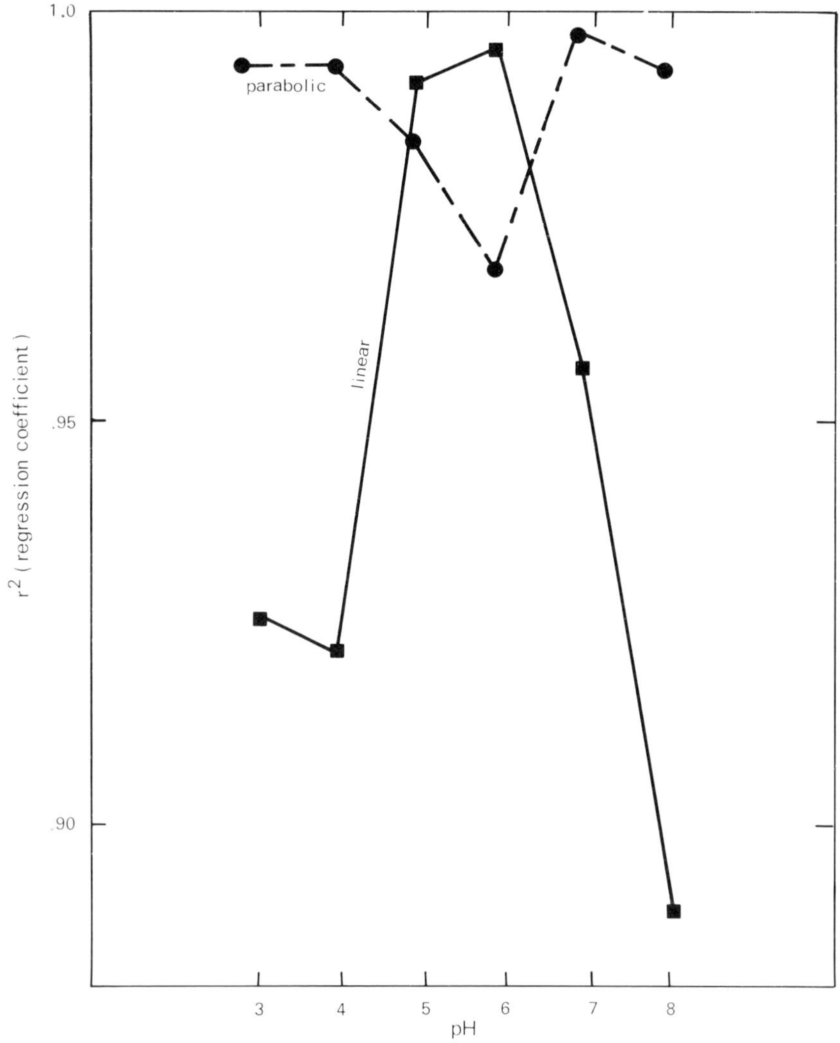

Figure 8. Comparison of the regression coefficients for parabolic and linear rate laws for plagioclase dissolution at 25°C and a pH range of 3–8

decrease in the parabolic rates for magnesium and silica with pH over the range 3.2-9.6.

The authors' work on several of natural glasses indicates that the pH effect is related to the valence state of the species diffusing from the silicate. For example, Figure 9 shows plots of the log of the parabolic rate constant, k_p, versus pH for silica (H_4SiO_4), sodium, calcium, and magnesium during dissolution of the vitric phase of the southern Nevada Rainer Mesa, member of the Timber Mountain Tuff. As expected, the rate constant for silica, diffusing presumably as an uncharged species, is independent of pH. The rate constant for monovalent sodium decreases slightly with increasing pH. The rate constants for bivalent calcium and magnesium show significant decreases with increasing pH. The pH valence state dependence can be explained by local equilibrium between hydrogen ions and cations in the silicate structure (29).

Charge balance requires that the outward flux of cations from the silicate be balanced by the inward flux of hydrogen ions. For example, the apparent sodium diffusion coefficient in equation 5 is therefore really the interdiffusion coefficient, D_{Na^+,H^+}, which describes the mobility of the sodium-hydrogen ion pair and is defined as

$$D_{Na^+,H^+} = \bar{D}_{Na^+} \cdot \bar{D}_{H^+} \cdot \frac{C_{Na^+} + C_{H^+}}{C_{Na^+}\bar{D}_{Na^+} + C_{H^+}\bar{D}_{H^+}} \qquad (7)$$

where $\bar{D}_{Na^+}$ and $\bar{D}_{H^+}$ are the constant self-diffusion coefficients of sodium and hydrogen ions, and C_{Na^+} and C_{H^+} are the concentration of diffusible sodium and hydrogen ions. As indicated by Helfferich (32), the interdiffusion coefficient is generally not a constant but varies continuously with changing concentrations of the interdiffusing species. Equation 7 is strictly true only if sodium is the sole species diffusing from the silicate. In the case of the vitric tuff previously modeled, the diffusion of sodium is also related to other cations diffusing from the glass by the common flux of hydrogen ions. The use of such an interdiffusion coefficient results in an exceedingly complex solution to equation 5. A simplified solution was established which permits the apparent sodium diffusion coefficient to be an empirical function of pH. Figure 10 shows four experiments conducted at an average pH of 7.13 and 25°C. Table III lists the experimental data. The solid lines represent the linear regression fit to the data. Although the surface-to-volume ratios are identical to data in Figure 3 (pH 4.75), the relative slopes are much gentler, indicating a decrease in the parabolic rates with increasing pH. The two variables in the numerical solution which define the rates of mass transfer in experiments of equal surface-to-volume ratios are the surface adsorption isotherm and the apparent diffusion coefficient. Exchange experiments were conducted in which glasses reacted with aqueous solutions at pH 6.0 and 7.0 for several weeks were rapidly

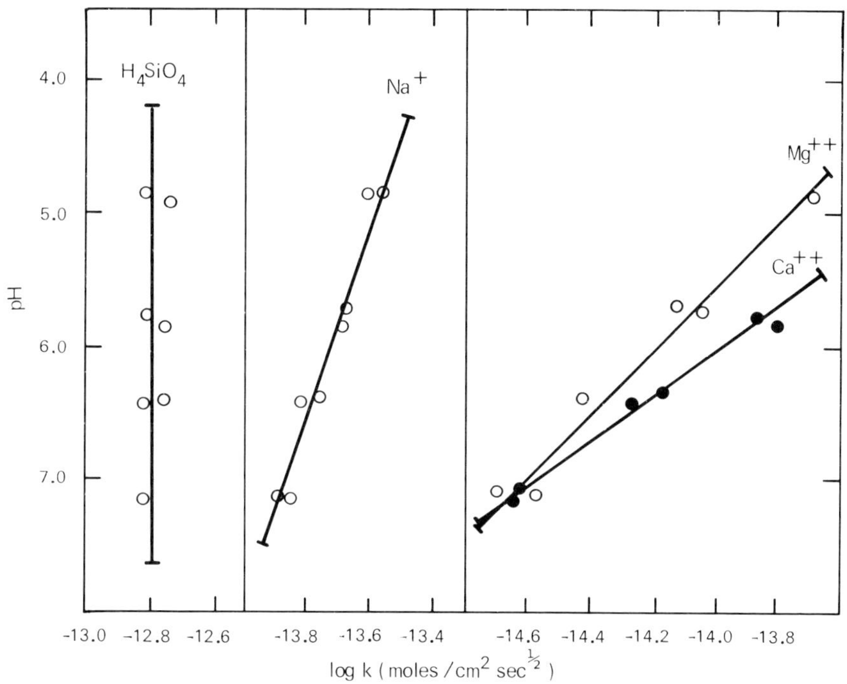

Figure 9. Effect of pH on the parabolic reaction coefficients for species of different valency during dissolution of a rhyolitic tuff at 25°C

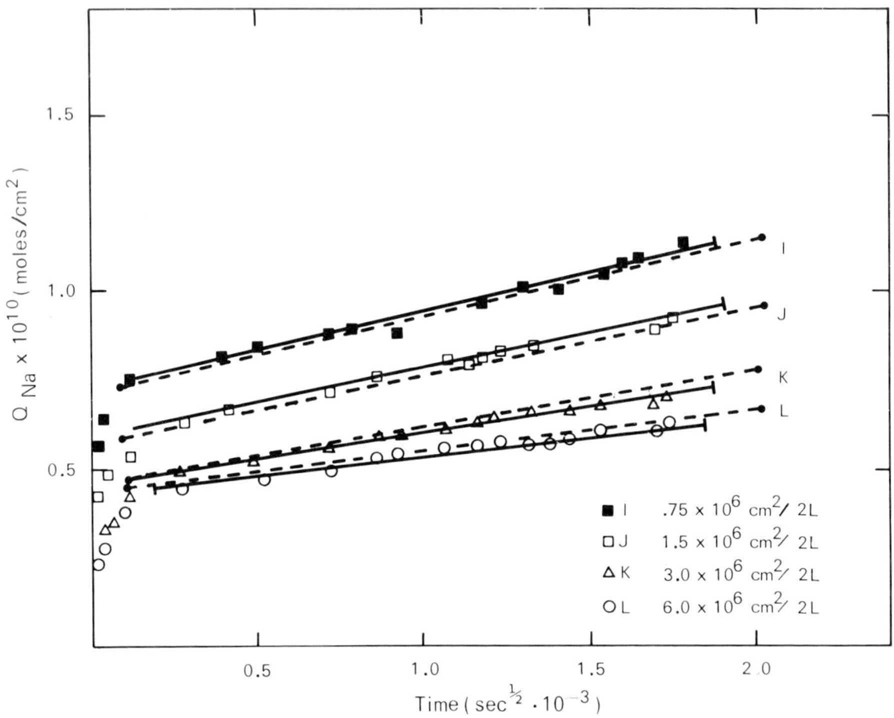

Figure 10. Effect of surface-to-volume ratios on inhibition of sodium mass transfer from a rhyolitic tuff at an average pH of 7.13 and 25°C. Solid lines are statistical fits to data points. Dashed lines predict inhibition based on diffusion model.

Table III.--Experimental Data Listed As Time, t ($s^{\frac{1}{2}}.10^{-3}$), Mass Transfer Of Sodium, Q_{Na} (mol/cm^6 X 10^{10}), And pH, For Experiments Of Different Total Surface Areas

Experiment	I			J			K			L		
Solid surface area - solution volume ratio	(0.38 X 10^6 cm^2/L)			(0.75 X 10^6 cm^2/L)			(1.5 X 10^6 cm^2/L)			(3.0 X 10^6 cm^2/L)		
	t	Q_{Na}	pH	t	Q_{Na}	pH	t	Q_{Na}	pH	t	Q_{Na}	pH
Sample No.												
1	0.0078	0.59	6.35	0.024	0.43	6.52	0.025	0.25	6.85	.025	0.26	7.08
2	.030	.60	6.62	.051	.49	6.67	.052	.35	6.77	.052	.29	7.00
3	.052	.64	6.49	.094	.60	*	.088	.37	6.85	*	*	*
4	.124	*	6.77	.130	.55	6.63	.120	.41	6.79	.109	.39	6.95
5	.292	*	6.80	.291	.63	6.69	.286	.47	7.01	.281	.46	7.16
6	.407	.83	6.70	.420	.67	6.74	.416	*	7.00	.411	*	6.90
7	.502	.85	6.69	.526	*	6.75	.523	.52	6.98	.523	.48	7.32
8	.715	.88	6.73	.732	.72	6.80	.730	.57	7.15	.730	.50	7.48
9	.785	.88	6.70	.879	.77	7.01	.887	.60	7.20	.877	.54	7.44
10	.926	.88	6.60	.939	.84	7.10	.937	.60	7.22	.939	.55	7.42
11	1.10	*	6.61	1.06	.81	6.91	1.06	.60	7.20	1.07	.57	7.54
12	1.18	.96	6.80	1.17	.81	7.00	1.17	.63	7.18	1.17	.57	*
13	1.32	1.01	6.70	1.22	.85	7.07	1.22	.65	7.20	1.22	.58	7.42
14	1.41	1.01	6.70	1.32	.85	6.91	1.32	.66	7.25	1.32	.57	7.50
15	*	*	*	1.39	.84	6.91	1.38	.65	7.12	1.38	.57	7.28
16	1.57	1.05	6.79	1.44	.87	6.91	1.44	.66	7.11	1.44	.59	*
17	1.62	1.09	*	1.53	.93	6.82	1.53	.68	7.06	1.53	.61	7.00
18	1.64	1.09	6.85	1.71	.89	6.95	1.71	.69	7.13	1.71	.62	7.29
19	1.79	1.15	6.89	1.74	.92	6.96	1.74	.70	7.06	1.74	.63	7.28

* Data not available.

titrated to pH 4.0. The lack of sodium release into solution strongly suggested that the surface sodium concentration was unaffected by aqueous hydrogen concentrations. A match to the data (Figure 10) was sought assuming an identical isotherm to that which defined the surface sodium concentrations at pH 4.75. A new pH-dependent apparent diffusion was then calculated. The dashed lines in Figure 10 are the solutions generated by the model. Comparison is good between rate constants based on the linear regression fit of the data and those based on the numerical model, using an apparent diffusion coefficient of 1.7×10^{-22} cm^2/s. As expected for an interdiffusion process involving available hydrogen ions, the apparent sodium diffusion coefficient decreases with increasing pH.

Conclusions

Although simple rate expressions can be utilized to explain short term, laboratory-controlled dissolution of many silicate phases, their application to modeling natural weathering processes is much more difficult. As indicated, the form of the rate expression, whether linear or parabolic, can depend on the type of silicate structure, the pH of reaction, and the time span of the weathering process itself. Perhaps the best method of determining the type of rate expression applicable to a specific natural situation is to compare natural water compositions to those predicted by laboratory-derived rate constants. In the case of long-term parabolic kinetics, the ratio of dissolved constituents should be directly proportional to the ratios of the parabolic rate constants. In the case of linear kinetics, the ratio of constituents should be stoichiometrically equal to the composition of the dissolving phase. In the case of a system in which more than one silicate phase is dissolving, the ratio of the constituents will reflect the relative reaction rates of individual phases, whether they are linear or parabolic. The above scheme applies only to those constituents which are substitutional components of major dissolving mineral phases, and which are conserved in the aqueous solution.

In addition to the form of the rate expression, the magnitude of the rate constant must be ascertained. As indicated in the discussion, the composition of the aqueous media can exert a strong control on the rate constant. Modeling of kinetic rates, therefore, can be considerably complicated, as in closed groundwater recharge systems where the aqueous composition would be expected to continually vary along the flow path.

In the case of the simple Rainier Mesa aquifer system, a quantitative model can be derived to explain such variations. As indicated in the case of sodium, the rate of mass transfer can, at any point in the reaction path, be related to the aqueous sodium concentration by use of an adsorption isotherm. This isotherm appears to be a separable variable independent of other aqueous constituents over the range of solution composition investigated by the authors. The rate of mass transfer can also be independently related to the aqueous pH by use of an apparent sodium-diffusion coefficient which is empirically determined and which allows for the codiffusion of hydrogen ions. Thus, the reaction rate for sodium dissolution for the vitric tuff can be modeled in a closed system with a variable pH and dissolved Na aqueous composition, by summing the results of small reaction increments, assuming constant source-rock composition. An example of this method is presented elsewhere in this volume.

Abstract

Experimentally determined dissolution kinetics are applicable to natural weathering processes of silicate rocks. Mass transfer from the mineral to the aqueous phase was determined to be incongruent under a range of experimental conditions. Transfer rates of individual species (Q) at times (t) can usually be described by one of two rate expressions;

$$Q = Q_o + k_1 t \text{ or } Q = Q_o + k_p t^{1/2}$$

where k_1 is a linear rate constant, k_p is a parabolic rate constant, and Q_o is the mass transferred during an initial surface exchange with hydrogen ions. The linear rate constant, k_1, represents continued surface exchange coupled to incongruent dissolution of the silicate framework whereas the parabolic rate constant, k_p, represents cation diffusion through a relatively undisturbed silicate framework. Detailed investigation of dissolution of a vitric tuff indicates that the rate of mass transfer of a species is described by a parabolic expression and is inversely dependent on the concentration of that species in aqueous solution. A numerical solution to the one-dimensional diffusion equation is presented using a Freundlich isotherm to relate the aqueous ion concentration and the ion density on the surfaces of the vitric tuff. This ion density on the surface in turn determines the concentration gradient within the glass. Results indicate that the apparent diffusion coefficient for sodium decreases from 1.3×10^{-21} at pH 4.8 to 1.7×10^{-22} at pH 7.0 and that the surface isotherm remains constant.

Literature Cited

1. Garrels, R. M., and Mackenzie, F. T. "Evolution of Sedimentary Rocks," 397 p. W. A. Norton and Co., New York, 1971.
2. Daubreé, A. Experiences sus des les decomposition chimiques provoquies par les action mecaniques dans divers mineraux tels que le feldspart, Compt. Rend., 339-345 (1867).

3. Rana, M. A., and Douglas, R. W. The reaction between glass and water Part 2. Discussion of the results, Phys. Chem. Glasses 2, 196-205 (1961).
4. Luce, R. W., Bartlett, R. W., and Parks, G. A. Dissolution kinetics of magnesium silicates, Geochim. Cosmochim. Acta 30, 35-50 (1972).
5. Tamm, Olof Expermentelle studien über die verwitterung und tonbildung von feldspäten, Chemie der Erde 4, 420-430 (1930).
6. Garrels, R. M. and Howard, P. Reactions of feldspar and mica with water at low temperature and pressure, Clays Clay Min. 6, 68-88 (1957).
7. Lagache, M., Wyart, J., and Sabatier, G. Dissolution des feldspaths alcalins dan l'eau pure ou chargee de CO_2 a 200°C, C. R. Acad. Sci. Paris, 253, 2019 (1961).
8. Lagache, M. Contribution à l'etude de l'altération des feldspaths den l'eau 100 et 200°C sous diverses pressions de CO_2 et application à la synthese des minéraux argileax, Bull. Soc. Fr. Mineral. Cust. 88, 223-253 (1965).
9. Aagaard, P., and Helgeson, H. C. Thermodynamic and kinetic constraints on the dissolution of feldspars, Geol. Soc. Amer. Abstr. 9 (7), 873, 1977.
10. Marshall, C. E. Reactions of feldspars and micas with aqueous solutions, Econ. Geol. 57, 1219-1227 (1962).
11. Paces, T. Steady-state kinetics and equilibrium between ground water and granitic rock, Geochim. Cosmochim. Acta 37, 2641-2663 (1973).
12. Boksay, Z., Bouquet, G., and Dobos, S. Diffusion processes in the surface layer of glass, Phys. Chem. Glasses 8, 140-144 (1967).
13. Boksay, Z., and Bouquet, G. On the reaction of water molecules with the silicate network in the glass phase, Phys. Chem. Glasses 16, 81-83 (1975).
14. Correns, C. W., and van Engelhardt, W. Neue untersuchungen über die verwitterung des kalifeldspates, Chem. Erde 12, 1-22 (1938).
15. Helgeson, H. C. Kinetics of mass transfer among silicates and aqueous solutions, Geochim. Cosmochim. Acta 35, 421-469 (1971).
16. Crank, J. "The Mathematics of Diffusion," 347 p. Oxford Press, London, 1955.
17. Correns, C. The experimental chemical weathering of silicates, Clay Mineral. Bull. 4, 249-265 (1961).
18. Csakvari, B., Boksay, Z., and Bouquet, G. Investigation of surface layers on electrode glass for pH measurement, Anal. Chim. Acta 56, 279-284 (1971).
19. Baucke, F. Investigation of surface layers, formed on glass electrode membranes in aqueous solutions, by means of an ion sputtering method, Jour. Non-Crystalline Solids 14, 13-31 (1974).

20. Tchoubar, C., and Oberlin, A. Altération de l'albite par action d'eau. Etude en microscopie et microdiffraction électroniques de la precipitation et de l'evolution des fibres de boehmite formée, Jour. Microsc. 2, 415-432 (1963).
21. Tchoubar, C. Formation de la kaolinite à partier d'albite por l'eau à 200°C. Etude en microscopie et diffraction électroniques, Bull. Soc. Fr. Mineral. Crist. 88, 483-518 (1965).
22. Parham, W. E. Formation of halloysite from feldspar: low temperature artificial weathering versus natural weathering, Clays Clay Min. 17, 13-22 (1969).
23. LaIglesia, A., Martin-Caballero, J. L., and Martin-Vivaldi, J. L. Formation de kaolinite par precipitation hologene à temperature ambiante, Emploi de feldspaths potassiques, Compt. Rend. D279, 1143-1145 (1974).
24. Petrović, R. Rate control in feldspar dissolution-II. The protective effect of precipitate, Geochim. Cosmochim. Acta 40, 1509-1521 (1976).
25. Petrović, R., Berner, R. A., and Goldhaber, M. B. Rate control in dissolution of alkali feldspars I. Study of residual grains by X-ray photoelectron spectroscopy, Geochim. Cosmochim. Acta 40, 537-548 (1976).
26. Busenberg, E., and Clemency, C. V. The dissolution kinetics of feldspars at 25°C and 1 atm-CO_2 partial pressure, Geochim. Cosmochim. Acta 40, 41-49 (1976).
27. Loughnan, F. "Chemical Weathering of Silicate Minerals", 155 p. Elsevier, Amsterdam, 1969.
28. Wollast, R. Kinetics of the alteration of K feldspar in buffered solutions at low temperature, Geochim. Cosmochim. Acta 31, 635-648 (1967).
29. White, A. F., and Claassen, H. C. Kinetic model for the dissolution of a rhyolitic glass, Geol. Soc. Am. Abst. 9, 1223 (1977).
30. Halsey, G., and Taylor, H. S. The adsorption of hydrogen on tungsten powder, Jour. Chem. Physics 15, 624-630 (1947).
31. Crank, J., and Nicolson, P. A practical method for numerical evaluation of solutions of partial differential equations of the heat type, Proc. Cambridge Phil. Soc. 43, 50-67 (1947).
32. Helfferich, F. "Ion Exchange", 327 p. McGraw-Hill, New York, 1962.

Disclaimer: The reviews expressed and/ or the products mentioned in this article represent the opinions of the author(s) only and do not necessarily represent the opinions of the U.S. Geological Survey.

RECEIVED November 16, 1978.

23

Calcium Phosphates—Speciation, Solubility, and Kinetic Considerations

G. H. NANCOLLAS, Z. AMJAD, and P. KOUTSOUKOS

Chemistry Department, State University of New York at Buffalo, Buffalo, NY 14214

The calcium phosphates are widely dispersed in the environment. They occur as constituents in all classes of rock and the apatites also occur naturally in pegmatites and other veins of hydro-thermal origin. In more recent times, the increase in phosphate concentrations in lakes and rivers near heavily populated areas has been a major factor in the resurgence of interest in physico-chemico processes such as precipitation and dissolution of phosphate salts. Despite the continuous entry of phosphate ions into a lake, the phosphate concentration does not increase proportionally (1, 2) indicating that at least some phosphate is removed by precipitation. Since the calcium concentration is also relatively high, perhaps through the use of lime additions for the removal of phosphates from sewage,calcium phosphate precipitation may be of particular importance. In the lime addition process, an induction period precedes the formation of precipitate and, typically, 85-90% of the phosphates can be removed in a recycling reactor at pH 8.0 (3). In addition to its precipitation in natural water systems, calcium phosphate deposits, normally attributed to hydroxyapatite ($Ca_5(PO_4)_3OH$ hereafter HAP), are also formed as deposits in boilers. One of the earliest studies of this phenomenon was made by Clark and Gerrard (4) who examined a number of other deposits where phosphate control was practiced. They observed a wide variation in chemical composition but concluded that HAP was the most likely calcium phosphate phase to be formed under the boiler conditions. Recently, the formation of calcium orthophosphates in industrial cooling water systems has become increasingly important. Higher phosphate levels are being encountered in cooling waters due to increased water reuse, use of lower quality make-up water such as tertiary sewage treatment plant effluent and the use of organic phosphonate scale and corrosion inhibitors which are degraded to orthophosphate. The increased phosphate levels combined with alkaline operating conditions can lead to the formation of highly insoluble calcium phosphate scale deposits (5).

Despite the importance of the precipitation of calcium

0-8412-0479-9/79/47-093-475$05.75/0
© 1979 American Chemical Society

phosphates, there is still considerable uncertainty as to the nature of the phases formed in the early stages of the precipitation reactions as a function of supersaturation, pH, and temperature. In the environment, complicating factors such as the presence of ions other than calcium and phosphate, and the heterogeneous nucleation of calcium phosphate phases on foreign substances have to be taken into account. Thus it has been shown (6, 7, 8, 9) that the nature of the calcium phosphate phase which forms during precipitation from aqueous solution is markedly dependent upon the calcium and phosphate concentrations, pH, and the nature and extent of the surface phase on which precipitation is occurring.

In this paper we discuss the chemistry of aqueous calcium phosphate systems from the point of view of both equilibrium and kinetic considerations. It will be shown that the chemical composition of the calcium phosphate precipitated under any given set of conditions, may be determined kinetically rather than simply on the basis of thermodynamic driving forces.

Equilibrium Considerations

In discussions of the precipitation of calcium phosphate, the phase which is usually emphasized is the thermodynamically most stable, HAP. However, most calcium phosphate solutions in which precipitation experiments are made, are initially supersaturated with respect to four additional phases. The solubility isotherms are shown in Figure 1 as a function of pH. Thus, at a pH of 7.0, in order of increasing solubilities, it is necessary to take into account tricalcium phosphate ($Ca_3(PO_4)_2$ hereafter TCP), octacalcium phosphate ($Ca_4H(PO_4)_3.2\ 1/2\ H_2O$, hereafter OCP), anhydrous dicalcium phosphate ($CaHPO_4$, hereafter DCPA), and dicalcium phosphate dihydrate ($CaHPO_4.2H_2O$, hereafter DCPD). The corresponding thermodynamic solubility products, K_{so}, at 25°, are:-

for HAP, $[Ca^{2+}]^5\ [PO_4^{3-}]^3\ [OH]y_2^5y_3^3y_1 = 4.7 \times 10^{-59} (mol\ l^{-1})^9$ (10)

for TCP, $[Ca^{2+}]^3\ [PO_4^{3-}]^2y_2^3y_3^2 = 1.20 \times 10^{-29} (mol\ l^{-1})^5$ (11)

for OCP, $[Ca^{2+}]^4\ [PO_4^{3-}]^3[H^+]y_2^4y_3^3y_1 = 1.25 \times 10^{-47} (mol\ l^{-1})^8$ (12)

for DCPA, $[Ca^{2+}]\ [HPO_4^{2-}]y_2^2 = 1.26 \times 10^{-7}\ (mol\ l^{-1})^2$ (13)

and

for DCPD, $[Ca^{2+}]\ [HPO_4^{2-}]y_2^2 = 2.49 \times 10^{-7}\ (mol\ l^{-1})^2$ (14)

In these expressions, the square brackets represent the molar concentrations of the species indicated, and y_z is the activity coefficient of a Z-valent species. In order to calculate the free ionic concentrations in these expressions, it is necessary to take into account ion-pair and complex formation. The equilibria in pure calcium phosphate solutions are:-

$H^+ + H_2PO_4^- \rightleftarrows H_3PO_4$, K = 164.1 $lmol^{-1}$, (15),
$H^+ + HPO_4^{2-} \rightleftarrows H_2PO_4^-$, K = 1.58 x 10^7 $lmol^{-1}$, (15)
$H^+ + PO_4^{3-} \rightleftarrows HPO_4^{2-}$, K = 2.33 x 10^{12} $lmol^{-1}$, (16)
$Ca^{2+} + H_2PO_4^- \rightleftarrows CaH_2PO_4^+$ K = 31.9 $lmol^{-1}$, (17)
$Ca^{2+} + HPO_4^{2-} \rightleftarrows CaHPO_4$ K = 681 $lmol^{-1}$, (17)
$Ca^{2+} + PO_4^{3-} \rightleftarrows CaPO_4^-$ K = 2.9 x 10^6 $lmol^{-1}$, (17)
$Ca^{2+} + OH^- \rightleftarrows CaOH^+$ K = 32.4 $lmol^{-1}$, (18)

Values of K, the thermodynamic association constants are given at 25°. The concentrations of ionic species in the solutions at any time can be determined from mass balance, electroneutrality, and the appropriate equilibrium constants as described previously (19, p. 85-92) by successive approximations for the ionic strength. The activity coefficients of Z-valent ionic species may be calculated from an extended form of the Debye-Huckel equation such as that proposed by Davies (20, p. 34-53).

$$-\log y_z = Az^2[I^{1/2}/(1 + I^{1/2}) - 0.3I], \qquad (1)$$

which has been shown to represent the activity coefficients of multiple charge ions up to ionic strengths of about 0.2 mol l^{-1} to within ± 1%. This and other activity coefficient equations have been discussed elsewhere (20, p. 34-53; 21, p. 242-279; 22, p. 229-252). In some cases, it is possible to study a system at a series of constant ionic strengths and to determine the best values of the parameters to be used in equations for the calculation of activity coefficients (23).

In typical environmental situations, the liquid phase may contain a number of ions which can form complexes with the multiple charge cations present. In order to be able to calculate the thermodynamic driving force for the precipitation of a particular solid phase, it is necessary to determine the concentrations of free lattice ions in the solutions. The commonly used procedures for the calculations of solute components in a homogeneous solution are the "equilibrium constant" and the "free energy minimization" methods. The former utilizes an approach wherein first, a "basis" is selected from the species, usually that having the highest concentration at equilibrium. The other solution species are formed from this "basis" by a series of chemical reactions, and their concentrations can be expressed through the use of equilibrium constants, in terms of the concentration of the chosen "basis". The resulting set of nonlinear simultaneous equations consists of as many unknowns as there are elements and can be solved by conventional numerical methods. The "free energy minimization" method utilizes only free energy criteria for chemical equilibria making no distinction among the constituent species and is essentially a constraint non-linear minimization problem. A number of search methods have been proposed in order to effect the desired minimization (24, 25, 26).

The efficiency of the "equilibrium constant" computational method is critically dependent upon the choice of the "basis" set and the initial estimate of the equilibrium concentrations of the components. In contrast, the "free energy minimization" method can be used for chemical systems containing a large number of elements or species without the need for a large number of memory locations in computer calculations. The method always converges to the solution, no matter how poor the starting estimates, but considerable time may be required to reach a solution, making it unsuitable for routine equilibrium calculations. The solution of the non-linear simultaneous equations are usually based upon techniques that change a many-variable iteration problem into a set of single-variable problems, treating the iterations as a nested set, each of which iterates on one variable only. Sillen (27) used a simple search technique whereas the program of Perrin and Sayce (28) used iteration formulae involving the ratios of the observed and calculated total concentrations of metal and ligand. Although these methods do not require explicit matrix inversion, they suffer from a rather slow convergence rate. The widely used Newton-Raphson method was employed in the development of EQUIL in order to effect rapid quadratic conversion for the calculation of the equilibrium concentrations of species in mixed electrolytes (29). The procedure, based on the "stability constant" method states the problem with simplicity and generality using a numerical technique based on a modified Newton-Raphson method in order to solve the non-linear simultaneous equations. The modifications made in the Newton-Raphson procedure included scaling the matrix, eigenvector analysis, development of an iteration matrix, and adopting a convergence forcer. In typical environmental situations, in the presence of relatively large concentrations of background electrolytes which are assumed not to form complexes with the reacting species, the activity coefficients can be assumed to remain constant. The EQUIL computer program provides a much faster convergence for the calculations of ionic species, and overcomes numerical shortcomings such as the ill-condition of the matrix and over-corrections to the estimates for each iteration.

It can be seen in Figure 1 (2) that although HAP is the most stable salt under many conditions, it becomes less stable than DCPA and DCPD if the solution is sufficiently acid. The position of the singular points in Figure 1 at which two solid phases are in equilibrium with the aqueous solution, is clearly dependent upon pH and upon the ionic strength of the solution. Where multiple charge ions are involved, the importance of introducing activity coefficient corrections becomes increasingly apparent. Thus, at ionic strengths of about 0.1 mol l^{-1}, HAP ionic products may be in error by as much as a factor of 10^6 if activity coefficients are neglected (6). A calcium phosphate solid phase

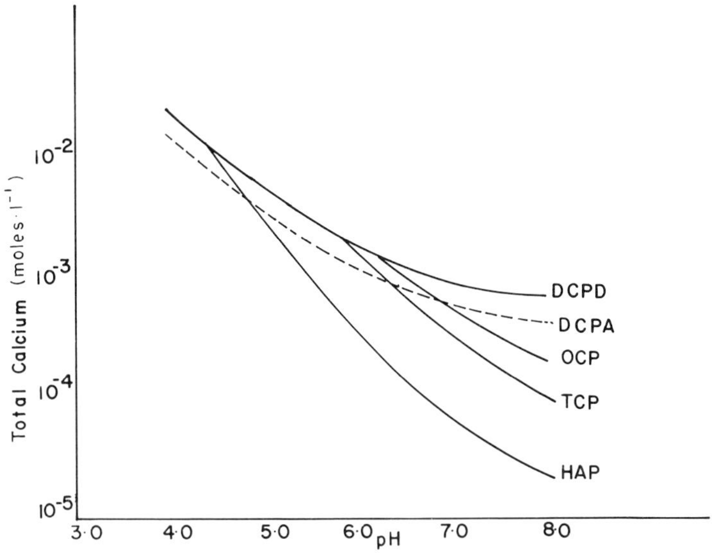

Wiley and Sons

Figure 1. Calcium concentrations and pH values of solutions saturated with respect to various calcium phosphate phases in the ternary system. $Ca(OH)_2$–H_3PO_4–H_2O at 25°C (2).

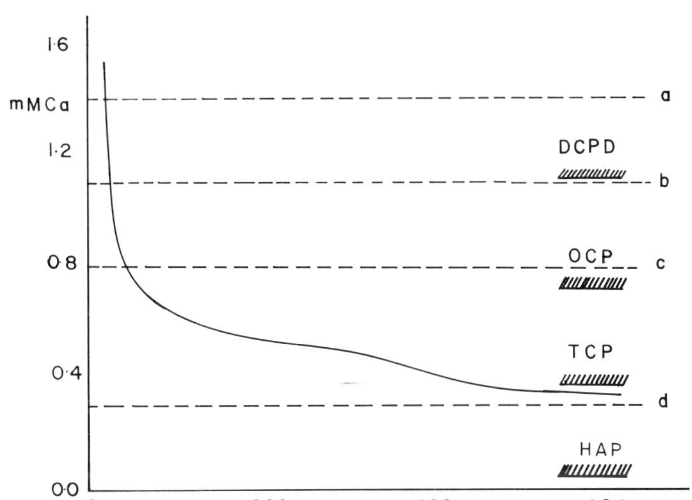

Figure 2. Growth of HAP seed crystals. Plots of total calcium as a function of time. Approximate saturation levels with respect to the various calcium phosphate phases also are shown.

which is exposed to a solution more acid than the corresponding singular point may be expected to become covered by a surface coating of a more acid calcium phosphate. The apparent solubility behaviour will then be quite different from that of the phase itself, and measurement of solution concentrations might lead to apparent high solubilities (2). Since calcium phosphate solutions are usually very dilute and solubilities low, the surface phases may persist for long periods of time. As pointed out by Brown (2), this may have led many workers to the conclusion that the properties of calcium phosphates in general, and HAP in particular, are rather variable (30, 31, 32). Clearly when working with a diagram such as that in Figure 1, it is essential to establish that the systems are at equilibrium in order to be able to use thermodynamic solubility products which refer to the free energy of a single solid phase. It is now quite well established that kinetic factors may be considerably more important in determining the nature of the solid phase present than are considerations based solely upon equilibrium solubility data.

Kinetic Considerations

Numerous kinetic studies have been made of the spontaneous precipitation of calcium phosphates from solutions containing concentrations of lattice ions considerably in excess of the solubility values (33, 34). Although attempts are usually made to control the mixing of reagent solutions, it is difficult to obtain reproducible results from such experiments since chance nucleation of solid phases may take place on foreign particles in the solution. Many of these difficulties can be avoided by studying the crystal growth of well-characterized seed crystals in metastable supersaturated solutions of calcium phosphate. Such solutions are stable for periods of days and the kinetics of growth of the added seed material may then be studied by following the concentration changes as a function of time. Not only are such methods capable of yielding highly reproducible results, but this model may be much closer to the environmental situations in which precipitations may take place preferentially on solid phases already present. During the precipitation reaction the possible formation of a number of calcium phosphate phases may be readily seen from Figure 1. Indeed, one might expect, following the Ostwald-Lussac empirical rule that the least stable phase would precipitate first, and experience has shown that the two hydrated salts DCPD and OCP are the ones that precipitate most easily, particularly at ambient temperatures (2). The eventual transformation of precursor phases to the thermodynamically stable HAP is also a problem which is little understood. The stoichiometry of the initially precipitated solid is invariably less than the 1.67 for the calcium:phosphate (Ca:P) molar ratio required for HAP (34, 35). However, the errors associated with these ratios, calculated from small differences in the measured total molar concentrations of calcium T_{Ca} and phosphate T_p, are very large unless special precautions are taken to increase the precision of the analytical methods; $\pm$ 0.1% precision is routinely required for stoichiometry determinations to 1%. In many experimental studies, the Ca:P ratio of the precipitating phases has been found to be 1.45 $\pm$ 0.05. This has led to the proposal of TCP as the precursor phase which is converted to the thermodynamically stable HAP by an autocatalytic mechanism (36). The conversion process has been studied as a function of solution environment (37), the presence of stabilizing agents such as pyrophosphate and organic phosphonates (38, 39, 40) and inorganic ions such as magnesium (41, 42). Recent studies of the maturation of freshly precipitated amorphous calcium phosphate at pH 7.4 (43) have shown that the ripening of the crystals is accompanied by inflexions in the time profiles of calcium and phosphate concentration in the solutions. These were attributed to phase changes during crystal maturation probably through an intermediate OCP-like, crystalline phase. The crystallographic similarity between OCP and HAP had already prompted Brown (44) to propose the phase as a precursor in the precipitation of apatite.

The problems associated with the irreproducibility of the results of spontaneous precipitation studies were overcome with the development of seeded growth techniques (45, 46) which enable the effects of factors such as temperature, supersaturation, and ionic strength to be studied quantitatively (6, 47, 48). The further development of a pH-stat method (49) enables studies to be made of calcium phosphate crystal growth on well characterized seed material under conditions of constant hydrogen ion activity. The simplest calcium phosphate phase for such studies is DCPD and, as can be seen in Figure 1, at pH values below about 6.0, DCPD, DCPA, and HAP are the only thermodynamically stable calcium phosphate phases. The kinetics of crystal growth of well-characterized DCPD seed crystals in metastable supersaturated solutions of calcium phosphate under these conditions has been studied and a typical growth curve following the addition of seed crystals is shown in Figure 3. In Figure 3 the volume of base required for constant pH, controlled by the pH-stat, is also plotted. Crystal growth takes place immediately upon inoculation with seed crystals and the rate curves are consistent with the kinetic equation

$$\text{Rate of crystal growth} = \frac{dT_{Ca}}{dt} = -ksN^2. \qquad (2)$$

In equation 2, k is the specific rate constant for growth, s is a term proportional to the number of DCPD growth sites and N is the number of moles of DCPD to be precipitated before equilibrium is reached (50). The applicability of equation 2 in describing DCPD crystal growth is demonstrated in Figure 4 in which the

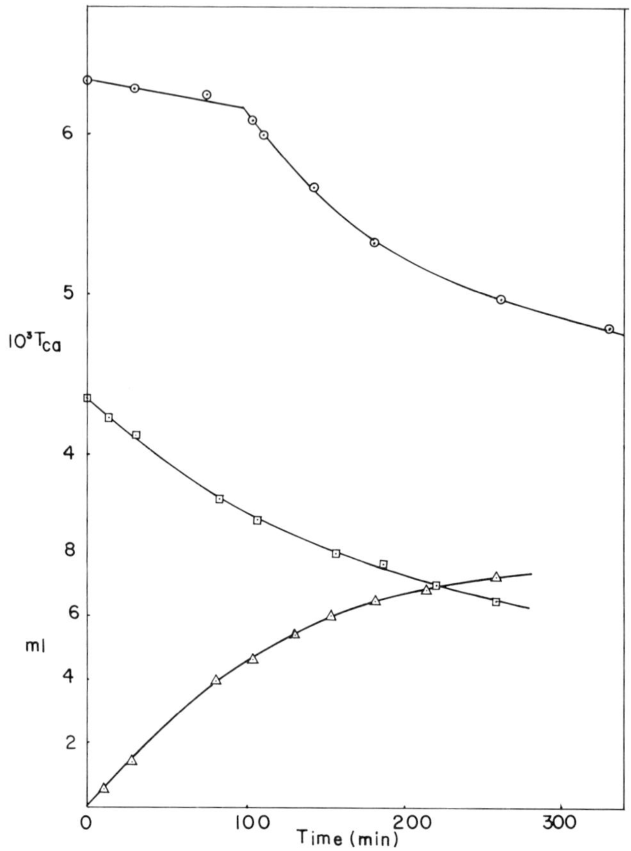

Figure 3. Crystal growth of DCPD. DCPD seeds (□) initial T_{Ca} = *4.37 m*M, T_p = *10.50 m*M, *pH* = *5.60, 25°C; (△) uptake of 0.05*M *potassium hydroxide. HAP seed (○)* T_{Ca} = *6.346 m*M, T_p = *6.317 m*M, *pH 5.59, 37°C.*

integrated form is plotted. The observed independence of the rate of crystallization upon the fluid dynamics and an activation energy of 48.9 ± 4 kJ mol^{-1}, obtained from the temperature coefficient of the rate constant, point to a surface-controlled process. The crystallization of a large number of other 2-2 symmetrical electrolytes has also been shown to follow the same kinetic equation (46, 47, 48).

As the pH is increased (see Figure 1), other calcium phosphate phases may be stabilized as precursors to the formation of apatite and the growth of HAP seed crystals in stable supersaturated solutions is much more complex (6, 7, 49). Measurements of the specific surface area, SSA, of the products grown at various times indicate that the initial formation of a microcrystalline or amorphous precursor leads to a rapid increase in SSA. The development of these phases is also observed by scanning electron microscopy, and dissolution kinetic studies of the grown material have indicated the formation of OCP as a precursor phase (6, 7). The overall precipitation reaction appears to involve, therefore, not only the formation of different calcium phosphate phases, but also the concomitant dissolution of the thermodynamically unstable OCP formed rapidly in the initial stages of the reaction. In the presence of magnesium ion the overall rate of crystallization is reduced and lower Ca:P ratios are observed for the first formed phases (51). It appears that the magnesium ion stabilizes a precursor phase, the stoichiometry of which appears to correspond to a mixture of DCPD and OCP. The possibility of DCPD, the simplest calcium phosphate phase, as a precursor to apatite precipitation was proposed by Francis (52).

Although the seeded crystallization experiments yield highly reproducible results and, in the case of the pH-stat method, the hydrogen ion activity could be held constant during the reaction, they suffer from the disadvantage that the calcium and phosphate ionic concentrations decrease appreciably as the reaction proceeds towards solubility equilibrium. At pH values above about 6.0 (see Figure 1), at each stage the supersaturated solutions are metastable with respect to various calcium phosphate phases which can form and subsequently dissolve as the concentrations in the supersaturated solutions decrease. Since the concentration changes become very small as the reaction proceeds, a relative analytical error of only a few percent in the total calcium or total phosphate concentration can preclude differentiation between the possible crystalline phases.

To overcome the problems associated with the changing of solution concentrations during precipitation, a new method has been developed (53) in which the chemical potentials of the solution species are maintained constant during the reaction. Following the addition of well-characterized seed material to stable supersaturated solutions of calcium phosphate at the required pH, the concentrations of lattice ions are maintained constant by the simultaneous addition of the reagent solutions containing calcium and phosphate ions, controlled by a glass electrode probe. The compositions of the reagent solutions are calculated from the results of exploratory measurements and, during the course of the reactions, their constancy is verified by analysis. Typical data are shown in Table I at a supersaturation corresponding to level c in Figure 2. At 37° and pH = 7.40, it can be seen that the use of reagent solution concentrations with a Ca:P ratio, R, of 1.45 results in the formation of a phase on the surface of the HAP seed crystals in which all solution parameters remain constant to within ± 0.3%. In addition, it is found that the rate of reaction is directly proportional to the inoculating seed concentration, thus confirming that the growth of the crystals takes place without interference from secondary nucleation. To have obtained a kinetic precipitation stoichiometry to this precision by techniques previously used would have required concentration analysis to at least ± 0.03%. Although, as mentioned previously, it had been assumed that the ratio, R, was close to that for TCP, thereby invoking TCP as the precursor, it is clear from the results in Table I that the ratio is significantly lower than the Ca:P = 1.50 required for TCP.

Results of a typical experiment at higher supersaturation are shown in Table II in which the initial supersaturation is above that for OCP (Figure 2, level b) but still below DCPD. It can be seen in Table II that the choice of reagent solutions with Ca:P of 1.33 yielded the required constancy of lattice ion concentrations for at least the first 10 min of reaction following the addition of HAP seed crystals. The results clearly indicate the formation of OCP in the early stages of the reaction and this phase was confirmed by x-ray and infra-red analysis. At longer times (15-20 min) the expected hydrolysis to a more basic phase takes place with a Ca:P ratio of approximately 1.45, the value observed in so many calcium phosphate precipitation studies. It is significant that, using the constant composition method, more than twice the original seed material could be grown as OCP in the early stages of the reaction. The results of these studies support a model for calcium phosphate precipitation in which OCP, formed as a precursor phase, hydrolyzes either partially or completely to HAP depending upon the rate of precipitation. The normally observed higher Ca:P values can be accounted for by assuming that only a fraction of the OCP molecules transforms to HAP. If the overall rate of reaction is such that only one in every three molecules is capable of transforming into HAP, the observed stoichiometry of the precipitating phase would be 1.44, or within 0.01 of the normally observed 1.45 ± 0.05. The hydrolysis probably takes place a layer at a time as has been proposed from unit cell x-ray analysis of the phases (54). The results of a preliminary experiment at even higher supersaturation corresponding to level a in Figure 2

(T_{Ca} = 1.20 mM, Tp = 1.20 mM, [KCl] = 30 mM, pH = 7.40) indicate that reagent solutions with a 1:1 calcium to phosphate ratio, corresponding to DCPD, produce an initial transient phase for approximately 10 min of reaction. Infrared analysis indicates the presence of DCPD and OCP in the grown material and this is the first time that DCPD has been grown on HAP seed material at pH levels above 7.0. The results suggest that DCPD is the ultimate apatite precipitation precursor. As will be shown later, the epitaxial growth of DCPD on HAP and other calcium phosphate at pH levels below 6.0 follows the characteristic DCPD growth kinetics.

Table I

Constant composition seeded growth of HAP crystals at pH 7.40, 37°C*

Time (min)	T_{Ca} (mM)	T_p (mM)	R**
0	0.800	1.160	1.450
20	0.797	1.161	1.457
34	0.796	1.160	1.457
48	0.797	1.156	1.450
60	0.794	1.157	1.457
75	0.800	1.170	1.462
90	0.797	1.161	1.457

* Initial solution: 150 ml of 0.800 mM $CaCl_2$, 0.552 mM KH_2PO_4, 0.457 mM KOH, 8.40 mM KCl and 5.0 mg of HAP seed. Titrant solutions: 10.00 mM $CaCl_2$ and 6.90 mM KH_2PO_4 with 11.91 mM KOH.

** R is the molar ratio of calcium to phosphate in solution at each time.

One of the advantages of the constant composition method is the ability to study growth rates, even at very low supersaturation, with a precision hitherto unobtainable. Experiments made at a concentration corresponding to level d in Figure 2 provide the opportunity for the formation of only a single phase, HAP. Typical data for an experiment in which HAP seed was added to such supersaturated solutions are shown in Table III from which it can be seen that HAP is grown by direct precipitation at pH = 7.40 and 37° (55). The stoichiometry of the precipitated phase is constant, 1.66 $\pm$ 0.01 for more than 6 hours of reaction and HAP was confirmed by infrared and x-ray diffraction studies. Experiments at a higher pH of 8.50, of significance in cooling tower applications, also indicate the

Table II

Constant composition seeded growth of HAP crystals at pH 7.40, 37°C*.

Time (min)	T_{Ca} (mM)	T_p (mM)	R
0	1.200	1.596	1.33
1	1.197	1.580	1.32
3	1.208	1.595	1.32
5	1.214	1.602	1.32
7	1.234	1.616	1.31
10	1.248	1.622	1.30

* Initial Solution: 150 ml of 0.800 mM $CaCl_2$, 0.575 mM KH_2PO_4, 0.854 mM KOH and 5.0 mg of HAP seed. Titrant solutions: 10.00 mM $CaCl_2$ and 7.19 mM KH_2PO_4 with 12.81 mM KOH.

exclusive formation of HAP at low supersaturation. Typical data are shown in Table IV from which is can be seen that more than 100% of the original seed material is grown directly at HAP. The results confirm previous suggestions based upon approximate stoichiometry determinations using conventional precipitation experiments that at sufficiently low supersaturation HAP can precipitate without the need for precursor formation both at the surface seed crystals (6) and spontaneously from solution (56).

Epitaxial Considerations

It is clear from the foregoing that the epitaxial growth of one calcium phosphate phase either upon another phase or upon a foreign substrate may constitute an important step in the precipitation of apatites. Crystal growth studies of the influence of different calcium phosphate seed material (9) have shown that all calcium phosphate phases are good nucleators for DCPD under solution conditions (see Figure 1, pH = 5.60) in which DCPD and HAP are the likely calcium phosphate phases. At higher pH (> than about 6) only OCP would readily induce growth in metastable supersaturated solutions (9). A more detailed study was made (8) in which HAP seed crystals were added to metastable supersaturated solutions at pH values between 4.5 and 5.1. As shown in Figure 3, the growth curve, following the addition of a relatively low concentration of HAP seed crystals (24 mg HAP l^{-1}) is characterized by a brief induction period (13-15 min) followed by the growth of DCPD with the required molar Ca:P ratio of 1.01 ± 0.01. The entire growth phase consists of DCPD which was

Table III

Constant composition seeded growth of HAP crystals at pH 7.40, 37°C*.

Time (min)	T_{Ca} (mM)	T_P (mM)	R	Newly formed HAP as % of original seed
0	0.300	0.180	1.67	0.0
60	0.303	0.178	1.69	8.0
120	0.300	0.180	1.67	15.0
180	0.298	0.176	1.69	21.0
240	0.295	0.176	1.67	27.0
360	0.300	0.180	1.67	37.0

* Initial solution: 500 ml 0.300 mM $CaCl_2$, 0.180 mM KH_2PO_4 2.000 mM KCl and 20.0 mg of HAP seed. Titrant solutions: 3.900 mM $CaCl_2$, 2.340 mM KH_2PO_4 and 4.390 mM KOH.

confirmed by x-ray analysis and which appeared to grow epitaxially from the surface of the HAP seed material. The formation of DCPD on the surface of the added HAP seed crystals involves a nucleation step and the subsequent crystal growth of these DCPD nuclei. It can be seen in Figure 4 that, following the induction period the reaction follows the same kinetics as that observed for the growth of DCPD crystals themselves. No evidence was found for the formation of a more basic calcium phosphate phase despite the fact that the solutions were supersaturated with respect to both HAP and DCPD. It is clear that the rate of nucleation, which is proportional to $(\ln T_{Ca}/T_{Ca^\circ})^{-2}$ where T_{Ca° is the solubility value, is markedly dependent upon supersaturation, sharply decreasing as the supersaturation decreases. Estimations based upon the relative rate of heterogeneous nucleation (8) indicate that heterogeneous nucleation of DCPD on the added HAP seed of low concentration is essentially complete within the induction period falling to zero at the commencement of normal DCPD growth.

At higher seed concentrations (about 230 mg HAP l^{-1}) a more basic calcium phosphate with Ca:P = 1.52 ± 0.04 crystallizes on the growth sites of the HAP seed material and no evidence is found for the presence of DCPD. The dependence of the growth phase on solid/solution ratio is of particular importance not only for the interpretation of the results of biological precipitation studies but also for the formation of calcium phosphate in environmental systems. It may be explained by the competition between heterogeneous nucleation of DCPD and the growth of the active sites already present on the seed substrate surface. The latter process occurs more extensively in the initial stages of reaction when the seed concentration is

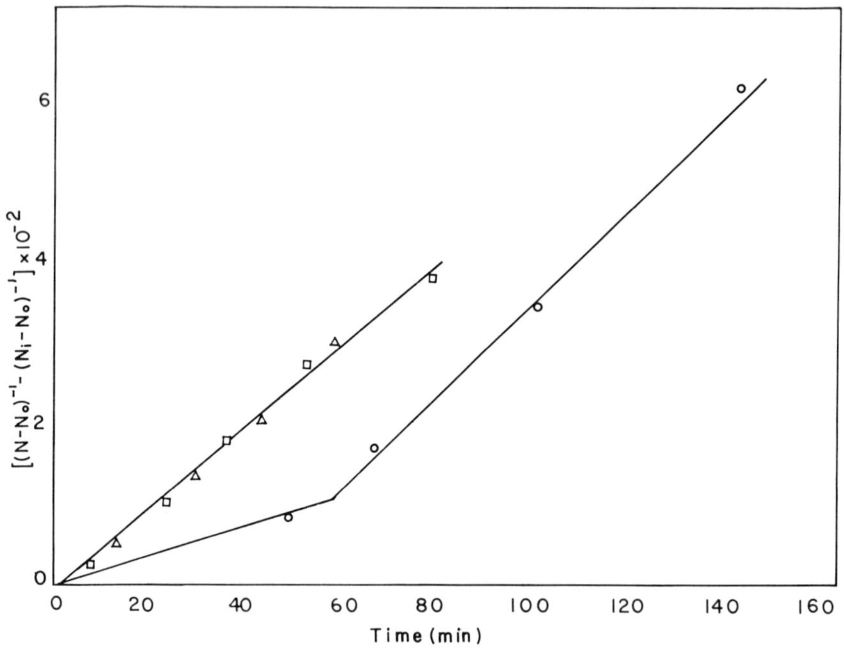

Figure 4. Plots of the integrated form of rate Equation 2. N_i *and* N_o *are the values of* N *corresponding to the instant of seed addition and equilibrium respectively.* (△) T_{Ca} = *9.314 m*M, T_p = *3.935 m*M, *pH 5.62, 37°C;* (○) T_{Ca} = *10.20 m*M, T_p = *5.861 m*M, *pH 5.62, 37°C.*

Table IV

Constant composition seeded growth of HAP crystals at pH 8.50, 37°C*.

Time (min)	T_{Ca} (mM)	T_P (mM)	R	Newly formed HAP as % of original seed
0	0.300	0.180	1.67	0.0
25	0.300	0.178	1.68	13.0
55	0.297	0.178	1.64	27.0
95	0.308	0.182	1.69	43.0
140	0.307	0.182	1.68	61.0
165	0.303	0.180	1.68	74.0
240	0.306	0.180	1.69	104.0

* Initial solution: 500 ml of 0.300 mM $CaCl_2$, 0.180 mM KH_2PO_4, 2.000 mM KCl and 20.0 mg HAP seed. Titrant solutions: 3.900 mM $CaCl_2$, 2.340 mM KH_2PO_4, 4.39 mM KOH.

relatively high. In these circumstances, the calcium and phosphate concentrations in the solution are rapidly lowered below that required for effective DCPD nucleation.

The results of these studies confirm the suggestions made earlier that the nature of the calcium phosphate phase which forms in the precipitation process is dependent not only upon the supersaturation, ionic strength, and pH, but also upon the rate at which reactions proceed.

Influence of Other Substances

As might be expected for a reaction which is controlled largely by a surface process, the presence of other components in the solution may have a marked influence upon the rate of precipitation. Studies have been made of the precipitation of calcium phosphate in the presence of a number of tricarboxylic acids of the Kreb's Cycle, citric acid, isocitric acid, cis-aconitic acid, trans-aconitic acid and tricarballylic acid (57). It was shown that those acids having an hydroxyl group, (citric and iso-citric) are effective precipitation inhibitors. Moreover, by computing the activities of calcium citrate complexes and following the uptake of citrate using carbon-14 labelled material, it was shown that the inhibiting effect of citric acid cannot be explained simply in terms of the formation of solution calcium complexes of the ligands. A relatively small degree of surface adsorption is sufficient to markedly retard the crystal growth reaction. Thus, in a recent study of the crystallization of barium sulfate (58), it was shown that the inhibition by an organic phosphonate additive,

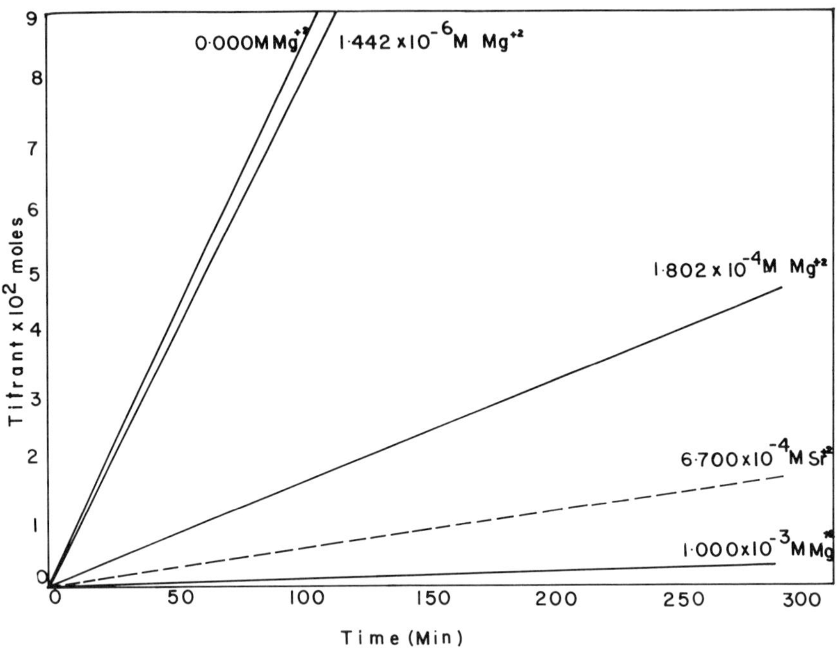

Figure 5. Constant composition growth of HAP seed at low supersaturation (T_{Ca} = 0.30 mM, T_p = 0.18 mM, [KCl] = 2 mM, 37°C, pH 8.50). Plots of titrant added as a function of time. Experiments in the presence of magnesium and strontium.

nitrilotri(methylenephosphonic acid), was sufficiently large that only about 4% of the surface of the barium sulfate crystals needed to be covered by inhibitor molecules for complete inhibition of crystal growth. This provides striking confirmation of the suggestion that precipitation may proceed through the development of a relatively small number of active sites. The inhibition of calcium phosphate crystal growth (initial conditions T_{Ca} = 1.66 mM, T_p = 1.00 mM, pH = 7.0, 25°) by typical growth inhibitors such as the organic phosphonate, hydroxyethylidene di(phosphonic acid), HEDP, has also been demonstrated (57). Precipitation is completely inhibited at a concentration of additive of 10^{-5}M and the results of adsorption experiments indicate that a relatively small fraction (10-30%) of the seed surface must be covered for complete inhibition of precipitation. The rate is markedly reduced even when only 1-10% of the surface is covered by HEDP.

At pH values below about 6.0 the influence of HEDP and fluoride, an important environmental constituent is particularly interesting. It has been shown that the nature of the precipitating calcium phosphate phase can be controlled not only by the concentration of inoculating HAP seed (8)but also by the presence of additives. Thus fluoride ion accelerates the crystallization of calcium phosphate probably through the formation of fluorapatite ($Ca_5(PO_4)_3F$, FAP) (59, 60). Under the conditions of the experiments, (T_{Ca} = 6.275 mM, T_p = 6.275 mM, 37°, pH = 5.59) the addition of HEDP, which retards the rate of mineralization, increases the nucleation of DCPD and the amount of this phase formed. Fluoride ions, on the other hand, in increasing the rate of mineralization, rapidly reduce the concentration levels of calcium and phosphate below those required for effective nucleation of DCPD. The resulting growth phase then consists of a more basic calcium phosphate with no evidence for the presence of DCPD (61). These results may have significant implications in discussions of the influence of other environmental components on the precipitation of calcium phosphates. The design of model experiments aimed at elucidating the mechanism of calcium phosphate precipitation under specific natural water conditions therefore requires very careful planning together with a knowledge of the specific influences of the individual components.

It is interesting to note that at pH values of about 7.0, there is a marked influence upon the crystallization of HAP seed crystals by changing the nature of the alkaline metal cation in the background electrolyte (6). The greatest differences are found between potassium and sodium chlorides, of particular importance in the environment. The possibility that sodium ion can replace calcium in the calcium phosphate solid is well known and it has been found that from 5.5 to 6.0 mol % of sodium is incorporated in the growing HAP crystals in the presence of sodium chloride but almost no potassium is observed in the presence of potassium chloride (6). The possibility of studying the influence of traces of metal ions using the new constant composition approach is particularly interesting. Studies have recently been made at concentrations corresponding to level d in Figure 2 at a higher pH of 8.5. Exclusive formation of HAP without the need for precursor phase was confirmed (P. Koutsoukos and G. H. Nancollas, unpublished data, 1968), the precipitating phase having a Ca:P molar ratio of 1.67 ± 0.02. A typical plot of the rate of mixed titrant addition, reflecting the rate of crystallization, is shown in Figure 5 and it is seen that low concentrations of magnesium ions have a marked inhibiting effect upon the HAP crystallization. It is interesting to note that the magnesium ion is not incorporated into the HAP crystals during precipitation. The inhibiting influence of magnesium ion upon the crystallization of calcium phosphate at supersaturation levels corresponding to level a in Figure 2, has been attributed to its adsorption at the surface of the crystals (51). The presence of magnesium ion also appears to change the stoichiometry of the first-formed surface phase as discussed earlier. It can be seen in Figure 5 that the presence of strontium ion also markedly retards the rate of HAP crystallization. In this case, however, an observed increase in calcium concentration as the reaction proceeds, indicates some substitution of Sr^{2+} for Ca^{2+} ions. This has been reported by other workers (41, 62, 63, p. 22-25) and may be ascribed to the similarity of ionic radii. The ions. This has been reported by other workers (41, 63, 64) and may be ascribed to the similarity of ionic radii. The constant composition technique enables more detailed studies to be made of the influence of magnesium, strontium, and other cations on the precipitation of calcium phosphates. As far as anions, typical of the environment, are concerned, the effect of carbonate is of particular significance (64, p. 514-525). Our preliminary studies show that low levels of carbonate ion (∿ 1 mg L^{-1}) significantly reduce the rate of calcium phosphate precipitation (P. Koutsoukos and G. H. Nancollas, unpublished data, 1968) while phosphate ions have a similar effect upon the constant composition growth of calcium carbonate crystals (T. Kazmierczak and G. H. Nancollas, unpublished data, 1978). Studies of these reactions under conditions in which the activities of all ion species are maintained constant are continuing.

Acknowledgements

We wish to acknowledge partial support of this work through research grants from the National Institutes of Health, National Institutes of Dental Research, (Grant # DE03223) and the National Science Foundation (Grant # ENG74,15486).

Abstract

The calcium phosphate phases which precipitate in solutions containing calcium and phosphate ions are markedly dependent upon factors such as pH, supersaturation, ionic strength, temperature and the nature and extent of the solid phases already present. Although thermodynamic considerations will yield the driving forces for precipitation of particular phases, their formation under a specific set of conditions may be determined much more by kinetic factors. Metastable phases may persist for considerable periods in the supersaturated solutions. In making the thermodynamic calculations, it is essential to take into account both activity coefficient terms and the formation of ion-pairs and complexes in the solutions. Methods are described for the highly reproducible kinetic study of precipitation reactions in the presence of seed materials. A recently developed constant composition techniques throws new light on the mechanism of calcium phosphate precipitation and enables solid stoichiometries and rates of reaction to be determined with a precision hitherto unobtainable even at the very low supersaturations of significance in the environment.

Literature Cited

1. "Phosphates in Detergents and the Eutrophication of America's Waters", 23rd Report, Comm. on Govt. Operations, US Printing Office, Washington, p. 27 (1970).
2. Brown, W.E. Solubilities of phosphates and other sparingly soluble compounds, p. 203-239, in Griffith, E.J., Beeton, A., Spencer, J.M., and Mitchell, D.T., ed., "Environmental Phosphorus Handbook," Wiley and Sons, New York, 1973.
3. Ferguson, J.F., Jenkins, D., and Eastman, J. Calcium phosphate precipitation, in 44th Ann. Conf. Water Pollution Control Fed., Ser. 5 (2) Oct. 4, (1971).
4. Clark, L.M. and Gerrard, W.E. The precipitation of hydroxyapatite, $3[Ca_3(PO_4)_2]$, $Ca(OH)_2$ Part I, J. Soc. Chem. Ind. 57, 295-297 (1938).
5. Varsanik, R.G. The nature and control of calcium orthophosphate depositions in cooling water systems. Paper 142, "Corrosion '75," Nat. Assoc. Corr. Eng., Toronto, Can., 1975.
6. Nancollas, G. H. and Tomazic, B. Growth of calcium phosphate on hydroxyapatite crystals. Effect of supersaturation and ionic medium. J. Phys. Chem. 78, 2218-2225 (1974).
7. Tomazic, B. and Nancollas, G.H. The seeded growth of calcium phosphates. Surface characterization and the effect of seed material. J. Coll. Interface Sci. 50, 451-461 (1975).
8. Barone, J.P. and Nancollas, G.H. The seeded growth of calcium phosphates. The effect of solid/solution ratio in controlling the nature of the growth phase. J. Coll. Interface Sci. 62, 421-431 (1977).

9. Nancollas, G.H, and Wefel, J.S. The seeded growth of calcium phosphates: Effect of different calcium phosphate seed material , J. Dental Res. 55, 617-624 (1976).

10. McDowell, H, Gregory, T.M, and Brown, W.E. Solubility of $Ca_5(PO_4)_3OH$ in the system $Ca(OH)_2$-H_3PO_4-H_2O at 5, 15, 25 and 37°C. J. Res. Nat. But. Stand. 81A, 273-281 (1977).

11. Gregory, T.M, Moreno, E.C, Patel, J.M, and Brown, W.E. Solubility of $Ca_3(PO_4)_2$ in the system $Ca(OH)_2$-H_3PO_4-H_2O at 5,15,25 and 37°C. J. Res. Nat. Bur. Stand. 78A, 667-674 (1974).

12. Moreno, E.C, Brown, W.E, and Osborn, G. Stability of dicalcium phosphate dihydrate in aqueous solutions and solubility of octacalcium phosphate. Soil Sci. 24, 99-102 (1960).

13. McDowell,H. "Solubility of $CaHPO_4$ and Ion-Pair Formation", Ph.D. Thesis, Howard University, Washington, D.C., 1968.

14. Patel, P.R, Gregory, T.M, and Brown, W.E. Solubility of $CaHPO_4$-$2H_2O$ in the quaternary system $Ca(OH)_2$-H_3PO_4-NaCl-H_2O at 25°C. J.Res.Nat.Bur.Stand. 78A, 675-681 (1974).

15. Bates, R.G, and Acree, S.F. ΔH values of certain phosphate-chloride mixtures and the second dissociation constant of phosphoric acid from 0° to 60° J.Res.Nat.Bur.Stand. 30, 129-155 (1943).

16. Vanderzee, C.E. and Quist, A.S. The third dissociation constant of orthophosphoric acid. J. Phys. Chem. 65, 118-123 (1961).

17. Chughtai, A., Marshall, R., and Nancollas, G.H. Complexes in calcium phosphate solutions. J. Phys. Chem. 72 208-211 (1968).

18. Davies, C.W. The extent of dissociation of salts in water. Part VI. Some calcium salts of organic acids. J. Chem. Soc. 277-281 (1938).

19. Nancollas, G.H. "Interactions in Electrolyte Solutions," 215 p. Elsevier, Amsterdam, 1966.

20. Davies, C.W. "Ion Association," 190 p. Butterworths, London, 1962.

21. Lewis, G.N. and Randall, M. "Thermodynamics" Revised by Pitzer, J.S. and Brewer, L., 2nd Ed. 723 p. McGraw-Hill, New York, 1961.

22. Robinson, R.A. and Stokes, R.H. "Electrolyte Solutions," 2nd Ed. Rev., 571 p. Butterworths, London, 1965.

23. Brannan, J.R. and Nancollas, G.H. Ion association in solutions of nickel malonate and n-butyl malonate. Trans. Faraday Soc. 58, 354-358 (1962).

24. Storey, S.H. and Van Zeggern, F. Computation of chemical equilibrium compositions. Can. J. Chem. Eng. 42, 54-55 and 48 591-593 (1970).

25. White, W.B., Johnson, S.M., and Dantzig, G.B. Chemical equilibrium in complex mixtures. J. Chem. Phys. 28, 751-755 (1958).

26. Warga, J. A convergent procedure for solving the thermo-classical equilibrium problem. J. Soc. Ind. Appl. Math. 11, 594-606 (1963).

27. Ingri, N., Kakolowicz, W., and Sillen, J.G. High speed computers as a supplement to graphical methods-V HALTAFALL A general program for calculating the composition of equilibrium mixtures. Talanta 14, 1261-1286 (1967).
28. Perrin, D.D. and Sayce, I.G. Computer calculation of equilibrium concentrations in mixtures of metal ions and complexing species. Talanta 14, 813-842 (1967).
29. I, Ting-Po and Nancollas, G.H. EQUIL-A general computational method for the calculation of solution equilibria. Anal. Chem. 44, 1940-1950 (1972).
30. Dietz, V. R., Rootare, H.M., and Carpenter, F.G. The surface composition of hydroxyapatite derived from solution behaviour of aqueous suspensions. J. Coll. Sci. 19, 87-101 (1962).
31. Rootare, H.M., Dietz, V.R., and Carpenter, F.G. Solubility product phenomena in hydroxyapatite water systems. J. Coll. Sci. 17, 179-206 (1962).
32. LaMer, V.K. The solubility behaviour of hydroxyapatite. J. Phys. Chem. 66, 973-978 (1962).
33. Newmann, W.F. and Newmann, M.W. The nature of the mineral phase of bone. Chem. Rev. 53, 1-45 (1953).
34. Eanes, E.D., Gillessen, I.H., and Posner, A.S. Intermediate states in the precipitation of hydroxyapatite. Nature 298, 365-367 (1965).
35. Walton, A.G., Bodin, W.J., Furedi, H., and Schwartz, A. Nucleation of calcium phosphate from solution. Can. J. Chem. 45, 2695-2701 (1967).
36. Eanes, E.D., Gillessen, I.H., and Posner, A.S. Mechanism of conversion of non-crystalline calcium phosphates to crystalline hydroxyapatite, p. 373-375, in "Proc. Int. Cong. on Crystal Growth, Boston, 1966." Pergamon Press, Oxford, 1967.
37. Termine, J.D., Peckauskas, R.A., and Posner, A.S. Calcium phosphate formation in vitro. II. Effect of environment on amorphous crystalline transformation. Arch. Biochem. Biophys. 140, 318-325 (1970).
38. Fleisch, H., Russel, R.G.G., Bisaz, S., Termine, J.D., and Posner, A.S. Influence of the pyrophosphate in the transformation of amorphous to crystalline calcium phosphate. Calcif. Tiss. Res. 2, 49-59 (1968).
39. Francis, M.D. The inhibition of calcium hydroxyapatite crystal growth by polyphosphonates and polyphosphates. Calcif. Tiss. Res. 3, 151-162 (1969).
40. Termine, J.D. and Conn, K.M. Inhibition of apatite formation by phosphorylated metabolites and macromolecules. Calcif. Tiss. Res. 22, 149-157 (1976).
41. Eanes, E.D. and Posner, A.S. Intermediate phases in the basic solution preparation of alkaline earth phosphates. Calcif. Tiss. Res. 2, 38-48 (1968).
42. Boskey, A.L. and Posner, A.S. Magnesium stabilization of amorphous calcium phosphate. Mat. Res. Bull. 9, 907-916 (1974).

43. Eanes, E.D. and Meyer, J.L. The maturation of crystalline calcium phosphate in aqueous suspensions at physiological pH. Calcif. Tiss. Res. 23, 259-269 (1977).
44. Brown, W.E. Crystal growth of bone mineral. Clin. Orthop. 44, 205-220 (1966).
45. Davies, C.W. and Jones, A.L. The precipitation of silver chloride from aqueous solutions. Part I. Faraday Discuss. Chem. Soc. 5, 103-111 (1949).
46. Nancollas, G.H. and Purdie, N. The kinetics of crystal growth. Q. Rev. Chem. Soc. 18, 1-20 (1964).
47. Nancollas, G.H. and Reddy, M.M. The kinetics of crystallization of scale forming minerals. J. Soc. Pet. & Eng., 117-126 (1974).
48. Nancollas, G.H. and Reddy, M.M. The crystallization of calcium carbonate II. Calcite growth mechanism. J. Coll. Interface Sci. 37, 824-830 (1971).
49. Nancollas, G.H. and Mohan, M.S. The growth of hydroxyapatite crystals. Arch. Oral Biol. 15, 731-745 (1970).
50. Marshall, R.W. and Nancollas, G.H. The kinetics of crystal growth of dicalcium phosphate dihydrate. J. Phys. Chem. 73, 3838-3844 (1969).
51. Tomazic, B., Tomson, M., and Nancollas, G.H. Growth of calcium phosphates on hydroxyapatitie crystals: The effect of magnesium. Arch. Oral Biol. 20, 803-808 (1975).
52. Francis, M.D. Solubility behavior of dental enamel and other calcium phosphates. Ann. N.Y. Acad. Sci. 131, 694-712 (1965).
53. Tomson, M.B. and Nancollas, G.H. Mineralization kinetics: A constant composition approach. Science 200, 1059-1060 (1978).
54. Brown, W.E., Smith, J.P., Lehr, J.R., and Frazier, W.A. Crystallographic and chemical relations between octacalcium phosphate and hydroxyapatite. Nature 196, 1050-1055 (1962).
55. Amjad, Z., Koutsoukos, P., Tomson, M.B., and Nancollas, G.H. The growth of hydroxyapatite from solution. A new constant composition method. J. Dental Res. (in press).
56. Boskey, A.L. and Posner, A.S. Formation of hydroxyapatite at low supersaturation. J. Phys. Chem. 80, 40-45 (1976).
57. Nancollas, G.H., Tomson, M.B., Battaglia, G., Wawrousek, H., and Zuckermann, M. Precipitation of calcium phosphate. The influence of tricarboxylic acid, magnesium, and phosphonate, in Rubin, A.J., "Chemistry of Wastewater Technology," Ann Arbor Science, (in press).
58. Leung, W. and Nancollas, G.H. Nitrilotri (methylenephosphonic acid) adsorption on barium sulfate crystals and its influence on crystal growth. J. Crystal Growth 44, 163-167 (1978).
59. Spinelli, M.A., Brudevold, F. and Moreno, E.D. Mechanism of fluoride uptake by hydroxyapatite. Archs. Oral Biol. 16, 187-203 (1971).
60. Kuyper, A.L. and Kultnerian, K. Mechanism of incorporation of fluoride into bone salt. J. Dental Res. 41, 345-350 (1962).

61. Barone, J.P. and Nancollas, G.H. The growth of calcium phosphates on hydroxyapatite crystals. The effect of fluoride and phosphonate. J. Dent. Res. (in press).
62. Duff, E.J. Interactions of divalent cations with apatites. Col. Int. CNRS 230, 419-421 (1977).
63. McConnel, D.C. "Apatite." 111 p. Springer Verlag, New York, 1973.
64. Stumm, W. and Morgan, J.J. "Aquatic Chemistry." 583 p. Wiley-Interscience, New York, 1970.

RECEIVED November 16, 1978.

24

Chemistry of Calcium Carbonate in the Deep Oceans

JOHN W. MORSE
Division of Marine and Atmospheric Chemistry, Rosenstiel School of Marine and Atmospheric Science, University of Miami, 4600 Rickenbacker Causeway, Miami, FL 33149

ROBERT C. BERNER
Department of Geology and Geophysics, Yale University, New Haven, CT 06520

Recent pelagic sediments containing over 30% calcium carbonate, by dry weight, cover a quarter of the surface of the earth (see Figure 1). These sediments make up a vast and chemically reactive carbonate reservoir which has a major influence on the chemistry of the oceans and atmosphere. In order to have a predictive understanding of the natural carbon dioxide system and the influence of man on it, the chemical dynamics of calcium carbonate deposition in the deep ocean basins must be known in detail.

The dominant source of calcium carbonate for deep sea sediments is biogenic, being derived from the remains of small pelagic organisms living near the ocean surface. Only a small portion of the calcium carbonate produced by these organisms is preserved in deep sea sediments. Current estimates indicate that between 75 and 95 percent of the calcium carbonate being produced in the open ocean is subsequently dissolved (2, 3). Most of the dissolution is thought to occur at or very near the sediment-water interface, rather than in the water column during settling (4).

The balance between calcium carbonate production and dissolution is the major pH buffering mechanism of seawater over periods of time at least on the order of thousands of years (5). The atmospheric carbon dioxide reservoir is less than 2 percent the size of the seawater reservoir (6) and there is active exchange between these two reservoirs across the air-water interface. Consequently, the carbon dioxide content of the atmosphere and accumulation of calcium carbonate in the deep oceans are closely coupled.

In order to understand the chemistry of calcium carbonate accumulation in the deep oceans, the sources of calcium carbonate, its distribution in recent pelagic sediments, the saturation state of seawater overlying deep-ocean sediments with respect to calcite and aragonite, and the relation between saturation state and dissolution rate must be known. These aspects of calcium carbonate chemistry are examined in this paper.

0-8412-0479-9/79/47-093-499$09.25/0
© 1979 American Chemical Society

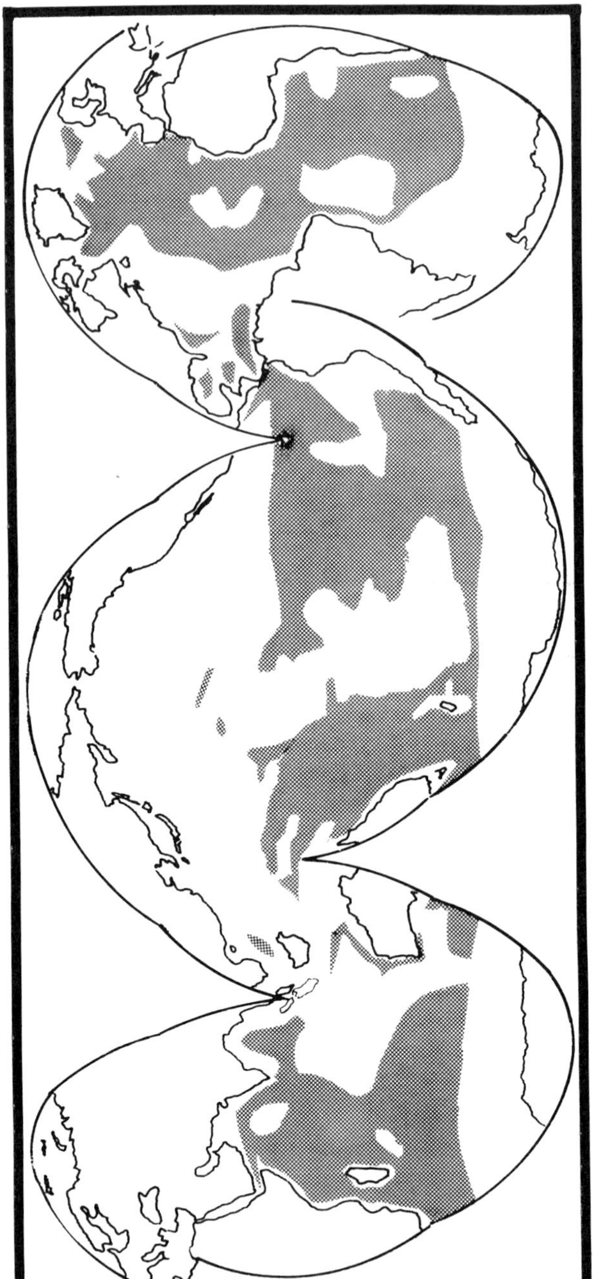

Figure 1. Distribution of recent pelagic sediments containing greater than 30 wt % calcium carbonate by dry weight (after Ref. 1).

Sources and Depositional Patterns of Calcium Carbonate in Deep Sea Sediments

Biogenic calcium carbonate, derived from the tests (shells) of pelagic organisms, is the dominant source for calcium carbonate presently being deposited in deep ocean basins. Foraminifera (animals) and coccolithophores (plants) are the most important sources of calcitic calcium carbonate. Pteropods (animals) are the most important source for aragonitic calcium carbonate. Photomicrographs of typical representatives of these organisms are presented in Figure 2. The relative abundance of these major sources and species distributions, within a given type of organism, are dependent on the environmental conditions at a given location. Consequently, the calcium carbonate arriving at the sediment-water interface is geographically and, perhaps, seasonally variable. Current estimates of the average pelagic aragonite to calcite production ratio range from 1 (7) to 0.25 (8). The relative abundance of aragonite to calcite and the size distribution of the biogenic calcite arriving at the sediment-water interface have a strong influence on dissolution kinetics and are important factors in calcium carbonate preservation.

The general pattern of calcium carbonate accumulation in deep ocean basins has been known for approximately one hundred years. In most areas the abundance of calcium carbonate in sediments is found to decrease with increasing water depth (see Figure 3). From this general pattern two important sediment marker levels have been identified. The first is the calcium carbonate compensation depth (CCD). Below the CCD calcium carbonate becomes a minor (less than 10%) component of the sediment. At the CCD close to 100 percent of the calcium carbonate is being dissolved and that the rate of input and dissolution closely balance. The second is the aragonite compensation depth (ACD), which frequently occurs at depths on the order of 3 km shallower than the CCD. Below the ACD aragonite is not preserved in the sediments. These early findings suggested that temperature and pressure are important controls on calcium carbonate deposition and, hence, that calcium carbonate accumulation in the deep oceans is chemically controlled.

Berger (10) defined an important sediment marker level which occurs above the CCD called the foraminiferal lysocline (FL). It is the depth at which there is a maximum change in the ratio of "easily dissolved" to "resistant" pelagic foraminifera. The classification of pelagic foraminifera into these categories is based on observations of foraminiferal assemblages suspended at different depths in the Pacific Ocean and the changes in foraminiferal assemblages in sediments with increasing water depth. At present, there is considerable controversy over how much dissolution of calcium carbonate must occur to produce the FL. Adelseck (11) has experimentally determined that approximately 80 percent of the calcitic fraction must be dissolved. This roughly

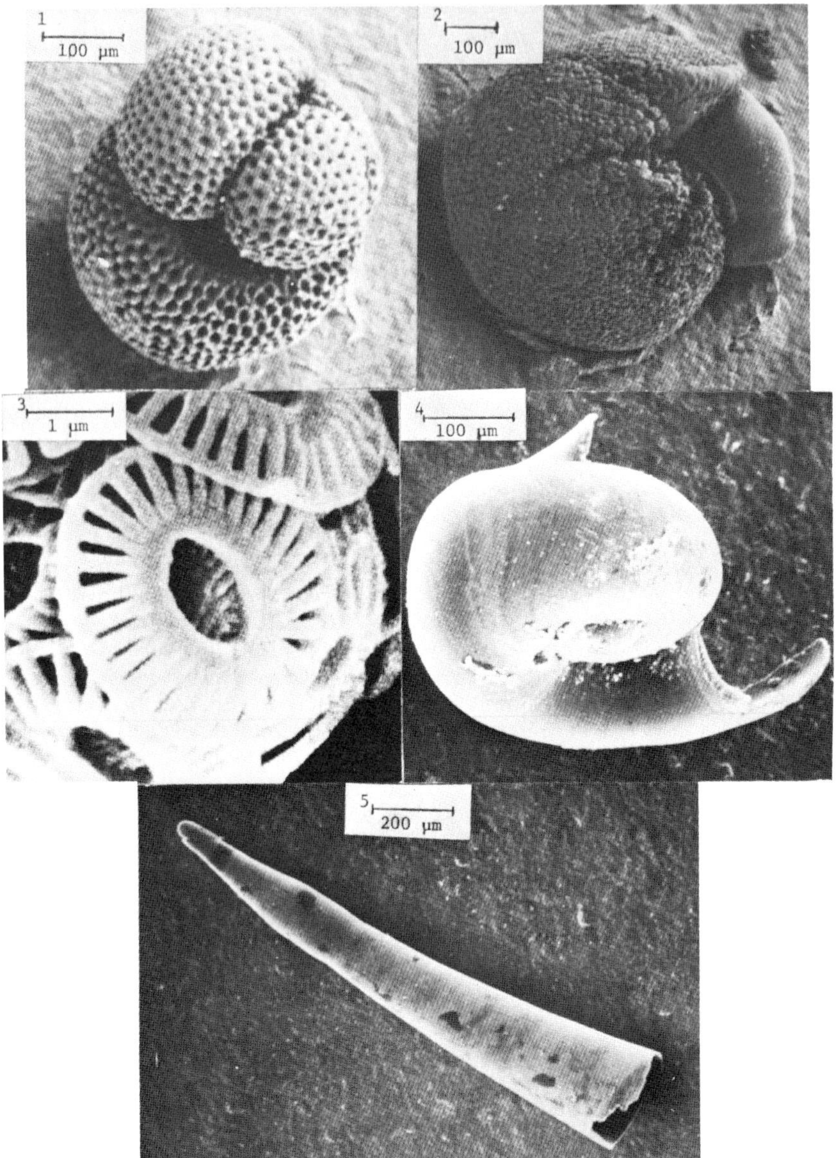

Figure 2. Examples of typical calcium carbonate tests which are deposited in pelagic sediments. (1) and (2) are foraminifera (calcite); (3) is a coccolithophore (calcite); (4) and (5) are pteropods (aragonite). Note the difference in size as shown on scales.

agrees with Berger's (12) minimum loss value, based on direct observations, of approximately 50 percent. The difference in depth between the FL and the CCD varies considerably. Under central subtropical water masses it is frequently about 800 m.

Berger (12) has identified another sediment marker level in the sediments. This is the R_o level at which the proportion of "resistant" planktonic foraminifera show the first significant increase. Berger's estimate of minimum dissolution loss at the R_o level is 10 percent. Adelseck's (11) experimentally determined amount of dissolution necessary to produce the R_o level is approximately 50 percent dissolution.

Calculation of Calcium Carbonate Saturation State

General Considerations. In order to facilitate the discussion of methods for calculating the saturation state of seawater with respect to calcium carbonate, initial consideration will be given to pure calcium carbonate phases. The method most frequently used expressing the saturation state of a solution with respect to solid phase is as the ratio (Ω) of the ion activity (a) product to the thermodynamic solubility constant (K). For the calcium carbonate phase calcite, the expression for the saturation state is defined as (e.g., 13):

$$\Omega_c = \frac{a_{Ca^{++}} \, a_{CO_3^{--}}}{K_c} . \qquad (1)$$

Where the subscript "c" denotes calcite. A subscripted "a" will be used in the same manner to denote aragonite. If Ω is greater than 1 the solution is supersaturated, less than 1 undersaturated, and equal to 1 in equilibrium with respect to calcite.

Because of difficulties in precisely calculating the total ion activity coefficient (γ) of calcium and carbonate ions in seawater, and the effects of temperature and pressure on the activity coefficients, a semi-empirical approach has been generally adopted by chemical oceanographers for calculating saturation states. This approach utilizes the apparent (stoichiometric) solubility constant (K'), which is the equilibrium ion molal (m) product. Values of K' are directly determined in seawater (as ionic medium) at various temperatures, pressures and salinities. In this approach:

$$\Omega_c = \frac{m_{Ca^{++}} \, m_{CO_3^{--}}}{K_c'} . \qquad (2)$$

The ratio of K_c to K_c' is then:

$$\frac{K_c}{K_c'} = \gamma_{Ca^{++}} \gamma_{CO_3^{--}} \quad (3)$$

Calculation of Calcium and Carbonate Total Molalities. Because of the constancy of composition of seawater, the total ion calcium concentration in seawater can be calculated, in "open" ocean seawater samples, directly from high precision salinity measurements, using the relationship (14):

$$m_{Ca^{++}} = \left(\frac{.010282}{1 - \frac{S}{1000}} \right) \quad (4)$$

where S is salinity in parts per thousand. This relationship can be expected to be constant under "open" ocean conditions to within at least 0.3% and is independent of temperature and pressure.

The calculation of the total ion molal carbonate ion concentration is more complex than for calcium because it is part of the carbonic acid system. The following reactions take place between CO_2 and carbonic acid in seawater:

$$CO_2(gas) \rightleftharpoons CO_2(aq) + H_2O \rightleftharpoons H_2CO_3 \rightleftharpoons HCO_3^- + H^+ \rightleftharpoons CO_3^{--} + 2H^+ . \quad (5)$$

Generally only one equilibrium constant is used to describe the reaction between gaseous carbon dioxide and H_2CO_3. The thermodynamic and apparent equilibrium constants for the above reactions are:

$$\alpha_o = \frac{a_{H_2CO_3}}{P_{CO_2}} \; ; \quad \alpha_s = \frac{m_{H_2CO_3}}{P_{CO_2}} \quad (6)$$

where α_o and α_s are the Henry's law coefficients for infinite dilution and seawater, respectively, and P_{CO_2} is the partial pressure of CO_2 gas, and:

$$K_1 = \frac{a_{HCO_3^-} \, a_{H^+}}{a_{H_2CO_3}} \; ; \quad K_1' = \frac{m_{HCO_3^-} \, a_{H^+}}{m_{H_2CO_3}} \quad (7)$$

$$K_2 = \frac{a_{CO_3^{--}} \, a_{H^+}}{a_{HCO_3^-}} \; ; \quad K_2' = \frac{m_{CO_3^{--}} \, a_{H^+}}{m_{HCO_3^-}} \quad (8)$$

where H_2CO_3 here represents the sum of dissolved CO_2 and true

H_2CO_3.

Using either the above thermodynamic or apparent (stoichiometric) equations, it is possible to calculate $m_{CO_3^{--}}$ if any two of the four possible measurable quantities for the carbonic acid system are known. These four measurable quantities are:

1. Total CO_2 defined as:

$$\Sigma\ CO_2 = m_{H_2CO_3} + m_{HCO_3^-} + m_{CO_3^{--}} \tag{9}$$

2. Carbonate alkalinity defined as:

$$A_c = m_{HCO_3^-} + 2m_{CO_3^{--}} \tag{10}$$

3. P_{CO_2}

4. pH

Carbonate alkalinity is not a directly analyzable quantity, but it can be derived from the titration alkalinity (A_t) by a relationship which is temperature and pressure dependent. pH is also variable with pressure and temperature. In addition, pH can be defined several ways depending on which buffer system is used, whether liquid junction potentials are considered, and what definition of ionic strength is used. The values of the apparent dissociation constants and calcium carbonate solubility constants are dependent on exactly what definition of seawater pH is used and what standardization technique is used (15, 16, 17).

Although every combination of the four carbonic acid system-analyzable variables has been used, the one pair frequently used because of its relative ease from both an analytical and calculation standpoint is pH and alkalinity. When this pair is used, the equation relating total carbonate molality to pH and carbonate alkalinity is:

$$m_{CO_3^{--}} = \frac{A_c}{2 + \frac{10^{-pH}}{K_2'}} \ . \tag{11}$$

The *in situ* pH can be calculated by making the appropriate temperature and pressure corrections on the measured pH. Temperature corrections are made using the equation of Harvey (18)

$$pH_{t_2} = pH_{t_1} + x(t_1 - t_2) \ . \tag{12}$$

For the general case where the pH is measured at 25°C (t_1) and the in situ temperature ($t_2 = t_{is}$) is less than 20°C, the values of x given by Harvey (18) yield Equation 13.

$$pH_{t_{is}} = pH_{25} - 0.0485 + 0.0105\ (t_{is} - 20). \tag{13}$$

Ben-Yaakov (19) has presented a more precise and complex method for calculating the change in pH of seawater with temperature, based on the changes in the apparent dissociation constants of boric and carbonic acid in seawater with temperature, pressure, and salinity. He found that the temperature coefficient depended on pH and salinity. However, the results obtained by his method and Equation 13 are in close agreement when the salinity is near 35°/oo and the pH is close to 8. Once the pH at the in situ temperature has been calculated, the actual in situ pH can be calculated by correcting the pH for the effect of pressure by use of Equation 14 which was derived from the data presented by Culberson and Pytkowicz (20, Table 4)

$$pH_{is} = pH_{t_{is}} - [(P-1) \times 10^{-4}]$$

$$[4.28 - 0.04\ t_{is} - 0.4\ (pH_{t_{is}} - 7.8)]\ . \tag{14}$$

where P is the absolute pressure in atmospheres.

In order to calculate the in situ carbonate alkalinity, it is first necessary to calculate the in situ alkalinity contribution of boric acid to the analyzed quantity, titration alkalinity. This is done by first calculating the concentration of boric acid via the boron to chlorinity ratio of Culkin (21) and a general salinity-molecular weight conversion formula

$$m_{B(OH)_3} = \frac{(1.21 \times 10^{-5})S}{\left(1 - \frac{S}{1000}\right)}\ . \tag{15}$$

The next step is the calculation of the first apparent dissociation constant of boric acid in seawater (K_B') under in situ conditions. K_B' can be calculated at the in situ temperature using the equation of Edmond (22) which is based on the data of Lyman (23)

$$-\log K_{B_{t_{is}}}' = \frac{2291.90}{(t_{is} + 273)} + 0.01756\ (t_{is} + 273)$$

$$-3.3850 - 0.32051 \left(\frac{S}{1.80655}\right)^{1/3}\ . \tag{16}$$

The change in -log K_B' due to pressure (generally designated in oceanographic literature as pK_B') can be calculated using the equation of Culberson and Pytkowicz (20)

$$\Delta(-\log K_B') = (1.809 \times 10^{-3})(P-1) - (4.515 \times 10^{-6})(t_{is} + 273)(P-1) - (1.69 \times 10^{-7})(P-1)^2 + (1.759 \times 10^{-12})(t_{is} + 273)^2(P-1)^2 . \quad (17)$$

From these considerations the in situ first apparent dissociation constant for boric acid in seawater is:

$$K_{B_{is}}' = 10^{-[(-\log K_{B_{t_{is}}}') - \Delta(-\log K_B')]} . \quad (18)$$

The carbonate alkalinity is calculated by the relationship of Skirrow (24):

$$A_{c_{is}} = A_t - \left[\frac{K_{B_{is}}' \, m_{B(OH)_3}}{K_{B_{is}}' + 10^{-pH_{is}}}\right] . \quad (19)$$

It is also necessary to calculate the in situ value for K_2'. Its value at the in situ temperature and salinity can be calculated using the equation of Mehrbach *et al.*, (17):

$$-\log K_{2_{t_{is}}}' = 5371.9645 + 1.671221(t_{is} + 273) + 0.22913S + 18.3802\log(S) - 128375.28/(t_{is} + 273) - 2194.3055\log(t_{is} + 273) - (8.0944 \times 10^{-4})(t_{is} + 273)S - 5617.11[\log(S)]/(t_{is} + 273) + 2.136S/(t_{is} + 273) . \quad (20)$$

The change in -log K_2' due to pressure, near 34.8°/oo salinity, is given by Culberson and Pytkowicz (20) as:

$$\Delta(-\log K_2') = -0.015 + (8.39 \times 10^{-4})(P-1) - (1.908 \times 10^{-6})(t_{is} + 273)(P-1) + (1.82 \times 10^{-7})(t_{is} + 273)^2 . \quad (21)$$

From these considerations the in situ value of K_2' is:

$$K'_{2_{is}} = 10^{-[(-\log K'_{2t_{is}}) - \Delta(-\log K_2')]}. \quad (22)$$

The total in situ carbonate ion molal concentration can now be determined using Equation 11. At this point, the only parameters lacking for calculation of the saturation state of calcium carbonate in seawater are the in situ values of the apparent solubility products of calcite and aragonite.

Calculation of Apparent Calcium Carbonate Solubility Products. The precise determination of the solubility of calcite and aragonite in seawater has proven to be one of the more elusive quests of low temperature geochemists. The two most recent determinations of apparent calcite solubility (25, 26) differ by approximately 40 percent at 2°C. The following discussion will be directed at the differences in the recently reported values for K_c' of Ingle et al. (25) and Berner (26).

In the study of calcite solubility by Ingle, et al. (26) both reagent grade and Iceland spar calcite were studied in seawater (35°/oo salinity) at different temperatures. An airtight saturometer (27) was used for solubility meaurements. Starting undersaturations were small enough so that pH changes of less than 0.2 were observed. However, in the experiments started in supersaturated seawater, pH changes as large as 0.8 occurred.

Berner (26) took a new approach to the determination of K_c' in seawater. His basic argument was that it is extremely difficult, if not impossible, to obtain good solubility data on calcite in seawater because, in precipitation from supersaturation, pure calcite does not form. Instead, a Mg-enriched calcite is formed. The solubility of the magnesian calcite, not pure calcite is, therefore, measured. This argument has previously been advanced by Weyl (28) from his saturometry results and was clearly demonstrated by Berner (29) in his study of the role of Mg in the crystal growth of calcite and aragonite. To avoid this problem, Berner (26) measured the solubility of calcite and aragonite in distilled water at various temperatures. The solubility of aragonite, which does not suffer from the Mg problem, in seawater was then determined at various temperatures. K_c' was obtained by assuming that the ratio of solubility of calcite and aragonite is a constant independent of solution composition. His experimental method for the determination of K_a' was different from that which Ingle et al. (25) used to determine K_c'. Instead of using a saturometer, a sample of the calcium carbonate to be studied was stirred in a solution kept at constant P_{CO_2} by bubbling a gas of constant composition through it. He then measured pH, alkalinity and $m_{Ca^{++}}$.

There appear to be two primary possibilities for the differences in the K_c' values. The first is that they are a result of the experimental techniques used to determine K_c'. The second possibility is that Ingle et al. (25) measured the solubility of calcite enriched by Mg substituting for Ca in the crystal structure. The main difficulty with the latter argument is that the K_c' values obtained from undersaturation agree quite well with those obtained from supersaturation. It is possible that, while pure calcite was dissolving, a Mg-enriched calcite of lower solubility was forming. If a guess is made that the least soluble phase contains 5 mole percent Mg, and the value for K_c' of Ingle et al. (25) at 25°C is used, then $K'_{Mg-c} = 5.43 \times 10^{-7}$. If $\gamma_{Ca^{++}} = 0.20$, $\gamma_{Mg^{++}} = 0.25$ and $\gamma_{CO_3^{--}} = 0.030$ the difference between the calculated free energy of formation and that determined by Berner (29) is only approximately 0.1 kcal/mole. This is probably well within the accuracy of the data. Although this is certainly not conclusive evidence that a magnesian calcite of approximately 5 mole percent Mg is being formed in the experiments of Ingle et al. (26), it certainly reinforces the possibility.

If a magnesian calcite forms as a surface phase which controls the solubility of calcite in seawater, it may be incorrect to use the solubility of pure calcite in trying to understand the deposition of calcitic carbonates in the sea. It could turn out that Ingle et al. (25) and Berner (26) are both correct in their way. Berner's values could be correct in a thermodynamic sense (i.e., the solubility of calcite if no different surface phase forms), while the results of Ingle et al. (25) could be correct in the empirical sense (i.e., calcite will always come to its equilibrium value due to formation of a Mg-enriched calcite surface phase). This possibility has recently been reinforced by the experimental work of Morse (30) and the deep sea observations of Broecker and Takahashi (31).

Berner (26) found a much larger change in the solubility of aragonite with temperature in seawater, than Ingle (32) found for calcite. The size of the change in the solubility of aragonite in seawater was close to that observed in distilled water. Recent calculations by Millero (33) on the effects of temperature and pressure on activity coefficients in seawater, as well as earlier studies on the activity coefficients of carbonate (34) and calcium ions (see 14) in seawater, allow a calculation of the expected change with temperature in the apparent solubilities of calcite and aragonite to be made. The results of such a calculation are summarized in Table I. They indicate that Berner's results are in much better agreement with theoretical values than are Ingle's. Also, the value calculated from Berner's aragonite measurements for the carbonate ion activity coefficient (0.031) is in better agreement with that determined by Pytkowicz (34) from apparent and thermodynamic values for the dissociation constants of carbonic acid in seawater (0.031) than is the value calculated from Ingle et al.'s (25) value for the apparent solubility of calcite in seawater (0.037). These calculations reinforce the idea that the phase that comes to equilibrium with calcite in seawater is not

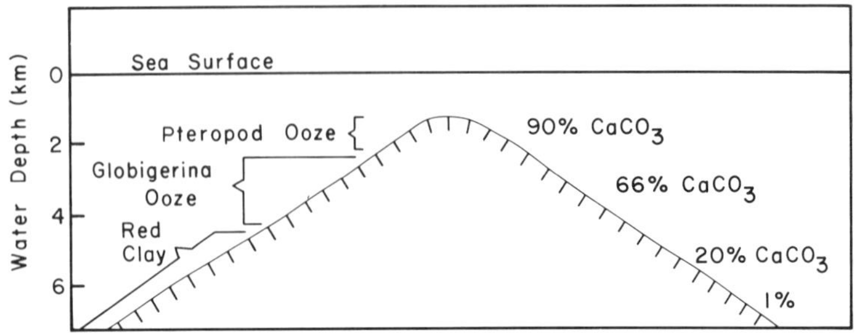

Figure 3. General distribution of calcium carbonate with increasing water depth in deep ocean basins (after Ref. 9)

Table I

The Change in Apparent Calcium Carbonate Solubility Constants with Temperature (After Morse et al. (35))

	$\frac{\partial K'_a}{\partial t}$ $mol^2\ kg^{-2}\ °C^{-1}$ (x 10^7)	$\frac{\partial K'_c}{\partial t}$ $mol^2\ kg^{-2}\ °C^{-1}$ (x 10^7)
Calculated	-0.060	-0.053
Berner (26)	-0.079	-0.064
Ingle (32)		-0.011

pure calcite. As a result at this time it is probably prudent to use both sets of directly measured changes in apparent solubility constants with temperature, i.e. the data of Ingle *et al.* (25) for calcite and Berner (26) for aragonite.

The equation given by Ingle (32) for the change in the apparent solubility constant of calcite in seawater with temperature and salinity is:

$$K'_{c_{t_{is}}} = [-34.452 - 39.866S^{1/3} + 110.21\log(S) - (7.5752 \times 10^{-6})(t_{is} + 273)^2] \times 10^{-7} \quad (23)$$

where K_c' is in mole^2 kg^{-2} of seawater.

If the apparent dissociation constants for carbonic acid in seawater of Mehrbach *et al.* (17) are used and the apparent solubility product for aragonite is expressed in mole^2 kg^{-2} (for consistancy with Ingle), the resulting change in the apparent solubility product for aragonite in seawater with temperature is:

$$K'_{a_{t_{is}}} = [8.15 + 0.079\ (25 - t_{is})] \times 10^{-7}. \quad (24)$$

The equation developed by Ingle (32) for determining the effect of pressure on the apparent solubility product of calcite in seawater is:

$$R(t_{is} + 273)\ \ln(K'_{is}/K'_{t_{is}}) = -\Delta\bar{V}_c(P-1) + (1/2)\Delta\bar{K}(P-1)^2 \quad (25)$$

where $\Delta\bar{V}$ is the change in the partial molal volume for dissolution of calcite in seawater and $\Delta\bar{K}$ is the compressibility, $-(\partial\Delta\bar{V}/\partial P)_{t_{is}}$.

Both $\Delta\bar{V}$ and $\Delta\bar{K}$ change with temperature and they must be calculated for the *in situ* temperature before use of Equation 25. The appropriate temperature corrections, based on the data of Ingle (32), are:

$$\Delta\bar{K}_{t_{is}} = [2.529 + 0.369\ (25 - t_{is})] \times 10^{-3} \quad (26)$$

$$\Delta\bar{V}_{c_{t_{is}}} = -35.5 - 0.53\ (25 - t_{is})\ . \quad (27)$$

Hawley and Pytkowicz (36) determined the partial molal volume change for aragonite in seawater at 22 and 2°C. At 22°C

$\Delta\overline{V}_a$ was -28.7 ± 0.2 cm^3/mole at 500 atm and -27.8 ± 0.1 cm^3/mole at 1000 atm. At 2°C the determination of $\Delta\overline{V}_a$ was not as precise and yielded an average value of -33.1 ± 2 cm^3/mole. Ingle (32) determined $\Delta\overline{V}_a$ at 2°C with a result of -31.8 cm^3/mole. When the same equations are used for calculating the $\Delta\overline{V}$ values of calcite and aragonite at 2°C (neglecting $\Delta\overline{K}$), Ingle's (32) value for calcite is -42.3 cm^3/mole. Thus, the difference in the $\Delta\overline{V}$ values between aragonite and calcite determined by Ingle (32) is 10.5 cm^3/mole. The dissolved phase volumes are the same for calcite and aragonite. The difference in $\Delta\overline{V}$ values should, therefore, be close to the difference in the solid phase volumes, which is 2.7 cm^3/mole. Consequently, the difference in $\Delta\overline{V}$ values for calcite and aragonite, determined by Ingle (32) is approximately 4 times larger than it theoretically should be.

We have chosen to calculate the effect of pressure on the apparent solubility of aragonite using the $\Delta\overline{V}_a$ values of Hawley and Pytkowicz (36), and equations 28 and 29

$$\Delta\overline{V}_{a_{t_{is}}} = -27.0 - 0.265\ (25 - t_{is}) \tag{28}$$

$$\ln(K'_{is}/K'_{t_{is}}) = \frac{-\ \Delta\overline{V}(P-1)}{R(t_{is} + 273)}\ . \tag{29}$$

From the above considerations all calculations necessary to determine the *in situ* saturation state of seawater can be carried out. A sample calculation using the methods described is presented in Table II.

Estimation of Uncertainty in Ω. The determination of the total ion molal concentration of calcium from salinity measurement is relatively precise with a probable error of less than 0.3% under open ocean conditions. Dickson and Riley (37) have recently discussed the effect of analytical errors on the evaluation of the components of the aquatic carbon-dioxide system for seawater at 25°C and 1 atmosphere total pressure. Their conclusions indicate that if alkalinity and total carbon dioxide are the measured parameters a probable combined uncertainty in the total carbonate ion molal concentration from 3 to 6 percent results, depending on P_{CO_2}. If pH and alkalinity are the measured parameters the uncertainty is approximately 4 percent. In addition to the probable error introduced by analytical precision, the absolute accuracy of the measurements introduces an error which is difficult to evaluate. The results of the GEOSECS intercalibration study (38) were indicative of this problem. A conservative guess is that accuracy introduces at least a one percent further uncertainty. It is also difficult to determine exactly what error is introduced through temperature and pressure corrections to *in situ* conditions. For the deep sea this may introduce a further uncertainty of at least

Table II

Sample Calculation of Calcium Carbonate Saturation State in Seawater (After Morse et al. (35))

Step 1. Data: The pH at 25°C is 7.85. The titration alkalinity at 25°C is 2.38 meq/kg. The sample was collected at 3400 m water depth where the in situ temperature was 2.0°C. The salinity is 34.6°/oo.

Step 2. Pressure = (3400 m)/(10 m/atm) = 341 atm.

Step 3. Using Equation 13 the pH at 2°C is calculated to be 8.09.

Step 4. An in situ pH = 7.95 is calculated by use of Equation 14.

Step 5. The molality of boric acid calculated by Equation 15 is 4.35×10^{-4}.

Step 6. The value of $K'_{B_{is}}$ calculated from Equations 16, 17, and 18 is $10^{-8.73}$.

Step 7. The in situ carbonate alkalinity determined by use of Equation 19 is 2.32 meq/kg.

Step 8. The in situ value of the second apparent dissociation constant of carbonic acid in seawater calculated by Equations 20, 21, and 22 is $10^{-9.33}$.

Step 9. The total carbonate ion molal concentration can now be calculated via Equation 11 to be 9.04×10^{-5}.

Step 10. The total calcium ion molal concentration can be calculated from Equation 4 to be 1.015×10^{-2}.

Step 11. From steps 9 and 10 the calcium carbonate total molal product is 9.18×10^{-7}.

Step 12. The value of K_c' at t_{is} is calculated using Equation 23 to be 4.69×10^{-7} and the value K_a' at t_{is} is calculated using Equation 24 to be 9.67×10^{-7}.

Step 13. The values of $\Delta\overline{V}_c$ and $\Delta\overline{K}$ are corrected to t_{is} by use of Equations 26 and 27. Using these values in Equation 25 yields $K'_{c_{is}} = 9.87 \times 10^{-7}$. $\Delta\overline{V}_a$ is calculated from Equation 28. Then by Equation 29, $K'_{a_{is}} = 15.92 \times 10^{-7}$.

Step 14. From Equation 2, $\Omega_c = 0.93$ and $\Omega_a = 0.58$.

one percent. Therefore, the total uncertainty in IMP in the deep oceans is at least 6 percent. Since the change in saturation state with depth in the deep oceans is generally about 20 percent/km this uncertainty in the IMP would result in an uncertainty of about 300 m in the depth of a given saturation state. In Figure 4 the uncertainty in the dissolution rate of biogenic deep sea calcite from the Pacific Ocean (30) which results from the uncertainty in the saturation state (Ω) is presented. The uncertainty in the IMP results in an uncertainty in the dissolution rate of approximately 40 percent.

At this time it is not possible to make an accurate estimate of the absolute uncertainty in the *in situ* values of the apparent solubility constants for calcite and aragonite. An approximate maximum uncertainty can be determined for K_c' at 5000 m water depth and 2°C by using two different data sets. The maximum value for K_c' is obtained by using K_c' value and temperature coefficient calculated by Berner (26) and the effect of pressure of Ingle (32). The value for K_c' obtained by this method is 19.8×10^{-7} mole^2 kg^{-2}. The minimum value is obtained by using the K_c' value and temperature coefficient of Ingle *et al.* (25), and the $\Delta\overline{V}_a$ at 2°C of Hawley and Pytkowicz (36) corrected for the difference in molar volumes between calcite and aragonite (-35.8 cm^3/mole). The value obtained by this method is 10.7×10^{-7} mole^2 kg^{-2}. There is approximately a factor of 2 difference in the value of K_c' obtained by these methods. Even if the uncertainty is a tenth of this value, it results in an approximate uncertainty of a factor of 2 in calculated dissolution rates, and indicates that the uncertainties in the *in situ* values for the apparent solubility constants of calcite and aragonite are much larger than in the *in situ* values for the calcium carbonate ion molal product.

The Distribution of Calcium Carbonate Saturation States and Their Relation to Sediment Marker Levels

The Geochemical Ocean Section Program (GEOSECS) has produced data from which it is possible to profile the saturation state of seawater with respect to calcite and aragonite in the Atlantic and Pacific oceans. Representative north-south calcite saturation profiles for the Western Atlantic and Central Pacific oceans are presented in Figures 5 and 6 (based on 39). It was observed that the saturation state of seawater with respect to calcite at the CCD was close to constant ($\Omega = 0.70 \pm 0.05$) except in the southern extremes (39). Broecker and Takahashi (31) have recently found that the carbonate ion concentration is close to constant at the FL, when appropriate corrections are made for pressure. The saturation state of seawater at the FL, calculated by the method presented in this paper, is 0.80 ± 0.05. Berger (40) has presented profiles for R_o, FL, CCD and CSL (calcite saturation level) in the eastern and western Atlantic ocean (see

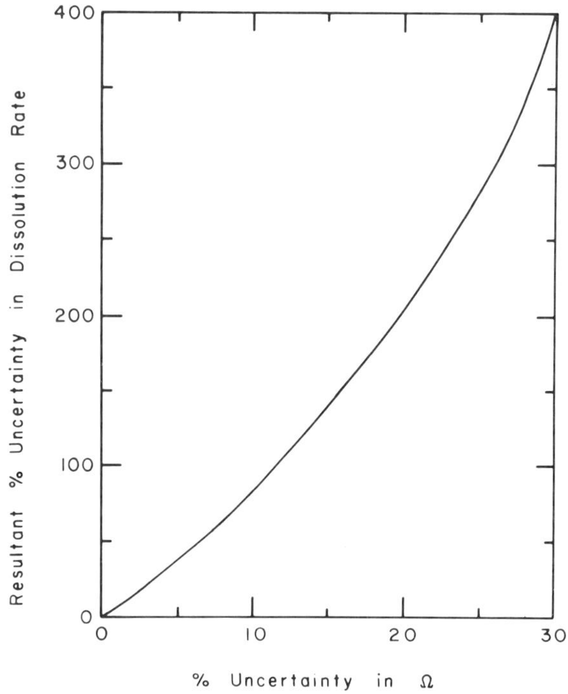

Figure 4. Relationship between the uncertainty in saturation state (Ω) and the resultant uncertainty in dissolution rate

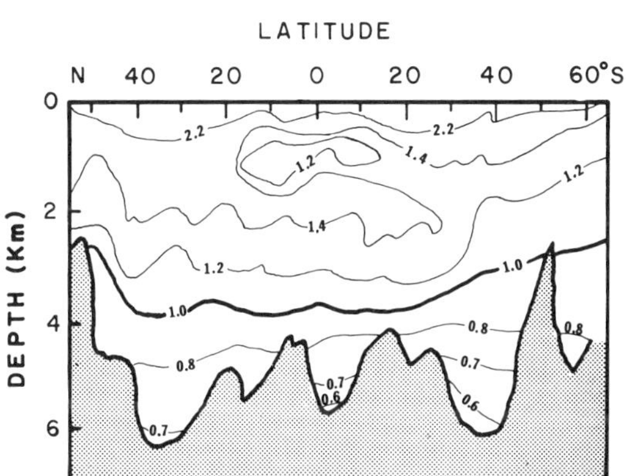

Figure 5. Distribution of calcite saturation states in the Western Atlantic Ocean (after Ref. 39)

Figure 7). His results indicate that R_0 and CSL are close to coincident with the probable uncertainty of their determination. A detailed profile showing the relations among the sediment marker levels and the saturation state of seawater with respect to calcite and aragonite in the Northwest Atlantic Ocean is presented in Figure 8.

An approximate relationship between the degree of undersaturation of seawater with respect to calcite and the extent of dissolution can be established by comparing the saturation state at the various sediment marker levels with estimates of the amount of dissolution required to produce these levels. In Figure 9 the "distance" from equilibrium ($1 - \Omega$) has been plotted against the estimated percent dissolution of the calcitic sediment fraction. Within the large uncertainties that exist in the amount of dissolution required to produce the FL and R_0 levels, a linear relation between the degree of undersaturation and extent of dissolution can be established. The intercept of the linear plot with the FL and R_0 levels indicates that approximately 15 percent more material has been lost than Berger's (12) minimum loss estimate of 50% and 10%, respectively.

The Dissolution Kinetics of Calcium Carbonate in Seawater

Calcium carbonate is accumulating in deep ocean sediments, in which the overlying water is undersaturated with respect to both aragonite and calcite, and sediment marker levels closely correspond to unique saturation states. This indicates that dissolution kinetics play an important role in determining the relation between seawater chemistry and calcium carbonate accumulation in deep ocean basins. It is, therefore, necessary to have knowledge of the dissolution kinetics of calcium carbonate in seawater if the accumulation of calcium carbonate is to be understood.

Two experimental approaches have been used to determine calcium carbonate dissolution kinetics in seawater. The first is suspension of different carbonates in the ocean at various depths. After a given period of time, the samples are recovered and the rate of dissolution determined by weight loss. The second experimental approach is the determination of dissolution kinetics in the laboratory at different undersaturations. A detailed discussion of the findings of these studies is presented in this section.

Open Ocean Dissolution Experiments. The first direct studies of calcium carbonate dissolution in deep seawater were made by Peterson (41) and Berger (42). Peterson suspended spheres of Iceland spar calcite, held in pronged plastic containers, at various depths in the Central Pacific Ocean for four months. The amount of dissolution was determined by weight loss, which was small relative to the total weight of the spheres. On the same mooring Berger suspended sample chambers, which consisted of

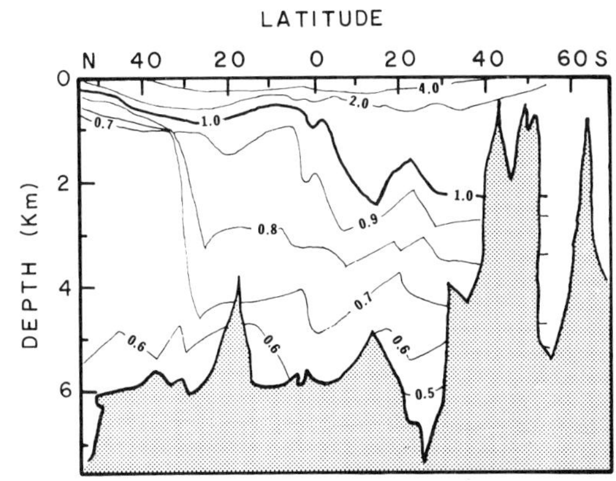

Figure 6. Distribution of calcite saturation states in the Central Pacific Ocean (after Ref. 39)

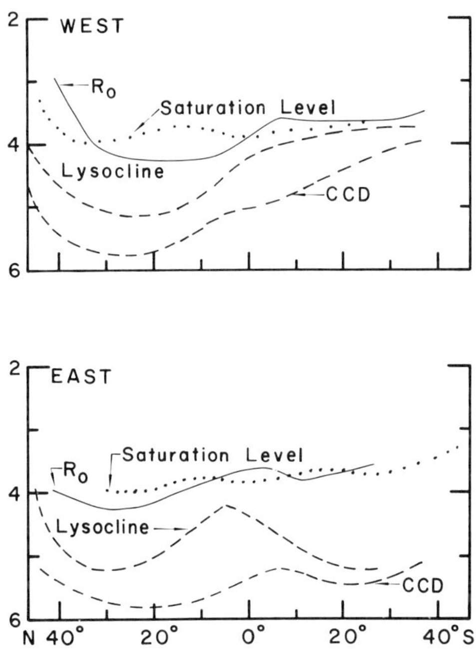

Figure 7. The depth distribution of the R_0 and calcite saturation levels, the foraminiferal lysocline and the calcium carbonate compensation depth in the Western and Eastern Atlantic Ocean (after Ref. 40)

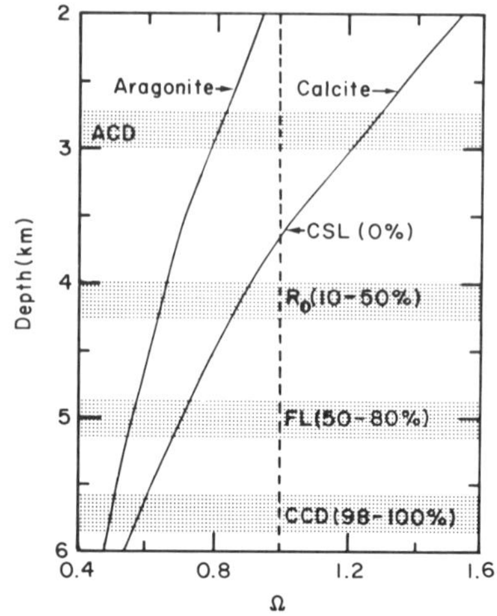

Figure 8. A detailed profile of calcite and aragonite saturation states and sediment marker depth in the Northwestern Atlantic Ocean. (Percentages are estimates of the amount of calcite dissolution which must occur to produce a given marker level.)

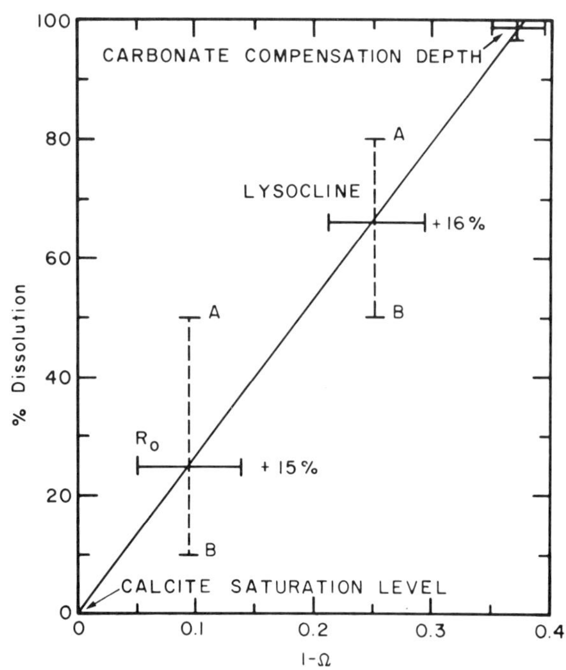

Figure 9. Relationship between the distance from equilibrium (1 − Ω) of seawater with respect to calcite and the amount of dissolution which is estimated to have occurred. (A) Adelseck's (11) experimentally determined amount of dissolution; (B) Berger's (12) minimum loss estimate for the amount of dissolution at the R_0 level and foraminiferal lysocline.

cylinders whose ends were covered with fine netting, containing an assortment of foraminifera and pteropods. Both freshly collected and H_2O_2- treated speciments were used. Berger determined the rate of dissolution for calcite and aragonite by the same method used by Peterson. The relative susceptibility of different species of foraminifera to dissolution was also determined.

Both Peterson (41) and Berger (42) found that dissolution started at approximately 0.5 km water depth and the rate of dissolution increased slowly with increasing water depth until a depth of approximately 3.8 km was reached. Below this depth the rate of dissolution rapidly increased with increasing water depth. The change in the saturation state of seawater, with respect to calcite, in the deep water of this region is close to linear with depth (43). Consequently, the results of these experiments indicated that the rate of dissolution was not simply related to saturation state. Edmond (44) proposed that the rapid increase in dissolution rate could be attributed to a change in water velocity. Morse and Berner (45) pointed out that this could be true only if the rate of dissolution was transport controlled. Their calculations indicated that the rate of dissolution measured by Peterson (41) was over 20 times too slow for diffusion controlled dissolution, this being the slowest transport process. It was, therefore, suggested that the rate of dissolution was surface controlled and the change in the relationship between increasing water depth and dissolution rate could be attributed to a change in surface-controlled reaction mechanisms.

Berger (10) found that the different species of foraminifera used in the water column dissolution experiment dissolved at different rates. Based on these findings, he constructed a solubility index for the relative resistance of different species of foraminifera to dissolution. From this index the different species of foraminifera were placed in the general categories of easily soluble and resistant. When the ratio of soluble-to-resistant species, occurring in surface sediments, is plotted against water depth, a distinct maximum in the rate of change in the ratio , with increasing water depth, is found. The depth at which this occurs is called the foraminiferal lysocline. In the area of the Pacific Ocean where the water column dissolution experiment was performed, the depth of the rapid increase in dissolution rate and the foraminiferal lysocline are close to coincident.

Milliman (46) has recently conducted an experiment in the Sargasso sea (NW Atlantic Ocean) similar in design to the Peterson (41) and Berger (42) experiments. In his experiment, preweighed samples of biogenic calcite, magnesium calcite and aragonite were suspended at different depths in sacks made of nylon panty hose. The results which he obtained for biogenic calcite were similar to those of Peterson and Berger, in that a gradual increase in dissolution rate with increasing water depth was found above a critical depth. Below this depth the rate of dissolution increased rapidly with increasing water depth. The critical depth was found to closely correspond with the depth of the FL in the area of the experiments. The same general pattern of dissolution rate versus water depth was found for the biogenic aragonite and magnesian calcite. An interesting finding for the aragonite was that its critical depth was approximately 700 m below the ACD and close to the calcite R_0 level in the area of the experiments. Milliman's results for calcite and aragonite are summarized in Figure 10.

Honjo and Erez (47) have attempted to overcome the problem of nonuniform water circulation through sample chambers by placing samples in chambers through which water is pumped at a constant rate. Their results from 5500 m in the Sargasso sea yield dissolution rates for biogenic calcite 3 to 4 times faster than observed by Milliman (46) and approximately 4 times greater for biogenic aragonite. However, even their rates may represent minimum values as the samples tended to "pond" in the bottom of the sample chambers and were thus not uniformly exposed to seawater of constant undersaturation (Honjo, S., Woods Hole Oceanogr. Inst., personal communication, 1978).

The dissolution pattern with increasing water depth found in all of the water column experiments are in general agreement, indicating that the dissolution kinetics of both synthetic and biogenic calcite and aragonite are not simply proportional to the degree of undersaturation. The results also indicate that the nature of the assemblage of biogenic tests should have a significant effect on the overall dissolution rate of a given calcium carbonate phase.

Laboratory Dissolution Experiments. The development of a high precision pH-stat method (48) for the determination of calcium carbonate reaction kinetics has made accurate laboratory measurement of near-equilibrium calcite and aragonite dissolution rates possible. The basic concept in the pH-stat technique is the maintenance of a constant degree of disequilibrium. The carbonate molality is kept constant by bubbling a gas of constant P_{CO_2} through the solution and maintaining a constant pH with a pH-stat. The rate at which acid must be added to transform the carbonate produced by dissolution to carbon dioxide and water is proportional to the rate of calicum carbonate dissolution. The concentration of calcium in seawater is approximately 1000 times that of carbonate. If only a small amount of calcium carbonate is allowed to dissolve in a relatively large volume of seawater, the change in calcium concentration is negligible. It is presently possible to make accurate dissolution rate determinations at saturation states that are constant to within better than 2%.

Using the pH-stat technique, Morse and Berner (45) found that the dissolution rate of synthetic and pelagic biogenic calcite did not increase linearly with increasing undersaturation. Beyond a critical undersaturation a very rapid increase in the rate of dissolution with increasing undersaturation was found. This change was attributed to a change in surface-controlled reaction mechanisms. By comparing the undersaturations used in the laboratory with those found in the water column at the site of the Peterson (41) and Berger (42) experiments, it was possible to construct a curve which correctly predicted the observed relation between water depth and dissolution rate. This is shown in Figure 11, along with results of the Peterson and Berger experiments.

A major problem encountered in comparing the rates of dissolution measured in the laboratory with rates measured in the water column was that the rates measured in the laboratory were 10 to 100 times faster. It has subsequently been shown (30) that the difference in rates for the biogenic calcite can be largely attributed to the fact that Berger (42) used only large foraminifera, while Morse and Berner (45) used a total carbonate sediment sample. The reason for the difference in rates for the Iceland spar and synthetic calcites is less certain. One possibility is that surfaces exposed in seawater experience biofouling. Peterson (41) observed macro-biofouling (e.g. barnacle attachment) on some of his spheres. Unfortunately, no scanning electron microscope examinations of the sphere surfaces were carried out. It is very likely that micro-biofouling occurred which may have substantially reduced the surface area available for dissolution. Also, Peterson (41) pretreated the spheres with HCl which may have reduced the surface concentration of active sites.

Berner and Morse (50) determined the dissolution rate of synthetic calcite in artificial seawater and $NaCl-CaCl_2$ solutions over a wide range of saturation states and different P_{CO_2} values. Their interpretation of the sudden increase in dissolution rate with increasing undersaturation in the near-equilibrium region of concern in the oceans was that it is necessary to reach a critical undersaturation for the retreat of monomolecular steps on the calcite surface. The degree of undersaturation necessary for the supposed change in reaction mechanism was found to depend on the concentration of orthophosphate in the solution. Recent studies of non-biogenic calcite (51) and biogenic magnesian calcites (52) have confirmed the importance of phosphate as an inhibitor of calcite dissolution. The effect of phosphate was explained by Berner and Morse (50) as a change in critical undersaturation necessary to enable monomolecular steps to penetrate, via curved embayments, between strongly adsorbed phosphate ions. The measured adsorbed phosphate concentrations and those predicted by theory agreed well. (A general discussion of calcite dissolution kinetics is presented elsewhere in this volume (53)).

A classical approach to reaction kinetics is to determine the empirical order of a reaction and its rate constant. The equation for this is:

$$R = k(1-\Omega)^n \qquad (30)$$

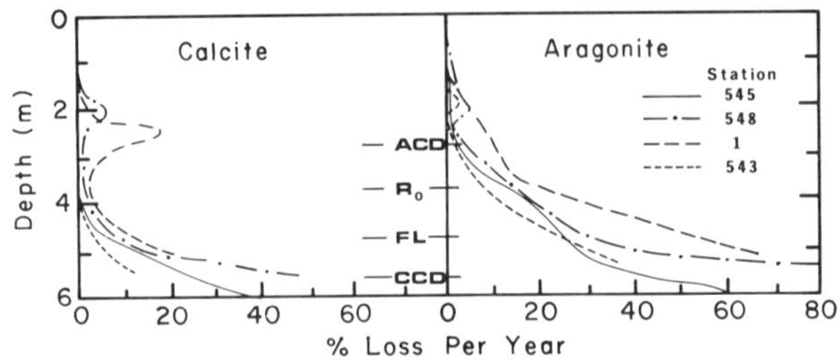

Figure 10. Rate of calcite and aragonite dissolution as a function of depth as determined by Milliman (46) in water column experiments in the Sargasso sea

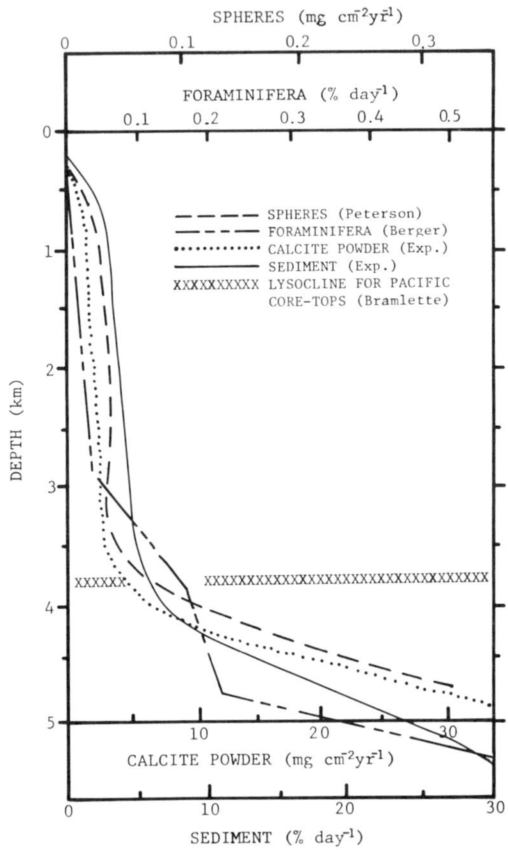

Figure 11. Plot of the Peterson (41) and Berger (42) results for their water-column dissolution experiments in the Central Pacific Ocean, and the Morse and Berner (45) laboratory experiments as a function of equivalent depth. The depth of the lysocline was calculated from the data of Bramlette (49) (after Ref. 45).

where R is the dissolution rate, k the rate constant, and n the empirical reaction order. In logarithmic form this equation is:

$$\log R = \log k + n \log(1-\Omega) \qquad (31)$$

The value of the rate constant and empirical reaction order can be determined by the intercept and slope, respectively, of a plot of log R versus log(1-Ω), if the plot is linear.

The data for synthetic calcite dissolution in seawater containing different phosphate concentrations (50) has been plotted in this manner in Figure 12. For a given phosphate concentration, the plots are linear over the range of undersaturations studied. This indicates that it is unlikely that a change in reaction mechanism occurs in the range of undersaturations of concern in the oceans. The concept of a "critical" undersaturation at which a near-equilibrium change in reaction mechanism occurs does not, therefore, appear to be strictly valid. The concept, which originated in the open ocean experiments (41, 42) and was furthered by early laboratory experiments (45, 50), is most probably an artifact of the manner in which the data were plotted. However, the concept is still quite useful (see below).

The linear plots in Figure 12 steepen with increasing phosphate concentration. In Figure 13 the log of the rate constant and the empirical reaction order have been plotted against the dissolved phosphate concentration. Both the empirical reaction order and the log of the rate constant change in a linear manner with phosphate concentration.

de Kanel and Morse (54) made a detailed study of the uptake kinetics of orthophosphate on the surfaces of calcite and aragonite in seawater. Their results indicate that the data are best fit to the Elovich equation for chemisorption (see 55, 56). This indicates that the uptake kinetics can be explained by an exponential decrease during reaction in available surface reaction sites and(or) a linear increase in the activation energy associated with the chemisorption process. Their conclusions strongly imply that carbonate surfaces are energetically heterogeneous.

The Berner and Morse (50) model for the influence of phosphate ions on dissolution was implicitly for a surface with a homogeneous distribution of adsorption sites or kinks. If energetic heterogeneity is accompanied by a spacially heterogeneous distribution of adsorbed ions on the surface of calcite, then the findings of de Kanel and Morse (54) suggest that at any given phosphate concentration there will be a range of "critical" undersaturations, each corresponding to a different portion of the calcite surface. Consequently, it is expected that as the surface phosphate concentration increases and surface sites are filled up, there would be both a decrease in the total dissolution rate and an increase in the sensitivity of the dissolution rate to the degree of undersaturation. Ultimately, a limit would be

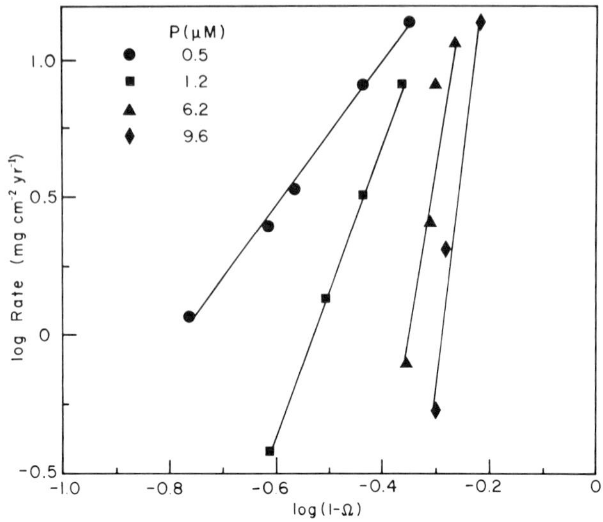

Figure 12. Log of the rate of dissolution of synthetic calcite in seawater containing different phosphate concentrations vs. the log of 1 — Ω. Based on the data of Berner and Morse (50). The two lowest dissolution rates measured have been excluded as they are highly anomalous.

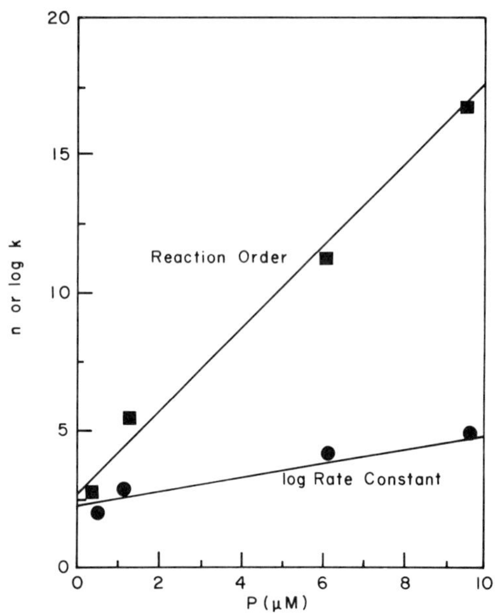

Figure 13. Empirical reaction order and log of the rate constant for synthetic calcite dissolution in seawater as a function of phosphate concentration based on the data presented in Figure 12

approached whereby all portions of the surface are equally covered by adsorbed phosphate ions and a true discontinuity in rate as predicted by the Berner and Morse (50) model, attained. Because of the present lack of knowledge of the distribution of surface sites and their energies, and of the dependence of and dissolution rate upon site energy and spacing, it is not possible to construct a quantitative model for the relationship between phosphate adsorption and dissolution rate.

Morse (30) carried out an examination of the near-equilibrium dissolution kinetics of calcium carbonate-rich deep sea sediments. His results are summarized in Figure 14. The sediment samples from different ocean basins have distinctly different reaction orders and empirical rate constants. The dissolution rate equations for the different sediment samples are:

$$R(\%/\text{day})_{\text{Indian}} = 10^{4.3}(1-\Omega)^{5.2} \quad (32)$$

$$R(\%/\text{day})_{\text{Pacific}} = 10^{2.7}(1-\Omega)^{3.0} \quad (33)$$

$$R(\%/\text{day})_{\text{Atlantic}} = 10^{3.1}(1-\Omega)^{4.5} \quad (34)$$

It was suggested that different particle size distributions combined to give different functionalities to the total rate or that different surface histories existed for the samples from different locations. Although the samples were taken from different ocean basins, it was Morse's (30) opinion that they should not be taken as representative of these basins, as variability within a given basin could probably be at least as large as that found in samples from different ocean basins.

Morse (30) found that the rate of dissolution per gram of calcium carbonate decreased with extent of dissolution in an irregular manner (see Figure 15). It was also found that the dissolution rates in deep sea sediment pore waters and seawater were the same at equivalent phosphate concentrations. This indicated that no important unknown dissolution inhibitor was present in the pore waters.

The rate of dissolution of synthetic and deep sea biogenic (pteropods) aragonite in seawater have also been determined in the laboratory by the pH-stat method (57). The results of the experiments to determine the change in the rate of dissolution as a function of undersaturation are presented in Figure 16. The pteropods were found to dissolve at only about 3% the rate, per unit surface area, of the synthetic aragonite. The results also indicate a change in the empirical reaction order from 2.92 to 7.37 at $\Omega = 0.44$. The rate equations for pteropod dissolution are:

$$R(\%/\text{day}) = 10^{2.04}(1-\Omega)^{2.92} \quad (35)$$

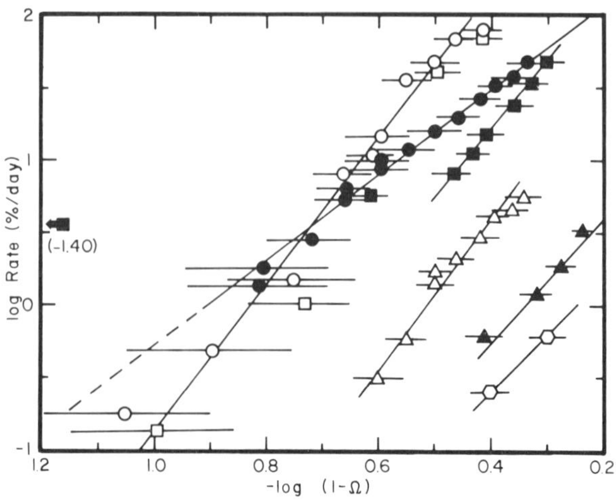

Figure 14. Log of the rate of dissolution in percent per day vs. the log of (1 — Ω). (○)Whole Indian Ocean sediment dissolved in Atlantic deep-seawater, (□) whole Indian Ocean sediment dissolved in deep-sea sediment pore water; (△) the greater than 62 μm size fraction of the Indian Ocean sediment dissolved in Atlantic deep-seawater; (●) whole Pacific Ocean sediment dissolved in Atlantic deep-seawater; (▲) the 125 —500 μm size fraction of Pacific Ocean sediment dissolved in Atlantic deep-seawater; (■) whole Atlantic Ocean sediment dissolved in Long Island Sound seawater (46); (⬡) 150 to 500 μm foraminifera dissolved in the Pacific Ocean water column. Error bars are based on an uncertainty in pH of ±0.01 (after Ref. 30).

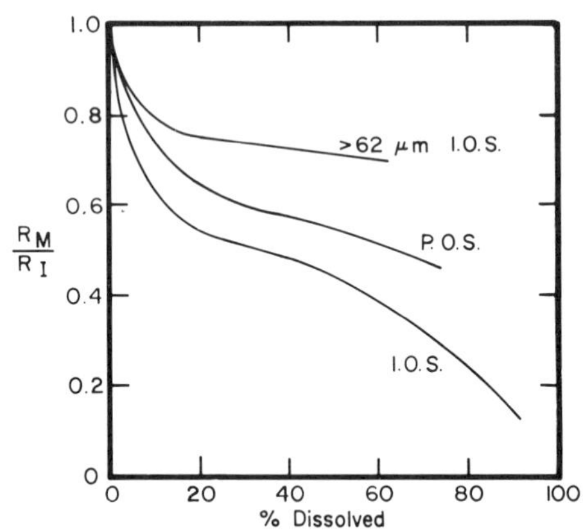

Figure 15. The measured rate of dissolution (R_M) per gram of calcium carbonate, divided by the initial rate of dissolution (R_I) vs. the percent of the calcium carbonate which has dissolved. Runs were carried at Ω = 0.40. I.O.S. = Indian Ocean sediment, P.O.S. = Pacific Ocean sediment (after Ref. 30).

$$R(\%/day) = 10^{3.14}(1-\Omega)^{7.37} . \quad (36)$$

Equation 35 applies over the undersaturation range Ω = 1 to 0.44. At greater undersaturations Equation 36 applies.

In Figure 17 the log the the dissolution rate of synthetic aragonite, pteropods, total calcitic Pacific Ocean deep sea sediment, and the 125 to 500 μm size fraction of the same sediment have been plotted against the seawater total carbonate ion concentration. It is interesting that at approximately 50% saturation, with respect to calcite (~22μM $CO_3^{=}$), the pteropod dissolution rate is not much greater than the total calcitic sediment dissolution rate. This is probably the result of the pteropods being much larger than the average biogenic calcite component and having a less reactive surface.

Morse et al. (57) found two major differences between the behavior of aragonite and calcite dissolution kinetics in seawater. The first was that the rate of pteropod dissolution per unit mass increased with increasing extent of dissolution (see Figure 18). The second was that phosphate, instead of inhibiting dissolution, caused an initial increase in dissolution rate and then had almost no effect (see Figure 19). The size of the short-term catalytic effect increased with increasing phosphate concentration and decreasing undersaturation (see Figure 20).

Comparison of Laboratory and Water Column Dissolution Rates. There are several problems in comparing calcium carbonate dissolution rates measured in the water column and laboratory experiments. These problems can be briefly summarized as: 1) samples of different origin and surface history have been studied, 2) different size fractions have generally been used, 3) in order to compare results it is necessary to be able to accurately calculate Ω of the water at the locations and depths where the water column experiments were conducted, and 4) dissolution rates for the water column experiments have generally been incorrectly reported as average dissolution rates (percent weight loss of original sample weight divided by time of exposure).

Morse et al. (57) suggested two possible approaches to these problems. One was to make an approximate calculation of the Ω in the water column experiment chambers. They used the following assumptions: 1) the dissolution kinetics, for a given sample type and size fraction, are the same in the laboratory and water column, 2) the Ω in the sample chambers was close to constant during the experiments and 3) that there was little change in the rate of dissolution per unit mass with extent of dissolution. The approximate saturation state in the sample chambers can then be compared with the saturation state of water at the sample chamber site calculated by the methods presented in this paper. In order to calculate the saturation state in the chambers, it is necessary to integrate Equation 30 and solve for Ω. The resulting equation is:

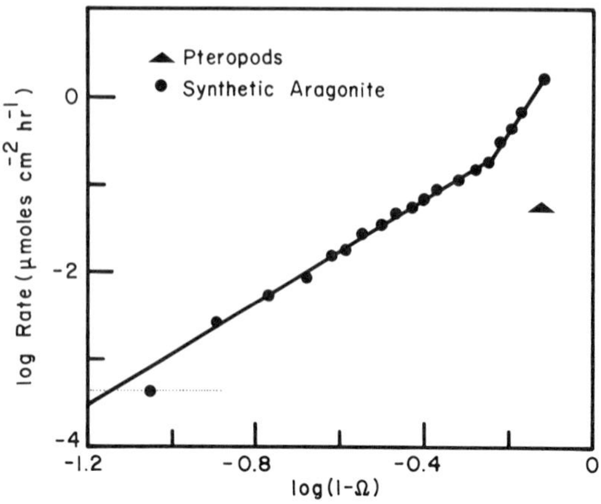

Figure 16. Log of the dissolution rate vs. the log of (1 — Ω) for synthetic aragonite and pteropods (after Ref. 57)

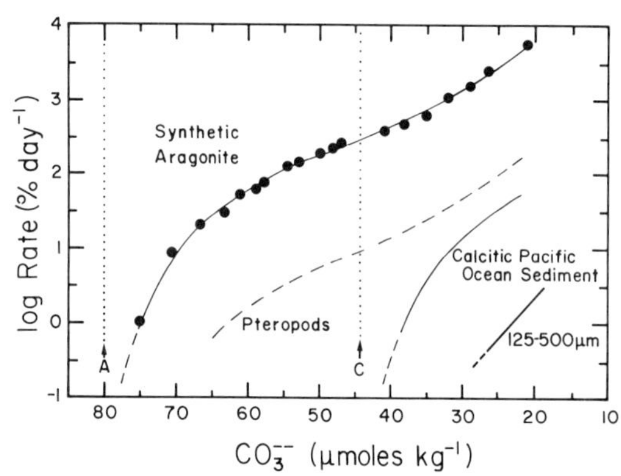

Figure 17. Log of the dissolution rate vs. total carbonate ion concentration for synthetic aragonite, pteropods, calcitic Pacific Ocean sediment, and foraminifera in the 125–500 μm size fraction. (A) indicates the aragonite equilibrium total carbonate ion concentration at 25°C, 1 atm (26). (C) indicates the calcite equilibrium total carbonate ion concentration at 25°C, 1 atm (25).

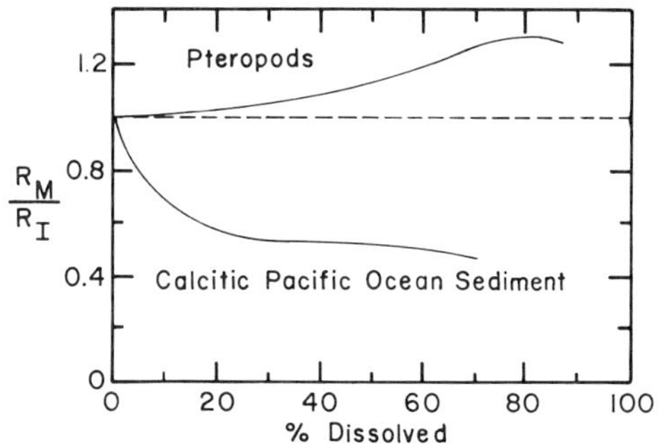

Figure 18. Ratio of measured rates of dissolution per gram to the initial dissolution for pteropods and calcite Pacific Ocean sediment (57)

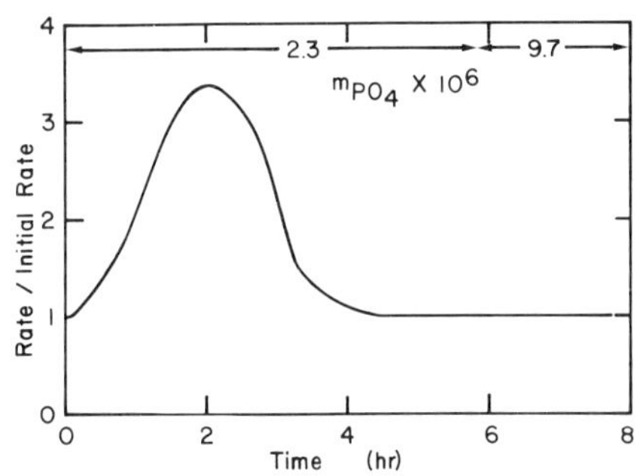

Figure 19. Change in the rate of synthetic aragonite dissolution, relative to dissolution in very low phosphate seawater, as a function of time (57)

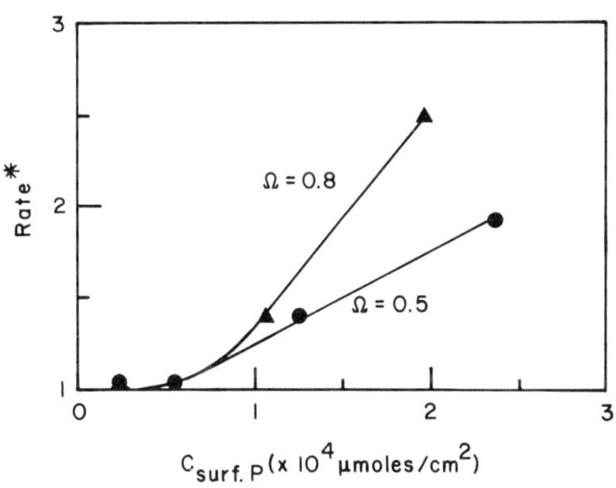

Figure 20. Rate of dissolution relative to the rate of dissolution in very low phosphate seawater (Rate) after a half hour of reaction vs. surface phosphate concentration at Ω = 0.8 and Ω = 0.5 (57)*

$$\Omega = 1 - \left[\frac{-100}{kt} \ln\left(1 - \frac{f}{100}\right)\right]^{\frac{1}{n}} . \qquad (37)$$

The second approach was to calculate the rate constant (k) for the water column experiments. The major assumptions for this approach were: 1) that Ω in the sample chamber was the same as that of the surrounding water, 2) the empirical reaction order (n) is the same in the water column and laboratory and 3) that there was little change in the rate of dissolution per unit mass with extent of dissolution. In order to calculate k, Equation 37 must be solved for k

$$k = \frac{100}{t} (1-\Omega)^{-n} \ln\left(1 - \frac{f}{100}\right) \qquad (38)$$

where f is the weight percent dissolved (for derivation see 57).

In comparing their laboratory dissolution rates for pteropods with water column results, Morse *et al.* (57) found that the major problem was the differences in the samples. Berger (43) used pteropods both freshly collected from net tows and cleaned with H_2O_2 to remove organic matter. Neither sample was truely representative of pteropods arriving at the sediment-water interface. Berger is of the opinion that the H_2O_2 treated samples are more representative of pteropods arriving at the sediment-water interface based on observations of pteropod tests from deep water net tows and box cores. Honjo and Erez (47) used only pteropods greater than 831 µm in size. This is a much larger size fraction than used in the laboratory experiments (greater than 125 µm). Based on experience with foraminifera (Morse, 30) it is reasonable to expect that the results of Honjo and Erez (47) should give significantly slower dissolution rates than found in the laboratory. Milliman (46) used ooids (roughly spherical fine-grained aragonite aggregates) making any intercomparison with other measurements virtually meaningless.

Another problem in comparing Berger's (42) results to those of other experiments is that all H_2O_2 treated pteropod samples suspended below the aragonite saturation depth completely dissolved. Morse *et al.* (57) found that if the slowest possible dissolution rate for Berger's sample from immediately below the aragonite saturation level was used to calculate the saturation state, in the sample chamber, that the calculated value of Ω in the chamber was only 0.08 higher than that of the surrounding water. This is equivalent to a difference of a factor of 2 in k. Since the assumption maximizes the difference, this result was considered to be very good agreement between their laboratory measurements and Berger's water column measurements.

Morse *et al.* (57) found that the calculated Ω in the sample chambers used by Honjo and Erez (47) was always higher than that of the surrounding water with the differences being 0.22 at

3600 m, and 0.25 at 4800 and 5500 m or factors of approximately 100, 12 and 9 in k, respectively. The differences were primarily attributed to the fact that Honjo and Erez (47) used a larger size fraction of ptetopods than Morse et al., (57). When the average amount of aragonite dissolution at 5500 m water depth in each of Milliman's (46) four experiments was calculated, the Ω in the chambers was higher than that of the surrounding water by 0.36, 0.36, 0.35, and 0.33 or factors of approximately 55, 55, 37 and 28 in k, respectively. Again the consistant difference was primarily attributed (57) to the different material used in the water column and laboratory experiments.

We have taken the first approach in comparing foraminifera dissolution rates. Because sample differences are less severe, this should provide a better comparison of laboratory and water column results. The Ω in Berger's (42) deepest chambers is calculated to be high by only 0.01 and 0.09, at 5000 m and 6000 m depth, respectively. The Ω in the Honjo and Erez (47) sample chamber is only 0.02 lower than the surrounding water. The Ω in Milliman's (46) sample chambers gives a larger disagreement with the Ω of the water column, with average calculated Ω being 0.18 higher than the surrounding water. This may be due to restricted flow of water through his chambers.

Conclusions

The accumulation of calcium carbonate in deep ocean sediments is a complex process. It is primarily governed by the interplay between biological production of calcium carbonate in the near-surface ocean and the chemistry of deep ocean waters. After over 100 years of study, the major problem of determining the saturation state of deep ocean water remains largely unresolved. It is currently possible, using recent laboratory measurements, to arrive at saturation states that differ by as much as a factor of 2. Both laboratory and water column experiments indicate that calcium carbonate dissolution kinetics are not simply related to saturation state. It is our opinion that the saturation state problem must be resolved and considerably more detail added to our present knowledge of calcium carbonate dissolution kinetics and accumulation patterns before attempts to model the accumulation of calcium carbonate in deep ocean sediments can be truly successful.

Acknowledgements

The authors wish to thank Drs. Frank J. Millero and John R. Southam, for their extremely helpful discussions and advice, and Catherine Tuttle for her patience and perseverance in preparing the figures and typing this paper. Support was provided by National Science Foundation grants (Marine Chemistry Program) OCE77-08134, J.W. Morse, P.I. and OCE77-03473, R.A. Berner, P.I.

Abstract

The balance between calcium carbonate infall and dissolution in deep ocean basins is the main control of both atmospheric CO_2 and the pH of seawater. Calcium carbonate is being deposited at depths 1 to 2 km below the depth at which seawater becomes undersaturated with respect to calcite. This indicates that the primary control of calcium carbonate deposition in the deep oceans is the kinetics of dissolution. The dissolution kinetics of pelagic biogenic calcite and aragonite in seawater are complex. The empirical reaction order and rate constant are dependent on the source of the calcium carbonate. The rate of dissolution changes in a nonlinear manner with extent of dissolution. Only 3 to 5% of the total surface area is available for reaction. Phosphate is a strong inhibitor of calcite dissolution, but not of aragonite dissolution. Modelling calcium carbonate deposition is difficult due to the nonlinear character of the depositional processes.

Literature Cited

1. Sverdrup, H.U., Johnson, M.W., and Fleming, R.H. "The Oceans," 1087 p., Prentice Hall, Englewood Cliffs, New Jersey, 1942.
2. Pytkowicz, R.M. The carbon dioxide-carbonate system at high pressure in the oceans, p. 83-135, in Barnes, H., ed. "Oceanogr. Mar. Biol. Ann. Rev. 6," George Allen and Unwin., London (1968).
3. Broecker, W.S. "Chemical Oceanography," 214 p. Harcourt Brace Jovanovich, New York, 1974.
4. Adelseck, C.G., Jr. and Berger, W.H. On the dissolution of planktonic formaminifera and associated microfossils during settling and on the sea floor, p. 70-81, in Sliter, W.V., Be, A.W.H., and Berger, W.H., eds., Cushman Found. Foram. Res. Spec. Publ. No. 13, 1975.
5. Pytkowicz, R.M. Carbonate cycle and the buffer mechanism of recent oceans, Geochim. Cosmochim. Acta 31, 63-73 (1967).
6. Woodwell, G.M., Whittaker, R.H., Reiners, W.A., Likens, G.E., Delwiche, C.C., and Botkin, D.B. The biota and the world carbon budget, Science 199, 141-145 (1978).
7. Berner, R.A. Sedimentation and dissolution of pteropods in the ocean, p. 243-260, in Andersen, N.R. and Malahoff, A., eds., "The Fate of Fossel Fuel CO_2 in the Oceans," Plenum Press, New York, 1977.
8. Berger, W.H. Deep-sea carbonate: pteropod distribution and the aragonite compensation depth, Deep Sea Res. (in press).
9. Murray, J. and Hjort, J. "The Depths of the Oceans," 831 p. Macmillan, London, 1912.
10. Berger, W.H. Planktonic foraminifera: selective solution and the lysocline, Mar. Geol. 8, 111-138 (1970).

11. Adelseck, C.G., Jr. Dissolution of deep-sea carbonate: preliminary calibration of preservational and morphological aspects, Deep Sea Res. 25, 1167-1185 (1978).
12. Berger, W.H. Sedimentation of planktonic foraminifera, Mar. Geol. 11, 325-358 (1971).
13. Edmond, J.M. and Gieskes, J.M. On the calculation of the degree of saturation of sea water with respect to calcium carbonate under in situ conditions, Geochim. Cosmochim. Acta 34, 1241-1292 (1970).
14. Millero, F.J. Seawater as a multicomponent electrolyte solution, p. 3-81, in Goldberg, E.D. ed., "The Sea, 5, Marine Chemistry," John Wiley and Sons, New York, 1974.
15. Bates, R.G. pH scales for seawater, p. 313-338, in Goldberg, E.D., ed., "The Nature of Seawater," Dahlem Konferenzen, Berlin, 1974.
16. Hansson, I. Determination of the acidity constant of boric acid in synthetic sea water media, Acta Chem. Scand. 27, 924-930 (1973).
17. Mehrbach, C., Culberson, C.H., Hawley, J.E., and Pytkowicz, R.M. Measurement of the apparent dissociation constants of carbonic acid in seawater at atmospheric pressure, Limnol. Oceanogr. 18, 897-907 (1973).
18. Harvey, H.W. "The Chemistry and Fertility of Seawaters," 240 p., Cambridge Univ. Press, Cambridge, 1963.
19. Ben Yaakov, S. A method for calculating the in situ pH of seawater, Limnol. Oceanogr. 15, 326-328 (1970).
20. Culberson, C. and Pytkowicz, R.M. Effect of pressure on carbonic acid, boric acid, and the pH in seawater, Limnol. Oceanogr. 8, 403-417 (1968).
21. Culkin, F. The major constituents of sea water, p. 121-161, in Riley, J.P. and Skirrow, G., eds., "Chemical Oceanography," Vol. 1, Academic Press, London, 1965.
22. Edmond, J.M. "The Carbonic Acid System in Seawater," 174 p., Ph.D. Thesis, Univ. of Calif., San Diego, Calif., 1970.
23. Lyman, J. "Buffer Mechanism of Seawater," 196 p., Ph.D. Thesis, Univ. of Calif., Los Angeles, Calif., 1957.
24. Skirrow, G. The dissolved gases - carbon dioxide, p. 227-322, in Riley, J.P. and Skirrow, G., eds., "Chemical Oceanography,", Vol. 1, Academic Press, London, 1965.
25. Ingle, S.E., Culberson, C.H., Hawley, J., and Pytkowicz, R.M. The Solubility of calcite in seawater at atmospheric pressure and 35‰ salinity, Mar. Chem. 1, 295-307 (1973).
26. Berner, R.A. The solubility of calcite and aragonite in seawater at one atmosphere and 34.5 parts per thousand, Amer. Jour. Sci. 276, 713-730 (1976).
27. Weyl, P.K. The carbonate saturometer, Jour. Geol. 69, 33-44 (1961).
28. Weyl, P.K. The solution behavior of carbonate minerals in seawater, p. 178-228, in Proc. Internat. Conf. Tropical Oceanogr., Miami Beach, Florida, 1965.

29. Berner, R.A. The role of magnesium in the crystal growth of calcite and aragonite from sea water, Geochim. Cosmochim. Acta 39, 489-504 (1975).
30. Morse, J.W. Dissolution kinetics of calcium carbonate in sea water: VI. The near equilibrium dissolution kinetics of calcium carbonate-rich deep sea sediments, Amer. Jour. Sci. 278, 344-353 (1978).
31. Broecker, W.S. and Takashashi, T. The relationship between lysocline depth and the in situ carbonate ion concentration, Deep Sea Res. 25, 65-95 (1978).
32. Ingle, S.E. Solubility of calcite in the ocean, Mar. Chem. 3, 301-319 (1975).
33. Millero, F.J. Effects of pressure and temperature on activity coefficients, Ch. 13, in Pytkowicz, R.M., ed. "Activity Coefficients in Electrolyte Solutions," CRC Press, W. Palm Beach, Florida (in press).
34. Pytkowicz, R.M. Activity coefficients of bicarbonates and carbonates in seawater, Limnol. Oceanogr. 20, 971-975 (1975).
35. Morse, J.W., de Kanel, J., and Craig, H.L., Jr. A literature review of the saturation state of seawater with respect to calcium carbonate and its possible significance for scale formation on OTEC heat exchangers, Ocean Engineering (in press).
36. Hawley, J.E. and Pytkowicz, R.M. Solubility of calcium carbonate in seawater at high pressures and 2°C, Geochim. Cosmochim. Acta 33, 1557-1561 (1969).
37. Dickson, A.G. and Riley, J.P. The effect of analytical error on the evaluation of the components of the aquatic carbon-dioxide system, Mar. Chem. 6, 77-85 (1978).
38. Takahashi, T., Weiss, R.F., Culberson, C.H., Edmond, J.M., Hammond, D.E., Wong, C.S., Li, Y.H., and Bainbridge, A.E. A carbonate chemistry profile at the 1969 GEOSECS intercalibration station in the Eastern Pacific Ocean, Jour. Geophys. Res. 75, 7648-7666 (1969).
39. Takahashi, T. Carbonate chemistry of seawater and the calcite compensation depth in the oceans, p. 11-26, in Sliter, W.V., Be, A.W.H., and Berger, W.H., eds., Cushman Found. Foram. Res. Spec. Publ. No. 13, 1975.
40. Berger, W.H. Carbon dioxide excursions in the deep sea record: Aspects of the problem, p. 502-542, in Andersen, N.R. and Malahoff, A., eds., "The Fate of Fossil Fuel CO_2 in the Oceans," Plenum Press, New York, 1977.
41. Peterson, M.N.A. Calcite: rates of dissolution in a vertical profile in the Central Pacific, Science 154, 1542-1544 (1966).
42. Berger, W.H. Foraminiferal ooze: solution at depths, Science 156, 383-385, (1967).
43. Berner, R.A. and Wilde, P. Dissolution kinetics of calcium carbonate in seawater: I. Saturation state parameters for kinetic calculations, Amer. Jour. Sci. 272, 826-839 (1972).

44. Edmond, J.M. An interpretation of the calcite spheres experiment [abst.], Amer. Geophys. Union 52, 256 (1971).
45. Morse, J.W. and Berner, R.A. Dissolution kinetics of calcium carbonate in sea water: II. A kinetic origin for the lysocline, Amer. Jour. Sci. 272, 840-851 (1972).
46. Milliman, J.D. Dissolution of calcium carbonate in the Sargasso sea (Northwest Atlantic), p. 641-654, in Andersen, N.R. and Malahoff, A., eds., "The Fate of Fossel Fuel CO_2 in the Oceans,", Plenum Press, New York, 1977.
47. Honjo, S. and Erez, J. Dissolution rates of calcium carbonate in the deep ocean: an in situ experiment in the North Atlantic Ocean, Earth Planet Sci. Lett. (in press).
48. Morse, J.W. Dissolution kinetics of calcium carbonate in sea water: III. Effects of natural inhibitors and the position of the chemical lysocline, Amer. Jour. Sci. 274, 638-647 (1974).
49. Bramlette, M.N. Pelagic sediments, p. 345-366, in Sears, M., ed., "Oceanography," Am. Assoc. Adv. Sci. Washington, D.C., 1961.
50. Berner, R.A. and Morse, J.W. Dissolution kinetics of calcium carbonate in seawater IV. Theory of calcite dissolution, Amer. Jour. Sci. 274, 108-134 (1974).
51. Sjoberg, E.L. Kinetics and mechanism of calcite dissolution in aqueous solutions at low temperatures, Acta Univ. Stockholmiensis, Stockholm Contrib. Geol. 32, 92 p. (1978).
52. Walter, L.M. and Hanor, J.S. Effect of orthosphosphate on the dissolution of biogenic magnesian calcites, Geochim. Cosmochim. Acta (in press).
53. Plummer. L.N. and Wigley, T.N.L. Critical review of the kinetics of calcite dissolution and precipitation, in Jenne, E.A., ed., "Chemical Modeling. Speciation, Sorption, Solubility and Kinetics in Aqueous Systems," Amer. Chem. Soc. Symp. Series, Washington, D.C. (this volume).
54. de Kanel, J. and Morse, J.W. The chemistry of orthophosphate uptake from seawater onto calcite and aragonite, Geochim. Cosmochim. Acta 42, 1335-1340 (1978).
55. Aharoni, C. and Ungarish, M. Kinetics of activated chemisorption Part 1 - The non-Elovichian part of the Isotherm, Faraday Trans. I 72, 400-408 (1976).
56. Aharoni, C. and Ungarish, M. Kinetics of activated chemisorption Part 2 - Theoretical models, Faraday Trans. I 73, 456-464 (1977).
57. Morse, J.W., de Kanel, J., and Harris, K. The dissolution kinetics of calcium carbonate in seawater VII. The dissolution kinetics of synthetic aragonite and pteropods, Amer. Jour. Sci. (in press).

RECEIVED November 16, 1978.

25

Critical Review of the Kinetics of Calcite Dissolution and Precipitation

L. N. PLUMMER and D. L. PARKHURST
U.S. Geological Survey, Reston, VA 22092
T. M. L. WIGLEY
University of East Anglia, Norwich NR47TJ, England

The rate of calcite dissolution is known to depend on the hydrodynamic conditions of the environment and on the rate of heterogeneous reaction at the mineral surface. Numerous laboratory studies demonstrate transport and surface-controlled aspects of calcite reactions in aqueous solutions, but until recently, no study has been comprehensive enough to enable comparison of kinetic results among differing hydro-chemical systems.

We have studied the dissolution kinetics of calcite in stirred CO_2 - water systems at CO_2 partial pressures between 0.0003 and 0.97 atm and between 5° and 60°C, using pH-stat and free drift methods (1). Our results suggest a mechanistic model for reactions at the calcite-aqueous solution interface that has broad implications to the controls on calcite dissolution and precipitation under diverse chemical and hydrodynamic conditions.

This paper reviews the subject of the kinetics of calcite dissolution and precipitation by comparing predictions made by our mechanistic model with published laboratory results.

Summary of the Results of Plummer et al.(1)

Experimental. We studied the dissolution of semi-optical grade crystals of Iceland spar (44.5 cm^2g^{-1} and 96.5 cm^2g^{-1}) in dilute solutions as a function of pH, PCO_2 and temperature. The "pH-stat" method was used to identify forward reactions far from equilibrium (in the near absence of backward reaction). The "free drift" method was used to study the reaction near equilibrium where both forward and backward rates must be considered. Details of the experimental procedures are given elsewhere (1).

Figure 1 summarizes our pH-stat results for three CO_2 partial pressures at 25°C as a function of pH. In general, the pattern shown in Figure 1 demonstrates three controls on the forward rate. 1) At low pH, dissolution rate shows little dependence on PCO_2. The slope of the log rate vs pH plot is approximately -1.0 and rate is proportional to the bulk fluid activity of H^+. 2) At intermediate pH, forward rate depends on PCO_2 and pH and the slope

0-8412-0479-9/79/47-093-537$09.25/0
This chapter not subject to U.S. copyright
Published 1979 American Chemical Society

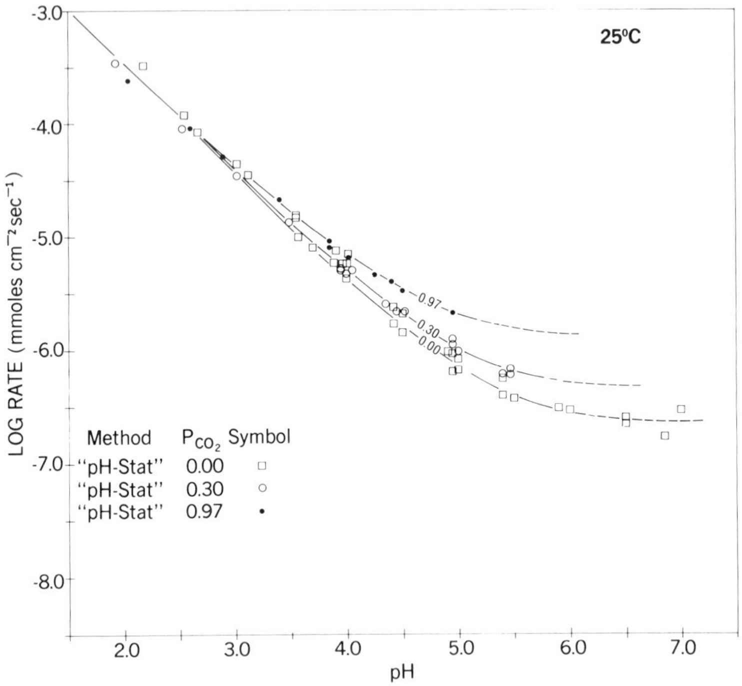

American Journal of Science

Figure 1. Log rate of calcite dissolution as a function of bulk fluid pH and P_{CO_2} *in the pH-stat experiments of Plummer et al. (1). All data are far from equilibrium and are described in the text by three simultaneous reactions (1)* on a log rate vs pH plot becomes less negative. 3) At higher pH, the rate becomes independent of bulk fluid pH, and at CO_2 partial pressures below 1 to 3 percent, the forward rate is, for all practical purposes, independent of both PCO_2 and pH at constant temperature. The boundaries between these three regions of control of the forward rate are functions of pH and PCO_2, occurring at higher pH as PCO_2 decreases.

Analysis of our rate data leads to the following expression for the forward rate

$$R_f = k_1 a_{H^+} + k_2 a_{H_2CO_3^*} + k_3 a_{H_2O} \quad (1)$$

where "a" denotes activity in the bulk fluid, $H_2CO_3^* \equiv H_2CO_3^o + CO_2^o$, and k_1, k_2 and k_3 are rate constants dependent on temperature. Expressions for these constants derived from the data are $\log k_1 = 0.198 - 444./T$, and $\log k_2 = 2.84 - 2177./T$. At T less than 298.15°K, $\log k_3 = -5.86 - 317./T$ and at temperatures greater than 298.15°K, $\log k_3 = 1.10 - 1737./T$. The forward rate ($R_f$) is in mmol of calcite dissolved per cm^2 per sec.

The expression given for k_1 (above) was obtained at stirring rates of 1800-2300 rpm but is in general a function of stirring. The activation energy of k_1 is 2.0 kcal/mol and we have concluded that the forward rate due to H^+ attack is a transport-controlled process. The CO_2 and water dependence of the forward rate do not appear to be significant functions of stirring. The activation energy of k_2 is 10.0 kcal/mol. k_3 has an activation energy of 1.5 kcal/mol below 25°C, while at higher temperatures, the activation energy of k_3 is 7.9 kcal/mol.

Calculations using equation 1 locate boundaries where each term of equation 1 balances the other two terms (Figure 2). Hydrogen ion attack is the dominant forward reaction at pH values less than those of line 1 (Figure 2) and at higher pH values the forward rate becomes increasingly independent of pH. Carbonic acid attack is the dominant forward mechanism at PCO_2 values above line 2; and at PCO_2 values below line 3, the water reaction becomes the dominant forward mechanism (Figure 2).

As the forward rate becomes increasingly independent of H^+ attack, the transition from $H_2CO_3^*$ to H_2O dominance of the forward rate approaches the limit $k_2 a_{H_2CO_3^*} = k_3 a_{H_2O}$, or

$$PCO_2 = \frac{k_3 a_{H_2O}}{k_2 K_o}, \quad (2)$$

where K_o is the Henry's law constant for CO_2. Equation 2 places the transition from $H_2CO_3^*$ dominance to water dominance of the forward rate near 0.10 atm CO_2 at 25°C. At pH greater than approximately 7, the forward rate becomes increasingly independent of PCO_2 at CO_2 partial pressures less than that given by line 2 (Figure 2). Forward reaction depends significantly on more than

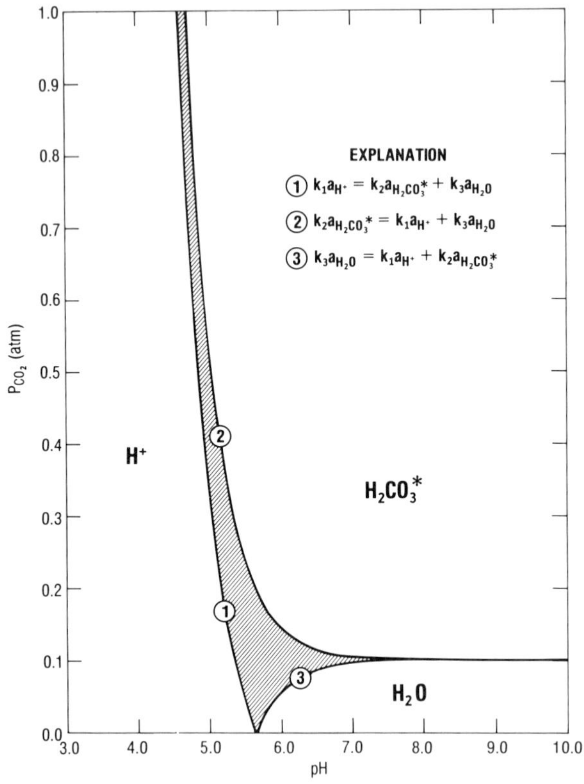

American Journal of Science

Figure 2. Reaction mechanism contributions to the forward rate of reaction as a function of pH and P_{CO_2} *at 25°C. Although* H^+, $H_2CO_3^*$, *and water reaction with calcite occur simultaneously throughout (far from equilibrium, as well as at equilibrium), the forward reaction is dominated by reaction with single species in the fields shown. More than one species contributes significantly to the forward rate in the stippled area, and along the lines labeled 1, 2, and 3, the forward rate attributable to one species balances that of the other two (1).*

one mechanism in the stippled area of Figure 2.

Data from free drift experiments at constant PCO_2 and temperature show a linear relationship between the terms ($R - k_1 aH^+$) and ($aCa^{2+} \cdot aHCO_3^-$) (Figure 3). The intercept in Figure 3 (at $aCa^{2+} \cdot aHCO_3^- = 0$) is the contribution to the forward rate far from equilibrium from CO_2 and water, that is, the intercept is equal to the term ($k_2 aH_2CO_3^* + k_3 aH_2O$). Regression analysis of free drift experiments similar to that shown in Figure 3 allows further estimation of k_2 and k_3 as a function of temperature, and these values agree well with those determined by the pH-stat method (1). The slope observed in plots of ($R - k_1 aH^+$) vs ($aCa^{2+} \cdot aHCO_3^-$) is a function of PCO_2 (Figure 3) and temperature (Figure 4). This slope can be used to describe the contribution of backward reaction to the net rate. Net rate, R, is determined by the difference in forward and backward rates, as given by the expression

$$R = k_1 a_{H^+} + k_2 a_{H_2CO_3^*} + k_3 a_{H_2O} - k_4 a_{Ca^{2+}} a_{HCO_3^-} \qquad (3)$$

which describes all of our results at constant temperature and PCO_2 using individual ion activities in the bulk fluid. The manner in which k_4 depends on temperature and PCO_2 is complex, and we have resorted to theoretical considerations in deriving an expression for k_4.

Theoretical. In deriving a theoretical expression for k_4, we have developed a reaction mechanism model for calcite dissolution which expands on the adsorption layer heterogeneous reaction model of Mullin (2). We assume that a thin (possibly only a few molecules thick) "adsorption layer" (or "surface layer") exists adjacent to the crystal surface, between the crystal surface and the hydrodynamic boundary layer. Species in the adsorption layer are loosely bound to the crystal surface and have relatively low mobility, particularly in comparison with species mobility in the boundary layer. The crystal surface is believed to be sparsely covered by reaction sites at discontinuities in the surface (3). To distinguish between species activities in the bulk fluid, at the base of the boundary layer (near the crystal surface), and in the adsorption layer, we use the subscripts (B), (o), and (s), respectively.

The observed forward and backward rate dependence (equation 3) suggests that three reactions may be occurring simultaneously on the surface

$$CaCO_3 + H^+ = Ca^{2+} + HCO_3^- \qquad (4)$$

$$CaCO_3 + H_2CO_3^o = Ca^{2+} + 2HCO_3^- \qquad (5)$$

$$CaCO_3 + H_2O = Ca^{2+} + HCO_3^- + OH^- \quad . \qquad (6)$$

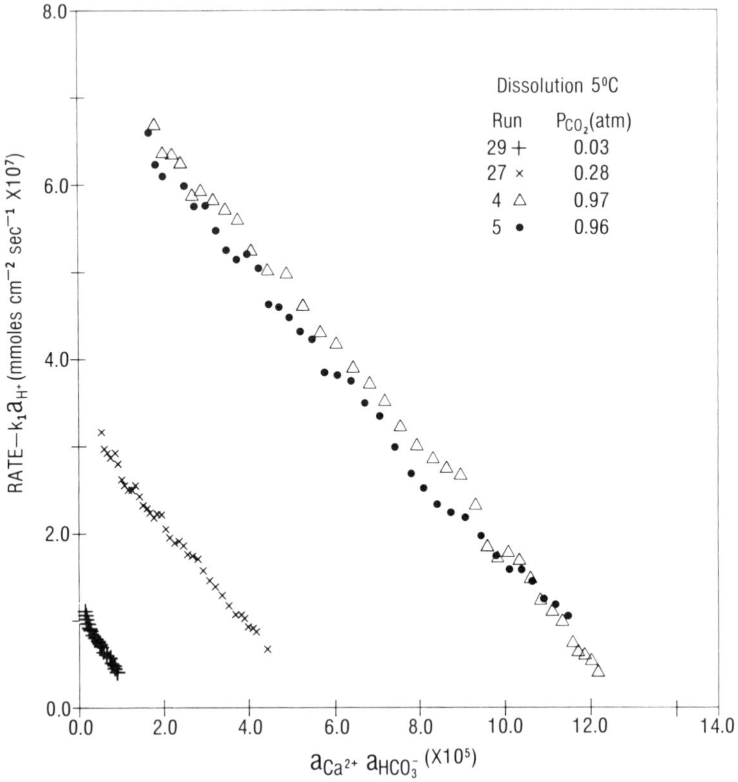

American Journal of Science

Figure 3. Rate minus k_1aH^+ as a function of the bulk fluid product $aCa^{++} \cdot aHCO_3^-$ at 5°C for P_{CO_2} values of 0.03–0.97 atm. The relationship is linear and is described by Equation 3. The intercept at $aCa^{++} \cdot aHCO_3^- = 0$ is the quantity $(k_2aH_2CO_3{}^* + k_3)$ and the slope is $-k_4$ (1).

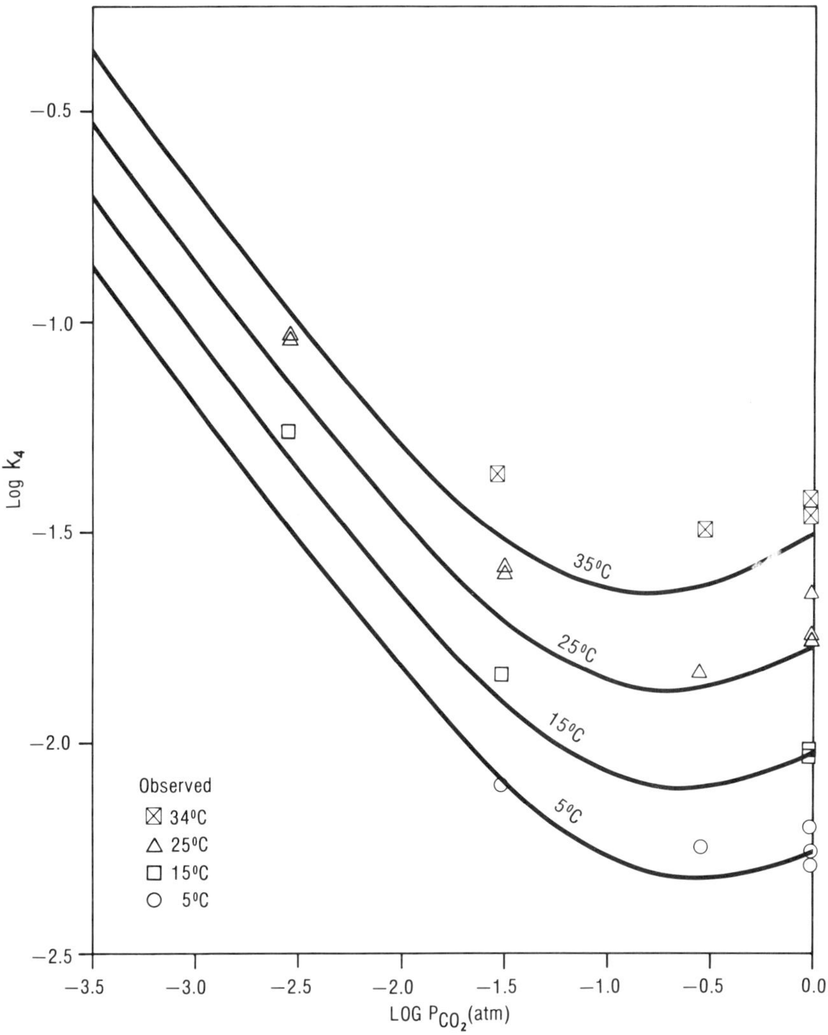

Figure 4. Comparison of observed values of k_4 *with theoretical values calculated from Equation 13 using values of* k_1 *rather than* k_1' *(1).*

In deriving a mechanistic model for calcite dissolution, we assume that reaction 4 is fast relative to reactions 5 and 6. This explains the dissolution rate dependence on H^+ transport and the lack of transport dependence at pH $\geq$5. If reactions 5 and 6 are slow, then, for our experimental conditions, the supply of $H_2CO_3^*$ and H_2O across the boundary layer will ensure that $aH_2CO_3^*(s) = aH_2CO_3^*(o)$ and $aH_2O(s) = aH_2O(o)$. In contrast, if 4 is fast, $aH^+(s)$ (and the activities of all species on the surface other than $H_2CO_3^*$ and H_2O) will be determined by calcite and carbonate species equilibria within the adsorption layer. In other words, $aH^+(s)$ will be the calcite saturation value for $aH_2CO_3^*(s)$ (= $aH_2CO_3^*(o)$) etc.

Secondly, we assume that net forward and backward reaction requires kinetic interaction of boundary layer reactant and product species (adjacent to the adsorption layer) with the adsorption layer species at reaction sites. As the empirical equation 3 shows, the rate of backward reaction is proportional to the bulk fluid (= boundary layer) activity product of the species Ca^{2+} and HCO_3^-. We conclude that net backward reaction involves simultaneous interaction (collision) of one boundary layer Ca^{2+} and one boundary layer HCO_3^- with the adsorption layer speciation at a reaction site. We are unable, however, to make the mechanistic distinction between the ion pair $CaHCO_3^+$ and individual Ca^{2+} and HCO_3^- collision at reaction sites; both mechanistic models are proportional to the boundary layer product $aCa^{2+} \cdot aHCO_3^-$.

Because the association of Ca^{2+} and HCO_3^- upon entering backward reaction has a net positive charge (as does the ion pair $CaHCO_3^+$) we assume, from electrostatic considerations, that backward reaction occurs at negatively charged reaction sites, that is, sites occupied by anions. For our experimental conditions, three anions are present (on the surface): CO_3^{2-}, HCO_3^-, and OH^-.

We now write the reactions 4 - 6 showing their relationships to boundary layer (o) and surface layer (s) species. The forward reactions are

$$CaCO_3 + H^+_{(o)} \rightarrow Ca^{2+}_{(s)} + HCO^-_{3(s)} \quad (4a)$$

$$CaCO_3 + H_2CO^o_{3(o)} \rightarrow Ca^{2+}_{(s)} + 2HCO^-_{3(s)} \quad (5a)$$

$$CaCO_3 + H_2O_{(o)} \rightarrow Ca^{2+}_{(s)} + HCO^-_{3(s)} + OH^-_{(s)}, \quad (6a)$$

and the backward reactions are

$$|Ca^{2+} - HCO_3^-|^+_{(o)} + CO^{2-}_{3(s)} \rightarrow CaCO_3 + HCO^-_{3(s)}$$

$$+ \quad HCO^-_{3(s)} = H^+_{(s)} + CO^{2-}_{3(s)}$$

$$\overline{|Ca^{2+} - HCO_3^-|^+_{(o)} \rightarrow CaCO_3 + H^+_{(s)} \quad (4b)}$$

$$|Ca^{2+} - HCO_3^-|^+_{(o)} + HCO^-_{3(s)} \rightarrow CaCO_3 + H_2CO^o_{3(s)} \quad (5b)$$

and

$$|Ca^{2+} - HCO_3^-|^+_{(o)} + OH^-_{(s)} \rightarrow CaCO_3 + H_2O_{(s)}, \quad (6b)$$

where HCO_3^- on the surface maintains equilibrium with the surface species and the notation $|Ca^{2+} - HCO_3^-|^+(o)$ points to our uncertainty as to the physical nature of the $Ca^{2+} - HCO_3^-$ association during backward reaction.

Recalling our assumption that the surface species are in equilibrium with calcite, forward and backward rate terms for surface species have a net rate of zero, and thus cancel. The rates of reactions 4,5,6 are then given by:

$$R_4 = k_1' \, a_{H^+(o)} - k_4' \, a_{Ca^{2+}(o)} \cdot a_{HCO_3^-(o)} \quad (7)$$

$$R_5 = k_2' \, a_{H_2CO_3^o(o)} - k_4'' \, a_{Ca^{2+}(o)} \cdot a_{HCO_3^-(o)} \cdot a_{HCO_3^-(s)} \quad (8)$$

$$R_6 = k_3' \, a_{H_2O(o)} - k_4''' \, a_{Ca^{2+}(o)} \cdot a_{HCO_3^-(o)} \cdot a_{OH^-(s)} , \quad (9)$$

and the net rate of dissolution is given by

$$R = R_4 + R_5 + R_6. \quad (10)$$

Flux calculations (4) show that boundary layer activities are approximately equal to bulk fluid activities for all species other than H^+; thus $k_2' \, aH_2CO_3^o(o)$ is replaced by $k_2 aH_2CO_3^*(B)$ and $k_3' aH_2O(o)$ becomes $k_3 aH_2O(B)$. Because the forward rate dependence of reaction 4 is transport-controlled,

$$R_{4,f} = k_1(a_{H^+(B)} - a_{H^+(o)}) \simeq k_1 \, a_{H^+(B)} \quad (11)$$

at low pH. At higher pH, the forward rate of 4 is more difficult to describe but is always small relative to the other terms in 10 and can be neglected.

Equation 10 is then identical with the empirical equation 3, and defines k_4 as

$$k_4 = k_4' + k_4'' \, a_{HCO_3^-(s)} + k_4''' \, a_{OH^-(s)} . \quad (12)$$

From the principle of microscopic reversibility, forward and backward rates of all reactions must balance at equilibrium, and thus, equations 7 - 9 give

$$k_4' = \frac{k_1' K_2}{K_c}, \quad k_4'' = \frac{k_2 K_2}{K_c K_1}, \text{ and } k_4''' = \frac{k_3 K_2}{K_c K_w},$$

where K_1, K_2, K_c, and K_w are equilibrium constants for $H_2CO_3^* = H^+ + HCO_3^-$, $HCO_3^- = H^+ + CO_3^{2-}$, $CaCO_3$ (calcite) $= Ca^{2+} + CO_3^{2-}$, and $H_2O = H^+ + OH^-$. Substitution into equation 12 defines K_4 as

$$k_4 = \frac{K_2}{K_c} \left\{ k_1' + \frac{1}{a_{H^+(s)}} \left[k_2 \, a_{H_2CO_3^*(s)} + k_3 \, a_{H_2O(s)} \right] \right\} . \quad (13)$$

By replacing k_1' in 13 with the experimentally determined values of k_1, we can compare the theoretical dependence of k_4 on PCO_2 and temperature with the observed PCO_2 and temperature dependence (Figure 4). Because $k_1' > k_1$, the computed values of k_4 should show a systematic bias (with the computed values below the experimental values). However, the PCO_2 dependence should remain unaltered by using k_1 instead of k_1', and, if the k_1' term is relatively small, the temperature dependence will also be largely unchanged.

Figure 4 shows that the PCO_2 and temperature trends in computed and observed values of k_4 are similar. Computed values are below the experimental values as expected, implying a k_1' value some ten to twenty times k_1. This qualitative and quantitative agreement between theory and experiment gives further support for our mechanistic model.

In terms of the saturation ratio, Ω ($\Omega = IAP/K_c$, where IAP is the ion activity product of calcite in solution; Ω is <1 during dissolution, Ω is >1 during precipitation, and $\Omega = 1$ at equilibrium), equations 3 and 13 define rate as

$$R = \alpha + \beta - \left(\alpha + \frac{a_{H^+}}{a_{H^+(s)}} \beta\right) \Omega \quad (14)$$

where $\alpha = k_1 a_{H^+}$, and $\beta = k_2 a_{H_2CO_3^*} + k_3 a_{H_2O}$. At low PCO_2 (<0.03 atm) and pH greater than 7.0, reaction 5 dominates and the rate is approximated by

$$R = k_3 a_{H_2O} \left(1 - \frac{a_{H^+}}{a_{H^+(s)}} \Omega\right) . \quad (15)$$

It is apparent that in order to use our rate model, a thermodynamic evaluation of the bulk fluid and surface speciation is required. Thermodynamic speciation in the bulk fluid can be calculated through the use of thermodynamic models of the aqueous phase, such as the computer program WATEQF (5). A parameter more difficult to assess in our rate model is the surface activity of H^+. At relatively high CO_2 partial pressures, as in our experiments (1), we have concluded that surface PCO_2 and aH_2O are near the bulk fluid values and surface pH is then determined by calcite equilibrium owing to the rapid reaction with H^+. But as shown in a later section of this paper, this conclusion (that surface PCO_2 = bulk fluid PCO_2) may be only approximately true in our own experiments, and it becomes crucial to any analysis of calcite kinetics at low CO_2 partial pressures.

Comparison With Other Studies

For purposes of discussion, relevant studies in the published literature can be grouped under four topics: 1) dissolution in acids, 2) CO_2-dependence, 3) effect of impurities and 4) precipitation. These topics cover a diverse literature in methods and conditions of experimental study, and no single set of rate measurements is complete enough for direct comparison with all other studies. In addition, rate equations derived from experiments are usually of limited applicability beyond the experimental results. To date, only the theoretical model of Plummer et al. (1) described above is comprehensive enough to allow predictions covering the range of experimental results in the literature. In some cases we are unable to make direct comparisons of observed rates. But where sufficient data are not available for direct comparison, we have either attempted to cast the results of others into the framework of our model, or we have compared observations with predictions based on our model. Our analysis of the literature is not exhaustive and is in part limited to reports giving sufficient data to enable direct or indirect comparison.

In order to cast results of other studies in terms of our model, some transformation of data has been necessary. In some cases we have had to estimate activity coefficients in solutions, and this has been done via the Davies equation (6),

$$\log \gamma_i = -AZ_i^2 \left(\frac{I^{1/2}}{1 + I^{1/2}} - 0.3I \right) \qquad (16)$$

where A is a constant dependent on temperature, z is ion charge, and I is ionic strength ($I = 1/2 \sum_i m_i z_i^2$, where m_i is molality of the ith ion). In other cases we have relied on an aqueous model and our rate model to predict rates under experimental conditions, or we have used our rate equation to predict concentration-time curves for comparison with observations. The aqueous model used to make predictions is similar to that of MIX2 (7) for the chemical system CaO-MgO-K_2O-Na_2O-HCl-H_2SO_4-H_2CO_3-H_2O. The aqueous model includes 23 ion pairs in addition to the major species and uses the constraints of mass action, mass balance and charge balance in solution. Using a computer program, RATECALC, reaction progress and rate are followed as a function of time by way of equations 3 and 13.

Dissolution in Acids. Most kinetic work with calcite has dealt with dissolution in acids. King and Liu (8) measured the initial rate of solution of rotated marble cylinders in dilute acids between 15° and 35^oC. They found that rate of dissolution increased with increased acid concentration, increased speed of rotation and increased temperature. Rate was inversely related to viscosity, supporting their conclusion of a transport-controlled reaction. Figure 5 shows some of their results in dilute HCl solutions at 4000 rpm. King and Liu's rates can be described by a linear relation in the activity of H^+ (Figure 5), where the slope is .022, .032, and .042 at 15°, 25°, and 35^oC, respectively, and comparable to our k_1, which is 0.051 at 25^oC (1) for suspended particles stirred at 1800-2300 rpm. The data of King and Liu (8) give an activation energy of 5.6 kcal/mol, which is nearly three times our value (2.0 kcal/mol). We do not expect close agreement in experimental values of k_1, owing to large differences in hydrodynamic conditions between experimental methods.

Figure 6 compares log rate as a function of pH from various sources. The slope of a best fit line between pH 2 and 5.5 is -0.95 confirming the first order reaction in aH^+. The data of King and Liu (8) and Plummer et al. (1) in dilute HCl solutions, and Berner and Morse (3) in artificial sea water solutions at low pH show close agreement. Table I summarizes experimental methods used by various workers leading to the rates given in Figure 6, and compares our estimates of k_1 from these data with the temperature dependence of k_1.

Clearly, there are real differences in k_1 between experiments. The highest value of k_1 is estimated from the data of Weyl (9), who directed a jet of CO_2-saturated water (pH ≃3.9) onto the surface of calcite. Weyl found that the rate of solution varied with the jet velocity. His rates imply that k_1 varies from 0.11 to 0.23 when velocity of the jet increases from 18 to 35 m sec^{-1}. The smallest value of k_1 (.0073) is derived from the data of Tominaga et al. (10). These authors rotated a disk of marble in HCl solutions (0.1750 - 0.5317N) at 485 rpm. Rate of dissolution was followed by the volume of CO_2 evolved. After an initial period for saturation of the acid with CO_2, rate of gas evolved becomes linear in the cumulative amount of CO_2 produced. Because the acid concentration decreased as calcite dissolved, we extrapolated the observed linear relation in CO_2 production back to the initial condition to estimate initial rates under known acid concentrations. Correction to pH via the Davies equation leads to the rates shown for these authors in Figure 6.

Most of the variation in k_1 (Table I) is probably due to

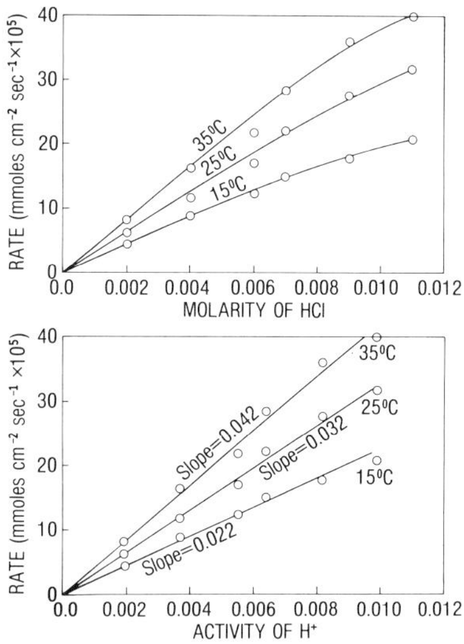

Figure 5. Dissolution of rotated marble cylinders in dilute HCl solutions between 15° and 35°C at 4000 rpm. Rate is shown as a function of concentration (upper) *and activity of H^+* (lower). k_1 *is defined by slope on plots of rate vs. aH^+.*

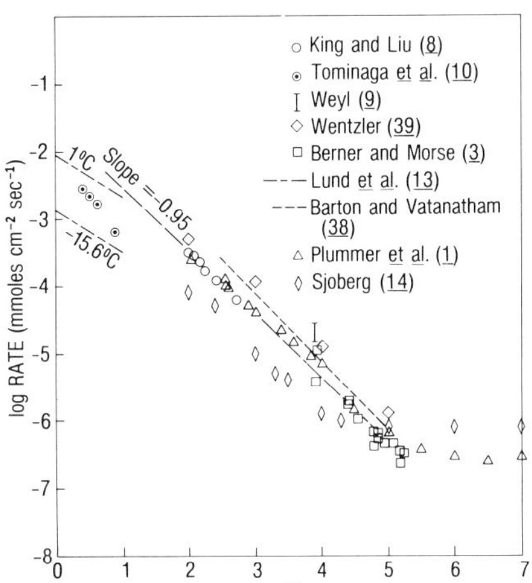

Figure 6. Summary of log rate as function of pH in acids at 25°C. All data are far from equilibrium and independent of P_{CO_2}. *The slope is —0.95 and rate may be assumed, within the uncertainties of the data, to be first order in aH^+.*

Table I. Dissolution in Acids

Description of Solution	pH Range	P_{CO_2} Range (atm)	Temperature °C	Experimental Method	Material	Stirring	Estimated k_1 cm/sec (25°)	Activation Energy kcal/mole	References
0.1750-0.5317 N HCl	0.41-0.88	0-1.0 (not controlled)	25	Rotating disk	Marble	485 rpm	0.0073	5.4 1/	Tominaga et al. (10)
0.002-0.011 N HCl	2.00-2.72	not controlled	15-35	Rotating Cylinder	Marble	4000 rpm	0.032	5.6	King and Liu (8)
0.01 N H_2SO_4	2.5-6.0	low	0-50	Free drift	Crushed seized limestone	Batch reactor "vigorously agitated"	0.078	3.6	Barton and Vatanatham (38)
0.1-9 N HCl	1.0	?	-15.6-25	Rotating disk	Marble	100-500 rpm	0.016	15.	Lund et al. (13)
HNO_3	2-5	?	25	Rotating disk	Limestone	?	0.053	—	Wentzler (39)
Pseudo-sea water, HCl	3.9-5.2	0-1.0	25	pH-stat	Calcite powder	Magnetic Stirrer	0.038	—	Berner and Morse (3)
HCl	2-4	0-1.0	0-60	pH stat	Iceland Spar	Impellor 1800-2300 rpm	0.051	2.0	Plummer et al. (1)
CO_2 - water	3.9	1.0	25	Solution jet	Iceland Spar	18-35 m/sec	0.11-0.23	—	Weyl (9)
0.7 M KCl	2-4	$10^{-2.5}$	25	pH stat	Single Crystal	Magnetic Stirrer 276-480 rpm	0.012	2.5	Sjoberg (14)
CO_2 - water	?	52.5 p.s.i.a.	97-282	Single crystal by weight	Natural clear calcite	?	?	2.5	Ellis (40)

1/ Recalculation of the data of Spring (41) as taken from Moelwyn-Hughes (42) corrected for temperature effects in viscosity.

differences in hydrodynamic conditions of the experiment, being the lowest under the approximately laminar boundary layer conditions at the surface of a slowly rotating disk (10), and highest at the impact of a high velocity jet on the calcite surface (9). Transport dependence of the reaction in HCl solutions is also documented by the work of Kaye (11) and Nierode et al. (12).

The temperature dependence of k_1 shows considerable variation (2.0 - 5.6 kcal/mole), and may also be a function of the hydrodynamic conditions of the experiment. Lund et al. (13) obtained an activation energy of 15 kcal/mol between -15.6 and +1.0°C under conditions where they concluded both surface reaction and transport processes influenced the rate. They found rate proportional to the 0.63 power of H^+ rather than the first order reaction observed for conditions of transport control.

The data shown in Figure 6 give the variation in log rate as a function of pH far from equilibrium. We have shown that below pH 4, CO_2 partial pressures between 0 and 1.0 atm do not influence the rate significantly. Above pH 4, all data shown in Figure 6 are at CO_2 partial pressures less than $10^{-1.5}$ atm and can thus be compared. Only the data of Plummer et al. (1) and Sjoberg (14) are at pH greater than 5.5 far from equilibrium, and both confirm a plateau in rate as a function of pH (Figures 1 and 6). We have concluded that this plateau defines the rate of the forward reaction in the near absence of both CO_2 and H^+ and is controlled by reaction with H_2O (1). It is not understood, however, why Sjoberg's rates are approximately 2.8 times greater than our rates far from equilibrium in the near absence of H^+ and CO_2, and are significantly lower than most observed rates at pH less than 5.0 (Figure 6).

The rates far from equilibrium in Figure 6 were obtained in a variety of aqueous solutions including HCl, HNO_3, H_2SO_4, NaCl-$CaCl_2$ and KCl to 0.7 molar at 25°C, and suggest that at least for these conditions, the forward rate is not significantly dependent upon the presence of these constituents in solution.

CO_2-dependence. There are relatively few investigations that deal with the effects of CO_2-dependence on rate of dissolution. Erga and Terjesen (15) followed calcite dissolution by the free drift method to near equilibrium in CO_2-water solutions at 25°C by measurement of dissolved calcium as a function of time. P_{CO_2} was maintained constant by bubbling gas mixtures. The calcite used was of "natural origin" and surface area estimated from particle size was 125 cm^2g^{-1}. Experiments at .95 - .97 atm CO_2 showed that rate was directly proportional to surface area of the particles, but independent of gas velocity, leading to their conclusion that the reaction rate is independent of 1) the transfer of carbon dioxide from the gas to liquid phase, and 2) chemical reaction in the bulk fluid. Rate of dissolution was proportional to stirring rate to the power (stirring coefficient) 0.22 (between 280 and 555 rpm). Because stirring coefficients of .5 to

1.0 are expected in transport-controlled reactions (16), the low stirring coefficient of Erga and Terjesen (15) further supports their conclusion that for their experimental conditions, "the rate determining steps are localized at the solid-liquid interface or in the liquid film in immediate contact with it".

Erga and Terjesen (15) published two sets of data that we can use for comparison with our rate model. The first is a plot of dissolved Ca vs time in a free drift run at 0.954 atm CO_2, and the second is a plot of dissolution rate vs total calcium in the solution for four free drift runs at CO_2 partial pressures of 0.135, 0.392, 0.664 and 0.952 atm (15).

Figure 7 compares two concentration-time curves calculated from equation 3 (using the program RATECALC mentioned earlier) with that observed by Erga and Terjesen. The dashed curve (Figure 7) is calculated using the reported surface area (1250. $cm^2\,l^{-1}$) and the solid curve is a simulation of their experiment assuming their surface area was 1/2 the reported value. Figure 7 shows that the results of Erga and Terjesen (15) differ from ours by a constant factor of two. That is, the form of our rate equation predicts the observed shape of the concentration-time curve with an uncertainty of a factor of two in the rate constants. This uncertainty is similar to the uncertainty in the measurement of k_3 (1). Surface area adjustments within a factor of two may also reflect uncertainties in estimation of area based on particle size, and (or) differences in reaction site density among differing material. What is important is that the shape of the free drift data of Erga and Terjesen (15) is closely matched by the form of our rate equation (Figure 7). The match may be even closer if one considers the expected (small) decrease in surface area during reaction, which was not accounted for in the computer simulation. Erga and Terjesen used 100g of calcite in 10 liters of distilled water, so that at 0.95 atm CO_2, nearly 10 percent of the weight of their starting material was dissolved by the end of their run.

Using the surface area correction of 1/2, we also closely match concentration-time curves for similar experiments reported by these authors in subsequent publications (17, 18) as shown for the results of Terjesen *et al*. (17) in Figure 8.

Figure 9 shows that if we use the surface area correction of 1/2, our rate equation also closely simulates the observed rates of Erga and Terjesen (15) at other CO_2 partial pressures. Because it is necessary to consider both forward and backward reaction in analysis of the results of Erga and Terjesen (15), we have made the calculations of Figure 9 assuming surface PCO_2 is equal to the bulk fluid value and the surface pH is determined by calcite equilibrium, identical to the calculations of Plummer *et al*. (1). As before, we have not corrected for expected decreases in surface area during their experiments. But even without this correction the match in computed and observed rates strongly supports the PCO_2 dependence of forward and backward rate predicted

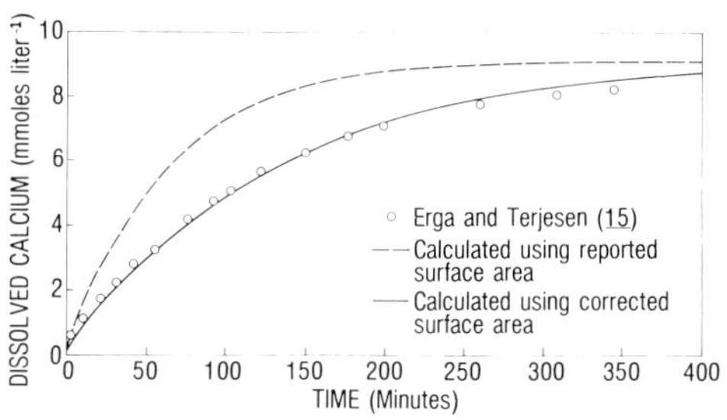

Figure 7. Comparison of observed calcium in solution (15) during free drift experiment at 0.954 atm CO_2 and 25°C with simulated reaction. The dashed line is calculated using the reported surface area, and the solid line is calculated assuming the area is half the reported value.

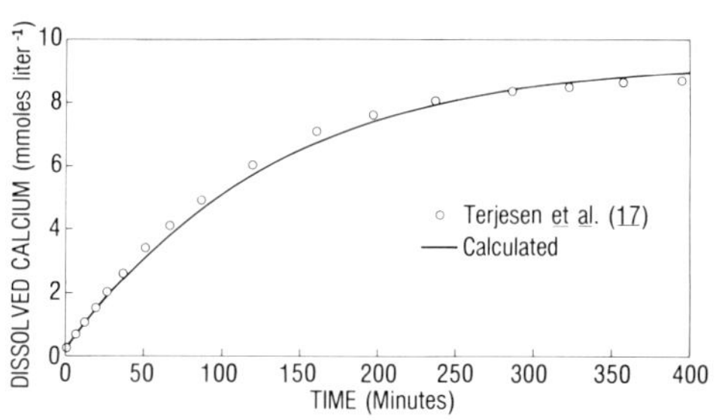

Figure 8. Comparison of simulated and observed calcium–time curve of Terjesen et al. (17) in a free drift run at 0.97 atm CO_2 and 25°C. The solid line represents the simulated reaction using the same surface area correction as found for the free drift experiments of Erga and Terjesen (15).

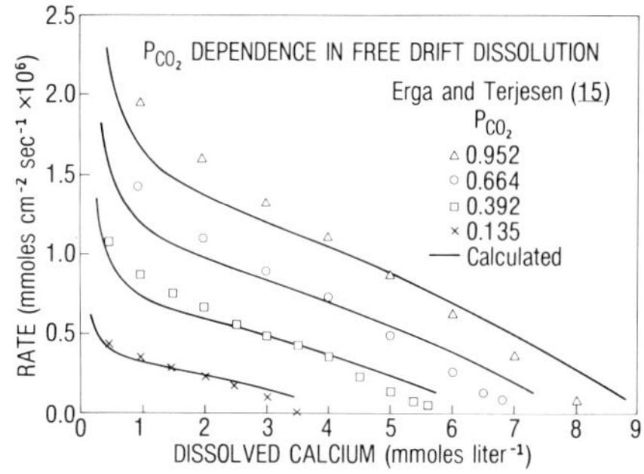

Figure 9. Comparison of observed and simulated rate for four CO_2 partial pressures at 25°C in the free drift experiments of Erga and Terjesen (15). The variation in rate as a function of P_{CO_2} is closely predicted by the rate model of Plummer et al. (1).

by Plummer et al. (1).

Figure 10 compares the CO_2 and H^+ dependence of dissolution rate observed by Berner and Morse (3) far from equilibrium in pseudo-sea water (a $CaCl_2$ - NaCl solution of the ionic strength and total calcium content of sea water) with the PCO_2 - aH^+ dependence given by equation 1. We have found that the rate of dissolution far from equilibrium becomes independent of PCO_2 below approximately $10^{-1.5}$ atm. Most of Berner and Morse's rates far from equilibrium are at CO_2 partial pressures equal to or below $10^{-1.5}$ atm, and as expected, show little or no CO_2 dependence far from equilibrium (Figure 10). According to equation 1, the difference in forward rate at constant pH between 0 and 1 atm CO_2 should be approximately 1.2×10^{-6} mmol cm^{-2} sec^{-1}, but the one value of Berner and Morse (3) at 1 atm CO_2 that can be plotted on Figure 10 shows no CO_2 dependence. Although most of the rates of Berner and Morse (3) shown in Figure 10 are within a factor of two of our rates, the expected relation of increased rate with increased PCO_2 is even reversed in some of their low PCO_2 data (Figure 10). Because most of the rates of Berner and Morse (3) far from equilibrium are at low PCO_2, their data do not provide an adequate test of PCO_2 dependence.

Almost all of the data of Sjoberg (14) are for very low CO_2 partial pressures. However, two rates far from equilibrium at 0.97 atm CO_2 (in 0.7*M* KCl at pH 4.4 and 5.0) are approximately twice the measured values at $10^{-2.5}$ atm CO_2 (14). Below pH 4, Sjoberg's rates far from equilibrium are dominated by reaction with H^+ and become independent of PCO_2, as we have found (1). Although there are significant differences between the absolute values of Sjoberg's rates and ours (Figure 6), the PCO_2 dependence observed by Sjoberg agrees reasonably well with that observed in our experiments.

Effect of Impurities. In the previous sections we have considered calcite dissolution 1) far from equilibrium in various solutions, and 2) both far from and near equilibrium but only in pure CO_2-water solutions. We now investigate calcite dissolution in solutions where backward reaction must be considered in the presence of impurities. Impurities can be defined as constituents present in the aqueous phase that are not part of the original stoichiometric composition of the reactant (2). The presence of impurities can have a profound effect on the rate of dissolution or precipitation, but, unfortunately, there is no simple way of predicting, *a priori*, the total effect that an impurity or combination of impurities can have on the rate of reaction.

The effect of impurities can be partially evaluated through thermodynamic analysis. Certainly, there are many other ways that impurities may affect the rate of reaction (2), and by accounting for the thermodynamic effect of impurities, other effects may be identified. Two classes of thermodynamic problems are evident.

The first is concerned with what thermodynamic changes will take place in a heterogeneous system upon addition of an impurity. This problem can be described in terms of changes in mineral solubility. If an impurity contains a constituent that is present in the reactant (common ion effect), the solubility of the reactant is usually decreased and precipitation may occur. Impurities may also cause changes in activity coefficients. Decreasing activity coefficients cause solubility to increase (salting-in). Increasing activity coefficients can cause precipitation (salting-out). These and other thermodynamic effects of impurities on solubility are discussed elsewhere (2, 19, 20).

The second type of thermodynamic problem is concerned with comparing parallel reactions in pure and impure systems. In the case of calcite dissolution we ask, how will rate change at constant pH and PCO_2 owing to the presence or absence of an impurity? Equation 14 shows that thermodynamic effects of impurities on the rate of calcite dissolution are accounted for by calculation of the bulk fluid saturation (Ω) and the equilibrium activity of H^+ in the adsorption layer ($aH^+(s)$). The following example demonstrates one possibility.

Consider the dissolution of calcite at 25°C and 0.95 atm CO_2 in separate starting solutions of pure water and 0.2 M $CaCl_2$ solution. Thermodynamic calculations indicate that the surface equilibrium pH (assuming surface PCO_2 is equal to the bulk fluid value) is lower in the $CaCl_2$ solution (5.51) than in the pure solution (6.02). After some initial dissolution, pH 5.36 is reached, at different times, in both solutions. In the pure solution Ω is lower (0.025) and the ratio, $aH^+/aH^+(s)$, is higher (4.57) than comparable values in the 0.2 M $CaCl_2$ solution, 0.479 and 1.41, respectively. The increase in Ω from the pure solution to the $CaCl_2$ solution is nearly 6 times larger than the accompanying decrease in $aH^+/aH^+(s)$ ratio. Equation 14 then shows that rate of calcite dissolution at constant pH and PCO_2 should be slower in the $CaCl_2$ solutions than in pure solutions.

These thermodynamic effects are complex and almost every impurity is a special case. Resolution of the thermodynamic effects of impurities usually requires a computer program, such as RATECALC mentioned earlier. One can only properly address the question as to what other effects impurities may have on rate of reaction, after the thermodynamic effects of impurities have been evaluated. This is the approach we have followed in our evaluation of the data of Berner and Morse (3), Morse (21) and Sjoberg (14) in various salt solutions of the ionic strength of sea water ($I = 0.7$). No attempt has been made to treat rate data of Berner and Morse (3) or others for which trace impurities (such as PO_4) cause large changes in rate without significantly altering the thermodynamic properties of the system. Clearly, other mechanisms must be considered in interpreting these results.

Table II summarizes our calculations of dissolution rate in pseudo-sea water and sea water for comparison with the pH-stat measurements of Berner and Morse (3). Berner and Morse tabulate measured pH, PCO_2 and rate. Because many of these data depend significantly on backward reaction, our calculations of the rates expected in their experiments have employed a thermodynamic model for their aqueous solutions. We have had to correct their measured pH values for the effects of residual liquid junction potential. The residual liquid junction potential in sea water has been estimated by Hawley and Pytkowicz (22) to be -3.2 mv, or +.054 pH (added to the measured value). As Berner and Morse report measured pH to two decimal places, we have added 0.05 pH to their measured values in making our thermodynamic calculations. For comparison, dual calculations are presented in Table II for uncorrected and corrected pH. In order to obtain different pH values in pseudo-sea water and artificial sea water at constant PCO_2, Berner and Morse added various amounts of HCl or NaOH to their starting solutions. Because the amounts of these additions are not given in their paper, we have calculated them using the reported pH, PCO_2, composition of the initial solution (pseudo-sea water or artificial sea water), and the charge balance criteria for the aqueous model in RATECALC. Our estimates of the amounts of these additions of HCl and NaOH have been considered in calculation of rate and are shown in Table II. Table II also gives the calculated surface equilibrium pH (assuming surface PCO_2 = bulk fluid PCO_2), bulk fluid saturation index (SI = log Ω) and predicted rate.

Far from equilibrium the correction for residual liquid junction potential is not significant, but the residual liquid junction potential becomes extremely important close to equilibrium (specifically for rates less than approximately 100 mg cm^{-2} yr^{-1}). Figure 11 shows a close correlation between calculated rates based on the junction potential corrected pH and the observed rate at various CO_2 partial pressures. Most of our calculated rates are within a factor of two of the observed. Some of the data (Figure 11) show a trend with the computed rates larger than the observed as rate decreases. This trend in departure of computed and observed rates is not understood. Sources of uncertainty include the following: 1) uncertainty in the (thermodynamic) pH of the bulk fluid, 2) uncertainty in the surface equilibrium pH due to some uncertainty in surface PCO_2, 3) possible inhibiting effects in their solutions, and 4) one must be aware of the possibility of other reaction mechanisms occurring which are presently unidentified. If higher pH values are not due entirely to the addition of NaOH, but due, in part, to the dissolution of small amounts of $CaCO_3$, the bulk fluid saturation with respect to calcite would be higher than that given in Table II and the calculated rate would be slower. One rate near equilibrium at one atmosphere CO_2 (pH 6.00, Table II) is calculated to be 15 times faster than the observed value. This difference could be accounted for by accompanying dissolution of a small amount of $CaCO_3$.

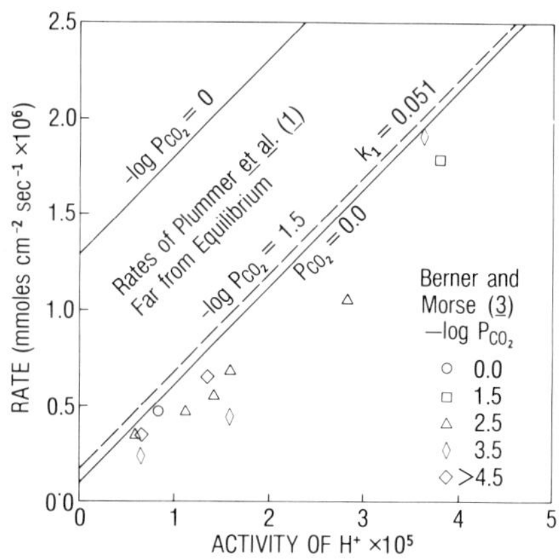

Figure 10. Dissolution rates from pH-stat experiments of Berner and Morse (3) in pseudo-sea water at 25°C. No P_{CO_2} dependence is observed at CO_2 partial pressures $< 10^{-1.5}$ atm, as expected. Range of rates observed by Plummer et al. (1) in dilute solutions shows that most rates at low P_{CO_2} far from equilibrium agree with our rate model. One rate measurement of Berner and Morse (3) at 1 atm CO_2 shows little or no P_{CO_2} dependence.

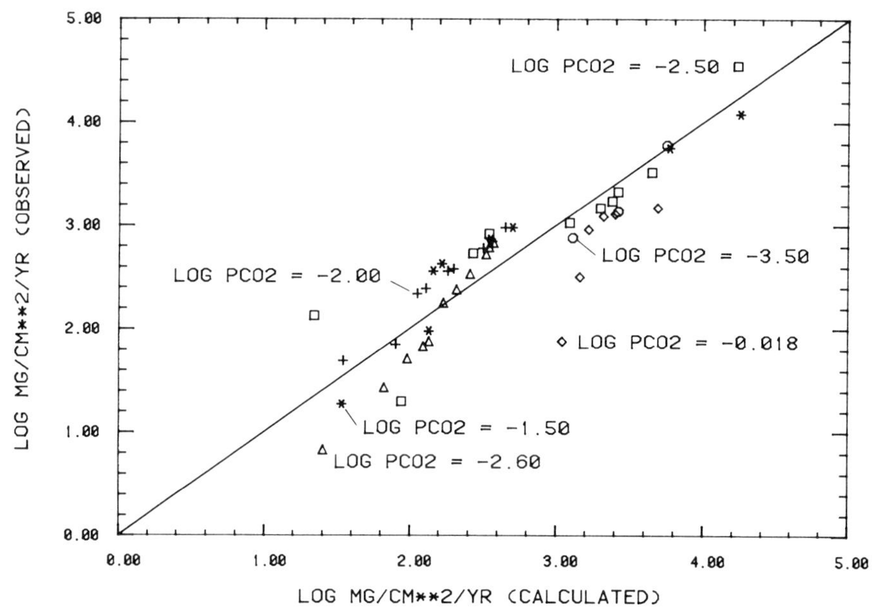

Figure 11. Comparison of observed rates of Berner and Morse (3) in pseudo-sea water and sea water at 25°C with the predicted rates. Most rates are within a factor of two of the observed.

Table II. Calculation of the Rates of Berner and Morse (3)

OBSERVED			COMPUTED USING UNCORRECTED pH					COMPUTED USING CORRECTED pH				
pH	Log P_{CO_2}	Rate mg cm^{-2} yr^{-1}	Computed adjustments to initial solution in mmole/kg H_2O		Log Ω	Equilibrium pH	Rate mg cm^{-2} yr^{-1}	Computed adjustments to initial solution in mmole/kg H_2O		Log Ω	Equilibrium pH	Rate mg cm^{-2} yr^{-1}
			HCl	NaOH				HCl	NaOH			
Pseudo - Sea Water, Far From Equilibrium												
5.08	-0.018	1479.	-0-	3.095	-2.11	5.987	5031.	-0-	3.475	-2.01	5.989	4854.
3.92	-1.5	12023.	0.153	-0-	-5.91	6.823	19892.	0.135	-0-	-5.81	6.823	17780.
4.42	-1.5	5623.	0.028	-0-	-4.91	6.824	6621.	0.020	-0-	-4.81	6.824	5952.
3.94	-2.5	35481.	0.152	-0-	-6.87	7.353	18914.	0.135	-0-	-6.77	7.353	16897.
4.55	-2.5	3311.	0.034	-0-	-5.65	7.354	4930.	0.030	-0-	-5.55	7.354	4434.
4.80	-2.5	1238.	0.016	-0-	-5.15	7.354	2937.	0.013	-0-	-5.05	7.354	2658.
4.82	-2.5	1738.	0.013	-0-	-5.05	7.354	2658.	0.010	-0-	-4.95	7.354	2410.
4.95	-2.5	1479.	0.007	-0-	-4.85	7.355	2189.	0.005	-0-	-4.75	7.355	1992.
5.23	-2.5	1072.	-0-	0.007	-4.29	7.355	1327.	-0-	0.009	-4.19	7.355	1224.
4.44	-3.5	6026.	0.048	-0-	-6.87	7.866	6230.	0.043	-0-	-6.77	7.866	5592.
4.80	-3.5	1380.	0.021	-0-	-6.15	7.866	2927.	0.018	-0-	-6.05	7.866	2649.
5.19	-3.5	759.	0.007	-0-	-5.37	7.866	1410.	0.006	-0-	-5.27	7.866	1297.
Pseudo - Sea Water, Near Equilibrium												
6.00	-0.18	75.	-0-	25.957	-0.29	6.091	1513.	-0-	29.144	-0.19	6.108	1089.
5.96	-0.18	318.	-0-	23.661	-0.37	6.079	1807.	-0-	26.565	-0.27	6.094	1439.
5.93	-0.18	920.	-0-	22.074	-0.42	6.071	2005.	-0-	24.782	-0.33	6.085	1665.
5.87	-0.18	1240.	-0-	19.214	-0.54	6.058	2354.	-0-	21.569	-0.44	6.069	2067.
5.79	-0.18	1280.	-0-	15.969	-0.70	6.042	2743.	-0-	17.926	-0.60	6.051	2509.
6.80	-1.5	18.8	-0-	5.409	-0.15	6.867	93.	-0-	6.075	-0.05	6.872	34.
6.71	-1.5	96.	-0-	4.390	-0.33	6.858	183.	-0-	4.930	-0.23	6.863	135.
6.70	-1.5	360.	-0-	4.289	-0.35	6.857	191.	-0-	4.817	-0.25	6.862	145.
6.68	-1.5	430.	-0-	4.095	-0.39	6.856	209.	-0-	4.598	-0.29	6.860	164.
6.42	-1.5	740.	-0-	2.244	-0.91	6.841	381.	-0-	2.519	-0.81	6.843	353.
6.12	-1.5	950.	-0-	1.122	-1.51	6.832	526.	-0-	1.259	-1.41	6.833	503.
7.11	-2.0	0.85	-0-	3.518	-0.03	7.123	15.	-0-	3.954	+0.07	7.127	-39.
7.09	-2.0	2.9	-0-	3.357	-0.07	7.122	35.	-0-	3.773	+0.03	7.126	-17.
7.04	-2.0	49.	-0-	2.987	-0.17	7.118	80.	-0-	2.597	-0.07	7.115	35.
6.99	-2.0	71.	-0-	2.659	-0.27	7.115	121.	-0-	3.357	-0.17	7.122	80.
6.95	-2.0	220.	-0-	2.422	-0.35	7.113	150.	-0-	2.987	-0.25	7.118	113.

6.93	-2.0	245.	-0-	2.312	-0.39	7.112	164.	-0-	2.721	-0.29	7.116	128.
6.85	-2.0	360.	-0-	1.920	-0.55	7.109	213.	-0-	2.156	-0.45	7.111	183.
6.82	-2.0	380.	-0-	1.791	-0.61	7.108	229.	-0-	2.011	-0.51	7.110	201.
6.57	-2.0	600.	-0-	1.003	-1.11	7.101	337.	-0-	1.126	-1.01	7.102	319.
6.45	-2.0	710.	-0-	0.760	-1.35	7.099	378.	-0-	0.853	-1.25	7.100	362.
6.19	-2.0	950.	-0-	0.416	-1.87	7.096	463.	-0-	0.468	-1.77	7.096	446.
7.30	-2.5	20.	-0-	1.665	-0.15	7.371	64.	-0-	1.618	-0.21	7.369	86.
7.22	-2.5	135.	-0-	1.439	-0.31	7.368	121.	-0-	1.880	-0.05	7.373	22.
6.87	-2.5	540.	-0-	0.145	-1.01	7.360	283.	-0-	0.222	-0.91	7.361	266.
6.58	-2.5	840.	0.156	-0-	-1.59	7.357	362.	0.116	-0-	-1.49	7.358	350.
Sea Water, Near Equilibrium												
7.37	-2.6	6.7	-0-	1.801	-0.16	7.444	67.	-0-	2.029	-0.06	7.447	25.
7.32	-2.6	27.	-0-	1.600	-0.26	7.442	103.	-0-	1.801	-0.16	7.444	67.
7.28	-2.6	51.	-0-	1.456	-0.34	7.441	130.	-0-	1.635	-0.24	7.443	96.
7.24	-2.6	68.	-0-	1.325	-0.42	7.440	154.	-0-	1.491	-0.32	7.441	123.
7.22	-2.6	76.	-0-	1.264	-0.46	7.439	165.	-0-	1.422	-0.36	7.441	136.
7.16	-2.6	176.	-0-	1.098	-0.57	7.438	196.	-0-	1.234	-0.48	7.439	170.
7.08	-2.6	242.	-0-	0.910	-0.73	7.436	231.	-0-	1.023	-0.63	7.437	210.
6.96	-2.6	336.	-0-	0.687	-0.97	7.434	275.	-0-	0.773	-0.87	7.435	258.
6.71	-2.6	530.	-0-	0.381	-1.48	7.433	342.	-0-	0.432	-1.37	7.432	330.
6.65	-2.6	610.	-0-	0.334	-1.59	7.431	356.	-0-	0.376	-1.49	7.431	335.
6.51	-2.6	675.	-0-	0.242	-1.87	7.430	386.	-0-	0.271	-1.77	7.430	375.

Figure 11 shows that our rate equation, with appropriate thermodynamic corrections, satisfactorily predicts observed rate in sea water over a wide range of pH and PCO_2 for values of Ω between 0.0 and 0.6. Nearer equilibrium, uncertainties in the calculations and the experimental data (3) are too large for a reliable test of our model.

We have a great deal more difficulty in interpreting the results of Sjoberg (23, 14). Most of our problems stem from Sjoberg's experimental design which involved dissolution in 0.7 M KCl solution using both pH-stat and free drift methods. Except for a few pH-stat measurements at 0.97 atm, most measurements were made at very low PCO_2. Unfortunately, PCO_2 was not actually controlled and we only know that PCO_2 was low owing to bubbling of "CO_2-free" nitrogen. Although there was probably very little transfer of CO_2 from the reaction system, and thus the reaction was essentially closed to CO_2 as Sjoberg assumed, the actual PCO_2 in solution depends on a number of factors including bubbling rate and dissolution rate. Our kinetic results show that surface PCO_2 is extremely important in controlling rate because CO_2 equilibrium with calcite determines surface pH. Although Sjoberg was correct in assuming a reaction system closed to CO_2 and in assuming PCO_2 was very low, surface PCO_2 may have varied significantly. Our calculations show that if surface PCO_2 varied by an order of magnitude in Sjoberg's free drift experiments, surface equilibrium pH would vary by 0.66 pH.

Sjoberg found that rate in 0.7 M KCl solutions was described by

$$d\,[Ca^{2+}]/dt = kA\left(C - [Ca^{2+}]^{\frac{1}{2}}\,[CO_3^{2-}]^{\frac{1}{2}}\right) \qquad (17)$$

where the brackets denote concentration, k is a rate constant, A is surface area, and C is the value of $[Ca^{2+}]^{\frac{1}{2}}\,[CO_3^{2-}]^{\frac{1}{2}}$ at equilibrium. Taking values of k and C from Sjoberg (23), equation 17 becomes, at 20°C,

$$R = 1.60 \times 10^{-6}\,(1 - \Omega^{\frac{1}{2}}) \qquad (18)$$

where R is in mmol cm^{-2} sec^{-1}. At low PCO_2 and high pH, our rate equation (equation 14) reduces to

$$R = 1.14 \times 10^{-7}\,\left(1 - \frac{a_{H^+}}{a_{H^+(s)}}\,\Omega\right) \qquad (19)$$

at 20°C. Ignoring, for the moment, the obvious difference in rate constants, we first compare the form of equations 18 and 19 correlating $\Omega^{\frac{1}{2}}$ with $(aH^+/aH^+(s))\Omega$. Using RATECALC, we reconstructed a free drift dissolution run in 0.7 M KCl in a closed system at 25°C. Figure 12 shows that these two terms behave similarly, but that $\Omega^{\frac{1}{2}}$ is slightly larger than the term

$(aH^+/aH^+(s))\Omega$. In our calculation, though, we assumed a closed system and that surface PCO_2 was equal to the bulk fluid value, which is calculated near $10^{-6.2}$ atm. Bubbling of pure N_2 gas will lower this value in the bulk fluid and thus lower the surface PCO_2. For example, referring to Figure 12, at $\Omega^{\frac{1}{2}} = 0.5$, our calculations show that $\Omega^{\frac{1}{2}}$ and our term $(aH^+/aH^+(s))\Omega$ are identical if surface PCO_2 is $10^{-6.4}$ rather than $10^{-6.2}$. Thus, considering the uncertainties in Sjoberg's PCO_2 and the problems associated with determining low values of surface PCO_2, there may be no fundamental difference between the form of Sjoberg's rate equation and ours under comparable experimental conditions.

Comparing rate constants in equations 18 and 19 shows that Sjoberg's initial rates in free drift experiments are 14 times faster than ours. There may be considerable uncertainty in Sjoberg's initial free drift rates as they are based on slopes of calcium vs time curves calculated from the measured pH. We found in our free drift experiments (1) that rate was unreliable for the first several hours of reaction at a PCO_2 of $10^{-2.5}$ owing to non-equilibrium of the CO_2-water system. Furthermore, Sjoberg's free drift rate far from equilibrium is 3.2 times faster than his initial rate determined by pH-stat.

Sjoberg (23, 14) also shows that initial rate decreases in a supposedly CO_2-free system by addition of $CaCl_2$. According to our rate equation, added calcium should have little effect on rate for a CO_2-free system. The decrease in rate observed by Sjoberg for addition of calcium may point to the presence of small amounts of CO_2 in his starting solutions.

Despite problems in interpreting Sjoberg's results, there are several points of general agreement. Sjoberg's pH and PCO_2 dependence far from equilibrium is similar to ours, as shown earlier. Forward rate in the near absence of CO_2 and H^+ appears constant (Figure 6) which also agrees with the form of our forward rate. We also find a similar temperature dependence in acids (Table I), and at low PCO_2 and high pH, Sjoberg's (14) rate constant has a temperature dependence of 6.8-8.4 kcal/mol which compares with 7.9 kcal/mol measured by us. Because of our uncertainties in estimating PCO_2 in Sjoberg's experiments, further speculation regarding the compatibility of his results and ours is not warranted.

Morse (21) recently reported near-equilibrium pH-stat rates in sea water at 25°C and $10^{-2.53}$ atm CO_2. His rates are all less than 8 mg cm^{-2} yr^{-1}. Table III compares our calculations (using equation 15 and assuming surface PCO_2 is equal to the bulk fluid value) of a set of Morse's rates near equilibrium, and shows generally poor agreement. In calculation of Ω and $aH^+(s)$, no correction for liquid junction potential error in measured pH was necessary, as in our earlier calculations for pseudo-sea water. In sea water one can use the apparent constant approach to solve thermodynamic problems in the carbonate system directly. We have calculated surface equilibrium pH using the CO_2 solubility data

Table III.

Observed[1]			Calculated			
pH	Ω^2	Rate mg $cm^{-2}yr^{-1}$	$pH(s)^3$ (open system)	Rate[4] mg $cm^{-2}yr^{-1}$	Implied[5] pH(s)	$pH(s)^6$ (closed system)
7.32	0.620	7.9	7.418	81.5	7.518	7.509
7.33	0.649	6.8	7.419	74.3	7.510	7.501
7.34	0.680	4.7	7.419	67.4	7.502	7.493
7.35	0.712	3.7	7.419	60.4	7.493	7.484
7.36	0.745	1.4	7.420	52.8	7.486	7.476
7.37	0.780	0.79	7.420	45.6	7.477	7.468
7.383	0.828	0.12	7.421	35.2	7.465	7.457
7.395	0.876	0.049	7.422	24.8	7.452	7.447
7.405	0.917	0.017	7.422	17.0	7.442	7.439

1/ Morse (21).

2/ Recalculated to show third significant figure using the apparent constants of Weiss (24), Lyman (25), Ingle *et al.* (26), and the measured pH and reported PCO_2.

3/ Theoretical calcite equilibrium pH at bulk fluid PCO_2 (open system).

4/ Calculated using equation 15 and the open system equilibrium pH.

5/ Value of pH(s) implied by the observed rate and equation 15.

6/ Theoretical calcite equilibrium pH in a sea water system closed to CO_2.

of Weiss (24), the apparent constants of Lyman (25), and the apparent calcite solubility constant of Ingle et al. (26). Also shown in Table III are values of surface pH necessary for our rate equation to reproduce Morse's rates. The implied surface pH is 0.10 to 0.02 higher than the calculated equilibrium value at $10^{-2.53}$ atm CO_2 (open system equilibrium). This difference in pH implies a surface PCO_2 slightly lower than the bulk fluid value, i.e., $10^{-2.68}$ to $10^{-2.56}$ during Morse's pH-stat rate measurements.

The surface equilibrium pH values implied by our rate equation and Morse's rates are all within 0.01 pH of the theoretical pH for calcite equilibrium in sea water closed to CO_2 (Table III). In a closed system, calcite equilibrium determines both surface pH and PCO_2, and rate depends, in part, on the flux of CO_2 to the surface. Sjoberg (23) noted a stirring dependence of rate at pH 8 and very low CO_2 partial pressures, where calcite dissolution has previously been attributed to surface reaction alone.

The material used by Morse was "whole Indian Ocean sediment" which is largely of biogenic origin. It is expected that the reacting surface area is considerably less than the BET surface area used by Morse to normalize rate. This may also explain some of the disparity between calculated and observed rates (Table III).

Calculation of dissolution rate using equations 14 and 15 becomes extremely sensitive to values of Ω and pH as equilibrium is approached. Calculated rates less than 100 mg cm^{-2} yr^{-1} must be open to considerable uncertainty owing to the difficulty in estimating Ω and surface pH with sufficient accuracy.

Precipitation. The rate model of Plummer et al. (1), although derived from dissolution experiments, accounts for both forward and backward reaction. If no other mechanisms occur, this rate model should also describe the kinetics of crystal growth of calcite. There are several studies of the crystal growth of calcite (27 - 34), all using the seeded growth technique of Reddy and Nancollas (27, 28). By this method, well characterized seed crystals are introduced into a stable supersaturated solution of $NaHCO_3$-$CaCl_2$. Crystal growth begins immediately and is followed by measurement of pH and total dissolved calcium. The partial pressure of CO_2 is not controlled and tends to increase in solution as crystallization proceeds. Our calculations show that some CO_2 is also outgassed from solution during crystal growth. Most crystal growth experiments have been conducted between pH 8 and 10, and at CO_2 partial pressures between 10^{-3} and 10^{-5} atm, the only exception being the work of Nancollas et al. (34) which is near pH 6 and 1.7 atm CO_2.

At low PCO_2, Reddy and Nancollas (27, 28) found that the rate of crystal growth was described by an equation of the form

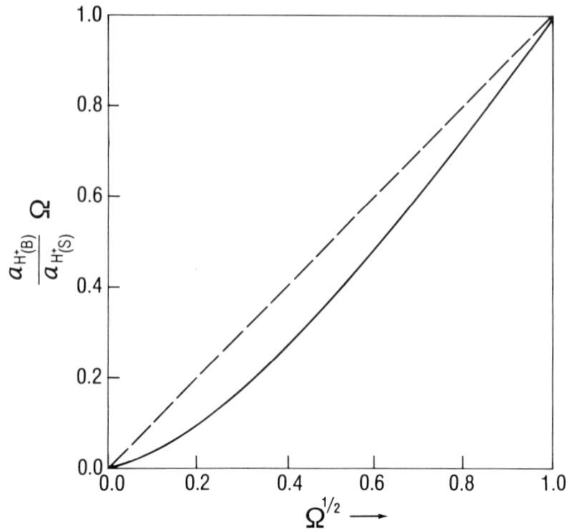

Figure 12. Correlation of $\Omega^{1/2}$ with the term $aH^+/aH^+(s)\ \Omega$ showing similarity of the terms, indicating that the form of Sjoberg's rate equation (14, 23) is similar to ours

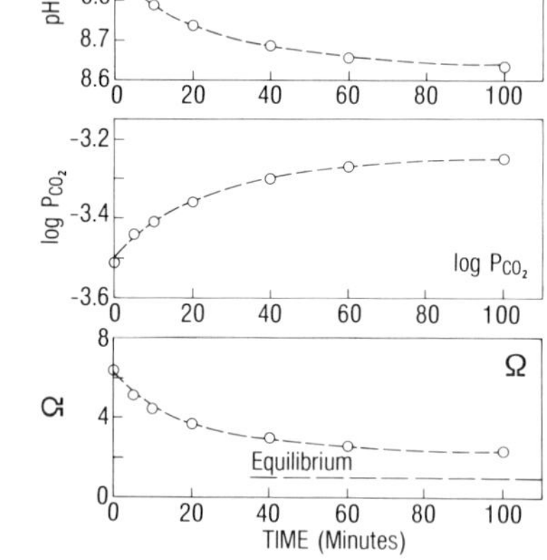

Figure 13. Typical variations in calcium, pH, log P_{CO_2}, and calcite saturation (Ω) during crystal growth experiments of Reddy (30, 35) in dilute $NaHCO_3$–$CaCl_2$ solutions

$$R = -k_G \left(m_{Ca^{2+}} m_{CO_3^{2-}} - \frac{K_c}{\gamma_\pm^2} \right) \tag{20}$$

where k_G is the crystal growth rate constant, m denotes molality, K_c is the calcite equilibrium constant, and $\gamma_\pm^2$ is the product of activity coefficients γCa^{2+} γCO_3^{2-}. This result was later confirmed by the experiments of Wiechers et al. (31). Equation 20 can be rearranged to the form

$$R = k_G \frac{K_c}{\gamma_\pm^2} (1 - \Omega) \tag{21}$$

At low PCO_2 our rate model reduces to

$$R = k_3 a_{H_2O} \left(1 - \frac{a_{H^+}}{a_{H^+(s)}} \Omega\right). \tag{22}$$

Calculation of the term $aH^+/aH^+(s)$ in several low PCO_2 crystal growth experiments gives values near 0.93 initially and increasing to 0.97 at the termination of the growth experiments. Thus the form of equations 21 and 22 is similar because the term $aH^+/aH^+(s)$ is nearly constant and near unity.

More recently a rate equation which is second order in the instantaneous calcium yield has been proposed (29) and applied to low PCO_2 crystal growth data (30, 32, 33). These studies describe rate of crystal growth by an equation of the form

$$R = k\,(C - C_s)^2 \tag{23}$$

where R is rate of crystal growth, k is a crystal growth rate constant, C is the concentration of dissolved calcium in solution as a function of time and C_s is the equilibrium calcium concentration at that time.

As a means of testing our rate model during crystal growth, we examined details of total calcium and pH during four comparable crystal growth experiments (30, 35). Figure 13 compares total calcium, pH, log PCO_2 and calcite saturation (Ω) in the bulk fluid during a typical run. Rate of crystal growth was calculated using the difference in successive calcium measurements, and varies from about 80 to 2 x 10^{-9} mmol cm^{-2} sec^{-1} in the four selected crystal growth experiments. When we assume that surface PCO_2 is equal to that calculated in the bulk fluid, and assume that calcite equilibrium determines surface pH, the calculated rates of crystal growth vary from 4 to 68 times faster than the observed values, and generally average ten to twenty times faster than the observed.

These differences between computed and observed rates are

significant and indicate either that our model is wrong, or that surface PCO_2 values differ from those in the bulk fluid. We have used these differences to predict the chemical properties of the surface during Reddy's crystal growth experiments (30, 35) assuming our reaction mechanism model describes the reaction. Figure 14 shows the ratio of the implied surface PCO_2 to the bulk fluid value as a function of reaction progress (Ω). Early in the reaction, when crystal growth is most rapid, our rate model and the data of Reddy (30, 35) imply that surface PCO_2 is approximately 5 times larger than the bulk fluid value. This ratio decreases as the rate of crystal growth decreases, with the implied surface PCO_2 nearly twice the bulk fluid value at the termination of the run (which is still quite far from equilibrium).

In Figure 15 we compare the theoretical open system equilibrium surface pH (assuming surface PCO_2 = bulk fluid PCO_2) with the surface pH implied by the PCO_2 imbalance shown in Figure 14. If our reaction mechanisms can be applied to crystal growth, our rate model predicts that surface pH is initially 0.6 pH less than the theoretical open system equilibrium value and the difference decreases to about 0.3 pH at termination of the experiment. These differences diminish as the reaction approaches equilibrium.

These calculations neither prove nor disprove our mechanism model, since the implied PCO_2 (and pH) differences are qualitatively consistent. That is, during the crystal growth experiments, CO_2 produced by $CaCO_3$ precipitation increases in the bulk fluid and some is lost to the atmosphere. This net flux of CO_2 is consistent with the calculated higher surface CO_2 partial pressures.

We have made a parallel calculation, similar to that given above for crystal growth at low PCO_2, using crystal growth data (34) near pH 6 and 1.7 atm. CO_2. During this experiment, of 110 minutes duration, most of the crystal growth occurred in the first 40 minutes, during which Ω varied from 17.3 (initially) to 7.4 (at 40 minutes), with the final value of Ω near 5.8 in the bulk fluid. Calculated rates (assuming surface PCO_2 is equal to the bulk fluid values, and that calcite equilibrium then determines surface pH) are all somewhat faster than the observed rates, but during the first 40 minutes of reaction, computed and observed rates differ by only a factor of two or less. After 40 minutes, the observed rates decrease faster than the calculated rates; the final calculated rate being approximately 12.4 times faster than the observed.

It is apparent from our calculations of the rate of calcite crystal growth that the agreement in computed and observed rate is far more satisfactory at higher PCO_2 than low PCO_2.

Summary and Discussion

This review has been limited to studies in which solid surface areas were reported. We have not considered specific exper-

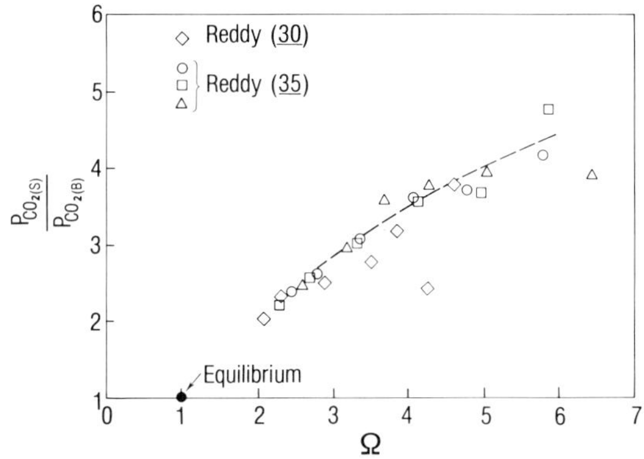

Figure 14. Ratio of surface P_{CO_2} *to bulk fluid* P_{CO_2} *during crystal growth experiments of Reddy (30, 35), as implied by the observed rates and the rate model of Plummer et al. (1)*

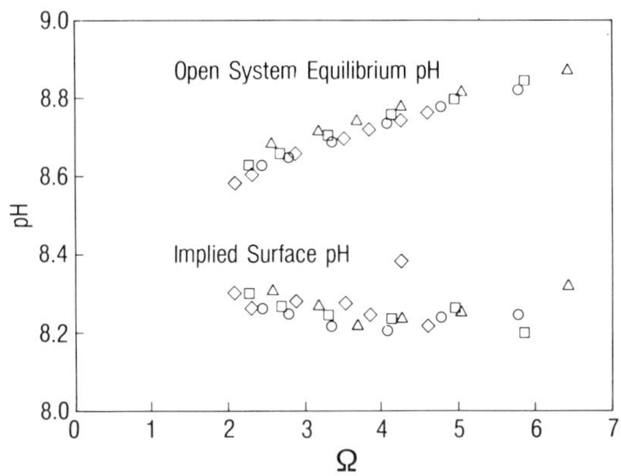

Figure 15. Comparison of calculated pH in equilibrium with calcite at the bulk fluid P_{CO_2} *during crystal growth experiments of Reddy (30, 35) with the surface pH implied by our rate model and the observed rate as a function of calcite saturation (Ω)*

iments involving inhibition. From the remaining kinetic studies with calcite, rates range from -1 x 10^{-5} mmol cm^{-2} sec^{-1} (precipitation) to about +1 x 10^{-2} mmol cm^{-2} sec^{-1}, while the results of various workers represent a range in pH of about 0 to 10, PCO_2 from 0.000001 to 1.7 atm, and temperature between -15 and 60°C. Solution compositions vary from the relatively simple CO_2 - H_2O system to the CO_2 - sea water system. Hydrodynamic conditions range from stirred batch experiments to experiments with rotating disks and cylinders and solution "drilling" experiments with high velocity jets.

Most rates predicted using equations 3 and 13, or 14 and 15 are within a factor of 20 of the observed and many are within a factor of 2 or better. The form of the forward rate as a function of PCO_2 and pH seems to agree with that observed by us. In addition, the shape of concentration - time curves predicted by equations 3 and 13 are similar to those observed in free drift dissolution experiments. Uncertainties greater than a factor of 2 are probably significant and point to three problem areas:
1) At low pH, rate depends significantly on the hydrodynamic transport constant for H^+ which is not well defined. For example, at 25°C, our calculations from observed rates show that k_1 may vary from ∿0.007 cm sec^{-1} under approximately laminar boundary layer conditions at the end of a rotating disk (10) to about 0.23 cm sec^{-1} at the impact of a jet (at ∿35 m sec^{-1}) on the calcite surface (9). Under the turbulent conditions of the stirred batch experiments of Plummer et al. (1), k_1 is near 0.05 cm sec^{-1}.
2) Accurate comparison of results requires knowledge of reaction site density per unit surface area. Calcite materials used for kinetic study have included natural marbles, limestones, hydrothermal crystals of Iceland spar, tests of calcareous organisms and laboratory and commercial precipitates. Surface areas, estimated by BET methods and graphical methods (based on particle size distribution) range from about 0.005 to 2 m^2g^{-1}. There are apparent discrepancies between graphical and BET surface areas and the question is raised as to which type of surface area estimate is most representative of the reacting surface area.

One could argue that BET surface areas overestimate the reacting surface area. The reacting surface area may coincide with a smoothed aqueous film on the surface which can have much less surface area than that determined by BET methods. The surface area of interstices may be essentially non-reactive owing to contact with solutions which are nearer to equilibrium. The uncertainty in reacting surface area is most significant in reactions with biogenic material, (36, 21). Even if the reacting surface area can be defined, it is still necessary to determine the number of reaction sites per unit reacting area. The number of reaction sites may depend on the number of imperfections in the crystal structure, so that less perfect crystals may dissolve faster than more perfect crystals under otherwise identical conditions. Clearly, an intercalibration is required where rates are measured for differing material of known BET surface area under identical hydrochemical conditions.

3) The third general area of uncertainty concerns the controls of surface pH. We have shown (1) that at relatively high bulk fluid PCO_2 (>0.03 atm), surface PCO_2 and aH_2O are near the bulk fluid values and surface pH is controlled by calcite equilibrium at the bulk fluid PCO_2 and aH_2O. Calcite equilibrium at the bulk fluid PCO_2 controls surface pH throughout most of our dissolution experiments, and accounts for our close agreement with the results of Erga and Terjesen (15) (PCO_2 = 0.135 - 0.952) and Terjesen (17) (PCO_2 = 0.97 atm). When we assume surface PCO_2 is equal to bulk fluid PCO_2 during calcite dissolution in sea water, calculated rates greater than 100 $mg/cm^2/yr$ are within a factor of two of those observed by Berner and Morse (3). Even calculated rates of calcite crystal growth at high PCO_2 are within a factor of two of the observed (34).

At low PCO_2 and nearer equilibrium, however, discrepancies in calculated rates of dissolution and precipitation become significant, with calculated rates, based on the assumption of surface PCO_2 = bulk fluid PCO_2, generally faster than the observed rate. Calculations show that near equilibrium, calculated rates of dissolution and precipitation are extremely sensitive to surface pH. It seems likely that our earlier prediction that surface PCO_2 is equal to the bulk fluid value (1) is only valid at relatively high PCO_2 (PCO_2 >0.03 atm).

Surface PCO_2 depends in part on the flux of CO_2 to the surface during dissolution and from the surface during precipitation. Possible kinetic problems related to the CO_2 flux are discussed elsewhere (37, 4). During dissolution, surface PCO_2 is expected to be slightly less than the bulk fluid value, and during precipitation, surface PCO_2 should be greater than the bulk fluid value. This effect is most noticable when PCO_2 of the bulk fluid is low (<.03 atm), and causes the surface pH to be greater in dissolution and less in precipitation than that expected for calcite equilibrium at the bulk fluid PCO_2.

Examination of equation 15 shows that these predicted departures in surface pH from values calculated at the bulk fluid PCO_2 are at least qualitatively significant for both dissolution and precipitation. For example, very low rates of dissolution in sea water at 0.0025 atm CO_2 (21) (Ω = 0.62 - 0.92) can be calculated from equation 15 if surface pH is only 0.02 to 0.10 pH units higher than that calculated at equilibrium with the bulk fluid PCO_2. The calculated values of surface pH, implied by the observed rate (21) and equation 15, are all within 0.01 pH units of the theoretical value for calcite equilibrium in a system closed to CO_2. During crystal growth experiments (30, 35), bulk fluid PCO_2 varies from 0.0003 to 0.0006 atm. while the implied surface PCO_2 is near 0.0015 atm., some 5.0 (initially) to 2.0 times the bulk fluid value.

According to our mechanistic model, the rate of calcite surface reaction far from equilibrium should be described by equation 1. At CO_2 partial pressures less than 0.03 atm., the forward rate becomes independent of PCO_2 (1). Because the activity of water is near unity in most natural waters, and pH is generally too high and PCO_2 too low for reactions 4 and 5 to contribute substantially to the forward rate, the rate of calcite dissolution far from equilibrium in most natural water systems should be dominated by reaction 6 and depend only on temperature.

Most natural water systems in contact with calcite (oceans, rivers, lakes, carbonate rock aquifiers) are, however, near equilibrium, and PCO_2 dependence cannot be ignored. According to our model, the rate of backward reaction is a significant function of surface pH, and surface pH is determined by calcite equilibrium at the surface PCO_2. At the relatively high pH, low PCO_2 conditions of most natural waters, the surface pH is least well defined and may depend, in part, on the flux of CO_2 between the surface and bulk fluid.

Acknowledgments

We thank Dr. M.M. Reddy of the New York State Department of Health for providing details of crystal growth experiments. Constructive criticism of an earlier version of this paper was provided by M.M. Reddy, and our colleagues at the U.S. Geological Survey, E.T. Sundquist and D.C. Thorstenson.

Abstract

Kinetic results of published laboratory studies of calcite dissolution and precipitation in the system CaO-MgO-Na_2O-K_2O-CO_2-H_2SO_4- HCl-H_2O are used to test the surface mechanism model of Plummer, Wigley, and Parkhurst (1978; Amer. Jour. Sci. 278, 179-216). This mechanism model includes: 1) three simultaneous reactions at the mineral-aqueous solution interface, 2) heterogeneous equilibrium within the adsorption layer, 3) forward reaction first order in the boundary layer activities of H^+, $H_2CO_3^*$ and H_2O, and 4) backward reaction controlled by interaction of boundary layer Ca^{2+} and HCO_3^- species with the adsorption layer speciation. Concentration - time curves predicted by this model have shapes similar to those observed in dissolution experiments. Most calculated rates are within a factor of 10 of the observed and many are within a factor of 2 or better. Discrepancies larger than a factor of 2 are probably significant and identify three problem areas: 1) quantification of transport aspects of the rate of dissolution under differing hydro-chemical conditions, 2) uncertainties in reaction surface area and reaction site density per unit surface area, and 3) difficulties in defining surface (adsorption layer) pH, which is essential in determining rate of backward reaction.

Literature Cited

1. Plummer, L.N., Wigley, T.M.L., and Parkhurst, D.L. The kinetics of calcite dissolution in CO_2-water systems at 5° to $60^{o}C$ and 0.0 to 1.0 atm CO_2. Amer. Jour. Sci. 278, 179-216 (1978).
2. Mullin, J.W. "Crystallization", 480 p. Chemical Rubber Co. Press, Cleveland, 1972.
3. Berner, R.A., and Morse, J.W. Dissolution kinetics of calcium carbonate in sea water, Part IV. Theory of calcite dissolution. Amer. Jour. Sci. 274, 108-134 (1974).
4. Plummer, L.N., and Wigley, T.M.L. The dissolution of calcite in CO_2-saturated solutions at $25^{o}C$ and 1 atmosphere total pressure. Geochim. Cosmochim. Acta 40, 191-202 (1976).
5. Plummer, L.N., Jones, B.F., and Truesdell, A.H. WATEQF - A FORTRAN IV version of WATEQ, a computer program for calculating chemical equilibrium of natural waters. U.S. Geol. Survey Water Resour. Invest. 76-13, 71 p. (1976).
6. Davies, C.W. "Ion Association", 190 p. Butterworths, Washington, D.C., 1962.
7. Plummer, L.N., Parkhurst, D.L., and Kosiur, D.R. MIX2: A computer program for modeling chemical reactions in natural waters. U.S. Geol. Survey Water Resour. Invest. 75-61, 73 p. (1975).
8. King, C.V., and Liu, C.L. The rate of solution of marble in dilute acids. Amer. Chem. Soc. Jour. 55, 1928-1940 (1933).
9. Weyl, P.K. The solution kinetics of calcite. Jour. Geol. 66, 163-176 (1958).
10. Tominaga, H., Adzumi, H., and Isobe, T. Viscosity effect on the rate of solution of calcium carbonate in hydrochloric acid. Chem. Soc. Japan Bull. 14, 348-352 (1939).
11. Kaye, C.A. The effect of solvent motion on limestone solution. Jour. Geol. 65, 34-46 (1957).
12. Nierode, D.E., and Williams, B.B. Characteristics of acid reaction in limestone formations. Soc. Petrol. Eng. Jour. Trans. 251, 406-418 (1971).
13. Lund, K., Folger, H.C., McCune, C.C., and Ault, J.W. Acidization - II. The dissolution of calcite in hydrochloric acid. Chem. Eng. Sci. 30, 825-835 (1975).
14. Sjoberg, E.L. Kinetics and mechanism of calcite dissolution in aqueous solutions at low temperatures. Stockholm Contrib. in Geol. 32, 92 p. (1978).
15. Erga, O., and Terjesen, S.G. Kinetics of the heterogeneous reaction of calcium bicarbonate formation, with special reference to copper ion inhibition. Acta Chem. Scand. 10, 872-875 (1956).
16. Bircumshaw, L.L., and Riddiford, A.C. Transport control in heterogeneous reactions. Quart. Rev. Chem. Soc. London 6, 157-185 (1952).

17. Terjesen, S.G., Erga, O., Thorsen, G., and Ve, A. Phase boundary processes as rate determining steps in reactions between solids and liquids: the inhibitory action of metal ions on the formation of calcium bicarbonate by the reaction of calcite with aqueous carbon dioxide. Chem. Eng. Sci. 14, 277-289 (1969).
18. Nestaas, I., and Terjesen, S.G. The inhibiting effect of scandium ions upon the dissolution of calcium carbonate. Acta Chem. Scand 23, 2519-2531 (1969).
19. Runnels, D.C. Diagenesis, chemical sediments, and the mixing of natural waters. Jour. Sed. Petrology 39, 1188-1201 (1969).
20. Wigley, T.M.L., and Plummer, L.N. Mixing of carbonate waters. Geochim. Cosmochim. Acta 40, 989-995 (1976).
21. Morse, J.W. Dissolution kinetics of calcium carbonate in sea water: VI. The near-equilibrium dissolution kinetics of calcium carbonate-rich deep sea sediments. Amer. Jour. Sci. 278, 344-353 (1978).
22. Hawley, J.E., and Pytkowicz, R.M. Interpretation of pH measurements in concentrated electrolyte solutions. Mar. Chem. 1, 245-250 (1973).
23. Sjoberg, E.L. A fundamental equation for calcite dissolution kinetics. Geochim. Cosmochim. Acta 40, 441-447 (1976).
24. Weiss, R.F. Carbon dioxide in water and sea water: the solubility of a non-ideal gas. Mar. Chem. 2, 203-215 (1974).
25. Lyman, J. Buffer mechanism of sea water. Ph. D. Thesis, 196 p., Univ. California, Los Angeles, 1956.
26. Ingle, S.E., Culberson, C.H., Hawley, J., and Pytkowicz, R.M. The solubility of calcite in sea water at atmospheric pressure and $35^{o}/oo$ salinity. Mar. Chem. 1, 295-307 (1973).
27. Reddy, M.M., and Nancollas, G.H. The crystallization of calcium carbonate. I. Isotopic exchange and kinetics. Jour. Colloid Interface Sci. 37, 824-830 (1971).
28. Nancollas, G.H., and Reddy, M.M. The crystallization of calcium carbonate. II. Calcite growth mechanism. Jour. Colloid Interface Sci. 37, 824-830 (1971).
29. Reddy, M.M., and Nancollas, G.H. Calcite crystal growth inhibition by phosphonates. Desalination 12, 61-73 (1973).
30. Reddy, M.M., Kinetics of calcium carbonate formation. Proc. Internat. Assoc. Theoret. Appl. Limnol. 19, 429-438 (1975).
31. Wiechers, H.N.S., Sturrock, P., and Marais, G.V.R. Calcium carbonate crystallization kinetics. Water Res. 9, 835-845 (1975).
32. Reddy, M.M. Crystallization of calcium carbonate in the presence of trace concentrations of phosphorous-containing anions. I. Inhibition by phosphate and glycerophosphate ions at pH 8.8 and $25^{o}C$. Jour. Crystal Growth 41, 287-295 (1977).

33. Reddy, M.M. Kinetic inhibition of calcium carbonate formation by wastewater constituents, p. 31-56, in Rubin, A.J., ed., "Chemistry of Wastewater Technology." Ann Arbor Science, Ann Arbor, Michigan, 1978.
34. Nancollas, G.H., Reddy, M.M., and Than, A. Crystallization of calcium carbonate in aqueous solutions at pH 6 (in press).
35. Reddy, M.M. unpublished data, 1978.
36. Honjo, S. Biogenic carbonate particles in the ocean; Do they dissolve in the water column? p. 269-294, in Anderson, N.R., and Malahoff, A., ed., "The Fate of Fossil Fuel CO_2 in the Oceans", Plenum Press, New York, 1977.
37. Curl, R.L. Solution kinetics of calcite, p. 61-66, Proc. 4th Internat. Cong. Speleology, Yugoslavia (1965).
38. Barton, P. and Vatanatham, T. Kinetics of limestone neutralization of acid waters. Environ. Sci. Technol. 10, 262-266 (1976).
39. Wentzler, T.H. A study of the interactions of limestone in acid solutions. M. S. Thesis, 190 p., Pennsylvania State Univ., University Park, 1971.
40. Ellis, A.J. The solubility of calcite in carbon dioxide solutions. Amer. Jour. Sci. 257, 354-365 (1959).
41. Spring, W. Uber den einfluss der temperatur auf die geschwindigkeit der einwirkung der mineralsauren auf marmor. Z. Physik, Chem. 1, 209-220 (1887).
42. Moelwyn-Hughes, E.A. "The kinetics of reactions in solution", 284 p., Oxford Univ. Press, Oxford, U. K., 1933.

Disclaimer: The reviews expressed and/ or the products mentioned in this article represent the opinions of the author(s) only and do not necessarily represent the opinions of the U.S. Geological Survey.

RECEIVED November 16, 1978.

Interfacing with Biologic Models

26

Trace Metal Bioavailability: Modeling Chemical and Biological Interactions of Sediment-Bound Zinc

S. N. LUOMA
U.S. Geological Survey, 345 Middlefield Road, Menlo Park, CA 94025
G. W. BRYAN
The Laboratory, Citadel Hill, Plymouth, England PL1 2PB

The processes controlling the availability of trace elements to aquatic organisms are poorly understood. Simple relationships are seldom found in natural systems between metal concentrations in organisms and total metal concentrations in either the food or water to which the organisms are exposed. A growing body of evidence suggests the effects of total metal concentrations may be strongly modified by both the chemical partitioning of metals in food components (1,2), and the chemical speciation of metals in solution (3,4). In this paper, we review the modeling approaches used to date to describe the biological, physical and chemical interactions which influence metal uptake by organisms, and we present a statistical model describing the effects of physicochemical form on the uptake of Zn from sediments by deposit feeding clams.

Two general modeling approaches have been employed to assist studies of trace metal impacts in aquatic environments: one approach has emphasized biological processes while the other has emphasized chemical controls on metal availability to organisms.

Biologically oriented models of trace metal behavior have been useful in assessing the relative importance of food and water as sources of metals for animals (5-9) and in simulating temporal dynamics of metals in systems subject to pulse inputs (10,11). Such models usually rely upon physiological transport constants determined in laboratory studies of metal exchange rates. Difficulties have been reported in the determination of such constants (12,13), especially when they are used to model long-term metal dynamics (14). The most important weakness of the biological approach, however, is the necessity of specifying the concentration of biologically available metal to which organisms are exposed. Norstrom *et al.* (15) successfully predicted methylmercury concentrations in a population of fish from a model employing bioenergetic parameters and estimates of uptake constants. Physicochemical factors appear to have a

0-8412-0479-9/79/47-093-577$08.25/0
This chapter not subject to U.S. copyright
Published 1979 American Chemical Society

minimal effect on methylmercury uptake by fishes, however. Defining the fraction of total metal available to organisms is a much more difficult task for other metals.

Chemical models of metal speciation have been used to assess the biological availability of different solute metal forms. Pagenkopf (4) and Andrew (16) used equilibrium models to suggest that the availability of Cu to fishes was controlled by the concentration of the free Cu ion. Equilibrium models were also used to show that the toxicity of Cu to phytoplankton followed the activity of metals rather than total metal concentrations (3,17, 18) and that the concentration of free Zn ion plus additional factors (e.g. competition from Ca and Mg) may affect the availability of solute Zn to fishes (19,20).

Use of solute speciation models to help predict metal availability to organisms in natural systems has not been attempted to date. Statistical comparisons of calculated Cu activities in various natural waters with concentrations of Cu in resident phytoplankton populations would provide a relatively simple test of the laboratory-based models of Cu availability. For organisms other than phytoplankton, however, the problem of modeling the bioavailability of metals in natural systems will be more complex.

Experiments employing chemical speciation models have dealt only with metal uptake from solution, but food may also be an important source of some metals (5,6,7,8). The availability of metals associated with solids is considerably lower than the availability of solute metal forms. However, the concentrations of metals bound to solids are usually orders of magnitude higher than metal concentrations in solution. Models of metal uptake by animals which include both solid and solute vectors could be used to show where the low availability of the particulate-bound metal forms is offset by the higher concentrations of these forms.

Models which include metal uptake from food must deal with the effect on uptake of metal partitioning among different forms within the solids. Some evidence suggests partitioning among ligands may affect the bioavailability of metals from organic food components. Carnivorous fish take up ^{65}Zn more efficiently from synthetic foods than from prey organisms (6,21). The availability of ^{65}Zn to fish ingesting labelled snails was lower from snails which were losing their burden of ^{65}Zn than from snails freshly labelled with the nuclide (21). Slowly exchanging compartments of ^{65}Zn would form a larger proportion of the nuclide in the former group of snails than in the latter (12,22), suggesting the availability of Zn to the predator varied with compartmentalization of the metal within the prey. Similarly, the retention of Hg by plaice varied when the fish were fed four different species of labelled invertebrates (7). Mercury apparently took a different biochemical form in the fish when originating from different prey organisms. Little is known about metal partitioning among organic ligands, especially within biological tissues. However, the effect of such partitioning on metal availability to higher trophic-level organisms appears to deserve further study.

The affect of physicochemical form on uptake from solids has been more carefully studied where sediments are the source of food to animals. When deposit-feeding clams were fed various types of sedimentary substrates (oxides of iron and manganese; organic detritus; inorganic and biogenic carbonates) labelled with ^{110m}Ag, ^{109}Cd, ^{60}Co and ^{65}Zn, uptake of the nuclides varied widely among substrates (1,2). The availability of the nuclides to the clam was inversely related to variations in the strength of metal-particulate binding among sediment types. The form of the sorbed metal also affected sediment-water distribution coefficients. If these results obtain in natural systems then the physicochemical form of particulate-bound metals may affect both the availability of metals ingested by particulate-feeding organisms, and (where contact times between water and particulates is long) the concentrations of solute metal to which the organisms are exposed.

Quantifying the influence of physicochemical form on metal uptake from solids is an important prerequisite to developing realistic models of metal dynamics in animals. In this paper we statistically assess the effect of selected chemical characteristics of estuarine sediments on the concentration of Zn in bivalves which ingest those sediments while feeding. Our objective is to determine what proportion of the variation in Zn concentrations of deposit-feeding bivalves from natural systems may be explained by the physicochemical partitioning of Zn in the sedimentary food of the animals.

Experimental Approach

Methodology for modeling the partitioning of Zn in sediments containing a mixture of sorption substrates has not been developed. To assist our statistical assessments we have made several simplifying assumptions about the chemical and biological interactions of sediment-bound Zn:

1) The relatively thin oxidized layer of sediment at the sediment-water interface is chemically different and biologically more important than are subsurface sediments. Many benthic organisms feed primarily at the sediment-water interface (including the bivalves used in this study). Sulfides dominate metal partitioning in the reducing environment of most estuarine subsurface sediments; however, oxides of iron, oxides of manganese, various organic materials and perhaps carbonates should all be important substrates binding metals under oxidized conditions. Our sediment samples were carefully collected from the oxidized layer of sediment, where, we assumed, sulfides were not important substrate.

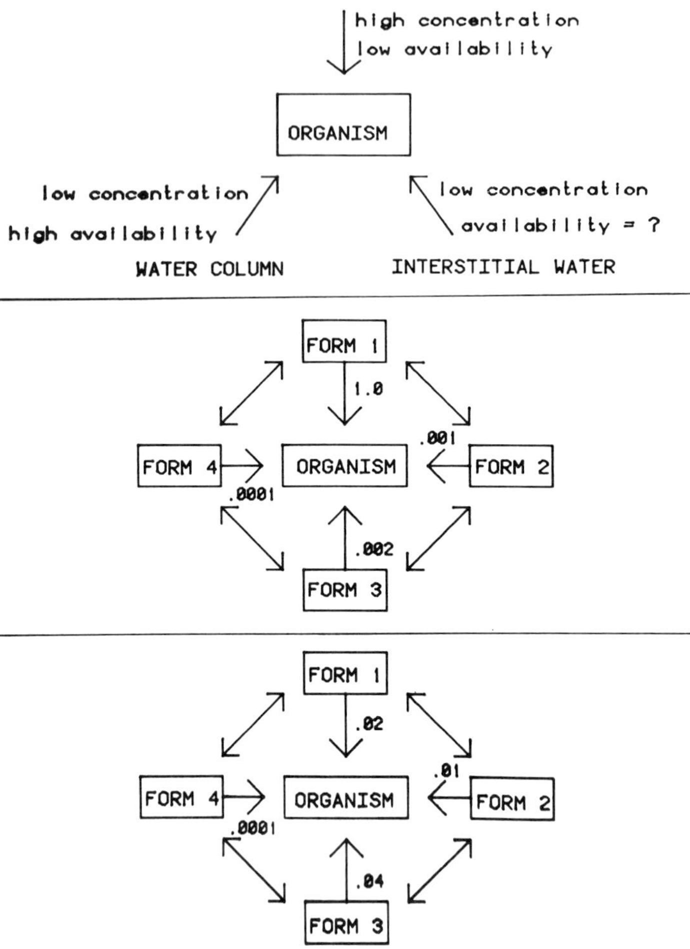

Figure 1. (a) Alternative vectors for trace metal uptake. Uptake from food components (solid forms) is characterized by metal forms of low availability occurring in high concentrations (approximately 3 × 10^4 to 3 × 10^6 μg/kg). The solute vector is characterized by metal forms in low concentrations (0.01–30 μg/L) but of potentially high availability. (b) Hypothetical uptake of different metal forms from a vector where a single form (e.g. the free ion) dominates uptake. Forms 2, 3, and 4 are of very low availability relative to Form 1. Availability numbers represent proportion of a given form which is assimilated when it reaches the environmental interface of the organism. (c) Hypothetical uptake among metal forms of low availability when highly available forms are not present in significant quantities (as is often true of the free ion). Significant uptake will occur when low availability forms are present in high concentrations, and uptake will be affected by relative abundance of low availability forms.

2) Total concentrations in the sediment do not necessarily reflect either the biologically or chemically reactive fraction of metals or substrates. Thus, we have used partial extraction techniques to characterize different phases of both metal and substrate. Under different estuarine conditions, hydrous oxides may vary in crystallinity and in their associations with other substrates (e.g., organics). The nature of the organic materials may also vary greatly. Different extractants remove different quantities of these substrates, in response to differences in substrate form (23). We may not assume that extractants selectively remove trace metals from any single sorption substrate (24), but differences in trace metal solubility among extractants may be useful to empirically separate metal forms susceptible to different treatments. Statistical association of the substrate forms defined by the partial extractants, with either bioavailability or the empirically defined phases of sediment-bound trace metals should reflect the forms of both substrate and metal which are the most reactive in chemical and/or biological interactions within estuarine sediments.

3) The partitioning of adsorbed metals in sediments is controlled by processes analogous to those controlling the speciation of metals in solution. Thus, the quantity of metal sorbed to competing substrates will be determined by metal-substrate stability constants and the relative abundance of the different substrates. Where physicochemical conditions change we would expect partitioning to change.

4) Too few stability constants are available presently to permit calculation of partitioning in multi-substrate systems. In lieu of stability constants we assume metal partitioning in sediments will change solely in response to changes in the relative abundance of the substrates which bind the metals. Since the availability of Zn to bivalves varies among substrates, changes in the relative abundance of substrate concentrations should correlate with changes in the biological availability of the metal to organisms whose metal burden is controlled by sediments. The biological availability of sediment-bound Zn in this case is defined as the concentration of Zn in the bivalves relative to that in the sediment,

$$B = \frac{Zn_{\text{organism}}\ (\mu g/g)}{Zn_{\text{sediment}}\ (\mu g/g)} \quad . \qquad (1)$$

The effect on bioavailability of changes in the chemical partitioning of Zn in sediments is assessed from correlations of B with ratios of the concentrations of different substrates removed from sediments by different chemical extractants.

Methods and Materials

Sediments and organisms were collected in two separate studies. In the first study sediments and the deposit-feeding bivalve Macoma balthica were collected intertidally from 19 stations in San Francisco Bay, California. Several stations were sampled during both the rainy and dry seasons. Sediment samples were scraped from the oxidized surface layer of the sediment, and immediately washed through 250 µm polyethylene mesh with seawater (salinity $32^{o}/oo$). Within 24 h of collection, subsamples (0.10 - 0.35 g dry wt.) were taken from a slurry (using a 0.5 ml micropipette) and extracted in duplicate using methods shown in Table I. Duplicate subsamples from each slurry were washed with distilled water and dried at $80^{o}C$ to determine the dry weight of solids. Subsample sediment weights varied less than 5 mg between duplicates (<5 percent). After extraction each sample was filtered (0.45 µm), and the filtrate was analyzed for Fe, Mn, and Zn by atomic-absorption spectrophotometry. To reduce contamination from filter membranes in the ammonium acetate extraction, 10 ml of ammonium acetate extract was passed through the filter before any filtrate was collected for analysis. Contamination levels were insignificant in other extractants relative to Zn concentrations in the samples. All extracts were stored at pH<2 prior to analysis. Total carbon and inorganic carbon were determined from solid samples using a Leco carbon analyzer. Organic carbon was determined by difference between inorganic carbonate and total carbon.

At least 20 individual Macoma were collected in each sample. In most situations, animals of median size (≈100 mg soft tissue dry wt.) were pooled for the analyses. Where variations in size were unavoidable, the tissues of individual animals were analyzed separately; however, no strong correlation (r<0.50) between Zn concentration and dry tissue weight was found at any station. The Macoma were kept in clean seawater for 24 h after collection to rid their alimentary tracts of undigested sediment; then the soft tissues were removed and dissolved in hot concentration H_2SO_4 with excess HNO_3. The acid solutions were evaporated just to dryness and reconstituted in 25 percent HCl.

Table I

Extractants applied to separate subsamples of sediment.

Extractant	Conditions[1] pH	Conditions[1] Comments	Comments
1 N ammonium acetate	--	---	dissolves manganese; removes exchangeable metals
0.4 N ammonium oxalate[2] in 0.4 N oxalic acid	3.3	---	extracts "amorphic iron oxides
25 percent acetic acid	2.2	---	
1 N hydrochloric acid	0.1	---	
1 N ammonia[2]	10	extract 1 week	extracts "humic substances"[3]
con nitric acid		hot digest	"total" metal[4]
hydroxylamine hydro-chloride in 0.01 N nitric acid	2.0	extract ½ h	poor buffering capacity

[1]All extractions were conducted for two hours at room temperature except where noted, using extractant ratios of approximately 30:1 (v/w) or greater.
[2]Extractants used in southwest England, but not San Francisco Bay.
[3]Alkali-soluble organics comprised less that 5% of the total organic carbon and were not dependent upon total organic concentrations (Luoma and Bryan, unpublished data, 1977).
[4]Some highly crystalline metal forms, soluble only in concentrated hydrofluoric acid are not extracted by concentrated nitric acid (Bryan, unpublished data, 1977).

In the second study, sediments and the deposit feeding clam Scrobicularia plana were collected from 36 stations in 17 estuaries in southwest England. Methods for collecting and extracting sediments in the English study were similar to those employed in San Francisco Bay, except that the English sediments were sieved through 100 µm mesh, were extracted with a greater variety of techniques (Table I) and were subjected to particle size analysis (percent particles less than 14 µm, as determined by density). Concentrations of Zn in the seaweed, Fucus vesiculosus, were used as an indicator of the concentration of biologically available solute Zn.

Scrobicularia and Macoma are morphologically and ecologically very similar species. Scrobicularia is larger (median dry wt. of soft parts approximately 300 mg) and longer-lived than the Macoma found in San Francisco Bay (Bryan and Luoma, unpublished data, 1977). Significant positive correlations were found between concentrations of Zn and soft tissue dry weight in all Scrobicularia samples (25). For our purposes mean concentrations of Zn were taken only from animals with lengths within 50 percent of that of the largest animal. This method does not remove all the biases introduced by the size correlation (25) but it should allow comparisons of animals of roughly similar ages. The Scrobicularia were kept in clean seawater for one week after collection, then soft tissues were dissolved in concentrated HNO_3 and reconstituted in HCl before atomic absorption analysis.

Regression analyses were conducted with both log-transformed and untransformed data. Only the linear regression of untransformed data is shown for the San Francisco Bay data, since the values covered a relatively narrow range and the linear equation yielded the highest correlation coefficient. Sedimentary variables were regressed against the bioavailability ratio in regressions with Macoma. While intuitively satisfying, the use of the bioavailability ratio as the dependent variable may lead to interpretative difficulties if the denominator of the ratio contributes significantly to the regression. To simplify interpretation of the multiple linear regression techniques employed with the more complex English data (and because Zn levels in the sediment correlated with those in the clam), the concentration of Zn in Scrobicularia alone was employed as the dependent variable and Zn concentrations in the sediment were included in the independent variable. The data from England were log-transformed for all statistical analyses because of: 1) the wide range of values, 2) the log-normal distribution of most variables, and 3) the higher correlation coefficients obtained for the log-log regressions than for linear regressions. F-values were calculated for all variables in the multiple regression model.

Results

Sediments.

Differences Between Study Areas. The physical differences (e.g., morphology, hydrodynamics, watershed characteristics and temperatures), between San Francisco Bay and the estuaries of southwest England contributed to major differences in the chemistry of the sediments in the two study areas. San Francisco Bay has a large watershed and is a large, physiographically complex estuary, over half of which forms a slowly circulating embayment. The English estuaries were generally much smaller in size than San Francisco Bay with small watersheds, and relatively straight channels to the sea which result in much shorter residence times for water. Sediments from several of the English estuaries were enriched with Zn from mine drainage, and the range of Zn concentrations found in both the bivalves and sediments in southwest England was considerably greater than the range of concentrations in San Francisco Bay (Figure 2a). The English sediments also showed a wider range of Fe concentrations (but a higher frequency of low concentrations) than San Francisco Bay, plus, generally higher concentrations of organic carbon, and lower concentrations of Mn (Figure 2b,c,d). Carbonates were consistently low in San Francisco Bay (less than 0.1 percent carbonate carbon; <1 percent $CaCO_3$) but were as high as 3.0 percent carbonate carbon (18 percent $CaCO_3$) in some English estuaries. Likewise, the concentrations of humic substances were much higher throughout the English study than were observed at the few stations for which we have data in San Francisco Bay. The significant chemical differences between study areas could result in differences in the partitioning of sediment-bound Zn. On the basis of mass balance we would expect sediment-bound Zn to be more closely associated with the oxides of iron and manganese in San Francisco Bay, while total organic materials, the humic fraction of the organics and, perhaps, carbonates are likely to be of greater importance in the estuaries of southwest England.

Extraction of Sedimentary Constituents. The proportion of the total concentration of Fe, Mn and Zn removed from the estuarine sediments by the different extractants varied widely among sampling stations (Table II). Hydrochloric acid, acetic acid and ammonium oxalate removed a similar proportion of total Zn from all sediments. Concentrations of Zn well above detection limits were also found in the ammonia, hydroxylamine and ammonium acetate extracts. The quantity of Zn removed from oxidized sediments by the latter two extractants was significantly greater than reported in previous studies in which reduced sediments were extracted (26).

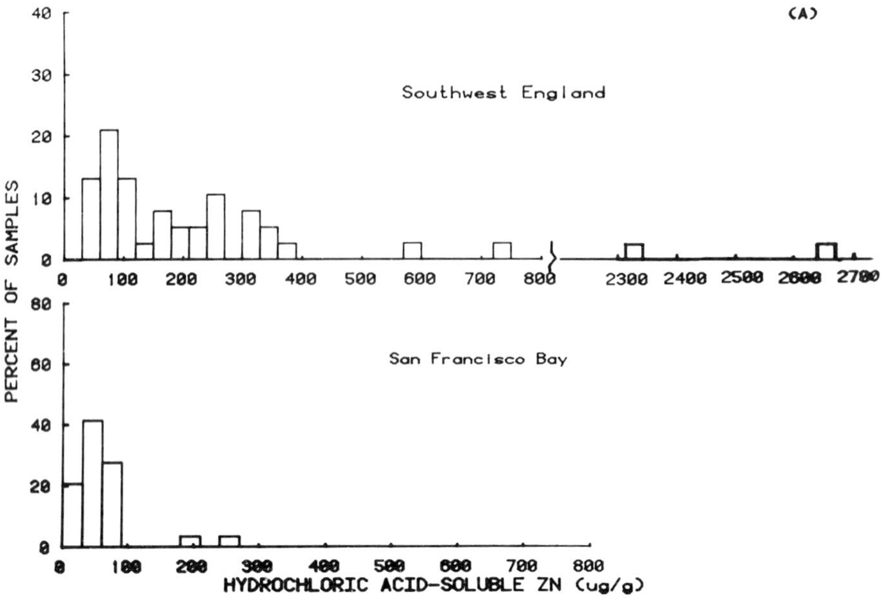

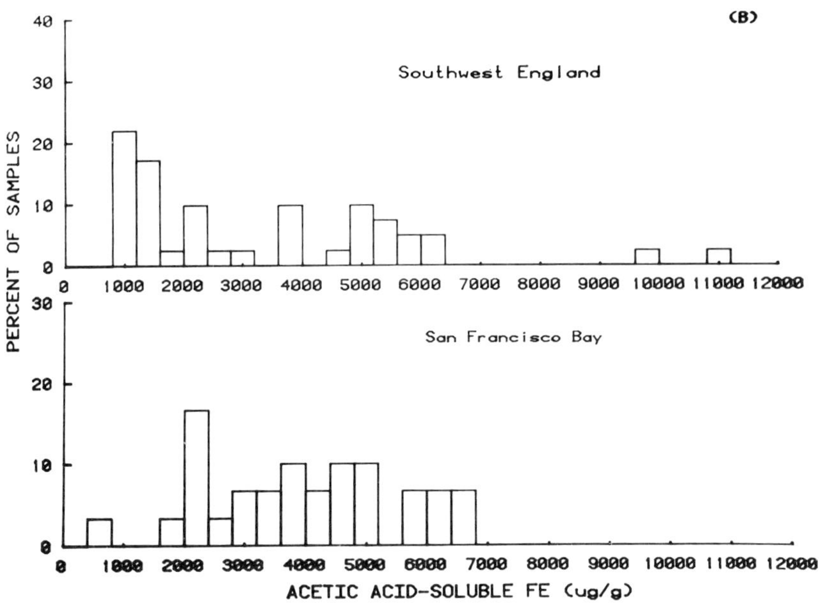

Figure 2. Histograms showing the percent distribution of (a) zinc, (b) acetic acid–soluble iron, (c) organic carbon, and (d) acetic acid–soluble manganese among sediment samples collected from southwest England (n = 40) *and San Francisco Bay* (n = 28). *Concentrations on dry weight basis.*

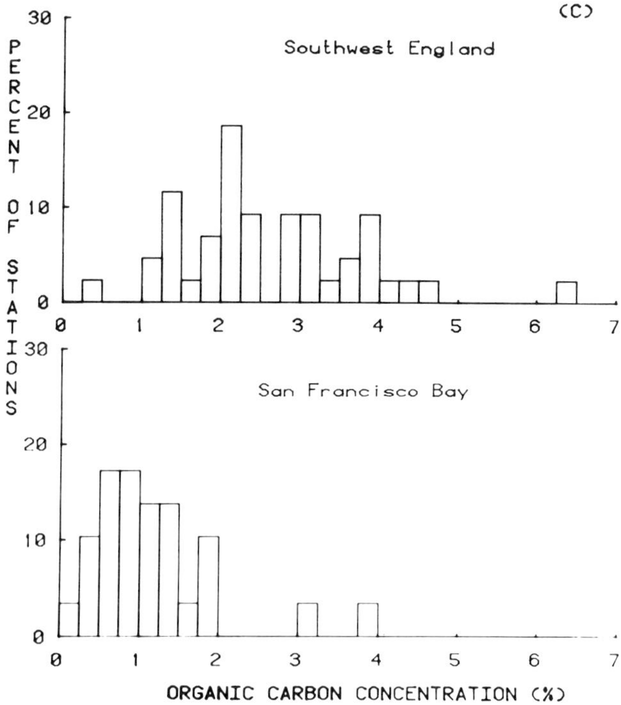
(C)
Southwest England
San Francisco Bay
PERCENT OF STATIONS
0
10
20
30
0 1 2 3 4 5 6 7
ORGANIC CARBON CONCENTRATION (%)

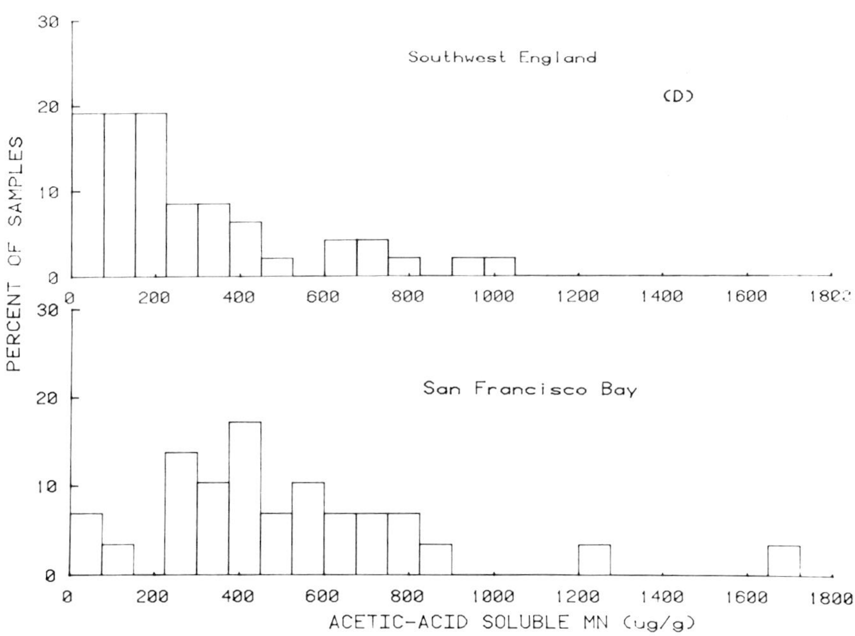
Southwest England
(D)
San Francisco Bay
PERCENT OF SAMPLES
0
10
20
30
0 200 400 600 800 1000 1200 1400 1600 1800
ACETIC-ACID SOLUBLE MN (ug/g)

Table II

Proportion of total Zn, Fe and Mn removed from oxidized sediments by various extractants. The first value is the mean; the values in parentheses represent the range of values observed.

Extractant	Fe	Mn	Zn
Hydrochloric acid[1]	0.200 (0.07-0.40)	0.463 (0.18-0.79)	0.624 (0.35-0.92)
Ammonium oxalate[1]	0.244 (0.08-0.55)	0.369 (0.14-0.68)	0.541 (0.28-0.83)
Acetic acid[1]	0.107 (0.03-0.22)	0.477 (0.14-1.00)	0.575 (0.29-0.91)
Hydroxylamine[2]	0.089 (0.02-0.193)	0.530 (0.13-0.70)	0.260 (0.09-0.38)
Ammonium acetate[1]	nd	0.084 (0.025-0.195)	0.029 (0.010-0.082)
Ammonia[1]	0.0024 (0.0002-0.0090	0.0023 (0.0001-0.0082)	0.0094 (0.0007-0.0391)

[1] Data from English estuaries. SF Bay values fall within this range. nd = Fe concentrations less than 5 μg/g in most samples.

[2] Data from San Francisco Bay; due to carbonate interference with this extractant a full set of data was not available from the English estuaries.

About half of the total Mn in sediments was removed by the low pH extractants. Ammonium acetate removed a smaller, but significant, quantity of Mn. Less than 1 percent of the Mn was solubilized with the humic substances dissolved by ammonia.

The largest proportion of the total Fe was removed by ammonium oxalate, which attacks the amorphic fraction of iron oxide in the sediments (23). Among the low pH extractants, hydroxylamine was the least efficient in extracting Fe. The difference between the extraction of Fe by acid and extraction by hydroxylamine was related to the crystallinity of the hydrous iron oxide. As pure iron oxides aged (and crystallized) in the laboratory, Fe solubility in hydroxylamine declined relative to solubility in acetic acid (Table III). In San Francisco Bay sediments, the ratio of hydroxylamine-soluble Fe to acetic acid-soluble Fe increased during the period of maximum runoff to the estuary (27) suggesting the proportion of the Fe in the sediments that was freshly precipitated varied seasonally. This was expected, since periods of heavy runoff are also times of maximum Fe movement from the watershed to the tributaries of the estuary (28).

Table III

Extractability of Fe from laboratory-prepared iron oxide as a function of the age of the precipitate.

Age	Extracted Fe		Ratio
	Hydroxylamine	Acetic Acid	Hydroxylamine/ acetic acid
(d)	(mol Fe)	(mol Fe)	
6	2.77	5.87	0.47
21	0.39	2.94	0.13

The concentrations of Zn extracted from sediments in both study areas correlated most closely with concentrations of Fe (Table IV). Correlations with amorphic iron were stronger than correlations with total iron. This was not surprising, since the number of surface sites available for binding metals declines substantially as iron oxides age from amorphic to crystalline forms (23). Extractable Zn also correlated significantly with Mn and, to a lesser extent, with total organic carbon in San Francisco Bay sediments. In the English sediments extractable Zn was also weakly correlated with total Mn and with organic carbon, but showed no significant correlation with particle size.

Statistical correlations are a somewhat ambiguous method of assessing metal form in sediments. A statistically significant regression may imply a chemical relationship within sediments although such a correlation may also occur if metals and substrates are deposited simultaneously (such similarities seem unlikely over the wide range of conditions in the English study), or if a third variable (e.g., particle size) correlates with both variables. In San Francisco Bay, significant correlations were observed between extractable Fe and extractable Mn ($r=0.77$, $p<0.01$, in acetic acid), but neither substrate was correlated with organic carbon. Thus, we may speculate that inorganic oxides are highly important in binding Zn in San Francisco Bay sediments but we cannot separate the influence of Mn from that of Fe. Separation of the importance of Fe from that of organic carbon in the English estuaries is also ambiguous, since these two substrates were positively correlated, and both correlated negatively with particle size. However, the strength of the correlation between Zn and Fe, and the significant correlation of Mn with only Zn, suggests the oxides are also of some importance as binding substrates in the estuaries of southwest England. The weak correlation of Zn with both organic carbon and particle size in the English estuaries does not necessarily mean these are not chemical and physical variables of significance within samples. Among samples, however, the wide range of Zn enrichment appears

Table IV
Correlation coefficients showing the relation of Zn to Fe, Mn, organic carbon and particle size in sediments extracted with either acetic acid, hydroxylamine hydrochloride in 0.01 N nitric acid, ammonium oxalate, ammonium acetate, or concentrated nitric acid. English data were log transformed for statistical analysis.

Zinc extractant	Fe			Mn			Organic carbon	Particle size
	Oxalate	Total	Hydroxyl-amine	Acetic acid	Total	Ammonium acetate		
Con nitric acid	0.80**	0.58**		0.14	0.37*	0.26	0.36*	-0.12
Acetic acid	0.78**	0.51**		0.14 [a]0.66**	0.36*	0.30	0.41** [a]0.56**	0.04
Ammonium oxalate	0.78**	0.50**		0.15	0.38*	0.27	0.42**	-0.02
Hydroxylamine			[a]0.81**	[a]0.74**			[a]0.52**	
Ammonium acetate	0.62**	0.41**		0.21	0.38*	0.30 [a]0.72**	0.19 [a]0.47**	-0.19

[a] Data from San Francisco Bay; unmarked coefficients are from English study.
** $p<0.01$. At n=40, r >0.39 (English study). At n=28, r >0.46 (San Francisco Bay).
* $p<0.05$. At n=40, r >0.30. At n=28, r >0.36.

to overwhelm any control of organics and particle size on Zn concentrations in the sediments.

Biological Availability.

Direct Extraction of Available Form. Concentrations of Zn in Scrobicularia correlated at a high level of significance with the wide range of extractable Zn concentrations in the sediments of the English estuaries (Table V). Within a single estuary (San Francisco Bay) there was no significant

Table V

Correlation coefficients describing the relationship between the concentration of Zn extracted from sediments and the concentration of Zn observed in *Macoma* (San Francisco Bay) and *Scrobicularia* (Southwest England).

Extractant	Correlation with Zn concentrations in clam: San Francisco Bay	Correlation with Zn concentrations in clam: Southwest England
Con nitric acid	0.09	0.48**
Hydrochloric acid	0.10	0.48**
Ammonium oxalate	nd	nd
Hydroxylamine hydrochloride	0.33	
Ammonium acetate	0.20	0.62**
Ammonia	nd	0.37*

** $p<0.01$. For English data (n=40), $r > 0.39$. For San Francisco Bay (n=28) $r > 0.46$.
* $p< 0.05$. For English data $r > 0.30$; for San Francisco Bay $r > 0.36$.
nd not determined.

correlation between Zn extracted from the sediments and Zn in the animals. The strongest correlations with the animals were observed for the ammonium acetate-soluble fraction of Zn in the sediments (English data) and for the two extractants which dissolve amorphic oxides of iron and manganese (oxalate in England and hydroxylamine hydrochloride in San Francisco Bay). The direct extraction methods explain less than 12 percent of the variance in the Zn concentrations of *Macoma*, and, despite their statistical significance, less than 40 percent of the variance in *Scrobicularia*.

Influence of Substrate Concentrations.

San Francisco Bay. To assess the influence of Zn partitioning in the sediments on uptake, regressions were calculated between the ratio describing Zn bioavailability (Equation 1) and concentrations of total organic carbon, iron oxides and manganese oxides (the latter two substrates as extracted by various techniques). Among all combinations of variables, the strongest correlation was observed for the regression

$$y_1 = a + b\,X_1 \qquad (2)$$

where y_1 = the bioavailability ratio (Equation 1), and

$$X_1 = \frac{Fe_{Hydam}}{Mn_{AmAc}}$$

where Fe_{Hydam} is the concentration of Fe removed by hydroxylamine hydrochloride, and Mn_{AmAc} is the concentration of Mn removed by ammonium acetate (Figure 3). The correlation coefficient was highly sensitive to the forms of Mn and Fe employed in the regression. Hydroxylamine-soluble Mn and acetic acid-soluble forms of both Fe and Mn showed a weak relationship to the bioavailability of Zn. The variable X_1 increased as the fraction of Mn extracted by ammonium acetate declined toward the head of the estuary. The concentrations of Mn removed by other extractants did not decline rapidly upstream. At the upstream stations the concentrations of Zn in *Macoma* relative to that in the sediments was also high. Although this suggested the variations in X_1 may be related to salinity, no significant correlation was found between bioavailability and salinity. Moreover, both X_1 and bioavailability were highest at a station outside the mouth of the estuary, where the salinity was near that of seawater.

The inverse correlation between the bioavailability of Zn and Mn_{AmAc} was relatively weak (r=0.41). The positive residuals of this relationship were largely data collected during the winter. The winter influx of fresh water into San Francisco Bay was accompanied by an increase in the sediments of hydroxylamine-extractable Fe (presumably, freshly precipitated Fe) and humic substances (27). We have data from too few San Francisco Bay stations to include the humics in regression calculations. However, the increase in hydroxylamine-Fe generally coincided with increases in Zn concentrations in *Macoma*; thus, the combined Fe/Mn ratio in Equation 2 explained 60 percent of the temporal and spatial variance in the Zn concentrations of the bivalve when all the data were considered.

The largest negative residuals in Figure 3 represented stations with high concentrations of total organic carbon, suggesting Zn concentrations in *Macoma* were lower than expected from the Fe/Mn ratio where total organic carbon concentrations were elevated. Organic carbon concentrations were included in the independent variable of the regression.

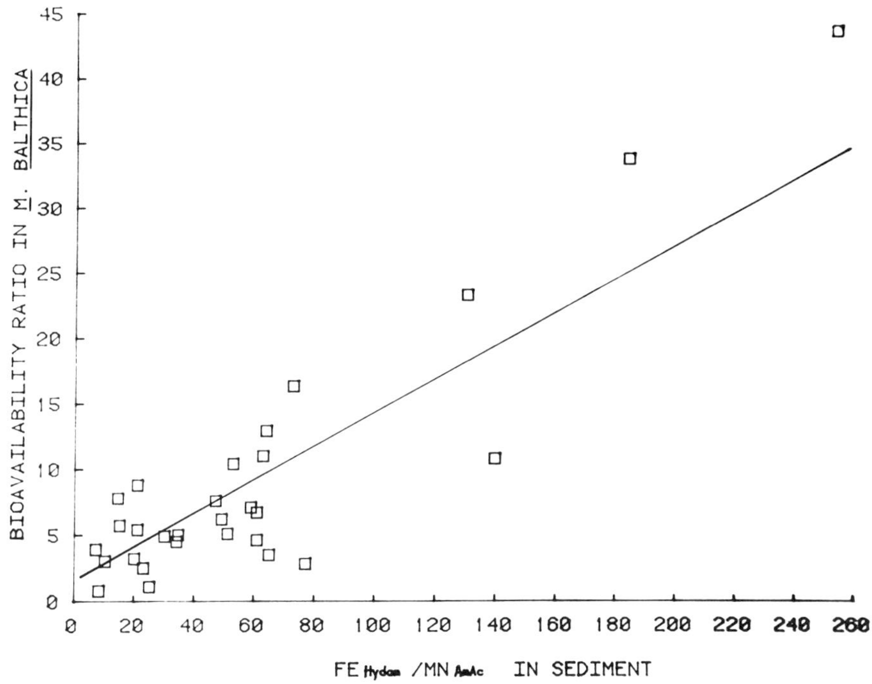

Figure 3. The bioavailability of zinc ($Zn_{clam}/Zn_{sediment}$) *to* Macoma balthica *as related to the ratio of hydroxylamine–soluble iron/ammonium acetate–soluble manganese in oxidized sediments of San Francisco Bay* (r = *0.77*).

$$y_1 = a + b\,X_2 \qquad (3)$$

where y_1 = the bioavailability ratio (Equation 1), and

$$X_2 = (Fe_{Hydam}/Mn_{AmAc}) \times (1/\text{organic carbon}).$$

The correlation coefficient of this regression was slightly higher (r=0.83 vs r=0.77) than when Fe/Mn alone was employed as the X_2 variable. The residuals of the regression suggested an overestimation of Zn bioavailability at the lowest carbon concentrations - i.e., the negative effect of carbon on bioavailability appeared to occur only at high levels of carbon. The effect of excluding low organic carbon concentrations from the relationship was tested by comparing correlation coefficients when carbon values below 0.6, 1.0, 1.5, or 3.8 percent were removed from calculation of the X_2 variable. This was accomplished by normalizing all carbon concentrations to one of the above and assuming concentrations lower than the chosen value were to equal one (i.e. all carbon values lower than the normalizing number had no effect on the value of the variable). The maximum correlation coefficient was observed when organic carbon was normalized to 1.0 percent (Figure 4). Normalizing to carbon levels lower than 1.0 percent progressively detracted from the relationship, while failure to include the higher carbon levels also resulted in a less-than-optimum correlation. When organic carbon was normalized to 1.0 percent the correlation coefficient (r=0.96) suggested 92 percent of the variance in the biological availability of Zn to _Macoma_ could be attributed to variations in concentrations of Fe, Mn and organic carbon in the sediments (Figure 5). The correlation between the denominator of the bioavailability ratio ($Zn_{sediment}$) and variable X_2 was weak (r=0.51); thus, most of the variance in the Zn concentrations of _M. balthica_ appeared to be due to variations in the ratio of iron, manganese and organic carbon. If this ratio is indicative of the partitioning of Zn in sediments, then nearly all the variation in Zn concentrations in _Macoma_ may be explained by the form of Zn in the sediments.

Southwest England. The effects of sedimentary variables on Zn concentrations in Scrobicularia were initially tested by multiple regression. Values of the F-statistic were calculated to determine the level of significance at which sedimentary variables explained Zn levels in the clam (Table VI). Several different regression equations were calculated to compare the different methods of extracting Zn, Mn and Fe. The multiple regression analyses employed only the 29 stations at which seaweed was found.

In the first run of the multiple regression, only the three variable (Fe, Mn, organic carbon) included in the San Francisco Bay study were analyzed (Table VI). The concentration of Zn, Mn and organic carbon in the sediments explained 49 percent of the

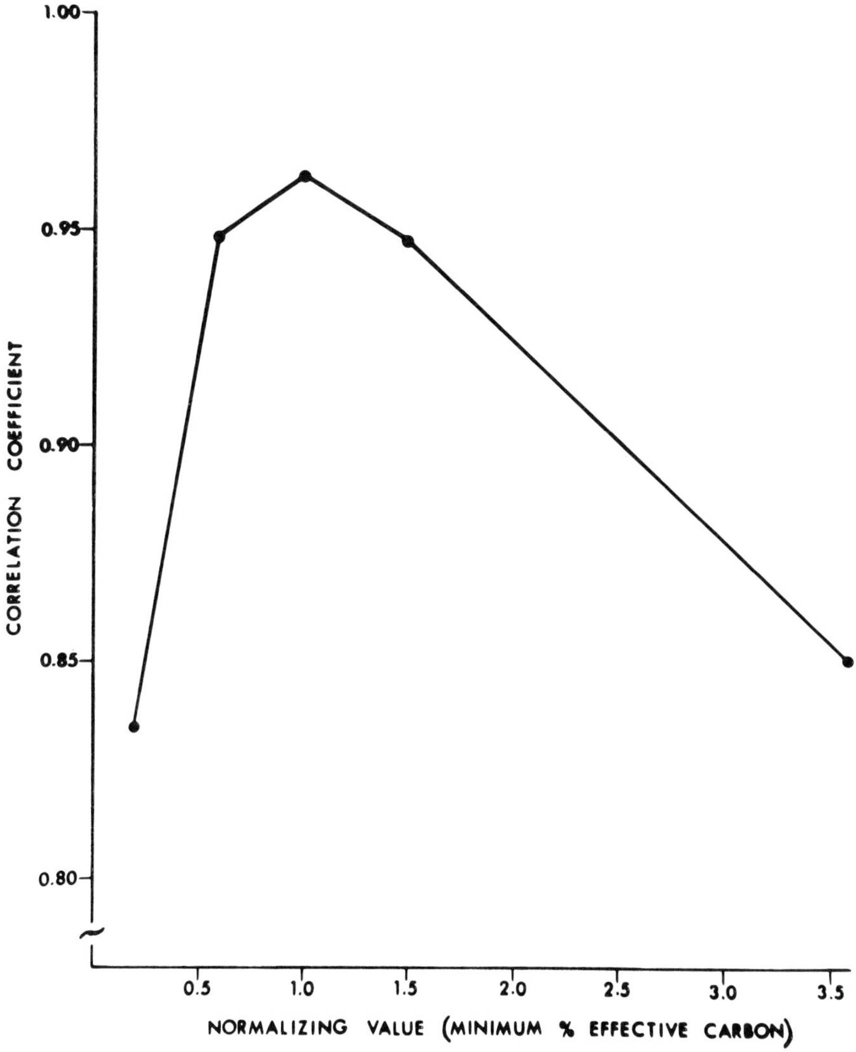

Figure 4. Correlation coefficients for the relationship between the bioavailability of zinc to M. balthica *and the sedimentary variable, $X_2 = (Fe_{Hydam}/Mn_{AmAc})$ (1/organic carbon), graphed as a function of the value to which organic carbon concentrations were normalized. All values less than the normalized value were assumed to equal one in the calculation.*

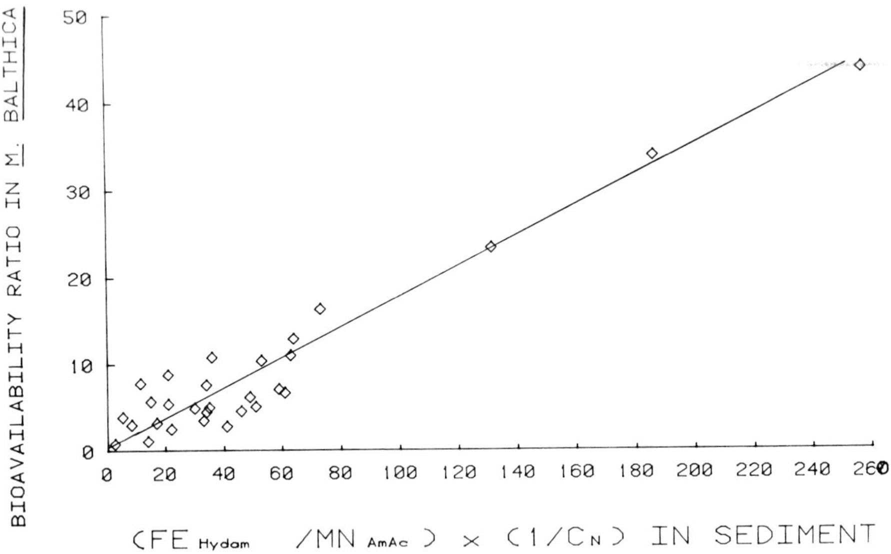

Figure 5. The bioavailability of zinc to Macoma balthica *($Zn_{clam}/Zn_{sediment}$) as related to the ratio in San Francisco Bay sediments of (Fe/Mn) × ($1/C_N$) where C_N signifies only organic carbon concentrations of greater than 1% are included in the calculation* (r = *0.96).*

Table VI

Values of the F-statistic calculated from a multiple regression of Zn concentration in Scrobicularia plana versus the combinations of variables listed. Percent variance (r^2) explained by the regression is also shown.

r^2	F-values												
	Zinc			Iron		Manganese			Organic carbon	Carbonate carbon	Humic substances	Extractable Cd	Solute Zn
	Ammonium acetate	Hydrochloric acid	Oxalate	Oxalate	Total	Ammonium acetate	Oxalate	Total					
	(+)	+	(+)	+	(+)	(+)	(+)	(+)	(-)	(-)	(+)	(+)	(+)
71.1	6.53**				0.01		3.66*		9.01**	3.13*	9.22**	2.21	5.80**
66.9	7.10**				0.05			0.40	6.00**	1.58	8.39**	0.91	4.42*
66.6	7.44**				0.03	0.20			5.04*	1.46	6.92**	0.67	5.03*
64.2			1.61		0.00		4.00*		6.09**	2.76	4.69*	2.44	7.50**
62.2			1.59	0.07			4.68*		5.65**	2.73	3.74*	2.78	6.76**
[a]49.0		11.2**			0.01		6.77**		5.44**				

* F-values >3.08= p< 0.05*

** F-value >5.29= p<0.01**

(+) and (-) indicate the sign of the slope taken by the variable.

a Only the variables used in San Francisco Bay study were included in this multiple regression run to facilitate comparison between the study areas.

variance in Scrobicularia. More of the variance in Scrobicularia was explained by ammonium acetate than when other extractants were employed (Table VI). Total organic carbon had a significant negative effect, the concentration of humic substances a positive effect, and solute concentrations of Zn a positive effect. Oxalate-soluble Mn had a weak positive effect, although neither total Mn nor ammonium acetate-soluble Mn significantly affected the regression. Carbonates had a weakly significant effect (negative), in one combination of variables. Neither Cd nor Fe showed any effect detectable by multiple regression. The regression equation yielding the highest correlation coefficient (r=0.843) was

$$\log Zn_{\underline{Scrob.}} = 0.03 + 0.30 (\log H) - 0.59 (\log C) + 0.19 (\log Zn_{AmAc}) + 1.22 (\log Zn_{solute}) + 0.17 (\log Mn_{oxal.}) - 0.08 (\log CO_3) \quad (4)$$

where H = the concentration of humic substances (absorbance/g sediment) and C = total organic carbon (percent).

Using data from all 40 stations, simple regressions were also calculated for the equation

$$\log y_2 = \log a + (b) (\log X_3) \quad (5)$$

where y_2 = the concentration of Zn in Scrobicularia and

$$X_3 = (Zn_{sediment}) \times (Substrate_1/Substrate_2)$$

for all combinations of the Fe, Mn, organic carbon, carbonate and humic material substrates. These calculations employed all 36 data points. The results were similar to those calculated by multiple regression (Table VII). The correlation was strongest when the fraction of Zn extracted by ammonium acetate was employed (Figure 6a). When the product of ammonium acetate-soluble Zn and the ratio of humic substance concentrations to total organic carbon was employed as X_3, i.e.

$$X_3 = (Zn_{AmAc}) \times (H/C) \quad (6)$$

67.1 percent (r=0.819) of the variance in Scrobicularia was explained (Figure 6b). When the ratios Mn/C and $MnCO_3$ were included with Zn_{AmAc} in X_3 the correlation was also improved over that between Zn_{AmAc} and Zn in the bivalve (Table VII). Including Mn with the organic ratio in a more complex independent variable

$$X_4 = (Zn_{AmAc}) \times (H/C) \times (Mn_{oxal}) \quad (7)$$

Table VII
Correlation coefficients showing relation of Zn concentrations in Scrobicularia to combinations of sedimentary variables, in the equation

$$\log y = \log a + (b)(\log X_3)$$

where y = Zn concentrations in Scrobicularia and $X_3 = (Zn_x)(substrate_w/substrate_z)$.

Headings are factors which make up X_3 variable.

Substrate w / Substrate z	Zn_1	Zn_4	Zn_3	Zn_6
Fe_4/Mn_1	0.54	0.36	0.39	0.37
Fe_4/Mn_3	0.44	0.22	0.25	0.25
Fe_6/Mn_6	0.58	0.36	0.42	0.32
Fe_4/Carbon	0.61	0.46	0.50*	0.46
Fe_6/Carbon	0.62	0.48	0.53*	0.44
Mn_1/Carbon	0.59	0.47	0.51*	0.43
Mn_4/Carbon	0.66*	0.55*	0.59*	0.52*
Mn_6/Carbon	0.62	0.51	0.55*	0.48
Fe_4/CO_3	0.61	0.48	0.51*	0.45
Fe_6/CO_3	0.60	0.46	0.49*	0.44
Mn_1/CO_3	0.64*	0.52	0.56*	0.50*
Mn_4/CO_3	0.68*	0.55*	0.58*	0.54*
Mn_6/CO_3	0.64*	0.52	0.55*	0.51*
humics/organic carbon	0.82*	0.60*	0.68*	0.50*
humics/carbonate	0.55	0.40	0.44	0.37

*Coefficients exceeding those between Scrobicularia and sedimentary Zn alone (from Table II). Subscript identification numbers are: 1=ammonium acetate; 3=hydrochloric acid; 4=oxalate; 6=concentrated nitric acid.

did not affect the correlation coefficient. Concentrations of Zn in Scrobicularia were higher than predicted by the independent variable in Equation 6 at several sites with very high Mn concentrations. A slight improvement in the correlation coefficient (r=0.842; r = 0.709; Figure 6c) was observed when only the high Mn values were inserted in Equation 5 by normalizing Mn concentrations to 350 μg/g to exclude the lower values (assuming all concentrations less than 350 μg/g took the value of 1 in variable X_4, Equation 7).

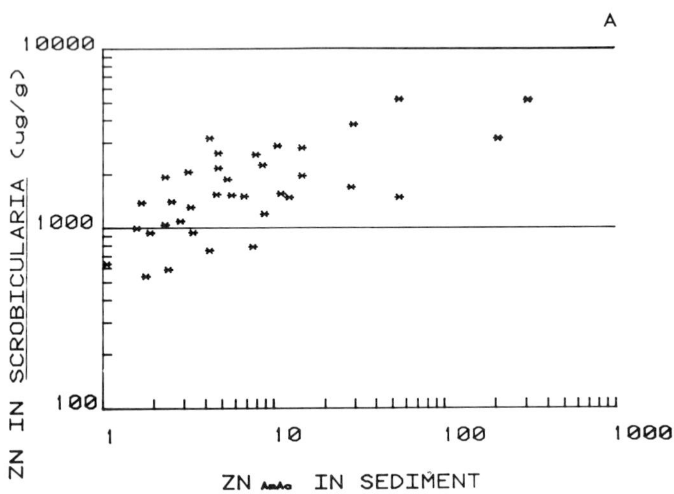

Figure 6. Zinc concentrations in Scrobicularia plana *from southwest England as related to: (a,* above*) ammonium acetate-soluble Zinc in sediments from southwest England* (r = *0.62); (b,* top right*) the product of ammonium acetate–soluble zinc and the ratio of humic substance concentrations (H) (absorbance in a 1N ammonia extract) to total organic carbon (C)* (r = *0.82); and (c,* bottom right*) the product of* $(Zn_{AmAc}) \times (H/C) \times (Mn_{oxal})$ *where* Mn_{oxal} = *oxalate soluble manganese where present at concentrations* > *350* $\mu g/g$ (r = *0.84).*

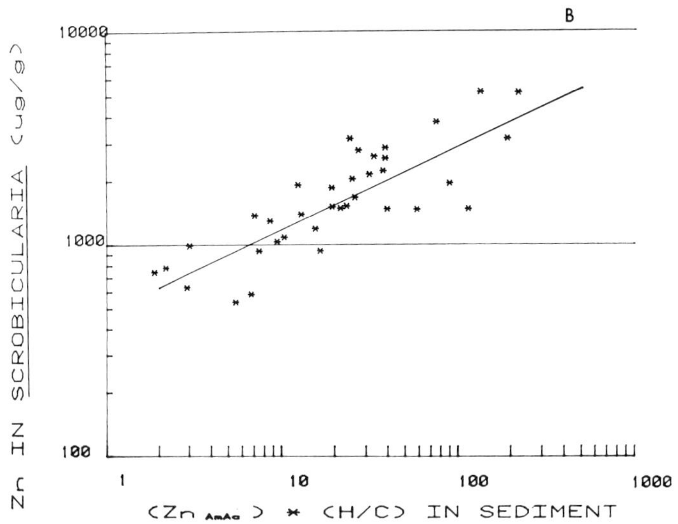
B
10000
1000
100
1
10
100
1000
Zn IN SCROBICULARIA (ug/g)
(Zn AmAc) * (H/C) IN SEDIMENT

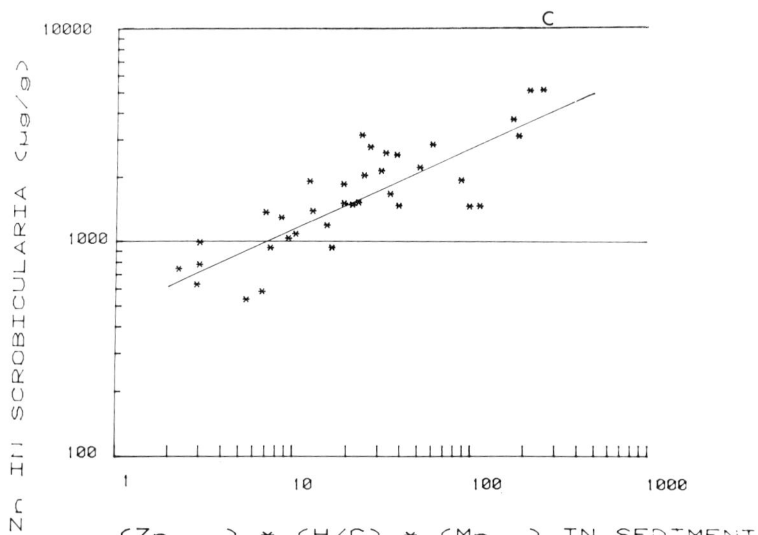
C
10000
1000
100
1
10
100
1000
Zn IN SCROBICULARIA (μg/g)
(Zn AmAc) * (H/C) * (Mn N) IN SEDIMENT

The residuals of the relationship in Figure 6c suggested salinity also had some effect on Zn concentrations in Scrobicularia. Stations from near the heads of the estuaries (with one exception) had high positive residuals, while those stations from near the mouths of the estuaries generally had negative residuals. Because of evaporation and the transient nature of salinities in the small English estuaries, it is difficult to reliably determine the salinity Scrobicularia was exposed to on the mudflat. Therefore, the salinity effect will require further investigation.

Discussion

To date, attempts to define the biologically available fraction of particulate-bound metals have largely bypassed any direct consideration of metal partitioning among physicochemical forms. Instead, extractants have been sought that simulate biological processes, directly removing the biologically available fraction of metals. Some correlation between metal concentrations extracted from soils and concentrations in plants have been observed within soil series (e.g. ammonium acetate extraction of Zn; DTPA extraction of Cu (30)), but among different types of soils these relationships break down. The search for a universal extractant for the biologically available fraction of metals has recently been extended to rationalize the development of complex extraction schemes for aquatic sediments (31,32). Luoma and Jenne (24) used less involved methodology to show some correlation between extractable nuclide concentrations and the uptake by Macoma of ^{65}Zn, ^{109}Cd, ^{60}Co and ^{110m}Ag from laboratory-prepared sediments. A close correlation between ammonium acetate-solubility and ^{65}Zn uptake was observed. In nature, ammonium acetate also reflected the bioavailability of Zn to Scrobicularia in the estuaries of southwest England better than did other extractants. The correlation was not sufficiently strong to have predictive value, however, and within the narrower range of Zn concentrations in San Francisco Bay no correlation was observed. The exchangeability of Zn, implied by ammonium acetate extraction, may influence the availability of the metal, but it is not the only factor determining uptake.

To directly assess the effect of Zn partitioning in sediments on uptake, we have assumed partitioning, and, in response, availability will change with changes in the concentrations of substrates which bind Zn. Despite this obvious oversimplification we have found ratios of sedimentary substrate concentrations that explain a very large proportion of the variance in either the bioavailability or the concentration of Zn in two species of bivalves. We have reported similar results for lead (33). Although the ratios of substrates which predicted Zn uptake differed between southwest England and San Francisco Bay, most of those differences were consistent with chemical differences between the areas. Substrate concentrations suggested inorganic oxides were more dominant than organic materials in the sediments of San Francisco Bay, while organic materials and carbonates occurred in high concentrations in the English estuaries. Distribution of two oxide forms in the sediments explained most changes in the availability of Zn to Macoma in the former estuary, with a negative effect on availability occurring only at higher concentrations (<1 percent) of total organic carbon. The ratio of humic substance concentrations to total organic carbon was an important control on availability in southwest England, with some indications that manganese oxides, solute Zn and low salinities also enhanced Zn uptake. The concentration limit on the effect of total organic carbon in San Francisco Bay (and, possibly, Mn in southwest England) was consistent with the participation of such substrates in partitioning only when present at sufficiently high concentrations to be competitive, although it is also possible the nature of these substrates was different in samples containing low concentrations.

The negative effect of total organic carbon on Zn availability to the bivalves was strongly evident in both San Francisco Bay and southwest England. If our assumptions are correct about the relationship between substrate concentration, Zn partitioning and the availability of Zn, then the Zn associated with the bulk of organic matter in the sediments may be of low availability to deposit feeders. Alternatively, organic materials could produce a similar statistical effect if total organic carbon concentrations influenced feeding rates in the animals. Whatever the mechanism, the negative effect of total organic carbon on Zn concentrations in the bivalves is a somewhat surprising result and warrants further investigation.

The statistical evidence also suggested that increased concentrations of humic substances enhanced the availability of Zn. The extractable humic materials represent a small, but extremely concentrated pool of trace metals in the sediments. Complexation with humic substances generally reduces the availability of metals in solution (34), but the possibility that Zn bound to humic substances may be of high availability relative to other bound forms deserves further investigation.

The role of inorganic oxides in determining the availability of sediment-bound Zn is not entirely clear from our comparisons of the two study areas. Within the sediments extractable Zn correlated most strongly with extractable Fe in both studies. Increased availability of Zn to Macoma correlated with increased concentrations of hydroxylamine-soluble Fe; but no influence of Fe on Zn levels in Scrobicularia was evident (although hydroxylamine-soluble Fe was not determined in the English study due to carbonate interferences with the extraction). In contrast, we have previously shown that extractable Fe inhibits the uptake of Pb by Scrobicularia (33). Ammonium acetate-soluble Mn strongly reduced the availability of Zn to Macoma, while oxalate-soluble Mn had a small positive effect on Zn availability to Scrobicularia. Manganese concentrations in the two study areas differed greatly, and a difference in the form of Mn in the sediments of the different estuarine environments is also possible. Moreover, the controls on Mn dissolution by either oxalate or ammonium acetate are unclear. One of the extractions may reflect undefined factors which correlate with Mn removal, or the forms of Mn removed by the two extractants may actually affect the availability of Zn differently. Not only the concentration of hydrous Fe and Mn oxides, but also the characteristics of the oxides themselves (as reflected in their solubility in different extractants) affected Zn uptake. An understanding of the role of oxide form in the physical, chemical and biological interactions in oxidized sediment awaits further study.

The concentration of Zn extracted from sediments by oxalate correlated with Zn levels in Scrobicularia more strongly than did Zn-solubility in other extractants (except, of course, ammonium acetate). Since oxalate dissolves amorphous Fe and Mn oxides (23), and presumably any associated Zn, the correlation is consistent with the positive effect of the oxides on Zn uptake suggested previously. Neither the concentration of Zn extracted by ammonia, nor the ratio of Zn removed by hydroxylamine and ammonium acetate showed any relationship to Zn levels in Scrobicularia or Macoma, however. Such extractants may extract trace metal-reactive fractions of the substrates involved in binding Zn in the sediments, but it is unlikely that they are selective for only the Zn associated with that fraction of those substrates (24). For example, the fraction of Fe removed from sediments by hydroxylamine may be highly amorphic, with a strong affinity for Zn. However, the concentration of Zn extracted by hydroxylamine includes not only that associated with the Fe, but also Zn associated with Mn oxides and quite possibly some organically-bound Zn (24), all of which may vary widely in their bioavailability. Thus, the concentration of Fe extracted by hydroxylamine may be a better indicator of the partitioning of Zn to the highly amorphic oxides of Fe (apparently a high-availability substrate) than is the concentration of Zn in hydroxylamine. Similarly, reactions occurring during the extraction may affect the concentration of Zn removed more strongly than the concentration of substrate removed. Both precipitation of $ZnOH_2$ and readsorption of Zn to other substrates are likely at the high pH used to extract humic substances. Thus, again, the concentration of humic substances extracted by alkaline reagents may more closely reflect Zn partitioning to this substrate than does the concentration of Zn extracted. Multi-substrate extraction of trace metals by most extractants and interferences from reactions occurring during the extraction process may help explain why the search for universal extractants which remove the biologically available fraction of the metals has met with little success.

There are inherent limitations in any statistical approach to studying controls on trace metal concentrations in organisms. Cross-correlations with undefined variables or spurious correlations within a data set are always possible. We have minimized those possibilities by studying two very different systems and a wide range of conditions within one of the systems. Nevertheless, precise determinations of the interacting effects of various substrates and the effects of other controls on Zn concentrations in deposit feeders must await further study in both the field and laboratory. We have attempted to point out specific questions such studies might address.

The methodology we have used to calculate our independent variables illustrates a possible method for statistically interfacing sophisticated chemical models of metal speciation or partitioning, and biological data from natural systems. In our calculations we have assumed ratios of abundance of substrates alone control Zn partitioning in sediments. If the relative abundance of substrates were weighted with stability constants, indices similar to those used as our independent variables could be derived that would vary as the calculated form of metal varied (the analog of a quantified speciation diagram). Regression of such partitioning indices against metal levels in various species might then point to common (at least within species) regression equations describing metal bioavailability. Such an approach is possible for solute metals and, at least, aquatic plants at the present level of knowledge, but will depend upon the development of realistic stability constants and a better understanding of substrate chemistry before it will be useful with solid metal forms. This method is more complex than is direct extraction of biologically available metals. In lieu of success in the search for such extractants, however, it represents one of many possible first steps toward realistically modeling, in natural systems, the complex interactions that occur at the interface of chemical and biological systems. Both the search for realistic bioavailability models and the search for direct extraction methodology should be facilitated by further study of the factors controlling the bioavailability of sediment-bound metals to animals.

Summary

1) The fraction of sediment-bound Zn extractable in ammonium acetate correlates significantly with Zn concentrations in bivalves when a broad range of Zn concentrations is considered. The relationship is not detectable, however, for a more narrow range of Zn concentrations. While solubility in ammonium acetate reflects a major control on the bioavailability of Zn, other processes are also important.

2) We have assumed the partitioning of Zn among substrates within aquatic sediments is controlled by changes in the relative abundance of the substrates. If so, then the availability of Zn to Macoma in San Francisco Bay is controlled by the competitive partitioning of Zn between two forms of hydrous oxide:hydroxylamine hydrochloride-soluble iron oxide and ammonium acetate-soluble manganese oxide. Organic materials affect the bioavailability of Zn when present in concentrations greater than 1 percent. Concentrations of organic carbon are higher and concentrations of the hydrous oxides are generally lower in the estuaries of southwest England than in San Francisco Bay. The availability of Zn to Scrobicularia appears to be primarily controlled by the ratio of humic substance concentrations to total organic carbon. Solute Zn and low salinities may enhance Zn levels in Scrobicularia at some stations, as may oxalate-soluble Mn when present in high concentrations.

3) The primary value of this study is its introduction of a number of new, specific questions about controls on the availability of Zn to deposit feeders. We have suggested a somewhat new approach to studying chemical and biological interactions in natural systems, and have proposed such an approach might be useful as a statistical interface between more sophisticated chemical and biological models.

Acknowledgements

We would like to thank Mr. G. Burt and Mr. L. G. Hummerstone for their assistance. Part of the study was supported by the Department of the Environment, U.K., under contract DGR 480/51 and part was supported by a National Science Foundation NATO Postdoctoral Fellowship.

Abstract

Extractable concentrations of sediment-bound Zn, as modified by the physicochemical form of the metal in the sediments, controlled Zn concentrations in the deposit-feeding bivalves Scrobicularia plana (collected from 40 stations in 17 estuaries in southwest England) and Macoma balthica (from 28 stations in San Francisco Bay). Over a wide range of concentrations, a significant correlation was found between ammonium acetate-soluble concentrations of Zn in sediments and Zn concentrations in Scrobicularia. This correlation was insufficiently precise to be of predictive value for Scrobicularia, and did not hold for Macoma over the narrower range of Zn concentrations observed in San Francisco Bay. Strong correlation of Zn concentrations in Scrobicularia and the bioavailability of sediment-bound Zn to Macoma with ratios of sorption substrate (oxides of iron and manganese, organic carbon, carbonates, humic materials) concentrations in sediments were found in both the English and San

Francisco Bay study areas. These correlations were attributed to substrate competition for sorption of Zn within sediments, assuming: 1) competition for sorption of Zn was largely controlled by the relative concentrations of substrates present in the sediments and 2) the bioavailability of Zn to the deposit feeders was determined by the partitioning of Zn among the substrates. The correlations indicated that the availability of Zn to the bivalves increased when concentrations of either amorphous inorganic oxides or humic substances increased in sediments. Availability was reduced at increased concentrations of organic carbon and, in San Francisco Bay, ammonium acetate-soluble Mn. Concentrations of biologically available Zn in solution and low salinities may also have enhanced Zn uptake, although the roles of these variables were less obvious from the statistical analysis.

Literature Cited

1. Luoma, S. N. and Jenne, E. A. Factors affecting the availability of sediment-bound cadium to the deposit-feeding estuarine clam Macoma balthica, p. 283-291 in Cushing, C. E., ed., "Radioecology and Energy Resources," Ecol. Soc. Amer. Spec. Pub. 1. Dowden, Hutchinson, and Ross. Stroudsberg, Pennsylvania, 1976.
2. Luoma, S. N. and Jenne, E. A. The availability of sediment-bound cobalt, silver and zinc to deposit-feeding clam, p. 213-231, in Wildung, R and Drucker, H., ed., "The Biological Implications of Metals in the Environment." NTIS CONF-750929, 1977.
3. Sunda, W. G. and Guillard, R. R. Relationship between cupric-ion activity and the toxicity of copper to phytoplankton, J. Mar. Res. 34, 511-529 (1976).
4. Pagenkopf, G. K., Russo, R. C. and Thruston, R. V. Effect of complexation on toxicity of copper to fishes, J. Fish. Res. Bd. Can. 31 (4), 462-465 (1974).
5. Benayoun, G., Fowler, S. W. and Oregioni, B. Flux of cadium through euphasiids, Mar. Biol. 27, 205-212 (1974).
6. Pentreath, R. J. Accumulation and retention of ^{65}Zn and ^{54}Mn by the plaice, Pleuronectes platessa L., J. Exp. Mar. Biol. Ecol. 12 (1), 1-18 (1973).
7. Pentreath, R. J. The accumulation of mercury from food by the plaice Pleuronectes platessa L., J. Exp. Mar. Biol. Ecol. 25, 51-61 (1976).
8. Fowler, S. W. and Benayoun, G. Selenium kinetics in marine zooplankton, Mar. Sci. Commun. 2, 43-51 (1976).
9. Huckabee, J. W., Goldstein, R. A., Janzen, S. A and Woock, S. E. Methylmercury in a fresh-water food chain, p. 199-211, Internat. Conf on Heavy Metals in Environment, Vol. II, Part I, Toronto, Canada, 1975.
10. Luoma, S. N. The dynamics of biologically available mercury in a small estuary, Est. Coast. Mar. Sci. 5, 643-652 (1977).

11. Price, A. H. II, Hess, C. T. and Smith, C. W. Observations of Crassostrea virginica cultured in the heated effluent and discharged radionuclides of a nuclear power reactor, Proc. Natl. Shellfish Assn. 66, 54-66 (1976).
12. Luoma, S. N. Physiological characteristics of mercury uptake by two estuarine species, Mar. Biol. 41, 269-273 (1977).
13. Fowler, S. W., Heyraud, M. and LaRosa, J. Factors affecting methyl and inorganic mercury dynamics in mussels and shrimp, Mar. Biol. 46, 267-277 (1978).
14. Elwood, J. W. and Eyman, L. D. Test of a model for predicting the body burden of trace contaminants in aquatic consumers, J. Fish. Res. Bd. Can. 33, 1162-1166 (1976).
15. Norstrom, R. J., McKinnon, A. E. and DeFreitas, A. S. W. A bioenergetics-based model for pollutant accumulation by fish. Simulations of PCB and methylmercury residue levels in Ottawa River yellow perch (Perca flavesceus), J. Fish. Res. Bd. Can. 33, 248-267 (1976).
16. Andrew, R. W. Toxicity relationships to copper forms in natural waters, p. 123-145, in Andrew, R. W., Hodson, P. V. and Konasewich, D. E., ed., "Toxicity to Biota of Metal Forms in Natural Waters," Great Lakes Res. Advisory Bd. Publ., Windsor, Ontario, Canada, 1975.
17. Jackson, G. A. and Morgan, G. A. Trace metal-chelator interactions and phytoplankton growth in seawater media: Theoretical analysis and comparison with reported observations, Limnol. Oceanogr. 23, 268-283 (1978).
18. Anderson, D. M. and Morel, F. M. Copper sensitivity of Gonyaulax tamarenisis, Limnol. Oceanogr. 23, 283-296 (1978).
19. Pagenkopf, G. K. Zn speciation and toxicity to fishes, p. 77-93, in Andrew, R. W., Hodson, P. V. and Konasewich, D. E., ed., "Toxicity to Biota of Metal Forms in Natural Waters," Great Lakes Res. Advisory Bd. Publ., Windsor, Ontario, Canada, 1975.
20. Zitko, V. Structure-activity relationships and the toxicity of trace elements to aquatic biota, p. 9-33, in Andrew, R. W., Hodson, P. V. and Konasewich, D. E., ed., "Toxicity to Biota of Metal Forms in Natural Waters," Great Lakes Res. Advisory Bd.Publ., Windsor, Ontario, Canada, 1975.
21. Merlini, M., Pozzi, G., Brazzelli, A. and Berg, A. The transfer of ^{65}Zn from natural and synthetic foods to a freshwater fish, p. 226-230, in Cushing, C. E., ed., "Radioecology and Energy Resources," Ecol. Soc. Amer. Publ. 1, Dowden, Hutchinson and Ross, Stroudsberg, Pennsylvania, 1976.
22. Cutshall, N. Turnover of ^{65}Zn in oysters, Health Physics 26, 327-331 (1974).
23. Jenne, E. A. Trace element sorption by sediments and soils, p. 425-553, in Chapel, W. and Peterson, K., ed., "Molybdenum in the Environment," Vol. 2, Marcel Dekker, New York, 1976.

24. Luoma, S. N. and Jenne, E. A. Estimating bioavailability of sediment-bound trace metals with chemical extractants, p. 343-551, in Hemphill, D. D., ed., "Trace Substances in Environmental Health," Vol. X, Univ. Missouri Press, 1977.
25. Bryan, G. W. and Hummerstone, L. G. Heavy metals in the burrowing bivalve Scrobicularia plana from contaminated and uncontaminated estuaries, J. Mar. Biol. Assn. U.K. 58, 401-419 (1978).
26. Serne, R. J. Geochemical distribution of selected trace metals in San Francisco Bay sediments, p. 280-288, in Drucker, H. and Wildung, ed., "The Biological Implications of Metals in the Environment," NTIS CONF-750929, 1977.
27. Luoma, S. N. and Cain, D. J. Temporal dynamics of copper, zinc and silver in South San Francisco Bay as related to freshwater discharge, in Conomos, T. J., ed., "San Francisco Bay: The Urban Estuary," Amer. Assoc. Adv. Sci. (in press).
28. Elder, J. F., Osborn, and Goldman, C. R. Iron transport in a Lake Tahoe tributary and its potential influence upon phytoplankton growth, Water Research 10, 783-689 (1976).
29. Shaw, E. and Dean, L. A. Use of dithizone as an extractant to estimate the zinc nutrient status of soils, Soil Sci. 73, 341-347 (1952).
30. Kelling, K. A., Kenney, K. A., Walsh, L. M. and Ryan, J. A. A field study of the agricultural use of sewage sludge III. Effect on uptake and extractability of sludge-borne metals, J. Environ. Qual. 6, 352-358 (1977).
31. Khalid, R. A., Gambrell, R. P., Patrick, W. H., Jr. Sorption and release of mercury by Mississippi River sediment as affected by pH and redox potential, p. 297-395, in Drucker, H., and Wildung, R. E., ed., "The Biological Implications of Metals in the Environment". NTIS CONF-750929, 1977.
32. Engler, R. M., Brammon, J. M., Rose, J. and Bigham, G. A practical selective extraction procedure for sediment characterization, Presented at 168th Amer. Chem. Soc. Meeting, Atlantic City, September 1974.
33. Luoma, S. N. and Bryan, G. W. Factors controlling the availability of sediment-bound lead to an estuarine bivalve, J. Mar. Biol. Assn. U.K. (in press).
34. Jenne, E. A. and Luoma, S. N. Forms of trace elements in soils, sediments and associated waters: An overview of their determination and biological availability, p. 110-142, in Wildung, R. and Drucker, H., ed., "The Biological Implications of Metals in the Environment," NTIS CONF-750929, 1977.

Disclaimer: The reviews expressed and/or the products mentioned in this article represent the opinions of the author(s) only and do not necessarily represent the opinions of the National Oceanic and Atmospheric Administration.

RECEIVED November 16, 1978.

Effect of the Physicochemical Form of Trace Metals on Their Accumulation by Bivalve Molluscs

FLORENCE L. HARRISON

Lawrence Livermore Laboratory, University of California, Environmental Sciences Division, P.O. Box 5507, Livermore, CA 94550

Bivalve molluscs effectively concentrate many trace elements (1-13). They are filter feeders and as such maintain a flow of water through their gills for feeding, respiration, and the removal of metabolic wastes. Trace metals occur in many physicochemical forms in water, and thus can enter animals by their ingestion of living and nonliving particulate material suspended in the water and from the sorption of substances dissolved in the water. We know neither the rates of accumulation nor the effects of physicochemical form on accumulation of many of the critical elements in most animals.

Many coastal ecosystems have elevated levels of metals and radionuclides (14). Anthropogenic sources of stable isotopes of metals include sewage disposal plants, electroplating plants, and mining and dredging operations; sources of radioactive isotopes include effluents from nuclear power plants and submarines, medical establishments, and uranium ore mining. The pollution from most of these operations results from routine or accidental discharges and are either continuous or episodic.

Models have been developed to predict concentration changes in bivalve molluscs after increased amounts of these pollutants are discharged into their environment. These models can be used to determine the conditions needed to maintain healthly populations of these animals and to minimize adverse effects on man from their consumption. The latter is important because certain metals are implicated in acute health problems in man, and a continued concern exists about the dose to man from radionuclides released into aquatic environments.

In the presentation that follows, we will consider the mathematical models that have been developed and the model parameters required to predict concentration changes in the animals. In addition, experimental data will be provided that indicate the sensitivity of model parameters to differences in physicochemical form of the elements in the water and to differences in metabolic responses among species. These kinds of

0-8412-0479-9/79/47-093-611$06.00/0
© 1979 American Chemical Society

information will provide an indication of the reliability of model predictions and of the areas in which additional data are needed.

Mathematical Models and Model Parameters

Mathematical Models. The accumulation of an element by any pathway can involve a number of different processes. If the rate-determining process can be described mathematically, a model can be developed to predict changes in concentration with time and location. A considerable effort has been made to develop models to predict the distribution of radionuclides released into the environment (15). The types of models developed to predict concentrations of radionuclides in aquatic organisms include equilibrium (16, 17, 18) and dynamic models (19, 20).

The type of model to be used in a given situation depends on the nature of the release and on the properties of the ecosystem. When releases of metals or radionuclides are continuous and steady-state conditions are present, an equilibrium model such as a concentration factor model can be used. In this case, the important parameter needed for the model is the concentration factor, the ratio of the concentration in the animal to that in the water. The concentration in the animal is determined then by multiplying the concentration in the water by the concentration factor. In the case of accidental episodic releases, a dynamic situation exists in which organisms accumulate the material for a relatively short period and then lose it with a characteristic time constant. In this case, the important parameters needed for the model are biological turnover rate constants and concentration factors.

Model Parameters. Concentration factors were determined for large numbers of bivalve molluscs. Values were obtained by determining the concentration of stable and/or radioactive nuclides in animals and water that were collected directly from the environment and that were sampled in laboratory experiments. These data were compiled for use in models to predict radionuclide concentrations in whole organisms or their tissues. In general, for bivalve molluscs a single value is given for each element.

The turnover of trace metals in organisms depends on dynamic processes of exchange with elements in the environment. Compartments of elements are identified from a mathematical analysis of the changes in concentration during accumulation or loss. The resolution of compartments is limited by experimental error, and the compartments that can be identified are those whose concentrations differ significantly in their exponential change. These compartments may be physiological, structural, or chemical entities, and their metabolic roles may not be known.

The turnover time of an element in an organism depends upon the organism and element. In small organisms with a large surface-to-volume ratio, the turnover time of monovalent elements such as Na or Cs may be minutes, whereas in large organisms and multivalent elements, months may be required. The turnover is also a function of the metabolism of the element by the organism. The quantities accumulated by organisms when the concentrations are increased in the water differ greatly for those elements that are and are not under homeostatic control.

Turnover rates were determined by following the accumulation or loss of radionuclides from animals in the field or in the laboratory. They were determined also by following the increased quantities of trace metals in animals that were exposed to increased quantities of the element in the water.

The change in concentration of a trace element in an organism at any time may be described by:

$$dC/dt = k_i W(t) - k_o C(t), \qquad (1)$$

where

$W(t)$ = the concentration in the water at time, t,
C = the concentration in the organism,
k_i = the biological accumulation rate constant,
k_o = the biological loss rate constant.

At steady-state conditions, $dC/dt = 0$, and

$$k_i W = k_o C(s), \qquad (2)$$

where

$C(s)$ = the concentration in the organism at steady state conditions.

Assuming first order kinetics and a constant concentration in the water, equation 1 upon integration becomes:

$$C(t) = \frac{k_i W}{k_o} [1 - e^{-k_o t}] \quad . \qquad (3)$$

Substituting $C(s) = k_i W/k_o$ into equation 3 , we have:

$$C(t) = C(s) [1 - e^{-k_o t}]. \qquad (4)$$

In those situations where the concentration in the water is known, concentration factors (CF) can be substituted for concentrations in the animal to give:

$$CF(t) = CF(s) [1 - e^{-k_o t}], \qquad (5)$$

where

CF(t) = the concentration factor in the organism at time, t,
CF(s) = the concentration factor in the organism at steady-state conditions.

The loss of stable or radioactive nuclides may be described by:

$$C(t) = C(i)\ [e^{-k_o t}], \quad (6)$$

where

C(i) = the initial concentration in the organism.

From k_o, the biological half-life ($T^{1/2}$) of a trace element in an organism may be determined from the relationship:

$$T^{1/2} = \frac{0.693}{k_o}\ . \quad (7)$$

Variations in Model Parameters Resulting from Species Differences

Whole Body Radionuclide Concentration Factors and Turnover Rates. A series of experiments were performed to determine the turnover rates and concentration factors of Co, Cs, Mn, and Zn in the marine clam *Mya arenaria* and the oyster *Crassostrea gigas*. The results published previously (21, 22) are included for comparison.

Radionuclide accumulation was followed in laboratory systems in which the concentrations of stable elements were kept constant (23). Changes in concentrations were followed in critical tissues as well as the entire body. The loss of radionuclides was followed in animals that had accumulated the radionuclides for 48 d in the laboratory. After exposure to the stable and radioactive nuclides, they were transferred to nonradioactive, unfiltered, circulating seawater (at the Marine Laboratory of California State University, Humbolt) in which biological loss of the radionuclides was followed.

In the oyster *C. gigas*, large differences in concentration factors and turnover rates of Co, Cs, Mn, and Zn were found in the soft tissues (Table I). The highest concentration factor measured was for Zn and the lowest for Cs.

In the clam *M. arenaria*, large differences in concentration factors and turnover rates of these elements were also found (Table I). Cobalt had the highest concentration factor and Cs had the lowest.

In both animals, two compartments were identified for most elements. Differences in the percentages of the element in each compartment were found for those elements which had two identifiable compartments. Comparison of the concentration factors for Mn and Co shows more than an order of magnitude difference between the values for oysters and clams. Greater similarity was found in turnover rates than concentration factors; Co, Mn, and Zn had turnover times of 2 to 3 mo in both

TABLE I
Concentration Factors and Half-Lives in Soft Tissues

		Half-life,d	
Element	Concentration Factor	Compartment 1	Compartment 2
		A. Oysters	
Zinc	1200	98 (100)[a]	—
Cobalt (22)	50	6 (38)	130 (62)
Manganese	35	4 (24)	98 (76)
Cesium (22)	10	2 (37)	6 (60)
		B. Clams	
Cobalt (21)	790	120 (100)	—
Manganese	590	8 (30)	70 (70)
Zinc	320	30 (33)	110 (67)
Cesium (21)	5	4 (75)	60 (25)

[a]Value in parenthesis is the percent of the total body element in the compartment.

species, whereas for Cs it was < 1 wk. These data on concentration factors, turnover rates, and compartment sizes suggest different metabolic pathways of these elements in both animals.

Tissue Turnover of Radionuclides. Oysters and clams that had accumulated either ^{54}Mn and ^{65}Zn or ^{60}Co and ^{137}Cs for about 1½ mo in the laboratory were transferred to unfiltered oceanic water, and changes in concentrations of the radionuclides in the tissues were followed for about 5 mo. Large differences were found in the rates of radionuclides loss from the tissues of both the oysters (Figure 1) and the clams (Figure 2). Three patterns of loss during the experimental period were observed. From many tissues, the loss was monophasic, from others it was biphasic, but in some there was an increase or a period of little change before the loss occurred. The last pattern was seen for ^{65}Zn in some tissues of oysters and clams and indicates that zinc may be mobilized from some tissues for accumulation in others.

The accumulation of ^{54}Mn, ^{60}Co, ^{65}Zn, and ^{137}Cs, was followed in M. arenaria under controlled laboratory conditions. The changes in concentration factors of the radionuclides during accumulation differed in each tissue, and, in a given tissue, with each radionuclide (Figure 3). Only for ^{137}Cs were steady-state conditions approached in all the tissues, even though the accumulation was followed for more than 5 mo. The scatter in the data during the sampling period was greatest in

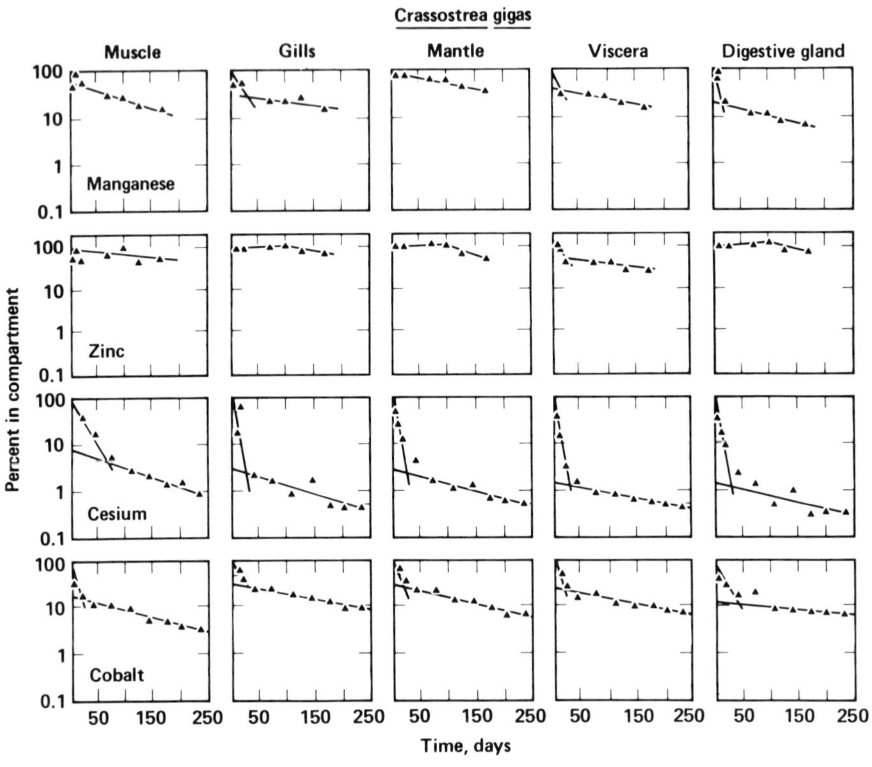

Figure 1. Percent of initial radioactivity in tissues of Crassastrea gigas *during loss period*

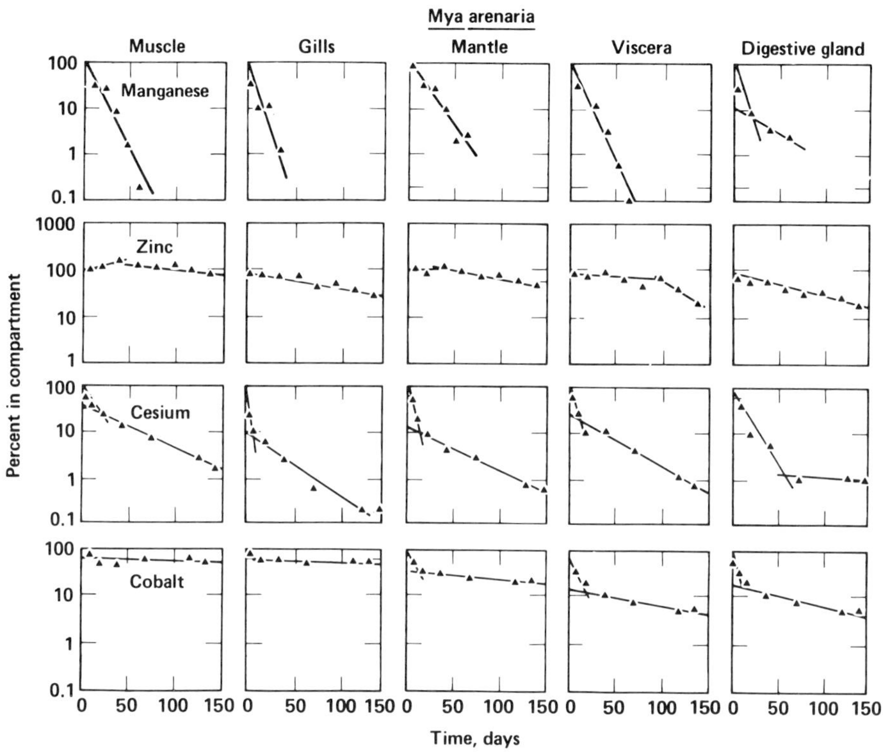

Figure 2. Percent of initial radioactivity in tissues of Mya arenaria *during loss period*

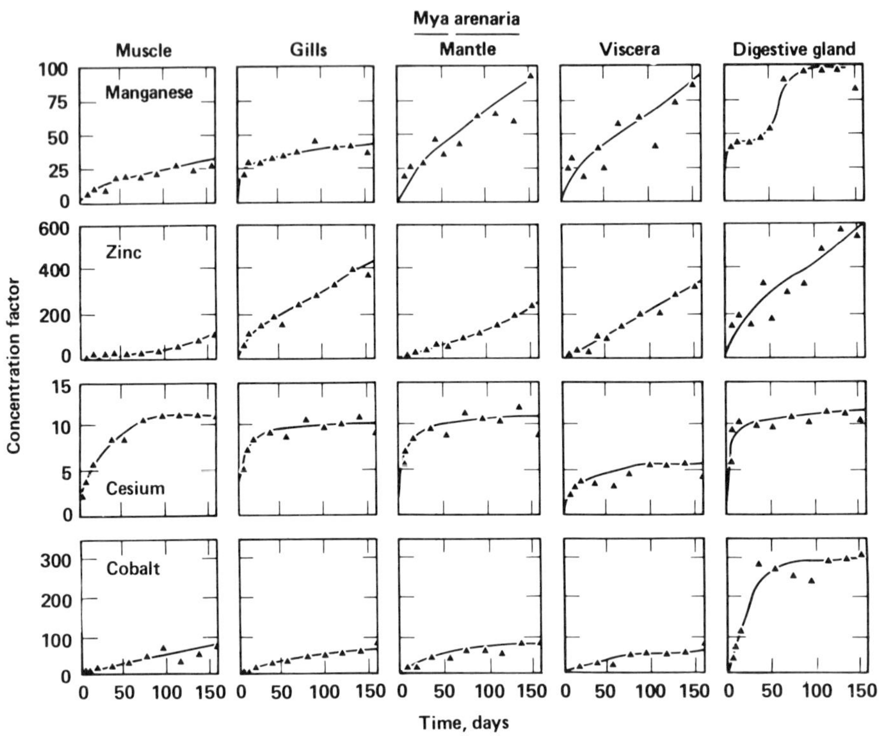

Figure 3. Concentration factors in tissues of Mya arenaria *during accumulation period*

the digestive gland and stomach and the viscera. Also in these tissues higher fractional standard deviations in the concentrations were found in the eight animals sacrificed at each sampling time.

The turnover rates of radionuclides were measured in bivalve molluscs both in the field and in the laboratory (21-30). Seymour (27) obtained a biological half-life of 300 d for ^{65}Zn in *C. gigas* and Wolfe (7) an ecological half-life of 347 d in *C. virginica*. Frazier (10) calculated a half-life on the order of 40 to 50 d for Zn in *C. virginica*, a value that is closer to the 100 d reported here for *C. gigas*.

Some results indicate that the rates obtained in the field vary seasonally; such variation is suggested by the turnover rates of ^{65}Zn in *C. gigas* (26) and of Zn and Cu in *C. virginica* (11). Frazier (10) found that the uptake of metals by oysters transferred to contaminated environments depends upon the season. Uptake by *C. virginica* was rapid in the summer and fall, but was low in the early spring.

Rates of accumulation and loss may differ significantly. George and Coombs (31) followed the accumulation and loss of Cd in *Mytilus edulis*. The rate of Cd uptake was 18 times faster than that of excretion. They concluded that the slower elimination is a consequence of a need to detoxify and store Cd by an immobilization mechanism.

Whole Body Stable Element Concentration Factors. Concentrations of some stable elements in populations of oyster *C. gigas* and the clam *Saxidomus nuttalli* were monitored to determine the normal seasonal changes in concentrations. For each sample, the soft tissues were separated from the shell, rinsed in seawater, and pooled to give a composite sample. Ten to twelve clams and 50 to 100 oysters were dissected at each sampling. The wet tissues were weighed and dried in an oven at 103 °C for at least 48 h. Samples were weighed after ashing to constant weight at 450 °C in a muffle furnace. Elemental analyses were performed by neutron activation (32).

The concentration in the soft tissues of oysters on an ash weight basis was Zn $\simeq$ Fe > Mn > Co (Table II). The concentration factors of the stable elements, Co, Mn, and Zn were higher than those determined by radionuclide studies in the laboratory (cf. Tables I and II).

The differences between the radioactive and stable nuclide concentration factors may be due to the presence in the organism of slowly exchangeable or nonexchangeable compartments. Steady-state conditions of stable and radioactive isotopes would not be reached in such compartments during the period of most laboratory experiments, and the contribution of these compartments to the concentration factor would be overlooked. Because most stable element analyses do not distinguish compartments but give the total amount in all compartments, the concentration factor of a stable isotope would be larger than

TABLE II
Trace Metal Concentrations and Concentration Factors in the Oyster *Crassostrea gigas*

Element	Concentration (µg/g wet weight) Mean	Range	Seawater (µg/l)	Concentration Factor
Zinc	100	73-140	6	17,000.
Manganese	10	3.1-21.	1.2	8,300.
Cobalt	0.06	0.008-0.21	0.1	600.
Iron	100	38. -380	3	33,000.

that of the radioisotope. Ophel (33) indicated that in some biota such a situation exists for Sr and perhaps for Co.

Differences in concentration factors between stable and radioactive isotopes may result also from other factors. If more than one chemical form of the element are present in the environment, they may be subject to different turnover and concentration processes. Seasonal effects could result if different physicochemical forms dominate during parts of the year.

Large differences in concentrations of Co, Fe, Mn, and Zn during the sampling period were found in oysters collected at near-monthly intervals over a number of seasons (Fig. 4). Concentrations of Zn ranged from 3300 to 9000, of Fe from 2000 to 9100, of Mn from 270 to 780, and of Co from 2.5 to 5.5 µg/g ash weight during the 17 mo period. Seasonal fluctuations in concentrations of these metals is clearly evident. The pattern of changes in Fe, Zn, and Co were similar; concentrations were high in the spring and late fall, and low in the summer and part of winter. The fall peak was seen in both 1972 and 1973. The pattern of change of Mn was opposite from the other three, i.e., its concentration was low when the others were high.

Large differences in concentration between some of the samples were measured. The concentration of Zn on October 1, 1973, was 4100, and on November 10, 1973, it was 9000 µg/g ash weight. This approximate doubling of the concentration in about 6 wks indicates a very rapid rate of change. The differences in rates of change in concentration suggest that the dynamics of accumulation and loss may differ during the year.

Large differences in concentrations of Co, Fe, and Zn with time were observed in clams also (Figure 5). Concentrations of Fe ranged from 600 to 15,000, of Zn from 250 to 500, and of Co from 6.5 to 15.5 µg/g ash weight. Seasonal trends were not as obvious in clams as in oysters, but cyclic changes during the 18-mo interval were apparent.

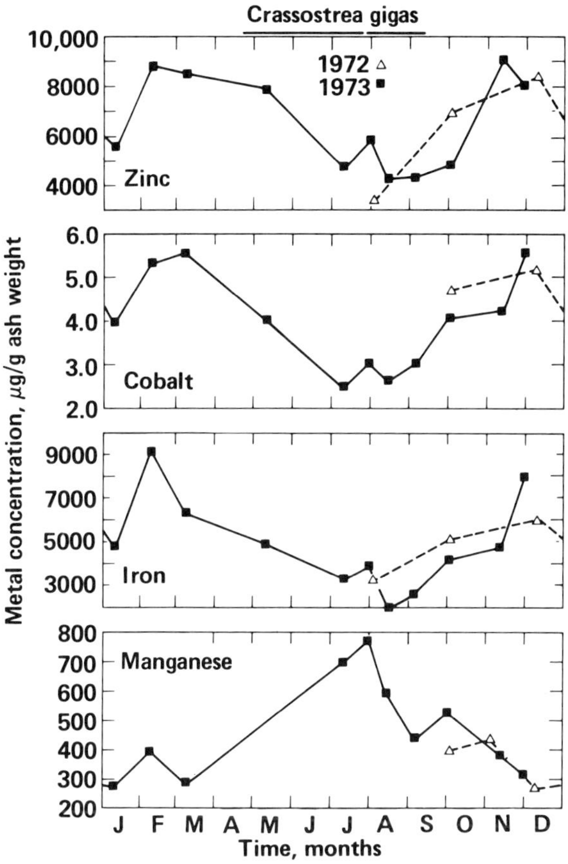

Figure 4. Concentration of metals in soft tissues of the oyster Crassostrea gigas

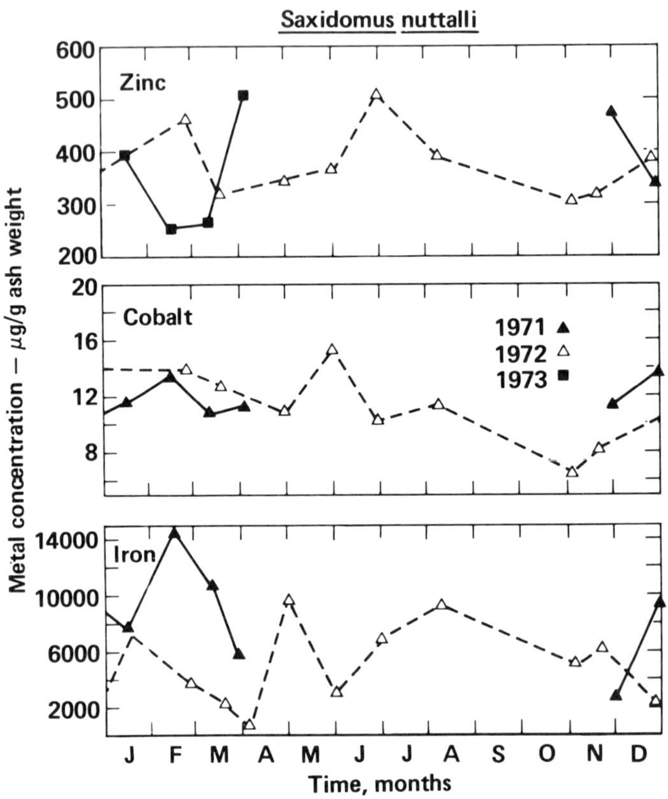

Figure 5. Concentration of metals in soft tissues of the clam Saxidomus nuttalli

Seasonal fluctuations in metal concentrations were described for the oyster C. virginica (1, 10); elevated levels in the summer are followed by decreased levels in the winter. Frazier (10) found that Mn concentrations in C. virginica were positively correlated only with those of Fe; Fe concentrations correlated with Cd and Zn; and Cd, Cu, Fe, and Zn concentrations were highly correlated with each other. We found Co, Fe, and Zn to behave similarly in C. gigas and Mn to vary inversely. Frazier (10), in reviewing the hypotheses suggested for the control of metal uptake, categorized these hypotheses into four groups related to: 1) shell growth and metabolism, 2) feeding, 3) spawning, and 4) physical chemistry of the metal either in the water or at the water-membrane interface. He identified two fundamentally different patterns of behavior in C. virginica exemplified by Mn and Zn. The concentrations of Mn in soft tissue are correlated significantly with shell growth, whereas Zn concentrations dramatically decrease before the period of maximum shell growth. The changes in Zn seem to be closely related to gonadal development and spawning. Differences in the patterns of changes in Mn and Zn were also evident in the results reported here on C. gigas. However, no data are available that relate the changes directly to shell growth or to reproductive changes.

Seasonal fluctuations in metal concentrations were reported for the mussel Mytilus edulis (34). Mussels from the Elbe/Cuxhaven site in Germany have highest concentrations normally in late winter and spring, and the lowest in late summer and autumn.

Bryan (35) reported annual fluctuations in the content of Co, Cu, Fe, Mn, Ni, Pb, and Zn in scallops. In general, highest concentrations were found in the winter and lowest in the summer and autumn.

The data available demonstrate seasonal changes in element concentration in a number of bivalve molluscs. The maximum and minimum concentrations do not occur at the same time of the year in all ecosystems. This is expected because seasonal changes are not concurrent in all ecosystems. Identification has not been made of the relative contribution to these changes of variations in the amounts and physicochemical forms of the elements in the water and in the metabolic processes in the organisms. However, seasonal changes can be significant and should be taken into consideration in model development.

Effect of Physical and Chemical Factors

Presence and Absence of Particles. Turnover of Co and Cs in Mya arenaria were measured by following both the accumulation and loss of radionuclides. However, the conditions under which the accumulation were measured differed from those of loss. During the accumulation period, the animals were maintained in filtered water in a closed system, whereas during the loss period, they were maintained in once-through, unfiltered seawater that contained the food organisms and particulate material to which they are exposed normally. Fluxes of Cs and Co in the clam during the accumulation and loss periods were calculated (Table III). The flux of an element in the tissues is the product of

TABLE III
Fluxes in *Mya arenaria* Determined During Accumulation and Loss Experiments (21)

Body part	Cesium μg/d/g ($1x10^4$) Accumulation	Cesium μg/d/g ($1x10^4$) Loss	Cobalt μg/d/g ($1x10^4$) Accumulation	Cobalt μg/d/g ($1x10^4$) Loss
Mantle	8.1	8.8	5.5	1.2
Gills and labial palps	14.	14.	3.5	14.
Muscle	3.0	5.1	3.7	0.4
Digestive gland and stomach	16.	9.3	47.	100.
Viscer[a]	3.4	7.1	4.4	21.

the loss rate constant and the concentration in the animal. Relatively good agreement for ^{137}Cs between the calculated values of fluxes from laboratory accumulation data and those from field loss data were obtained; the rate constants determined by the two methods were similar. These results suggest that the chemical form of the radionuclide was similar and that the major source of the ^{137}Cs for the clams is from the water; no significant differences were detected between the two situations. Greater differences were found in the flux of Co during the two periods than in that of Cs. Much larger fluxes of Co were measured during loss than during uptake,in the gills, the digestive gland and stomach, and the viscera. These differences may reflect the availability of food or may indicate that cobalt follows a different metabolic pathway during accumulation and loss. Thus, cobalt may be deposited more rapidly in a compartment than it can be mobilized from it.

We investigated the effects of particulate material in another series of experiments (25). Oysters were introduced into a discharge canal of a nuclear plant before scheduled releases of radioactivity. The quantities of the radionuclides accumulated during the release were measured during a period of normally high (July) and a period of normally low biological productivity (December). During each period, the animals were divided into two groups: one received untreated seawater, and the other filtered seawater.

During the July experiment, the oysters were placed in the canal at 1700 h the evening before the release. Some oysters were removed from the water at 0850 h (prerelease); the release occurred at 0900 h and terminated at 1100 h. Only ^{65}Zn was detected in the prerelease animals maintained in nonfiltered seawater (Table IV). The relatively higher concentrations in the prerelease oysters maintained in filtered water compared to those in nonfiltered water may be because of increased availability resulting from the leaching of particles that had been deposited on the filter during the 8-h interval during which the particles were collected. Desorption of radionuclides from the particles on the filter would occur any time the water flowing through the filter had a lower specific activity than that in which the particles were originally suspended.

In the animals sampled at 1100 h (postrelease), concentrations of all radionuclides were elevated. Concentrations in postrelease oysters in nonfiltered seawater were higher than those in filtered seawater for all radionuclides. The amounts accumulated per hour in animals in filtered water compared to those in nonfiltered water were lower by 95% for ^{60}Co, by 78% for ^{54}Mn, by 52% for ^{65}Zn, and by 40% for $^{134,137}Cs$.

On December 4, 1973, the prerelease oysters were removed at 0900 h, and the postrelease were removed at 1230 h; animals were placed in the discharge canal water at 1530 h on December 3. Compared with the expected concentrations in the water, the amounts of most radionuclides accumulated per hour during the release were lower in December than in July (Table IV).

The differences between the quantities accumulated in oysters held in filtered vs. nonfiltered water suggest that particles play an important role in the accumulation of elements. Particles in the water may be living microorganisms, organic detritus, inorganic material, or any combination of the three and may vary in quantity, both with time and location. The role of particles in radionuclide accumulation differed with each radionuclide. In oysters, ^{60}Co appears to be accumulated primarily from the suspended particulate fraction, whereas $^{134,137}Cs$ appears to be accumulated primarily from the soluble fraction. The differences between the amounts of radionuclides accumulated in July and December probably resulted primarily from differences in the composition of the suspended particles; particle loads were 25 and 15 mg/l dry weight, respectively (25).

Lack of correlation was reported in the levels of Cu and Zn in filtered water and those in oysters (36). This poor degree of correlation may be related to the ingestion of suspended particulate material in the seawater (2, 37). Because the particles may vary in quantity and composition both with time and location in an environment, the quantities of those elements with high affinities for particles that are accumulated by organisms should vary correspondingly.

Effect of Physicochemical Form. The physicochemical form of the element may affect the quantities accumulated. To assess the

TABLE IV

Radionuclide Concentrations in Oysters (25)

	^{54}Mn	^{60}Co	^{65}Zn	^{134}Cs	^{137}Cs
July 31, 1973					
Nonfiltered seawater					
Prerelease (pCi/kg)	ND[a]	ND	150 ± 25	ND	ND
Postrelease (pCi/kg)	120 ± 4.7	260 ± 7.7	750 ± 30	160 ± 4.8	340 ± 6.7
Rate (pCi/kg h)[b]	60	130	300	80	170
Filtered seawater					
Prerelease (pCi/kg)	46 ± 5.1	1.0 ± 1.0	200 ± 20	1.0 ± 1.0	8.0 ± 2.7
Postrelease (pCi/kg)	72 ± 3.6	14 ± 3.8	490 ± 20	99 ± 4.0	210 ± 6.3
Rate (pCi/kg h)	13	7	145	49	101
Seawater concentration (pCi/1)[c]	4.85	12.8	9.05	258.	423.
December 4, 1973					
Nonfiltered seawater					
Prerelease (pCi/kg)	70 ± 10	30 ± 8	130 ± 30	ND	ND
Postrelease (pCi/kg)	60 ± 12	40 ± 8	200 ± 40	324 ± 15	500 ± 22
Rate (pCi/kg h)	0	3	23	108	162
Filtered seawater					
Prerelease (pCi/kg)	80 ± 7	ND	60 ± 25	ND	ND
Postrelease (pCi/kg)	157 ± 11	ND	200 ± 30	310 ± 12	498 ± 20
Rate (pCi/kg h)	26	0	14	103	166
Seawater concentration (pCi/1)[c]	6.28	9.31	2.97	349.	467.

[a]ND, not detected.

[b]Amount accumulated per hour during the release

[c]Radionuclide concentration was calculated from the known dilution of the source material.

effects of the physicochemical form of Cu and Zn on accumulation, groups of twelve oysters were held in a flow-through system in seawater containing ^{64}Cu, ^{65}Zn, and either glycine (1 x 10^{-6} M), ethylenediaminetetraacetate (EDTA, 1 x 10^{-6} M), "yellow stuff" (2 mg/l), or clay (5 mg kaolin/l). Animals were removed from the test solutions, dissected, weighed, and the radionuclides quantified. After a 24 h exposure, the concentration factors in the oysters differed both with the test material and with the element (Table V). In the tissues, concentration factors of both elements generally were high in the gills and the digestive gland and stomach and low in the muscle and the blood.

The data on physicochemical forms suggest that this is an important factor in the accumulation of elements. Concentration factors both increased and decreased. George and Coombs (31) reported that complexation to EDTA, humic or alginic acids, or pectin doubled the rates of accumulation and final tissue concentrations in *M. edulis* and eliminated a lag period.

Whether the changes in the quantities accumulated are due to differences in rates of transport of the different forms of metals across biological membranes, to changes in feeding rates stimulated by the presence of organic material in the water, or to interactions of metals and ligands in the water that alter the quantities of the biologically available forms of the metal is not known. However, differences in turnover rates from laboratory and field studies may be explained in part by this factor.

Effect of Concentration. To predict the amounts of radioactive or stable nuclides of an element that may be accumulated when increased quantities are present in the water, we must know if the concentration of the stable nuclide is under metabolic control. To test for a homeostatic mechanism for Zn and Mn in oysters and for Co and Cs in clams, groups of oysters and clams were exposed to increased concentrations of these elements in an attempt to saturate any regulatory mechanisms. The results reported earlier for Co and Cs on clams (21) are included for comparison to those for Mn and Zn in oysters.

The accumulation of ^{54}Mn and ^{65}Zn from the water by oysters was followed in groups of animals held in different concentrations of Mn and Zn in the water (2, 5, or 10 μg/l) and the same concentration of ^{65}Zn and ^{54}Mn. Concentration factors of both ^{54}Mn and ^{65}Zn were lower in animals maintained in water containing 5 or 10 g/l (Figure 6). This reduction was not linear, however, as a five-fold increase in concentration did not result in a five-fold decrease in concentration factor.

The accumulation of ^{60}Co and ^{137}Cs from the water by clams was followed at concentrations of 0.5, 2.5, and 12.5 μg/l of stable Co and Cs (21). Clams do not seem to regulate Cs in the range of concentrations tested, but do seem to

TABLE V
Copper-64 and Zinc-65 Concentration Factors in Oyster Soft Tissues After 24-h Exposure to Test Materials

Body Part	None (control)	Glycine	Clay	Yellow Stuff	EDTA
		A. ^{64}Copper			
Mantle	100	160	120	120	30
Gills and labial palps	210	340	280	170	90
Shell muscles	60	60	30	20	7
Digestive gland and stomach	160	230	180	120	50
Viscera	100	220	80	90	30
Body fluid	10	10	5	4	2
Remains	30	30	30	14	6
Whole body	80	110	80	60	20
		B. ^{65}Zinc			
Mantle	16	23	82	60	1.1
Gills and labial palps	31	37	170	153	3.0
Shell muscles	4	4	23	37	0.1
Digestive gland and stomach	10	13	82	86	1.0
Viscera	9	12	64	61	0.9
Body fluid	0.8	0.8	4	4	0.1
Remains	2	2	24	11	0.1
Whole body	10	12	57	49	0.1

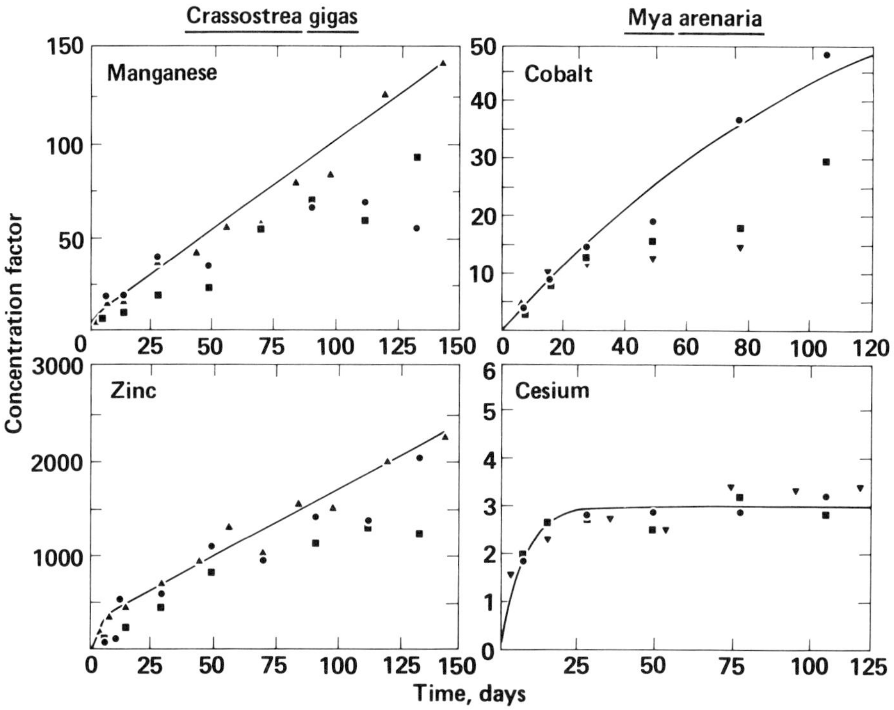

Figure 6. Concentration of metals in soft tissues of Crassostrea gigas *and* Mya arenaria *exposed to different levels of metals in the water. Manganese and zinc:* (▲) *2;* (■) *5; and* (●) *10 µg/L. Cobalt and cesium:*(●) *0.5;* (▼) *2.5; and* (■) *12.5 µg/L.*

regulate Co at the higher concentrations; the Co concentration factors were lower (Figure 6).

Shuster and Pringle (5) exposed C. virginica to increased levels of metals in the water under controlled laboratory experiments. They reported an approximate doubling of metal concentrations in the tissues upon doubling the metal ion concentration in the water.

The uptake of Cd by M. edulis at low concentrations was directly proportional to the concentrations in the seawater with a maximum concentration factor of 167 at 0.7 mg Cd/l (31). At higher concentrations, the concentration factor decreased, which was considered to indicate a saturation of the available binding sites.

Boyden (38) reports that Cd concentration factors of Patella were 10 times greater in a Cd-contaminated environment than could be expected from a proportional increase related to environmental concentrations. On the other hand, Cu concentration factors of Ostrea edulis, C. gigas, and M. edulis were greatly decreased in a grossly Cu-contaminated environment.

Although regulation of turnover of some elements at elevated concentrations is suggested from the data, the mechanisms are not clearly established, and the response is species dependent. The changes in concentration factors may not be a cellular but instead may be an organismic response to the toxic effects of the increased levels of metals. Oysters decreased the time the shell was opened at increased concentrations of Cu (13). A closure response was also found in mussels (F.L. Harrison and D.W. Rice, unpublished data, 1978). The toxic responses may include shell closure, changes in pumping and feeding rates, and changes in respiration.

Conclusion

The turnover of stable and radioactive nuclides of trace metals differs greatly with animal species, element, time, and physicochemical form of the metal in the water. Concentration factors and turnover rates of a given element can range several orders of magnitude in value in different bivalve molluscs. The reliability of models developed to predict concentration change is dependent on the selection of input parameters appropriate for the situation under consideration.

Equilibrium models are used to assess the environmental impact of power plant siting. The use of a single, maximum concentration factor for bivalve molluscs as input into the model in this situation is appropriate for screening purposes, i.e., to determine whether the maximum credible value would impact the environment. However, when more realistic estimates are required, selection of concentration factors applicable to the site, species, and situation is necessary.

Dynamic models are used in the case of accidental releases where time is an important consideration. The use of appropriate rate constants and concentration factors even in simplified dynamic models can result in better approximations of the concentration maxima and concentration changes that can be expected than the use of equilibium models.

Acknowledgements

The author thanks John W. Dawson and David W. Rice, Jr. for their technical assistance.

Work performed under the auspcies of the U.S. Department of Energy by the Lawrence Livermore Laboratory under contract number W-7405-ENG-48.

Abstract

Bivalve molluscs effectively concentrate numberous elements. To maintain healthy populations of molluscs and to minimize adverse effects on man from their consumption, we must be able to predict changes in concentration after stable and radioactive nuclides are released into their environment. Both equilibrium and dynamic predictive models have been developed. Model parameters needed include concentration factors and turnover rate constants.

Concentration factors and rate constants determined experimentally in the oyster *Crassostrea gigas* and the clam *Mya arenaria* differ widely with species and element. The physical form of the element in the water affects turnover also; accumulation of radionuclides of Co, Cs, Mn, and Zn is greater in water containing suspended particles. The chemical form of the element in the water affects its accumulation. When glycine, ethylenediaminetetraacetate (EDTA), "yellow stuff", or clay are added to seawater, the accumulation of Cu and Zn by the oyster differs with each test material; glycine increases and EDTA decreases the accumulation of both elements.

The concentrations of stable Co, Fe, Mn, and Zn in oysters and of Co, Fe, and Zn in clams fluctuate. In oysters, seasonal changes were evident. Concentrations of Co, Fe, and Zn seem to vary together, whereas that of Mn varied inversely.

Before accurate predictions can be made of the accumulation of stable and radioactive nuclides of elements by bivalve molluscs, we need concentration factors, rate constants, and information on the effects of metabolic and environmental factors on these parameters for each animal species of interest.

Literature Cited

1. Galtsoff, P.S. The American oyster, *Crassostrea virginica gemlin*. *U.S. Fish. Wildl. Serv. Fish Bull.* *64*, 480 p. (1964).
2. Brooks, R.R. and Rumsby, M.G. The biogeochemistry of trace element uptake by some New Zealand bivalves. *Limnol. Oceanogr.*

10, 521-527 (1965).

3. Pringle, B.H., Hisson, D.E., Katz, E.L., and Mulawka, S.T. Trace metal accumulation by estuarine mollusks. J. Sanit. Eng. Div. Amer. Soc. Civil Eng. 94, 455-475 (1968).
4. Kopler, F.C. and Mayer, J. Studies on trace metals in shell fish, p. 67-80, in Proc. Gulf and South Atlantic Shellfish Sanit. Res. Conf., March 21-22, 1967. U.S. Public Health Serv. Publ. No. 999 UIH-9, 1969.
5. Shuster, C.N., Jr. and Pringle, B.H. Trace metal accumulation by the American Eastern oyster Crassostrea virginica. Proc. Nat. Shellfish Assoc. 59, 91-103 (1969).
6. Roosenburg, W.H. Greening and copper accumulation in the American oyster, Crassostrea virginica, in the vicinity of a steam electric generating station. Chesapeake Sci. 10, 241-252 (1969).
7. Wolfe, D.A. Levels of stable Zn and ^{65}Zn in Crassostrea virginica from North Carolina. J. Fish. Res. Bd. Canada 27, 47-57 (1970).
8. Huggett, R.J., Bender, M.E., and Stone, H.D. Utilizing metal concentration relationships in the Eastern oysters (Crassostrea virginica) to detect heavy metal pollution. Water Research 7, 451-460 (1973).
9. Windom, H.L. and Smith, R.G. Distribution of iron, magnesium, copper, zinc, and silver in oysters along the Georgia coast. J. Fish Res. Bd. Canada 29, 450-452 (1972).
10. Frazier, J.M. The dynamics of metals in the American oyster, Crassostrea virginica. I. Seasonal effects. Chesapeake Sci. 16, 162-171 (1975).
11. Frazier, J.M. The dynamics of metals in the American oyster, Crassostrea virginica. II. Environmental effects. Chesapeake Sci. 17, 188-197 (1976).
12. Okazaki, R.K. Copper toxicity in the Pacific oyster, Crassostrea gigas. Bull. Environ. Contam. Toxicol. 16, 658-664 (1976).
13. Harrison, F.L. and Rice, D.W., Jr. Copper sensitivity of adult Pacific oysters, in Thorp, J.H., III, and Gibbons, J.W., ed., "Symp. Energy and Environmental Stress in Aquatic Ecosystems," (in press).
14. Phillips, D.J. The use of biological indicator organisms to monitor trace metal pollutants in the marine and estuarine environments. Environ. Pollut. 13, 281-317 (1977).
15. Hoffman, F.O. Chm, "The Evaluation of Models Used for the Environmental Assessment of Radionuclide Releases," 131 p., Proc. Workshop, Oak Ridge National Laboratory, Oak Ridge, Tennessee, CONF-770901, 1978.
16. Ng, Y.C., Burton, C.A., Thompson, S.E., Tandy, R., Kretner, H.K., and Pratt, M. "Prediction of the Maximum Dosage to Man from the Fallout of Nuclear Devices. IV. Handbook for Estimating the Maximum Internal Dose to Man from Radionuclides Released to the Biosphere." Lawrence Livermore Laboratory

Report UCRL-50163 (Part IV), 1968.

17. Soldat, J.K., Baker, D.A., and Corley, J.P. Application of a general computation model for composite environmental radiation doses, p. 483-498, in "Environmental Behavior of Radionuclides Released in the Nuclear Industry," IAEA, Vienna, 1973.
18. Lyon, R.B. "RAMM: A System of Computer Programs for Radionuclide Pathway Analysis Calculations," Atomic Energy of Canada Limited, AECL-5527, 1976.
19. Hijama, Y. and Shimizu, M. Uptake of radioactive nuclides by aquatic organisms; the application of the exponential model, p. 463-476, in "Environmental Contamination by Radioactive Materials," Proc. Seminar, IAEA, Vienna, 1969.
20. Cutshall, N. Turnover of Zinc-65 in oysters. Health Phys. 26, 327-333 (1974).
21. Harrison, F.L. Accumulation and loss of cobalt and caesium by the marine clam *Mya arenaria*, under laboratory and field conditions, in Symp. Interaction of Radioactive Contaminants with the Constituents of the Marine Environment, Seattle, Washington, July 10-14, 1972. IAEA, Vienna, 1973.
22. Cranmore, G. and Harrison, F.L. Loss of ^{137}Cs and ^{60}Co from the oyster, *Crassostrea gigas*. Health Phys. 28, 319-333 (1975).
23. Harrison, F.L. Accumulation and distribution of ^{54}Mn and ^{63}Zn in freshwater clams, p. 198-220, in Proc. Second Nat'l. Symp. on Radioecology, Ann Arbor, Michigan, May 15-17, 1967. AEC Symp. Series, CONF-670503, 1969.
24. Salo, E.O. and Leet, W.L. The concentration of ^{65}Zn by oysters maintained in the discharge canal of a nuclear power plant, p. 363-371, in Proc. Second Nat'l. Symp. on Radioecology, Ann Arbor, Michigan, May 15-17, 1967. AEC Symp. Series, CONF-670503, 1969.
25. Harrison, F.L., Wong, K.M., and Heft, R.E. Role of water and particulates in radionuclide accumulation in the oyster *Crassostrea gigas*, p. 9-20, in Cushing, C.E., Jr., ed., "Radioecology and Energy Resources," Dowden, Hutchinson, and Ross, Stroudsberg, Pennsylvania, 1976.
26. Seymour, A.H. and Nelson, V.A. Biological half-lives for zinc and mercury in the Pacific oyster, *Crassostrea gigas*, p. 849-856, in Proc. Third Nat'l. Symp. on Radioecology, Oak Ridge, Tennessee, May 10-12, 1971. CONF-710501, 1972.
27. Seymour. A.H. and Nelson, V.A. Decline of ^{65}Zn in marine mussels following the shutdown of Hanford reactors in Symp. Interaction of Radioactive Contaminants with the Constituents of the Marine Environment, Seattle, Washington, July 10-14, 1972. IAEA, Vienna, 1973.
28. Pentreath, R.J. The accumulation from water of ^{65}Zn, ^{54}Mn, ^{58}Co, ^{59}Fe by the mussel, *Mytilus edulis*. J. Mar. Biol. Ass. U.K. 53, 127-143 (1973).

29. Pentreath, R.J. The roles of food and water in the accumulation of radionuclides by marine teleost and elasmobranch fish, p. 421-436, in Symp. Interaction of Radioactive Contaminants with the Constituents of the Marine Environment, Seattle, Washington, July 10-14, 1972. IAEA, Vienna, 1973.
30. Van Weers, A.W. Uptake and loss of ^{65}Zn and ^{60}Co by the mussel Mytilus edulis L., p. 385-401, in Symp. Interaction of Radioactive Contaminants with the Constituents of the Marine Environment, Seattle, Washington, July 10-14, 1972. IAEA, Vienna, 1973.
31. George, S.G. and Coombs, T.L. The effects of chelating agents on the uptake and accumulation of cadmium by Mytilus edulis, Mar. Biol. 39, 261-268 (1977).
32. Heft, R.E. Absolute instrumental neutron-activation analysis at Lawrence Livermore Laboratory in Third Intern'l Conf. on Nuclear Methods in Environmental and Energy Research, October 10-13, 1977.
33. Ophel, I.L., Fraser, C.C., and Judd, J.M. Strontium concentration factors in biota and bottom sediments of a fresh water lake, p. 509-530, in Intern'l Symp. Radioecology Applied to the Protection of Man and His Environment, Vol. 1/2, Rome, Italy, Sept. 7, 1971, Eur-4800, 1972.
34. Karbe, L., Schnier, C.H., and Siewers, H.O. Trace elements in mussels (Mytilus edulis) from coastal areas of the North Sea on the Baltic. Multielement analyses using instrumental neutron activation analysis. J. Radioanal. Chem. 37, 927-943 (1977).
35. Bryan, G.W. The occurrence and seasonal variation of trace metals in the scallop Pecten maximus (L.) and Chlamys opercularis (L.). J. Mar. Biol. Ass. U.K. 53, 145-166 (1973).
36. Cronin, L.L., Pritchard, D.W., Schubel, J.R., and Sherk, J.A. "Metals in Baltimore Harbor and Upper Chesapeake Bay and their Accumulation by Oysters," Chesapeake Bay Institute, The Johns Hopkins University, Univ. Maryland, 72 p., 1974.
37. Kopler, F.C. and Mayer, J. Concentrations of 5 trace metals in the waters and oyster (Crassostrea virginica) of Mobile Bay, Alabama. Proc. Nat. Shellfish Assoc. 63, 27-34 (1973).
38. Boyden, C.R. Effect of size upon metal content of shellfish. J. Mar. Biol. Ass. U.K. 57, 675-714 (1977).

RECEIVED November 16, 1978.

28

Relationships of Activities of Metal-Ligand Species to Aquatic Toxicity

V. R. MAGNUSON, D. K. HARRISS, M. S. SUN, and D. K. TAYLOR
Department of Chemistry, University of Minnesota, Duluth, Duluth, MN 55812

G. E. GLASS
U.S. Environmental Protection Agency, Environmental Research Laboratory, 6201 Congdon Boulevard, Duluth, MN 55804

Much has been published concerning toxicity of metals to aquatic life although only during the past decade have there been interpretations of the toxicity data in terms of the relative toxicity of particular speciation forms, e.g., Cu^{2+}, $CuOH^{+}$, $CuCO_3^{o}$ (1-7). The specific objective of this paper is to illustrate the use of Factor Analysis in discriminating between toxic and non-toxic species. The procedure to be followed is: determination of equilibrium aqueous speciation, calculation of appropriate factors, correlation of toxicity with these factors, and interpretation of the correlation analysis in terms of particular species activities.

The analysis of toxicity data in terms of speciation products is a difficult task since the variables, species activities, usually are numerous and often are interrelated. Factor Analysis allows one to determine a small number of statistically independent, linear combinations of activities (factors). Correlation of these combinations (factors) with toxicity allows discrimination to be made between toxic and non-toxic species.

Conclusions drawn from a statistical study of this type are only as valid as the data upon which the study is based. The published toxicity study (Andrew, Biesinger, and Glass (2)) which we use for illustration is scientifically sound, however, the number of experimental points and limited ranges of some of the experimental variables do restrict the ability to discriminate between species.

Andrew *et al.* (2) studied the effects of carbonate, orthophosphate, and pyrophosphate on the toxicity of copper(II) to *Daphnia magna* at constant pH and total hardness. They reported mortality rates and reciprocal survival times to be directly correlated with cupric and copper-hydroxo ion activities as determined by equilibrium calculations. They also found toxicity to be negatively related to activities of soluble copper carbonate ($CuCO_3^{o}$), and independent of total or dissolved copper concentration. Data for their set of experiments involving addition of orthophosphate are included in Table I but were not used in some

0-8412-0479-9/79/47-093-635$05.50/0
© 1979 American Chemical Society

of our later calculations due to uncertainty as to the stability constants for phosphate complexes. Calculations when all points were used are referred to as 30 point and those in which the orthophosphate points are excluded are referred to as 26 point. The calculations described below are of two types; first, recalculation of metal-ligand speciation and comparison with previous results, and secondly, interpretation of the toxicity data in terms of speciation using factor analysis and multiple regression. The speciation calculations were performed using REDEQL2 (8) and SPSS (9) was used for factor analysis and multiple regression.

Sources of Data and Limitations

Stability Constants. The choice of equilibria to include and the accuracy of the related stability constants have, of course, a major effect upon the predicted speciation and upon inferences drawn from correlations of activities of the species with toxicity. A measure of the effects of limited accuracy in the stability constants on predicted speciation concentrations can be obtained through repetitive calculations on a system while allowing for small, random variations in the stability constants. Such calculations have been carried out using a three-metal, three-ligand model (31 complexes) with concentrations ranging from 1 m*M* to 0.01 μ*M*. Random variations of 0.05 to -0.05 units in the mantissas of the pK's led to concentration changes ranging from 6% to 30% with a mean change of 14%. In general, the percentage changes were greatest for the species whose concentrations were small fractions of the total concentration. Use of smaller random variations of 0.01 to -0.01 units produced proportionately smaller changes, about 1/5 as great.

Toxicity Studies. It is difficult to find published toxicity studies which are well documented and which also contain a sufficient number of data points to allow meaningful statistical analyses. Minimal necessary documentation requires pH, alkalinity, measured concentrations of the particular metals or ligands under study, and a complete analytical background analysis. This presumes of course that: equilibria involving the background species are important and will be considered; total inorganic carbon will be derived from alkalinity and knowledge of other acids present; and the nominal added metal or ligand does not necessarily equal the measured dissolved metal or ligand.

With regard to including all equilibria, for the experiment referred to in line 7 of Table I, the calculated activity of $CuSO_4^0$ is twice that of $Cu_2(OH)_2^{2+}$ and five times that of $Cu(OH)_3^-$. The relative magnitudes of these activities lead one to believe that the amount of copper combined with sulfate and other ligands must be taken into account if calculated activities of species such as $Cu(OH)_3^-$ and $Cu_2(OH)_2^{2+}$ are to be meaningful.

With regard to calculation of total inorganic carbon (TIC), knowledge of the analytical background and the alkalinity allow the calculation of inorganic carbon from (for alkalinity expressed as mg ℓ^{-1} $CaCO_3$)

$$TIC = \frac{1+10^{-pH+6.35}+10^{-10.33+pH}}{(1+2x10^{-10.33+pH})} \left(\frac{Alk}{50,000} - 10^{-14+pH} + 10^{-9.5} \right) \quad (1)$$

when other acids are not present, with appropriate adjustments when other acids are present. Empirical equations expressing TIC in terms of pH and hardness have been used but seem not to be satisfactory over a wide range of pH (10).

With regard to calculated versus measured amounts of metals, in the work of Andrew, Biesinger and Glass (2), the measured dissolved copper varied between 53 and 100% of the nominal copper added to the system (see Table I) and their data is representative of several other published reports when like comparisons are made.

Analysis of Data

The issue to be addressed in a subsequent section is the attribution of the toxicity of copper in aquatic systems to particular speciation forms, e.g., Cu^{2+}, $Cu_m(OH)_n^{2m-n}$, $Cu_j(CO_3)_k^{2j-2k}$. Analysis of published experimental data in this area normally is difficult for several reasons: statistically small numbers of data points in relation to the number of variables, lack of independence of the variables with correlation coefficients often of the order of 0.8 or 0.9 (see Table II), and small ranges and scatter of points due to the usual experimental practice of varying as few parameters as possible in a particular run with the intent of determining bivariate relationships.

In the first set of experiments of Andrew *et al.* (2), for example, pH, hardness and total alkalinity are held constant while copper is added to solutions with fixed added amounts of $NaHCO_3$, or Na_2HPO_4, or $Na_4P_2O_7$, in addition to copper added to the background, with no subset of experiments including more than six cases. In our speciation calculations related to this set of nineteen experiments, some twenty-five complexes of copper including three with carbonate, two with phosphate, six with hydroxide, and six with pyrophosphate are involved.

It is possible to create new variables, e.g., as linear combinations of the activities, intended to characterize a particular aspect of the system and then interpret the data in terms of this smaller set of new variables. One might, for example, use the sum of the activities of Cu^{2+}, $CuOH^+$ and $Cu_2(OH)_2^{2+}$ as a variable rather than the three species separately (10). In this way a large number of bivariate correlations can be made, but

TABLE I

Calculated Equilibrium Values for Activities of Selected Species Involved in the Andrew's Experiments. Values shown are the negative logarithms of the activities

Nominal Added Cu(μM)	Measured Dissolved Cu(μM)	Cu^{2+}	$CuOH^{+}$	$Cu_2(OH)_2^{2+}$	$Cu(OH)_2^{o}$	$Cu(OH)_3^{-}$	$Cu(OH)_4^{2-}$	$CuCO_3^{o}$	$Cu(CO_3)_2^{2-}$
Lake Superior Water Only, pH 7.85-8.05									
Control	~0.02	10.06	9.80	14.57	7.85	12.99	17.84	8.77	11.06
0.20	0.20	9.00	8.72	12.41	6.75	11.87	16.71	7.69	9.96
0.32	0.33	8.78	8.50	11.97	6.54	11.65	16.49	7.48	9.74
0.51	0.41	8.69	8.41	11.79	6.44	11.56	16.40	7.38	9.64
0.80	0.61	8.51	8.24	11.44	6.27	11.39	16.22	7.21	9.47
1.27	1.02	8.29	8.01	10.99	6.05	11.16	16.00	6.99	9.25
2.01	1.54	8.11	7.83	10.64	5.87	10.98	15.82	6.81	9.07
Lake Superior Water + 48 μM $NaHCO_3$, pH 7.85-8.05									
Control	~0.02	10.11	9.83	14.62	7.85	12.97	17.80	8.77	11.01
0.80	0.60	8.50	8.24	11.44	6.28	11.41	16.26	7.19	9.44
1.27	0.98	8.27	8.02	11.00	6.07	11.21	16.07	6.97	9.23
2.01	1.45	8.08	7.83	10.64	5.90	11.05	15.92	6.78	9.06
Lake Superior Water + 210 μM Na_2HPO_4, pH 7.85-8.05									
Control	~0.02	10.12	9.84	14.64	7.86	12.98	17.82	8.71	10.88
0.80	0.54	8.58	8.30	11.57	6.34	11.46	16.31	7.18	9.35
1.27	0.76	8.42	8.15	11.27	6.19	11.32	16.16	7.03	9.20
2.01	1.07	8.27	8.00	10.97	6.04	11.17	16.03	6.88	9.06

TABLE I Cont'd.

Nominal Added Cu(μM)	Measured Dissolved Cu(μM)	Cu^{2+}	$CuOH^{+}$	$Cu_2(OH)_2^{2+}$	$Cu(OH)_2^{o}$	$Cu(OH)_3^{-}$	$Cu(OH)_4^{2-}$	$CuCO_3^{o}$	$Cu(CO_3)_2^{2-}$
Lake Superior Water + 110 μM $Na_4P_2O_7$,pH 7.85-8.05									
Control	~0.02	10.85	10.54	16.06	8.55	13.64	18.45	9.51	11.74
0.80	0.77	9.16	8.86	12.69	6.87	11.97	16.78	7.83	10.06
1.27	1.24	8.95	8.65	12.28	6.67	11.76	16.58	7.62	9.86
2.01	1.95	8.74	8.45	11.86	6.46	11.57	16.39	7.41	9.66
Lake Superior Water + (0,1,2,4,10 mM) added $NaHCO_3$,pH 7.3-7.5									
5	3.38	6.77	7.07	9.01	5.67	11.35	16.76	6.03	8.86
5	3.82	6.83	7.12	9.21	5.72	11.40	16.80	5.78	8.29
5	4.13	6.88	7.17	9.31	5.77	11.44	16.85	5.66	7.99
5	4.21	7.01	7.29	9.56	5.89	11.56	16.97	5.57	7.67
5	4.37	7.24	7.52	10.02	6.12	11.78	17.20	5.47	7.23
Lake Superior Water + (0,2.5,5.0,10.0,50.0,500 μM) added $Na_4P_2O_7$, pH 7.4-7.6									
5	***	6.75	6.96	8.88	5.47	11.07	16.39	5.93	8.67
5	***	6.90	7.09	9.15	5.59	11.18	16.48	6.06	8.79
5	***	7.03	7.21	9.39	5.70	11.27	16.56	6.18	8.89
5	***	7.24	7.40	9.76	5.86	11.42	16.69	6.36	9.06
5	***	7.89	8.04	11.05	6.50	12.04	17.30	7.00	9.67
5	***	9.75	9.91	14.78	8.38	13.92	19.21	8.80	11.41

*** Values Assumed To Be 5.

TABLE II

Correlation Coefficients Between the Activities of Copper Species in the 26 point Calculations

	Cu^{2+}	$CuOH^{+}$	$Cu_2(OH)_2^{2+}$	$Cu(OH)_2^{o}$	$Cu(OH)_3^{-}$	$Cu(OH)_4^{2-}$	$CuCO_3^{o}$	$Cu(CO_3)_2^{2-}$
Cu^{2+}		0.99	0.94	0.89	0.34	-0.21	0.65	0.22
$CuOH^{+}$	0.99		0.96	0.95	0.45	-0.12	0.60	0.17
$Cu_2(OH)_2^{2+}$	0.94	0.96		0.91	0.41	-0.14	0.46	0.05
$Cu(OH)_2^{o}$	0.89	0.95	0.91		0.70	0.18	0.47	0.07
$Cu(OH)_3^{-}$	0.34	0.45	0.41	0.70		0.83	0.06	-0.12
$Cu(OH)_4^{2-}$	-0.21	-0.13	-0.14	0.18	0.83		-0.26	-0.21
$CuCO_3^{o}$	0.65	0.60	0.46	0.47	0.06	-0.26		0.84
$Cu(CO_3)_2^{2-}$	0.22	0.17	0.05	0.07	-0.12	-0.21	0.84	

direct comparisons between the new variables are difficult since interrelationships of the new variables are not clear and they may be correlated with each other.

What is desired is a logical, systematic procedure to generate a small set of uncorrelated variables which allow direct comparison, make chemical sense, and are amenable to easy interpretation. A mathematical procedure which meets these criteria is the use of factor analysis (9, 11, 12, 13) to generate statistically independent new variables (factors) followed by correlation of toxicity with the generated factors.

Factor analysis has two general uses; to see whether underlying patterns of relationships exist among sets of variables, and to rearrange or reduce the data (variables) to a smaller set of factors or components that may be taken as primary or source variables. One often is cautioned about the first use, employment of factor analytic methods to determine relationships among the variables, in that spurious patterns may be indicated. In our applications, however, the desired patterns, i.e. combinations of activities, normally will be known or can be predicted and the use of factor analysis will be to generate a reduced set of statistically independent components (factors), e.g. three factors replacing eight individual activities. In cases where the patterns are not clear, one retains the criterion that the results must be chemically rational and may reject those that fail to meet this criterion. These factors then are used in multiple regression analyses to attempt to identify toxic species.

Toxicity of Copper to Aquatic Forms

General. A variety of conclusions have been published in recent years on the relative toxicity of free copper ion (Cu^{2+}), hydroxo copper complexes, ($CuOH^+$, $Cu(OH)_2^o$, $Cu(OH)_3^-$, $Cu(OH)_4^{2-}$, $Cu_2(OH)_2^{2+}$), and carbanato copper complexes ($CuCO_3^o$, $Cu(CO_3)_2^{2-}$). The statements have been, in general, qualitative in nature rather than quantitative. Shaw and Brown (14) concluded that $CuCO_3^o$ is as toxic as Cu^{2+}; Pagenkopf *et al.* (15) find the major toxic species to be Cu^{2+} with a possible contribution from $CuOH^+$: Andrew *et al.* (2) state that copper toxicity is directly related only to the activities of Cu^{2+}, $CuOH^+$, and $Cu_2(OH)_2^{2+}$; Chakoumakos *et al.* (16) report Cu^{2+}, $CuOH^+$, $Cu(OH)_2^o$ and $Cu_2(OH)_2^{2+}$ to be toxic forms and found $CuHCO_3^+$, $CuCO_3^o$ and $Cu(CO_3)_2^{2-}$ not toxic under their conditions; Hawarth and Sprague (10) found the smoothest response surface, of those tried, for copper toxicity to be generated by $[Cu^{2+}]$ + $[CuOH^+]$ + $[Cu_2(OH)_2^{2+}]$and, on the basis of this, claim these three to be the toxic forms and, on the basis on non-smooth response surfaces, that $Cu(OH)_2^o$ and the carbonato copper species are not toxic.

Most of the experiments reported upon in the above referenced papers involved systems with pH in range 6.5-8.5, hardness up to 350mg ℓ^{-1} as $CaCO_3$, initial alkalinity as high as 300mg ℓ^{-1} as $CaCO_3$, and total copper up to 5mg ℓ^{-1}.

In the following sections, factor analysis and multiple regression are used in an attempt to determine the relative toxicity of the copper species discussed above. The activities of eight species (Cu^{2+}, $CuOH^{+}$, $Cu_2(OH)_2^{2+}$, $Cu(OH)_2^{o}$, $Cu(OH)_3^{-}$, $Cu(OH)_4^{2-}$, $CuCO_3^{o}$, $Cu(CO_3)_2^{2-}$) are included in the calculations.

Factor Determination. One is strongly tempted to generate "universal" factors for the species of interest, valid over typical ranges of the parameters involved. The problem is that the points for a given experiment are unlikely to be representative of the points used to generate the "universal" factors and the factors lose their statistical independence. One has little choice then, when using published data, but to use the experimental points directly in the determination of the factors as well as in the correlations and we have done so in the systems reported here although we later discuss the structure of the experimental design which would be most useful to this type of analysis. The quality of species separation obtained through use of previously published experimental points is good however, since the parameter selections made by the experimenters often were for the purpose of studying toxicity as a function of speciation.

A comparison of factors determined with 100 random points and with 30 and 26 experimental points can be made from the data shown in Table III. A short description of the data may be useful to those unfamiliar with factor analysis. The values shown are "factor loadings" for the species on the factors, e.g. the top line under the 100 random point heading states that

$$\text{activity}(Cu^{2+}) = 0.88\text{xFactor } 1 - 0.19\text{xFactor } 2 + 0.27\text{xFactor } 3 \quad (2)$$

and the activity of Cu^{2+} can be calculated from values of the factors. The "factor loadings" are correlation coefficients and the relative importance of the factors in determining the activity of Cu^{2+} is given as the square of the coefficients, implying that Factor 1 is sufficient to explain 77% of the variance in the activity of Cu^{2+}, Factor 2 explains 4%, and Factor 3 explains 7%. The fact that these add up to 0.88 rather than 1.00 states that these three factors are sufficient to explain only 88% of the variation in the activity of Cu^{2+} in the 100 points. In comparison, three factors are sufficient to explain 98% of the variation of the activity of Cu^{2+} in Andrew's 30 point and 26 point cases.

The actual activities of the species are not used in the determination of factors, but instead Kaiser normalized values of the activities are used,

TABLE III

Three-Factor Case for Random Points and the Andrew *et al.* Experiments

Varimax Rotated Factor Matrices After Rotation with Kaiser Normalization

	100 Random Points			Andrew's 30 Points			Andrew's 26 Points		
Species	Factor 1	Factor 2	Factor 3	Factor 1	Factor 2	Factor 3	Factor 1	Factor 2	Factor 3
Cu^{2+}	0.88	-0.19	0.27	0.96	-0.05	0.24	0.96	-0.05	0.24
$CuOH^{+1}$	0.97	0.06	0.22	0.98	0.05	0.20	0.98	0.05	0.19
$Cu_2(OH)_2^{2+}$	0.96	0.04	-0.03	0.97	***	0.05	0.97	0.02	0.04
$Cu(OH)_2^{o}$	0.59	0.73	0.08	0.92	0.36	0.11	0.92	0.35	0.11
$Cu(OH)_3^{-}$	0.04	1.00	-0.01	0.39	0.92	-0.04	0.41	0.91	-0.04
$Cu(OH)_4^{2-}$	-0.12	0.95	0.02	-0.16	0.97	-0.13	-0.15	0.97	-0.13
$CuCO_3^{o}$	0.38	0.04	0.88	0.46	-0.07	0.87	0.45	-0.09	0.87
$Cu(CO_3)_2^{2-}$	-0.18	0.70	0.56	0.01	-0.08	0.95	***	-0.08	0.95
Percent of overall variance in the points accounted for by the 3 factors	48.6	39.0	12.4	57.6	27.5	14.9	57.5	27.5	14.9

*** Value less than precision shown

$$X' = \frac{X - \bar{X}}{\sigma_X} \tag{3}$$

where X is the experimental value, X' is the normalized value, $\bar{X}$ the mean of the X and σ_x the standard deviation of the X. Kaiser normalized variables have nominal means of zero and standard deviations of one. Equation 2 should be written instead as

$$A'(Cu^{2+}) = 0.88\ F_1 - 0.19\ F_2 + 0.27\ F_3 \tag{2'}$$

where $A'(Cu^{2+})$ is the Kaiser normalized activity of Cu^{2+} and the factors are Kaiser normalized variables.

From known values of activities, one can calculate the value of the factor (factor score) *via* factor score coefficients,

$$F_1 = c_{11}X'_1 + c_{12}X'_2 + \ldots . +c_{1n}X'_n \tag{4}$$

where the c_{pq} are the factor score coefficients and the X'_q are the normalized values of the activites. The means and standard deviations of the activities of the species, plus the factor score coefficients for the 26 point case are given in Table IV. The second factor would be written

$$F_2 = -1.09\ A'(Cu^{2+}) + 7.19\ A'(CuOH^+) + \ldots \tag{5}$$

TABLE IV

Means, Standard Deviations, and Factor Score Coefficients for Andrew's 26 Points with Three Factors. (Means and Standard Deviation in Molarities)

Species	Mean Activity	Standard Deviation	Factor Score Coefficients Factor 1	Factor 2	Factor 3
Cu^{2+}	4.28×10^{-8}	6.10×10^{-8}	-19.55	-1.09	3.11
$CuOH^+$	2.65×10^{-8}	3.33×10^{-8}	48.73	7.19	-7.89
$Cu_2(OH)_2^{2+}$	1.91×10^{-10}	3.36×10^{-10}	0.23	0.29	0.06
$Cu(OH)_2^o$	9.39×10^{-7}	9.03×10^{-7}	-38.18	-14.37	4.95
$Cu(OH)_3^-$	3.62×10^{-12}	2.85×10^{-12}	16.76	14.23	-1.05
$Cu(OH)_4^{2-}$	3.62×10^{-17}	3.81×10^{-17}	-5.61	-7.48	-0.09
$CuCO_3^o$	5.75×10^{-7}	9.26×10^{-7}	-1.34	-0.01	1.14
$Cu(CO_3)_2^{2-}$	4.14×10^{-9}	1.20×10^{-8}	0.49	0.05	0.18

Using multiple regression, biological response (BR) can then be correlated with the factors (factor scores) as

$$BR = aF_1 + bF_2 + cF_3 + \ldots \tag{6}$$

thereby generating a relationship between biological response and activities, e.g. Equations 7 and 8.

The factor score coefficients (c_{pq}) are not correlation coefficients for independent variables and the concept of relative weight of the activities of the species does not hold, i.e. the squares of the coefficients do not give the relative importance of the activities in the factor.

The quality of species separation appears to be better in the 30 and 26 experimental point cases, that is to say that the loadings generally are more restricted to single factors, especially for $Cu(OH)_2^o$ and $Cu(CO_3)_2^{2-}$ although $Cu(OH)_3^-$ and $CuCO_3^o$ are somewhat less well separated (Table III). The boxes drawn are to indicate the principal species dominating a factor, e.g. under Andrew's 30 point case Factor 1 is primarily (Cu^{2+}, $CuOH^+$, $Cu_2(OH)_2^{2+}$, $Cu(OH)_2^o$) with Factor 2 represented well by ($Cu(COH)_3^-$, $Cu(OH)_4^{2-}$) and Factor 3 by ($CuCO_3^o$, $Cu(CO_3)_2^{2-}$). Note in the 100 random point case that $Cu(OH)_2^o$ distributes 35% on Factor 1 and 53% on Factor 2 and $Cu(CO_3)_2^{2-}$ distributes 49% on Factor 2 with 31% on Factor 3, relatively poor separations. An often used rule-of-thumb is that loadings greater than 0.3 should be considered significant (12, p.10). As will be evident from the discussion below on separability of species, the 100 random-point-factors are a more accurate representation of the behavior of the hydroxo complexes over a wide pH range than are the 30 and 26 point factors. The experiments of Andrew *et al.* are restricted to three pH values (7.4, 7.5, 7.95) thereby limiting the relative variation of hydroxo complex activities, creating a set of data more easily represented by three factors and possibly giving the false impression that the separations achieved are better. Further attention has not been given to factors generated from random points due to problems of statistical independence earlier discussed.

Separation of species by factor analysis is limited in several ways. In order of decreasing importance the separability is controlled by: speciation behavior over the ranges of parameters used, experimental point distribution over the ranges of parameters, number of species involved in the calculation, and number of factors determined.

With regard to speciation behavior: in Figure 1 are displayed the speciation curves for the activities of free copper ion and the hydroxo-complexes of copper over a range of pH. Inspection of the curves reveals that $CuOH^+$ and $Cu_2(OH)_2^{2+}$ have the same pattern of behavior over the entire range and cannot be separated, $Cu(OH)_3^-$ and $Cu(OH)_4^{2-}$ are similar enough that separation is not to be expected, and over the range of pH values used by Andrew *et al.* 7.4-8.0, one should not expect separations other than those observed. One might, upon first glance, feel that the forms of the factors are completely described by the correlation coefficients

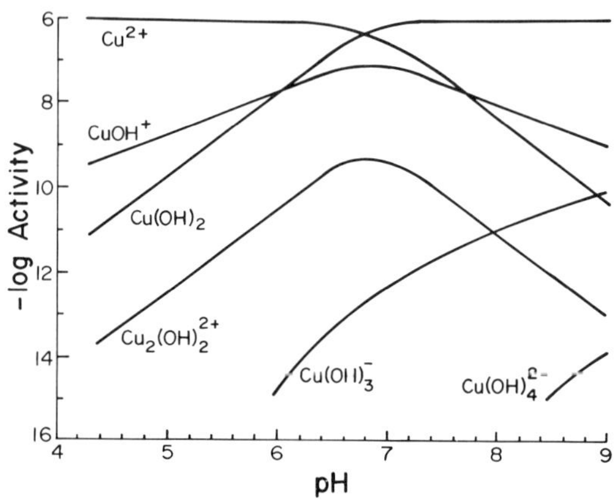

Figure 1. Speciation curves for the activities of free copper ion and the hydroxo complexes of copper vs. pH

given in Table II. The boxes drawn around sets of correlation coefficients correspond to the boxes in Table III and indeed, the largest correlation coefficients lie within the boxes although quite substantial values do lie without. No mathematical manipulation can obviate the data in Table II and of course the factors are derived from the correlation coefficients. The usefulness of factor analysis lies in its ability to determine a small set of statistically independant variables which retain as much information as possible about the system.

With regard to point distribution: it should be clear from Figure 1 that the pH values must be representative of the range 6.0-8.0 for the above discussed separation to be achieved.

With regard to the number of species involved: the number of factors is restricted to no more than one-half the number of variables if convergent, unique values are to be obtained for communalities of the variables (11, p.200), (the communality, of a variable is the amount of the variance of that variable accounted for by the common factors). Therefore, to use a five factor decomposition of a set of species one must have at least 10 species included in the calculation.

With regard to the number of factors: it should be obvious that to obtain a separation of the four groups mentioned later, at least four factors must be used. If other species which have quite different patterns of behavior over the set of points are included in the calculation, sufficient additional factors must be included to account for each speciation pattern or the patterns will smear over the factors.

Addition of more species or factors for Andrew's data would not have proven fruitful due to the limited pH range involved in the experiment.

Finally, after all of the above have been taken into account, it is possible to "rotate" a set of factors with different types of rotation yielding somewhat different species loading (9). If the speciation patterns for the points are not distinct however, rotation of the factors cannot achieve a separation. We have found Varimax rotation, which maximizes the squared loadings of variables in each factor while retaining the orthogonality of the factors to be most useful.

Copper Toxicity to Daphnia Magna

Lake Superior water was used as the dilution water for all experiments by Andrew (2). A copy of the current analytical background for Lake Superior water, Table V, was obtained from the Environmental Research Laboratory-Duluth and was used in our speciation calculations. The water has a total hardness of ~45mg ℓ^{-1} as $CaCO_3$, alkalinity of ~42mg ℓ^{-1} as $CaCO_3$, and a pH from 7.4 to 8.2.

TABLE V

Concentrations used to Simulate the Analytical Background of Lake Superior Water (Molarities)

Calcium	$3.26x10^{-4}$	Zinc	$1.10x10^{-8}$
Magnesium	$1.10x10^{-4}$	Cobalt	$3.40x10^{-9}$
Sodium	$5.60x10^{-5}$	Barium	$1.00x10^{-7}$
Potassium	$1.29x10^{-5}$	Strontium	$1.80x10^{-7}$
Cadmium	$1.80x10^{-10}$	Silver	$1.00x10^{-10}$
Chromium	$3.80x10^{-9}$		
Copper	$1.60x10^{-8}$		
Iron	$5.40x10^{-8}$	Carbonate	$8.81x10^{-4}$
Mercury	$5.01x10^{-11}$	Chloride	$3.61x10^{-5}$
Methyl Mercury	$5.01x10^{-12}$	Sulfate	$3.65x10^{-5}$
Manganese	$5.50x10^{-9}$	Silicate	$4.47x10^{-5}$
Nickel	$3.40x10^{-9}$	Phosphate	$3.24x10^{-8}$
Lead	$4.84x10^{-10}$	Nitrate	$1.66x10^{-5}$

The pH was adjusted in each of their experiments by regulating CO_2-air mixtures bubbled through the solution, e.g. $Na_4P_2O_7$ was added to the solution followed by bubbling sufficient CO_2 through the solution to bring the pH back to 7.95. This process increases the total inorganic carbon. REDEQL2 does allow one to determine the amount of $2H^+ + CO_3^{2-}$ necessary to add to bring the pH back to 7.95. The total carbonate concentrations used in our calculations are listed in Table VI.

TABLE VI

Total Carbonate Values used in Recalculation of Andrew's Data

Andrew's Table	Experiments	Total Carbonate (mM)
2	1 - 7	0.881
	8 - 11	0.929
	12 - 15	1.114
	16 - 19	0.889
3	1	0.935
	2	1.936
	3	2.938
	4	4.932
	5	10.940
4	1 - 4	0.920
	5	0.931
	6	1.119

Speciation Calculations. Although REDEQL2 was used to determine speciation in Andrew's work as well as ours, considerable differences exist in predicted equilibrium activities due to modifications which have been made to the program and the associated data base. The major contribution to the differences is the inclusion of more complexes, e.g. $Cu(OH)_2^o$. Complexation by uncharacterized organic ligands and absorption of species or suspended matter were not considered in these speciation calculations. (Both organic complexation and absorption should be less serious problems in this Lake Superior study as compared to a non-laboratory system.) In Table VII some comparative data are listed and in Table I are shown the calculated activities for the free copper ion plus the hydroxo and carbonato complexes.

TABLE VII

Comparisons of Some Calculated Copper Species Activities from this work (M-G) with that of Andrew, *et al* (ABG)

Measured dissolved copper (μM)	Calculated Activity of Copper Species (μM): Cu^{2+} ABG	Cu^{2+} M-G	$CuOH^+$ AbG	$CuOH^+$ M-G	$Cu(OH)_2^o$ ABG*	$Cu(OH)_2^o$ M-G	$CuCO_3^o$ ABG	$CuCO_3^o$ M-G
~0.02	0.001	***	0.001	***		0.014	~0.02	0.002
0.20	0.011	0.001	0.009	0.002		0.178	0.18	0.020
0.33	0.016	0.002	0.015	0.003		0.288	0.30	0.033
0.41	0.020	0.002	0.019	0.004		0.363	0.36	0.042
0.61	0.031	0.003	0.028	0.006		0.537	0.54	0.062
1.02	0.034	0.005	0.047	0.010		0.891	0.89	0.102
1.54	0.076	0.008	0.071	0.015		1.349	1.35	0.155

* $Cu(OH)_2^o$ was not considered in ABG calculations
*** Value less than precision shown

Single Ion Correlations with Toxicity. The data given in the first fifteen cases, excluding those with added Na_2HPO_4, were used to determine correlations of inverse median survival time with species activity for free Cu^{2+} and the seven listed complexes. The values for r ranged from 0.97 to 0.98 with significance 0.00001 for all species but one, $Cu_2(OH)_2^{2+}$, which exhibited an r of 0.95.

Five of the points (1, 8, 9, 10, 12 excluding Na_2HPO_4 points) are "zero-points", i.e. median survival time exceeded the life of the experiment, and a better test of the hypothesis of toxicity due to copper species results from discarding the zero-points. When this is done, the correlation coefficients decrease by 0.01 or less and the level of significance remains at the 10^{-5} level.

In Figure 2 are shown the inverse median survival time versus activity plots for Cu^{2+} and $CuCO_3^o$, both for the 15 point cases and the 10 point cases.

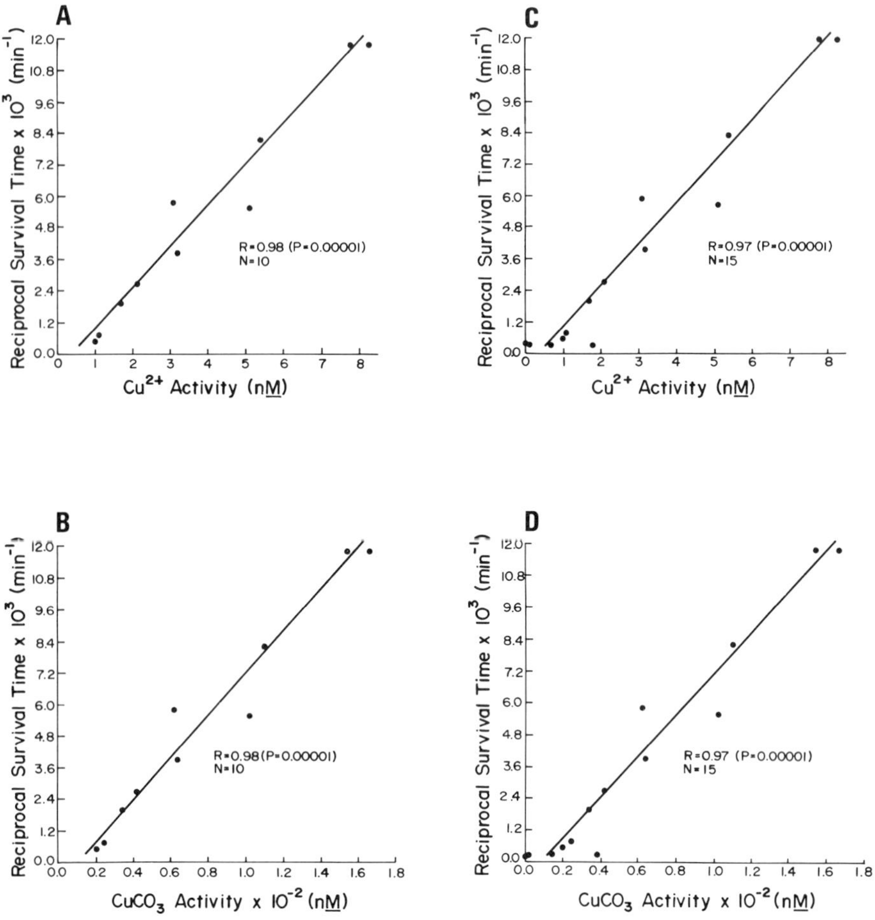

Figure 2. Reciprocal median survival time vs. activity for Cu^{2+}: (A) 10 point case, (C) 15 point case; for $CuCO_3{}^{\circ}$: (B) 10 point case, (D) 15 point case

Factor Analysis and Multiple Regression. Factor scores for the 26 and 30 point cases, generated using the factor score coefficients in Table IV, were correlated with inverse median survival time, t^{-1}.

The 26-point case yielded the following relationship

$$\frac{1}{t} = 0.0046\ F_1 + 0.0027\ F_2 - 0.00013\ F_3 + 0.0061 \quad (7)$$

while the 30-point case yielded

$$\frac{1}{t} = 0.0051\ F_1 + 0.0024\ F_2 - 0.00011\ F_3 + 0.0055 \quad (8)$$

Standard errors, 95% confidence intervals, and levels of significance are tabulated in Table VIII for the coefficients in Equations 7 and 8.

TABLE VIII

Standard Errors, 96% Confidence Intervals, Levels of Significance and Changes in R^2 for the 26-Point and 30-Point Regression Equations

26-Point Case	Coefficient x 10^3	Standard Error x 10^3	95% Confidence Interval x 10^3	Level of Significance	Change in R^2
F_1	4.6	0.57	(3.4,5.8)	<0.001	0.70
F_2	2.7	0.60	(1.4,3.9)	<0.001	0.15
F_3	-0.13	0.62	(-1.4,1.2)	0.84	***
Constant	6.1	0.60	(4.9,7.4)		
30-Point Case					
F_1	5.1	0.46	(4.1,6.1)	<0.001	0.61
F_2	2.4	0.46	(1.4,3.4)	<0.001	0.18
F_3	-0.11	0.47	(-1.1,0.86	0.82	***
Constant	5.5	0.46	(4.6,6.5)		

*** Value less than precision shown

The constant terms are the mean values of inverse median survival time in each case since the factor scores (F_1,F_2, F_3) have nominal means of zero and standard deviations of 1.

The relative contributions to the variance in 1/t for each factor are given in the last column of Table VIII. The values are the squares of the correlation coefficients for each variable in Equations 7 and 8 since the factors are orthogonal variables. The sums of contributions are 85% in the 30-point case and 80% in the 26-point case. That these values are less than 100% states that the three factors are sufficient to account for 85% and 80%, respectively, of the variance in 1/t.

Although single ion correlations differ substantially between the 26 and 30 point cases (r decreases by about 0.1 to 0.2 units in going from 26 to 30 points), the derived factors and multiple regression coefficients are in good agreement for the two calculations.

Conclusions for *Daphnia magna* - Needed Experiments. Conclusions drawn from the above data treatment are necessarily tentative due to the quality of species separation, the numbers of data points involved, and the narrow pH range involved in the calculations.

If the numbers were accepted at face value, the data indicate that: the species involved account for 80-85% of the toxicity, the carbonato complexes do not contribute to toxicity, the anionic hydroxo copper complexes contribute 15-18% to the total toxicity, and the free copper ion and/or the neutral and cationic hydroxo complexes of copper are responsible for 60-70% of the toxicity of copper to aquatic life.

If the carbonato complexes are accepted as being non-toxic, a set of experiments are needed in which the toxicity of copper to Daphnia magna is measured at equal intervals over the pH range 6.0-8.0. The speciation curves shown in Figure 1 would predict that this set of experiments plus inclusion of all copper complexes involved in the speciation calculations should allow separation of the species of concern into four groups, (Cu^{2+}), ($Cu(OH)_2^o$), ($CuOH^+$,$Cu_2(OH)_2^{2+}$), ($Cu(OH)_3^-$, $Cu(OH)_4^{2-}$) and a determination of the relative toxicity of these species. Underlying these statements is the assumption that the mechanism of interaction of copper at the environmental interface remains the same over this pH range.

Sensitivity of Conclusions to Errors in Stability Constants

As one measure of the sensitivity of the conclusions concerning toxic species to errors in the input data, the 26 point calculations have been repeated with a drastic change in the value of the logarithm one stability constant, β for $Cu(OH)_2^o$. The correct value of β_2 for $Cu(OH)_2^o$ has been the subject of much discussion. A value of 14.3 was used in the species calculations in the first part of this paper. Log β_2 was changed to 11.8 and the speciation calculation was repeated with all other input data held constant. Since $Cu(OH)_2^o$ was the dominant copper species in the first calculation, the effect of changing the stability constant by a factor of ∿300 has a marked effect on the calculated activities (Table VII and Table IX). The resulting activities were used as input for factor analysis. In Table X are shown the factors for the 26 point case. Decreasing the stability constant of $Cu(OH)_2^o$ ∿300 fold causes a shift in the loading of $Cu(OH)_2^o$ from Factor 1 to Factor 2, of $CuCO_3^o$ from Factor 3 to Factor 1 and of $Cu(OH)_3^-$ from Factor 1 to Factor 2, with minimal changes for the other species. Examination

TABLE IX

Calculated Copper Species Activities with log (β_2) for $Cu(OH)_2$ of 11.8

Measured dissolved copper (μM)	Activity of Copper Species (μM)			
	Cu^{2+}	$CuOH^+$	$Cu(OH)_2^o$	$CuCO_3^o$
∿0.02	0.001	0.001	***	0.013
0.20	0.008	0.016	0.004	0.166
0.33	0.014	0.026	0.008	0.275
0.41	0.017	0.032	0.010	0.347
0.61	0.026	0.048	0.014	0.513
1.02	0.043	0.079	0.023	0.851
1.54	0.064	0.120	0.035	1.288

*** Value less than precision shown

of speciation curves versus pH, Figure 1, would lead one to predict these shifts. It should be noted that the species that shifted from one factor to another have moderate or high loadings in two factors.

TABLE X

Comparison of the Varimax Rotated Factor Matrices for Three-Factor 26-Point Case

Species	Factor 1		Factor 2		Factor 3	
	A	B	A	B	A	B
Cu^{2+}	0.96	0.97	-0.05	-0.05	0.24	0.17
$CuOH^+$	0.98	0.97	0.05	0.23	0.19	0.12
$Cu_2(OH)_2^{2+}$	0.97	0.94	0.02	0.07	0.04	0.01
$Cu(OH)_2^o$	0.92	0.61	0.35	0.79	0.11	0.02
$Cu(OH)_3^-$	0.41	0.10	0.91	0.99	-0.04	-0.04
$Cu(OH)_4^{2-}$	-0.15	-0.10	0.97	0.99	-0.13	-0.05
$CuCO_3^o$	0.45	0.60	0.09	-0.01	0.87	0.77
$Cu(CO_3)_2^{2-}$	***	-0.02	0.08	-0.05	0.95	0.91

*** Value less than precision shown

A = Calculations with $\beta_2 = 10^{14.3}$ for $Cu(OH)_2^o$

B = Calculations with $\beta_2 = 10^{11.8}$ for $Cu(OH)_2^o$

Multiple regression analysis utilizing factor scores for the above factors gives the following:

$$\frac{1}{t} = 0.0052F_1 + 0.0021\ F_2 + 0.00084\ F_3 + 0.0061 \qquad (9)$$

with changes in R^2 in the regression analysis due to F_1, F_2, and F_3 of 0.68, 0.10, and 0.02, respectively. As before, the Cu species involved account for 80% of the toxicity. Since only the carbonato species dominate the third factor (which accounts for 2% of the variance in t^{-1}), the carbonato complexes contribute, at worst, only slightly to the toxicity. The anionic hydroxo copper complexes, along with $Cu(OH)_2^o$, loaded in factor 2, can contribute only moderately to copper toxicity. Free copper ion and/or the cationic hydroxo complexes, along with $Cu(OH)_2^o$, again are predicted to be responsible for the major portion of the toxicity of copper to aquatic life.

Summary

A procedure is developed through which the relative toxicity of metal speciation products can be determined. The procedure is illustrated utilizing data from a published study on the toxicity of copper to *Daphnia magna* and tentative conclusions are drawn with respect to the relative toxicity of certain combinations of species. The experiments necessary to obtain a better description of the relative toxicity of copper species to toxicity are described.

Some limitations of the procedure are discussed. A thorough error analysis of the procedure is not given but is in progress and will be the subject of a later paper.

Acknowledgement

The authors wish to express their appreciation to the U.S. Environmental Protection Agency (Grant # R804996) for support of this work, and to the University of Minnesota, Duluth Computer Center for their assistance.

Abstract

The use of factor analysis and multiple regression in determining relative toxicities of species is described. Procedures are illustrated using the data of Andrew, Biesinger and Glass from a study of toxicity of copper to *Daphnia magna* following a recalculation of the equilibrium speciation of copper. The results yield tentative conclusions that the carbonato copper complexes are not toxic, the anionic hydroxo copper complexes contribute 15-18% to the toxicity of copper, and that free copper and/or the neutral and/or cationic hydroxo copper complexes are responsible for 60-70% of the toxicity in this set of experiments.

The limitations of this analysis are discussed and a set of experiments which should provide more definitive answers to the relative toxicity of these species is described.

Literature Cited

1. Quality Criteria for Water, U.S. Environmental Protection Agency, Washington, D.C. (July 1976).
2. Andrew, R.W., Biesinger, K.E., and Glass, G.E. Effects of inorganic complexing on the toxicity of copper to Daphnia magna, Water Res. 11, 309-315 (1977).
3. Glass, G.E. Identification and distribution of inorganic components in water: What to measure? Ann. N.Y. Acad. Sci. 298, 31-46 (1977).
4. Lee, G.F. Graduate education in environmental chemistry, J. Chem. Educ. 51, 772-774 (1974).
5. Doudoroff, P. Recent advances in fish toxicology, U.S. Environmental Protection Agency, GP/3-77-085, 1977.
6. Mount, D.I. Research to develop heavy metal standards for fresh water, p. A19-20, in Hutchinson, T.C., ed., "International Conference on Heavy Metals in the Environment," Vol. 1, Toronto, Ontario, Canada, October 27-31, 1975.
7. "Proceedings of Workshop, Toxicity to Biota of Metal Forms in Natural Water," October 7-8, 1975, Duluth, Minnesota, sponsored by the Standing Committee on the Scientific Basis for Water Quality Criteria of the International Joint Commission's Research Advisory Board.
8. Morel, F., and Morgan, J. A numerical method for computing equilibria in aqueous chemical systems, Environ. Sci. Tech. 6, 58-67 (1972).
9. Nie, N.H., Hull, D.H., Jenkins, J.G., Steinbrenner, K., and Bent, D.H. "Statistical Package for the Social Sciences," 675 p., McGraw-Hill, New York, 1975.
10. Howarth, R.S., and Sprague, J.B. Copper lethality to rainbow trout in waters of various hardness and pH, Water Res. (in press).
11. Cattel, R. Factor analysis: An introduction to essentials. I. The purpose and underlying models, Biometrics 21, 190-215 (1965).
12. Bennett, S., and Bowers, D. "An Introduction to Multivariate Techniques for Social and Behavioral Sciences," 156 p., John Wiley & Sons, New York, 1976.
13. Harman, H.H. "Modern Factor Analysis," Second Edition, 487 p., Univ. of Chicago Press, 1967.
14. Shaw, T.L., and Brown, V.M. The toxicity of some forms of copper to rainbow trout, Water Res. 8, 377-382 (1974).
15. Pagenkopf, G.K., Russo, R.C., and Thurston, R.V. Effect of complexation on toxicity of copper to fishes, J. Fisheries Res. Bd. Can. 31, 462-465 (1974).

16. Chakoumakos, C., Russo, R.C., and Thurston, R.V. The toxicity of copper to rainbow and cutthroat trouts under different conditions of alkalinity, pH and hardness, Environ. Sci. Tech. (in press).

RECEIVED November 16, 1978.

29

Critical Assessment of the Relationship between Biological Thermodynamic and Electrochemical Availability

M. WHITFIELD and D. R. TURNER

Marine Biological Association of the United Kingdom, The Laboratory, Citadel Hill, Plymouth, England PL1 2PB

Much of the current interest in the B-subgroup trace metals (e.g. Zn, Cu, Cd, Pb, Hg) in natural waters centers around their influence on the biota. It is clear from the work published so far (1, 2, 3) that the biologically available fraction of a particular element (i.e. the fraction of the total concentration of that element that is available for biological uptake) is intimately dependent on the chemical form of the element in solution. Where the solution composition is clearly defined and all the relevant conditional stability constants are known it is possible to calculate the equilibrium speciation of a metal from thermodynamic principles (2). By suitable manipulation of the solution chemistry the response of organisms to particular chemical species can then be studied (1, 3). Such detailed knowledge of the medium composition is rarely available for natural waters which can exhibit considerable variations in the concentrations of the major ions and of the suite of trace metals present. In addition, significant and variable quantities of unidentified organic compounds, possibly with appreciable complexing capacities, might be present and exert a considerable influence on the biological availability of trace metals (3, 4). Not only do these variations make the equilibrium calculations difficult but the presence of organic matter, and also the presence of solid phases, might make the equilibrium concept itself untenable (5). In response to such difficulties a number of operational procedures have been developed to determine the fraction of 'available' metal in solution. The most direct are the bioassay procedures in which the response of a test organism to changes in the solution chemistry (usually simply to changes in the total concentration of the metal) is monitored. Frequently, problems have arisen because of variability in the test organism itself and because inadequate attention has been given to ensuring consistency in the solution chemistry. Recent studies suggest that these problems can be successfully circumvented (1,3) and that useful techniques may soon be available for detailed studies of sublethal effects (6, 7). Such procedures are time

0-8412-0479-9/79/47-093-657$06.00/0
© 1979 American Chemical Society

consuming, however, and their usefulness rests heavily on the aptness of choice of the test organism. Direct chemical measurements of 'available' metal are more suitable for a rapid assessment of metal availability in the field and in routine water samples. Where metal concentrations are sufficiently high, ion selective electrodes (ISE's) can be used to give a direct measure of conventional metal ion activities. In some instances such measurements have been shown to provide a direct correlation with the response of organisms to changes in solution chemistry (8, 9 and Crisp, D.J., Marine Science Laboratories, Menai Bridge, North Wales, personal communication, 1978). Since potentiometric techniques require relatively high concentrations of the selected ion (not less than 10^{-6}M unless the metal is well buffered in solution) they are rarely sufficiently sensitive for the direct analysis of natural waters. Consequently, considerable interest has recently been shown in anodic stripping voltammetry (ASV) which is able to make selective, direct and non-destructive measurements of amalgam forming metals (e.g. Cu, Pb, Cd, Zn) at concentrations down to 10^{-10}M (10, 11). ASV is sensitive to the chemical form of the metal in solution and it has been suggested that the fraction of metal determined by this procedure might provide a useful guide to the biological availability of that metal in solution (3). When we use chemical procedures measuring trace metal availability we are, in effect, assuming that these procedures provide useful models of biological uptake. The significance of the results obtained will therefore depend on the suitability of the uptake mechanism implied by the experimental measurement. Since electrochemical procedures involve the transfer of metal ions to an electrode surface in direct contact with the solution we will confine our theoretical analysis to the uptake of trace metals by phytoplankton cells which might be expected to utilise a similar mechanism. The uptake of ionic lead from sea water will be used as the model process. We will first clarify the chemical definitions of the available metal fraction and then investigate the relationship of these chemically defined fractions to the fraction of metal available for uptake by a model phytoplankton cell.

Chemical Definitions of Trace Metal Availability

Thermodynamically available fraction (TAF). Early studies of the influence of non-electrolyte solutes on aquatic organisms identified two kinds of toxicity - physical toxicity (or narcosis) and chemical toxicity (12). Narcosis is caused by a wide variety of substances (including the atmospheric gases) and seems to arise because essential pathways are physically blocked by an excess of inert molecules that have entered the organism *via* an equilibrium distribution across an outer membrane. At equilibrium the activities of the toxic compound are the same in the organic phase and in the aqueous phase. Consequently, the thermodynamic activity of the narcotic compound in the aqueous phase is directly related to its biological availability. This thermodynamic scale of narcotic potency was first proposed by Ferguson (12) and has since been confirmed for a wide range of compounds (13). This scale implies that all compounds that exert narcotic effects will have the same influence on a given organism if they have the same thermodynamic activity. Chemical toxicity, which is caused by chemically reactive agents (such as heavy metals and organometallic compounds), should not show such a simple overall correlation with thermodynamic activity since the toxicity of each component will depend on its own unique ability to interfere with vital chemical processes. Nonetheless, the continued development of the theory of biological membranes (14, 15) and particularly the recent intense interest in artificial ion-selective membranes (16, 17, 18) has emphasised the significance of gradients in thermodynamic activity rather than concentration in controlling the transport of chemical components across membranes. As a result, it has been assumed, on the basis of very little direct evidence, that the biological availability of an element is related to the activity in solution of the particular chemical form that is taken up preferentially by the organism. This definition of the thermodynamically available fraction of the element (TAF) is consistent with the thermodynamic scale of narcotic potency but it does not imply a common activity threshold for all elements. The parameter relevant to our model uptake process is the activity of lead in sea water, a_{Pb}, which can be written as

$$a_{Pb} = [Pb]f_{Pb} \qquad (1)$$

where f_{Pb} is the conventional free single-ion activity coefficient and [Pb] is the concentration of free lead. The natural concentration of lead in sea water is so low ($\simeq 10^{-10}$M) that the complexes formed by the metal with inorganic anions will have a negligible effect on the concentration of the free ligands which can be calculated from a conventional chemical model for sea water (19). Against this background of constant ligand concentration the equilibrium speciation of lead can be calculated using the side reaction concept of Ringbom (20) which can be summarised in the equations

$$\sigma_{L,j} = \beta_j^*(L)[L]^j = [PbL_j]/[Pb] \qquad (2)$$

$$\Sigma_{Pb} = \Sigma(\sigma_{L,j}) \qquad (3)$$

$$[Pb]_T = [Pb](1+\Sigma_{Pb}) = [Pb]\,\bar{\alpha}_{Pb} \qquad (4)$$

$$[PbL_j] = [Pb]_T\,\sigma_{L,j}/\bar{\alpha}_{Pb} \qquad (5)$$

where $\beta_j^*(L)$ is the overall stoichiometric equilibrium constant describing the formation of PbL_j from the free components, $[Pb]_T$

is the total concentration of lead and $\bar{\alpha}_{Pb}$ is termed the overall side reaction coefficient. The calculated equilibrium speciation of lead in sea water, using the above equations, is shown in Table I.

In defining $\bar{\alpha}_{Pb}$ a number of assumptions have been made concerning the conventional values of f_X ascribed to the individual chemical species involved in the various equilibria (see footnotes to Table I). In particular it was assumed that, at constant ionic strength, f_X is also constant and all changes in the activity of X associated with changes in solution composition are attributed to changes in [X]. Consequently, the activity of lead in sea water of constant salinity is proportional to [Pb]. In view of the number of assumptions involved in the determination of $\bar{\alpha}_{Pb}$ and f_{Pb} (21, 23, 25, 26) it would be foolhardy to go beyond this and to place too much significance on the value of a_{Pb} that can be calculated. In establishing our model for biological uptake we will consider conditions where activity coefficients remain constant throughout. The thermodynamically available fraction (TAF) will therefore be equal to $\bar{\alpha}_{Pb}^{-1}$ (equation 4).

If lead also forms a complex with an organic ligand (J) it is readily shown that

$$-\log [Pb] = \log (\bar{\alpha}_{Pb} + \sigma_{Pb,J}) - \log[Pb]_T \qquad (6)$$

so that a plot of - log [Pb] versus log $\sigma_{Pb,J}$ will give a curve common to all organic ligands (21). At low ligand concentrations $\bar{\alpha}_{Pb} \gg \sigma_{Pb,J}$ so that the curve is a horizontal straight line and [Pb] is unaffected by increases in the ligand concentration although the buffer capacity for the free metal is increased slightly. At high ligand concentrations $\sigma_{Pb,J} \gg \bar{\alpha}_{Pb}$ and the curve is a straight line of unit slope. Here the free metal concentration drops off rapidly with increasing ligand concentrations. Equation 6 therefore provides a simple basis for rationalising many of the observations, summarised by Siegel (4) and Mancy and Allen (3) on the influence of organic complexing agents on the biological availability of metals and thus strengthens the correlation with the TAF.

$\bar{\alpha}_{Pb}$ can be calculated directly if the concentrations of all ligands and of all competing cations are known (21). Where this information is not available, ISE's can in principle enable conventional single-ion activites to be measured directly. The limited sensitivity of present-day ISE's precludes their use in natural waters, although they can be used in experimental systems involving elevated concentrations of trace metals. Provided that the salinity remains constant, a cell without liquid junction, composed of perfectly selective chloride and lead ISE's could be used. The difference between the emfs measured in the sample (E_x) and in a standard solution with the same temperature and major ion composition (E_s) would be given by

Table I.
The equilibrium speciation of lead in sea water at pH 8 (21)

Species	log β^{*a}	$\sigma_{L,j}$ [b]	%$[Pb]_T$
$PbCl^+$	0.90	4.50	7
$PbCl_2^0$	1.32	6.70	11
$PbCl_3^-$	1.20	2.87	5
$PbCl_4^{2-}$	1.06	1.18	2
$PbOH^+$	6.21	2.62	4
$Pb(OH)_2^0$	10.35	5.88×10^{-2}	-
$Pb(OH)_3^-$	13.08	5.10×10^{-5}	-
$PbCO_3^0$	6.12	34.7	55
$Pb(CO_3)_2^{2-}$	9.32	1.44	2
$PbSO_4^0$	1.4	3.69×10^{-1}	1
$Pb(Cl,OH)^0$	6.14	1.25	2
$Pb(Cl,CO_3)^-$	5.62	6.21	10
$Pb(OH,CO_3)^-$	10.14	5.8×10^{-1}	1
Pb^{2+}	-	-	2

a Values corrected to I = 0.72 using the Davies equation (22) to calculate f_i values.

b The following free ligand concentrations were calculated for an ion-pair model based on f_i values calculated from the MacInnes convention (23) using the sea water recipe of Millero (24).

Ion	Cl^-	OH^-	SO_4^{2-}	CO_3^{2-}	HCO_3^-
pX	0.247	5.79	1.833	4.58	2.80

$$E_x - E_s = (\underline{k}/2)\ \log[Pb]_x/[Pb]_s \tag{7}$$

where $\underline{k}$ = 0.1986 TmV (T = temperature in °K). For measurements in sea water the standard solution could be prepared in an organic-free sea water adjusted to pH 5 to 6 where lead speciation depends only on the concentration of chloride and sulphate (21).

There is a little evidence in the literature suggesting that both calculated values of free metal concentration (1) and values measured with ISE's (8, 9) can be correlated with the biological availability of copper. Additionally, Sunda *et al.* (27) have presented evidence that the toxicity of cadmium in seawater containing NTA can be correlated with the observed and calculated concentration of free cadmium ions.

Electrochemically available fraction (EAF). The significance of the electrochemically available fraction of a trace metal, as measured by ASV, can best be appreciated by comparison with measurements of the thermodynamic availability made using an ISE. We will consider in each case an electrode in a stirred solution with a diffusion layer of thickness δ at its surface. All changes induced in the equilibrium speciation by reactions at the electrode surface will be assumed to take place within this diffusion layer. Potentiometric measurements are carried out at equilibrium so that there is no concentration gradient and no net metal flux across the diffusion layer. In contrast, during the plating step of an ASV analysis, metal is deposited electrolytically at a mercury electrode from a rapidly stirred solution. This deposition process generates a gradient of metal concentration within the diffusion layer and the only species that will be sensed by the method will be those that can contribute to the metal flux at the electrode surface. The time scale of the plating process is defined by the diffusion layer thickness (δ) and by the diffusion coefficients of the individual species (D_i). If the electrode process is the reversible reduction of metal ions, only complexes that dissociate sufficiently rapidly to release the free metal within the diffusion layer will contribute to the metal flux (J) at the electrode surface. In a recent study (28, 29) we have described a general theory which enables J to be calculated for a multi-ligand system. From this we obtain

$$J = M_0[Pb]_T - M_1[Pb]^0 \tag{8}$$

where M_0 and M_1 are complicated matrix functions of δ and D and also of the kinetics and thermodynamics of the metal-ligand interactions. $[Pb]^0$ is the concentration of the free metal at the electrode surface. Under current limiting conditions (i.e. on the plateau of the peak stripping current versus plating potential curve) $[Pb]^0 = 0$ and the limiting flux in the presence of kinetic control can be written as

$$J_k = M_O[Pb]_T \quad . \tag{9}$$

If there were no kinetic control (i.e., if all species could dissociate sufficiently rapidly to contribute fully to the metal flux) then, under the same conditions, the diffusion limited flux of the metal would be

$$J_d = D[Pb]_T/\delta \tag{10}$$

if we assume that the metal and its complexes all have the same diffusion coefficient. The <u>electrochemically available fraction</u> (<u>28</u>, <u>29</u>) will therefore be

$$J_k/J_d = M_O\,\delta/D$$
$$= I_k/I_d \tag{11}$$

where I_k and I_d are the stripping peak currents measured under kinetically and diffusionally limited conditions, respectively. ASV and potentiometry therefore impose quite different limiting conditions for the measurement of trace metal availability (Figure 1, (a) and (b), and Table II).

Table II
Comparison of potentiometric and voltammetric techniques

	Potentiometry	Voltammetry
Measured quantity	Electrode potential	Metal flux at the electrode surface
Measure of trace metal obtained	Thermodynamic Activity	Electrochemical availability
Concentration of free metal at the electrode surface $[Pb]^o$	Bulk concentration $[Pb]$	Zero
Trace metal flux (J) at the electrode surface	Zero	Kinetically limited flux J_k
Mathematical summary	$[Pb]^o/[Pb] = 1$ $\xi = J/J_k = 0$	$[Pb]^o/[Pb] = 0$ $\xi = J/J_k = 1$

Combining equations 8 and 9 we find that

$$J/J_k = 1 - M_1[Pb]^o/M_O[Pb]_T \quad . \tag{12}$$

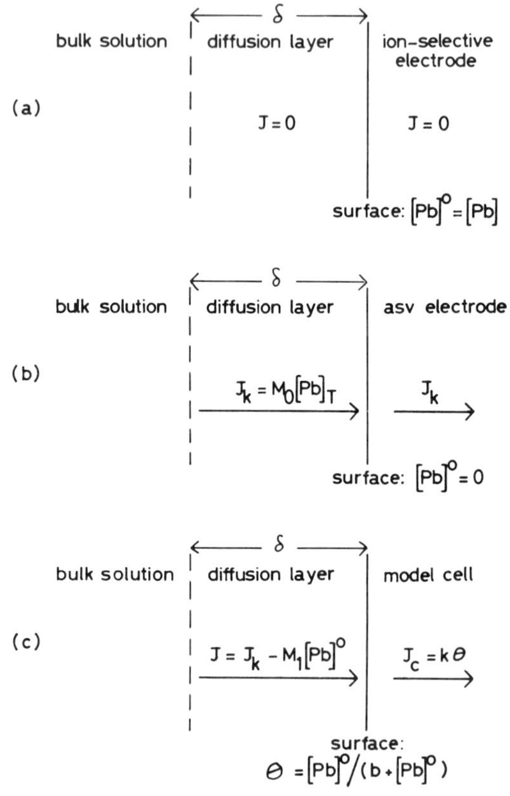

Figure 1. Comparison of conditions imposed for the sensing of trace metal by (a) an ion-selective electrode, (b) a mercury film asv electrode, and (c) the model cell. In each case the system is divided into four zones: bulk solution, diffusion layer of thickness (δ), cell/electrode surface, and cell/electrode interior.

Since $[Pb]^{o} \longrightarrow [Pb]$ as $J/J_k \longrightarrow 0$ we also find that $M_1/M_0 = \bar{\alpha}_{Pb}$ (equation 4) so that

$$J/J_k + [Pb]^{o}/[Pb] = 1 \; . \tag{13}$$

Under current limiting conditions $[Pb]^{o} = 0$ so that $J/J_k = 1$. To simplify matters we will set $J/J_k = \xi$ so that $[Pb]^{o}/[Pb] = 1-\xi$.

By estimating ξ for a model cell we should be able to decide whether the uptake of metal will reflect the thermodynamic or the electrochemical availability of lead in the solution (Table II).

Definition of the Model Cell

Description. The model organism is a free-floating unicellular sphere with characteristics selected, where possible, to match those of a phytoplankton cell. The organism and its environment (Figure 1c) are divided into four concentric zones - the bulk solution, the diffusion layer, the containing membrane and the cell contents. We will assume that the species taken up by the cell is the free metal ion since most of the studies of the uptake of B-subgroup metals by organisms support this hypothesis (1, 2, 3, 7, 27, 30). Two steady-state transport processes are considered, namely (i) transport of trace metal to the cell surface described by equation 8 and (ii) assimilation of the trace metal into the interior of the cell described by

$$J_c = k\theta \tag{14}$$

where k is a constant and θ is the fraction of the available surface adsorption sites occupied by the trace metal ions. The form of equation 14 assumes that metal already present in the cell has no effect on the assimilation rate, and is thus best considered as a description of the early stages of metal uptake. The fractional coverage θ is described by the Langmuir isotherm

$$\theta = [Pb]^{o}/(b + [Pb]^{o}) \tag{15}$$

which assumes that adsorption and desorption processes are rapid compared to transport processes. From equations 14 and 15 we have

$$J_c = k[Pb]^{o}/(b + [Pb]^{o}) \; . \tag{16}$$

At steady state $J_c = J$ so that, eliminating $[Pb]^{o}$ from equations 8 and 16 results in,

$$J^2 - J(M_1 b + M_0[Pb]_T + k) + kM_0[Pb]_T = 0 \; . \tag{17}$$

Only one root of this equation lies in the range $0 \leqslant J \leqslant J_k$. Combination of this root with equation 9 enables ξ to be calculated.

Before we can consider the response of our model organism to lead in sea water we must define k (equation 14) and b (equation 15) and estimate the range of diffusion layer thicknesses (δ) that are characteristic of phytoplankton cells. Since we have no direct experimental evidence on which to base an estimate of k we will treat it as a variable in the calculations.

Estimation of half saturation constant (b). Values of b determined at the surfaces of phytoplankton cells, corrected to refer to [M] in the sea water ionic medium, are shown in Table III. Similar values are found for lead and zinc and a somewhat lower value for cadmium. The value estimated for mercury is unrealistically low and suggests that here the species adsorbed is not the free metal. Initially we will take b = 50 nM as an order of magnitude figure to describe the surface or our model cell.

Estimation of diffusion layer thickness (δ). δ for a moving particle is related to the velocity of motion (u) of the particle through the water. For a sphere of radius a moving through the water at a constant velocity it can be shown using the equations given by Levich (34, p. 84-85) that the average diffusion layer thickness δ_{AV} is given by,

$$\delta_{AV} = \pi D^{1/3} (u/a^2)^{-1/3}/2 \quad . \qquad (18)$$

All quantities in equations, including numerical constants, are expressed in c.g.s. units. With $D = 10^{-5}$ cm^2 sec^{-1} equation 18 becomes

$$\delta_{AV} = 0.034 \, (u/a^2)^{-1/3} \quad . \qquad (19)$$

The problem of estimating the diffusion layer thickness therefore reduces to that of estimating u and a. We have considered three possible mechanisms for the movement of phytoplankton through the water: gravitational sinking, convective water movements and the swimming of flagellates.

Gravitational sinking. Phytoplankton cells are generally denser than the surrounding sea water and therefore sink unless kept in suspension by convective water movements (35). δ_{AV} for sinking cells was estimated from a table of sinking rates for a variety of phytoplankton cells given by Smayda (35). A value of a was estimated from the cell volume assuming a spherical cell. Only data for living cells were used and a histogram of δ_{AV} (Figure 2) indicates values closely grouped in the range 2 to 5 x 10^{-3} cm.

Convective water movements. Wind-driven Langmuir cells are considered to be among the most important convective mechanisms which keep phytoplankton in the euphotic zone (35). Phytoplankton cells which are denser than the water will not be fully entrained by these motions and will have a resultant net velocity through the water. Levich (34, p. 182) gives an approximate expression for this motion,

$$u/a^2 \simeq 0.39\,(\Delta\rho/\rho_0)(u_L^9/\nu^5 L^3)^{\frac{1}{4}}$$

where $\Delta\rho/\rho_0$ is the relative density excess of the cell over sea water, L and u_L are the scale and velocity of the convective motion respectively and ν is the kinematic viscosity. Taking $\rho_0 = 1.024$ and $\nu = 10^{-2}$ cm^2 s^{-1} (s^{-1} is defined below) we obtain,

$$u/a^2 \simeq 119\,\Delta\rho\,(u_L^3/L)^{\frac{3}{4}}. \quad (20)$$

Investigations into Langmuir cells have resulted in the approximate expressions $u_L \simeq 0.008W$ (36) and $L \simeq 4.8W$ (37) where W is the wind speed in cm s^{-1}. Using these expressions equation 20 becomes

$$u/a^2 \simeq 0.00077\,\Delta\rho\,W^{3/2}. \quad (21)$$

A histogram of $\Delta\rho$ values for phytoplankton cells (Figure 3) indicates that most values lie in the range $0 < \Delta\rho < 0.08$. A plot of δ_{AV} versus W for several values of $\Delta\rho$ (Figure 4) indicates that even at high wind speeds the diffusion layer thickness will not be less than the value of 10^{-2} cm normally maintained by chance convective motion (39, p. 2).

Gavis (40) has also considered the effect of open water turbulence, which may be characterised by a rate of shear S, with a maximum value of 6 s^{-1}, where S is given by

$$S = u/a.$$

The extreme effect of open water turbulence can then be characterised by

$$u/a^2 = 6/a. \quad (22)$$

Substitution of equation 22 in equation 19 shows that the maximum shear rate of 6 s^{-1} will result in values of δ_{AV} in the range 2×10^{-3} to 4×10^{-3} cm for cells of radius 10^{-3} to 10^{-2} cm, similar to that resulting from gravitational sinking (Figure 2).

Swimming of motile cells. When expressed in body lengths, s^{-1}, the speed attained by animals from paramecia to tuna in water varies little with size (41) and most organisms have a maximum speed of 10 lengths, s^{-1}. Taking $u = 20a$ as a maximum for motile

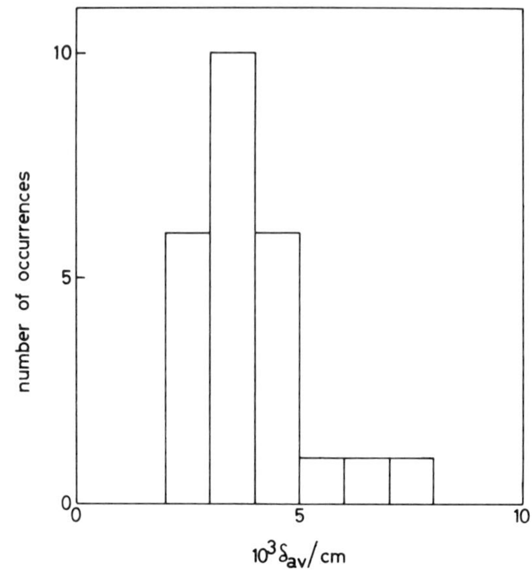

Figure 2. Histogram of estimated values of $(\delta)_{AV}$ for gravitational sinking of phytoplankton cells. Estimates obtained from data given for live cells by Smayda (35) using Equation 19.

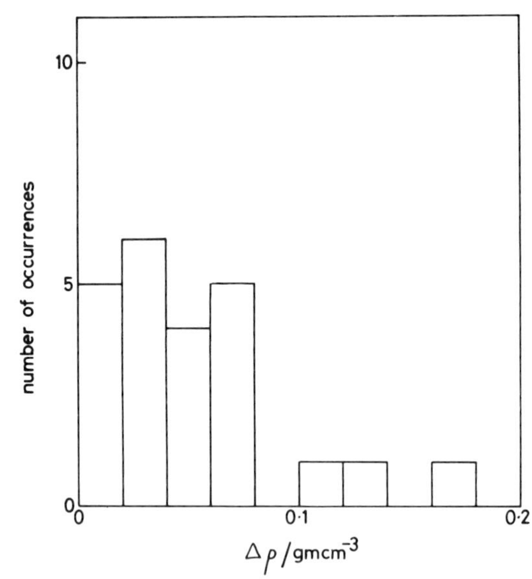

Figure 3. Histogram of excess density ($\Delta\rho$) for phytoplankton cells; data taken from Ref. 38

Table III.
Measured values of half-saturation constant (b) for metal adsorption at phytoplankton cell surfaces

Species	Metal	b/nM measured using total metal concentrations	b/nM corrected to free metal concentration	Reference
Phaeodactylum tricornutum	Zn	91 [a]	43 [c]	(30)
Phaeodactylum tricornutum	Pb	2100 [b]	42 [d]	(31)
Platymonas subcordiformis	Pb	3800 [b]	76 [d]	(31)
Isochrysis galbana	Cd	400 [a]	8 [e]	f
Isochrysis galbana	Hg	97 [a] 50 [a]	$\sim 10^{-13}$ [g]	(32)
Dunaliella tertiolecta	Hg	23 [a] 25 [a]	$\sim 10^{-13}$ [g]	(32)

a determined using Langmuir isotherm.
b determined using Freundlich isotherm: the figure given is the metal concentration corresponding to half saturation of the surface.
c assuming 47% free zinc in sea water (33)
d assuming 2% free lead in sea water (21)
e assuming 2% free cadmium in sea water (33)
f Davies, A.G., Marine Biological Association, Plymouth, personal communication 1978.
g estimated using data given in Mantoura *et al.* (33).

cells results in

$$\delta_{AV} = 0.013\ \underline{a}^{1/3} \ . \qquad (23)$$

This equation gives δ_{AV} values in much the same range as gravitational sinking and open water turbulence (Figure 5).

The values of δ_{AV} obtained by these three procedures are order of magnitude estimates only since the relationships used to calculate $(\underline{u}/\underline{a}^2)$ are approximate and e uation 19 is not strictly valid when $\delta_{AV} > \underline{a}$ (34, p. 84-85). Nonetheless, the values obtained suggest that the movement of phytoplankton cells through the water is unlikely to produce a diffusion layer thickness less than 10^{-3} cm. This is of the same order as the diffusion layer thickness during an ASV analysis. At the rotating disc electrode, rotation speeds in the range 25 to 2500 rpm give diffusion layer thicknesses of 10^{-2} cm to 10^{-3} cm, and diffusion layer thicknesses at conventional stationary electrodes in stirred solutions will also be in this range. Measurements of the electrochemically available fraction made by ASV are therefore likely to be relevant to situations where biological uptake is controlled by transport of trace metal to the cell surface.

Trace Metal Availability for the Model Cell - Calculation and Discussion.

The calculation of ξ (= J/J_k, Table II) under different conditions will enable us to define the parameters that dictate whether the model cell responds to the thermodynamically or to the kinetically available fraction of the trace metal in solution. Values of M_0 and M_1 (Table IV) were calculated for lead in sea water at pH 8 using the speciation picture given in Table I.

Table IV. Matrix terms (M_0 and M_1) for the calculation of the electrochemically available fraction of lead in sea water (28,29)

Log	$10^3\ M_0$ (cm s^{-1})	M_1 (cm s^{-1})
-2.00	0.98	0.060
-2.25	1.72	0.105
-2.50	2.97	0.182
-2.75	5.05	0.309
-3.00	8.33	0.510

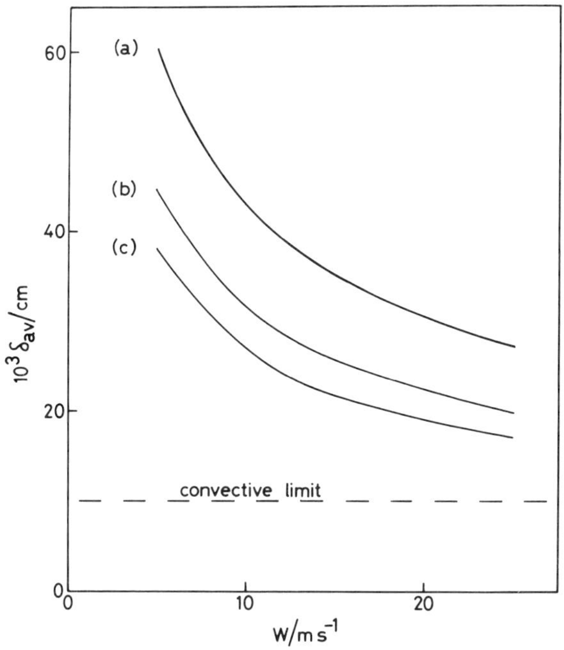

Figure 4. Estimated average diffusion layer thickness $(\delta)_{AV}$ resulting from Langmuir convection as a function of wind speed W. (a) $(\Delta\rho) = 0.02$ g cm^{-3}, (b) $(\Delta\rho) = 0.05$ g cm^{-3}, (c) $(\Delta\rho) = 0.08$ g cm^{-3}.

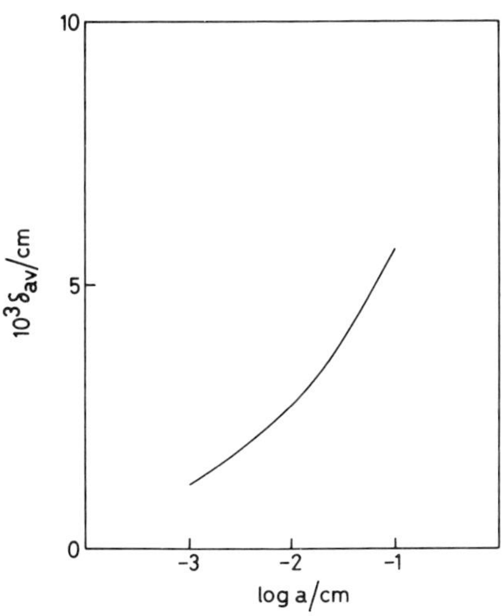

Figure 5. Estimated average diffusion layer thickness $(\delta)_{AV}$ for motile cells of radius a swimming at maximum rate (Equation 23)

Influence of assimilation rate constant (k). Values of ξ were calculated as a function of k using fixed values for $[Pb]_T$ (10^{-10}M), b (50 nM) and δ ($\log \delta = -2.5$). As the assimilation rate increases (higher values of k, Figure 6) the uptake of metal at the cell surface begins to be controlled by dissociative reactions in the diffusion layer so that $\xi \longrightarrow 1$ and the cell begins to sense the electrochemically available fraction of the metal. Under these limiting conditions the transport of metal across the diffusion layer is the rate determining process. Conversely at low values of k, corresponding to a slow assimilation of lead, $\xi \longrightarrow 0$ (i.e., $[Pb]^o/[Pb] \longrightarrow 1$) so that the cell begins to sense the thermodynamically available fraction of the metal. In this case the transport of metal from the surface to the interior of the cell is the rate determining process. There is no sharp division between these limiting cases and in this instance (Figure 6) there is a range of two to three orders of magnitude in k over which neither describes the trace metal availability. To see how these limits depend on other factors we will assume arbitrarily that when $\xi > 0.9$ the cell responds to the EAF and when $\xi < 0.1$ (i.e., when $[Pb]^o/[Pb] > 0.9$) the cell responds to the TAF. We can rearrange equations 9 and 17 to give

$$k = \xi M_o[Pb]_T - (\xi/(\xi - 1))M_1 b \qquad (24)$$

The value of k at the 'electrochemical limit' (when $\xi = 0.9$) will be referred to as k_e and the value at the 'thermodynamic limit' (when $\xi = 0.1$) as k_t (Figure 6).

The k-values used in the calculations so far can be compared with experimental observations (Davies, A.G., Marine Biological Association, Plymouth, personal communication, 1978) if we remove all reference to the cell surface and work simply in terms of the assimilation flux (J_c). The change-over from thermodynamic to kinetic control will occur in the region $0.1 J_k < J_c < 0.9 J_k$. In the present system M_o lies in the range 0.98 to 8.33×10^{-3} cm s^{-1} so that, from equation 9, the change-over for lead will occur when $J_c \simeq 10^{-3} [Pb]_T$ (J_c in mol cm^{-2} s^{-1}, $[Pb]_T$ in mol cm^{-3}). Unfortunately no data are available for lead but the fluxes have been measured for the initial stages of zinc uptake by *Phaeodactylum tricornutum* (Table V). The values of J_c and $M_o[Zn]_T$ are within an order of magnitude of one another so that the uptake is close to the change-over region.

Table V. Assimilation fluxes for the initial uptake of zinc by Phaeodactylum tricornutum

$[Zn]_T$ ($mol\ cm^{-3}$)	$M_o[Zn]_T$ [a] ($mol\ cm^{-2}\ s^{-1}$)	J_c (observed) [b] ($mol\ cm^{-2}\ s^{-1}$)
7.3×10^{-11}	7×10^{-13} to 7×10^{-14}	4×10^{-15}
7×10^{-10}	7×10^{-12} to 7×10^{-13}	10^{-14}

a Assuming that for zinc $I_k = I_d$ (i.e. there is no kinetic control) and hence $M_o = D/\delta$ (equation 11). $D = 10^{-5}\ cm^2\ s^{-1}$ and $\delta = 10^{-3}$ to 10^{-2} cm.

b Davies, A.G., Marine Biological Association, Plymouth, personal communciation, 1978.

Influence of diffusion layer thickness (δ). Equation 9 defines J_k as $J_k = M_o[Pb]_T$, and since M_o increases with decreasing δ (see e.g., Table IV) the effect of a reduction in δ is to increase J_k. This increase causes a shift of the change-over region to higher values of k, as evidenced by the dependences of k_e and k_t on δ shown in Figure 7.

Influence of the fractional surface coverage (θ). The dependence of k_e and k_t on $[Pb]_T$ for various half saturation values of the cell surface is shown in Figure 8. When θ is small, k_e and k_t are independent of $[Pb]_T$. As the surface coverage exceeds about ten per cent ($\theta > 0.1$), the values of k_e and k_t, and hence the nature of the cell's response to lead in solution, become concentration dependent. This is particularly important in considering the extrapolation of the results of laboratory experiments at high metal concentrations to the lower levels found in natural waters.

If the surface of the cell is close to saturation at the higher concentration, the nature of the cell's response might be quite different to that found at natural metal levels.

Influence of organic complexation. Ligands whose lead complexes dissociate only partially within the diffusion layer during the ASV plating process (i.e., complexes partially labile to ASV) (29) have little effect on k_t and k_e. For example, the presence of 10^{-4} M 'humic' acid (21), which is partially labile to ASV, reduces the TAF for lead by 31% and the EAF by 5-25% over the range of diffusion layer thicknesses considered here. The

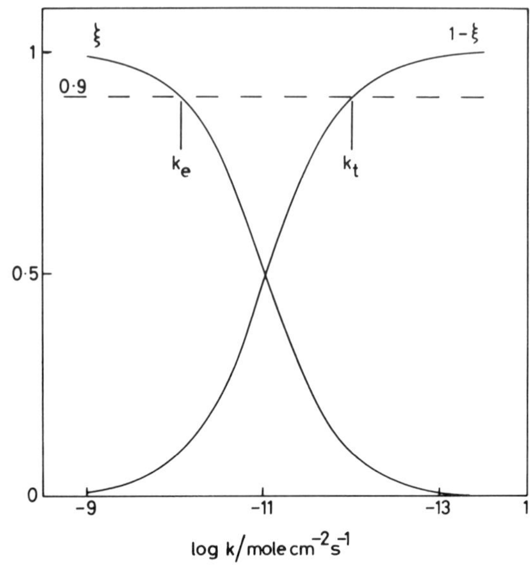

Figure 6. Calculated dependence of (ξ) *and* ($1 - \xi$) *on log* k *for the model phytoplankton cell. Other parameters* $[Pb]_T = 10^{-10}$M, b = 50 nM, *and log* $\delta = -2.5$.

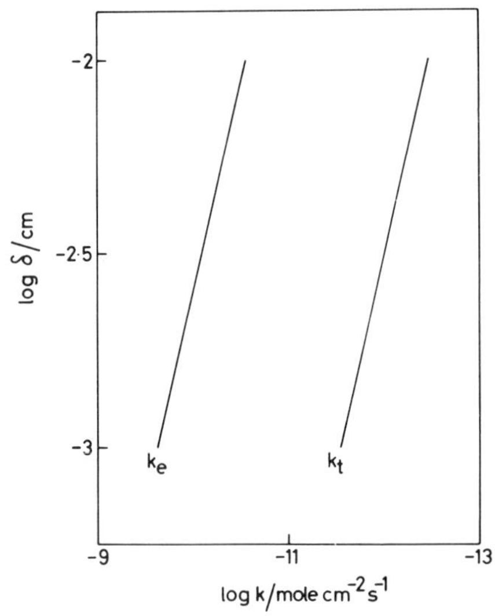

Figure 7. Calculated dependence of log k_e *and log* k_t *on log* δ *for the model phytoplankton cell. Other parameters* $[Pb]_T = 10^{-10}$M, $b = 50$ nM.

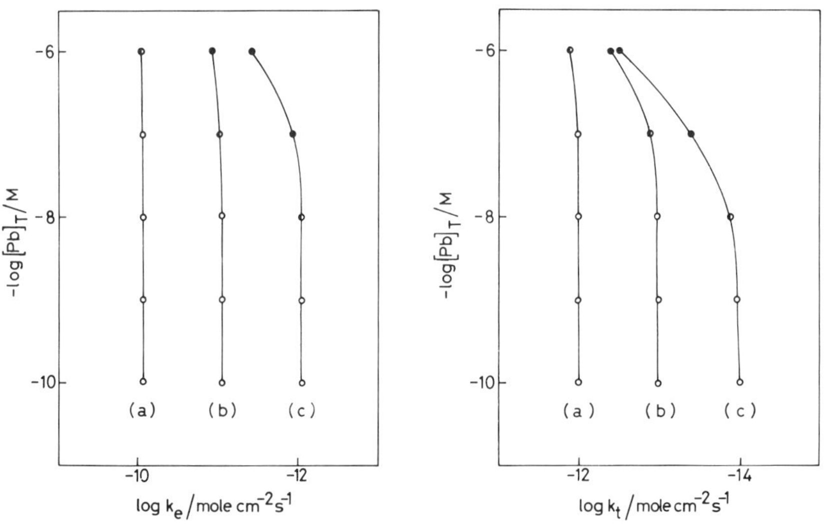

Figure 8. Calculated dependences of log k_e *and log* k_t *on* $[Pb]_T$ *for the model phytoplankton cell. Other parameters: log* $\delta = -2.5$; *(a)* b = 50 *n*M, *(b)* b = 5 *n*M, *(c)* b = 0.5 *n*M. *Fractional surface coverage* (θ) *indicated by:* (○) $\theta < 0.1$; (◐) $0.1 < \theta < 0.5$; (●) $\theta > 0.5$.

corresponding changes in log k_t and log k_e were only of the order of 0.1 units. The addition of ligands forming complexes with lead that are non-labile to ASV (29) will reduce both the EAF and the TAF by a factor corresponding to the change in chemical speciation. The decrease in trace metal availability on strong complexation is illustrated by the results of Shulz-Baldes and Lewin (31) who showed that the uptake of lead by *Phaeodactylum tricornutum* was strongly suppressed by the addition of EDTA to the culture medium. The change-over region will be unaffected unless the cell surface is close to saturation (Figure 8). In this case the change-over will depend on the concentration of metal *not* bound in complexes non-labile to ASV.

Summary and Conclusions

For a particular diffusion layer thickness δ the thermodynamic availability, as measured with an ISE or calculated from a speciation model, and the electrochemical availability, as measured by ASV, represent limiting cases of a continuum of trace metal availability. The nature of this continuum is most simply defined by considering the flux of the free metal ion across the diffusion layer to a surface which senses the metal availability. The ratio ξ of the observed flux (J) to the limiting flux (J_k) is unity for ASV measurements under current limiting conditions and zero for ISE measurements.

Several imperfections remain, both in our understanding of the chemistry of trace metals in natural waters and in the sophistication of our experimental techniques, that prevent an exact determination of the thermodynamically available fraction ($\bar{\alpha}_{Pb}^{-1}$, equation 4) and the electrochemically available fraction (I_k/I_d, equation 11). The stability constants used in calculating the individual σ-values (equation 2) are subject to considerable uncertainty (2, 21, 42) and the conventional f_i-values used in their adjustment to sea water conditions are based on a multiplicity of conventions. For many complexes that may be important in natural samples the stability constants are unknown and, frequently, the ligands have not been identified. The ISE's available at the present time are neither sensitive enough nor selective enough for direct measurements of [Pb] in most natural systems. No procedure has yet been developed for the direct determination of I_k/I_d under well-defined conditions. None of the ASV data published so far have included a determination of δ and none have used a well defined rotating disc electrode for which δ can be calculated as a function of rotation speed (see discussion by Turner and Whitfield (28, 29)). The normal procedure for estimating the fraction of electrochemically available metal involves a standard addition analysis of an untreated sample and is therefore dependent on the kinetics of the reactions controlling the assimilation of the ionic metal 'spike'. In addition, the theory used in the present paper (28, 29) assumes that the only process occurring at the electrode during the plating step is the reversible reduction of metal ions. In natural samples, adsorption phenomena may be observed (43) and other metal species may be reduced irreversibly at the electrode surface (29). The electrode process would then no longer correspond to the uptake of ionic metal that appears, on the basis of current evidence, to be the most common model for biological uptake in the absence of lipid soluble species (e.g. methylated metals). Despite these limitations the flux ratio ξ provides a useful starting point for discussing trace metal availability for a simple model cell.

The cell used in the present paper is designed, as far as possible to resemble a phytoplankton cell and its behaviour is described by a surface half-saturation constant (b) and an assimilation rate constant (k). This model has been used to identify the change-over region between response to the thermodynamically and electrochemically available fractions of lead in sea water. The most significant control is exerted by k (Figure 6) and direct experimental measurements of this parameter would be of considerable interest. The natural range of diffusion layer thicknesses causes a shift of one order of magnitude in both k_e and k_t and has no effect on the width of the change-over region (Figure 7). The width of this region becomes strongly concentration dependent when $b < [Pb]$ and this effect becomes more pronounced as surface coverage increases (Figure 8).

These effects, and the nature of chemical measurements of trace metal availability should be borne in mind when planning or interpreting experiments on the uptake of trace metals by phytoplankton cells.

Acknowledgement

The authors wish to thank Dr. A.G. Davies for valuable discussions.

Abstract

It is seldom possible to relate the biological availability of a particular element (i.e. its ability to influence biological processes) simply to changes in its stoichiometric concentration. This paper considers two concepts that have been used to provide chemical analogues of the biological availability of trace metals. The thermodynamic availability of an element in solution is defined as the activity of the particular chemical form that is taken up preferentially by the organism. The electrochemical availability is defined as the fraction of the total metal concentration that is available for electro-deposition at an electrode immersed in the solution during the timescale of the plating process. The implications of these definitions when related to conventional measurement techniques (using ion-selective electrodes and anodic stripping voltammetry respectively) are considered for the determination of 'available' lead in artificial sea water. The analysis is also extended to consider the uptake of ionic lead from solution by a spherical, unicellular 'model' organism. Discussion of the balance struck between the rate processes associated with uptake release and assimilation are used to define conditions under which the chemical analogues are relevant to biological availability.

Literature Cited

1. Sunda, W. and Guillard, R.R.L. The relationship between cupric ion activity and the toxicity of copper to phytoplankton. J. Mar. Res. 34, 511-529 (1976).
2. Millero, F.J. Thermodynamic models for the state of metal ions in sea water, p. 653-693, in Goldberg, E.D., McCave, I.N., O'Brien, J.J., and Steele, J.H. ed., "The Sea." Vol. 6, Wiley Interscience, New York, 1977.
3. Mancy, K.H. and Allen, H.E. "A Controlled Bioassay System for Measuring Toxicity of Heavy Metals." U. S. Environ. Protect. Agency, Duluth, Minnesota, EPA-600/3-77-037, 1977.
4. Siegel, A. Metal-organic interactions in the marine environment. p. 265-295, in Faust, S.D. and Hunter, J.V., ed., "Organic Compounds in Aquatic Environments." Marcel Dekker, New York, 1971.
5. Lerman, A. and Childs, C.W. Metal-organic complexes in natural waters: control of distribution by thermodynamic, kinetic and physical factors, p. 201-236, in Singer, P.C., ed., "Trace Metals and Metal-Organic Interactions in Natural Waters". Ann Arbor Science, Ann Arbor, Michigan, 1973.
6. Stebbing, A.R.D. The effects of low metal levels on a clonal hydroid. J. Mar. Biol. Ass. UK 56, 977-994 (1976).
7. Stebbing, A.R.D. and Pomroy, A.J. A sublethal technique for assessing the effects of contaminants using Hydra Littoralis. Water Research (in press).
8. Davey, E.W., Morgan, M.J., and Erickson, S.J. A biological measure of the copper complexation capacity of sea water. Limnol. Oceanogr. 18, 993-997 (1973).
9. Manahan, S.E. and Smith, M.J. Coppper micronutrient requirement for algae. Environ. Sci. Tech. 7, 829-833 (1973).
10. Whitfield, M. The electroanalytical chemistry of sea water, p.1-154, in Riley, J.P. and Skirrow, G., ed., "Chemical Oceanography." Vol. 4, Academic Press, London, 1975.
11. Davison, W. and Whitfield, M. Modulated polarographic and voltammetric techniques in the study of natural water chemistry. J. Electroanal. Chem. 75, 763-789 (1977).
12. Ferguson, J. The use of chemical potentials as indices of toxicity. Proc. R. Soc. B 197, 387-404 (1939).

13. Crisp, D.J., Christie, A.O. and Ghobashy, A.F.A. Narcotic and toxic action of organic compounds on barnacle larvae. Comp. Biochem. Physiol. 22, 629-649 (1967).
14. Lakshminarayanaiah, N. Membrane phenomena, Ch. 5, p. 203-286, in Hills, G.J., ed., "Electrochemistry a Specialist Periodical Report." Vol. 2. The Chemical Society, London, 1972.
15. Lakshminarayanaiah, N. "Membrane Electrodes." 368 p. Academic Press, London, 1976.
16. Koryta, J. Theory and application of ion-selective electrodes. Anal. Chim. Acta 61, 329-411 (1972).
17. Koryta, J. "Ion-Selective Electrodes." 207 p. Cambridge Univ. Press, Cambridge, 1975.
18. Koryta, J. Theory and applications of ion-selective electrodes. Part II. Anal. Chim. Acta 91, 1-85 (1977).
19. Whitfield, M. Sea water as an electrolyte solution p. 43-171, in Riley, J.P. and Skirrow, G., ed., "Chemical Oceanography," Vol. 1, Academic Press, New York, 1975.
20. Ringbom, A. "Complexation in Analytical Chemistry." 395 p. Interscience, New York, 1963.
21. Whitfield, M. and Turner, D.R. in "Lead-Occurrence, Fate and Pollution in the Marine Environment". International Experts Discussion, Rovinj, Yugoslavia, Pergamon Press, London (in press).
22. Davies, C.W. "Ion Association." 190 p. Butterworths, London, 1962.
23. Whitfield, M. The ion-association model and the buffer capacity of the carbon dioxide system in sea water at 25°C and 1 atmosphere total pressure. Limnol. Oceanogr. 19, 235-248 (1974).
24. Millero, F.J. Sea water as a multi-component electrolyte solution, p. 3-80, in Goldberg, E.D., ed., "The Sea." Vol. 5, Wiley Interscience, New York, 1975.
25. Whitfield, M. A chemical model for the major electrolyte component of sea water based on the Brønsted-Guggenheim hypothesis. Mar. Chem. 1, 251-266 (1973).
26. Whitfield, M. The extension of chemical models for sea water to include trace components at 25°C and 1 atmosphere pressure. Geochim. Cosmochim. Acta 39, 1545-1557 (1975).
27. Sunda, W.G., Engel, D.W. and Thuotte, R.M. Effect of chemical speciation on toxicity of cadmium to grass shrimp, Paleomonates pugio: importance of free cadmium ion. Environ. Sci. Tech. 12, 409-413 (1978).
28. Turner, D.R. and Whitfield, M. The electrodeposition of trace metal ions from multi-ligand systems. I. Theory. J. Electroanal. Chem. (submitted).
29. Turner, D.R. and Whitfield, M. The electrodeposition of trace metal ions from multi-ligand systems. II. Calculations on the electrochemical availability of lead at trace levels in sea water. J.Electroanal. Chem. (submitted).

30. Davies, A.G. The kinetics of and a preliminary model for the uptake of radio-zinc by Phaeodactylum tricornutum in culture, p. 403-420, in "Radioactive Contamination of the Marine Environment." Proceedings of a symposium, Seattle, 1972. International Atomic Energy Agency, Vienna, 1973.
31. Shulz-Baldes M. and Lewin, R.A. Lead uptake in two marine phytoplankton organisms. Biol. Bull. 150, 118-127 (1976).
32. Davies, A.G. An assessment of the basis of mercury tolerance in Dunaliella Tertiolecta. J. Mar. Biol. Assoc. UK 56, 39-57 (1976).
33. Mantoura, R.F.C., Dickson, A. and Riley, J.P. The complexation of metals with humic materials in natural waters. Est. Cstl. Mar. Sci. 6, 387-408 (1978).
34. Levich, V.G. "Physicochemical Hydrodynamics." 700 p. Prentice-Hall, Englewood Cliffs, New Jersey, 1962.
35. Smayda, T.J. The suspension and sinking of phytoplankton in the sea. Oceanogr. Mar. Biol. Ann. Rev. 8, 353-414 (1970).
36. Scott, J.T., Myer, G.E., Stewart, R. and Walther, E.G. On the mechanism of Langmuir circulations and their role in epilimnion mixing. Limnol. Oceanogr. 14, 493-503 (1969).
37. Faller, A.J. and Woodcock, A.H. The spacing of windrows of Sargassum in the ocean. J. Mar. Res. 22, 22-29 (1964).
38. Eppley, R.W., Holmes, R.W. and Strickland, J.D. Sinking rates of marine phytoplankton measured with a fluorometer. J. Exp. Mar. Biol. Ecol. 1, 191-208 (1967).
39. Albery, W.J. and Hitchman, M.L. "Ring-disc Electrodes." 175 p. Clarendon Press, Oxford, 1971.
40. Gavis, J. Munk and Riley revisited: nutrient diffusion transport and rates of phytoplankton growth. J. Mar. Res. 34, 161-179 (1976).
41. Harkness, R.D. Locomotion, p.133-162, in Bligh, Cloudesley-Thompson, and Macdonald, ed., "Environmental Physiology of Animals." Blackwell, Oxford, 1976.
42. Whitfield, M. Activity coefficients in natural waters, in Pytkowicz, R.M., ed., "Activity Coefficients in Electrolyte Solutions." CRC Press, Cleveland (in press).
43. Heyrovský, M. and Hyerovská, R. Surface and volume redox processes: difference between potentiometry and polarography. J. Electroanal. Chem. 52, 141-143 (1974).

RECEIVED November 16, 1978.

Comprehensive Models and Applications

30

Equilibrium Chemistry of Heavy Metals in Concentrated Electrolyte Solution

ABRAHAM E. VAN LUIK
6700 South Cass Ave., Bldg. 8, Argonne National Laboratory, Argonne, IL 60439
JEROME J. JURINAK
Soil Science and Biomediology, Utah State University, Logan, UT 84322

Multicomponent chemical equilibria in solutions have been successfully modeled in a variety of ways (1). Mathematical models based on thermodynamic equilibrium, mass balance, and charge balance equations, and free energy minimization models have been developed and shown applicable to systems for which the given thermodynamic data are valid (2, 3). For dilute solutions, infinite dilution based thermodynamic constants apply, while for more concentrated solutions up to and including the ionic strength of sea water, appropriate published constant ionic medium stability constants may be used (1). For brines, typically of ionic strengths several times that of sea water, constant ionic medium approaches would require treating each individual brine as a unique solvent, meaning that the necessary thermodynamic stability constants would need to be experimentally determined for each individual system (1).

Recent applications of thermodynamic modeling to brines have included empirical measurements of solute mean ionic activities at or near the ionic strengths under investigation (4, 5), and extensions of infinite-dilution based treatments by the addition of non-ideality terms as required by theory (6,7,8). These applications were concerned with answering specific problems connected with the detailing of brine composition relationships to geological formations, and were not attempts at formulating a general theory of electrolyte solution structure. Work is in process, however, on developing a self-consistent, quantitative theory comprehending the structure of aqueous electrolyte solutions from the infinitely dilute to the hydrated salt state (1).

One of the newer theoretical treatments, based on the pioneering statistical thermodynamic work of McMillan and Mayer (6), as mathematically formulated by Friedman (9), does appear to hold significant promise as a theory of sufficient generality that it may eventually embody other working theories as demonstrated special cases. This theory, known as the cluster integral expansion theory (1) or simply as cluster theory (9), has been developed to the point where applications have been made to calculating

0-8412-0479-9/79/47-093-683$07.00/0
© 1979 American Chemical Society

osmotic and activity coefficients for sea water (10) as well as osmotic coefficients for a quaternary brine system up to an ionic strength of 6 molal (11).

The consequences of this theory are used in this model to predict activities and solubilities of the brine constituents for which necessary thermodynamic data were available in the literature. Brine constituents for which such data were not available were treated using other models.

Theory

The solubility product equation (12) may be written:

$$(a_+)^i \ (a_-)^j = K_{so} \ (a_{solid}) \ (a_{H_2O})^{(ih_+ + jh_-)} \quad (1)$$

in which the left hand side may be called the ion activity product (IAP), and where a_+ is the cation activity in solution, a_- is the anion activity in solution, h_+ is the cation hydration number, h_- is the anion hydration number, a_{solid} is the activity of the solid phase, a_{H_2O} is the solvent (water) activity, i, j are the stoichiometric coefficients for the cation and anion respectively, and K_{so} is the thermodynamic solubility product constant.

In this study, for lack of available data, the following assumptions were made with respect to the use of thermodynamic solubility product constants: 1) $a_{solid} = 1$, 2) K_{so} is not significantly affected by the co-solutes of the brines to be studied, 3) solute and solvent activities can be accurately calculated for the systems to be considered, 4) hydration numbers, or the number of water molecules bound to some extent by each ion in solution may be treated as adjustable parameters, and may be estimated as a function of ionic strength and solvent activity as detailed elsewhere (16).

Solute and Solvent Activity Calculations. For the purposes of this study, the derivations necessary to the calculation of the solute and solvent activities will begin with the equation for the prediction of the excess free energy of a single electrolyte solution based on the work of Friedman (9).

For any imaginary ideal solution of an electrolyte, at any given T and P, in which all activity and osmotic coefficients are unity, we can write for the chemical potential of a solute s,

$$\mu_s^{ideal} = \mu_s^\circ + RT \ln m_s \quad (2)$$

where μ_s° is the standard state chemical potential of the solute, R is the gas content, T is the absolute temperature, and m is the molality of the solute. The difference between μ_s of a real solution and μ_s^{ideal} is designated as an excess function, μ_s^{EX}, and is written

$$\mu_s^{EX} = \mu_s - \mu_s^{ideal} = RT \ln \gamma_s \tag{3}$$

where the excess free energy in the real solution is equated to the activity coefficient of the solute γ_s.

Equation 3 defines the excess chemical potential for one solute; if more than one solute exists in the solution, a weighted sum of the excess chemical potentials of the solutes may be called the mean partial molal Gibbs free energy of solutes, $\mu^{EX}{}_{\pm}$:

$$\mu^{EX}{}_{\pm} = \sum_{s=1}^{n} x_s \cdot \mu_s^{EX} = \sum_{s=1}^{n} x_s \ln \gamma_s \tag{4}$$

where $x_s = m_s/m$, $m = \sum_{s=1}^{n} m_s$, and n = the number of solutes.

The chemical potential of the solvent μ_w, is given by

$$\mu_w = \mu_w^\circ - RTM_w m\phi/1000 \tag{5}$$

where M_w is the molecular weight of water and ϕ is the osmotic coefficient defined by

$$\phi = 1 + \frac{1}{m} \int_{m=0}^{m} \sum_{s=1}^{n} m_s d \ln \gamma_s \tag{6}$$

Similarly, for a real solution the excess solvent chemical potential μ_w^{EX}, may be shown as a function of the osmotic coefficient, ϕ:

$$\mu_w^{EX} = \mu_w^\circ - \frac{RTM_w m}{1000} \phi - \mu_w^\circ + \frac{RTM_w m}{1000} = \frac{RTM_w m}{1000} (1 - \phi) \tag{7}$$

The total excess free energy, which corresponds to the part of the free energy function which arises from ionic interactions in the real mixture is therefore

$$G^{EX} = \mu_w^{EX} + \mu^{EX}{}_{\pm}$$

$$= \frac{RTM_w m}{1000} (1 - \phi) + RT \sum_{s=1}^{n} x_s \ln \gamma_s$$

$$= RT \left[\frac{M_w m}{1000} (1 - \phi) + \sum_{s=1}^{n} x_s \ln \gamma_s \right] \quad (8)$$

Equation 8 can be used to calculate excess free energies for single solute solutions (n = 1) using tabulated values such as those appearing in Robinson and Stokes (13).

If one kilogram of solution is produced by adding a solution containing electrolyte A and a solution containing electrolyte B, and if electrolytes A and B have a common ion (an assumption for the sake of simplicity), then the excess free energy for the final solution can be written (9):

$$G^{EX}(y,I) = \Delta m G^{EX}(y,I) + yG^{EX}(1,I) + [1-y]\, G^{EX}(0,I) \quad (9)$$

where $I = 1/2 \sum m_s Z_s^2$, the ionic strength, y = the ionic strength fraction of electrolyte A, $G^{EX}(1,I)$ applies to the solution containing only electrolyte A at ionic strength I and $G^{EX}(0,I)$ applies to the solution containing only electrolyte B at ionic strength I. The G^{EX} terms may be evaluated using equation 8.

The term $\Delta m G^{EX}(y,I)$ is the increase in excess free energy due to the mixing of the two solutions, and represents the changes in molecular configuration, such as the formation of ion-pairs, ion triplets, etc.

Since $\Delta m G^{EX}$ is sensitive to both the composition and concentration of electrolytes A and B, its calculation would be simplified considerably if the initial solutions and the final mixture were at the same stoichiometric ionic strength based on 1 kg of solvent in the final mixture. This suggests that for a given mixture of electrolytes in solution sufficient sets of common ion mixtures at the same $\bar{I}$ should be found which in sum describe the total solution, allowing the excess free energy of the mixture to be found using a series of calculations shown by equation 9. Basically, this is the approach used to calculate $\Delta m G^{EX}$ in this study.

For the three ion mixture solution, Friedman (9) showed that equation 9 could be expressed in terms of the $\Delta m G^{EX}(y,I)$ dependence on y by the following expansion:

$$\Delta m G^{EX}(y,I) = I^2 RTy(1-y)[g_0 + g_1(1-2y) + g_2(1-2y)^2 + \ldots] \quad (10)$$

where the g_i terms are coefficients characteristic of the various ionic species (complexes) found in the final mixture.

In addition, the excess free energy of mixing term as derived by Friedman (9) from cluster theory can be written in the general form (14):

$$\Delta m G^{EX} = RT \sum_{u} B^u \ \Delta m C^u \quad (11)$$

where B^u is a constant for each specie u, and ΔmC^u is the change in the concentration product of the species u when the mixture comes to equilibrium. Reilly and Wood (14) derive the ΔmC^u terms for all possible ion pair and ion triplet formations, and show that the term $\{I^2y(1-y)\}$ may be factored out of each of the ion pair and ion triplet ΔmC^u terms. From this result it was shown that in equation 10 the g_0 term represented all ion pair and some ion triplet formation excess free energy contributions in the three ion solution. The ion triplet formations contributing to g_0 were shown (12) to be those composed of three ions of the same charge magnitude. Since in comparison to ion pairing the excess free energy contribution due to ion triplet formations is very small, usually, neglect of the g_1 and higher g_i terms in equation 10 was suggested as an appropriate simplifying assumption by Reilly and Wood (14).

If we define concentrations in terms of equivalents per kg of solvent, E, we may write equation 9 for three ions in solution E_1, E_2, and E_3, and their ionic charges Z_1, Z_2, and Z_3 as

$$\Delta mG^{EX} = \frac{RT}{4E} E_1E_2E_3(Z_1 - Z_3)(Z_2 - Z_3) g_0 \qquad (12)$$

where $E = E_1 + E_2 = E_3$ for the system being defined, which may be seen to be a normalizing parameter defining ΔmG^{EX} on a per equivalent basis.

Even though in this specific instance $E = E_3$, the importance of retaining the concept of E for the general case may be demonstrated by defining equation 12 for a system consisting of one cation and 2 anions where $E \neq E_3$, but $E = E_1 = E_2 + E_3$, where E_1 is the cation concentration.

A more complex solution, made by mixing two single electrolyte solutions, may be treated by combining the excess free energy of mixing terms for each binary common-ion mixture into which the final solution may be divided. For example, consider the mixing of two solutions of the same ionic strength to yield one kilogram of solvent in the final mixture. The first solution contained cations M_1 and M_2, and anion X_1 and X_2. The final solution contains cations M_i, i = 1,2,3, in concentration m_i, and anions X_j, j = 1,2, in concentration m_j. The possible binary common ion combinations for the hypothetical component solutions for which the theory has been developed, consists of six sets of two possible cations and one anion, and three sets of one cation, two anion combinations. These three-ion solutions may be described by two triple summations:

$$\sum_{k=2}^{k=1} \sum_{\ell=1}^{\ell=k-1} \sum_{m=1}^{m=j} \quad \text{where } i = 1,2,3 \text{ and } j = 1,2 \qquad \text{and}$$

$$\sum_{k=1}^{k=1} \quad \sum_{\ell=2}^{\ell=j} \quad \sum_{m=1}^{m=\ell-1} \qquad \text{where } i = 1,2,3 \text{ and } j = 1,2.$$

Each of the above combinations may have their excess free energy evaluated by a term such as given by equation 12.

To make the writing of the excess free energy of mixing relation for the mixture M_iX_j less laborious, the notation as given by Reilly and Wood (14) was adopted,

$$\Delta mG^{EX} = \frac{RT}{4E} \left\{ \sum_{k=2}^{k=1} \sum_{\ell=1}^{\ell=k-1} \sum_{m=1}^{m=j} E_k^M E_\ell^M Z_{km} Z_{\ell m} g_{M_k}{}^{M_\ell^{X_m}} + \sum_{k=1}^{k=i} \sum_{\ell=2}^{\ell=j} \sum_{m=1}^{m=\ell-1} E_k^M E_\ell^X E_m^X Z_{k\ell} Z_{km} g_{X_\ell}{}^{X_m^{M_k}} \right\} \tag{13}$$

where the superscripts M or X identify cations or anions respectively, and where the g_0 terms are labeled to emphasize their specificity for each of the nine hypothetical component solutes required by this construction. The equivalence normalizing term E is now

$$E = \sum_{k=1}^{k=i} E_k^M = \sum_{\ell=1}^{\ell=j} E_\ell^X .$$

Equation 8 gave the procedure for calculating the molar excess free energy for any pure, single salt solution at ionic strength I. To use equation 8 for the more complex mixed solution M_iX_j (i = 1,2,3; j = 1,2), the mixture must be divided into single salt solutions containing M_1X_1, M_1X_2, M_2X_1, M_2X_2, M_3X_1, and M_3X_2, respectively, all at molal ionic strength I.

The result of the application of equation 12 on each of these mixtures may be labeled $G^0_{M_iX_j}$ (i = 1,2,3; j = 1,2). The mixing fractions corresponding to each M_iX_j component solution must be a ratio of the moles contributed by the single salt solution to the total moles of the mixture.

In the general notation of our complex mixture, Reilly and Wood (14) showed this mixing fraction for each component solution to have the form

$$\frac{E_i^M E_j^X (Z_i^M - Z_j^X)}{2I\ E}$$

which is a concentration ratio of the component solution to the total solution or the kilogram of solvent in the component solution. A summation of all these fractional terms times their excess free energies calculated by equation 8 gives the total mixture excess free energy when added to the ΔmG^{EX} from equation 12:

$$G^{EX} = \sum_{\ell=1}^{\ell=i} \sum_{m=1}^{m=j} \frac{E_\ell^M \, E_m^X \, Z_{\ell m}}{2I \;\; E} G^0_{M_\ell X_m} + \Delta mG^{EX} \qquad (14)$$

where ΔmG^{EX} is defined by equation 13 and $Z_{\ell m} = (Z_\ell^M - Z_m^X)$.

Equation 14 corresponds to equation (A-1) of Reilly, Wood, and Robinson (15) for the total excess free energy for a general mixed electrolyte solution containing m_i^M moles of cation M_i with charges Z_i^M, m_j^X moles of anions X_j with charges Z_j^X, and one kilogram of solvent.

In terms of one of the single salt component solutions M_iX_j:

$$G^0_{M_iX_j} = RT\, m^0 \,(1 - \phi^0_{M_iX_j} + \ln \gamma^0_{\pm\, M_iX_j}) \qquad (15)$$

$$\text{where } m^0 = m_j^X = - \frac{2\, I}{Z_i^M \cdot Z_j^X}$$

If we differentiate equation 15 with respect to m^0, we have (9)

$$\frac{\partial}{\partial m^0} [G^{0\;EX}_{M_iX_j}] = RT \ln \gamma^0_{\pm M_iX_j} \qquad (16)$$

The total excess free energy expression (equation 14) for the general mixture $M_\ell X_m$ may be written in a form analogous to equation 15 (12).

$$G^{EX} = RT \left\{ m(1 - \phi) + \sum_{\ell=1}^{\ell=i} m_\ell^M \ln \gamma_\ell^M + \sum_{m=1}^{m=j} m_m^X \ln \gamma_m^X \right\} \qquad (17)$$

$$\text{where } m = \sum_{\ell=1}^{\ell=i} m_\ell^M + \sum_{m=1}^{m=j} m_m^X$$

Equation 17 relates the excess free energy of a mixed electrolyte solution to the osmotic coefficient of the solution and the activity coefficients of its component ions.

The Osmotic Coefficient of a Mixed Electrolyte Solution. It may be seen that differentiation of equation 17 with respect to m will leave a term including the osmotic coefficient of the solution, ϕ.

In order to evaluate ϕ, the total excess free energy expression given in equation 17 will be differentiated with respect to the solvent concentration. This involves no changes in the concentration ratios of any ionic specie m to the total concentration of all ionic species m. Differentiation of equation 17 with respect to m results in the working equation for the osmotic coefficient of a mixed electrolyte solution. Details of the differentiation are given elsewhere (15,16).

$$
\begin{aligned}
m(1-\phi) = &- \sum_{\ell=1}^{\ell=i} \sum_{m=1}^{m=j} \frac{E_\ell^M E_m^X Z_{\ell m}}{Z_\ell^M Z_m^X E} \left(1 - \phi^0_{M_\ell X_m}\right) \\
&- \sum_{k=2}^{k=i} \sum_{\ell=1}^{\ell=k-1} \sum_{m=1}^{m=j} \frac{E_k^M E_\ell^M E_m^X Z_{km} Z_{\ell m}}{4E} \left\{ g_{M_k M_\ell X_m} + I \frac{\partial}{\partial I}\left(g_{M_k M_\ell X_m}\right) \right\} \\
&- \sum_{k=1}^{k=i} \sum_{\ell=2}^{\ell=j} \sum_{m=1}^{m=\ell-1} \frac{E_k^M E_\ell^X E_m^X Z_{k\ell} Z_{km}}{4E} \left\{ g_{X_\ell X_m M_k} + I \frac{\partial}{\partial I}\left(g_{X_\ell X_m M_k}\right) \right\}
\end{aligned}
\qquad (18)
$$

The change of variable $\frac{\partial X}{\partial m}$ to $I \frac{\partial X}{\partial I}$ was made for convenience in applying the equation, as in Reilly, Wood, and Robinson (15). It can be shown that $m \frac{\partial X}{\partial I} = I \frac{\partial X}{\partial I}$ for any function of $X=f(m)$ (16).

The Activity Coefficient of Any Component Salt in the Aqueous Electrolyte Mixture. Analogous to the derivation of the osmotic coefficient expression, equation 17 was differentiated with respect to m to give a relationship between $\frac{\partial G}{\partial m}$ and the activity coefficient, and then that relationship was evaluated by substituting in the value of $\frac{\partial G}{\partial m}$ obtained by differentiation of G from equation 14 with respect to m, where m is now the result of a change in m, m*, added to the initial m, $\overline{m}$' as caused by a variation in m_p^M and m_q^X. Using m in this new expression,

$m = \overline{m} + m^*$, complicates the derivatives significantly. The steps are detailed elsewhere (15, 16).

The resulting working equation for the activity coefficient of a salt M_pX_q is

$$\begin{aligned}
\ln \gamma_{\pm M_pX_q} = {} & \frac{Z_p^M Z_q^X}{E - Z_{pq}} \sum_{\ell=1}^{\ell=i} \sum_{m=1}^{m=j} \Bigg\{ \frac{E_m^X Z_{pm}}{Z_p^M Z_m^X} \left(1 - \phi^0_{M_pX_m} + \ln \gamma^0_{M_pX_m}\right) \\
& + \frac{E_\ell^M Z_{\ell q}}{Z_\ell^M Z_q^X} \left(1 - \phi^0_{M_\ell X_q} + \ln \gamma^0_{M_\ell X_q}\right) \\
& + E_\ell^M E_m^X Z_{\ell m} \Bigg(\Bigg[\frac{-1}{Z_\ell^M Z_m^X} \cdot \left(\frac{1}{E} + \frac{Z_{pq}}{2I}\right) \cdot \left(1 - \phi^0_{M_\ell X_m} + \ln \gamma^0_{M_\ell X_m}\right) \Bigg] \\
& + \frac{Z_{pq}}{2I\, Z_\ell^M Z_m^X} \cdot \ln \gamma^0_{M_\ell X_m} \Bigg) \Bigg\} + \frac{Z_p^X Z_q^X}{4EZ_{pq}} \Bigg[\sum_{k=2}^{k=i} \sum_{\ell=1}^{\ell=k-1} \sum_{m=1}^{m=j} \Bigg\{ -E_\ell^M E_m^X Z_{pm} Z_{\ell m} \\
& \cdot g_{M_pM_\ell}^{\;X_m} - E_k^M E_m^X Z_{km} Z_{pm} \cdot g_{M_kM_p}^{\;X_m} - E_k^M E_\ell^M Z_{kq} Z_{\ell q} \cdot g_{M_kM_\ell}^{\;X_q} \\
& + \frac{E_k^M E_\ell^M E_m^X}{E} Z_{km} Z_{\ell m} \cdot \left(g_{M_kM_\ell}^{\;X_m} - \frac{Z_{pq}E}{2} \cdot \frac{\partial}{\partial I}\left[g_{M_kM_\ell}^{\;X_m}\right] \right) \Bigg\} \\
& + \sum_{k=1}^{k=i} \sum_{\ell=2}^{\ell=j} \sum_{m=1}^{m=\ell-1} \Bigg\{ -E_\ell^X E_m^X Z_{pm} Z_{p\ell} \cdot g_{X_\ell X_m}^{\;M_p} \\
& - E_k^M E_m^X Z_{kq} Z_{km} \cdot g_{X_qX_m}^{\;M_k} - E_k^M E_\ell^X Z_{kq} Z_{k\ell} \cdot g_{X_\ell X_q}^{\;M_k} \\
& + \frac{E_k^M E_\ell^X E_m^X}{E} Z_{km} Z_{k\ell} \cdot \left(g_{X_\ell X_m}^{\;M_k} - \frac{Z_{pq}E}{2} \cdot \frac{\partial}{\partial I}\left[g_{X_\ell X_m}^{\;M_k}\right] \right) \Bigg\} \Bigg]
\end{aligned} \qquad (19)$$

Equation 19 is equivalent to the final equation for $\ln \gamma_{\pm M_p X_q}$ as derived by Reilly, Wood, and Robinson (15), the form is slightly different due to different algebraic manipulations.

The data needed to evaluate equation 19 for $\ln \gamma_{\pm M_p X_q}$ may be catalogued as follows:

1. Osmotic and activity coefficients for pure aqueous solutions of salts $M_p X_m$, $m = 1, 2, \ldots, j$, and $M_\ell X_q$, $\ell = 1, 2, \ldots, i$.
2. Excess free energies of mixing for all the component binary common ion mixtures $M_p M_\ell X_m$, $M_p M_\ell X_q$, $M_p X_\ell X_m$, $M_p X_\ell X_q$, $M_k M_\ell X_q$, $M_k M_\ell X_m$, $M_k X_\ell X_m$, and $M_k X_\ell X_q$, where $m = 1, 2, \ldots, j$ and $\ell = 1, 2, \ldots, i$.

Activity Coefficient for a Trace Component in a Mixed Electrolyte Solution. If the salt $M_p X_q$ is present in the electrolyte mixture in only a very minute quantity, it may be seen that its contribution to I, E, and m are negligible. Applying equation 19, and treating the trace salt as any other component, it becomes apparent that any interaction terms with an M_p dependence will be vanishingly small. Also, when $k = p$, any terms multiplied by E_k^M may be neglected since they will approach zero. The result of such a treatment is that the trace salt activity coefficient will be largely determined by the activity coefficient of the pure salt in a solution at the same ionic strength as the mixture (15).

This result holds only for trace components with a common ion in the mixture, meaning that only one of the ions comprising the trace salt is actually present in a trace quantity.

It has been shown elsewhere (10) that the data for the single salt solutions are more important than the binary common ion solution data in calculations involving equations 18 and 19. Often, in fact, neglect of the interaction parameters causes little loss in predictive accuracy (10). Since for trace salts the interaction parameter contributions to equations 18 and 19 are naturally small, equations 18 and 19 will be used for trace salts, with one ion common to a non-trace salt, neglecting the $\Delta_m G^{EX}$ contributions.

Osmotic and Activity Coefficient Interpolation From Published Tables. Glueckauf (17) has developed equations for the osmotic and activity coefficients of single salt solutions which, when "tuned" to measured data, will accurately interpolate between measured values and predict values to ionic strengths well above those usually encountered in natural brines. For brines of ionic strengths above 2 molal, the following equations for the osmotic and activity coefficients were developed (17):

$$\phi = -\frac{2.303\ A}{9} |z^+z^-| (\frac{2I}{b})^{1/3} \cdot \frac{c}{d_0 m} + k'\ m \qquad (20)$$

$$\ln \gamma_\pm = -\ 2.303A \cdot |z^+z^-| (\frac{2I}{b})^{1/3} (\frac{3}{4} + \frac{1}{4} \frac{c}{d_0 m}) + k''\ m \qquad (21)$$

where A = the Debye-Hückel A parameter
z^+z^- = cation and anion charges, respectively
I = molal ionic strength
b = $\kappa a/I^{1/2}$ where a is the Debye-Hückel distance of closest approach, κ is the reciprocal length constant of the Poisson-Boltzmann equation
c = molar concentration
d_0 = density of water at 0°C
m = molal concentration
k',k"= adjustable parameter matching calculated osmotic and activity coefficients to measured values.

The parameters k' and k" are most easily determined by linear regression of the values calculated using the electrostatic portions of equations 20 and 21 as a function of concentration on corresponding measured values (13).

Only single ion molarities are given as input data, therefore, published (18) single salt solution data relating solute concentrations to solvent concentrations were necessary in the computing of the $\frac{c}{d_0 m}$ ratio for each major component salt.

One major component salt, $CaSO_4$, is sufficiently insoluble so that single salt solution data does not exist in the concentration range of interest. Osmotic and activity coefficient data for $CaSO_4$ were assumed, therefore, to be approximated well enough by arbitrarily assigning a value slightly below (90 percent of) corresponding $MgSO_4$ single salt solution values.

The Carbonate System. Since no osmotic or activity coefficient data are available for single salt solutions of carbonate and bicarbonate salts, estimates had to be made of the activities of these ligands in solution, as well as for the activities of carbonate and bicarbonate salt cations.

Single Ion Activity Coefficients. The cation single ion activity coefficients were estimated using the model of Bates, Staples, and Robinson (19). Their model was derived specifically for chloride solutions, and was shown to give reasonable results in solutions of ionic strengths up to and including 6 molal.

The following equations were used, incorporating the hydration number h (19).

(a) for the salt MCl:

$$\log \gamma_{M^+} = \log \gamma_{\pm MCl} + 0.00782\ hm\phi \quad (22)$$

$$\log \gamma_{Cl^-} = \log \gamma_{\pm MCl} - 0.00782\ hm\phi \quad (23)$$

(b) for the salt MCl_2:

$$\log \gamma_{M^{+2}} = 2 \log \gamma_{\pm MCl_2} + 0.00782\ hm\phi + \log [1 + 0.018\ (3-h)m] \quad (24)$$

$$2 \log \gamma_{Cl^-} = \log \gamma_{\pm MCl_2} - 0.00782\ hm\phi - \log [1 + 0.018\ (3-h)m] \quad (25)$$

Although the above equations are derived for pure solutions of MCl and MCl_2, the following technique was used to adapt their use to a mixed electrolyte solution in which chloride is the dominant anion:

a. The osmotic coefficient for the mixture, as given by equation 18 is related to the molality of an isopiestic pure NaCl solution by equation 20:

$$m = \frac{1}{k'} (\phi - d) \quad (26)$$

where $d = \frac{-2.303\ A|Z^+Z^-|}{9} (\frac{2I}{b})^{1/3}$

b. Using the osmotic coefficient and molality of the isopiestic NaCl solution, the activity coefficient for NaCl is given by using equations 20 and 21 to give:

$$\ln \gamma_{\pm NaCl} = \phi_{NaCl} - (k''_{NaCl} + k'_{NaCl})\ m + 3d \quad (27)$$

where d is as given for equation 26.

c. The activity of the chloride ion in this isopiestic NaCl solution was obtained using equation 23.

d. The monovalent cation activity coefficient of interest was calculated using equations 22 and 23 as follows:

$$\log \gamma_M + \log \gamma_{Cl} - .01564\ hm\phi \quad (28)$$

where h refers to the salt MCl.

e. The divalent cation activity, by manipulation of equations 24 and 25 is given by:

$$\log \gamma_{M^{+2}} = 4 \log \gamma_{Cl^-} + .02346\ hm\phi + 3 \log [1 + 0.018\ (3-h)m] \quad (29)$$

where h refers to the salt MCl_2.

Note that the most important assumptions made here are: 1) chloride activity coefficients in isopiestic, NaCl dominated solutions are equal, and 2) the mole fraction statistical hydration model for splitting mean activity coefficients yields reasonable single ion activities.

A comparison of mean activity coefficient splitting models and their results in the ionic strength range of interest to a brine study (1) quickly shows that, perhaps contrary to expectation, the second assumption is the more questionable.

The Solubility of Carbon Dioxide in Brines. Information required to describe the carbonate system in a brine includes the solubility of carbon dioxide, the extent of carbonic acid formation, and the distribution of the anionic carbonic acid dissociation products HCO_3^- and CO_3^{2-}.

Stewart and Munjal (20) determined the solubility of CO_2 in distilled water, a synthetic sea water, and in three and five-fold concentrations of the synthetic sea water. The Henry's Law constants for CO_2 computed for these systems at different temperatures were found to be linearly related to temperature and solution molality. The concentration of CO_2 (aq) in moles per total moles of solution (mole fraction) may be written

$$[CO_2]aq = K_H P_{CO_2} \quad (30)$$

where [] refers to concentration, K_H = Henry's Law constant, and P_{CO_2} = the partial pressure of CO_2 in atmospheres.

K_H, by statistical treatment of the data of Stewart and Munjal (20), was found to be described by the function:

$$K_H = \frac{[55.6 - (C_o - C_w) \cdot 0.0556] + M_s}{[657.6 + 38.69 \cdot t] \cdot 1.292 \cdot I^{0.07831}} \quad (31)$$

where K_H has the units of moles CO_2/(liter solution x atmospheres), 55.6 = moles of H_2O per liter of H_2O, (C_o-C_w) = grams per liter of water that has been displaced by the anhydrous solute, 0.0556 = moles of H_2O per gram of H_2O, M_s = moles of solute per liter of solution, t = temperature in degrees C, and where the denominator is the linear regression function which evaluated K_H in terms of the centigrade temperature for an aqueous solution where M_s and (C_o-C_w) both equal zero, multiplied by a power curve which corrects K_H for ionic strength. The quantities M_s and (C_o-C_w) were evaluated for each solution by the following routine:

$$M_s = \sum_{i=1}^{n} M_i$$

where M_i = the molarity of each major component solute salt i which was calculated using the molar inputs for each ion according to the following sequence of calculations where all concentrations are molarities, and where the subscript T refers to a total concentration input value for an ion, or a total calculated concentration value for a salt:

$$M_{CaCl_2} = [CaCl_2]_T = [Ca^{+2}]_T$$

$$M_{KCl} = [KCl]_T = [K^+]_T$$

$$[Cl^-] = [Cl^-]_T - 2[CaCl_2]_T - [KCl]_T$$

$$[NaCl] = \text{smallest of } [Na^+]_T \text{ or } [Cl^-]$$

$$[Cl^-] = [Cl^-] - [NaCl]$$

$$M_{MgCl_2} = [MgCl_2] = \text{smallest of } (1/2[Mg^+]_T \text{ or } [Cl^-])$$

$$[Mg^{+2}] = [Mg^{+2}]_T - [MgCl_2]_T$$

$$[Cl^-] = [Cl^-] - 2[MgCl_2]_T$$

$$M_{MgSO_4} = [MgSO_4]_T = \text{smallest of } [Mg^{+2}] \text{ or } [SO_4^{-2}]_T$$

$$[SO_4^{-2}] = [SO_4^{-2}]_T - [MgSO_4]_T$$

$$M_{Na_2SO_4} = [Na_2SO_4]_T = [SO_4^{-2}]$$

$$[NaCl] = [NaCl] - 2[Na_2SO_4]$$

$$[Cl^-] = [Cl^-] + 2[Na_2SO_4]$$

$$M_{NaCl} = [NaCl]_T = [NaCl] + [Cl^-] \qquad (32)$$

In the FORTRAN program, any inadvertant negative numbers were set equal to zero. It is noted that any inaccuracy in the data which results in a surplus of the dominant anion, Cl^-, will be assigned to the dominant cation, Na^+.

The next step in the calculation of CO_2aq is:

$$(C_o-C_w)s = \sum_{i=1}^{n} M_i \ (C_o-C_w)i + X_i \tag{33}$$

where X_i is the intercept and $(C_o-C_w)i$ is the slope of a straight line fit computed for each M_i versus $(C_o-C_w)i$ using published data (18).

After the aqueous CO_2 concentration in moles per liter has been calculated, it is converted to moles per kilogram by multiplying the molarity by a density based factor of the form:

$$m = \frac{1000.}{\rho(1000.- G)} \quad \text{where G represents } \frac{\text{kg solute}}{\text{kg solvent}} \tag{34}$$

For a brine of unknown density, the density may be approximated by the following:

$$\rho = \frac{1}{1000} \sum_{i=1}^{n} M_i Mw_i - (C_o-C_w)_s + 1000. \tag{35}$$

where Mw_i is the gram-molecular weight of each salt specie i.

The assumption was made that all $[CO_2]aq$ forms carbonic acid, that is

$$[CO_2]\,aq = [H_2CO_3^\circ] \tag{36}$$

Activity of HCO_3^- and CO_3^{2-}. The activity coefficient of CO_2aq in brine systems was defined by the following empirical equation based on data provided by Helgeson (7):

$$\gamma_{CO_2\,aq} = 0.93 + 0.17\ M_{Na}/\rho \tag{37}$$

The product of equations 36 and 37 gives the carbonic acid activity in solution $(H_2CO_3^\circ)$. If we write,

$$K_1 = \frac{(H^+)\ (HCO_3^-)}{(H_2CO_3^\circ)} \tag{38}$$

having computed $(H_2CO_3^\circ)$, then (H^+) and K_1 values are required to calculate (HCO_3^-).

It was assumed that brine pH values obtained using a conventional pH electrode standardized in distilled water based buffer solutions would provide pH values of sufficient accuracy (21).

Empirical curves were fit to data (22) giving values K_w, K_1, and K_2 over 5-30°C temperature and 2-4 percent salinity ranges. It was assumed that the extrapolation from 2 to 4 percent salinity to 20-30 percent salinity would not result in serious discrepancies in K_w, K_1, or K_2. The resulting equations for estimating K_w, K_1, and K_2 as functions of concentration and temperature were

$$- \log K_w = 1.005 - 0.04023 \cdot °C + 13.253 \cdot I^{-0.01736} \quad (39)$$

$$- \log K_1 = 0.2497 - 0.00996 \cdot °C + 5.896 \cdot I^{-0.02061} \quad (40)$$

$$- \log K_2 = 0.4078 - 0.01628 \cdot °C + 9.039 \cdot I^{-0.03630} \quad (41)$$

The activities of the HCO_3^- and CO_3^{2-}anions in the brine were approximated by using the temperature and concentration corrected constants:

$$(HCO_3^-) = K_1 \cdot (H_2CO_3^°)/(H^+) \quad (42)$$

$$(CO_3^{2-}) = K_2 \cdot (HCO_3^-)/(H^+) \quad (43)$$

The activity of the hydroxyl anion, necessary to the calculation of basic carbonate ion activity products, was calculated by the equation

$$(OH^-) = K_w/(H^+) \quad (44)$$

Temperature Effects. The temperature range for which this model was assumed to be valid was 0°C through 40°C, which is a range covering most natural surface water systems (28). Equilibrium constants were adjusted for temperature effects using the Van't Hoff relation whenever appropriate enthalpy data was available (23, 24, 25). Activity and osmotic coefficients were temperature corrected by empirical equations describing the temperature dependence of the Debye-Hückel parameters of equations 20 and 21. These equations, obtained by curve-fitting published data (13), were

$$A = 0.4916 \cdot e^{(1.488 \cdot 10^{-3} \cdot t)} \quad (45)$$

$$b = 0.3249 \cdot e^{(4.503 \cdot 10^{-4} \cdot t)} \quad (46)$$

where t is the temperature in °C.

Experimental

Validation of the model involved application of the model to 1) measurements of natural brine macro- and micro-component concentrations to determine whether predicted saturated phases corresponded with naturally occurring precipitated phases, and 2) measurement of equilibrium saturation concentrations of Cu, Pb, Cd, and Zn in 2, 4, and 6 molal artificial brines of known composition to validate the predictive ability of the micro-component portion of the model.

In these studies, thermodynamic equilibrium conditions were assumed. Constant pH monitoring was used as an estimator of the equilibration times involved, since a stable pH is an overall indicator of the internal chemical stability of pH dependent processes. In order to allow the assumptions of atmospheric O_2 and CO_2 partial pressures, water scrubbed compressed air was pumped into the reaction vessels under constant temperature and constant vigorous mixing conditions. The air was introduced by fritted glass bubblers which had been teflon coated while air flowed through them in order to minimize the glass-solution interface.

Sampling from the reaction vessels was done using a 50 ml Class A pipet. With reaction vessels containing initially 4000 ml of solution, this represented a withdrawal of 1.25 percent of the total solution per sampling, allowing the assumption that the solution was not significantly affected by any given sampling with regard to volume or composition. Samples were immediately filtered through Whatman #42 filter paper.

Macro-component cation analyses were done by distilled water dilution and subsequent flame emission or flame atomic adsorption.

Chlorides were determined by potentiometric titrations performed on distilled water dilutions. Carbonates and bicarbonates were either determined by titration or by calculation from pH measurements. Sulfates were determined by the gravimetric method utilizing $BaSO_4$ precipitation.

Trace metal analyses were performed using sample acidification, complex formation using ammonium dithiocarbamate (APDC), extraction with methylisobutylketone (MIBK), and analysis of extracts by flame atomic adsorption. Details of the method are given by Brooks, Presley, and Kaplan (29).

Results and Discussion

Validation of the Osmotic Coefficient Calculation. The results for the osmotic coefficients calculated for a series of brines representing an evaporative concentration sequence of sea water brines are given in Figure 1 and Table I. In Figure 1, the agreement between the measured (10) and calculated osmotic coefficients are compared. Table I also includes the coefficient of variation between the calculated and measured values. The maximum coefficient of variation encountered was 0.90 percent and the average coefficient of variation over the range considered was 0.26 percent.

The Mean Molal Activity Coefficient Calculation for the Macro-Component Salts. The mean molal activity coefficients for the macro-component chloride and sulfate salts are given in Figures 2a and 2b, respectively. Because of the unavailability of measured activity coefficient data in brines, validation of the results presented in Figures 2a and 2b was not possible.

The Solvent Activity Calculation. A series of measurements on a Great Salt Lake (GSL) brine concentration sequence showed that the saturation vapor pressure lowering over a GSL brine at or near the labile saturation point should be about 9.5 millibars at 25°C. Application of the model to comparable GSL brine resulted

Table I. Osmotic Coefficient for Sea Water Concentrates at 25°C

I (molal)	ϕ calc.*	ϕ meas.†	Coeff. of ‡ variation (%)
1.800	0.941	0.953	0.896
2.135	0.962	0.970	0.586
2.535	0.989	0.995	0.427
3.013	1.023	1.023	0.000
3.586	1.065	1.065	0.000
4.275	1.117	1.115	0.127
5.107	1.181	1.180	0.059
6.115	1.260	1.260	0.000

*Calculated from the model, using molar input data. Molar composition ratios normalized with respect to sodium: Na^+ = 1, K^+ = .021, Mg^{2+} = .136, Ca^{2+} = .0001, Cl^- = 1.174, SO_4^2 = -.056.

†Interpolated graphically from data presented by Robinson and Wood (10).

‡Coefficient of variation = $\frac{S_x}{\overline{X}} * 100$.

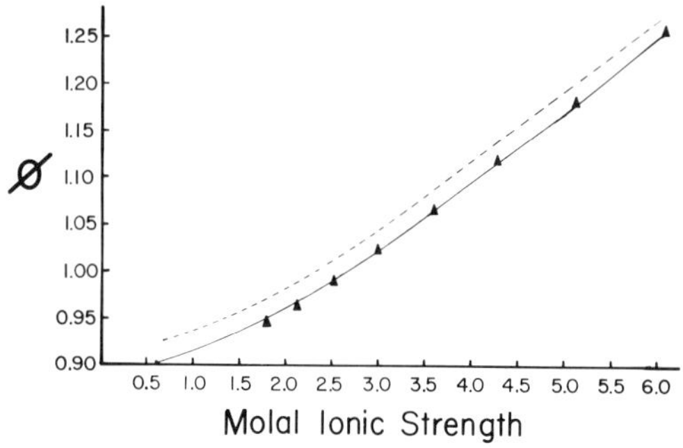

Figure 1. Comparison of osmotic coefficients calculated in the model with measured osmotic coefficients for sea water concentrates. (– – –) NaCl, measured; (——) sea water concentrates, measured; (▲) sea water concentrates, calculated using the model.

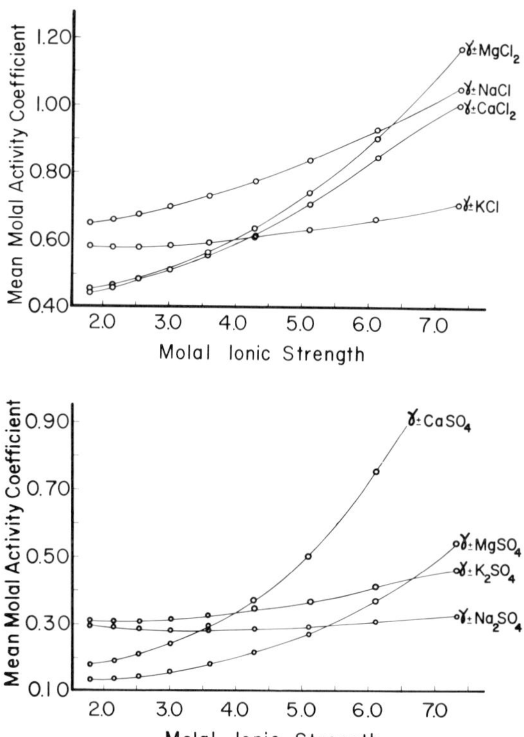

Figure 2. Mean molal activity coefficients of macrocomponent salts as a function of ionic strength for sea water concentrates. Normalized molar composition ratios: $Na^+ = 1.00$; $K^+ = 0.021$; $Mg^{2+} = 0.136$; Ca^{2+}, Cd^{2+}, Cu^{2+}, Pb^{2+}, $Zn^{2+} = 0.001$; $Cl_2^- = 1.174$; $SO_4^{2-} = 0.056$. (○) Computation points.

in a solvent activity value equivalent to a saturation vapor pressure lowering of 8.2 millibars (coefficient of variation 10.39%). To assess the meaning of such a discrepancy, field measurements by Turk (31) were compared to the laboratory results of Dickson et al. (30), for two brines similar in composition to GSL brine, at a specific gravity of 1.20 kg/liter. Turk (31) found that the saturation vapor pressure lowering for this brine was 9.5 millibars while the comparable value measured by Dickson et al. (30) was nearer 7.6 millibars (coefficient of variation 15.7%). Saturated vapor pressure lowering values for sea water concentrates presented by Turk (31) allowed further comparisons between the vapor pressure lowering as calculated by the model and experimental values. The data are shown in Table II. Since the solvent activity is strongly dependent on the solute composition, the agreement demonstrated in Table II is considered reasonable.

Table II. Saturation Vapor Pressure Lowerings for a Series of Sea Water Concentrates: Calculated Vs. Measured Values at 25°C

I molal*	ρ* g/ml	a_w*	$(1-a_w)e_s$* millibars	$(e_s-e'_s)$† millibars	Coeff. of variation (%)
1.80	1.064	.9522	1.51	1.17	17.94
2.14	1.075	.9424	1.83	1.68	6.04
2.54	1.087	.9301	2.22	1.90	10.98
3.01	1.103	.9148	2.70	2.28	11.93
3.59	1.120	.8955	3.31	2.60	16.99
4.73	1.142	.8834	3.70	3.80	1.89
5.11	1.166	.8400	5.07	5.39	4.33
6.11	1.196	.8004	6.33	6.97	6.81

*As calculated from molar ion concentration data using the model in this study, e_s is the saturated vapor pressure in millibars at 25°C.

†Measured values interpolated from graphically presented data giving vapor pressure lowering at 25°C as a function of brine density (31), e'_s is the saturated vapor pressure of the brine in millibars at 25°C.

Single Ion Activity Predictions. In lieu of an attempt to test the carbonate system predictive capability of the model, use was made of the fact that in most aerobic natural aquatic systems the solution concentration of heavy metal cations in the absence of interfacial phenomena is often controlled by the carbonate, basic carbonate or oxide (hydrous) solid phase form of the metal (22, 32, 33, 34).

Thus, the ability of the model to predict the chemistry of heavy metals in brine in a sense was used to test the validity of the carbonate subroutine. The general procedure was to assume that the trace metal solubility in brine was controlled by either the carbonate, basic carbonate or hydrous oxide form of the metal. The heavy metal and carbonate ion activities were determined by the model. The resultant calculated solubility of the heavy metals in brine was then compared with experimentally determined values.

It should be noted that the complexity and ill definition of carbonate salts in highly saline solutions is documented in literature by the complete absence of single salt osmotic and activity coefficient data. This situation requires the use of this less than satisfying technique to define heavy metal chemistry in brines.

Considerable effort was expended in attempts to validate the ability of the model to predict trace element solubilities in both natural and artificial brines of various compositions and ionic strengths in the laboratory, but after nine long term experiments it became apparent that the kinetics of trace metal carbonate precipitation would make this type of laboratory validation of the model extremely difficult. A search of the literature showed that metal carbonate precipitation kinetics were generally very slow, especially in complex solutions (35). In particular, the slowness of the kinetics of cadmium carbonate precipitation were described as notorious and as presenting a serious problem in wastewater treatment and corrosion control systems aimed at cadmium removal from solution (33, 34). Of the four metals being studied, copper was the only trace metal which was precipitated rapidly enough to be essentially at equilibrium with the artificial brines after a period of ten days. Characteristically, in both the natural and artificial brines, the reduction of copper in solution was accompanied by the establishment of algal growths and subsequent severe frothing in the stored, aerated, constant temperature reaction vessels.

The attempt to validate the model, therefore, shifted from the addition of metals to brine to measuring metals in natural brines which could be assumed to be near equilibrium with their constituent trace metals. North Arm Great Salt Lake, Utah (GSL) brine was chosen because its composition relative to South Arm brines is fairly constant (36, 37), allowing data from other workers to be meaningfully compared with this present study. Cu, Pb, Cd, and Zn were determined in the GSL North Arm Brines and compared with data on GSL North Arm brines given by Taylor et al. (38).

The filtered brine results for the two studies are in essential agreement, as shown in Table III. The disagreement between the unfiltered brine results of the two studies reflects sampling location differences.

Table IV shows the predicted and measured concentration of Cu, Pb, Cd, and Zn in the GSL North Arm brine. The measured values are compared with the predicted solubilities calculated for the basic carbonate, carbonate, and hydroxide phases. The lead sulfate form was also included. The measured and calculated values in most cases appear with coefficients of variation representing the six brines used to provide the average values reported.

In general, the agreement between predicted and measured was good for a first generation model. The data show that the concentration of Cd in the brine was controlled by the solid phase otavite, $CdCO_3$. The copper in the brine was supersaturated about seven times with respect to basic copper carbonate, malachite, $Cu_2(OH)_2CO_3$. Lead in the brine was found to be supersaturated about fourteen times with regard to either $PbCO_3$ or $Pb(OH)_2$. Zinc was shown to be slightly supersaturated with respect to hydrozincite, $Zn(OH)_{1.2}(CO_3)_{0.4}$.

The predicted values for the various solid phases shown in Table IV were calculated using the hydroxide and carbonate activity printout of the model. These values were used to calculate the equilibrium trace cation activities in the brine. The activity values were then divided by the predicted single ion activity coefficients and thereby converted to the molal concentration scale. These molal units were then converted to molar units, and then into the milligrams per kilogram of solution units reported.

Considering the assumptions incorporated into the carbonate subroutine and in using the mole-fraction statistical model for the partitioning of the mean molal activity coefficients into single ion activity coefficients, the agreement between predicted and measured trace metal concentrations in GSL brine was better than expected.

Analysis of X-ray diffraction patterns of the precipitates resulting from the addition of salts of the four metals to GSL North Arm brine showed that both copper and lead precipitated in the basic carbonate form. Zinc and cadmium compounds were not found in these X-ray diffraction traces, which may be due to these compounds comprising less than five percent of the solid material analyzed.

Summary and Conclusions

The objective of this study was to develop a model which would describe the chemistry of trace metals in highly concentrated electrolyte solutions.

Table III. Cu, Zn, Pb, and Cd concentrations in milligrams per kilogram of solution for GSL North Arm brine

	This Study		Tayler et al. (1977)	
Metal	Filtered*	Unfiltered	Filtered†	Unfiltered
Cu	.0006 ± .0002‡	.0010 ± .0006	<.001 ± ---#	.014 ± .0007
Zn	.0049 ± .0007	.0060 ± .0009	.005 ± .0007	.053 ± .003
Pb	.0026 ± .0002	.0028 ± .0012	.004 ± .001	.012 ± .002
Cd	.0025 ± .0001	.0066 ± .0015	<.001 ± ---	<.001 ± ---

*Using one Whatman GFA 13 cm glass fibre filter per liter of brine.

†Using a Millipore FHLP Teflon filter followed by a Whatman 540 paper filter efficient to .45 microns.

‡±Standard deviation for eight determinations for the unfiltered and six determinations for the filtered brines.

#±Standard deviations for the two separate North Arm brine analyses reported.

Table IV. Model solubility predictions for Cu, Pb, Cd, and Zn in GSL North Arm brine at 25°C

	Concentrations in mg per kg of solution				
	Predicted at equilibrium for the listed salts				
Metal	Basic carbonate	Carbonate	Hydroxide	Sulfate	Measured*
Cu	8.4 (± 4.0)† x 10^{-5}	1 x 10^{-2}#	1.4 (± 0.7) x 10^{-5}	--	60.0 (± 20.0) x 10^{-5}
Pb	0.72 (± 0.5) x 10^{-4}	1.9 (± 1.0) x 10^{-4}	1.9 (± 1.0) x 10^{-4}	1.26	26.0 (± 2.0) x 10^{-4}
Cd	--	2.7 (± 1.2) x 10^{-3}	30.7#	--	2.5 (± 0.1) x 10^{-3}
Zn	0.2 (± .05) x 10^{-2}	4.1 (± 2.0) x 10^{-2}	2.9 (± 2.0) x 10^{-2}	--	0.5 (± 0.07) x 10^{-2}

*Using one Whatman GFA 13 cm glass fibre filter per liter of brine.

†Average plus or minus the standard deviation for the sic brine samples. Large standard deviations related to pH dependence of the ligand activities. Measured pH range of the collected samples was from 7.70 to 8.05.

#Single values hand calculated for a pH 7.9 brine from hydroxide and carbonate activities as given by the model.

This was accomplished by expanding an existing model based on the cluster integral expansion theory of electrolyte solution structure into a comprehensive thermodynamic model describing the major and minor components of brines, including metals present in trace amounts.

The model was shown to predict osmotic coefficients to within the experimental error for the determination of osmotic coefficients from saturated vapor pressure measurements. Activity coefficient calculations appeared to be consistent with the available data for mixed electrolyte solutions.

The model predicts equilibrium concentrations for metals in concentrated electrolyte solutions which are in contact with a precipitated solid phase. An application of the model to a Great Salt Lake brine showed that predicted cadmium, zinc, and copper solubilities were in good agreement with measured dissolved cadmium, zinc, and copper levels in these same brines. Lead was supersaturated with respect to its basic carbonate in the Great Salt Lake brine according to the model prediction.

Laboratory validation studies showed the copper prediction to be accurate, but extremely slow kinetics for zinc, lead, and cadmium precipitation prevented validation for these three metal solubilities in the laboratory.

Limitations of the Model. The internal supporting routines for the otherwise completely general model restrict the apppli-cability of this model to sodium chloride dominated brines of ionic strengths from 2 to 6 molal. Temperature compensations were included to make the model applicable to solutions in the 10 to 35°C temperature range, approximately.

Acknowledgments

Partial funding for this project through USDI OWRT Contract No. 14-34-001-7094 is gratefully acknowledged.

Abstract

The equilibrium chemistry of lead, cadmium, copper, and zinc in concentrated electrolyte solutions (brines) in the concentration range of 2 to 6 molal is described using a comprehensive first generation model requiring molar analytical input data, the pH and temperature of the solution. Osmotic and activity coefficients for each component salt including salts of the heavy metals were computed on a pure single salt solution basis at the same ionic strength as the brine. Weighted combinations of the single salt solution data were used to predict the activity coefficients of the constituent salts. The theoretical basis for the equations used in the model was taken from the cluster integral expansion theory of electrolyte solution structure as presented in the literature. The carbonate and bicarbonate ion activities were estimated using empirical equations. The nature of these equations restricts the use of this model to sodium chloride dominated brines. The model predictions for the heavy metal concentrations in the North Arm of the Great Salt Lake were in reasonable agreement with measured values. The data imply that copper, cadmium, and zinc were essentially in equilibrium with $Cu_2(OH)_2CO_3$ (malachite), $CdCO_3$ (otavite), and $Zn(OH)_{1.2}(CO_3)_{0.4}$ (hydrozincite). Lead was oversaturated with respect to $PbCO_3$ and $Pb(OH)_2$, and undersaturated with respect to $PbSO_4$.

Literature Cited

1. Whitfield, M. Sea water as an electrolyte solution, p. 43-171, in Riley, J. P. and Skirrow, G., ed., "Chemical Oceanography." Vol. 1, 2nd Ed. Academic Press, New York, 1975.
2. Crerar, D. A. A method for computing multicomponent chemical equilibria based on equilibrium constants. Geochim. Cosmochim. Acta, 39, 1375-1384 (1975).
3. Morel, F., and Morgan, J. A numerical method for computing equilibria in aqueous chemical systems. Environ. Sci. Tech. 6, 58-67 (1972).
4. Polzer, W. L., and Roberson, C. E. Calculation of ion activity products for a brine from the Bonneville Salt Falts, Utah. U.S. Geol. Survey Prof. Paper 575-C, 116-119 (1967).
5. Brown, D. W., and Gulbrandsen, R. A. Chemical composition of a saline lake on Enderbury Island, Phoenix Island Group, Pacific Ocean. U.S. Geol. Survey J. Res. 1, 105-111 (1973).
6. Lerman, A. Model of chemical evolution of a chloride lake -- The Dead Sea. Geochim. Cosmochim. Acta. 31, 2309-2330 (1967).
7. Helgeson, H. C. Thermodynamics of hydrothermal systems at elevated temperatures and pressures. Am. J. Sci. 267, 729-804 (1969).
8. Wood, J. R. Thermodynamics of brine-salt equilibria - I. The systems NaCl - KCl - $MgCl_2$ - $CaCl_2$ - H_2O and NaCl - $MgSO_4$ - H_2O at 25 C. Geochim. Cosmochim. Acta. 39, 1147-1163 (1975).
9. Friedman, H. L. "Ionic solution theory." 265 p. Interscience Publishers, New York, 1962.
10. Robinson, R. A., and Wood, R. H. Calculation of the osmotic and activity coefficients of seawater at 25 C. J. Solution Chem. 1, 481-488 (1972).
11. Platford, R. F. Thermodynamics of the system H_2O - NaCl - Na_2SO_4 - $MgSO_4$ to the saturation limit of NaCl at 25 C. J. Solution Chem. 3, 771-780 (1974).
12. Lewin, S. "The solubility product principle." 116 p., Sir Isaac Pitman and Sons, LTD., London. 1960.
13. Robinson, R. A., and Stokes, R. H. "Electrolyte solutions." 512 p. Butterworths, London, 1955.
14. Reilly, P. J., and Wood, R. H. The prediction of properties of mixed electrolytes from measurements of common ion mixtures. J. Phys. Chem. 73, 4292-4297 (1969).

15. Reilly, P. J., Wood, R. H., and Robinson, R. A. The prediction of osmotic and activity coefficients in mixed-electrolyte solutions. J. Phys. Chem. 75, 1305-1315 (1971).
16. Van Luik, A. E., and Jurinak, J. J. A chemical model of heavy metals in the Great Salt Lake. Utah Agricultural Experiment Station, Research Report 34, Utah State Univ., Logan. 155 p., 1978.
17. Glueckauf, E. Electrostatic interactions in electrolyte solutions. Proc. Roy. Soc. Lond. A. 310, 449-462 (1969).
18. Wolf, A. V., Brown, M. G., and Prentiss, P. B. Concentrative properties of aqueous solutions: Conversion tables. in Weast, R. C., ed., "Handbook of Chemistry and Physics." CRC Press, Cleveland. D194-D236, 1974.
19. Bates, R. G., Staples, B. R., and Robinson, R. A. Ionic hydration and single ion activities in unassociated chlorides at high ionic strengths. Anal. Chem. 42, 867-871 (1970).
20. Stewart, P. B., and Munjal, P. K. The solubility of carbon dioxide in distilled water, synthetic sea water and synthetic sea water concentrates. Water Resour. Center, Univ. Calif. Sea Water Conversion Laboratory Report 69-2. Water Resour. Center Delineation Report 30. 1969.
21. Ben-Yaakof, S., and Sass, E. Independent estimate of the pH of Dead Sea brine. Limnol. Oceanogr. 22, 374-376 (1977).
22. Hansson, I. A new set of acidic constants for carbonic acid and boric acid in sea water. Deep Sea Res. 20, 461-478 (1973).
23. Truesdell, A. M., and Jones, B. F. WATEQ, a computer program for calculating chemical equilibria of natural waters. U.S. Geol. Survey J. Res. 2, 233-248 (1974).
24. Wagman, D. D., Evans, W. H., Parker, V. B., Halow, I., Bailey, S. M., Schumm, R. H., and Churney, K. L. Selected values of chemical thermodynamic properties. NBS Tech. Notes 270-3 through 270-5. 1968-1971.
25. Parker, V. B., Wagman, D. D., and Evans, W. H. Selected values of chemical thermodynamic properties. NBS Tech. Note 270-6. 1971.
26. Stokes, R. H., and Robinson, R. A. Ionic hydration and activity in electrolyte solutions. Am. Chem. Soc. Jour. 70, 1870-1878 (1948).
27. Stokes, R. H., and Robinson, R. A. Solvation equilibria in very concentrated electrolyte solutions, p. 75-94, in Kay, R. L., ed., "The Physical Chemistry of Aqueous Systems." Plenum Press, New York, 1972.
28. Stumm, W., and Morgan, J. J. "Aquatic chemistry." 583 p. Wiley-Interscience, New York, 1970.
29. Brooks, R. R., Presley, B. J., and Kaplan, I. R. APDC-MIBK extraction system for the determination of trace elements in saline waters by atomic-absorption spectrophotometry. Talanta 14, 809-816 (1967).
30. Dickson, D. R., Yepson, J. H., and Hales, J. V. Saturated vapor pressures over Great Salt Lake brine. J. Geophys. Res. 70, 500-503 (1965).

31. Turk, L. J. Evaporation of brine: A field study on the Bonneville Salt Flats, Utah. Water Resour. Res. 6, 1209-1215 (1970).
32. Santillan, J., and Jurinak, J. J. The chemistry of lead and cadmium in soil: Solid phase formation. Soil Sci. Soc. Am. Proc. 39, 851-856 (1975).
33. Weber, W. J., and Posselt, H. S. Equilibrium models and precipitation reactions for cadmium (II). Ch. 7, in Rubin, A. R., ed., "Aqueous-environmental chemistry of metals." Ann Arbor Science Publishers, Inc., Ann Arbor. 1974.
34. Posselt, H. S., and Weber, W. J. Studies on the aqueous corrosion chemistry of cadmium. Ch. 8, in Rubin, A. R., ed., "Aqueous-environmental chemistry of metals." Ann Arbor Science Publishers, Inc., Ann Arbor. 1974.
35. Nancollas, G. H., and Reddy, M. M. Crystal growth kinetics of minerals encountered in water treatment processes. Ch. 6, in Rubin, A. R., ed., "Aqueous-environmental chemistry of metals." Ann Arbor Science Publishers, Inc., Ann Arbor. 1974.
36. Handy, A. H., and Hahl, D. C. Great Salt Lake: Chemistry of the water, p. 135-151, in Stokes, W. L., ed., "The Great Salt Lake: Guidebook to the geology of Utah." Utah Geol. Mineralog. Survey Water Resour. Bull. 12 (1966).
37. Whelan, J. A. Great Salt Lake, Utah: Chemical and physical variations of the brine, 1966-1972. Utah Geol. Mineralog. Survey Water Resour. Bull. 17. 27 p. (1973).
38. Tayler, P. L., Hutchinson, L. A., and Muir, M. K. Heavy metals in the Great Salt Lake, Utah. Utah Ccology 4, 19-28 (1977).

RECEIVED November 16, 1978.

31

Heterogeneous Interactions of Arsenic in Aquatic Systems

THOMAS R. HOLM, MARC A. ANDERSON, DENNIS G. IVERSON, and ROBERT S. STANFORTH

University of Wisconsin, Water Chemistry Program, Madison, WI 53706

The chemistry of arsenic in soil and aquatic ecosystems can best be described as a complex array of homogeneous and hetergeneous chemical, biochemical, and geochemical reactions that together control the dissolved concentrations of arsenic in these systems. Although many of these reactions have been studied independently, the total description of the combined major reaction mechanisms that control the cyclic behavior of arsenic in environmental systems has largely been undeveloped. In the paper presented here, results obtained in laboratory and field studies are used to describe some of the major control mechanisms that affect the distribution as well as transformations of arsenic in an aquatic environment.

Arsenic can exist in several oxidation states, as both inorganic and organometallic species, and in dissolved and gaseous phases (Table I). Dissolved arsenic species can adsorb to suspended solids and be carried down to the sediments in an aquatic system. Since gaseous arsenic compounds can form, arsenic can be removed from the sediments as dissolved gas or in gas bubbles (e.g. CH_4). Thus, arsenic can cycle within aquatic ecosystems and this cyclic behavior has been reviewed by Ferguson and Gavis (1) and Woolson (2). In any given system, it is necessary to understand the behavior of a variety of different arsenic compounds as well as a variety of environmental compartments in order to totally characterize the cyclic behavior of this element.

Four arsenic species common in natural samples are arsenate, arsenite, methanearsonic acid (MMAA) and dimethylarsinic acid (DMAA). These species possess different chemical properties which affect the mobility of arsenic in natural systems. For example, methanobacterium form trimethyl arsine from DMAA faster than from MMAA or arsenate (3) and arsenate and MMAA are more strongly adsorbed than DMAA on alluvial soils (4). Transformation between the different oxidation states and species of arsenic may occur as a result of chemical or biochemical reactions (1, 5, 6, 2, 7, 8, 9). Inorganic chemical

0-8412-0479-9/79/47-093-711$06.50/0
© 1979 American Chemical Society

Table I
Arsenic Species Commonly Found in Environmental Samples

Species	Name(s)	Oxidation State
AsO_4^{3-}	Arsenate	+5
AsO_3^{3-}	Arsenite	+3
$CH_3AsO(OH)_2$	Methanearsonic Acid Monomethyl Arsonic Acid	+3
$(CH_3)_2AsOOH$	Hydroxydimethyl Arsine Oxide[a] Dimethyl Arsinic Acid Cacodylic Acid	+1
AsH_3	Arsine[b]	-3
$(CH_3)_2AsH$	Dimethyl Arsine[b]	-3
$(CH_3)_3As$	Trimethyl Arsine[b]	-3

[a] IUPAC name

[b] Gaseous

transformations may be thermodynamically predicted to be a function of Eh and pH (1). Geochemically, arsenic may form insoluble precipitates with calcium, sulphur, iron, aluminum, and barium compounds in natural waters (10). These precipitates are, however, slow in nucleating and show slow growth rates. Arsenic species are more likely to be adsorbed on the surface of organic and inorganic substrates than as crystalline precipitates. Indeed, it will be shown in this paper that not only do adsorption reactions affect the transport of arsenic but they additionally alter the apparent transformation rates of arsenicals in natural systems. Since most natural systems are heterogeneous, that is, they have solid and solution phases, one can immediately recognize the importance of understanding the role played by adsorption reactions in controlling dissolved elemental concentrations and transformation rates.

It is the purpose of this paper to describe some of the major mechanisms that control arsenic in aquatic systems. Particularly, this paper addresses the problem of arsenic speciation and compartmentalization in sediments. To this end, results obtained from speciation, compartmentalization, kinetic, and adsorption studies using both field and laboratory samples will be interfaced in a descriptive model for arsenic in heterogeneous systems. The model has particular significance with respect to the biochemical transformation studies that have previously been performed by other researchers (11, 12, 13, 14, 3, 15).

In order to investigate the complex array of mechanisms that control the transformations and distribution of arsenic in an aquatic system, it was advantageous to study a system that was stressed by high concentrations of arsenicals. Then analytical measurements could be performed at levels well above detection limits and the rates of processes such as biotransformations and sedimentation would be enhanced over corresponding rates in systems having background levels of arsenicals. The lower Menominee River (Figure 1) is an example of such a system and is the area chosen for this research. Due to the improper disposal of arsenical wastes, total arsenic concentrations are as high as 0.08 M in ground water, 1.6 x 10^{-5} M in river water, 0.08 M in sediment pore fluids, and 4000 mg kg^{-1} in sediment solids (16). This paper is based on ongoing research in the lower Menominee River (16) and represents results to date. Each of the following sections has its own methods, results, and discussion subsections.

Speciation and Compartmentalization

Methods of Speciation and Fractionation. It is apparent that in order to understand the mobility of arsenic and its availability for reactions, methods of speciation and fractionation must be applied to sediment samples in field and laboratory studies. In this paper speciation refers to the separation and quantitative determination of inorganic arsenic, methanearsonic acid, and cacodylic acid. Compartmentalization involves identifying the major compartments for arsenic in a heterogeneous system (e.g. aqueous, adsorbed, occluded,...) and determining the amounts of arsenic in each compartment. Fractionation involves the extraction of arsenic from operationally defined fractions of the solid phase of an aquatic system (e.g. sediment).

Several analytical methods for speciating arsenic have been reported. They include chromatographic techniques such as electrophoresis and ion-exchange (17), paper chromatography (18) and HPLC (19); selective volatilization of arsenic compounds to analogous arsines followed by GC-MES (20); boiling point separation/spectral emission (21); and atomic absorption (22). The above techniques have been applied to samples such as; commercial pesticides (20), coal and fly ash (23), rocks, sediments, soils and minerals (24, 22), plant tissue (18), bovine liver (23), and water samples (25).

Studies in this laboratory require the use of analytical and speciation procedures which can be routinely performed on a variety of samples. Most of the above mentioned speciation techniques require that a considerable amount of equipment be dedicated to these particular analyses. In addition, some of them involve experimental parameters which are difficult to reproduce, hence, making routine analysis difficult.

The speciation technique adopted for this study is a modification of Yamamoto's (25) chromatographic method which is simple to perform, requires a minimum amount of equipment and is reliable for separating inorganic-As (i.e. arsenite and arsenate), monomethyl-arsonic acid, and dimethyl-arsincic acid in river water and interstitial water. The procedure involves: 1) conditioning a Dowex AG 50W-X8 resin column to the acid form and adjusting its pH to 1.5-1.8; 2) placing a sample (at pH <2) on the column and eluting sequentially with a 0.2 M trichloroacetic acid and 1 M ammonium acetate; 3) collecting fractions and measuring arsenic by flameless atomic absorption (FAA) with matrix modification (26). The efficiency of separating inorganic arsenic (first eluted) from MMAA (second eluted) depends on the initial column pH. Lowering the pH from 4 (25) to 1.5 (16) results in complete separation of the two arsenical forms. The use of ammonium salts in the eluent solutions eliminates certain cation interferences for FAA analysis of the effluent. Matrix modification with 0.1% $NiNO_3$ for arsenic determination by FAA normalizes instrumental response for inorganic and methylated arsenic compounds (16).

Arsenic in soilds has been fractionated by Jackson's (28) procedure for soil phosphorus (15, 27). In this laboratory, a modification of Jackson's procedure is being used on sediment solids. A series of 1 molar solutions of NH_4Cl, NH_4OH, acid ammonium oxalate (29) and HCl are used in sequence. The chloride fraction, or exchangeable fraction, contains weakly adsorbed, coulombically bound arsenic. The hydroxide fraction, contains iron and aluminum arsenate precipitates and surface precipitates (i.e. adsorbed arsenic species having both chemical and coulombic bonding to oxide surfaces). The oxalate, or reductant soluble fraction, contains arsenic occluded in amorphous weathering products. The acid, or calcium, fraction contains arseno-apatites.

Results of Speciation and Fractionation. The arsenic species in the pore waters of a sediment core collected in September 1977 (Site 70A) are shown in Figure 2. Total arsenic concentrations in these samples ranged from about 0.013 M to 0.107 M and the samples had to be diluted to about 10 μM arsenic for chromatography. Although methylated arsenic is the main species in the ground water for that area, inorganic arsenic is abundant in the pore waters and particularly in the oxidized near-surface layer of the core. These data suggest that demethylation was occurring in the sediments, but little can be discerned about the mechanisms or rates of reactions. Table II shows similar data for a core collected in April, 1978 (Site 69A). MMAA is the most abundant species in this core

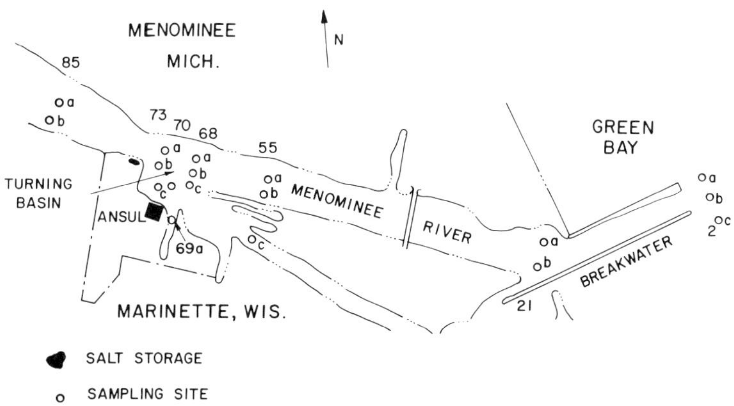

Figure 1. Map of the lower Menominee River showing sampling sites. Transect numbers are distances from the mouth of the breakwater in hundreds of feet.

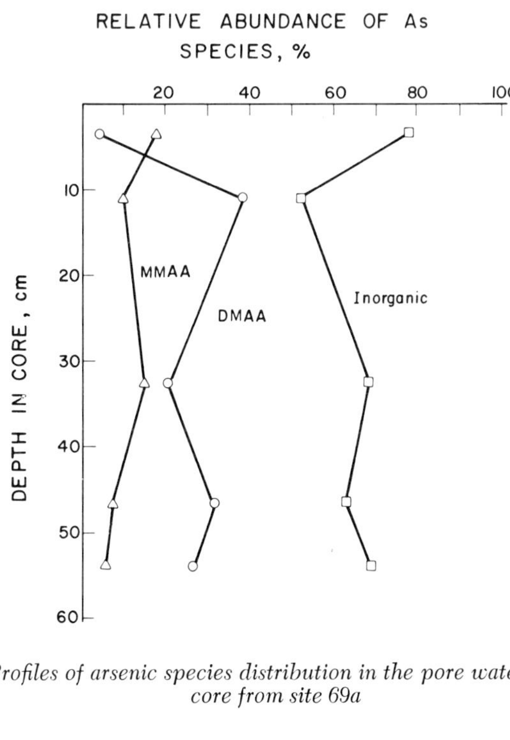

Figure 2. Profiles of arsenic species distribution in the pore waters of a sediment core from site 69a

collected closer to shore. The differences in species between the September, 1977 core and the April, 1978 core may be due to their locations with respect to the source or to seasonal effects.

Arsenic was distributed among the operationally defined fractions of the sediment solids of a sediment core collected in September, 1977 as shown in Figure 3. Generally, the order of abundance of As in the fractions was OH^- (Fe and Al) > oxalate (amorphous or occluded) >> Cl^- (exchangeable) >> H^+ (Ca or arsenoapatite). This order was greatly enhanced in the top section of the core, for which the redox potential in pore waters was -5 mV. All other redox potentials were less than -230 mV. Thus, the top section of the core was the only section in which amorphous $Fe(OH)_3$ was stable (30) and was the section in which the most iron and aluminum arsenic should be found. This data indicates that most of the solid phase arsenic was arsenate (11).

Table II
Arsenic Species in Pore Water From Sediment Core
(Site 69A, April, 1978)

Average Depth (cm)	Species (%)		
	Inorganic	MMAA	DMAA
4.5	6	91	3
12.6	9	82	8
19.8	6	83	11
27.0	4	96	--[a]

[a]Not detected.

Competitive Adsorption

Adsorption processes may be particularly important in influencing species concentrations, since the arsenic present in the pore waters will probably be in equilibrium with arsenic adsorbed on solid surfaces. Arsenic in any species measured in pore waters may be only a fraction of the total amount of that species present in the sediments, the rest being adsorbed to or incorporated into particulate matter. Thus, it is important to study the sorptive characteristics of each of the arsenic species in the sediments. In the Menominee River sediments studied, the four oxygenated arsenic species (arsenate, arsenite, monomethyl arsonic acid and cacodylic acid) are often present together and competing among themselves and with phosphate for the same sorption sites. The competitive adsorptive characteristics of the species could greatly influence

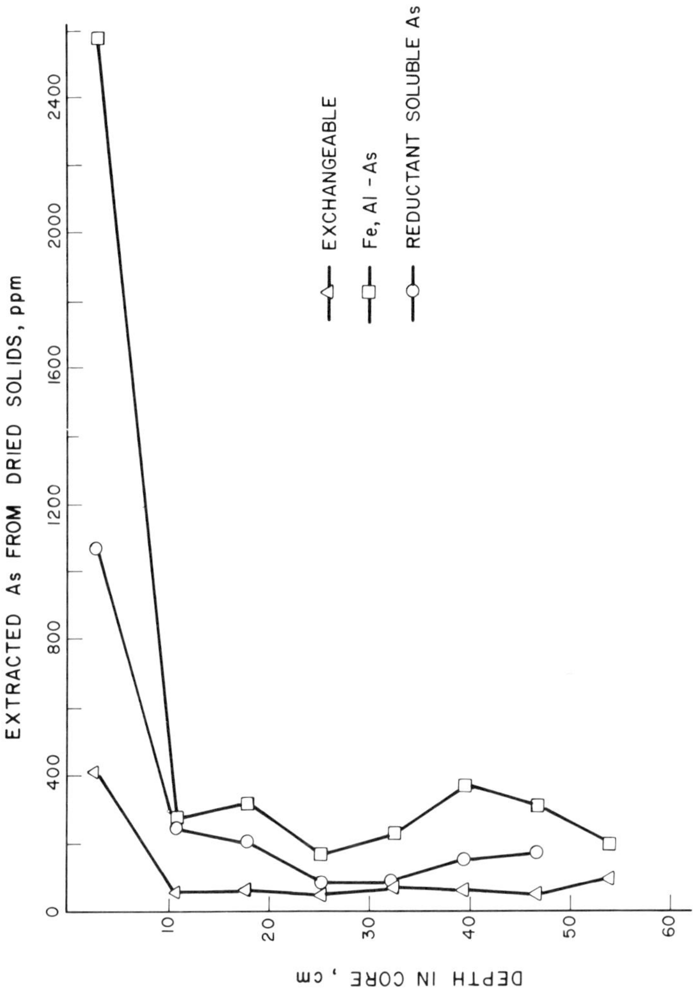

Figure 3. Profiles of arsenic fractionation in the solid phase of a sediment core from site 69a (same core as in Figure 2). The HCl fraction was much smaller than the other fractions and was not plotted.

the concentrations of each species, and hence any interpretations made from those concentrations.

To illustrate this point, assume that a sediment initially contains only cacodylic acid but that there are bacteria present which convert cacodylic acid to arsenate. Assume further that cacodylic acid is not adsorbed by the sediment, while arsenate is strongly adsorbed. As the cacodylic acid is converted to arsenate, arsenate is adsorbed by the sediment and arsenic disappears from solution. Assuming a 90% water content in the sediments, an initial cacodylic acid concentration of 1000 μM, and using the constants for arsenate given later in this paper, the amount of arsenic in solution and adsorbed on the sediments can be calculated. Results of such a calculation are listed in Table III.

If the loss of arsenate to the sediments through adsorption is ignored, an apparent large loss of arsenic occurs as the cacodylic acid is converted to arsenate. The significance of adsorption to the solution concentration is obvious from this example. In instances where both species in the reaction adsorb significantly the situation is more complicated. Again, however, the effects of adsorption on the observed concentrations must be understood if the system is to be adequately understood and described. Particularly interesting is the situation in which a weakly adsorbed species is converted to a strongly adsorbed one. The more strongly adsorbed species may replace the more weakly adsorbed species as the more strongly adsorbed species are formed. Thus, the concentration of the more weakly adsorbed species in solution will remain relatively constant until much of the conversion has occurred. To ignore the composition of the species adsorbed on the solids would lead to a serious misunderstanding of the actual dynamics of the system.

In order to obtain a general overview of the sorptive characteristics of the various arsenic species and phosphate on Menominee River sediments a short-term experiment was performed. A Menominee River sediment slurry was spiked with arsenic or phosphate and the loss from solution was measured. By comparing the relative losses of the various arsenic species and phosphate from solution an indication of the relative sorptive strengths was obtained.

Experimental. An uncontaminated sediment sample was collected from the Menominee River. The sediment had no distinctive odor, appeared to be a mixture of sand and clay, and was anaerobic (i.e., the pore waters had a redox potential of approximately -115 mV). The sample was stored in a plastic bag at $4^{o}C$ until use.

To faciliate working with the sediments, the solids content of the sediment was reduced by diluting the sediments with pore water. Pore water, obtained by centrifuging a

Table III
Results of Calculation of Conversion of Cacodylic Acid to Arsenate in Anaerobic Sediment

% of Cacodylic Acid Converted to Arsenate	Cacodylic Acid Concentration (mM)	Arsenate Concentration (μM)	Arsenate Adsorbed (%)	Fraction of Total Arsenic Remaining in Solution (%)
0	1000	0.0	0	100
10	900	0.405	99.7	90.0
50	500	2.13	99.6	50.2
90	100	4.04	99.6	10.4
99	10	4.50	99.5	1.5

portion of the sediments, was mixed with fresh sediment in approximately a 1:2 sediment:pore water ratio. A 2 ml sample of the sediment slurry was placed in a small glass tube and stoppered with a rubber septum. The sediment transfers were done under an N_2 atmosphere to avoid oxidation. A volume of standard arsenic or phosphate solution was injected with a syringe through the septum. The tubes were shaken for approximately 24 hours. The solids were separated by centrifugation and filtration through a 0.45 μm filter, and the pore water analyzed for arsenic or phosphorus. Blanks were run using arsenic and phosphorus in distilled water and pore water to investigate sorption to the glass tubes and precipitation in the pore waters. Both factors were found to be unimportant in removing arsenic or phosphorus from the experimental solution.

Results. The experimental results (Figure 4) show significant differences in the loss from solution of the various arsenic species and phosphate. Clearly, some species are removed from solution to a much greater extent than others, with the relative removals being as follows:

$$PO_4 > AsO_4 > MMAA > AsO_3 > CA$$

most strongly removed — least strongly removed

where MMAA = monomethyl arsonic acid and CA = cacodylic acid.

Phosphate and arsenate were both removed to a great extent, while cacodylic acid was removed only slightly and in some cases not at all. Since the loss from solution occurred only when the sediments were present, either sorption or surface precipitation must be controlling the concentrations. Since adsorption is probably the major process controlling the observed loss from solution, it is instructive to fit the data obtained to a Langmuir isotherm. Adsorption constants for the various species calculated from the data obtained are given in Table IV. Constants for arsenite were not calculated because the data indicated that arsenite loss from solution does not follow an adsorption isotherm. Inspection of the arsenite loss from solution isotherm shown in Figure 4 indicates that the loss is linearly dependent on concentration over the concentration range used.

The constants given in Table IV show the strong adsorption of phosphate and arsenate relative to cacodylic acid, with monomethyl arsonic acid intermediate. Constants for the adsorption of arsenate on amorphous iron and aluminum hydroxide are also included in Table IV to compare the sediments with well-characterized adsorbents. It was assumed that loss from

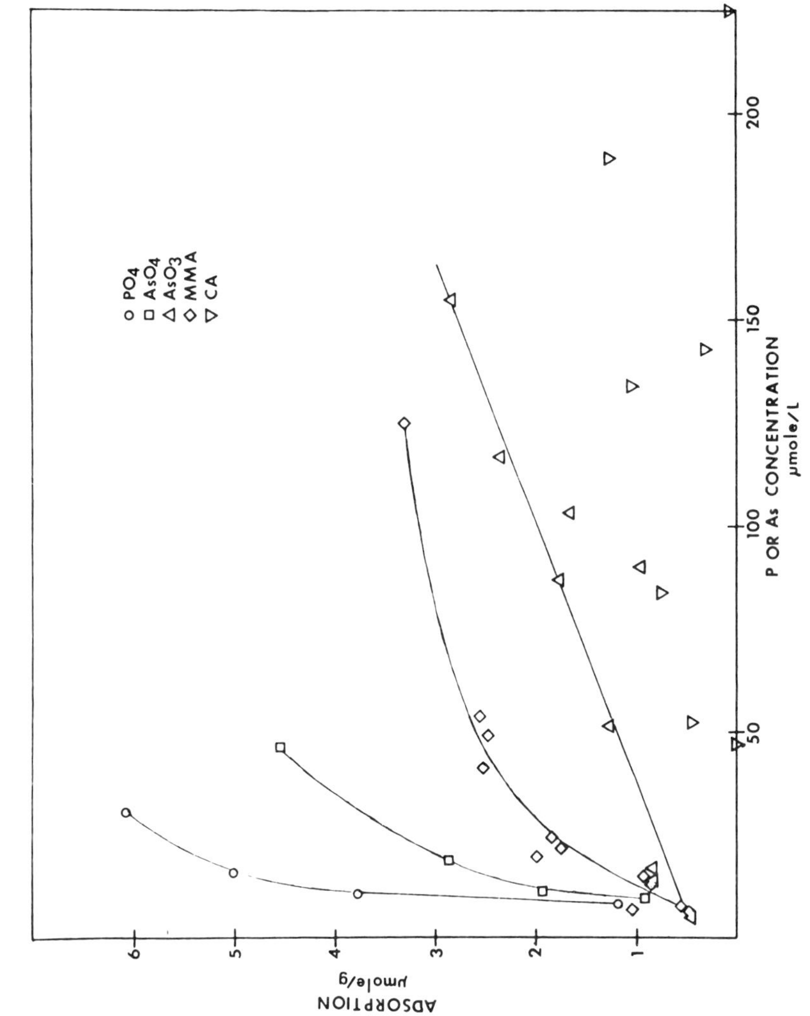

Figure 4. Sorption of phosphate and arsenic species by anaerobic Menominee River sediments

Table IV
Sorption of Phosphate and Arsenic Species by Anaerobic Sediments

Species	Γ^a $\mu mol\ g^{-1}$	K^a $\mu mol\ l^{-1}$	N^b	Correlation Coefficient[c]
PO_4	9.1	15	3[d]	0.86
AsO_4	7.7	32	3[d]	0.98
MMAA	3.6	28	11	0.38
AsO_4[f,h]	1250	137	--	--
AsO_4[g,h]	1700	51	--	--
CA	3.6	320	4[e]	0.85

[a] Langmuir coefficients.
[b] Number of points.
[c] Correlation of linearized form of data to a least squares approximation of the best straight line through the points.
[d] Ignoring one data point shown in Figure 4 due to possible background interference.
[e] Ignoring three points in which little or no adsorption was observed.
[f] Adsorption on amorphous $Al(OH)_3$.
[g] Adsorption on amorphous $Fe(OH)_3$.
[h] Data from Ferguson *et al.* (*31*).

solution followed a Langmuir isotherm. Since a small number of points were used and there was some scatter in the data, there is some uncertainty in the constants. However the constants are useful for calculations, as will be seen later in this paper.

Discussion. Wauchope (4) reported that adsorption of phosphate and various arsenic species on alluvial soils varied as follows:

$$AsO_4 \simeq MMAA > DMAA > PO_4$$

most strongly adsorbed   least strongly adsorbed

The arsenic species adsorbed in the same pattern in both studies, while phosphate changed from being the most strongly adsorbed in this study to the least strongly adsorbed in Wauchope's study. The soils used in Wauchope's study were presumably aerobic, with iron and aluminum hydroxides present. The sorption of phosphate and the arsenic species on the twelve soils used in Wauchope's experiment correlated well with both the clay content and the iron content of the soils. The Menominee River sediments, however, were anaerobic, so iron should have been present as $Fe(OH)_2$ rather than the $Fe(OH)_3$ which was, presumably, present in Wauchope's alluvial soils. The solubility of $Fe(OH)_2$ is greater than that of $Fe(OH)_3$ (30) and its surface properties may differ from those of $Fe(OH)_3$. Thus, differences in amount and speciation of iron may have accounted for differences in phosphate sorption.

Ferguson and Anderson (31) investigated the adsorption of arsenate and arsenite on iron and aluminum hydroxides and found that for both adsorbents the adsorption of arsenate followed a Langmuir isotherm while the sorption of arsenite varied linearly with concentration. In the concentration range used in the experiment described above, they found that arsenate was much more strongly adsorbed than arsenite. They did not explain the rather surprising difference in behavior between arsenate and arsenite; however, their results give support to the results found in this experiment.

Combining the results of the three studies, it appears that arsenate is more strongly adsorbed than monomethyl arsonic acid, which is more strongly adsorbed than cacodylic acid, and that all three probably follow Langmuir isotherms. Arsenite sorption, on the other hand, varies linearly with concentration over concentration ranges of environmental significance. Phosphate adsorption also follows a Langmuir isotherm, but the strength of adsorption relative to the arsenic species varies depending on the adsorbent.

Adsorption processes are very dependent on the solid surface characteristics and pH, as well as other factors (adsorbate and adsorbent identity and concentration, solution ionic strength, identity of other ions present, etc.). Since neither surface characteristics nor pH were measured or controlled, the numerical values obtained in the experiment need to be interpreted with care, as was discussed previously. The relative trends, however, will probably be valid for sediments similar to those used in the experiment, i.e., anaerobic river sediments where sediments control the concentration of the species in solution. Wauchope's findings indicate that the relative trends may not be valid for aerobic sediments or soils. Adsorption in sediments or soils occurs primarily on various clays and on amorphous iron and aluminum hydroxide particles and coatings on other particles. The concentration of the adsorbing species obviously varies greatly in different sediment samples. Thus the constants determined for one sediment sample will apply only to the particular concentrations and composition of adsorbing species found in that sample, i.e., the constants will be sample specific. As long as the adsorbing species remain the same in different sediment samples, although at different concentrations, the relative adsorbing patterns found for different species will remain the same, and the results from one set of experiments can be used as an aid to understanding the processes occurring in a similar sediment sample. If, however, the adsorbing species change between samples, then the relative adsorptive strengths of the sediments for different anions may change, and the results from experiments with one sediment cannot be applied to the different sediments. Further experiments are planned to better elucidate the relative adsorbing strengths of phosphate and arsenic species on aerobic sediments.

Rates of Species Transformations in Anaerobic Sediments

Kinetics and Adsorption. If microorganisms only take up dissolved arsenic species, then adsorption can affect rates of species transformations by lowering concentrations of reactants. In this section calculations of changes in concentrations of reactants and products of arsenic species transformations in sediments are presented. Rates of transformation are assumed to be proportional to species concentrations and the rate of another reaction, V, coupled to the transformation. That is, the transformation reaction is not assumed to be a source of energy or structural material for the microorganisms. Thus,

$$- \frac{d\,(\text{Species})}{dt} = kV\,(\text{Species})\ .$$

If V is constant, for example if V is described by the Michelis-Menten equation and the substrate concentration is high enough that V approaches V_{max}, then kV is a lumped constant and the transformation rate is described by a first order equation.

Calculating the effects of adsorption on kinetics necessarily involves mixing kinetic and equilibrium data. However, the time scales for adsorption and desorption reactions are much shorter than those for microbially mediated arsenic species transformations. Adsorption and desorption reactions reach equilibrium over a period of 24 hours or less (32, 33). On the other hand, Woolson (2) estimated conversion rates of 0.067 to 0.404 % day^{-1} for oxidative metabolism of cacodylic acid to arsenate in model aquatic systems. Assuming first order kinetics, these conversion rates translate to half lives of 5.8 to 34.7 months. Thus, Woolson's model ecosystems were probably at equilibrium at all times with respect to adsorption and desorption of arsenic species.

Examples of the Influence of Adsorption on Kinetics. Three cases of arsenic species transformations influenced by adsorption are considered in this section. The three cases considered are: 1) reactant not adsorbed, product adsorbed (e.g. demethylation of cacodylic acid to arsenate), 2) reactant adsorbed, product not adsorbed (e.g. methylation of arsenate to cacodylic acid), 3) both reactant and product adsorbed and competing for adsorption sites (e.g. demethylation of methanearsonic acid to arsenate). The equations for each case are shown in Tables V through VII, respectively.

Most of the symbols in Tables V, VI, and VII have their common meanings. The symbol $[B]_s$ stands for the concentration of adsorbed species B and is the product of Γ, the number of moles of B adsorbed per gram of adsorbent, and U, the number of grams of adsorbent per liter. Thus, $[B]_s$ has the units moles per liter.

An example of each case was calculated using the adsorption constants for anaerobic Menominee River sediments listed in Table IV. The first order rate constant used was 0.04 day^{-1}, corresponding to a half life of 17 days. Initial reactant concentrations of 0.001 M were chosen to approximate conditions in moderately contaminated sediments (16). A sediment solids concentration of 60 g l^{-1} (corresponding to 95% water) was used.

In the first two cases, setting up and solving the differential equation was straightforward. In case 2 a closed form solution could not be obtained, so the equation was solved numerically.

Setting up the differential equation for case 3 involved solving a cubic equation, so a different approach was taken. For case 3 the system of mass balance and Langmuir equations was solved for [A] and [B]. Then the system was allowed to

Table V

Case 1. First Order Reaction. Product Adsorbed According to Langmuir Isotherm. Reactant not Adsorbed

$$A \rightarrow B$$

$$\frac{-d[A]}{dt} = k\,[A]$$

$$C = [A] + [B]_s + [B]$$

$$[B]_s = \frac{U\,\Gamma_\infty\,[B]}{K + [B]}$$

at $t = 0$ $[A] = C, \quad [B] = 0$

$[A] = Ce^{-kt}$ (independent of adsorption)

$$B = \frac{-N + (N^2 + 4KC(1-e^{-kt}))^{1/2}}{2}$$

where $N = U\Gamma_\infty + K - C(1-e^{-kt})$

Example: demethylation of cacodylic acid (weakly adsorbed) to arsenate (strongly adsorbed).

Key to Symbols:

A	reactant, B product
[A]	aqueous concentration of [A]
U	concentration of adsorbent in g ℓ^{-1}
K	constant in Langmuir isotherm
Γ_∞	adsorption capacity of adsorbent in mol g^{-1}
$[B]_s$	concentration of adsorbed product $[B]_s = U\Gamma_B$

Table VI
Case 2. First Order Reaction. Reactant Adsorbed According to Langmuir Isotherm. Product Not Adsorbed.

$$-\frac{d[A]}{dt} = \frac{d[B]}{dt} = k\,[A]$$

$$C = [A] + [A]_s + [B] \qquad \text{at } t = 0 \quad C = [A]_{s,o} + [A]_o$$

$$[B] = 0$$

$$[A]_s = \frac{U\Gamma_\infty\,[A]}{K + [A]}$$

$$\frac{N_1}{\alpha+\kappa} + \ell n\,\frac{\alpha\,N_2}{(\alpha+\kappa)^{N_2-1}} = -kt + \text{constant}$$

where $N_1 = 2\,\kappa + \mu - 2$ $\qquad \alpha = [A]/C$

$N_2 = 2 + \kappa\mu$ $\qquad \kappa = K/C$

$\mu = \Gamma_\infty U/C$

Other symbols described in Table V.

Solve numerically using Newton iteration.

Constant calculated by substituting $\alpha_o = [A]_{o/C}$ at $t = 0$

Table VII
Case 3. First Order Reaction. Product and Reactant Adsorbed According to Competitive Langmuir Isotherm.

$$A_{tot} = [A] + [A]_s \qquad 1$$

$$[A]_s = \frac{U\,\Gamma_{\infty A}\,[A]}{K_A + [A] + [B]\,K_A/K_B} \qquad 2$$

$$B_{tot} = [B] + [B]_s \qquad 3$$

$$[B]_s = \frac{U\,\Gamma_{\infty B}\,[B]}{K_B + [B] + [A]\,K_B/K_A} \qquad 4$$

Symbols described in Table V.

Procedure:
1. Solve equations 1-4 for [A] and [B].
2. Compute $\Delta A = [A]\,(1-e^{-kt})$ for a short interval t. (Choose t such that $\Delta A << [A]$).
3. Subtract ΔA from A_{tot} and add ΔA to B_{tot}.
4. Go to step 1 and repeat the procedure.

react as if there were no solid present and [A] was the total concentration of A. In other words, an amount of A, ΔA, was changed to B according to the equation $\Delta A = [A](1-e^{-kt})$, where t was chosen to be less than 0.1 $t_{1/2}$. Then A was subtracted from C_A and added to C_B and [A] and [B] were recalculated. The entire process was repeated until the desired time interval was covered.

The method used for the solution of case 3 was tested on systems for which the differential equation could be set up. The approximate solution was within 0.1% of the exact solution in each case.

Demethylation of cacodylic acid to arsenate was chosen as an example of case 1, product adsorbing with reactant not adsorbing, because cacodylic acid is only weakly adsorbed by anaerobic sediments and arsenate is strongly adsorbed. For the same reason, methylation of arsenate to produce cacodylic acid was chosen as an example of case 2, reactant adsorbed and product not adsorbed. Demethylation of methanearsonic acid to arsenate was chosen as an example of case 3, reactant and product both adsorbing.

The predicted arsenate vs time curves for demethylation of cacodylic acid are shown in Figure 5. Both curves are for first order decay with a rate constant of 0.04 day^{-1} but with different amounts of adsorbent present. Adsorption of arsenate results in dissolved arsenate concentrations that are lower than total arsenate. An s-shaped curve results because arsenate produced during the first few days is almost entirely adsorbed.

Predicted arsenate and cacodylic acid vs time curves for methylation of arsenate to produce cacodylic acid are plotted in Figures 6 and 7, respectively. Adsorption of arsenate lowers the dissolved arsenate concentration and, therefore, results in a slower rate of methylation. The apparent rate constant for this reaction in sediment containing 60 g of solids per liter is 0.03 day^{-1} compared with 0.04 day^{-1} in the absence of sediment.

Demethylation of monomethylarsonic acid to produce arsenate with the two species competing for adsorption sites is depicted in Figure 8. Adsorption of MMAA results in slower conversion rates and the apparent rate constant in this case is 0.03 day^{-1}. Since arsenate is adsorbed more strongly than MMAA the final arsenate concentration is less than the initial MMAA concentration.

The apparent rate of decomposition of adsorbed reactant is quite sensitive to the product of the parameters U and Γ_∞. For MMAA demethylation using the parameters of case 3, for which $U\Gamma_\infty$ is 216 μM, the apparent rate constant is 0.04 day^{-1}. If $U\Gamma_\infty$ is increased to 500 the apparent rate constant drops to 0.03 day^{-1}. Dilution of sediment decreases U and, thus,

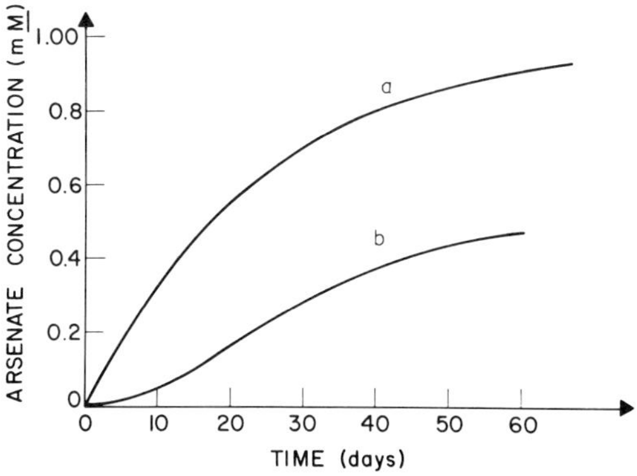

Figure 5. Dissolved arsenate from demethylation of cacodylic acid. k = *0.04 day*$^{-1}$*,* Γ_∞ = *7.7 μmol g*$^{-1}$*,* K = *32 μmol L*$^{-1}$*. (a) U = 0; (b) U = 60 g L*$^{-1}$*.*

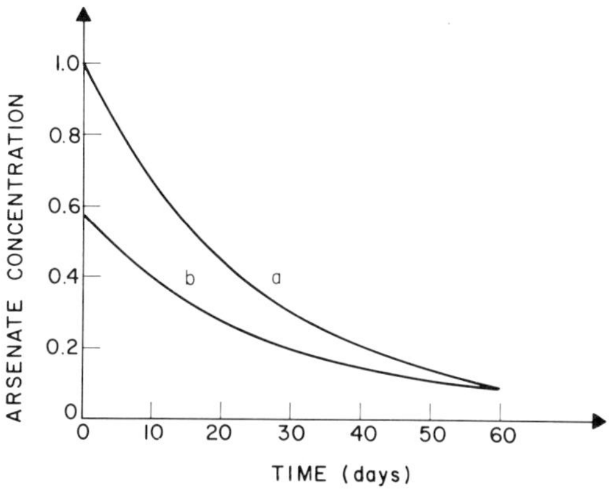

Figure 6. Dissolved arsenate remaining after methylation to cacodylic acid. (a) U = 0; (b) U = 60 g L^{-1}*.*

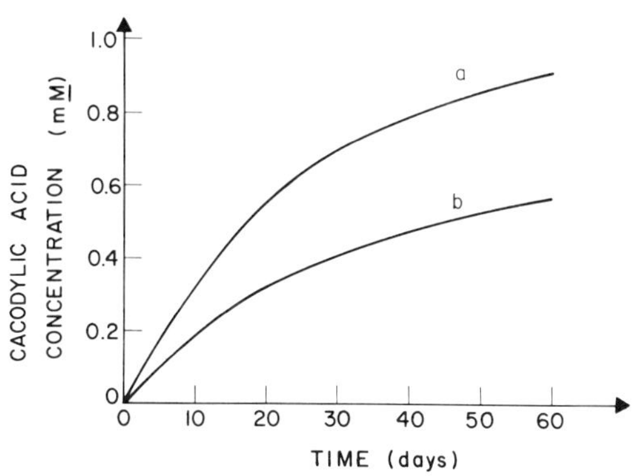

Figure 7. Dissolved cacodylic acid concentration from methylation of arsenate. (a) U = 0; (b) U = 60 g L^{-1}*.*

decreases $U\Gamma_\infty$. Oxidation of iron from FeS_2 or $Fe(OH)_2$ to amorphous $Fe(OH)_3$ may increase Γ_∞ and, thus, increase $U\Gamma_\infty$.

Since the apparent rate constant is sensitive to $U\Gamma_\infty$ in experiments in which reactant species are adsorbed, care must be taken to hold $U\Gamma_\infty$ constant. Sediment is commonly diluted to facilitate handling in incubation experiments (A. Zehnder, Univ. Wisconsin, privation communication, 1978). Such dilution could result in an increase in the apparent reaction rate as shown by the above example. Incubation experiments are also done using flooded soils (11) and if reactants in such experiments adsorb, $U\Gamma_\infty$ must be held constant. Thus, comparison of experiments done with different sediments or soils or with varying ratios of water to solids should be interpreted with care.

Rates of Arsenic Species Transformations in Incubated Sediments

Rates of changes in arsenic speciation were studied by spiking anaerobic sediments with arsenic standards, incubating the sediments, and monitoring arsenic speciation over a period of two months. The incubation experiments were modeled using the methods described in the previous section.

Experimental. Anaerobic sediments were obtained from an uncontaminated area of the Menominee River (site 55c, Figure 1). (Note: Sediments from a different site were used in the adsorption experiments.) Aliquots of 6 ml of sediment were transferred to 15 ml serum vials. The vials were stoppered with rubber septa. Sediment transfer operations were done under nitrogen to prevent oxidation of the sediments. Control vials were sterilized by autoclaving. Arsenic standards were added to the vials by injecting 4×10^{-5} mole of standard through the septa. The sediments were incubated at $9^{o}C$ for up to 65 days.

Results and Discussion. Arsenic speciation in the sediments spiked with MMAA is shown plotted as a function of time in Figure 9. The results of the DMAA experiment are shown in Figure 10. Error limits were estimated from scatter in FAA measurements. The lines connecting the experimental data points were drawn to make trends in concentrations easier to see and were not meant to imply linear changes in concentration with time between samples.

The concentrations of the spiked methylated arsenic species decreased with time while the concentrations of inorganic arsenic increased with time. There were no changes in arsenic speciation in the sterilized control sediments. Thus, the observed demethylation of both MMAA and DMAA was microbially mediated.

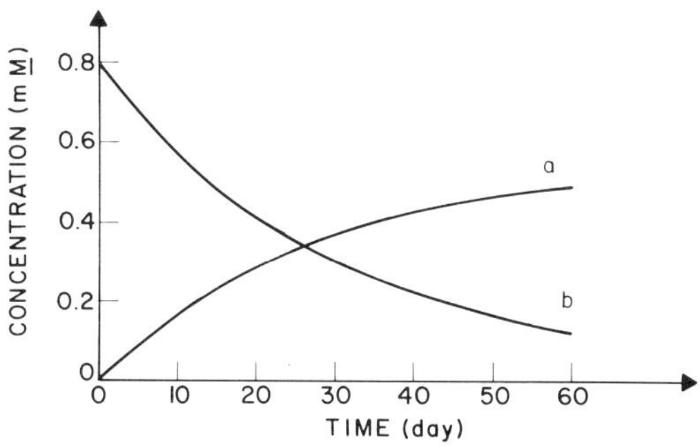

Figure 8. Dissolved arsenate and monomethyl arsonic acid (MMAA) concentrations from demethylation of MMAA to arsenate. k = 0.04 day^{-1}, Γ_{∞,AsO_4} = 7.7 mol g^{-1}, K_{AsO_4} = 32 mol L^{-1}, $\Gamma_{\infty,MMAA}$ = 28 mol L^{-1}. *(a) arsenate; (b) MMAA.*

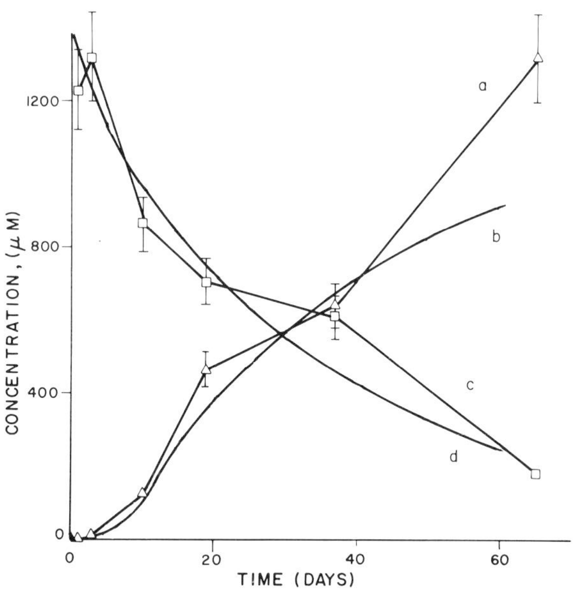

Figure 9. Dissolved arsenate and monomethyl arsonic acid (MMAA) concentrations in pore waters of sediments spiked with MMAA and predictions of sorption/kinetic model. Model parameters: k = 0.07 day^{-1}, $\Gamma_{\infty,MMAA}$ = 4.5 μmol g^{-1}, K_{MMAA} = 5.0 μmol L^{-1}, Γ_{∞,AsO_4} = 3.5 μmol g^{-1}, K_{AsO_4} = 5.0 μmol L^{-1}. *(a) Inorganic arsenic, measured; (b) arsenate, predicted; (c) MMAA, measured; (d) MMAA, predicted.*

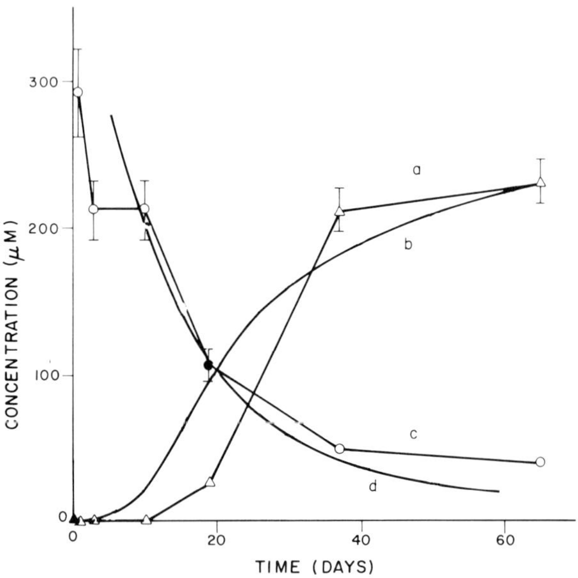

Figure 10. Dissolved arsenate and cacodylic acid (DMAA) concentrations in pore waters of sediments spiked with DMAA and predictions of sorption/kinetic model. Model parameters: k = *0.03 day*$^{-1}$, Γ_{∞,AsO_4} = *3.5 μmol g*$^{-1}$, K_{AsO_4} = *5.0 μmol L*$^{-1}$, $\Gamma_{\infty,DMAA}$ = *3.6 μmol g*$^{-1}$, K_{DMAA} = *320.0 μmol L*$^{-1}$. *(a) Arsenate, measured; (b) arsenate, predicted; (c) DMAA, measured; (d) DMAA, predicted.*

The initial concentration of MMAA in solution was less than the initial concentration of DMAA because MMAA is adsorbed more strongly than DMAA. The inorganic arsenic *vs* time curve was s-shaped, implying that the inorganic arsenic was adsorbed.

The experimental results were modeled using the adsorption/kinetic model and the parameters listed in Table VIII.

Table VIII
Parameters Used to Model Sediment Incubation Experiments

Parameter	MMAA	DMAA
k (day^{-1})	0.07	0.03
U ($g\ l^{-1}$)	60.0[c]	60.0
Γ_{∞}, r ($\mu mol\ g^{-1}$)[a]	4.5	3.6
K_r ($\mu mol\ l^{-1}$)	5.0	320.0
Γ_{∞}, p ($\mu mol\ g^{-1}$)[b]	3.5	3.5
K_p ($\mu mol\ l^{-1}$)	5.0	5.0

[a]Subscript r refers to adsorption of reactant.

[b]Subscript p refers to adsorption of product.

[c]Sediment is 95% water.

As can be seen in Figures 9 and 10, there was good agreement between the model predictions and the experimental data. The same arsenate adsorption constants were used for both the MMAA and DMAA models.

The adsorption parameters Γ_{∞} and K used for modeling the incubation experiments were different from but similar in size to the parameters listed in the adsorption section. However, the sediments used in the adsorption experiments were taken from a different site than the sediments used in the incubations and, therefore, may have had different surface properties. Thus, first order reactions with reactant and product concentrations controlled by adsorption describe the rates of demethylation of MMAA and DMAA in anaerobic sediments.

Summary

Arsenic speciation in anaerobic sediments is controlled by both microbially mediated transformations of species and by abiotic chemical processes including adsorption. The two classes of reactions probably interact and this interaction is important in understanding the environmental chemistry of arsenic. Adsorption can strongly affect the apparent rates of species transformations, so that if only dissoved species concentrations are monitored, true rates of conversion may not be measured.

Acknowledgements

The authors thank Ms. Susan Barta for assistance with field and laboratory work and Professor David E. Armstrong for critically reading the manuscript.

This work was supported by grants from the Environmental Protection Agency.

Abstract

Sorption of monomethyl arsonic acid (MMAA), dimethyl arsinic acid (DMAA), and arsenate on anaerobic bottom sediments from the Menominee River, Wisconsin are described by Langmuir isotherms. These results were incorporated into a kinetic model of arsenic species transforamtion which takes sorption into account. Model predictions were found to be sensitive to the sediment water content and Γ_{∞}, the adsorptive capacity of the sediment. Demethylation of MMAA and DMAA was observed in sediment incubation experiments. The predictions of the sorption/kinetic model were in good agreement with the results of the incubation experiments.

Literature Cited

1. Ferguson, J.F. and Gavis, J. A review of the arsenic cycle in natural waters. Water Res. 6, 1259-1274 (1973).
2. Woolson, E.A., "Generation of dimethyl arsine from soil," 91 p., Abstracts of 16th Meeting, Weed Science Society of America, 1976.
3. McBride, B.C. and Wolfe, R.S., Biosynthesis of dimethylarsine by methanobacterium, Biochem. 10, 4312-4317 (1971).
4. Wauchope, R.D. Fixation of arsenical herbicides, phosphates, and arsenate in alluvial soils, J. Environ. Qual. 4, 355-358 (1975).
5. Challenger, D.P., Higginbottom, C., and Ellis, L. Formation of organometalloidal compounds by microorganisms, J. Chem. Soc., 95-101 (1933).
6. Braman, R.S. Arsenic in the environment, p. 108-123, in Woolson, E.A. (ed.), "Arsenical Pesticides," Amer. Chem. Soc. Symp. Ser., 1975.
7. Peoples, S.A. Review of arsenical pestidices, p. 1-12, in Woolson, E.A. (ed.), "Arsenical Pesticides," Amer. Chem. Soc. Symp. Ser., 1975.

8. Crecelius, E.A. "The Geochemistry of Arsenic and Antimony in Puget Sound and Lake Washington, Washington," Ph.D. thesis, Univ. of Washington, Seattle, 1974.
9. Hiltbold, A.E. Behavior of organoarsenicals in plants and soils, p. 53-69, in Woolson, E.A. (ed.), "Arsenical Pesticides," Amer. Chem. Soc. Symp. Ser., 1975.
10. Wagemann, R. Some theoretical aspects of stability and solubility of inorganic arsenic in the freshwater environment, Water Res. 12, 139-145 (1978).
11. Woolson, E.A. and Kearney, P.C. Persistence and reactions of ^{14}C-cacodylic acid in soils, Environ. Sci. Tech. 7, 47-50 (1973).
12. Schuth, C.K., Isensee, A.R., Woolson, E.A., and Kearney, P.C. Distribution of ^{14}C arsenic derived from [^{14}C] cacodylic acid in an aquatic ecosystem, J. Agr. Food Chem. 22, 999-1003 (1974).
13. Woolson, E.A., Isensee, A.R., and Kearney, P.C. Distribution and isolation of radioactivity from ^{74}As-arsenate and ^{14}C-methanearsonic acid in an aquatic model ecosystem. Pestic. Biochem. Physiol. 6, 261-269 (1976).
14. Cox, D.P. and Alexander, M. Production of trimethylarsine gas from various arsenic compounds by three sewage fungi, Bull. Environ. Cont. Toxic. 9, 84-88 (1973).
15. Woolson, E.A., Axley, J.H., and Kearney, P.C. The chemistry and phytotoxicity of arsenic in soils: I. Contaminated field soils, Soil Sci. Soc. Am. Proc. 35, 938-943 (1971).
16. Anderson, M.A., Holm, T.R., Iverson, D.G., and Stanforth, R.S., "Mass balance and speciation of arsenic in the Menominee River, Wisconsin," Project report 6, Water Chemistry Program, Univ. Wisconsin, Madison, 1978.
17. Overby, L.R., Bocchieri, S.F., and Frederickson, R.L. Chromatographic, electrophoretic, and ion exchange identification of radioactive organic and inorganic arsenicals, J. Assoc. Off. Anal. Chem. 48, 17-22 (1965).
18. Sachs, R.M., Michael, J.L., Anastasia, F.B., and Wells, W.A. Determination of arsenical herbicide residues in plant tissues, Weed Sci. 19, 412-416 (1971).
19. Brinkman, F.E., Blair, W.R., Jewett, K.L., and Iverson, W.P. Application of a liquid chromatograph coupled with a flameless atomic absorption detector for speciation of trace organometallic compounds, J. Chrom. Sci. 15, 493-503 (1977).
20. Talmi, Y. and Bostick, T.D. Determination of alkylarsenic acids in pesticides and environmental samples by gas chromatography with a microwave emission spectrometric detection system, Anal. Chem. 47, 2145-21540 (1975).
21. Braman, R., Johnson, D., Foreback, C., Ammons, J., and Bricker, J. Separation and determination of nanogram amounts of inorganic arsenic and methylarsenic compounds, Anal. Chem. 49, 621-625 (1977).

22. Thompson, A.J. and Thoresby, P.A. Determination of arsenic in soil and plant materials by atomic absorption spectrophotometry with electrothermal atomization, Analyst 102, 9-16 (1977).
23. Talmi, Y. and Norvell, U.E. Determination of arsenic and antimony in environmental samples using gas chromatography with a microwave emission spectrometric system, Anal. Chem. 47, 1510-1516 (1975).
24. Terashima, S. Determination of arsenic in rocks, sediments and minerals by arsine generation and atomic absorption spectrometry, Anal Chim. Acta 86, 43-51 (1976).
25. Yamamoto, M. Determination of arsenate, methanearsonate, and dimethylarsinate in water and sediment extracts, Soil Sci. Soc. Am. Proc. 39, 859-861 (1975).
26. Ediger, R.D. Atomic absorption analysis with the graphite furnace using matrix modification, At. Abs. Newsletter 14, 127-139 (1975).
27. Jacobs, L.W., Syers, J.K., and Keeney, D.R. Arsenic sorption by soils, Soil Sci. Soc. Am. Proc. 34, 750-754 (1970).
28. Jackson, M. "Soil chemical analysis-advanced course," 89 p., 5th printing, Department of Soil Science, Univ. Wisconsin, Madison, 1969.
29. Saunders, W.M.H. Phosphate retention by New Zealand soils and its relationship to free seaquioxides, N.Z.J. Agr. Res. 8, 30-57 (1965).
30. Stumm, W. and Morgan, J.J. "Aquatic Chemistry," 583 p., Wiley-Interscience, New York, N.Y., 1970.
31. Ferguson, J.F. and Anderson, M.A. Chemical forms of arsenic in water supplies and their removal, p. 137-157, in Rubin, A.J. (ed.), "Chemistry of Water Supply, Treatment, and Distribution," Ann Arbor Science Publishers, Inc., Ann Arbor, Michigan, 1974.
32. Anderson, M.A., Ferguson, J.F., and Gavis, J. Arsenate adsorption on amorphous aluminum hydroxide. J. Coll. Interf. Sci. 54, 391-399 (1976).
33. Kafkafi, U., Posner, A.M., and Quirk, J.P. Desorption of phosphate from kaolinite. Soil Sci. Soc. Amer. Proc. 31, 348-353 (1967).

RECEIVED November 16, 1978.

32

Distribution, Transport, Adsorption, and Precipitation of Inorganic Phosphorus in the Genesee River

M. M. REDDY

State of New York, Department of Health, Albany, NY 12201

Understanding the fate of phosphorus in a watershed is critical for the development of strategies to cope effectively with eutrophication (1). Sediment-bound phosphorus is contributed by sewage, eroded soil, plant meterial, and urban runoff. These solids are transported as suspended sediment and as bedload. Transport of phosphorus associated with larger particle size sediment occurs only during periods of high river discharge. Soluble phosphorus entering a watershed as runoff or wastewater inputs, on the other hand, is apparently sorbed on suspended particles and/or reacts with other water column constitutents to form insoluble substances which are transported by water and eventually deposited in greater abundance in fine grained sediment. In lakes soluble phosphorus inputs are typically converted to biomass, recycled through the water column, and eventually deposited in bottom sediments. Establishment of realistic and successful strategies for controlling phosphorus inputs to the North American Great Lakes requires that periods of intense hydrological activity in a basin be carefully studied and characterized. In this paper, we shall examine several model chemical reactions in hard waters which may influence phosphorus concentrations in the water column and sediment. Chemical reactions influencing phosphate distribution in the Genesee River were identified using three separate techniques: 1) sediment samples collected during synoptic surveys were analyzed using a variety of selective phosphorus extraction procedures to characterize sediment-bound phosphorus; 2) water column chemical concentrations from samples obtained concurrently with the sediment samples were used to calculate equilibria for the determination of ion activity products of several mineral phases which may remove phosphorus from the water column; and 3) a seeded crystallization procedure was used to monitor the distribution of inorganic phosphate between solution and solid phases during calcite precipitation in simulated natural water. Genesee River sampling sites are listed in Table I and are shown in Figure 1.

0-8412-0479-9/79/47-093-737$05.75/0
© 1979 American Chemical Society

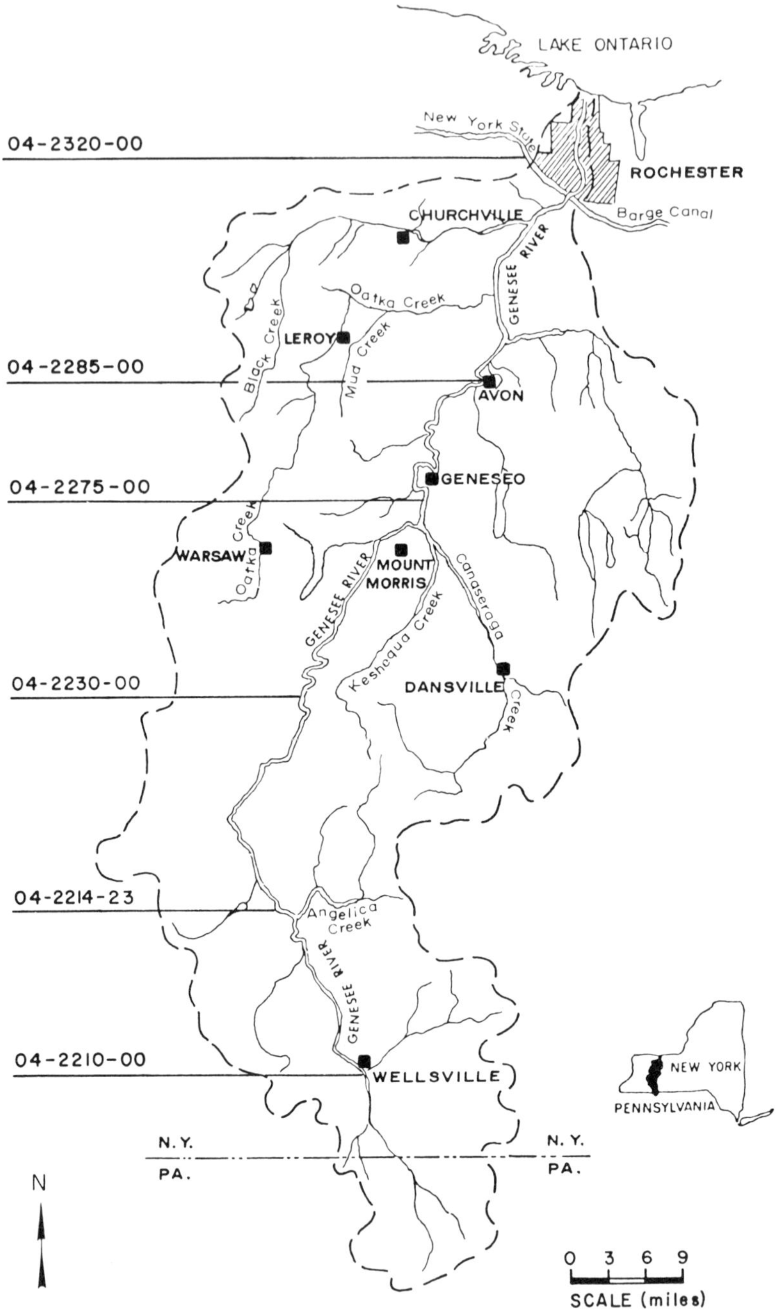

Figure 1. Genesee River sampling sites

Table I. Sampling Sites on the Genesee River, N.Y.

Site location	Miles from mouth of Genesee	U.S.G.S. station number	Latitude	Longitude
Wellsville	137	04-2210-00	42°07'20"	77°57'27"
Transit Br.	117	04-2214-23	42°19'46"	78°04'36"
Portageville	85	04-2230-00	42°34'13"	78°02'33"
Mt. Morris	62	04-2275-00	42°44'00"	77°50'21"
Avon	35	04-2285-00	42°55'04"	77°45'27"
Rochester	5	04-2320-00	43°10'50"	77°37'40"

Published reports discussing the influence of calcium carbonate solid surfaces on the phosphate ion concentration in natural water (2, 3) include: 1)solubility product-based calculations (4, 5); 2) spontaneous precipitation experiments from highly supersaturated solutions (6, 7, 8, 9); 3) adsorption of phosphate onto calcium carbonate (10, 11, 12, 13). These experiments have provided useful information, but several have not included activity coefficient or ion pair corrections in their analyses, seriously limiting the applicability of the results. In other studies, solutions of high supersaturation were employed which yielded complex, poorly defined solid phases. The multiplicity of complex crystal polymorphs and hydrates in the calcium phosphate system is now well recognized and must be considered in the interpretation of experiments involving the removal of phosphate from natural waters (14, 15).

Experimental Procedures

Sample Collection, Pretreatment, and Analysis. Sediment-bound phosphorus in the Genesee River was studied by sampling bottom sediment, fine-material washed from bottom sediment, suspended sediment, and water column particulate material at six stations on the river. The sampling program was planned to be synoptic with complete chemical and hydrological parameters recorded at each site. One kilogram surficial sediment samples were collected in midstream at most sites during six field trips from June 1975 to March 1977. Bottom sediment samples were obtained with nonmetallic equipment and most were wet-sieved immediately through a 2-mm polyethlyene sieve using river water. Sediment samples were frozen on site and stored frozen until analysis. Sample collection and pretreatment procedures (16, 17) have been adopted from techniques employed by the U. S. Geological Survey (B. Malo, U. S. Geological Survey, personal communication, 1975). Suspended sediment samples were obtained by continuous high-speed centrifugation of large volumes of river water, or by collecting the suspended sediment obtained in the field wet-sieving procedure. Particulate material is defined as that which is retained by filtering a river water sample through a 0.45-um Millipore filter. Typically, two liters of water were filtered for particulate analysis. Suspended sediment measurements made at the same time, by the U. S. Geological Survey, were used to convert the particulate concentrations to a dry weight basis.

Frozen bottom sediment samples were thawed for analysis, dried at 110°C, crushed and sieved through a 100-mesh nylon sieve and digested with HNO_3-H_2O_2 at 100°C for two hours (18), and/or with acid-alkaline persulfate (19). Since silicate minerals are not solubilized in these procedures the results are considered only as an estimate of the total available phosphorus. Suspended sediments were freeze-dried in the laboratory before analysis. Particulate material was digested directly on the filter used for collection, employing the digestion technique described above.

Fractional analysis procedures were chosen for high reproducibility and precision. A phosphorus extraction procedure employing NaOH (with NaCl) to determine occluded phosphorus was adopted from the work of Williams et al. (20, 21). The sequence in this procedure was first extraction with NaOH and then extraction with dilute hydrochloric acid, the residue being discarded. Work performed in this laboratory (M. M. Reddy and E. Shpirt, New York State Dept. of Health, unpublished data, 1976) and elsewhere (22) indicates that the NaOH extractable phosphorus is a measure of bioavailable sediment phosphorus. Hydrochloric acid extractable phosphorus, following NaOH extraction, measures apatite plus phosphorus incorporated in iron oxides. The NaOH extract obtained in this study was not digested and reflects only inorganic forms of phosphorus. Additional fractionation methods were used to identify elemental association between phosphorus and hydrous metal oxides in sediments. These extraction procedures are largely empirical. The procedure employed here consisted of one initial extraction by hydroxylamine hydrochloride-nitric acid reagent (23) followed by extraction with ammonium oxalate-oxalic acid solution (24), the residue being discarded. The hydroxyl amine extraction procedure was used to identify the manganese oxide-phosphate relationship, while ammonium oxalate-oxalic acid removes amorphous iron oxides and phosphate associated with these oxides. A detailed description of the procedures used in this study for water column analysis has been presented elsewhere (25,

26).

During each analysis, replicates of samples collected in the Genesee River were analyzed as quality-control check samples. The coefficients of variation for total phosphorus analysis of three sediments used as quality-control check samples were 0.12 (n = 12), 0.16 (n = 11), and 0.13 (n = 11), with means of 412, 487, and 758 ug P/g, respectively.

Seeded Crystallization Experiments. A detailed description of the seeded growth technique has been published recently (27). The following summarizes the experimental procedure employed. Reagent grade chemicals; distilled, deionized, filtered (0.22-um Millipore filter) water; and grade A glassware were used in all experiments. Supersaturated calcium carbonate solutions were prepared by drop-wise addition of 200 ml of 5 x 10^{-4} M calcium chloride solution to 200 ml of 8 x 10^{-3} M sodium bicarbonate solution in a thermostated double-walled Pyrex glass reaction vessel. Stability of the supersaturated solution was verified by the constancy of pH for at least one hour before the start of each experiment. Solution pH changes accompanying calcite growth after inoculation of the stable supersaturated solution with seed crystal were monitored with a Corning pH meter and a strip chart recorder. Calcium concentration in solution during crystallization was followed by analysis of solution filtrates. An EDTA titration procedure employing calcein indicator (28) with a micrometer burette was used to determine calcium concentration in the filtered solution. Total carbonate concentration was calculated from a titrimetric analysis of the filtrate using 0.01 N sulfuric acid and methyl Purple indicator (pH range 4.8 - 5.4). A Quantachrome Monosorb surface area analyzer was used to measure seed crystal surface area; a Phillips powder diffraction apparatus with copper Ka radiation and a nickel filter was employed for X-ray powder diffraction verification of seed crystal composition. Calcite seed crystals were prepared by rapidly adding 0.5 M $CaCl_2$ solution to 0.5 M Na_2CO_3 solution at 5°C. The viscous suspension formed was gradually warmed to 25°C, stirred overnight, then washed with distilled water. Seed was aged in distilled, deionized water 6 months before use. The seed consisted of uniform aggregates of flat crystals shown to be calcite by X-ray diffraction, with a surface area of 1.71 m^2/g. Seeded crystal growth experiments were performed in solutions resembling natural waters. Ionic species concentrations were calculated from measured solution pH and from total calcium and carbonate concentrations, using the mass action and mass balance equations in Table II. Calculations were performed using successive approximations for ionic strength, I, (34) with a Wang 720 C programmable calculator. Ion activity coefficients were obtained from the modified Debye - Hückel equation proposed by Davies (36). Bicarbonate ion was the predominant carbonate species in the experimental solutions, accounting for more than 95% of the total carbonate

Table II. Equations Used for Calculation of Ionic Species at 25°C and 1 Atm Pressure

Equation*	Reference for constant
Mass action	
1. $\frac{[H^+][HCO_3^-]}{[H_2CO_3]} = 10^{-6.35}$	(29)
2. $\frac{[H^+][CO_3^{2-}]}{[HCO_3^-]} = 10^{-10.33}$	(30)
3. $[H^+][OH^-] = 10^{-14.00}$	(31)
4. $\frac{[CaCO_3]}{[Ca^{2+}][CO_3^{2-}]} = 10^{3.2}$	(32)
5. $\frac{[CaHCO_3^+]}{[Ca^{2+}][HCO_3^-]} = 10^{1.3}$	(33)
6. $\frac{[CaOH^+]}{[Ca^{2+}][OH^-]} = 10^{1.4}$	(34)
7. $([Ca^{2+}][CO_3^{2-}])_{calcite} = 10^{-8.4}$	(35)
Mass balance†	
8. $T_{Ca} = \frac{[Ca^{2+}]}{f_2} + \frac{[CaOH^+]}{f_1} + \frac{[CaHCO_3^+]}{f_1} + \frac{[CaCO_3^0]}{f_0}$	
9. $T_{CO_3} = \frac{[CO_3^{2-}]}{f_2} \quad \frac{[HCO_3^{-1}]}{f_1} \quad \frac{[H_2CO_3]}{f_0} \quad \frac{[CaHCO_3^+]}{f_1} + \frac{[CaCO_3^0]}{f_0}$	

* Bracketed terms denote thermodynamic activities in solution.

† The term f_z denotes the activity coefficient for ion of charge z.

concentration. Calcium carbonate and bicarbonate ion-pair concentrations were considered in solubility calculations. The influence of phosphate ion-pair formation on the crystallization reaction was examined and found to be negligible (27) at the phosphate concentration levels employed.

Results and Discussions

Phosphorus Distribution. The major sediment phosphorus fraction is that extracted by hydrochloric acid (Table III). Ammonium oxalate-oxalic acid solution extracts somewhat less phosphorus than hydrochloric acid, while sodium hydroxide as well as hydroxylamine extract much less.

Phosphorus extracted from sediment by NaOH has been related to non-occluded, surface-exchangeable, bioavailable forms (22). Hydrochloric acid extraction yields occluded phosphorus incorporated in hydrous metal oxides, carbonate and phosphate minerals of sediment. Hydroxylamine reagent specifically removes hydrous manganese oxides, while amorphous hydrous oxides of iron and aluminum are removed by the oxalate reagent. Total available sediment phosphorus analyses includes sediment organic phosphorus components in addition to the inorganic portion determined by the selective extraction procedures.

The variation in total available sediment phosphorus concentration among the three sediment types shown in Table II is clear. A statistical analysis of this data shows that both the suspended sediment and particulate total phosphorus concentrations are greater than the bottom sediment value at the 99% confidence level. Phosphorus content increases in the sequence; bottom sediment, suspended sediment, and particulate material in accordance with the increase in surface area (M. M. Reddy, New York State Dept. of Health, unpublished data, 1977). High surface area sediment components may adsorb phosphorus-containing substances, from the water column, increasing their phosphorus concentration. Another possible explanation includes dilution of bottom sediment by relatively inert primary minerals in the sand and silt size fractions.

Phosphorus Transport. Sediments are recognized as a major transport medium for phosphorus to the North American Great Lakes. Phosphorus transport in watersheds such as the Genesee occurs in large part during rainfall and snow-melt discharge events. The transport of elements from a watershed can be expressed as instantaneous unit load. This quantity is defined as the amount of material carried by a river at a given point divided by the area drained by the river above that point. For the synoptic studies described here, the instantaneous unit loads are expressed as grams of phosphorus per second per acre (Figure 2). The major component of the phosphorus load transported by the Genesee River during the two sampling periods discussed here

Table III. Statistics for Phosphorus Analyses for Several Sediment Types Collected in the Genesee River Watershed, New York, (μg/g)

Sample	$\bar{x}$	Range	σ	CV	N
Bottom Sediment					
Total Available	560	330 - 980	140	0.25	99
NaOH Extractable	58	5 - 410	62	1.07	98
HCl Extractable	398	177 - 731	99	0.25	98
NH_2OH Extractable	74	6 - 313	63	0.86	98
$(NH_4)_2C_2O_4$ Extractable	185	49 - 453	93	0.50	83
Suspended Sediment[1]					
Total Available	770	390 - 2020	360	0.46	46
NaOH Extractable	163	19 - 1000	232	1.43	17
HCl Extractable	528	258 - 664	109	0.21	17
NH_2OH Extractable	70	3 - 385	102	1.46	17
$(NH_4)_2C_2O_4$ Extractable	474	119 - 1110	222	0.47	17
Particulate Analysis[2]					
Total Analysis	910	400 - 3000	640	0.70	61

[1]Suspended sediment samples were obtained by continuous high-speed centrifugation of large volumes of river water, or by collecting the suspended sediment obtained in the field wet-sieving procedure.

[2]Particulate material is defined as that which is retained by filtering a river water sample through a 0.45-μm Millipore filter.

(December 14, 1975 and March 13, 1976) is that associated with the suspended sediment. Dissolved phosphorus in the water column during these periods was typically less than half of the total water column concentration.

Unit loads varies widely (Figure 2). A flood control impoundment, located just upstream of the Mount Morris sampling station, markedly influences the Genesee River discharge and suspended sediment concentrations at the downstream stations. Three sampling stations upstream of the Mount Morris impoundment (Wellsville, Transit Br., and Portageville) show a smooth and systematic increase in the phosphorus unit load with increasing discharge. In contrast, unit loads at the mid-basin sites show much larger abolute values and fluctuations than the other stations. These stations exhibit large variations in discharge and therefore in suspended sediment concentrations.

Mineral Saturation in the Genesee River. The importance of heterogeneous equilibria in regulating dissolved inorganic phosphorus concentrations in the Genesee River was examined by calculating the ion activity products of several mineral phases using the WATEQF (37) chemical model and/or mass action and mass balance equations with a small laboratory computer. During high discharge periods in December and March there is extensive undersaturation in the water column with respect to calcium carbonate and phosphate phases, while during August 1976, a relatively lower-flow period, the first four downstream sampling stations in the Genesee showed saturation or supersaturation with respect to calcite (Figure 3). Thus, during the summer, it appears that in the lower reach of the Genesee River calcite precipitation limits the concentration of dissolved calcium. Saturation levels for the thermodynamically stable calcium phosphate mineral, hydroxyapatite, are 10^{10} below the equilibrium values for the Genesee River Stations during the high flow sampling periods of December and March. Precipitation of hydroxyapatite or other calcium phosphate phases does not occur so these solid phases do not regulate phosphate concentration in the subsaturated river water.

Fertilizer applied to calcareous soils produces minerals such as hydroxyapatite (13). When such soils are eroded, and subsequently carried to the Genesee River, this mineral phase will be in a markedly subsaturated solution and will tend to dissolve, releasing inorganic phosphate to the water column. Two major sub-basins in the Genesee River watershed have different calcite saturation levels. Oataka Creek sub-basin, a predominately agricultural region, had calcite saturation values much closer to equilibrium than those found for the Genesee River on the sampling dates. Canasaraga Creek sub-basin, an area less intensively cultivated than Oataka Creek sub-basin, exhibited saturation or supersaturation in August but was subsaturated in December 1975 and March 1976. From these data, it can be hypothesized that there are higher sediment phosphorus contents in

a

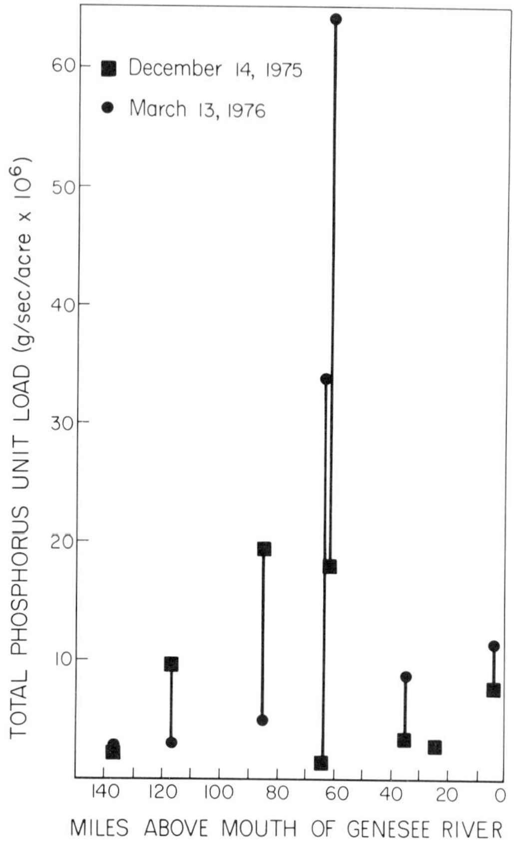

Figure 2. (a, above*) Total phosphorus unit load (g P/sec/acre), (B,* top right*) percent unit load as dissolved phosphorus, and (c,* bottom right*) discharge (cfs) plotted as a function of sampling point distance from Lake Ontario. Samples represent dates preceeding and following the snow melt runoff event in 1976.*

b

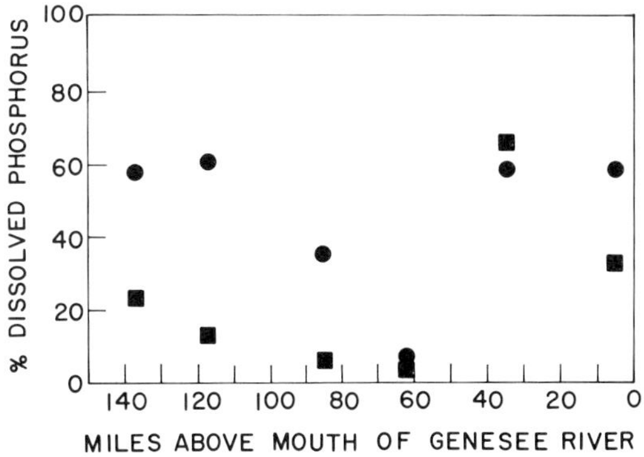
% DISSOLVED PHOSPHORUS
100
80
60
40
20
0
140 120 100 80 60 40 20 0
MILES ABOVE MOUTH OF GENESEE RIVER

c

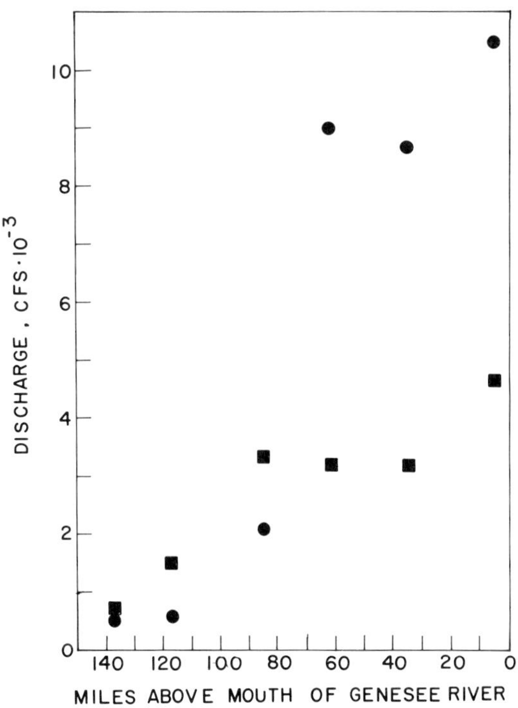
DISCHARGE, CFS·10⁻³
10
8
6
4
2
0
140 120 100 80 60 40 20 0
MILES ABOVE MOUTH OF GENESEE RIVER

the agricultural sub-basin because of carbonate mediated phosphorus mineralization. In support of this hypotheses, Oataka basin sediments are the Genesee watershed sediments with the greatest hydrochloric acid extractable phosphorus content. Release of phosphorus from agriculturally derived soils and sediments during high discharge periods may be counteracted by runoff dilution. Data reported by the U. S. Geological Survey for the St. Lawrence and Lake Ontario basins, for the period considered here, show phosphorus concentration in rainwater to be 0.010 mg P/1 (38). This concentration is much less than the soil water phosphorus concentration. Base flow Genesee River dissolved inorganic phosphorus concentrations are 0.004 mg P/1, while peak flow values are approximately 0.011 mg P/1. Since recent evidence (39) indicates that less than 25% of rainwater phosphorus is dissolved inorganic phosphorus, these results support the suggestion that some form of solid dissolution is involved in the regulation of the water column phosphorus concentration.

Dissolved metals other than calcium have a minor effect on the distribution of phosphorus between the water column and sediment in this fluvial system. The two principal metals of potential interest, iron and aluminum, are present in Genesee River water almost entirely in the particulate phase (40). Dissolved concentrations of these metals are below the detection limit (less than 50 ug/1). Iron and aluminum minimum detectable dissolved concentrations were used to estimate the saturation levels of the corresponding phosphate minerals. These calculations suggest that both iron and aluminum phosphate minerals are substantially below saturation levels. The solid surfaces exhibited by iron and aluminum hydrous oxides (as particulate material in the water column) undoubtedly serve as sites for phosphorus adsorption and incorporation in the fluvial system. Data presented for the oxalate extraction procedure in Table III demonstrate the importance of phosphorus binding by hydrous metal oxides.

Nriagu (Canadian Center for Inland Waters, unpublished data, 1975) has proposed that basic metal phosphates are important sinks for heavy metals in the environment. In most natural waters of New York State, dissolved basic metals including Pb, Cu, and Zn are at low concentrations (below 10 ug/1), and these metals would not be expected to be a major factor governing phosphorus distribution.

Phosphate Distribution During Calcite Crystallization. The crystallization rate data illustrated in Figure 4 follow a rate equation

$$dN / dt = ksN^2 \quad (1)$$

where N is calcium carbonate (mol / 1) at time t to be precipitated from solution before equilibrium is attained; k is the

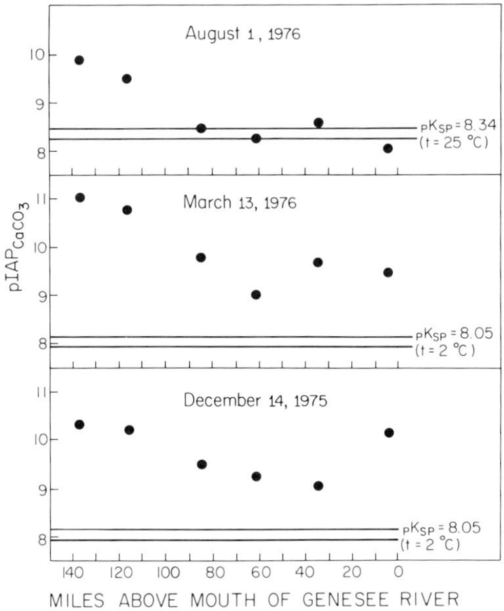

Figure 3. Ion activity product of calcite plotted as a function of sampling point distance from Lake Ontario

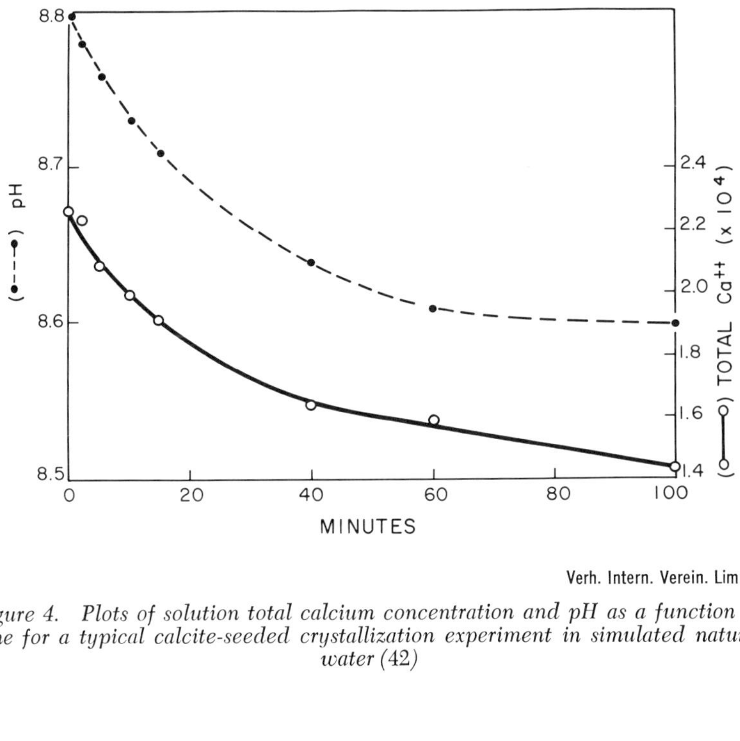

Verh. Intern. Verein. Limnol.

Figure 4. Plots of solution total calcium concentration and pH as a function of time for a typical calcite-seeded crystallization experiment in simulated natural water (42) crystal growth rate constant; and s is the seed crystal concentration (mg / 1), which is proportional to the surface area available for growth. Examination of calcite growth rate data is facilitated by presentation in the integrated form of Equation 1

$$N^{1} - N_{o}^{-1} = k\ s\ t \qquad (2)$$

where N_o is calcium carbonate to be precipitated from the supersaturated solution at zero time. The linear plot of $N^{-1} - N_o^{-1}$ as a function of time (Figure 5) confirms the validity of equation 2 for the interpretation of the experimental results. The inhibition of crystallization by phosphate is marked (Figure 6) and appears to be due to simple adsorption of the ions at growth sites on the crystal surface. This is demonstrated by plotting the rate constants, in the presence and absence of phosphate, in a form corresponding to a simple Langmuir - type adsorption isotherm (27)

$$k_o / (k_o - k) = 1 + k_2 / k_1 \cdot (P) \qquad (3)$$

where k_o is the crystal growth rate constant in the absence of phosphate ion, k_1 and k_2 are the adsorption and desorption rate constants, and P is the solution phosphate molarity. In Figure 7, $k_o / (k_o - k)$ is plotted against the reciprocal of the phosphate concentration. The linear relation with an intercept of unity indicates that the Langmuir isotherm satisfactorily describes the marked inhibitory effect of phosphate in terms of a monomolecular blocking layer of these ions at the growth sites on the crystal surface. There is no evidence for the formation of a second phase on the surface of the calcite seed even though these solutions are supersaturated with respect to the thermodynamically stable calcium phosphate phase, hydroxyapatite. This can be clearly seen in Figure 8 which shows solution phosphate concentration during a series of calcite crystallization experiments. Here, solution phosphorus concentrations changed only slightly. In experiments with low phosphate concentration (less than 10^{-5} M), and reaction times of 1 day there are significant reductions in solution phase phosphate concentration (10 to 20% removal). These phosphate changes showed little systematic behavior and may be related to the surface nucleation of solid calcium phosphate.

Conclusions Relative to Phosphate Distribution in Alkaline Natural Waters

Selective extractions, chemical equilibria calculations, and crystallization measurements presented here imply that the hydrous iron oxides, even in the carbonate dominated Genesee River, play a major part in inorganic phosphorus transport by sediments in the fluvial system. Saturation levels of inorganic phosphate and calcium carbonate minerals in the Genesee River

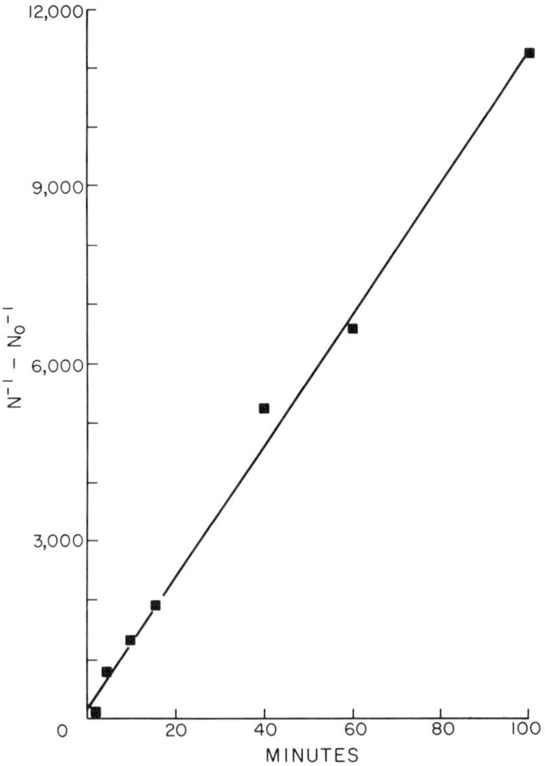

Verh. Intern. Verein. Limnol.

Figure 5. Calcite crystallization rate function N^{-1}–N_0^{-1} vs. time for data shown in Figure 4 (42)

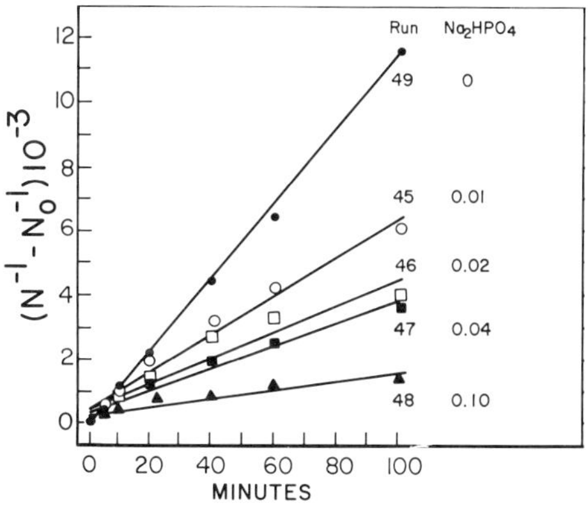

Figure 6. Calcite crystal growth in the presence and absence of phosphate ion, as expressed by the rate function N^{-1}–N_0^{-1} vs. time. The symbols and the numbers beside the curves indicate the phosphate concentration multiplied by 10^4. Adapted from Ref. 43.

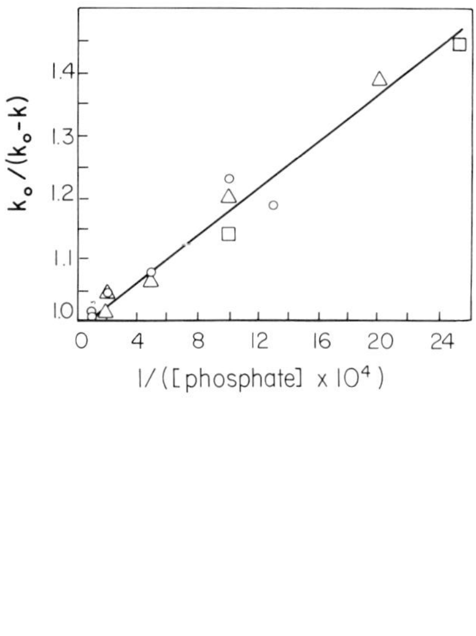

Figure 7. Langmuir isotherm plot of $k_0/(k_0 - k)$ against the reciprocal of the phosphate concentration, where k_0 is the calcite growth rate constant in the absence of phosphate and k is the rate constant in the presence of phosphate: (△) $k_0 = 0.824$; (□) $k_0 = 1.205$; (○) $k_0 = 0.790$. Adapted from Ref. 43.

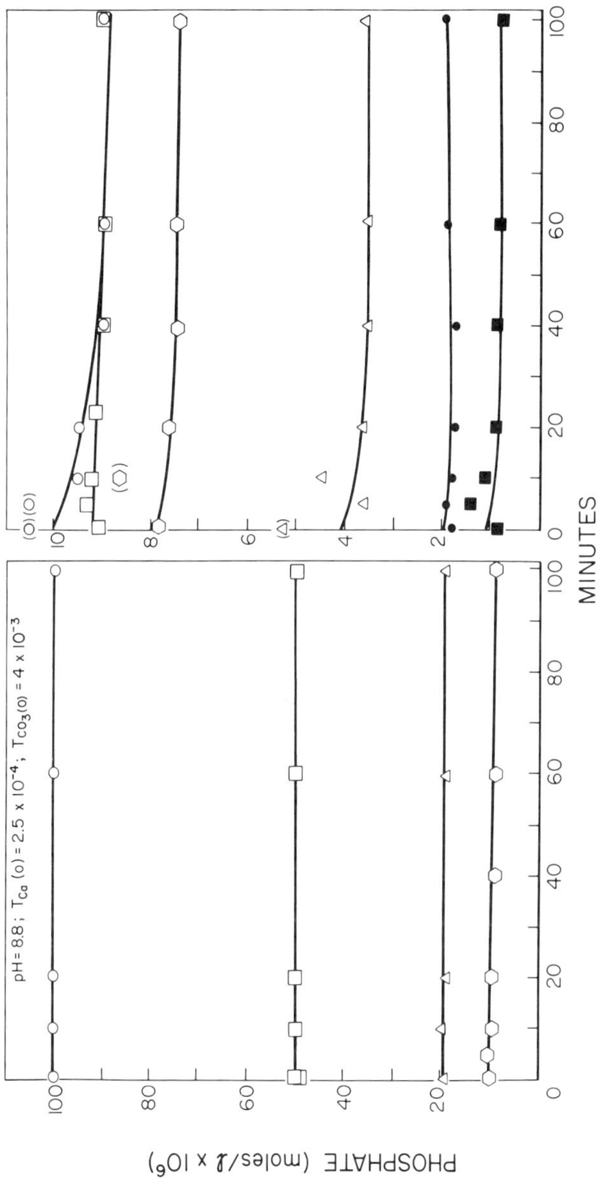

Figure 8. Phosphate concentration vs. time for several calcite seeded growth experiments in the presence of phosphate ion

indicate that phosphate mineral dissolution, and not precipitation, may be the predominant heterogeneous reaction during periods of high discharge. Elevated discharge and sediment transport occur primarily in March through May. Phosphate mineral dissolution occurring during fluvial transport then has an immediate impact on Lake Ontario. Dissolution of phosphorus-containing minerals transforms sediment phosphorus from a biologically unavailable form to a form that is readily taken up by microorganisms in the lake. Evidence of the relationship between lake water phosphorus contents and algal productivity is the positive correlation of spring-time dissolved phosphorus values with summer chlorophyll concentrations. In low flow periods, downstream sampling stations in the Genesee River show saturation or supersaturation with respect to calcite. During this interval, calcite can act as a substrate for the mineralization of phosphate, and may also regulate solution calcium ion concentration.

In the lower reaches of the Genesee River, the results of the extractions suggest that substances other than hydrous oxides are phosphorus sinks. This is evident where the amount of sediment phosphorus extracted by hydrochloric acid steadily increases down river, while the oxalate extractable phosphorus remains relatively constant. Schwertmann (24) emphasized that the results of such procedures are best considered as a measure of the relative amount of a phase or, more generally, a measure of an element's reactivity in a sediment under carefully controlled conditions. Laboratory experiments (Figure 8) show that phosphorus uptake by calcium carbonate, under simulated natural conditions, proceeds slowly. The large hydrochloric acid extractable component observed at Rochester may arise from slow uptake and subsequent mineralization of dissolved inorganic phosphorus by carbonate minerals.

From the selective extraction analysis of sediment, chemical equilibrium calculations, and seeded crystallization measurements, several conclusions can be reached concerning the behavior of phosphate in alkaline surface waters.

1) Inorganic phosphorus in bottom sediment appears to reside primarily in association with surface hydrous oxide coatings of sediments. Hydrous metal oxides, in particular those of iron, transported as suspended sediment, may scavenge phosphate from the water column in a fluvial system.

2) Inorganic phosphorus incorporated with easily reducible hydrous manganese oxides is typically less than that found for the hydrous oxides of iron and aluminum.

3) A fraction of the sediment bound phosphorus exists in a form which is not extracted with the hydrous metal oxides, but is removed by treatment with dilute hydrochloric acid. This suggests the occurrence of phosphorus in carbonate minerals and/or occurrence of phosphate minerals.

4) Phosphorus transport in the Genesee River, expressed as instantaneous unit load of total water column phosphorus, shows

moderate agreement for the stations reported here, with the exception of two midbasin sites. These locations exhibit discharge dependent bottom sediment remobilization which leads to inappropriately high unit loading.

5) During high discharge, many areas of the Genesee basin are subsaturated with respect to calcite, while during summer baseflow periods calcite saturation or supersaturation is widespread. Under these conditions calcite could mediate phosphorus mineral formation.

6) A detailed examination of phosphate distribution between solution and solid phase during calcite crystallization in a simulated natural water shows that phosphorus adsorbs as a monolayer, causing slight changes in the solution phosphorus concentration. It appears that under the conditions examined in this study, calcite-mediated phosphorus mineralization has a role in the movement of phosphorus from the water column to bed sediments, although the extent and rates of the process in natural systems remain to be determined.

Differences in river basin morphology, soil characteristics, rainfall, and land use in a watershed influence phosphorus transport in a fluvial system. However, the dominance of iron oxides as an inorganic phosphate sink and the discharge dependent behavior of calcium carbonate-phosphate minerals found in this study would be expected to exist in other calcareous agricultural regions of New York State as well. Mountainous terrain and areas of sand and muck soil would probably not exhibit the same behavior. It would seem that the results of this study could also apply to other agricultural watersheds adjacent to the North American Great Lakes.

Acknowledgment

This study was carried out as part of the Task C work of the Pollution from Land Use Activities Reference Group (PLUARG), International Joint Commission for the Great Lakes, and was funded through the United States Environmental Protection Agency and the State of New York

The author wishes to express his appreciation to G. W. Fuhs for his valuble suggestions and discussions during the course of his work; to K. V. Krishnamurty and E. Shpirt for assistance in the development of the Genesee River Watershed analysis program; and to K. Wang, C. Baggenstoss, N. Sopchak, C. Snyder, and the staff of the Laboratories for Analytical Chemistry of this Division for technical support.

Abstract

Several chemical reactions, including calcium carbonate and hydroxyapatite precipitation, have been studied to determine their relationship to observed water column and sediment phosphorus contents in hard water regions of New York State. Three separate techniques have been used to identify reactions important in the distribution of phosphorus between the water column and sediments: 1) sediment sample analysis employing a variety of selective extraction procedures; 2) chemical equilibrium calculations to determine ion activity products for mineral phases involved in phosphorus transport and; 3) seeded calcium carbonate crystallization measurements in the presence and absence of phosphate ion.

Typical results for Genesee River watershed bottom sediment phosphorus contents are: (phosphorus analysis, $\bar{x}$ (ug/g), σ (ug/g), n) total phosphorus, 560, 140, 99; NaOH extractable phosphorus, 58, 62, 98; HCl extractable phosphorus, 398, 99, 98; NH2OH extractable phosphorus, 74, 63, 98; $(NH_4)_2C_2O_4$ extractable phosphorus: 184, 93, 83. Bottom sediment, suspended sediment, and particulate total phosphorus (560 ug/g, 770 ug/g, and 910 ug/g respectively) increase as does specific surface area, across these fractions indicating that adsorption processes may be important in increasing sediment phosphorus concentrations. Sediment phosphorus extraction analyses show that hydrous iron oxides (extracted by $(NH_4)_2C_2O_4$) play a major role in the transport of sediment phosphorus. In northern areas of the Genesee River watershed calcium carbonate formation also appears to be involved in phosphorus fixation. Ion activity product calculations for water column samples from the Genesee River consistently exhibit subsaturation with respect to the stable calcium phosphate phase, hydroxyapatite. Calcium carbonate, which can serve as a substrate for phosphate mineralization, shows an ion activity product below the solubility product in the Genesee River except during the summer low-rainfall season.

Extensive seeded calcite growth experiments in the presence of phosphate ion indicate that the phosphate ion adsorbs onto the crystal surface as a monolayer. At a concentration of 10^{-5} M, phosphate ion can strongly inhibit calcite formation; however, short term experiments show that this monolayer adsorption removes insignificant amounts of phosphorus from solution. In experiments lasting several days a further decrease in solution phosphate concentration occurs, presumably caused by nucleation of a surface calcium phosphate phase on the calcite seed. Reaction of phosphate with calcite surfaces appears likely in hard water areas, but low adsorption capacity and slow kinetics of the phosphate-calcite reaction under natural conditions probably prevent calcite mediated phosphorus mineralization from becoming a greater phosphorus sink than binding to the amorphous iron oxides.

Literature Cited

1. Stumm, W. The acceleration of the hydrogeochemical cycling of phosphorus, Water Res. 7, 131-144 (1973).
2. Stumm, W. and Morgan, J. J. "Aquatic Chemistry", 583 p., Wiley Interscience Pub., New York, 1970.
3. Kamp-Nielson, L. Mud-water exchange of phosphate and other ions in undisturbed sediment cores and factors affecting the exchange rates, Arch. Hydrobiol. 73(2), 218-237 (1974).
4. Golterman, H. L. ed. "Proc. I. B. P. Symp." 297 p. Amsterdam, 1967.
5. Norvell, W. A. Insolubilization of inorganic phosphate by anoxic lake sediment, Soil Sci. Soc. Am. Proc., 38 441-445 (1974).
6. Leckie, J. and Stumm, W. p. 237-249. in: Gloyna, E. and Eckenfelder, W. eds., "Advances in Water Quality Improvement-Physical and Chemical Processes", Univ. Texas Press, Austin, Texas, 1970.
7. Bachra, B. N. Trautz, O. R. and Simon, S. L. Precipitation of calcium carbonate and phosphates, Arch. Biochem. Biophys 103, 124-138 (1963).
8. Simkiss, K., Phosphates of as crystal poisons of calcification, Biol. Rev. 39, 487-505 (1964).
9. Otsuki, A. and Wetzel, R. Coprecipitation of phosphate with carbonates in a marl lake, Limnol. Oceanogr. 17, 763-767 (1972).
10. Cole, C. V., Olsen, S. R. and Scott, C. O. The nature of phosphate sorption by calcium carbonate, Soil Sci. Soc. Am. Proc. 117, 352-256 (1953).
11. Kuo, S. and Lotse, E. G. Kinetics of phosphate adsorption by calcium carbonate and Ca-Kaolinite, Soil Sci. Soc. Am. Proc. 36, 725-729 (1972).
12. Griffin, R. A. and Jurinak, J. J. The interaction of phosphate with calcite, Soil Sci. Soc. Am. Proc. 37, 847-850 (1973).
13. Mattingly, G. E. G. Labile phosphate in soils, Soil Sci. 119(5), 369-375 (1975).
14. Tomson, M. B. and Nancollas, G. H. Mineralization kinetics: a constant composition approach. Science 200, 1059-1060 (1978).
15. Tomazic, B. and Nancollas, G. H. The seeded growth of calcium phosphates. Surface characterization and the effect of seed material, J. Colloid and Interface Sci. 50(3), 451-461 (1975).
16. Reddy, M. M. p. 244-251. in Golterman, H.L., ed., "Interactions Between Sediments and Fresh Water," The Hague, The Netherlands 1977.

17. Reddy, M. M. p. 500-506. in: "Conference Proceedings, 4th Joint Conference on Sensing of Environmental Pollutants." Amer. Chem. Soc., Washington, D. C., 1978.

18. Krishnamurty, K. V., Shpirt, E. and Reddy, M. M. Trace metal extraction of soils and sediments by nitric acid-hydrogen peroxide. Atomic Absorption Newslett. 15, 68-70 (1976).

19. Fuhs, G. W. Determination of particulate phosphorus by alkaline persulfate digestion, Inter. J. Environ. Anal. Chem. 1, 123-129 (1971).

20. Williams, J. D. H., Syers, J. K., Harris, R. F. and Armstrong D. E. Fractionation of inorganic phosphate in calcareous lake sediments, Soil Sci. Soc. Am. Proc. 35, 250-255 (1971).

21. Williams, J. D. H., Syers, J. K., Armstrong, D. E. and Harris R. F. Characterization of inorganic phosphate in noncalcareous lake sediments Soil Sci. Soc. Am. Proc. 35, 556-561 (1971).

22. Sagher, A. "Microbial Availability of Phosphorus in Lake Sediments", M. S. Thesis, Univ. Wisconsin, Madison, 1974.

23. Chao, T. T. Selective dissolution of manganese from soils and sediments with acidified hydroxylamine hydrochloride, Soil Sci. Am. Proc. 36, 764-768 (1972).

24. Schwertmann, V. Use of oxalate for Fe extraction from soils, Can. J. Soil. Sci. 53, 244-246 (1973).

25. Krishnamurty, K. V. and Reddy, M. M. "The Chemical Analysis of Water and Sediments in the Genesee River Watershed Study" on file in the library of International Joint Commission, Regional Office, Windsor, Ontario, Canada, 1975.

26. Krishnamurty, K. V. and Reddy, M. M. Tris (pyrrolidine dithiocarbamato) cobalt (III) chelate matrix for trace metal preconcentration from aqueous solution by coprecipitation, Anal. Chem. 49, 222-226 (1977).

27. Reddy, M. M. p. 31-58 in: Rubin, A., ed., "Chemistry of Wastewater Technology", Ann Arbor Science Publishers, Inc., Ann Arbor, Michigan, 1978.

28. Diehl, H. "Calcein, Calmogite and o,o' dehydroxyazobenzene titrimentric, colorimetric and fluorometric reagents for calcium and mangnesium", G. F. Smith Chemical Co. Columbus, Ohio, 1964.

29. Harned, H. S. and Davis, R. The ionization constant of carbonic acid in water and the solubility of carbon dioxide in water and aqueous salt solutions from 0 to 50°, J. Am. Chem. Soc. 65, 2030-2037 (1943).

30. Harned, H. S. and Scholes, S. R. The ionization constant of HCO_3^- from 0 to 50°, J. Am. Chem. Soc. 63, 1706-1712 (1941).

31. Harned, H. S. and Hamer, W. J. The ionization constant of water and the dissociation of water in potassium chloride

solutions from electromotive forces of cells without liquid junction, J. Am. Chem. Soc. 55, 2194-2206 (1933).

32. Garrels, R. M. and Thompson, M. E. Chemical model for sea water at 25°C and atmosphere total pressure, Am. J. Sci. 260, 57-66 (1962).

33. Greenwald, I. The dissociation of calcium and mangesium carbonates and bicarbonates, J. Biol. Chem. 141, 789-796 (1941).

34. Nancollas, G. H. "Interactions in Electrolyte Solutions" 214 p. Elsevier Publishing Co., Amsterdam, 1966

35. Langmuir, D. Stability of calcite based on aqueous solubility measurements, Geochim. Cosmochim. Acta. 32, 835-851 (1968).

36. Davies, C. W. "Ion Association", 190 p., Butterworths, Washington, D. C., 1963.

37. Plummer, L. N., Jones, B. F. and Truesdell, A. H. WATEQF- A Fortran IV version of WATEQ, a computer program for calculating chemical equilibrium of natural waters, U. S. Geol. Survey Water Resour. Invest. 76-13.

38. "Water Resources Data for New York Water Year 1976", Vol. 1, U. S. Geol. Survey Water Data Report NY761 .

39. Nicholls, K. H. and Cox, C. M., Atmospheric nitrogen and phosphorus loading to Harp Lake, Ontario, Canada, Water Resour. Res. 14, 589-592 (1978).

40. Reddy, M. M., in "Particulate Metal Transport in the Genesee River, New York", presented at the American Chemical Society National Meeting, Los Angeles, Calif. March 1978, E 42.

41. Lorenzen, M., p. 31-50, in Mitchell, R., ed., "Water Pollution Microbiology," Vol. 2, John Wiley and Sons, Inc., New York, 1978.

42. Verh. Intern. Verein. Limnol., 19, 433 (1975).

43. J. Crystal Growth, 41, 294 (1977).

RECEIVED November 16, 1978.

33

Equilibrium, Kinetic, and Chromatographic Controls of the Solution Composition Obtained during the in situ Leaching of a Uranium Orebody

R. W. POTTER II, J. M. THOMPSON, M. A. CLYNNE, and V. L. THURMOND

U.S. Geological Survey, Menlo Park, CA 94025

Extraction of uranium from uranium ore deposits in sandstones and arkosic sandstones is being accomplished economically by leaching the ores in place with the added advantage of minimal physical disturbance of the rocks (1). In practice the leaching solution (lixiviant) consists of groundwater to which has been added an oxidant (air, O_2, or H_2O_2) and a complexing agent, ($(NH_4)_2CO_3$, Na_2CO_3, $NaHCO_3$, etc.), for the uranyl ion produced by the oxidation of the uranium ores (1, 2). The focus of current research is towards optimizing the lixiviant so as to get the maximum recovery of uranium while at the same time minimizing any unfavorable impact on the environment. In order to attain these goals, efforts are being put forth to develop empirical and/or theoretical models which are capable of predicting changes in composition of the lixiviant as a function of time and distance travelled through the orebody (2).

The modelling that has been applied to date to the *in situ* leaching of uranium has primarily consisted of modelling the hydrologic behavior of the lixiviant as a function of time. A more sophisticated model has combined a chemical model with the hydrologic model to describe the dissolution of uranium and the pore volumes of lixiviant required to leach the orebody (2). This combined hydrologic-chemical model assumes equilibrium processes, although the basis for its construction was largely empirical data obtained from small scale pilot tests. Inherent in the construction of the simple hydrologic models or the combined hydrologic-chemical model is the assumption that the lixiviant moves as an essentially homogeneous mass (slug flow). Non-equilibrium effects on solution composition such as the kinetics of interaction of the minerals in the host rock with the lixiviant and chromatographic effects resulting from adsorption of ions from a moving solution onto mineral surfaces and/or ion exchange between the moving lixiviant and clays are generally disregarded or assumed to be minimal. These simplistic assumptions have in large part resulted from a lack of

0-8412-0479-9/79/47-093-761$05.00/0
This chapter not subject to U.S. copyright
Published 1979 American Chemical Society

available field and laboratory data on the variation of solution composition as a function of time and volume of rock traversed. The purpose of this paper is to present this type of data and to call attention to some of the difficulties involved in constructing viable models describing the variation of solution composition during an in situ leaching operation.

Initial Conditions and Predictions

A suite of 314 groundwater-lixiviant samples were collected from monitoring wells located in an uranium orebody before and while the orebody was being in situ leached. The collected samples cover a 61-day period. The major chemical species initially present in the groundwater were Na^+, K^+, Mg^{++}, Ca^{++}, SiO_2, Cl^-, $SO_4^=$, and $HCO_3^=$. In addition to the groundwater analyses, a suite of cores from the orebody was studied to ascertain the initial mineralogic composition of the host formation.

The host formation containing the orebody consists of a poorly consolidated, highly sorted sandstone composed chiefly of quartz with subordinate amounts of clay minerals (dominantly Ca-montmorillonite), potassium feldspars, chert, plagioclase, pyrite-marcasite, zeolites (dominantly clinoptilolite) and calcite. The clay minerals and zeolites are distributed rather uniformly as a fine matrix throughout the orebody. These phases contain readily available ion exchange sites. They therefore introduce a high potential for the chromatographic separation of chemical species by selective adsorption or ion exchange as the lixiviant flows through the formation. Chromatographic separation would invalidate the assumption that the flow is dominantly slug flow as used in the current models (2).

The lixiviant used at this site was manufactured by dissolving gaseous ammonia and carbon dioxide into groundwater that had previously been pumped from the orebody. This solution was then injected into the formation via injection wells and recovered from production wells after flowing through the orebody. Samples for this study were taken from monitor wells located between the injection and production wells at distances ranging from 2 to 25 meters from the injection wells. The sampling interval was generally once per day at each of the four wells although occasionally two samples per day were collected.

Using the initial groundwater pH, Eh, and compositions, calculations of mineral stabilities were made using the computer programs SOLMNEQ (3) and WATEQ (4). Both programs yielded results which showed that the groundwater composition was compatible with the observed mineral assemblages and with the state of alteration of the respective minerals. The pH of the groundwater prior to injection of the lixiviant ranged from 6.9 to 7.4 and was apparently buffered by equilibria involving calcite and aqueous CO_2 species. Pyrite was several orders of magnitude supersaturated based on the computer calculation as evidenced by the detectable Fe (0.12 ppm) and H_2S (0.3 ppm) present in the formation waters.

The initial pH of the lixiviant varied somewhat but in general was greater than 9. Using the average pH, Eh, and composition of the lixiviant, the host rock mineral stabilities were again calculated using SOLMNEQ and WATEQ. The calculations indicated that calcite, which had been in equilibrium with the groundwater, became greatly supersaturated. The potassium feldspar and plagioclase solution equilibria was shifted so that the feldspars became unstable and tended to alter with the amount of alternation being controlled by the applicable kinetic rate law. On the basis of the above equilibrium calculations and the properties of the lixiviant, one would predict that as the lixiviant arrived at a monitoring well the pH would increase, silica concentration most likely would be constant or would slightly increase owing to the increased pH; the Ca^{++} concentration would decrease owing to calcite precipitation; the Mg^{++} concentration would remain essentially unchanged; and NH_4^+ and $CO_3^=$ would simultaneously increase. The effects of precipitation and dissolution equilibria on the concentrations of dissolved Na^+ and K^+ would probably not be as significant as the ion-exchange reactions with the clays and zeolites. Based on the ion-exchange characteristics of the montmorillonites, the concentraion of K^+ and Na^+ would be expected to increase at approximately the same rate. The predictions made above assume slug flow and that strict equilibrium conditions control the solution composition.

There are many problems involved with making reliable predictions about the behavior of uranium in in situ leach environments. The first problem is specifying which uranium-bearing phases are present in the orebody. In the ore deposit we studied, the major phases appear to be amorphous UO_2 and uraninite, although a significant amount of uranium is present in an unidentified form in the rock matrix rich in clays and zeolites. A second problem is a lack of internally consistent thermodynamic data which makes it difficult to calculate the speciation of the uranium in the ammonium carbonate lixiviant. The most troubling aspect, however, centers on the dependence of the solubility and rate of solution of uranium upon the oxidation state of the lixiviant. At the site we studied there was an initial Eh gradient of over 200 mv across the travel path of the lixiviant. In modelling the in situ leaching at this site it is necessary to calculate or predict the capacity of the oxidant to overcome this pre-mining gradient. However, the data necessary to accomplish this are not available, hence only qualitative predictions can be made. Oxidized uranium previously mobilized by the lixiviant might be re-reduced as it is transported through the reduced portion of the formation. The uranium concentration would then be expected to decrease. The amount of decrease would be moderated both by the capacity and concentration of the oxidant. On the other hand, the concentration of mobilized uranium would remain unchanged, or it would increase as the lixiviant moved into the oxidized portion of the orebody.

Results

As predicted by the equilibrium calculations, the Ca^{++} concentration decreased owing to calcite precipitating from the lixiviant. However, the rate of decrease was markedly slower than predicted. The SiO_2 concentration did not behave as predicted. The initial concentration of silica was approximately 33 mg/L and decreased steadily as a function of time until it reached a value of 7 mg/L where it stabilized (Figure 1). This behavior could be the result of the precipitation of a silica-bearing phase not contained in the computer codes such as buddingtonite (NH_4 feldspar), some ammonia clays, or layered silicates. At the temperature and pH of the solution, a dissolved silica value of 7 mg/L is below the solubility of quartz, thus ruling out the precipitation of quartz as a control on the silica concentration. The uranium concentrations did not behave completely as predicted by the qualitative arguments given above. Uranium was mobilized near the center of the orebody; and as it migrated outwards towards the production wells, the concentration in solution decreased by two orders of magnitude in the reduced region (as predicted) and decreased in the oxidized portion by an order of magnitude.

The most striking departure from prediction, although expected, was the observation that the various species were chromatographically separated. The first species to arrive was H^+ (Figure 2) which apparently was generated by the reactions between ammonia, silica and silicates. Its arrival time is about 8 days prior to the arrival of NH_3, the last species to arrive. Closely following H^+ was HCO_3^- which arrived about 6 days prior to the arrival of the NH_3 (Figure 3). At approximately the same time as the HCO_3^- arrived at the monitor well, Mg^{++} was also arriving (Figure 4) most likely as $(MgHCO_3)^+$ or related species. Approximately 5 days before the NH_3, the first Cl^- began to arrive (Figure 5). The first arrivals of Na^+ (Figure 5) and K^+ (Figure 6) are approximately the same, i.e., about 3 days prior to the NH_3. The Ca^{++} concentration began to decrease in response to the HCO_3^- and $CO_3^=$ increase about half a day prior to its arrival (Figure 5). The supersaturated concentrations which were observed, as well as the slight increase of Ca^{++} 2 days prior to the NH_3 arrival, suggest that Ca^{++} was being ion exchanged by the clays at a rate slightly greater, or nearly equal to, the kinetic precipitation rate of calcite, thus explaining its slow decrease in concentration (Figure 5). The

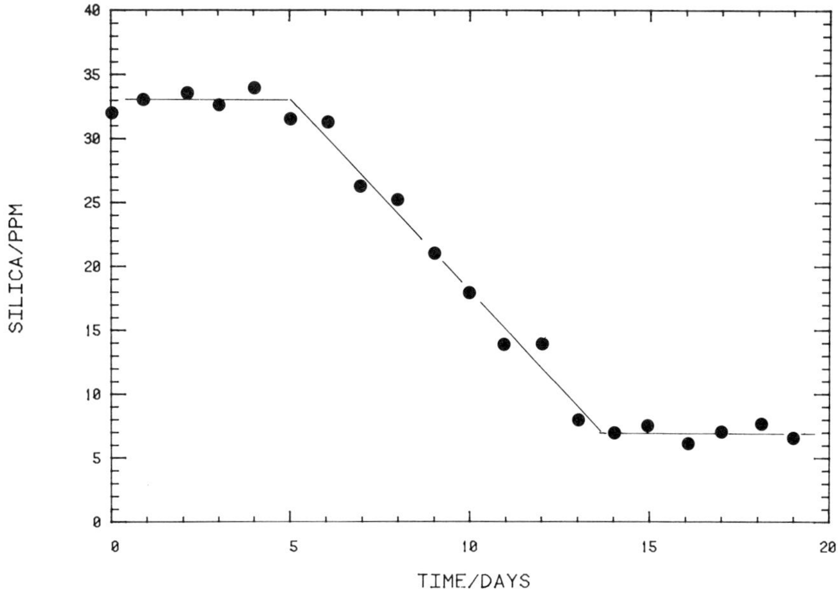

Figure 1. Variation of SiO_2 concentration with time

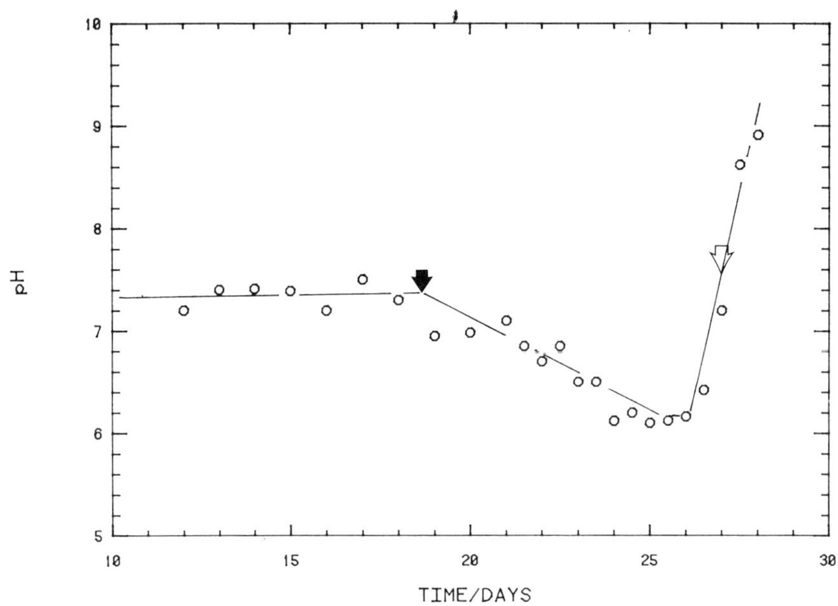

Figure 2. pH as a function of time. Solid arrow indicates the arrival of H^+ while the open arrow indicates the arrival of the first detectable NH_3.

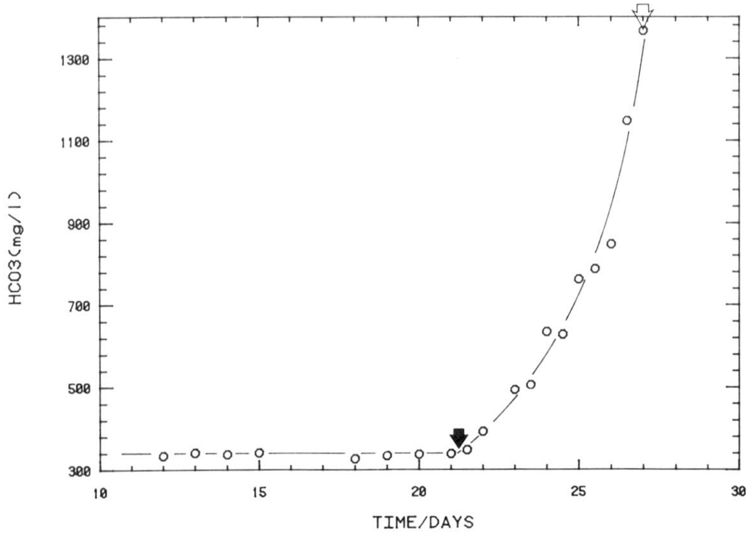

Figure 3. *HCO_3^- concentration as a function of time. Solid arrow indicates the arrival of HCO_3^- while open arrow indicates the arrival of the first detectable NH_3.*

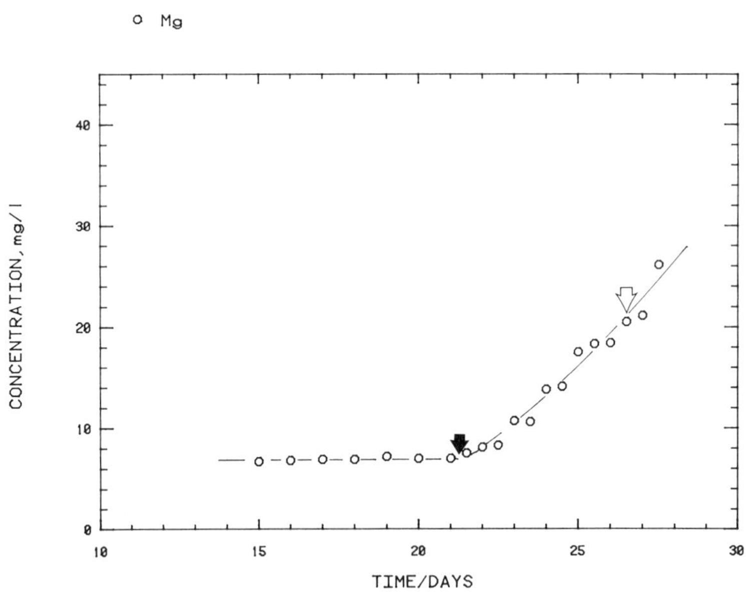

Figure 4. *Mg^{++} concentration as a function of time. The solid arrow indicates the arrival of Mg^{++} while the open arrow indicates the first arrival of detectable NH_3.*

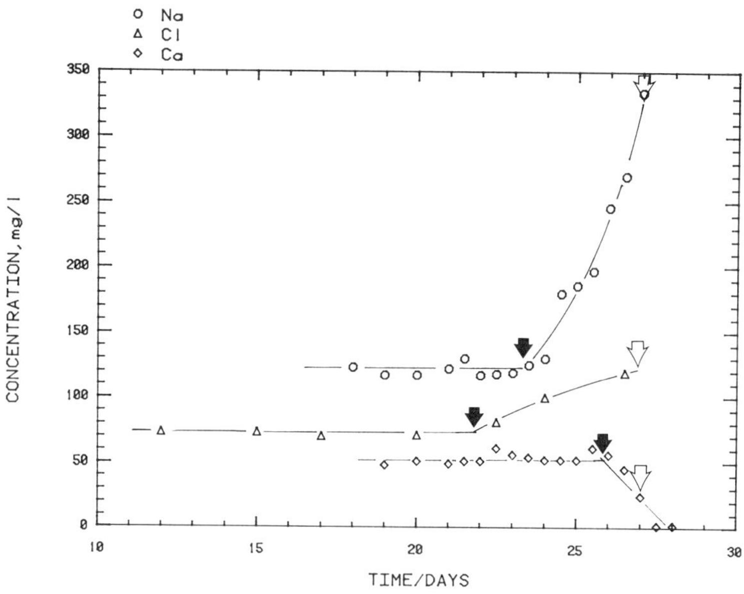

Figure 5. Na^{+}, Cl^{-}, and Ca^{++} concentration as a function of time. Solid arrows indicate the arrivals of Na^{+}, Cl^{-}, and Ca^{++} while the open arrow indicates the arrival of the first detectable NH_3.

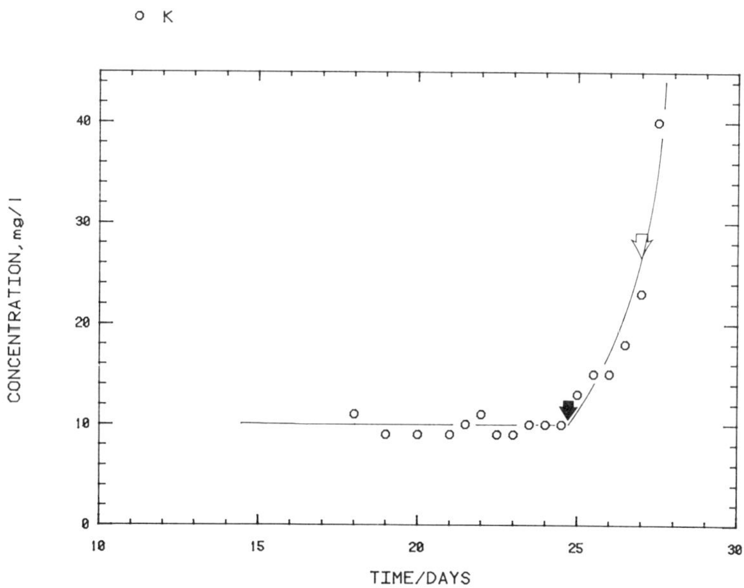

Figure 6. K^{+} concentration as a function of time. Solid arrow indicates the arrival of K^{+} while the open arrow indicates the first arrival of detectable NH_3.

mobilized uranium began to arrive about one day prior to the NH_3. This high retention time is suggestive of extensive interaction with the clays and zeolites. The explanation for the decrease in concentration of uranium as it moved into the oxidized region may result from extensive ion-exchange with the clays and zeolites, thus fixing the uranium in these phases.

Conclusions

Before computer models can reach the required sophistication to be truly capable of predicting the behavior of an orebody as it is *in situ* leached, there are several significant pieces of information which must be incorporated into the model. First and foremost is the effect of the mineralogic assemblage on the process. This necessitates the accurate characterization of the host formation prior to *in situ* leaching. Secondly, a better and internally consistent data base is needed for the kinetic and thermodynamic properties of uranium compounds and their speciation and oxidation. In addition, reliable kinetic data for ion exchange reactions are also required. Thirdly, the lixiviant can not be treated as a homogeneous mass as it moves through the formation but rather some corrections are required for chromatographic effects.

If all of these elements are incorporated into a basic hydrologic flow model, then detailed modelling of the behavior of the leaching system will be possible. The development of such comprehensive models will not only aid in the optimization of solution compositions for the most effective uranium recovery, but will also allow a more realistic environmental impact assessment and corrective measures if required.

Abstract

The process of *in situ* leaching of uranium ore bodies offers a method of extracting uranium that is economically viable and more environmentally acceptable than current surface mining technology. Models capable of predicting solution composition as a function of time and distance travelled are required to optimize uranium recovery as well as to evaluate environmental impact. The current approach is to treat the problems as being strictly controlled by equilibrium considerations and to either disregard or to consider negligible the kinetic and chromatographic effects.

We have collected 314 ground-water samples from wells located across an uranium ore body being leached. These samples reflect conditions before and during the leaching of an uranium ore body. The samples were analyzed by a variety of methods for different elements and species, Na, K, Mg, Ca, SiO_2, Cl, U, NH_3 , and HCO_3, in order to better understand the processes involved and how well the behavior is predicted by existing equilibrium models.

The observed data agrees well only with the prediction for Ca. The magnitude of the SiO_2 effect was significantly larger than expected, possibly due to the precipitation of an ammonium feldspar not considered in the equilibrium calculations. Even when corrected for the regional Eh gradient, the behavior of uranium differed substantially from the qualitative predictions.

The concentration of Na, K, and Mg were controlled by ion exchange reactions; however, these concentrations were also strongly affected by chromatographic effects which were observed for all of the species studied. The most striking was the 8-day separation between the arrival of NH_3 and H. These effects are significant and invalidate normal assumptions that the flow is slug flow.

In order to be truly predictive, models for the complex leaching process must contain equilibrium, kinetic, and chromatographic parameters. Development of such comprehensive models will not only aid in the optimization of solution compositions for the most effective uranium recovery, but will also allow a more realistic environmental impact assessment.

Literature Cited

1. White, L. In-situ leaching opens new uranium reserves in Texas. Eng. Mining J. July, 1975, 73-81 (1975).
2. Shock, D.A. The Vail Conference on in-situ leaching of uranium. In Situ 1, 103-113 (1977).
3. Kharaka, Y.K. and Barnes, Ivan. "SOLMNEQ: Solution-mineral equilibrium computations." U.S. Geol. Survey Comp. Contrib., PB-215899, NTIS, Springfield, Virginia, 82 p. (1973).
4. Truesdell, A.H. and Jones, B.F. WATEQ, A computer program for calculating chemical equilibria of natural waters. U.S. Geol. Survey J. Res. 2, 233-248 (1974).

Disclaimer: The reviews expressed and/ or the products mentioned in this article represent the opinions of the author(s) only and do not necessarily represent the opinions of the U.S. Geological Survey.

RECEIVED November 16, 1978.

Application of Geochemical Kinetic Data to Ground-water Systems

A Tuffaceous-Rock System in Southern Nevada[1]

HANS C. CLAASSEN and ART F. WHITE

U.S. Geological Survey, Denver, CO 80225

In principle, the chemical composition of water recharging a lithologically homogeneous aquifer will depend upon the mineral phase present in the aquifer, the amount of dissolved carbon dioxide (H_2CO_3), the amount of aquifer surface area (S) in contact with the hydraulically effective pore volume (V), the temperature at which reaction occurs (T), the contact time (t), and the reaction rate (k). Laboratory experiments may be carried out using specific lithologic media to determine the rates of reaction of these media with water containing dissolved carbon dioxide. If the interrelationships of the above variables can be sufficiently defined, a determination of one of the above aquifer properties can be made, if the others are known, and if a representative water sample from the aquifer is available.

Ordinarily, aquifer properties are measured using hydraulic-testing techniques. These techniques yield estimates of rates and quantities of water movement within a given aquifer and are not related to chemical reactivity. However, knowledge of the chemical reactivity of an aquifer is necessary, if meaningful predictions of the effects on water quality of wastes injected into the aquifer are to be made, because sorption of waste materials by the aquifer is usually relied upon to reduce any potential hazard. The amount of active aquifer surface in relation to the solution volume with which it is in contact (S/V or σ) must be known for a given aquifer system to predict the efficiency of the aquifer to mitigate the effect of waste introduction. This σ is unobtainable by hydraulic testing of an aquifer, but could be estimated from water-qualtiy data if values of the other required parameters were available.

Hydrologic Setting

To test the above hypothesis, a system in southern Nevada was chosen for which sufficient hydraulic and water-quality data were available. This sytem, Rainier Mesa, is located about 160 km northwest of Las Vegas, Nev. The lithology and hydrology of

[1] This is Part I of a series.

0-8412-0479-9/79/47-093-771$05.75/0
This chapter not subject to U.S. copyright
Published 1979 American Chemical Society

Rainier Mesa have been described by Thordarson (1) and are schematically illustrated on Figure 1.

The caprock of the mesa is predominantly devitrified tuff, the Rainier Mesa Member of the Timber Mountain Tuff. Its thickness varies, but is generally about 100 m, although in small areas the underlying Paintbrush Tuff crops out. The tuff of the Rainier Mesa Member has a vitric basal zone of chemical composition similar to the overlying devitrified tuff. Underlying the Rainier Mesa Member is the vitric Paintbrush Tuff with a chemical composition also similar to that of the overlying rocks. The Paintbrush is generally 200 m thick and is underlain by vitric rocks which have undergone extensive alteration to clay minerals and zeolites. These altered rocks are informally referred to as the "tunnel beds," so named because of the many drifts which have been mined into Rainier Mesa. Although chemically similar, these rocks are hydrologically quite dissimilar.

Recharge to the Rainier Mesa hydrologic system occurs mainly through fractures in the competent devitrified tuff; some recharge may also occur directly to the Paintbrush Tuff where it crops out on the surface of the mesa. The fracture (secondary) hydraulic conductivity may be locally very high, but the average conductivity of the devitrified rock is probably low. That water which passes through the fractures of the devitrified tuff enters the Paintbrush Tuff, which has a relatively high interstitial (primary) hydraulic conductivity as determined by measurements made on core samples (1). Continued downward movement of water in the Rainier Mesa system is retarded by the tunnel beds which have both low primary and secondary conductivities. Thus, recharged water probably travels rapidly through the fractured Rainier Mesa Member; perhaps more slowly through the underlying Paintbrush; and even more slowly through the tunnel beds. This constitutes one example of a perched water body: a zone of saturation above an unsaturated zone--in this case caused by an underlying zone of low conductivity, the zeolitized tuff of the tunnel beds. In Rainier Mesa, the top of the zone of saturation is irregular. Composite water levels measured in drill holes penetrating both the Paintbrush and the tunnel beds indicate that the top of the zone of saturation is in the lower part of the Paintbrush. Data from cores (Larry Benson, Lawrence Berkeley Laboratory, Berkeley, written communication, 1978; 1, 2), however, indicate that other parts of the Paintbrush may be nearly saturated. It appears that the devitrified Rainier Mesa Member is unsaturated (1).

An additional feature that may influence the hydrologic system is a gently east-northeast plunging syncline. Lithologic, hydrologic, and water-quality data suggest that flow in the Paintbrush, in addition to downward, may be lateral toward the synclinal axis.

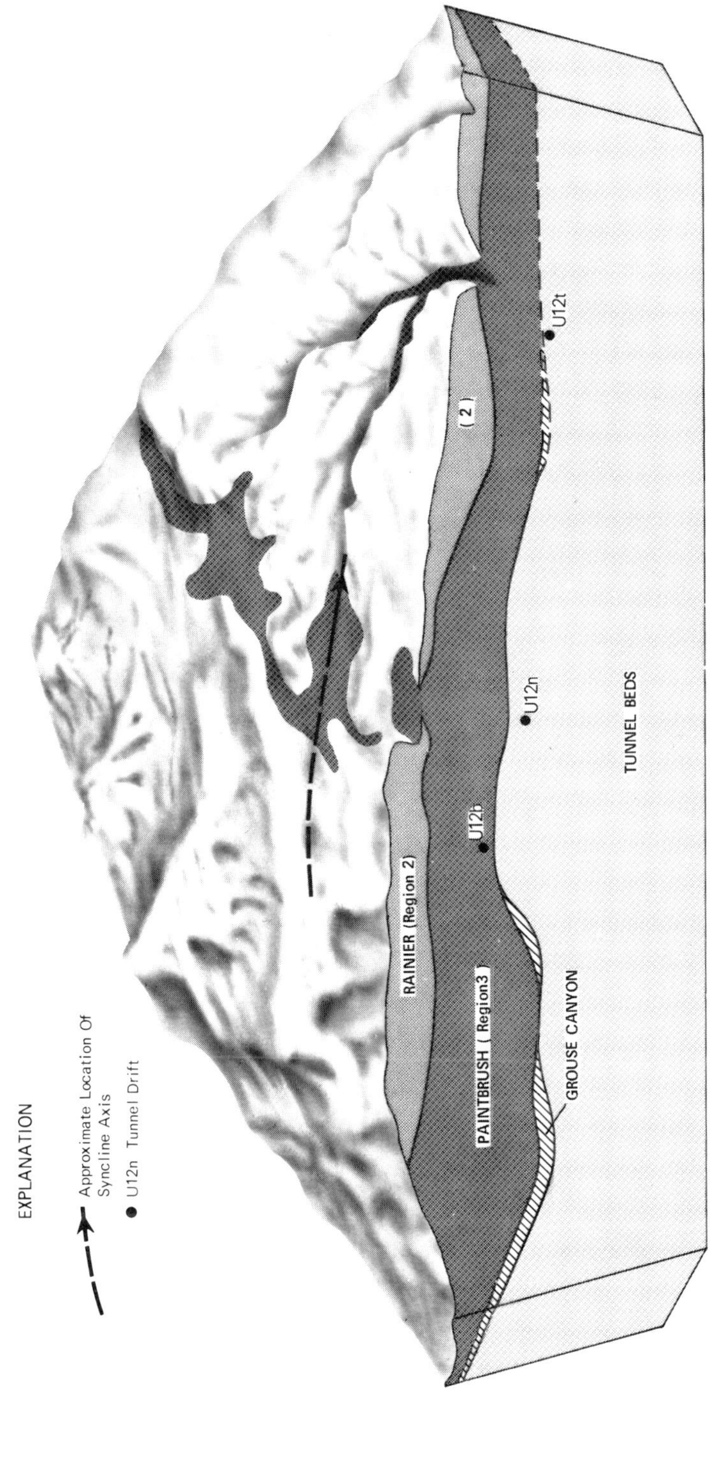

Figure 1. Iscmetric view of Rainier Mesa, Nevada, showing major geohydrologic features

Water Quality

For many years, water samples have been collected from seeps in tunnels, mined for the most part in the tunnel beds. These samples have shown one significant compositional similarity: high percent sodium. The B-tunnel system was driven into the Paintbrush Tuff, and considerable calcium was also present in the water samples collected. Some samples collected from the T-tunnel system contained high concentrations of sodium, typical of water from the tunnel beds; other samples contained significant calcium. These latter samples were collected from regions of T-tunnel quite near the synclinal axis. This proximity results in a lesser thickness of tunnel beds overlying locations nearer the axis, and the samples represented water which had traveled a shorter distance in the tunnel beds. The high-sodium water is believed to result from precipitation of, or ion exchange reactions with, zeolites which are present in the tunnel beds. Furthermore, these regions of T tunnel produced large water flows of more than 6 L/s, as contrasted with the more usual seeps of a few hundredths of a liter per second. Benson (LBL, Berkeley, written commun., 1978) reported analyses of a few samples centrifuged from cores taken in the Paintbrush in the vicinity of the T-tunnel complex; these showed significant calcium contents. Figure 2 shows some of the water-quality data from the Rainier Mesa system plotted on a Na-Ca-K trilinear diagram. Those samples from B- and T-tunnel systems with significant Ca contents are plotted as open circles, and the more typical tunnel water samples with low Ca are closed circles.

Laboratory experiments (3) with both vitric and devitrified tuff of the Rainier Mesa Member have disclosed the rates and mechanism of dissolution of these rocks. Many experiments were carried out to define the interrelationships among pH, S/V (or σ), and temperature (T), in determining the rates and cation products of the reactions. Figure 2 also shows the progressive changes in reaction products for both vitric and devitrified rocks at an experimental pH value similar to that at which most of the reaction occurs in the natural system. The reaction results have been extrapolated beyond the experimental reaction time using the relationship

$$Q_{M,t} = Q_o + k_M t^{\frac{1}{2}},$$

where

$Q_{M,t}$ is the total mass of cation M transferred from solid to solution in time, t,

Q_o is the mass transferred from solid to solution by ion exchange at fresh rock surface,

k_M is the rate constant associated with transfer of cation M at constant pH.

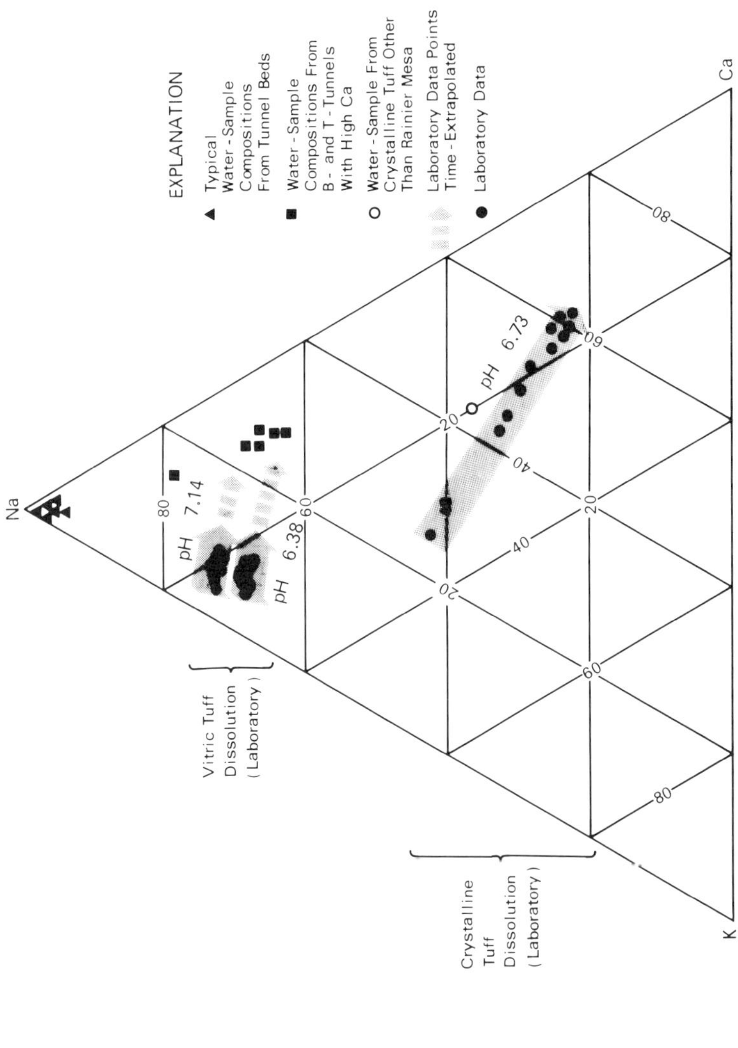

Figure 2. Composition relationships among naturally occurring and laboratory samples of various kinds

The relationship is derived from the previously mentioned laboratory experiments. Although the Q_o term is a significant determinant of total mass transferred in the experiments using material freshly disaggregated, the authors believe that $Q_o \simeq 0$ in the environment of a steady-state ground-water system. The fresh surface has associated cations, which are more easily removed than those within the rock itself. These cations are no longer available for removal in a natural ground-water system not undergoing large-scale physical disruption. Extrapolation of the laboratory data to much longer times further presumes a continuation of the same mechanism as at shorter times; for glasses, the assumption appears valid, as nonparabolic rates have not generally been observed (see (3) for a more complete discussion). Ratios of parabolic rate constants successfully predict the ion ratios observed in Rainier Mesa ground-water samples, whereas, if linear rates would become important at longer times in this system, congruent dissolution must take place, and the dissolved-ion ratios should reflect the composition of the solid. This is not observed in the field data (see (3) for a more detailed discussion of dissolution mechanisms). It is clear that reaction with the devitrified (crystalline) tuff of the Rainier Mesa Member results in water extremely high in calcium and magnesium, whereas reaction with the vitric tuff of Rainier Mesa of the same chemical composition produces water of a composition similar to that of the samples from B and T tunnels believed to have been little affected by the zeolitized tuff. Field-data confirmation of the laboratory results from reaction of the crystalline tuff is difficult, because no water samples have ever been obtained from the unsaturated Rainier Mesa Member. One spring sample obtained from devitrified Timber Mountain Tuff, however, is believed to be representative of a system similar to that of Rainier Mesa, but without the vitric phase; its composition is also indicated on Figure 2.

The location in the geologic section from which the B and T tunnel-complex water samples were obtained, comparison of their chemical composition with data obtained from laboratory experiments on both vitric- and crystalline-tuff, and hydrologic data from the Rainier Mesa system, combine to imply that the vitric material (Rainier Mesa Member and Paintbrush Tuff) of Rainier Mesa is responsible for the water quality observed above the zeolitized zone.

It is likely that because the recharging water must first pass through the fractures in the crystalline tuff of Rainier Messa, at least some of the dissolved material in the water must be derived from this material. The question, of course, is not "if," but "how much." The development that follows will assume that the effect is not significant, and the probable error introduced by this assumption will be dealt with later.

Chemical Kinetic Modeling Of The Rainier Mesa System

To model the chemical-kinetic evolution of the Rainier Mesa ground water, several simplifying assumptions and parameter estimates had to be made. Figure 1 diagrammatically illustrates the Rainier Mesa system, indicating regions of lithologic and hydrologic importance; these will be discussed separately.

Region 1 is the soil zone (or equivalent). This is the zone of aerobic biological activity and the region of high partial-pressure of carbon dioxide gas (P_{CO_2}) resulting from plant respiration and microbiological-decay mechanisms. Some of this gas is dissolved by the recharging water as it passes through region 1, according to the reaction $H_2O_l + CO_{2g} \rightleftharpoons H_2CO_{3\,aq}$. An estimate of H_2CO_3 in recharge water was made from six Rainier Mesa water samples for which reliable pH and HCO_3^- values were available. In simple silicate rock dissolution, all carbonate species result from reaction of H_2CO_3 with the rock. Thus, $HCO_3^- + CO_3^{2-} + H_2CO_3$ at any point in the flow path is assumed to be a constant. Calculations of concentrations of these species in the six samples previously mentioned allowed calculation of H_2CO_3 initially available for reaction. These data are presented in Table I. The arithmetic mean was used in subsequent calculations.

The reaction process was viewed as a series of simple steps, outlined hereafter. First, recharging water dissolves soil-zone CO_2 to yield a typical sample value of $H_2CO_3 + HCO_3^-$ of 2.16 x 10^{-3} mol/L (CO_3^{2-} is negligible). The distribution of major carbonate species is governed by the equilibrium relationship (4):

$$K_1^{25°} = \frac{a_{H^+}\, a_{HCO_3^-}}{a_{H_2CO_3}} = 4.31 \times 10^{-7} \simeq \frac{[H^+][HCO_3^-]}{[H_2CO_3]} \quad . \qquad (1)$$

Table I.--Data Used To Estimate Recharge Concentration of H_2CO_3 In Typical Rainier Mesa Ground Water

Sample designation	pH	Total carbonate [1/] species
U12t.03 UG1	7.18	2.14
U12t.03 UG3	7.40	2.36
U12n.07 Byp.	8.04	.89
U12n.02 UG1	7.31	2.51
U12t main 1805	7.83	2.53
U12t main	8.21	2.52
Arithmetic mean		2.16

[1/] $HCO_3^- + H_2CO_3 + CO_3^{2-}$ in mmol/L; CO_3^{2-} is negligible.

For a total carbonate species concentration of 2.16×10^{-3} mol/L, $[HCO_3^-] = 3.05 \times 10^{-5}$ and $[H_2CO_3] = 2.13 \times 10^{-3}$ mol/L at 25°C. The associated pH value was calculated to be 4.5. The actual temperature of recharge is not known; 25°C was chosen because temperatures within Rainier Mesa are believed to range from ~ 20° to 25°C and most of the reaction takes place under those conditions.

Once the recharged ground water leaves the mesa soil zone (region 1) where it received its charge of H_2CO_3, it enters rocks (region 2, Rainier Mesa Member, and region 3, Paintbrush Tuff), where it reacts with the tuffaceous aquifer utilizing only the H_2CO_3 derived from the soil zone. This is called a closed system reaction:

$$\text{(silicate rock)}\ (M^+) + H_2CO_3 \rightarrow \text{(silicate rock)}\ (H^+) + M^+ + HCO_3^-$$

thus, H^+ ions are consumed, increasing the pH and bicarbonate, and metal ions appear in solution. The reaction progress may be modeled as a series of steps, the beginning and end of each step defined by a specific pH change, which, in turn, fixes concentrations of all major carbonate species by equilibrium (equation 1) and mass-balance (equation 2) relationships:

$$[H_2CO_3] + [HCO_3^-] = 2.16 \times 10^{-3}. \qquad (2)$$

For the analysis presented here, an approximation to continuously varying pH was made by choosing 0.5-pH intervals, and the boundary values for HCO_3^- in Table II were calculated. Changes in HCO_3^- within each pH interval were then determined, these changes resulting from consumption of hydrogen ions from H_2CO_3 with an equivalent production of both HCO_3^- and cations, M^{n+}. Because 2.08 meq/L H_2CO_3 is converted to HCO_3^-, in a reaction which goes from pH = 4.5 to pH = 8.0, 2.08 meq/L cations are produced. Cation compositions of several Rainier Mesa ground-water samples are shown on Figure 3, with the cation composition of a sample having 2.08 meq/L total cations indicated by the arrow and by the numerical values listed. The straight lines drawn through the data points are to facilitate determination of a cation composition and are not intended to imply composition changes during reaction; this would only be true if all the samples represented points along a single flow path or along similar flow paths. In Rainier Mesa, it is more probable that the samples represent similar locations along a series of dissimilar flow paths, and therefore, in theory, any one of the actual samples could be modeled. Instead, a single median value was chosen as typical.

Associated with each pH interval then is a midpoint pH, a change in bicarbonate concentration ($\Delta[HCO_3^-]$), and a set of rate constants. These values are presented in Table II. White and

Table II.--Boundary Values and Kinetic Rate Constants Used in Simulation of Rainier Mesa Ground Water

pH interval boundary	Mean pH (reaction pH)	$[HCO_3^-]$ at boundary mol/L	$\Delta[HCO_3^-]$ mol/L	$-\log k_{Na^+}$	$-\log k_{K^+}$	$-\log k_{Ca^{2+}}$	$-\log k_{Mg^{2+}}$
				mol/cm^2 sec$^{1/2}$ (25°C)			
4.5	----	3.05×10^{-5}	---------	-----	-----	-----	-----
---	4.75	---------	5.75×10^{-5}	13.57	14.54	13.40	13.67
5.0	----	8.80×10^{-5}	---------	-----	-----	-----	-----
---	5.25	---------	1.68×10^{-4}	13.64	14.54	13.65	13.88
5.5	----	2.56×10^{-4}	---------	-----	-----	-----	-----
---	5.75	---------	3.86×10^{-4}	13.73	14.54	13.90	14.08
6.0	----	6.42×10^{-4}	---------	-----	-----	-----	-----
---	6.25	---------	5.87×10^{-4}	13.82	14.54	14.15	14.28
6.5	----	1.23×10^{-3}	---------	-----	-----	-----	-----
---	6.75	---------	5.00×10^{-4}	13.90	14.54	14.40	14.49
7.0	----	1.73×10^{-3}	---------	-----	-----	-----	-----
---	7.25	---------	2.56×10^{-4}	13.98	14.54	14.66	14.70
7.5	----	1.98×10^{-3}	---------	-----	-----	-----	-----
---	7.75	---------	9.73×10^{-5}	14.07	14.54	14.91	14.90
8.0	----	2.08×10^{-3}	---------	-----	-----	-----	-----

Claassen (3) determined kinetic rate constants ($k_{M^{n+}}$) for release of cations from vitric material of the Rainier Mesa Member. These rate constants are pH-dependent, because the mechanism of the reaction involves diffusion of hydrogen ions into the glass, and a corresponding (and equivalent) diffusion of cations out; thus, codiffusion determines the rate at which cations appear in solution.

The equation which determines the mass of any one species transferred during a given pH interval is:

$$q_{M^{n+}} = \sigma t^{\frac{1}{2}} n k_{M^{n+}} \quad (3)$$

where

$q_{M^{n+}}$ is mass of species M^{n+} transferred during any pH interval (at a constant pH equal to the mean pH of the interval), equivalents/L,

σ is the aquifer surface area in cm^2 in contact with 1 L of ground water,

t_i is the time that the ground water is in contact with the aquifer to produce a 0.5-pH change during reaction interval, i, in seconds,

$k_{M^{n+}}$ is the kinetic rate constant of production of species M^{n+} at the mean pH of the interval considered, in moles per square centimeter per second,

n is the formal charge of species M^{n+}.

The total mass of major cations transferred to solution during reaction interval, i, is:

$$T_i = \Sigma q_{M^{n+}},$$

or

$$T_i = \sigma t_i^{\frac{1}{2}} \left[k_{Na^+} + k_{K^+} + 2k_{Ca^{2+}} + 2k_{Mg^{2+}} \right] = \Delta[HCO_3]_i. \quad (4)$$

Because it was assumed at first that neither the σ nor time of reaction for any interval, i, was known, an estimate of the product $\sigma t_i^{\frac{1}{2}}$ was made, and $q_{M^{n+}}$ values determined for each species. These values were summed and compared with the $\Delta[HCO_3^-]$ value for that interval. A computer program was developed which allowed rapid convergence toward $T_i = \Delta[HCO_3^-]$. Each interval indicated in Table II was treated in the same manner. The resulting values for each $q_{M^{n+}}$ are shown in Table III, and the relative changes in ion composition with increasing concentration are plotted on Figure 4.

Table III.--Mass Transferred From Rainier Mesa Vitric Material Per Liter of Solution for Each pH Interval

(No allowance is made for precipitation of any solid phase)

Mean pH of interval	$\sigma t^{\frac{1}{2}}$ $cm^2/L\ sec^{\frac{1}{2}}$	$q_{M^{n+}}$, meq/L				$\Sigma q_{M^{n+}} = T_i$ meq/L
		Na^+	K^+	Ca^{2+}	Mg^{2+}	
4.75	3.777×10^8	0.0102	0.0011	0.0301	0.0162	0.0576
5.25	1.728×10^9	.0396	.0050	.0774	.0456	.1676
5.75	6.100×10^9	.1135	.0177	.1537	.1015	.3864
6.25	1.376×10^{10}	.2078	.0399	.1949	.1445	.5871
6.75	1.671×10^{10}	.2105	.0485	.1330	.1083	.5003
7.25	1.173×10^{10}	.1232	.0340	.0514	.0469	.2555
7.75	5.924×10^9	.0504	.0172	.0146	.0149	.0971

The agreement with the typical field sample is not good; the outstanding feature is that the Mg^{2+} values are predicted to be higher than the field data indicate. The Na^+ and Ca^{2+} values are too low; the reason becomes clear when lithologic data from the Paintbrush Tuff are examined in detail. Numerous workers have reported montmorillonite in tuffaceous vitric rocks in the vicinity of Rainier Mesa (5). Most recently, Larry Benson (LBL, Berkeley, written commun., 1978), in examination of the mineralogic changes with depth in the Rainier Mesa system, reported montmorillonite in the Paintbrush Tuff, as well as other clays and zeolites in the lower 30 m of Paintbrush and throughout the tunnel beds. Analysis of a sample of nearly pure montmorillonite from the Grouse Canyon Member of the Belted Range Tuff (5, p. 87-89), which underlies the Paintbrush in some locations at Rainier Mesa, has yielded the following molecular formula:

$$(Na_{.06}K_{.06}H_{.09})(Mg_{.43}Ca_{.17}Fe^{II}_{.03})Fe^{III}_{.15}Al_{1.44}Ti_{.01}Si_{3.92}O_{10}(OH)_2$$

If it is assumed that precipitation of montmorillonite of this composition is occurring contemporaneously with dissolution of vitric material in the Paintbrush Tuff, the ratios of major cations (in solution) consumed is approximately $Mg^{2+}:Ca^{2+}:Na^+:K^+ = 1:0.40:0.14:0.14$.

To correct the kinetic dissolution calculations for montmorillonite precipitation, it was assumed that, for each pH interval

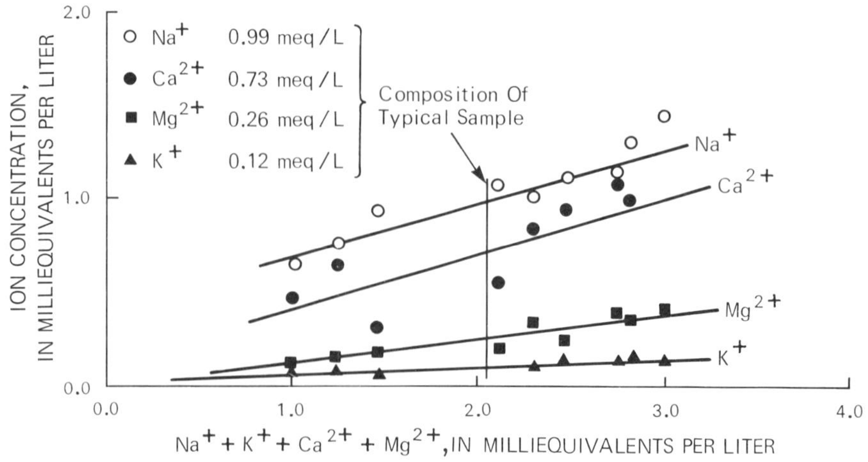

Figure 3. Rainier Mesa groundwater data

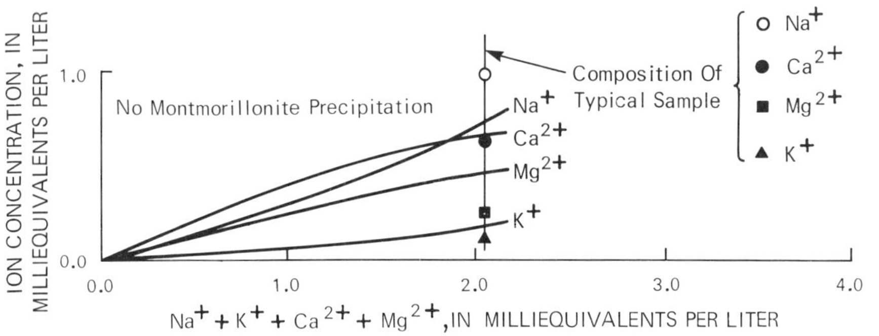

Figure 4. Change in water composition of typical sample from kinetic data (no montmorillonite precipitation)

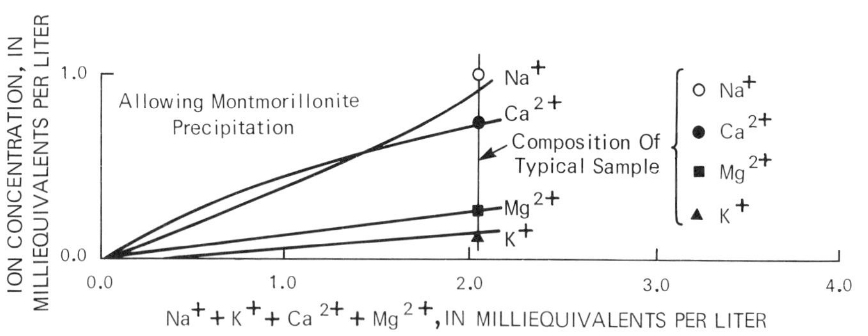

Figure 5. Change in water composition of typical sample from kinetic data (allowing montmorillonite precipitation)

of reaction, the amount of Mg^{2+} produced ($q_{Mg^{2+}}$) minus the amount precipitated (Mg^{2+}_{xs}) would equal the amount expected in the resulting solution, as defined by the Mg^{2+} curve of Figure 3. This assumes that a sample undergoes changes in Mg^{2+} along a flow path in a manner similar to all samples, regardless of flow path or variations in σ. As previously discussed, this assumption is not strictly true, but the gentle slope of the Mg^{2+} curve makes it a reasonable approximation. The results of the modified kinetic dissolution calculations are presented in Table IV and a plot of the changes in ion composition with increasing total concentration is shown on Figure 5. Since H^+ is consumed to produce ions used in precipitation of montmorillonite and released in the precipitation process, it has been assumed in the calculations that an equivalent number of H^+ are released when Mg^{2+}, Ca^{2+}, Na^+, K^+ are precipitated as clay. This is approximately true if a simple precipitation reaction takes place, although the actual mechanism is in doubt. Even though the fit to the Mg^{2+} value of the typical Rainier Mesa ground-water sample is a condition of the modified calculations, note the almost perfect prediction of the Ca^{2+} and K^+ values and the greatly improved prediction of the Na^+ value.

In a further attempt to confirm the hypothesis that montmorillonite precipitation initiates between pH 5.0 and 5.5 as required (see Table IV), ion activity products (IAPs) were determined for the solution compositions resulting from reaction at each half-pH interval. Values for Fe, Al, and Si were obtained from field data of Benson (LBL, Berkeley, written commun., 1978). These IAPs were divided by the equilibrium constant for Aberdeen montmorillonite (K_{Ab}) reported by Kittrick (6). This montmorillonite has a composition similar to that estimated for the Rainier Mesa system:

$$(Na,K,H)_{.42}(Mg)_{.45}Fe^{III}_{.34}Al_{1.47}Si_{3.82}O_{10}(OH)_2.$$

Although it is probable that the equilibrium constant for the montmorillonite being precipitated within the Rainier Mesa system is different from that of the Aberdeen montmorillonite, data presented by White and Claassen (7, 8) indicate that the deviation may not be great. The log IAP/K_{Ab} values were plotted vs log t to show progression from a solution composition which is undersaturated ($\log IAP/K_{Ab} < 0$) at pH 5.0, to one which is supersaturated ($\log IAP/K_{Ab} > 0$) at pH 5.5 and above (Figure 6). Note that the solution continues to oversaturate as the pH increases. Further discussion of this phenomenon and the methods used to determine the relation between IAP and t are offered in the following section.

Table IV.--Mass Transferred From Rainier Mesa Vitric Material Per Liter of Solution For Each pH Interval

(Allowance has been made for precipitation of montmorillonite of composition $Mg^{2+}:Ca^{2+}:Na^{+}:K^{+}$ = 1:.40:.14:.14)

Mean pH of interval	$\sigma t^{1/2}$ $cm^2/L\ sec^{1/2}$	$q_{M^{n+}}$, meq/L				$q_{(Mg^{2+}xs)}$ mmol/L	Amount of montmorill-onite precipitated mmol/L	$\Sigma q_{M^{n+}} = T_i$ meq/L
		Na^+	K^+	Ca^{2+}	Mg^{2+} [1/]			
4.75	3.889×10^{8}	0.0105	0.0011	0.0310	0.0166	------	------	0.0581
5.25	2.992×10^{9}	.0578	-----	.1053	.02	0.0378	0.0879	.1831
5.75	8.431×10^{9}	.1439	.0117	.1776	.05	.0452	.1051	.3832
6.25	1.791×10^{10}	.2551	.0367	.2122	.08	.0540	.1256	.5840
6.75	2.006×10^{10}	.2440	.0496	.1363	.07	.0300	.0698	.4999
7.25	1.361×10^{10}	.1394	.0360	.0500	.03	.0122	.0284	.2554
7.75	6.623×10^{9}	.0554	.0182	.0137	.01	.0034	.0079	.0973

[1/] Amount in solution limited to value determined from Mg^{2+} plot of Figure 3.

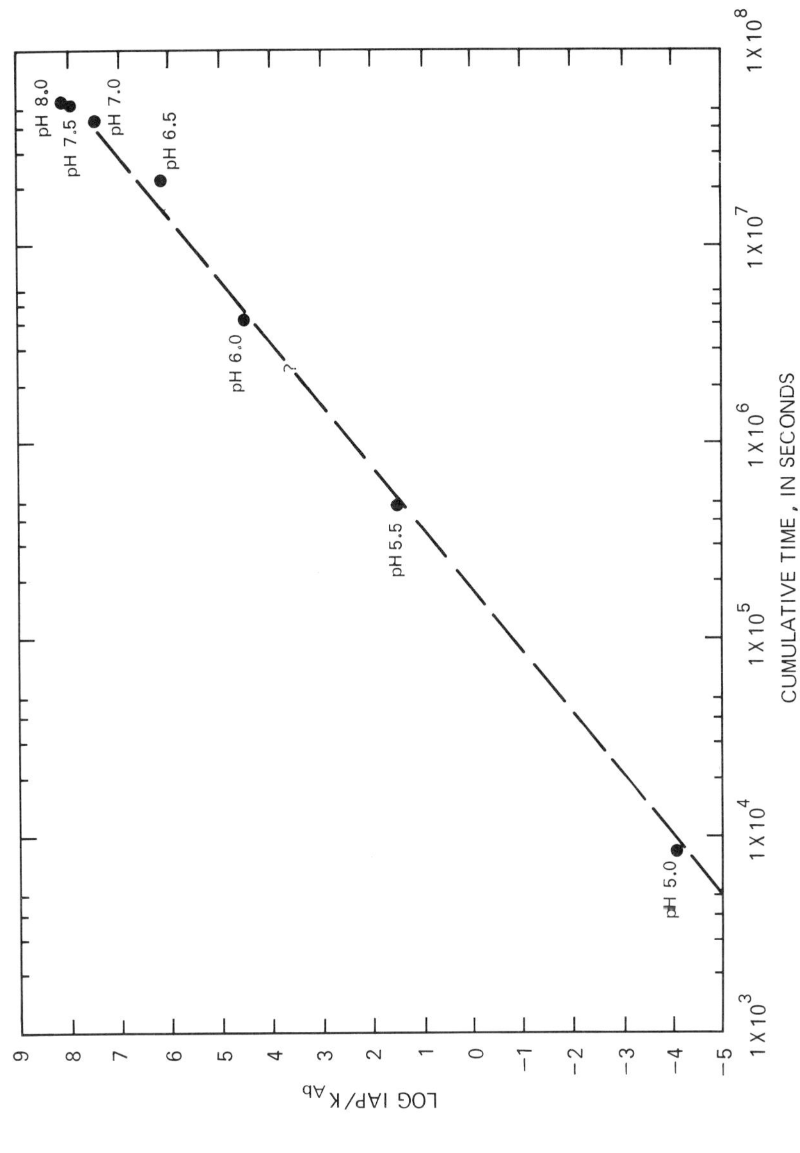

Figure 6. Changes in saturation index IAP/K_{Ab} with progress of the glass-dissolution reaction

Discussion of Results and Conclusions

The imperfect fit of the kinetic predictions to the field data may have its origin in any of several areas. The intervals of 0.5 pH units chosen are probably too large for constant kinetic rate constants to be assumed, especially for Ca^{2+} and Mg^{2+} in the pH range 4.75 - 6.75. This shortcoming can be eliminated simply by decreasing the intervals to, for example, 0.1 or 0.01 pH unit.

The authors have developed an expression to modify kinetic rate constants for surface-sorption effects. Corrections to rate constants have not been made in the calculations reported here, but the experimental values for rate constants are within a few percent of those which would be appropriate for the Rainier Mesa aquifer. The basic methodology reported here is believed valid; only minor changes in the computed values are expected from more precise calculations.

The assumption that montmorillonite of reported composition is being produced and is the only authigenic phase is probably only an approximation. The composition may be in error or may vary areally within Rainier Mesa as a result of areal variations in water quality. Although no zeolites or other clay minerals were reported in the bulk of the Paintbrush Tuff, very small amounts may remain undetected by either x-ray diffraction analysis or thin-section petrography and may affect the aqueous composition by precipitation or by ion exchange.

It has also been assumed that the devitrified Rainier Mesa Member has had insignificant effect on the water chemistry. This assumption is supported by comparison of the observed water composition with compositions resulting from reactions of both devitrified and vitric material with water containing dissolved carbon dioxide (see Figure 2). It is obvious that at least some of the water recharging Rainier Mesa must pass through the devitrified material and that there will be some interaction, but the results of this study indicate that the interaction is small, because significant interaction should preclude a reasonable match to field data.

Effective Aquifer Surface

Table IV and Figure 5 contain the best results obtainable within the assumptions and approximations outlined above. Included in the table are a set of $\sigma t^{\frac{1}{2}}$ product terms which contain the most significant application of the methodology presented here. As previously discussed, one of the purposes of this work is to determine the effective surface area of aquifer material in contact with ground water. Such a value is unobtainable through present hydraulic-testing techniques and is necessary for valid predictions to be made concerning the aquifer sorption properties from laboratory-developed sorption data. Only the water actually in contact with aquifer material, under natural conditions of σ, can reflect the nature of surface with which the waste products will come in contact. If we consider that the match to the typical Rainier Mesa ground-water sample presented on Figure 5 is adequate, an estimate of total residence time of the ground water within Rainier Mesa will allow determination of σ. Clebsch (9) presented a range for residence of Rainier Mesa ground water of 0.8 - 6 years based on its tritium content.

A plot of σ vs t was made using each of several values of σ and the following relationship: $t = \frac{\left(\sum_{i=1}^{7} \sigma t_i^{1/2}\right)^2}{\sigma^2}$ (see Table IV).

The resulting curve is presented in Figure 7. The residence time limits of 0.8 - 6 year (2.5×10^7 - 1.9×10^8 s) were then identified with the corresponding σ values of 6.4×10^6 to 2.3×10^6 cm^2/L.

Thordarson (1) and Benson (LBL, Berkeley, written commun., 1978) both describe the Paintbrush Tuff as partially saturated. Core samples obtained from the lower part of the Paintbrush in the tunnel complex were reported by Diment and others (2) to have saturation levels of 55-91 percent, with an average of 77 percent. Benson (LBL, Berkeley, written commun., 1978) reported an average value for water saturation in the Paintbrush of 90 percent. It would therefore be expected that the total aquifer-pore surface would not be involved in transport of water. As a matter of fact, assuming that all of the saturated-pore space is effective in transmitting water, one might predict that, on the average, somewhere between 77 and 90 percent of the aquifer surface is in contact with percolating water.

Estimation of aquifer σ through B.E.T. and porosity measurements of cores obtained from the Paintbrush Tuff in widely scattered locations in Rainier Mesa yielded values ranging from 8.0×10^7 to 2.3×10^8 cm^2/L. Even if the lower of the measured saturation values is used to estimate the fraction of surface area actually in contact with liquid, the σ values (6.2×10^7 to 1.8×10^7 cm^2/L) appear to be much too high when compared with the results of the geochemical kinetic analysis. The implication is that, on the average, only about 3 percent of the total pore space is effective in transmitting water, as determined by the ratio of measured σ to the σ derived from water-quality kinetics. Stating this differently, the rock-surface area apparently in contact with unit volume of liquid is only about 3 percent of that which would be predicted on the basis of measurements made on so-called natural-state material. This discrepancy may be magnified somewhat by the fact that we are dealing with partially saturated material, but none of the measurements made have indicated saturation at the 3 percent level. We must conclude that the major water-transporting pores constitute only a small fraction of the total interconnected (and probably saturated) pore space.

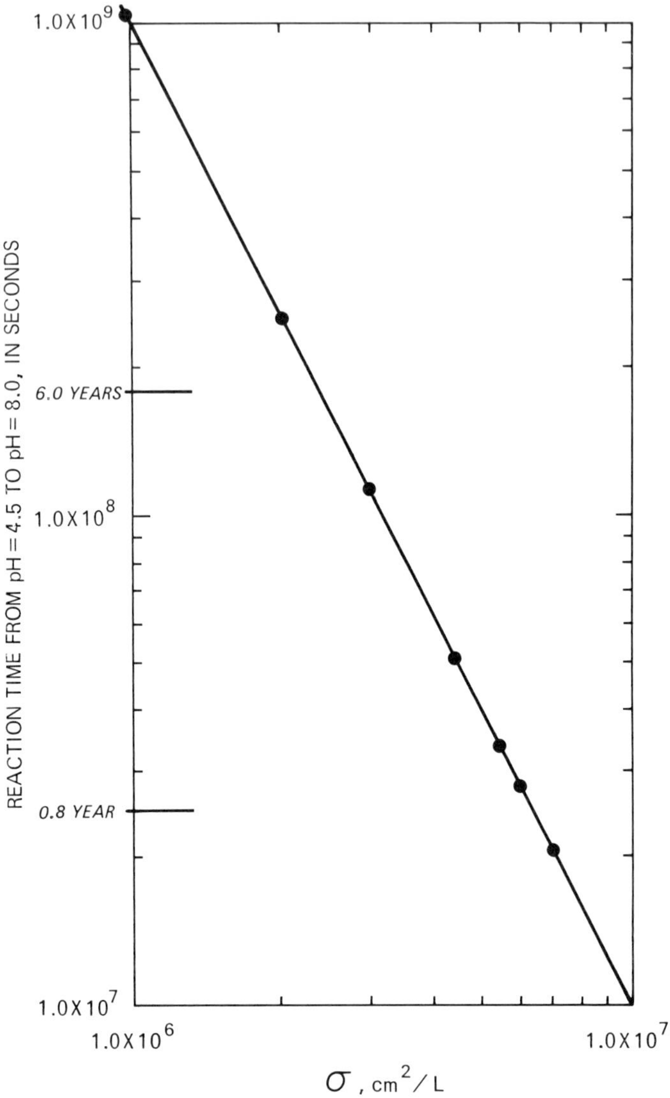

Figure 7. Relation between reaction time (t) *and surface to volume ratio* (σ) *for aqueous reaction of Rainier glass*

An important aspect of this finding lies in its contribution to understanding transport of waste materials in natural systems. Too often, predictions of the waste-sorbing potential of an aquifer are made on the basis of laboratory experiments utilizing crushed aquifer material which is intended to reach equilibrium with a waste-containing solution. The amount of waste removed from solution is measured, and the sorption capacity of the aquifer material is determined as weight, or equivalent-weight contaminant, per unit weight of aquifer material. More sophisticated experiments may measure surface area of crushed material and attempt to relate to grain size of in-situ material. Cores are utilized in such experiments only rarely. None of these techniques have the potential to predict the aquifer surface area effective in removing waste. The techniques used in this report will be refined and further tested in this and other systems.

Montmorillonite Supersaturation

The calculation of montmorillonite saturation index present at the end of each 0.5-pH interval from the kinetically generated solution composition and the equilibrium constant for the Aberdeen montmorillonite was presented on Figure 6. A rapid increase in saturation at lower values of pH slowing at higher pH values is indicated. This behavior suggests that the rate of production of soluble cations is greater than the rate at which species required for montmorillonite precipitation are removed from solution. Note that it has not been stated that montmorillonite precipitates in the classical sense; that is, as a simple crystalline substance. It is more likely that formation of an amorphous-aluminosilicate material precedes both the incorporation of the iron and magnesium and the subsequent rearrangement to form the final phase.

Figure 8 shows the changes in rate of production of the sum of all cations as the reaction proceeds. Because the reaction rate is dependent on pH, greater rates occur at lower pH values. It is, however, curious to observe the time-dependence on removal of Mg^{2+} from solution, and, presumably, also the rate of precipitation of montmorillonite. As shown on Figure 6, montmorillonite saturation is reached at about 1.9×10^5s. Prior to this time, the precipitation rate must be zero; this is indicated on Figure 8 by the vertical arrow. Once saturation has been reached, a small but finite rate of precipitation is anticipated and expected to increase as the saturation index increases. The rates of precipitation calculated from the data of Table IV are in direct opposition to this expectation; rates are initially high and decrease with time more rapidly than the rate at which production of all dissolved species decreases. This would indicate that the rate-controlling step in the formation of montmorillonite is one that involves a species undergoing significant change with time or reaction progress. The most likely species is the hydrogen ion, or the species whose solution concentration is controlled by

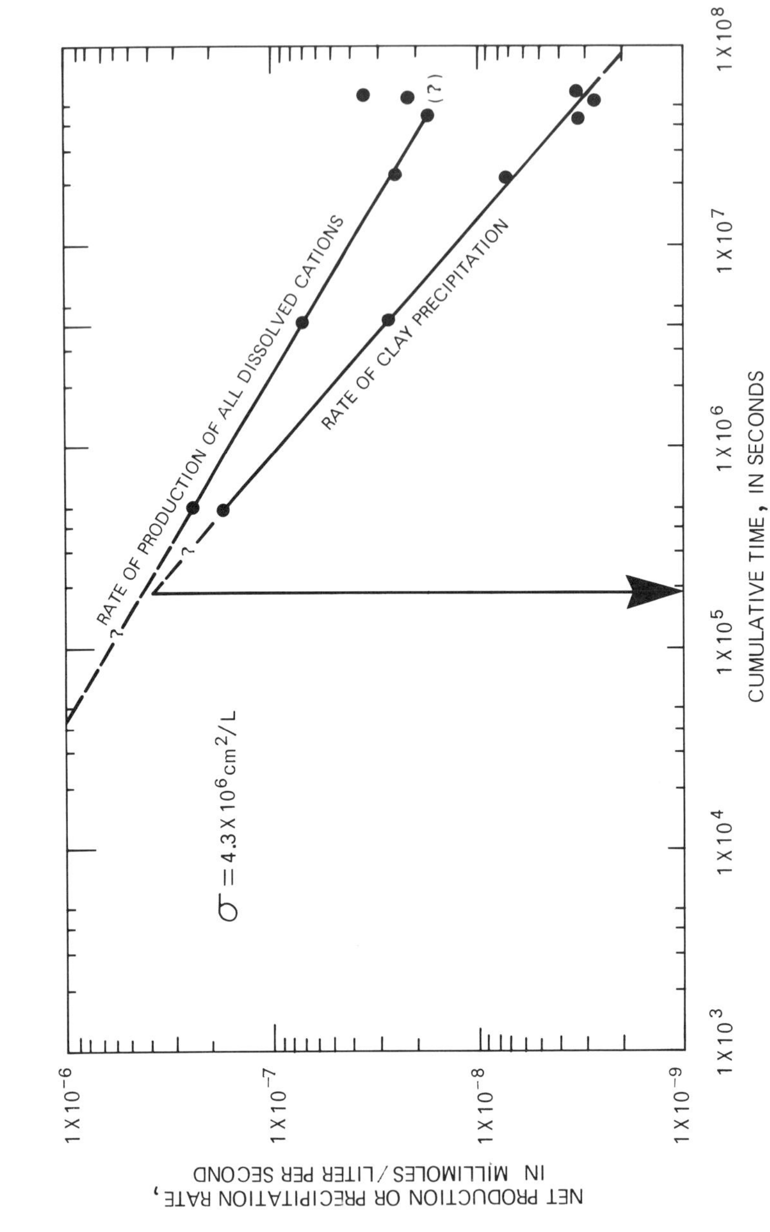

Figure 8. Changes in rate of production of both clay and total cations with time

hydrogen such as $Al(OH)_4^-$ (10) or Fe^{3+} (11). Further work is required to substantiate this hypothesis.

Summary

A technique for determination of effective aquifer surface area in contact with percolating ground water has been presented. The method utilizes laboratory-determined chemical kinetic dissolution data with geologic, hydrologic, and ground-water-quality data to yield an estimate of effective aquifer surface area, a parameter not obtained by other techniques.

Analysis of the results obtained for the Rainier Mesa groundwater system by application of the above method indicated that although measurements of water saturation on core samples from the Paintbrush Tuff evidenced a high degree of saturation (77 to 90 percent), the major water-transmitting regions of the aquifer constitute only about 3 percent of the total, if the porosity is equally distributed. If the regions of major hydraulic conductivity are fractures, measurements of σ on core samples will be meaningless, and the method presented here will provide the only means for an estimate of σ. Although the Paintbrush has been described as a hydraulic system of primary porosity, the calculations of σ suggest that it is a system of secondary porosity, which provides the major throughput. This allows for highly water-saturated rock, with little water movement, coupled to fractures or other conductive passages through which most of the water flows. The effect of such a system on the transport of waste can be significant if core data are used to predict sorptive behavior; in the Paintbrush Tuff, sorptive behavior is nearly two orders of magnitude lower than that predicted from core measurements.

Results of experiments utilizing crushed rock and equilibration with waste solutions to determine sorption behavior cannot be extrapolated to actual aquifer conditions, even if B.E.T. surface area is known. This technique has been commonly used in the past to assess waste-storage-site safety. As the primary hydraulic conductivity decreases and secondary conductivity becomes more prominent, this methodology becomes less and less viable for input to modeling of waste transport. The method presented in this report should result in more realistic waste-transport modeling.

Abstract

Kinetic modeling was used to estimate the effective surface area of aquifer in contact with a unit volume of ground water for a composite saturated-unsaturated groundwater system in southern Nevada. This aquifer property, not obtainable by other means, is necessary for realistic modeling of solute transport in groundwater systems. The results of the kinetic modeling indicate that only a small part of the total interconnected pore space is available for transport of water to the water table. The aquifer studied is composed of both vitric (glassy) and divitrified (crystalline) volcanic tuff of nearly identical chemical composition. Comparison of laboratory and field data indicated that only the vitric phase has a significant influence on ground-water composition. Laboratory determination of mass-transfer rates from the vitric material to solution as functions of pH allowed a simulation of the natural water's cation composition. Simulated results were improved considerably when the model was modified to take into account precipitation of the clay mineral, montmorillonite. Estimates of surface area per unit volume obtained from the kinetic model are about 3 percent of those obtained independently from surface area measurements using the Braunauer, Emmett, and Teller (B.E.T.) equation.

Literature Cited

1. Thordarson, William, Perched ground water in zeolitized-bedded tuff, Rainier Mesa and vicinity, Nevada Test Site, Nevada, U.S. Geol. Survey Open-file Rept., TEI-862, 93 p. (1965).
2. Diment, W. H., Wilmarth, V. R., McKeown, F. A., Dickey, D. D., Hinrichs, E. N., Botinelly, T., Roach, C. H., Byers, S. M., Jr., Hawley, C. C., Izett, G. A., Clebsch, Alfred, Jr. Geological Survey investigations in the U12b.03 and U12b.04 tunnels, Nevada Test Site, U.S. Geol. Survey Open-file Rept., TEM-996, 75 p. (1959).
3. White, A. F., and Claassen, H. C. Dissolution kinetics of silicate rocks, application to solute modeling, in Jenne, E.A., ed., "Chemical Modeling--Speciation, Sorption, Solubility, and Kinetics in Aqueous Systems," Am. Chem. Soc., 1978 (this volume).
4. Garrels, R. M., and Christ, C. L. "Solutions, Minerals and Equilibria", 450 p., Harper & Row, New York, 1965.
5. U.S. Geological Survey, Results of exploration of Baneberry Site, early 1971, U.S. Geol. Survey Rept. USGS-474-245, NTIS, Springfield, VA, 92 p. (1974).
6. Kittrick, J. A. Stability of montmorillonites II, Aberdeen Montmorillonite, Soil Sci. Soc. Amer., 35, 820-823 (1971).
7. White, A. F., and Claassen, H. C. Geochemistry of ground water associated with tuffaceous rocks, Oasis Valley, Nevada, Geol. Soc. Am. Abs. 7 (7), 1316-1317 (1976).
8. White, A. F., and Claassen, H. C. Kinetic model for the dissolution of a rhyolitic glass, Geol. Soc. Am. Abs. 9 (7), 1223 (1977).
9. Clebsch, Alfred, Jr. Tritium age of ground water at the Nevada Test Site, Nye County, Nevada, C122-C125, *in* "Short Papers in the Geologic and Hydrologic Sciences", U.S. Geol. Survey Prof. Paper 424-C, (1961).

10. Hem, J. D., Roberson, C. E., Lind, C. J., and Polzer, W. L. Chemistry of aluminum in natural water, U.S. Geol. Survey Water-Supply Paper 1827-E, 57 p. (1973).
11. Hem, J. D., and Cropper, W. H. Survey of ferrous-ferric equilibria and redox potentials, U.S. Geol. Survey Water-Supply Paper 1827-A, 55 p. (1959).

Disclaimer: The reviews expressed and/ or the products mentioned in this article represent the opinions of the author(s) only and do not necessarily represent the opinions of the U.S. Geological Survey.

RECEIVED November 16, 1978.

35

A Preliminary Comprehensive Model for the Chemistry of Sulfidic Marine Sediments

LEONARD ROBERT GARDNER

Department of Geology, University of South Carolina, Columbia, SC 29208

Diagenetic reactions in subaqueous sediments commonly produce systematic vertical changes in the composition of pore waters. In general these reactions involve the decomposition of sediment organic matter, the regeneration of nutrients and the precipitation and dissolution of various mineral phases. As a result of the vertical concentration gradients that develop in pore waters, diffusive exchange of dissolved substances between the sediment and overlying water can occur. In environments where the overlying water is oxygenated, decomposition of sediment organic matter is carried on by aerobic organisms that utilize dissolved oxygen diffusing into the sediment. However, if the oxygen demand is greater than that supplied by diffusion, the sediment will become anoxic at some depth. At this point the oxidation of organic matter can be continued by anaerobic organisms that utilize sulfate or nitrate as oxidants. In marine sediments the high concentration of sulfate ion in sea water enables sulfate reduction to dominate the process of anaerobic decomposition.

In recent years various workers (1-7) have successfully developed models based on the mathematics of diffusion (8) to describe vertical profiles of selected chemical parameters in marine sediments dominated by sulfate reduction. Several papers (9, 10) have also proposed models for nitrogen diagenesis in the upper aerobic zone of such sediments. Most of these models, however, deal with only one or two relatively well behaved parameters, such as SO_4^{2-} or CO_2, which do not interact strongly with other components of the sediment besides organic matter. A truly comprehensive model for such sediment should deal simultaneously with all of the major chemical parameters of the system and ideally should be formulated as an initial value prob-

0-8412-0479-9/79/47-093-795$05.00/0
© 1979 American Chemical Society

lem so that the depth at which the sediment becomes anoxic could be freely predicted rather than assumed as in the nitrogen diagenesis models previously cited. Formulation of such a model would require consideration of the kinetics of the many complex reactions that may occur near the oxidized-reduced boundary as a result of interactions between downward diffusing oxygen with upward diffusing reduced substances such as H_2S and NH_4^+. Furthermore the upper portion of most marine sediments are inhabited by burrowing organisms and thus the model would have to take into account the movement of substances resulting from bioturbation in order to accurately predict the depth to the oxidized-reduced boundary. The author has attempted to formulate such a model but because of the many ad hoc assumptions that were required and because of the difficulties encountered in obtaining reasonable numerical solutions this work is not ready for presentation. However if one is willing to ignore the upper oxidized portion of the sediment or to assume that the sediment is anoxic at the surface, then, as will be shown below, it is possible to develop a reasonably comprehensive and realistic model for sediment dominated by sulfate reduction. Although previous models for sulfidic marine sediments have successfully simulated profiles of sulfate, carbon dioxide and ammonia, development of a model for H_2S profiles has been stymied by the fact that H_2S may precipitate as FeS and/or FeS_2. This inability to predict H_2S profiles has in turn prevented prediction of pH profiles. Only recently have experimental studies of the kinetics of iron sulfide formation (11, 12) provided a basis for formulating models for the diagenesis of H_2S in the presence of iron oxides.

This paper proposes a system of 10 non-linear, simultaneous differential equations (Table I) which, upon further development and validation, may serve as a comprehensive model for predicting steady state, vertical profiles of chemical parameters in the sulfide dominated zones of marine sediments. The major objective of the model is to predict the vertical concentration profiles of H_2S, hydrotriolite (FeS) and pyrite (FeS_2). As with any model there are a number of assumptions involved in its construction that may limit its application. In addition to steady state, the major limiting assumptions of this model are the assumptions that the sediment is free of $CaCO_3$, that the diffusion coefficients of all dissolved sulfur species are equivalent and that dissolved oxygen does not penetrate into the zone of sulfate reduction.

Also it should be emphasized that the model developed here has not been validated by quantitative comparisons with comprehensive field data because of my lack of access to such data. However, as will be shown, the profiles of H_2S, FeS and FeS_2 generated by the model compare favorably in a qualitative fashion with such profiles as are available in the literature. Thus I feel that the work presented here represents a significant step towards the development of a truly comprehensive model for the chemistry of sulfidic marine sediments.

Construction of the Model

The formulation of this model is based on the mathematics of diagenesis developed by Berner (8). Using the chain rule of partial differentiation (13, p. 335) Berner showed that for any property of a sediment, P, that is a continuous differentiable function of depth, x, and time, t,

$$\frac{\delta P}{\delta t} = \frac{dP}{dt} - S\frac{\delta P}{\delta x} \qquad (1)$$

where S(= dx/dt) is the rate of sediment deposition minus the sum of the rates of erosion and compaction. Equation 1 is known as the general diagenetic equation and is the basis for the system of equations presented in Table I. It should be noted here that horizontal variations in P are assumed negligible. The total derivative dP/dt represents the rates of all the diagenetic processes affecting P. Omitting the usually negligible effect of the flow of water resulting from compaction, dP/dt can be represented in general by

$$\frac{dP}{dt} = D\frac{d^2P}{dx^2} + CR \qquad (2)$$

where D is the whole sediment diffusion coefficient (8) and CR represents the additive rates of all chemical reactions affecting P. Positive terms for CR imply production of P whereas negative terms imply consumption. In the case where P is a solid substance, D can be assumed to equal zero so that there is no effect due to diffusion. Also since steady-state conditions are assumed, $\delta P/\delta t = 0$ and thus the general diagenetic equation reduces to

$$D\frac{d^2P}{dx^2} - S\frac{dP}{dx} + CR = 0 \qquad (3)$$

In view of the fact that sulfate reduction is driven by the microbial decomposition of organic matter [(OC) in Table I] the diagenetic equation for this substance will be derived first. Although the mechanisms by which sulfate reducing bacteria decompose sediment organic matter are complex and not fully understood, Berner (3, 4) and others (7) have successfully modeled steady-state profiles of sulfate on the assumption that the kinetics of decomposition are first order and that the organic substrate is in solid form. Thus by substituting D = 0 and CR = -K (OC) in equation 3 the diagenetic equation for organic carbon is found as

$$\frac{d(OC)}{dx} = - \frac{K}{S} (OC) \tag{4}$$

where K is the first order reaction constant (Table II). Note that this is a first-order, ordinary differential equation for which a simple exponential solution is easily obtained.

In the absence of oxygen and the possible exchange of ammonia on clays, the only reaction affecting sulfate, carbon dioxide and ammonia is that of sulfate reduction whereby sulfate is consumed as a reactant and carbon dioxide and ammonia are liberated from the decomposing organic matter. Thus the expressions for CR for these substances will be the first order reaction term K (OC) modified by the appropriate stoichiometric coefficients. These coefficients are determined by the reaction chosen to describe the process of sulfate reduction. A reaction that has been used successfully by Berner (3, 4) and others (7) is

$$2H_2CO(NH_3)_\alpha(\text{organic matter}) + SO_4^{2-}(aq) + 2\alpha H^+ \rightarrow$$
$$2HCO_3^-(aq) + 2\alpha NH_4^+(aq) + HS^-(aq) + H^+ \tag{5}$$

According to equation 5 each mole of organic carbon consumed results in the consumption of 0.5 mole of sulfate and thus the CR term for sulfate equals -0.5 K(OC). Accordingly the CR term for total dissolved CO_2 equals K(OC) whereas the CR term for NH_4^+ is αK(OC). The reader can easily verify that substitution of these CR expressions in equation 3 with a non-zero value for D leads to the differential equations for ΣCO_2, sulfate and ammonia shown in Table I.

The CR expression for total dissolved H_2S will consist of three terms; a positive term, 0.5K(OC), resulting from production according to equation 5 and

Table I. System of equations for comprehensive model of sulfide diagenesis.

PARAMETER	DIFFERENTIAL EQUATIONS
Org. C	$(OC)' = -C_1(OC)$
ΣCO_2	$(CO_2)'' = C_2(CO_2)' + C_3(OC)$
ΣH_2S	$(H_2S)'' = C_2(H_2S)' + 0.5\ C_3(OC) - C_4(HFeO_2)(H^+)^2(H_2S)^{3/2} - C_5(H^+)(S^o)(FeS)^2(H_2S)$
SO_4^{2-}	$(SO_4)'' = C_2(SO_4)' - 0.5\ C_3(OC)$
NH_4^+	$(NH_4)'' = C_2(NH_4)' + \alpha C_3(OC)$
$HFeO_2$	$(HFeO_2)' = -C_6(H^+)^2(H_2S)^{3/2}(HFeO_2)$
FeS	$(FeS)' = C_6(H^+)^2(H_2S)^{3/2}(HFeO_2) - C_7(H^+)(S^o)(FeS)^2(H_2S)$
S^o	$(S^o)' = 0.5\ (FeS)'$
FeS_2	$(FeS_2)' = C_7(H^+)(S^o)(FeS)^2(H_2S)$
H^+	(see equation 17 in text)

Constants α, C_1, C_2, etc. are derived in Table II.

two negative terms resulting from consumption according to the following equations:

$$HFeO_2(s) + 2H^+ + HS^-(aq) + \tfrac{1}{2} S^{-2}(aq) \rightarrow FeS(s) + \tfrac{1}{2}S^o(s) + 2H_2O \quad (6)$$

$$FeS(s) + S^o(s) + H^+ + HS^-(aq) \rightarrow 2FeS_2(s) + H_2(aq) \quad (7)$$

The kinetics of reactions 6 and 7 have recently been investigated by Rickard (11, 12). According to Rickard the rate law for reaction 6 is

$$\frac{d(FeS)}{dt} = KFS\ (H^+)^2(H_2S)^{3/2}A_f \quad (8)$$

where

(FeS) = FeS in mol l^{-1}

KFS = rate constant (Table II)

(H_2S) = total dissolved H_2S in mol l^{-1}

(H^+) = hydrogen ion activity

A_f = surface area of $HFeO_2$ in cm^2.

If the reacting goethite is assumed to occur in spherical particles of uniform size, A_f can be expressed as

$$A_f = SHM\ (HFeO_2) \quad (9)$$

where SHM is the surface area of $HFeO_2$ per mole and is a function of sphere diameter and the density and molecular weight of goethite. As for reaction 8, pyrite formation, Rickard reports that the kinetics of this process obey the following equation:

$$\frac{d(FeS_2)}{dt} = KFS2\ (H^+)(H_2S)A_h^2\ A_s \quad (10)$$

where

(FeS_2) = FeS_2 in mol l^{-1}

KFS2 = rate constant (Table II)

A_h = surface area of FeS in cm^2

Table II. Primary and derived constants for model equations.

PRIMARY CONSTANTS:

Symbol	Description	Units	Value	Source
S	Sedimentation rate	cm sec^{-1}	4.8×10^{-9}	a
D	Diffusion coef.	cm^2 sec^{-1}	3.0×10^{-6}	assumed
K	Rate constant for sulfate reduction	sec^{-1}	5.0×10^{-10}	b
KFS	Rate constant for FeS precipitation	$mol^{-\frac{1}{2}}\ 1^{-\frac{1}{2}}\ cm^{-2}\ sec^{-1}$	1.5×10^{-7}	(11)
KFS2	Rate constant for FeS_2 precipitation	$mol^{-1}\ cm^6\ sec^{-1}$	1.0×10^{-13}	(12)
α	Ratio of N to C in organic matter	none	0.155	assumed
SHM	Molar surface area of $HFeO_2$	$cm^2\ mol^{-1}$	4.53×10^5	c
SOM	Molar surface area of S^o	$cm^2\ mol^{-1}$	2.29×10^5	c
STM	Molar surface area of FeS	$cm^2\ mol^{-1}$	2.87×10^5	c

a. Typical of coastal sediments, southeastern U.S.A.
b. Consistent with Berner's (22) empirical relation between S and K.
c. Based on spherical particles 2 μm in diameter.

DERIVED CONSTANTS:

$C_1 = K/S$ $C_2 = S/D$ $C_3 = -K/D$

$C_4 = -1.5\ (SHM)(KFS)/D$ $C_5 = -0.5\ (SOM)(STM)^2(KFS2)/D$

$C_6 = (SHM)(KFS)/S$ $C_7 = (SOM)(STM)^2(KFS2)/S$

A_s = surface area of S^o in cm^2.

Again assuming spherical particles A_h and A_s can be expressed as

$$A_h = STM\ (FeS) \tag{11}$$

$$A_s = SOM\ (S^o) \tag{12}$$

where STM and SOM are the surface areas per mole of FeS and S^o, respectively. Thus in summary the CR expression for H_2S is given by

$$CR = 0.5(OC) - 1.5\ KFS(H^+)^2(H_2S)^{3/2}SHM(HFeO_2)$$
$$-0.5\ KFS2(H^+)(H_2S)STM^2(FeS)^2SOM(S^o) \tag{13}$$

where the coefficients, -1.5 and -0.5, are derived from the stoichiometry of reactions 6 and 7. Again it is left to the reader to verify that substitution of the above result for CR in equation 3 leads to the differential equation for total dissolved H_2S shown in Table I. From this discussion the origin of the differential equations for $HFeO_2$, FeS, S^o and FeS_2 should also be obvious.

With regard to the stoichiometry of reactions 6 and 7 it should be mentioned that Rickard was unable to ascertain the exact mechanisms and reactions underlying his kinetic results. I have written reaction 6 as an overall description of the process of FeS formation because the coefficients of each reactant in equation 6 matches its corresponding exponent in equation 8. Furthermore reaction 6 is compatible with Rickard's interpretation of the kinetic mechanisms involved in this process and with his report of the production of native sulfur in his experiments. In the case of reaction 7, however, the achievement of harmony between kinetics and stoichiometry is more problematic. Again Rickard was unable to ascertain the exact mechanisms and reactions involved in pyrite formation. He did not attempt to write an overall reaction but suggested that the process involves dissolution of both FeS and S^o, the formation of polysulfide ions and their subsequent reaction with Fe^{2+} ions to form pyrite plus native sulfur. As with reaction 6 I have chosen to write reaction 7 in such a fashion that the stoichiometric coefficients of the reactants correspond with their respective exponents in equation 10. If one assumes that Rickard's reaction vessels were devoid, as seems likely, of strong oxidizing agents such as ferric iron or O_2 and that the N_2 carrier gas used in his experiments was not reduced to ammonia, then the constraints of reaction 7 force one to postulate the reduction of H^+ to H_2. This is a somewhat unattractive requirement in view of the lack of supporting experimental evidence for H_2 production but it is at least thermodynamically feasible in that the equilibrium concentration of aqueous H_2 would have to be at least 10^4 times greater than the concentration of H_2S. Given typical concentrations of H_2S in anoxic sediments it is unlikely that an equilibrium pressure of H_2 could develop at one atmosphere total pressure. If one is willing to admit the presence of O_2 in the system, then reaction 7 could be modified as follows:

$$2FeS(s) + S^o(s) + H^+ + HS^-(aq) + \tfrac{1}{2}O_2 \rightarrow 2FeS_2(s) + H_2O. \quad (14)$$

This, however, is undesirable because O_2 does not occur in the rate equation and because, as assumed, it is unlikely that there is sufficient O_2 present in sediments at depths where sulfate reduction and pyrite formation are active. Furthermore the often cited reaction between FeS and S^o,

$$FeS(s) + S^o(s) \rightarrow FeS_2(s) \quad (15)$$

is unappealing because it bears no obvious relation to the rate equation and because in conjunction with reaction 6 it would upon completion (i.e. at a sufficiently great depth) produce FeS and FeS_2 in at least a one to one mole ratio which contradicts field studies that commonly show almost total conversion of FeS to FeS_2 at depth. On the other hand the combination of reactions 6 and 7 allow but, as will be seen, do not demand complete conversion to FeS_2. Although other reactions involving siderite formation, methane formation, N_2 reduction and/or reactions with organic complexes have been proposed for pyrite formation (14) none of them have been studied kinetically and thus simply cannot be considered at this time. Finally although Rickard reports that framboidal pyrite was not produced during his experiments this does not necessarily mean that framboidal pyrite formation does not obey the stoichiometry and kinetics of equations 7 and 10.

At this point construction of the model is complete except for incorporation of H^+ into the system of equations. As discussed above, H^+ is an important variable in the kinetics of iron sulfide formation.

However because the rate law expressions for iron sulfide formation (equations 8 and 10) are non-linear the differential equations for H_2S and the iron sulfides are not amenable to explicit solution. Thus it is important to develop an equation for H^+ that can be incorporated in a numerical solution technique such as that of Runge-Kutta (15). Fortunately an appropriate differential equation for H^+ can be developed from charge balance considerations. Here it is assumed that dissolved substances other than those listed in Table I are not affected by diagenesis. If this is true, then a charge balance difference equation can be written (16):

$$\begin{aligned}(NH_4^+) - (NH_4^+)_o + (H^+) - (H^+)_o &= (OH^-) - (OH^-)_o \\ &+ 2(SO_4^{2-}) - 2(SO_4^{2-})_o + (HCO_3^-) - (HCO_3^-)_o \\ &+ 2(CO_3^{2-}) - 2(CO_3^{2-})_o + (HS^-) - (HS^-)_o \qquad (16)\end{aligned}$$

where the subscripted quantities represent the concentrations of these substances in the overlying sea water. Here it should be pointed out that in developing the equations in Table I it has been tacitly assumed that pH dependent species of CO_2, H_2S and NH_3 (such as HCO_3^-, HS^-, etc.) have equivalent diffusion coefficients. In principle a separate differential equation could have been written for each species. However each of these would incorporate a pH dependent distribution function. The programing complexity that this would generate is not justified given the preliminary nature of the model and the other rough assumptions involved. In equation 16 each of the carbonate species can be expressed in terms of total CO_2 and H^+ by means of the appropriate carbonate equilibrium constants. The other pH dependent species such as HS^- can be treated in a similar fashion. In this manner equation 16 can be expressed in terms of the variables (H_2S), (SO_4), (CO_2), (NH_4) and (H^+), the initial concentrations and the various equilibrium constants. Using the chain-rule of partial differentiation (13, p. 335) the total differential of (H^+) can be obtained and isolated on the left hand of the equation. Division of both sides of the resulting equation by dx yields a first order differential equation of the form

$$\frac{d(H^+)}{dx} = f[(H_2S), (SO_4), (CO_2), (NH_4), (H_2S)', (SO_4)', (CO_2)', (NH_4)'] \qquad (17)$$

which can be incorporated in a Runge-Kutta algorithm. Space limitations unfortunately do not permit a more explicit derivation of this equation than that sketched here.

Development of this model has neglected several substances that in principle possibly could have been included. The most obvious omissions are dissolved phosphate released by the decomposition of organic matter and the possible dissolution or precipitation of $CaCO_3$. Both of these substances would affect the pH profile. The effect of phosphate on pH would probably be small because its concentration in organic matter is low compared to N and C. The effect of $CaCO_3$ on pH however could be important and thus application of the model should probably be restricted to $CaCO_3$ free sediments.

The Simulation Algorithm

The equations of the system presented in Table I are both non-linear and simultaneous. Thus there is no obvious way of obtaining explicit solutions and one must therefore resort to numerical techniques. Because of its efficiency and wide use, I chose a Runge-Kutta method with fourth-order accuracy (15). The details of the algorithm developed for this problem will eventually be published in an appropriate journal. The algorithm can in principle handle the system of equations presented in Table I. In practice however problems arise because some of the equations are second order and thus the algorithm must be supplied with initial values (i.e. at x = o) for both concentration and the first derivative. As will be discussed below the algorithm may generate unrealistic concentrations if inappropriate values are chosen for the first derivative. It should be emphasized that this does not result from deficiencies in the algorithm, such as error buildup, because such effects can be minimized by choosing sufficiently small depth increments. The performance of the algorithm with respect to error buildup was evaluated by having the algorithm compute profiles for a hypothetical sediment devoid of reactive iron oxide. For such a case the equations in Table I either vanish or uncouple and become linear so that explicit solutions can be obtained for each (except H^+). Using explicitly determined initial values for the required first derivatives and a step size of 0.05 cm, all of the numerically computed values agreed with those computed from their corresponding explicit functions to at least four digits to a depth of 100 cm.

Furthermore at each step the ratio of the change in cationic charge to the change in anionic charge (see equation 16) was 1.0000. Thus it appears that the chief limitations of the algorithm are those due to possible deficiencies in the conceptual framework of the model and to the appropriateness of the initial conditions supplied to the algorithm.

The problems associated with correctly specifying the initial first derivative values became evident during early attempts to simulate profiles of H_2S and iron sulfides. As pointed out earlier the differential equation for organic carbon has an explicit solution. This solution can be substituted for (OC) in the equations for SO_4, NH_4 and CO_2 which are linear and can thus be solved analytically. Thus the explicit solutions obtained can be used to calculate the correct initial first derivatives for these parameters. The equation for H_2S, however, is non-linear and coupled to those for $HFeO_2$, FeS, S^o and FeS_2. Thus there is no obvious analytical way to correctly specify the initial derivative for H_2S. However since FeS precipitation can not occur at the zero concentration of H_2S assumed at $x = o$, it seemed reasonable to assume that at $x = o$ the derivative of H_2S should be equal in magnitude but opposite in sign to that of SO_4. Utilization of this assumption in the algorithm resulted in H_2S concentrations that greatly exceed the concentrations of H_2S that would be produced in a system devoid of reactive iron oxide but otherwise similar. Furthermore simulation runs showed that although the algorithm obeyed mass balance constraints on Fe it invariably converts at depth essentially all of the initially supplied reactive $HFeO_2$ to FeS_2. In cases where the initial concentration of organic carbon is less than four times the initial concentration of $HFeO_2$, this behavior of the algorithm is also unrealistic because according to reaction 5 the maximum number of moles of reduced sulfur in all forms (H_2S, FeS, FeS_2, S^o) that can be produced is equal to one half of the initial number moles of decomposable carbon. Thus one is faced with the problem of either finding an initial derivative for H_2S that gives acceptable results or incorporating into the algorithm the appropriate organic carbon substrate constraint on iron sulfide formation. The first alternative was attempted on a trial and error basis but soon proved futile. Fortunately however I was able to modify the algorithm by incorporation of an appropriate organic carbon constraint.

The carbon constraint on iron sulfide formation is based on reaction 5 and Berner's (17) demonstration that the total sulfate reduced during burial to depth x, SR, is given by

$$SR = \frac{(OC)_o}{2}[1 - \exp(\frac{-Kx}{S})] \quad (18)$$

where $(OC)_o$ = initial concentration of organic carbon. Since the solution of the differential equation for (OC) is

$$(OC) = (OC)_o[\exp(\frac{-Kx}{S})] \quad (19)$$

it follows that $SR = 0.5[(OC)_o - (OC)]$.

In the case of a sediment devoid of iron oxide, such that none of the reduced sulfate is incorporated in solid phases such as FeS, the concentration of total dissolved H_2S equals $(SO_4)_o - (SO_4)$ providing the diffusion coefficients for H_2S and SO_4 are equivalent. Thus it is reasonable to assume that in a system where half of the reduced sulfate has been incorporated in solid phases as a result of reactions with iron oxide, the concentration of H_2S should be equal to one half of H_2S concentration that would prevail in a system devoid of iron oxide but otherwise similar. This idea can be expressed mathematically as

$$\frac{(H_2S}{(SO)_o - (SO_4)} = \frac{SR - 2(FeS_2) - (FeS) - (S^o)}{SR} \quad . \quad (20)$$

Thus if no iron sulfide genesis occurs because of the absence of iron oxide, then $(H_2S) = (SO_4)_o - (SO_4)$. Otherwise (H_2S) will be some fraction of $(SO_4)_o - (SO_4)$ depending on the amount of SR that has not been incorporated into solid phases as reduced sulfur. This constraint on H_2S and iron sulfide genesis was incorporated into the original algorithm in the following manner. At the end of the Runge-Kutta calculations for each depth increment the following tolerance ratio was computed:

$$SRC = [(SR - 2(FeS_2) - (FeS) - (S^o) - \frac{SR(H_2S)}{(SO_4)_o - (SO_4)}]/SR \quad . \quad (21)$$

If the carbon constraint is obeyed exactly SRC equals zero. If SRC lies between -0.05 and +0.05 the algorithm accepts the results and proceeds to the next increment. If SRC lies outside this range, the algorithm repeats the calculations for the increment with appropriately adjusted values of $(H_2S)'$ until tolerance is satisfied.

Simulation of Iron Sulfide Genesis

In order to illustrate the qualitative agreement of the carbon constrained model developed here with field data cited in the literature, several simulations of H_2S, FeS, FeS_2 and pH profiles are presented in Figures 1 and 2. Because of their appearance in the rate equations for iron sulfide kinetics these simulations have also been designed to illustrate the possible effects of pH and reactant molar surface areas on iron sulfide genesis. In both figures sulfate reduction and iron sulfide formation begin about one cm below the sediment surface. The profiles of pH above this depth have been simulated on the basis of a model developed (but not discussed here) for the oxidized zone of the sediment. This model incorporates reactions pertaining to the aerobic decay of organic matter, bioturbation and nitrogen diagenesis. Since reactions in the aerobic zone are assumed to be uncoupled from those in the zone of sulfate reduction the model for the aerobic zone has been used here simply as a means of varying the initial pH at the top of the zone of sulfate reduction.

As can be seen, the H_2S profiles for all four simulations shown in Figures 1 and 2 show maximum concentrations in the submillimolar range and subsequent decreases with depth. In sediment hosting active sulfate reduction and pyrite formation, H_2S concentrations attain maximum values of about 10^{-3} moles liter^{-1} and commonly decrease thereafter with depth (18, 19). In general incorporation of the carbon constraint in the algorithm produces simulations that agree with qualitative predictions of H_2S profiles by Goldhaber and Kaplan (18, p. 638). The simulations of FeS shown by jobs 433, 1502 and 1673 are similar to published profiles (19, 20) in that they show abrupt peaks just below the oxidized-reduced boundary. Also, as is commonly the case in natural sediments, the simulated peak concentrations of FeS are about an order of magnitude lower than the ultimate concentration of FeS_2. Thus in both magnitude and shape the simulations shown here bear strong resemblance to those recently described in the literature, suggesting that the carbon constrained model has potential application to the study of sulfur diagenesis in marine sediments.

The simulations shown in Figure 1 are designed to illustrate the possible effect of the initial pH at the

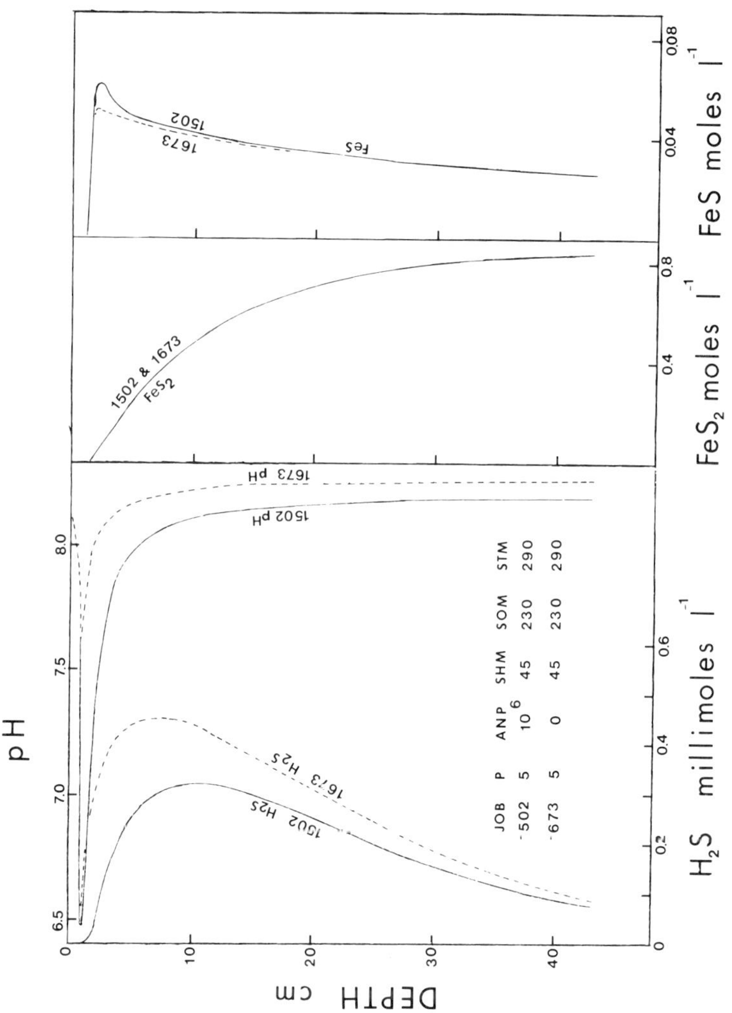

Figure 1. Simulations of sediment chemistry showing effects of differences in pH at the top of the zone of sulfate reduction. Values for model constants are listed in Table II unless otherwise on figure. P and ANP are parameters of the aerobic model that control the depth and pH at the oxidized-reduced boundary.

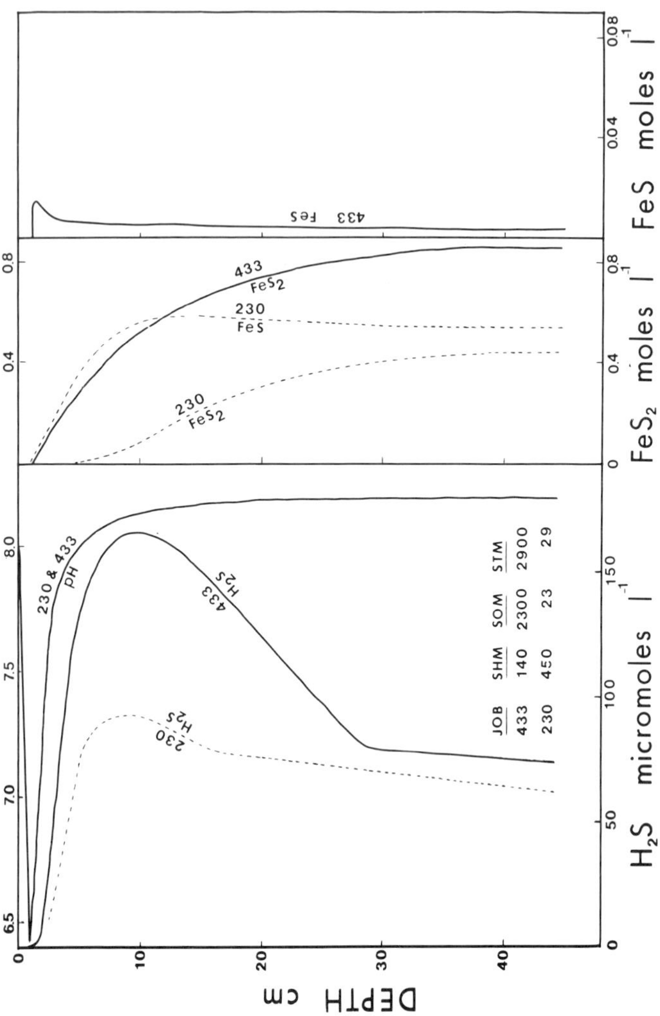

Figure 2. Simulations of sediment chemistry showing effects of differences in the molar surface areas of solid reactants. Values for model constants are listed in Table II unless otherwise on figure.

start of sulfate reduction on the course of iron sulfide genesis. The pH at the oxidized-reduced boundary must be controlled at least in part by reactions in the oxidized zone, such perhaps as ammonia oxidation. Thus it is conceivable that a comprehensive understanding of iron sulfide genesis might require an understanding of the factors that control the pH at this boundary. In both simulations the surface area parameters (SHM, STM, SOM) and the initial organic carbon concentration are identical. Thus the only difference between the simulations as far as iron sulfide genesis is concerned is that in job 1502 the starting pH is 6.47 whereas in job 1673 it is 7.75. As can be seen this difference in starting pH has no effect on the profile of FeS_2 and only slightly alters the maximum FeS concentration. Although the results of this simulation experiment are preliminary they suggest that the pH at the oxidized-reduced boundary is not likely to be an important factor in iron sulfide genesis. This results from the fact that the rapid depletion of SO_4^{2-} ion just below the oxidized-reduced boundary is not compensated by an adequate increase in dissolved H_2S. Thus charge balance consideration require a rapid rise in pH which soon dissipates the initial kinetic advantage of low pH.

The simulations shown in Figure 2 are designed to illustrate the effect of surface area. In both, the depth to the zone of sulfate reduction as well as the initial conditions at the beginning of sulfate reduction are identical except for the assumed molar surface areas. In job 433 the molar surface areas of S^o, SOM, and FeS, STM, are about an order of magnitude larger than the molar surface area of $HFeO_2$, SHM. Although the magnitudes of SOM, STM and SHM chosen for job 433 are larger than for jobs 1502 and 1673 (Figure 1) the ratios of SOM and STM to SHM are similar in all three jobs. As can be seen by comparing Figures 1 and 2 the profiles of FeS and FeS_2 in job 433 are very similar to those for jobs 1502 and 1673. On the other hand in job 230 (Figure 2) the molar area of $HFeO_2$ is about an order of magnitude larger than the molar areas of S^o and FeS so that the ratios of SOM and STM to SHM are about one hundred times smaller than they are in jobs 1502, 1673 and 433. This dramatic difference results in the precipitation of approximately equal amounts of FeS and FeS_2 at depth in job 230. Thus the size (and reactivity) of iron oxide particles supplied to the sediment could be an important factor in producing FeS enriched sediments such as those discussed by Berner (8) from the Black Sea.

Concluding Remarks

Although the above discussion does not constitute a rigorous verification of the model (21), the similarity of the simulations presented here to observed profiles of H_2S, FeS and FeS_2 in marine sediments indicates that the model may be of value in the study and interpretation of vertical patterns in sulfur diagenesis. Comprehensive multiparameter analyses of sediment profiles from a variety of sites will be required to validate the model. In this endeavor techniques will have to be devised to ascertain the molar surface areas of the various solid phase reactants. Eventually it may be possible to expand the model presented here to include processes in the aerobic zone so that the depth to the oxidized-reduced boundary can be predicted as well as the pH profile through this boundary. This achievement would constitute a truly comprehensive model.

Acknowledgements

The author gratefully acknowledges the assistance of R. Underwood and G. Johnson who guided the author in developing the Runge-Kutta algorithm. W. E. Sharp, W. S. Moore and M. Goldhaber reviewed the manuscript and provided helpful suggestions and discussions.

Abstract

This paper proposes a system of ten non-linear, simultaneous differential equations which, upon further development and validation, may serve as a comprehensive chemical model for sulfide dominated marine sediments. These equations were developed in accordance with R. A. Berner's mathematics of diagenesis and incorporate D. T. Rickard's work on the kinetics of iron sulfide formation. A Runge-Kutta algorithm has been developed to provide numerical solutions for the equations. When utilized with an organic carbon substrate limitation on the production of reduced sulfur, the algorithm generates profiles of H_2S, FeS and FeS_2 that agree qualitatively with measured profiles reported in the recent literature. Experiments with the algorithm suggest that the ratio of FeS to FeS_2 at depth depends strongly on the initial molar surface area of goethite but that the profiles of FeS, FeS_2, and H_2S are not greatly affected by the initial pH.

Literature Cited

1. Berner, R.A. Stoichiometric models for nutrient regeneration in anoxic sediments. Limnol. Oceanogr. 22, 781-786 (1977).
2. Berner, R.A. Diagenetic models of dissolved species in the interstitial waters of compacting sediments. Amer. Jour. Sci. 275, 88-96 (1975).
3. Berner, R.A. Kinetic models for the early diagenesis of nitrogen, sulfur, phosphorus, and silicon in anoxic marine sediments, p. 427-450, in Goldberg, E.D. (ed.), "The Sea," vol. 5, John Wiley & Sons, New York, 1974.
4. Berner, R.A. An idealized model of dissolved sulfate distribution in recent sediments. Geochim. Cosmochim. Acta 28, 1497-1503 (1964).
5. Martens, C.S. and Berner, R.A. Interstitial water chemistry of anoxic Long Island sound sediments. 1. Dissolved gases. Limnol. Oceanogr. 22, 10-25 (1977).
6. Lasaga, A.C. and Holland, H.D. Mathematic aspects of non-steady-state diagenesis. Geochim. Cosmochim. Acta 40, 257-266 (1976).
7. Toth, D.J. and Lerman, A. Organic matter reactivity and sedimentation rates in the ocean. Amer. Jour. Sci. 277, 465-485 (1977).
8. Berner, R.A. "Principles of Chemical Sedimentology," 204 p. McGraw-Hill Book Co., New York, 1971.
9. Vanderborght, J., Wollast, R., and Billen, G. Kinetic models of diagenesis in disturbed sediments. Part 2. Nitrogen diagenesis. Limnol. Oceanogr. 22, 794-803 (1973).
10. Vanderborght, J. and Billen, G. Vertical distribution of nitrate concentration in interstitial water of marine sediments with nitrification and denitrification. Limnol. Oceanogr. 20, 953-961 (1975).
11. Rickard, D.T. Kinetics and mechanism of the sulfidation of goethite. Amer. Jour. Sci. 274, 941-952 (1974).
12. Rickard, D.T. Kinetics and mechanism of pyrite formation at low temperature. Amer. Jour. Sci. 275, 636-652 (1975).
13. Hildebrand, F.B. "Advanced Calculus for Applications," 646 p. Prentice-Hall, Inc., Englewood Cliffs, New Jersey, 1965.
14. Berner, R.A. Sedimentary pyrite formation. Amer. Jour. Sci. 268, 1-23 (1970).

15. Froberg, C.D. "Introduction to Numerical Analysis," 433 p. Addison-Wesley Publishing Co., Reading, Massachusetts, 1969.
16. Gardner, L.R. Chemical models for sulfate reduction in closed anaerobic marine environments. Geochim. Cosmochim. Acta 37, 53-68 (1973).
17. Berner, R.A. Sulfate reduction, pyrite formation, and the oceanic sulfur budget, p. 347-361, in Dryssen, D. and Jagner, D., ed., "The Changing Chemistry of the Oceans," Nobel Symposium 20, Almquist and Wiksell, Stockholm, 1972.
18. Goldhaber, M.B. and Kaplan, I.R. The sulfur cycle, p. 569-655, in Goldberg, E.D., ed. "The Sea," vol. 5, John Wiley & Sons, New York, 1974.
19. Jorgensen, B.B. The sulfur cycle of a coastal marine sediment (Limfjorden, Denmark). Limnol. Oceanogr. 22, 814-831 (1977).
20. FOAM (Friends of Anoxic Mud): Goldhaber, M.B., Allen, R.C., Cochran, J.K., Rosenfeld, J.K., Martens, C.S., and Berner, R.A. Sulfate reduction, diffusion, and bioturbation in Long Island sound sediments: Report of the Foam Group. Amer. Jour. Sci. 277, 193-237 (1977).
21. Behrens, J.C., Beyer, J.E., Madsen, O.B.G. and Thomsen, P.G. Some aspects of modelling the long-term behaviour of aquatic ecosystems. Ecological Modelling 1, 163-198 (1975).
22. Berner, R.A. Sulfate reduction and the rate of deposition of marine sediments. Earth Planet. Sci. Lett. 37, 492-498 (1978).

RECEIVED November 16, 1978.

36

WATEQ2—A Computerized Chemical Model for Trace and Major Element Speciation and Mineral Equilibria of Natural Waters

JAMES W. BALL and EVERETT A. JENNE
U.S. Geological Survey, Water Resources Division, Menlo Park, CA 94025
DARRELL KIRK NORDSTROM
Department of Environmental Sciences, University of Virginia, Charlottesville, VA 22903

The protection of ecosystems, upon which our health and lives depend (1), requires that we understand natural processes and develop the capability to predict the effect of changes, such as the addition of pollutants, on these ecosystems. The prediction of trace-element behavior in ecosystems requires a multicomponent model by which one can: 1) calculate aqueous speciation of the trace elements among both natural organic and inorganic ligands; 2) evaluate solubility hypotheses; 3) account for sorption-desorption processes; and 4) incorporate chemical kinetics. This paper documents a chemical model that partially accomplishes the first two of these four goals. The present model has evolved from WATEQ, the earlier water-mineral equilibria model written in Pl/1 by Truesdell and Jones (2, 3), and from WATEQF, the Fortran version of Plummer *et al*. (4). These models, in turn, drew on the preceding model of Barnes and Clarke (5). The related PL/1 model, SOLMNEQ (6), drew on the models of Barnes and Clarke (5) and a prepublication version of Truesdell and Jones (2) as well as the thermodynamic data treatment of Helgeson (7) and Helgeson *et al*.(8).

The WATEQ program contains an extensive thermodynamic data base which was carefully selected for use with low-temperature natural waters (9, 2). Activity coefficients for the major ions are calculated from a computer fit of an extended Debye-Hückel equation containing two adjustable parameters (2, 3). These activity coefficients are considered more reliable than the standard Debye-Hückel equation or the Davies equation for high ionic strength solutions (up to 1-3 molal). The method of calculation in WATEQ is back-substitution for the cations and successive approximation for the anions with convergence on mass balance for anions. WATEQF changed to the more rapid back-substitution method for anion mass balance convergence. In addition, manganese speciation is included in WATEQF, and an option for calculating activity coefficients by either the Debye-Hückel or the Davies equation is provided. WATEQ2 retains most of these features, and additional modifications are explained below.

0-8412-0479-9/79/47-093-815$05.25/0
This chapter not subject to U.S. copyright
Published 1979 American Chemical Society

We have added ten additional elements (Ag, As, Cd, Cs, Cu, Mn, Ni, Pb, Rb, and Zn), complexes of Br and I, several metastable solids, some sparingly soluble salts, and several ion pairs of major constituents to the model. Other changes include reorganization of the computer code into a series of external subroutines and changing the mode of convergence to decrease the number of iterations required.

Because of the interactive nature of aqueous solute speciation calculations, it would be desirable to enter at once into the chemical model the reactions and thermodynamic data for all elements whose inclusion might affect the computed activity or equilibrium solubility of other solute species. However, our experience is that the greatest reliability is obtained by adding only the data for one element, or for one ligand group, at a time; then test data sets and real world water sample analyses are run before making further additions to or changes in the model.

Various associates, whom we have frequently called on for specialized knowledge and information, have materially assisted in this modeling effort. Collaborative studies have often provided the impetus to add some specific element, ligand group, or group of solid phases to the model. Apparent oversaturation with one or more solid phases of an element has often prompted us to seek out and add data for additional solute complexes or more soluble solid phases. The partitioning of an unexpectedly large portion of an element into a particular complex has led us to make an expanded compilation for the complex or to consult with colleagues to assist in selecting best values. Colleague criticisms (constructive and kind for the most part) of studies in press and in preparation have prompted us to make specific tests and proceed immediately with some change or addition, which would otherwise have awaited a "more opportune time." L. N. Plummer provided frequent consultation and supplied a prepublication copy of the reactions and associated thermodynamic data for the manganese section of the WATEQF chemical model (4). B. F. Jones and A. H. Truesdell have also been particularly helpful on many occasions.

In our effort to collect the appropriate data and develop the requisite understanding of geochemical processes, we have developed some adjunct computer programs. These include AACALC (Atomic Absorption and emission spectrometry CALCulation), EQLIST (EQuilibrium computation LISTing), and EQPRPLOT (EQuilibrium computation PRinting and PLOTing). AACALC (FORTRAN) reduces atomic absorption or emission spectrometry data to concentrations, EQLIST (PL/1) constructs tables from the WATEQ2 (input) card file, and EQPRPLOT (FORTRAN) constructs ratio and scatter plots of dissolved constituents, activity products (AP), or activity product to solubility product ratios (AP/K) via computer terminal printer or tape-driven plotter.

Additions and Modifications to the Model

The thermochemical reaction values vary according to the way the reaction is written. Therefore, all reactions in the present model which have been added or revised, together with the selected thermochemical values, and operational information to facilitate input and output of the data, are available in an adjunct report (10).

Elements. Rather large sets of solute complexes and mineral phases have been added for Ag, As, Cd, Cu, Mn, Ni, Pb, and Zn. Additionally, Cs and Rb have been added to the model.

The merit of including Cs and Rb in the model, in spite of the near absence of solute complexes or pure solids, is that in the future the activities of the uncomplexed alkali metal ions can be used to compute their probable substitution into certain silicate minerals and to examine ion-exchange processes. Alkali metals complex with OH^-, Cl^-, and NO_3^- only at such high ionic strengths that the basic assumptions of a multicomponent ion-association model are no longer valid. In addition, thermodynamic data for these complexes are highly uncertain. For these reasons, such complexes have been dropped from the model. Similarly, there are data on alkali metal compounds such as Rb_2S which might have been included. However, the pure alkali metal sulfides are known to be highly unstable and/or deliquescent (11) and are rarely found as minerals. Therefore, they have not been included in the model.

The manganese sections of WATEQF (4) were heavily utilized in preparing a similar section for WATEQ2 (10). The solute portion was utilized in its entirety, but the $MnOH^+$ and $Mn(OH)_3^-$ association reactions were expressed in terms of H_2O and H^+ instead of OH^-, and the $HMnO_2^-$ complex, a duplication of the $Mn(OH)_3^-$ complex, was excluded. The following subset of the mineral species was selected: pyrolusite, birnessite, nsutite, bixbyite, hausmannite, pyrochroite, manganite, rhodochrosite, $MnCl_2 \cdot 4H_2O$, MnS(green), $MnSO_4$, $Mn_2(SO_4)_3$, $Mn_3(PO_4)_2$ and $MnHPO_4$. The following subset was excluded: MnO, $Mn(OH)_3$, $MnCl_2$, $MnCl_2 \cdot H_2O$, $MnCl_2 \cdot 2H_2O$, Mn_2SiO_4 and $MnSiO_3$. To the selected set were added six minerals for which thermochemical data are unknown, in order to obtain log AP values of the individual minerals for different waters. The six minerals are: cryptomelane ($K_{0.8}Mn^{2+}_{0.6}Mn^{4+}_{8}O_{17}$), hollandite ($Ba_{0.78}Fe^{2+}_{0.57}Mn^{2+}_{0.59}Mn^{4+}_{7}O_{16}$), psilomelane ($Ba_{0.78}Ca_{0.19}K_{0.03}Mn^{2+}Mn^{4+}_{7}O_{16} \cdot 2.5H_2O$), todorokite ($Ca_{0.393}Mg_{0.473}Mn^{2+}_{1.134}Mn^{4+}_{5}O_{12} \cdot 2H_2O$), lithiophorite ($Li_2Al_8Mn^{2+}_{2}Mn^{4+}_{10}O_{35} \cdot 14H_2O$), and rancieite ($Ca_{0.44}Mn^{2+}_{0.56}Mn^{4+}_{4}O_9 \cdot 3H_2O$).

Aqueous Complexes. All solute reactions are written as association (formation) reactions whereas the solid reactions are written as dissociation (dissolution) reactions. For mass balancing purposes, all solute reactions are written in terms of the free form of the parent species, so that all constants are for overall rather than stepwise reactions.

Polysulfides and Sulfide. The polysulfide complexes of Ag and Cu have been added to the model in an attempt to reduce the apparent oversaturation with $Ag_2S(s)$ calculated for San Francisco Bay waters (12). Calculation of the activity of polysulfide ions requires the assumptions: 1) the quantity of S_8° (free sulfur) is not a limitation on its reaction with bisulfide (HS^-) to form polysulfides; and 2) polysulfides are in equilibrium with bisulfide.

In addition to the sulfide complexes of the added trace elements, the $Fe(HS)_2^\circ$ and $Fe(HS)_3^-$ complexes (13) have been included to increase the rigor of the sulfide speciation. The sulfide reactions have been rewritten in terms of HS^- rather than S^{2-} since HS^- is the dominant sulfide ion in most waters.

Sulfate. Various published values for the association constants and association enthalpies of metal sulfate ion pairs and triplets (14, 15, 16, 17) show good agreement (± 10%) except for $NaSO_4^-$. Log K values for the formation of $NaSO_4$ range from the 0.226 value of Lafon and Truesdell (18) to the 1.17 value of Pytkowicz and Kester (19), as cited by Fisher (20). If the one low value of Lafon and Truesdell (18) and the high values of Fisher and Fox (21) and Fisher (20) are dropped, the remaining four values average 0.70 ± 0.05 (22, 23, 24, 25) which is identical to the value selected by Smith and Martell (26), who may have used the same evaluation technique. G. M. Lafon (Johns Hopkins U., personal communication, 1978) has suggested that the low value should be discounted and that the formation of a sodium sulfate ion triplet is unlikely. Most of the other association constants were obtained from R. M. Siebert and C. L. Christ (Continental Oil Co., U. S. Geol. Survey, personal communication, 1976) after comparing their values with those reported in the literature. Enthalpy values were also selected from the preliminary data of R. M. Siebert and C. L. Christ which had been evaluated by the Fuoss equation (27). Careful checking with published literature values showed no serious discrepancies and it was felt that using data from one source would help maintain internal consistency. The ion triplet $Fe(SO_4)_2^-$ was one exception. The log K for this complex is the average of the results of Izatt *et al.* (25) and Mattoo (28) which differ by less than 1%. The enthalpy of association has not been published but it has been estimated by assuming that the difference between it and $FeSO_4^+$ is equal to the difference between $Al(SO_4)_2^-$ and $AlSO_4^+$. Although the reliability of this estimation cannot easily be determined, it certainly is better than assuming $\Delta H_r = 0$.

Fluoride. For equilibrium calculations in acid solutions, stability constants are needed for HF°, HF_2^-, and $(HF)_2^\circ$ species. These species become important when the pH drops below 4.5 and fluoride concentration rises above 5×10^{-4} *M*. Several measurements have been made on the dissociation of HF° at 298.15°K and the agreement is excellent (29-37). The weighted mean value of log K = 3.169 ± 0.010 (1σ, unweighted) given in Ball et al. (10) was calculated from these investigations after dropping the high value of Patel et al. (34) and the low value of Vasil'ev and Kozlovskii (37) which is necessary in order to maintain consistency with the kinetic data of Kresge and Chiang (38, 39).

For the reaction:

$$HF^\circ + F^- \rightleftarrows HF_2^- \qquad (1)$$

the stability constant has a larger uncertainty owing to the competing HF° association equilibrium. Reported log K values range from 0.49 to 0.70 (29, 30, 32, 33, 35, 36) and the weighted mean value is 0.58 ± 0.05.

Aqueous HF dimers have been shown to exist by Warren (35) who measured a log K of 0.43 ± 0.05 for the reaction:

$$2HF^\circ \rightleftarrows (HF)_2^\circ. \qquad (2)$$

Enthalpy values for the calculation of temperature dependence are not available for reaction 2. The log K for the dissociation of HF° and for reaction 1 have been measured between 0 and 100°C by Broene and DeVries (29), Ellis (30) and Hamer and Wu (33). Vasil'ev and Kozlovskii (37) have also obtained enthalpy data for these reactions by calorimetric titration. The average ΔH = -3.46 ± 0.75 kcal mol^{-1} for HF° dissociation and ΔH = 1.09 ± 0.30 kcal mol^{-1} for reaction 1.

Silicate minerals are more soluble in natural waters having high fluoride concentrations and low pH values than in other waters. High concentrations of dissolved silica may be maintained by the formation of hexafluorosilicic acid:

$$Si(OH)_4^\circ + 6F^- + 4H^+ \rightleftarrows SiF_6^{2-} + 4H_2O. \qquad (3)$$

The equilibrium constant for this reaction has been measured at 25°C by Roberson and Barnes (40) and the enthalpy is estimated from the data of Wagman et al. (41). Reaction 3 is important in many applications: a) chemical processes involving volcanic gases and condensates (40), b) chemical reactions in acid, halogen-rich hot springs (42, 43), c) waters receiving effluent from phosphate processing plants (44), d) fluoridation of water supplies (45), e) analytical chemistry, such as in the determination of either fluoride or dissolved silica (46) and f) the formation and hydrothermal alteration of ore deposits (47, 48, 49).

The log K and ΔH for the formation of the aqueous complexes CaF^+, FeF^{2+}, FeF_2^+, FeF_3°, $BF(OH)_3^-$, $BF_2(OH)_2^-$ and $BF_3(OH)^-$ have been evaluated by Nordstrom and Jenne (50) and are in good agreement with those selected by Smith and Martell (26) who used a different evaluation procedure. This addition to WATEQ2 makes a fairly complete inventory of aqueous fluoride complexes.

Neutral and Polymeric Aluminum and Iron. The association constants and enthalpies of aluminum and iron hydroxides have been evaluated by comparing the critically selected data of Baes and Mesmer (51) with that of R. M. Siebert and C. L. Christ (personal communication, 1976). Differences between the two data sets are negligible and the final selection was from Baes and Mesmer (51) because data on more complexes are found there. Important new species added to the model are the polynuclear complexes $Fe_2(OH)_2^{4+}$ and $Fe_3(OH)_4^{5+}$. Some controversy has arisen over the existence of $Fe(OH)_3^\circ$ and $Al(OH)_3^\circ$. Baes and Mesmer (51) have indicated that although the formation constant of $Al(OH)_3^\circ$ is only known from one measurement (52) and has a large uncertainty, it is real, with a log K $\leq$ -15.0 for the reaction

$$Al^{3+} + 3H_2O \rightleftarrows Al(OH)_3^\circ + 3H^+. \quad (4)$$

Baes and Mesmer (51) also suggest that the log K for

$$Fe^{3+} + 3H_2O \rightleftarrows Fe(OH)_3^\circ + 3H^+ \quad (5)$$

is less than -12 and this agrees with the generally accepted value of -13.6 (53). Recently, Byrne (54) and Kester et al. (55) have presented evidence for the existence of $Fe(OH)_3^\circ$ and reconfirmed the value of the log K. We have therefore included both neutral species in the model.

Others. Equilibrium association constants calculated from free energy data (41) for two aqueous arsenic fluoride species, AsO_3F^{2-} and $HAsO_3F^-$, were so high that the two species accounted for virtually all the As^{5+} in several water samples, practically irrespective of the fluoride concentration. The E^H calculated from the activities of As^{3+} and As^{5+} under these conditions was near -4 volts, i.e. well below that at which water decomposes (0 to -0.83 volts from pH 0 to 14). From the original data of Dutt and Gupta (56) the log K = 2.832 for

$$H_3AsO_4^\circ + F^- = HAsO_3F^- + H_2O \quad (6)$$

and log K = -3.037 for

$$H_3AsO_4^\circ + F^- = AsO_3F^{2-} + H^+ + H_2O. \quad (7)$$

Thus, there appears to have been a computational error in converting the stability data of Dutt and Gupta (56) to standard free energies of formation.

Equilibrium log K_r° values calculated from free energy data (41) for two lead hydroxychlorides (PbOHCl, $Pb_2(OH)_3Cl$) did not agree with those of the original authors (57). However, revised ΔG_f° data (10) from NBS (B. R. Staples, Nat'l Bur. Stand., personal communication, 1978) agree very well with the original data.

Organic Ligands. The model has been expanded to permit sensitivity analyses of naturally occurring organic ligands. These composite ligand groups are referred to as fulvate and humate. The model defaults to molecular weights of 650 and 2000, respectively, for these two ligand groups. Reported molecular weights for these substances vary widely (S. A. Jacobs and E. A. Jenne, unpub. data, 1978). Therefore, if a concentration for either substance is used as input data, without an accompanying analytically determined molecular weight, a warning message is printed, and all pertinent output data are flagged. Reported equilibrium constants for these metal-ligand complexes also vary widely and should therefore be user supplied. In the absence of supplied values, the model defaults to data from Smith and Martell (26) for oxalic acid (10).

Solid Phases. Sulfates. Solubility product constants and free energies of formation for the jarosite mineral group (jarosite, natrojarosite, and hydronium jarosite or carphosiderite, as the hydrogen form is termed in the older literature) have been compiled by Nordstrom (58). Considerable discrepancies occur between different investigations because the solution equilibria are very complicated: several strong complexes are formed and attempts are seldom made to account for the effect of hydroxide and sulfate complexation of the cations involved on apparent solubility. There is also a lack of consistency between values for the jarosite solubility product constant, partly because different complexes were used. Of the four investigations made on jarosite, Brown's results (59, 60) must be discounted because of very large uncertainties in the results. A mean value of -98.80 ± 1.1 for the log K_{sp} has been selected for WATEQ2 from the works of Zotov *et al.* (61), Vlek *et al.* (62) and Kashkai *et al.* (63). If the dissolution reaction is written as:

$$6H^{+} + KFe_3(SO_4)_2(OH)_6(s) \rightleftarrows K^{+} + 3Fe^{3+} + 2SO_4^{2-} + 6H_2O \qquad (8)$$

then the log K is -14.8 ± 1.1. The log K for natrojarosite, written in the same manner as reaction 8, is -11.2 ± 1.0 from the work of G. Clifton (Continental Material Co., personal communication, 1977) and agrees with the value obtained by Kashkai *et al.* (63). Only one log K value is available for hydronium jarosite (63). The ΔH for the reaction 8 has been estimated to be -31.28 kcal mol^{-1} by utilizing the data of Zotov *et al.* (61) for the entropy of jarosite, the previously cited investigations for the mean ΔG_f° value and Wagman *et al.* (41, 64) for the enthalpies of the ions. The enthalpy for natrojarosite dissolution is derived from the entropy and free energy values given by G. Clifton (personal communication, 1977) and for hydronium jarosite a linear correlation between free energies and enthalpies was assumed for the jarosite group since no data are available.

We have observed melanterite, $FeSO_4 \cdot 7H_2O$, to be one of the common sulfate minerals produced by the oxidation of pyrite during weathering. Unfortunately, its solubility and related thermodynamic properties are not well established. The log K for melanterite dissolution has been derived from the free energies of formation of the constituent species and the greatest source of uncertainty lies with the ΔG_f° for Fe^{2+}. We prefer the value of -21.8 ± 0.5 (65) which results in a log K of -2.47. The enthalpy of dissolution is 2.86 kcal mol^{-1} based on ΔH_f° = -22.1 kcal mol^{-1} for Fe^{2+} from Larson and Hepler (65).

The log K and ΔH° values in Ball et al. (10) for the dissolution of epsomite have been obtained from the free energy and enthalpy data given in Wagman et al. (41) and Parker et al. (66). For the log K and ΔH° of potassium alum solubility, values were obtained from the free energies and enthalpies of Wagman et al. (41) and Kelly et al. (67).

Fluorite. The solubility and related thermodynamic properties of fluorite have had large uncertainties, i.e. 2 to 3 orders of magnitude. Nordstrom and Jenne (50) utilized simultaneous multiple regression analysis (68) to evaluate these thermochemical data. The revised log K (10) agrees quite well with the upper limit of fluorite ion activity product calculations of many geothermal waters in the western United States. Although a total uncertainty of ± 0.5 was assigned to the log K (to include analytical and computational errors), more recent investigations indicate that the log K falls between -10.5 and -11.0 (69, 70, 71) so that the uncertainty in the log K at 298.15°K is ±0.25. At this level of deviation the analytical and computational uncertainties inherent in the calculations of the ion activity product are likely to be greater than those in the thermodynamic properties.

Other alkaline earth fluorides (BaF_2, SrF_2) have been added to the model. However, they are less likely than their respective sulfates or carbonates to be solubility limiting phases.

Others. Thermochemical data for the ferrous chlorite, greenalite ($Fe_3Si_2O_5(OH)_4$), and phlogopite ($KMg_3AlSi_3O_{10}(OH)_2$) dissolution were taken from Plummer et al. (4) who used the free energy values of Eugster and Chow (72) for greenalite and Bird and Anderson (73) for phlogopite.

Solubility calculations were added for two allophanes, for which the equilibrium constants and formulae are a function of pH. Paces (74) found cold ground waters collected from springs in granitic rocks of the Bohemian Massif of Czechoslovakia to be supersaturated with respect to kaolinite while being unsaturated with respect to amorphous silica. He interpreted this as an indication that a metastable aluminosilicate more soluble than kaolinite was controlling the concentrations of alumina and silica in these waters. This aluminosilicate was further hypothesized to be of varied chemical composition, controlled by the mole fraction of silica and dissolved by the reaction:

$$[Al(OH)_3]_{(1-x)}[SiO_2]_x(s) + 3(1-x)H^+ \rightleftarrows (1-x)Al^{3+} + xH_4SiO_4^\circ + (3-5x)H_2O \quad (9)$$

In equation 9, x is the mole fraction of silica and is equal to 1.24 - 0.135pH. This expression describes the linear variation between pure amorphous hydrous alumina and silica as a function of pH (75). The equilibrium constant for this substance was calculated by combining two endmember constants from the literature and incorporating the pH-dependence equation into the resulting expression, yielding an expression for the equilibrium solubility (75) of:

$$\log K = -5.7 + 1.68pH. \quad (10)$$

Under field conditions the solubility of this material should be lower due to the large difference in the speed of crystallization of amorphous alumina versus amorphous silica. In fact, a best-fit line to field samples from the Sierra Nevada is described by the equation (75):

$$\log K = -5.4 + 1.52pH. \quad (11)$$

Copper ferrites have been included in the model, but have as yet not been found to be equilibrium controls on copper or iron solubility. The calculated activity products for the two minerals, cuprous ferrite and cupric ferrite, are characteristically several orders of magnitude oversaturated when compared to their respective equilibrium constants in a wide variety of surface waters.

Ponnamperuma *et al.* (76) describe a ferrosoferric hydroxide ($Fe_3(OH)_8$), offering evidence that most iron (II) in reduced soils other than acid sulfate soils is present in this form. Using a log K of 17.56 from Ponnamperuma *et al.* (76) for the reaction:

$$2Fe(OH)_3(s) + Fe^{2+} + 2OH^- \rightleftarrows Fe_3(OH)_8(s) \quad (12)$$

and log K values for the ionization of water and ferric hydroxide dissolution, we calculate a log K of 20.222 for the reaction:

$$Fe_3(OH)_8(s) + 8H^+ \rightleftarrows 2Fe^{3+} + Fe^{2+} + 8H_2O \quad (13)$$

Biedermann and Chow (77) describe a ferric hydroxy-chloride ($Fe(OH)_{2.7}Cl_{0.3}$) which has been seen to precipitate from sea water, having a log K of 3.04 ± 0.05 (51) for the reaction:

$$Fe(OH)_{2.7}Cl_{0.3}(s) + 2.7H^+ \rightleftarrows Fe^{3+} + 2.7H_2O + 0.3Cl^-. \quad (14)$$

Chien and Black (78) calculate a log K of -114.4 for the fluorocarbonato apatite reaction:

$$Ca_{9.5}Na_{.36}Mg_{.144}(CO_3)_{1.2}F_{2.48}(PO_4)_{4.8}(s) \rightleftarrows 9.5Ca^{2+} + 0.36Na^{+} + 0.144Mg^{2+} + 1.2CO_3^{-} + 2.48F^{-} + 4.8PO_4^{3-} \quad (15)$$

These reactions have been added to the model.

Morey et al. (79) have calculated an equilibrium constant for amorphic silica which best fits their field data. The log K for this reaction of -2.71 has also been added.

Redox Couples. The model calculates the redox potential of the couples: H_2O_2/O_2, H_2O/O_2, Fe^{2+}/Fe^{3+}, NO_2^-/NO_3^-, S^{2-}/SO_4^{2-}, and As^{3+}/As^{5+}, given the requisite concentrations of the couple members. Dissolved oxygen is all that is required for calculation of both the H_2O/O_2 and H_2O_2/O_2 couples. The H_2O/O_2 couple is kinetically inhibited and is grossly out of equilibrium except at elevated temperatures (80). Therefore, the option of using pE from dissolved oxygen for redox speciation has been dropped from the model.

Recent studies (81) show that when the following three conditions are fulfilled, the platinum electrode provides a reliable and accurate estimate of the ferrous-ferric redox potential, $E^H_{Fe^{2+}/Fe^{3+}}$, in acid mine drainage waters. The conditions are: 1) large volumes of water must flow past the electrodes during emf measurement; 2) water samples must be properly filtered ($\leq$ 0.1 μm membrane) and preserved for ferrous-ferric iron analyses; and 3) activities of Fe^{2+} and Fe^{3+} rather than concentrations must be used to compute the potential of the ferrous-ferric couple. Other studies (R. E. Stauffer, E. A. Jenne, J. W. Ball, unpub. data, 1974) show that when a flow-through cell is used for emf measurements of geothermal waters, the platinum electrode potential mirrors decreasing dissolved sulfide and then increasing dissolved oxygen values in the downstream drainage of individual hot springs.

As alluded to above, input data for total iron, Fe(II) and/or Fe(III) are accepted by the model, with solute modeling calculations done using whatever data are input. If either Fe(II) or Fe(III) are present, Fe(total) is ignored; if Fe(II) only is present, speciation is done among Fe(II) complexes only, and likewise for Fe(III). To accomplish this, the reactions of the iron section have been extensively rewritten (10) and a procedure, named SPLIT_IRON, has been added, which performs the mass balance calculations separately for Fe(II) and Fe(III) when they are input separately. An E^H value is calculated from the computed activities of Fe^{2+} and Fe^{3+} and may, by user option, be used to distribute other redox species in lieu of an input E^H value. If only Fe(total) is input, the input E^H value is used to distribute all redox species including Fe^{2+} and Fe^{3+}; if there is only Fe(total) input, and no input E^H value, all Fe calculations are bypassed.

Standard Deviations. In order to evaluate the effect on the modeling calculations of errors in the analytical input data, propagated standard deviations are now computed for a subset of the solid phase activity products considered in the model. Arrangements have also been made to enter and output standard deviations for thermodynamic data.

A new procedure named ERRCALC was written to calculate the propagated standard deviation. The basic equation for the activity product of a solid phase is:

$$\log_{10}AP = Ax + By \tag{16}$$

where AP is the activity product, x and y are the log activities of the constituent ions in solution, and A and B are the number of ions of each species making up the solid. From Bevington (81, p. 60) we construct the following equation to calculate the standard deviation in the $\log_{10}$ of the activity product, which in practice is the sum of the $\log_{10}$ of the activities of the constituent ions:

$$\sigma^2_{\log_{10}AP} = A^2\sigma^2_x + B^2\sigma^2_y \;. \tag{17}$$

Unfortunately, the standard deviation in the $\log_{10}$ of the activity of the species is not known; rather the known quantity is the standard deviation in the input analytical value. Therefore, a second equation is required for x and y as the $\log_{10}$ of another number. From Bevington (82, p. 63-64),

$$\text{for:}\quad x = a\ln(\pm bu) \tag{18}$$

$$\sigma_x = a\,\frac{\sigma_u}{u} \tag{19}$$

where u and x are the activity and $\log_{10}$ of the activity, respectively, of the species, and a and b are weighting constants. If a is set equal to 1/2.303 (i.e., 1/ln (10)), equation 18 becomes:

$$x = \log_{10}(\pm bu) \tag{18a}$$

and equation 19 becomes:

$$\sigma_x = \frac{1}{2.303}\;\frac{\sigma_u}{u}\;. \tag{19a}$$

The standard deviation in the activity of the species is also unknown, but we have defined the propagated standard deviation to be strictly in terms of errors in analytical input data, so that the relative standard deviation of input data and computed activities are, under this definition, equal:

$$\frac{\sigma_v}{v} = \frac{\sigma_u}{u} \tag{20}$$

where u is the activity of the species, v is the analytical concentration of the species from which the activity of the species was derived and σ_u and σ_v are the respective standard deviations. Substituting from equation 20 to equation 19a we obtain:

$$\sigma_x = \frac{1}{2.303} \left[\frac{\sigma_v}{v}\right]. \qquad (19b)$$

Deriving an analogous expression for σ_y in terms of w and substituting it and the right side of 19b into 17 we obtain:

$$\sigma^2_{\mathrm{Log}_{10}AP} = A^2 \left(\frac{\sigma_v}{2.303v}\right)^2 + B^2 \left(\frac{\sigma_w}{2.303w}\right)^2 \qquad (17a)$$

where σ_w and w are the analogously obtained values for σ_y.

The sigma values thus computed are printed along with the results of the activity product calculations under the heading SIGMA(A), the uncertainty ascribed to the analytical values used. The uncertainties ascribed to the thermodynamic data are labeled SIGMA(T).

Sigma values are routinely obtained for analyses performed in our laboratory. However, we are utilizing a considerable amount of literature data to evaluate the role of various solid phases as effective solubility controls. These data rarely contain either sigma values or the data from which they can be calculated. Therefore, we have resorted to use of twice the uncertainty resulting from rounding of the analytical values as a minimum estimate of the analytical uncertainty. These are not distinguished in the model from true sigma values.

Other Modifications. All new reactions which include OH^- are written in terms of H_2O and H^+, following the convention of Baes and Mesmer (51). If the reactions are written in terms of OH^-, errors in the dissociation of water and in its temperature dependence are encountered. These errors are thus avoided since A_{H^+} is a directly measured quantity. This also has the advantage of making the propagated standard deviations for different solid phases more comparable than if some are written in terms of OH^- and some in terms of H_2O and H^+. Most of the preexisting reactions which include OH^- were similarly modified to conform to this convention.

Other features of WATEQF (4) which have been incorporated into WATEQ2 include: 1) revised set of analytical expressions for the effect of temperature on the stability constant (10); 2) printed table of log K values for which analytical expressions exist, following the listing of the input thermochemical data (10); 3) check on the charge balance computed from input data, with aborting of the calculations if the balance is off by >30%; 4) revised anion mass balance calculation, allowing for faster convergence; and 5) improved set of headings used in the printed results of the solute modeling calculations.

Specific conductance calculated from input major constituent data using the method of Laxen (83) has been added to the model as a check on analytical input data. Differing input and calculated specific conductances indicate that one or more errors may exist in the analytical input data.

A mass balance section for the hydrogen sulfide species was added to the anion mass balance calculations when we observed that strong HS^- complexing of some trace metals sometimes rendered cation mass balance convergence impossible.

Activity coefficients were originally calculated using the extended Debye-Hückel equation and whenever a new complex was added to the program it was necessary to estimate the å parameter. This problem was overcome by substituting the more general Davies equation which has adequate reliability at low ionic strengths and is usually more accurate at high ionic strengths (84). Since acid mine waters can have ionic strengths approaching that of sea water, it is desirable to use a theory for activity coefficients that can reach somewhat above 0.1 molal, the usual upper limit for extended Debye-Hückel calculations. The Davies equation is considered satisfactory to 0.5 molal. The extended Debye-Hückel equation with fit parameters (2, 3) has been retained for the major ions, Ca, Mg, Na, K, Cl and SO_4, and the Debye-Hückel equation is used to calculate the polysulfide activity coefficients, for which å parameters have been estimated by Cloke (85).

There are inconsistencies in the model for the calculation of activity products for the "clays." Exchangeable cations are disregarded for the low exchange capacity kaolinite, halloysite, chlorite, and moderate capacity illite. For certain expansible layer silicates and two zeolites, the $\log_{10}$ of the activity of selected cations is added into the sum of the activity products. The mineral phases treated in this manner, and the solute cations considered as exchangeable cations, are beidellite ($[A_{Mg^{2+}}]^{\frac{1}{2}} + A_{Na^+} + A_{K^+}$), clinoptilolite and mordenite ($A_{Na^+} + A_{K^+}$), Belle Fourche montmorillonite and Aberdeen montmorillonite ($A_{H^+} + A_{Na^+} + A_{K^+}$). Note that the square root of the divalent cation is used in the sum in keeping with the practice in the ion exchange literature (86). Revision of the calculation of exchangeable cation contribution to the activity product has been delayed pending the pertinent reviews of Kittrick (87), as well as that of Bassett et al. (88) presented at this symposium.

Modifications in the Code

The PL/1 language computer code has been extensively altered in the process of building it into WATEQ2; in fact, minor alterations are far too numerous to mention here. Several errors in the original code were corrected and some major changes, noted below, were made to improve program execution and ease of use and to broaden its usefulness. The input and output aspects of program operation are given, along with the thermodynamic data base, in a supplementary report (10).

Arrays which must be increased in size when species are added are now automatically adjustable in dimension merely by supplying appropriate input data. The results of solute and mineral calculations will not appear if the activity or activity product, respectively, has not been calculated. This eliminates extraneous non-information and shortens the listing considerably, an advantage especially when a simple laboratory solution is considered and/or a low-speed remote computer terminal is utilized.

Some data are entered into the model as "carried-only" data, primarily for plotting using a subsequent computer program. However, as the model evolves, some of these carried-only data become input to the model itself or to adjunct calculations. Specific conductance, which was initially carried-only data, is now compared to a computed "specific conductance" as a quality-of-analysis screening technique.

The listing of the results of the mineral equilibrium calculations has been drastically altered, with the deletion of ΔG_r, ΔG_r per equivalent cation, and all values in base 10 form. Information now printed for each species for which an activity product is calculated includes log AP/K, SIGMA (Analytical), SIGMA (Thermodynamic), log AP/K_{min}, and log AP/K_{max}. As discussed previously, SIGMA(A) is the propagated standard deviation in the analytical values and SIGMA(T) is the standard deviation in the thermodynamic data. The $\log_{10}K_{min}$ and $\log_{10}K_{max}$ values have been changed from $\pm$ 5% of the $\log_{10}K$ value (2) to experimentally determined values which may represent a less soluble or more soluble form of the solid phase than that selected as the "best" value.

WATEQ2 consists of a main program and 12 subroutines and is patterned similarly to WATEQF (4). WATEQ2 (the main program) uses input data to set the bounds of all major arrays and calls most of the other procedures. INTABLE reads the thermodynamic data base and prints the thermodynamic data and other pertinent information, such as analytical expressions for effect of temperature on selected equilibrium constants. PREP reads the analytical data, converts concentrations to the required units, calculates temperature-dependent coefficients for the Debye-Hückel equation, and tests for charge balance of the input data. SET initializes values of individual species for the iterative mass action-mass balance calculations, and calculates the equilibrium constants as a function of the input temperature. MAJ_EL calculates the activity coefficients and, on the first iteration only, does a partial speciation of the major anions, and performs mass action-mass balance calculations on Li, Cs, Rb, Ba, Sr and the major cations. TR_EL performs these calculations on the minor cations, Mn, Cu, Zn, Cd, Pb, Ni, Ag, and As. SUMS performs the anion mass action-mass balance calculations, and tests the results against input concentration values for the anions CO_3^{2-}, SO_4^{2-}, F^-, PO_4^{3-}, Cl^- and S^{2-}, and prints the results of each set of iterative calculations. MAJ_EL, TR_EL and SUMS are executed repetitively until mass balance to within 0.1% of the input concentrations is achieved for the six anions, or until 40 iterations have elapsed. If convergence is not reached in 40 iterations, a warning message is printed and execution continues just as through convergence had been reached. SOLUTES performs computations not related to the mass balance calculations, such as E^H, specific conductance, pO_2 and pCH_4 calculations, prints out all the solute data, and performs necessary logarithm conversions for use in subsequent calculations. RATIO calculates and prints mole ratios calculated from analytical molality and log activity ratios. APCALC calculates thermodynamic activity products for the various mineral species considered by WATEQ2. OUTPNCH generates a card deck of a subset of the calculated activities, activity products and input concentrations for subsequent use with plotting programs. ERRCALC, discussed previously, uses input analytical standard deviations to calculate the propagated standard deviation in the log of the activity products for a subset of minerals considered. PHASES prints the results of the activity product and error calculations, and computes and prints the saturation state of each mineral with respect to a thermodynamic equilibrium constant for each reaction considered.

Acknowledgements

We are pleased to acknowledge the efforts of L. N. Plummer and R. W. Potter II, both of the U. S. Geological Survey, for their careful reviews of this manuscript; to B. F. Jones, U. S. Geological Survey, for many helpful discussions; and to J. M. Burchard, whose help in many aspects of this work is especially appreciated.

Abstract

The computerized aqueous chemical model of Truesdell and Jones (2, 3), WATEQ, has been greatly revised and expanded to include consideration of ion association and solubility equilibria for several trace metals, Ag, As, Cd, Cu, Mn, Ni, Pb and Zn, solubility equilibria for various metastable and(or) sparingly soluble equilibrium solids, calculation of propagated standard deviation, calculation of redox potential from various couples, polysulfides, and a mass balance section for sulfide solutes. Revisions include expansion and revision of the redox, sulfate, iron, boron, and fluoride solute sections, changes in the possible operations with Fe (II, III, and II + III), and updating the model's thermodynamic data base using critically evaluated values (81, 50, 58) and new compilations (51, 26; R. M. Siebert and

C. L. Christ, unpublished data 1976). Mechanical revisions include numerous operational improvements in the computer code.

Literature Cited

1. Odum, E. P. The emergence of ecology as a new integrative discipline: Science 195, 1289-1293 (1977).
2. Truesdell, A. H., and Jones, B. F. WATEQ, a computer program for calculating chemical equilibria of natural waters: NTIS-PB2-20464, 77 p. (1973).
3. Truesdell, A. H., and Jones, B. F. WATEQ, a computer program for calculating chemical equilibria of natural waters: U. S. Geol. Survey J. Res. 2(2) 233-274 (1974).
4. Plummer, L. N., Jones, B. F., and Truesdell, A. H. WATEQF - A Fortran IV version of WATEQ, a computer program for calculating chemical equilibrium of natural waters: U. S. Geol. Survey Water-Resour. Invest. 76-13, 61 p. (1976).
5. Barnes, Ivan and Clarke, F. E. Chemical properties of ground water and their corrosion and encrustation effect on wells. U. S. Geol. Survey Prof. Paper 498-D, 58 p. (1969).
6. Kharaka, Y. K., and Barnes, Ivan. SOLMNEQ: Solution-mineral equilibrium computations: U. S. Geol. Survey Water Res. 73-002, 88 p. (NTIS PB-215 899) (1973).
7. Helgeson, H. C. Thermodynamics of hydrothermal systems at elevated temperatures and pressures. Amer. J. Sci. 267, 729-804 (1969).
8. Helgeson, H. C., Brown, T. H., Nigrini, A., and Jones, T. A. Calculation of mass transfer in geochemical processes involving aqueous solutions. Geochem. Cosmochem. Acta, 34, p. 569-592 (1970).
9. Nathenson, M., "Thermodynamic calculations," NTIS Tech. Report PB214372, 29 p. (1973).
10. Ball, J. W., Jenne, E. A., and Nordstrom, D. K. Additional and revised thermochemical data and computer code for WATEQ2-- A computerized chemical model for trace and major element speciation and mineral equilibria of natural waters. U. S. Geol. Survey Water Res. Invest. 78-116 (in press).
11. Weast, R. C., ed., "CRC Handbook of Chemistry and Physics" Chemical Rubber Co., Cleveland, Ohio, 51st Ed. 1970-1971.
12. Jenne, E. A., Girvin, D. C., Ball, J. W., and Burchard, J. M. Inorganic speciation of silver in natural waters - fresh to marine, Chapter 4, p. 41-61, in Klein, D. A., ed., "Environmental Impacts of Nucleating Agents Used in Weather Modification Programs," 256 p. Dowder, Hutchinson and Ross, Stroudsberg, Pa. 1978.
13. Naumov, G. B., Ryzhenko, B. N. and Khodakovsky, I. L. "Handbook of Thermodynamic Data," 328 p. U. S. Geol. Survey WRD-74-001, NTIS PB 226 722/AS. 1974.

14. Sillen, L. G., and Martell, A. E. Stability constants of metal ion complexes: Chem. Soc. Spec. Publ. No. 17, London, 754 p. (1964).
15. Sillen, L. G., and Martell, A. E. Stability constants of metal ion complexes. Suppl. No. 1: Chem. Soc. Spec. Publ. No. 25. 865 p. (1971).
16. Yatsimirskii, K. B. and Vasil'ev, V. P. "Instability Constants of Complex Compounds." 218 p. (Tr. from Russ., D. A. Paterson), Pergamon Press, New York, 1960.
17. Ashcroft, S. J., and Mortimer, C. T. "The Thermochemistry of Transition Metal Complexes." 478 p., Academic Press, New York, 1970.
18. Lafon, G. M., and Truesdell, A. H. Temperature dependence of sodium sulfate complexing in aqueous solutions. [abst.] Amer. Geophys. Union Trans. 52, p. 362 (1971).
19. Pytkovicz, R. M. and Kester, D. R. Harned's rule behavior of NaCl-Na_2SO_4 solutions explained by an ion association model. Amer. J. Sci. 267, 217-229 (1969).
20. Fisher, F. H. Dissociation of Na_2SO_4 from ultrasonic-absorption reduction in $MgSO_4$-NaCl solution. J. Sol. Chem. 4, 237-240 (1975).
21. Fisher, F. H. and Fox, A. P. $NaSO_4^-$ ion pairs in aqueous solutions at pressures up to 2000 atm. J. Sol. Chem. 4, 225-236 (1975).
22. Righellato, E. C., and Davies, C. W. The extent of dissociation of salts in water. Part II. Uni-bivalent salts. Trans. Faraday Soc. 26, 592-600 (1930).
23. Jenkins, I. L., and Monk, C. B. The conductances of sodium, potassium and lanthanum sulfates at 25°C. J. Amer. Chem. Soc. 72, 2695-2698 (1950).
24. Austin, J. M., and Mair, A. D. Standard enthalpy of formation of complex sulfate ions in water. I. HSO_4^-, $LiSO_4^-$, $NaSO_4^-$. J. Phys. Chem. 66, 519-521 (1962).
25. Izatt, R. M., Eatough, D., Christensen, J. J. and Bartholomew, C. H. Calorimetrically determined Log K, ΔH°, and ΔS° values for the interaction of sulfate ion with several bi- and tervalent metal ions. J. Chem. Soc. A, 47-53 (1969).
26. Smith, R. M., and Martell, A. E. "Critical Stability Constants. V. 4. Inorganic Complexes." 257 p., Plenum Press, New York, 1976.
27. Bockeris, J. O'M., and Reddy, A. K. N. "Modern Electrochemistry," 1432 p., Plenum Press, New York, 1970.
28. Mattoo, B. N. Stability of metal complexes in solution. III. Ion association in ferric sulfate and nitrate solutions at low Fe III concentrations. Z. Phys. Chem. [Frankfurt] 19, 156-167 (1959).
29. Broene, H. H., and DeVries, T. The thermodynamics of aqueous hydrofluoric acid solution. J. Amer. Chem. Soc. 69, 1644-1646 (1947).

30. Ellis, A. J. The effect of temperature on the ionization of hydrofluoric acid. J. Chem. Soc., 4300-4304 (1963).
31. Vanderborgh, N. E. Evaluation of the lanthanum fluoride membrane electrode response in acidic solutions. Talanta 15, 1009-1013 (1968).
32. Baumann, E. W. Determination of stability constant of hydrogen and aluminum fluorides with a fluoride-selective electrode. J. Inorg. Nucl. Chem. 31, 3155-3162 (1969).
33. Hamer, W. J., and Wu, Y.-C. The activity coefficients of hydrofluoric acid in water from 0 to 35°C. J. Res. Nat'l. Bur. Stand 74A, 761-768 (1970).
34. Patel, P. R., Moreno, E. C., and Patel, J. M. Ionization of hydrofluoric acid at 25°C. J. Res. Nat'l. Bur. Stand 75A, 205-211 (1971).
35. Warren, L. S. The measurement of pH in acid fluoride acid solutions and evidence for the existence of $(HF)_2$. Anal. Chim. Acta 53, 199-202 (1971).
36. Kresge, A. J., and Chiang, Y. Solvent isotope effects on the ionization of hydrofluoric acid. J. Phys. Chem. 77, 822-825 (1973).
37. Vasil'ev, V. P., and Kozlovskii, E. V. Thermochemistry of acid-base reactions in aqueous hydrofluoric acid solutions. Russ. J. Inorg. Chem. 18, 1544-1546 (1973).
38. Kresge, A. J., and Chiang, Y. Acid catalysis in hydrofluoric acid buffers. J. Amer. Chem. Soc. 90, 5309-5310. (1968).
39. Kresge, A. J., and Chiang, Y. Vinyl ether hydrolysis. V. Catalysis in dilute hydrofluoric acid solution. J. Amer. Chem. Soc. 94, 2814-2817 (1972).
40. Roberson, C. E., and Barnes, R. B. Stability of fluoride complex with silica and its distribution in natural water systems. Chem. Geol. 21, 239-256 (1978).
41. Wagman, D. D., Evans, W. H., Parker, V. B., Halow, I., Bailey, S. M., and Schumm, R. H. Selected values of chemical thermodynamic properties. Table for the first thirty-four elements in the standard order of arrangement. Nat'l Bur. Stand. Tech. Note 270-3, 264 p. (1968).
42. Ellis, A. J. Chemical processes in hydrothermal systems--a review, p. 1-26, in Ingerson, E., ed., "Proceedings of Symposium on Hydrogeochemistry," v. 1, Clarke, Washington, D. C. 1973.
43. Ozawa, T., Kamada, M., Yoshida, M., and Sanemasa, I. Genesis of acid hot spring, p. 105-121, in Ingerson, E., ed., "Proceedings of Symposium on Hydrogeochemistry," v. 1, Clarke, Washington, D. C. 1973.
44. Martin, D. F. and Taft, W. H. Occurrence and implication of sedimentary fluorite in Tampa Bay, Fla., p. 202-210, in Gibb, T. R. P., ed., "Analytical Methods in Oceanography." Amer. Chem. Soc. Mono. Ser. 147. 1975.

45. Crosby, N. T. Equilibria of fluorosilicate solution with special emphasis to the fluoridation of public water supplies. J. Appl. Chem. 19, 100-102 (1969).
46. Shapiro, L. Spectrophotometric determination of silica at high concentrations using fluoride as a depolymerizer. U. S. Geol. Survey J. Res. 2, 357-360 (1974).
47. Lovering, T. S., and Shepard, A. O. Hydrothermal alteration zones caused by halogen acid solutions, East Tintic District, Utah. Amer. J. Sci 258-A, 215-229 (1960).
48. Cadek, J., and Malkovsky, M. Transport of fluorine in natural waters and precipitation of fluorite at low temperatures. Acta Univ. Carolinae Geol. 4, 251-270 (1966).
49. Shawe, D. R., ed. Geology and resources of fluorine in the United States. U. S. Geol. Survey Prof. Paper 933, 99 p. (1976).
50. Nordstrom, D. K., and Jenne, E. A. Fluorite solubility equilibria in selected geothermal waters. Geochim. Cosmochim. Acta 41, 175-188 (1977).
51. Baes, C. F. Jr., and Mesmer, R. E. "The Hydrolysis of Cations," 458 p. John Wiley and Sons, New York, 1976.
52. Nazarenko, V. A., and Nevskaya, E. M. Chemistry of the reactions between the ions of multivalent elements and organic reagents. XVII. Interaction of aluminum and gallium ions with methylthymol blue. J. Anal Chim. USSR 24(6), 670-673 (1969).
53. Langmuir, D. The Gibbs free energies of substances in the system $Fe-O_2-H_2O-CO_2$ at 25°C. U. S. Geol. Survey Prof. Paper 650-B, B180-B184 (1969).
54. Byrne, R. H. Jr. "Iron speciation and solubility in seawater," Ph. D. Thesis, Univ. Rhode Island, Kingston, R.I., 1974.
55. Kester, D. R., Byrne, R. H. Jr., and Liang, Y.-J. Redox reactions and solution complexes of iron in marine systems, p. 56-79, in Church, T. M., ed., "Marine Chemistry in the Coastal Environment," Amer. Chem. Soc. Symp. Ser. 18, 1975.
56. Dutt, N. K. and Gupta, A. Fluorarsenates and their analogues with sulphates. Part III. Stability of fluorarsenate ion and its comparison with fluorphosphate ion. J. Indian Chem. Soc. 38, 249-252 (1961).
57. Deschamps, Pierre, and Charreton, Barthe. Determination of the solubility products of the hydroxide and basic chlorides of lead. Acad. Sci. Comptes Rendus 232, 162-163 (1951).
58. Nordstrom, D. K. "Hydrogeochemical and microbiological factors affecting the heavy metal chemistry of an acid mine drainage systems." Ph. D. Thesis, Stanford University, 1977.
59. Brown, J. B. A chemical study of some synthetic potassium-hydronium jarosites. Amer. Mineralog. 10, 696-703 (1970).
60. Brown, J. B. Jarosite-goethite stabilities at 25°C, 1 atm. Mineral. Deposita 6, 245-252 (1971).

61. Zotov, A. V., Mironova, G. D., Rusinov, V. L. Determination of the Gibbs free energy $\Delta G^{\circ}_{f,298}$ of jarosite synthesized from a natural solution. Geokhimiya, 739-745 (1973).
62. Vlek, P. L. G., Blom, Th. J. M., Beek, J., and Lindsay, W. L. Determination of the solubility product of various iron hydroxides and jarosite by the chelation method. Soil Sci. Soc. Amer. Proc. 38, 429-432 (1974).
63. Kashkai, C., Borovskaya, Y. B., and Babozade, M. A. $\Delta G^{\circ}_{f,298}$ determination of synthetic jarosite and its sulfate analogues. Geokhimya 5, 778-784 (1975).
64. Wagman, D. D., Evans, W. H., Parker, V. B., Halow, I, Bailey, S. M., and Schumm, R. H. Selected values of chemical thermodynamic properties. Tables for elements 35 through 53 in the standard order of arrangement. Nat'l Bur. Stand. Tech. Note 270-4, 141 p. (1969).
65. Larson, J. W., and Hepler, L. G. Calorimetric measurements on metal sulfates and their hydrates: electrode potentials and thermodynamic data for aqueous ions of transition elements, p. 195-201, in "Analytical Calorimetry," Plenum Press, New York, 1968.
66. Parker, V. B., Wagman, D. D., and Evans, W. H. Selected values of chemical thermodynamic properties. Tables for the alkaline earth elements (Elements 92 through 97 in the standard order of arrangement): Nat'l Bur. Stand. Tech. Note 270-6, 106 p. (1971).
67. Kelly, K. K., Shomate, C. H., Young, F. E., Naylor, B. S., Salo, A. E., and Huffman, E. H. Thermodynamic properties of ammonium and potassium alums and related substances with special reference to extraction of alumina from clay and alunite. U. S. Bur. Mines Tech. Paper 688. 104 p. (1946).
68. Haas, J. L. Jr., and Fisher, J. R. Simultaneous evaluation and correlation of thermodynamic data. Amer. J. Sci. 276, 525-545 (1976).
69. Stearns, R. I., and Brandt, A. F. Solubility product variability at constant temperature and pressure. J. Phys. Chem. 80, 1060-1063 (1976).
70. Macaskill, J. B., and Bates, R. G. Solubility product constant of calcium fluoride. J. Phys. Chem. 81, 496-498 (1977).
71. Brown, D. W., and Roberson, C. E. Solubility of natural fluorite at 25°C. U. S. Geol. Survey J. Res. 5, 509-517 (1977).
72. Eugster, H. P., and Chou, I.-M. The depositional environments of precambrian banded iron-formations. Econ. Geol. 68, 1144-1168 (1973).
73. Bird, G. W., and Anderson, G. M. The free energy of formation of magnesium cordierite and phlogopite. Amer. J. Sci. 273, 84-91 (1973).
74. Paces, Tomas. Active mineral surfaces: origin and possible effects on trace elements in natural water systems, p. 361-368, in Hemphill, D. D., ed., "Trace Substances in Environments Health-VI: Univ. Missouri, Columbia." 1973.

75. Paces, Tomas. Steady-state kinetics and equilibrium between ground water and granitic rock: Geochim. Cosmochim. Acta 37, 2641-2663 (1973).
76. Ponnamperuma, F. N., Tianco, E. M., and Loy, Teresita. Redox equilibria in flooded soils: I. The iron hydroxide system. Soil Sci. 103, 374-382 (1967).
77. Beidermann, G., and Chow, J. T. Studies on the hydrolysis of metal ions. Part 57. The hydrolysis of iron(III) ion and the solubility product of $Fe(OH)_{2.70}Cl_{0.30}$ in 0.5 M $(Na^+)Cl^-$ medium. Acta Chem. Scand. 20, 1376-1388 (1966).
78. Chien, S. H., and Black, C. A. Free energy of formation of carbonate apatites in some phosphate rocks. Soil Sci. Soc. Amer. J. 40, 234-239 (1976).
79. Morey, G. W., Fournier, R. O., and Rowe, J. J. Solubility of amorphous SiO_2 at 25°C. J. Geophys. Res. 69, 1995-2002 (1964).
80. Sato, Motoaki. Oxidation of sulfide ore bodes, 1. Geochemical environments in terms of Eh and pH. Econ. Geol. 55, 928-961 (1960).
81. Nordstrom, D. K., Jenne, E. A., and Ball, J. W. Redox equilibria of iron in acid mine waters, in Jenne, E. A., ed., "Chemical Modeling in Aqueous Systems. Speciation, Sorption, Solubility, and Kinetics," Amer. Chem. Soc., (this volume).
82. Bevington, P. R. "Data Reduction and Error Analysis for the Physical Sciences," 336 p., McGraw-Hill Book Co., New York, 1969.
83. Laxen, D. P. H. A specific conductance method for quality control in waters. Water Res. 11, 91-94 (1977).
84. Butler, J. N. "Ionic Equilibrium, A Mathematic Approach," 547 p., Addison-Wesley, New York, 1964.
85. Cloke, P. L. The geologic role of polysulfides - Part II. The solubility of acanthite and covelite in sodium polysulfide solutions. Geochim. Cosmochim. Acta 27, 1299-1319 (1963).
86. Helffrich, Friederich, "Ion Exchange." 624 p., McGraw-Hill, New York, 1962.
87. Kittrick, J. A. Ion exchange and mineral stability: Are the reactions linked or separate? in Jenne, E. A., ed., "Chemical Modeling in Aqueous Systems, Speciation, Sorption, Solubility, and Kinetics," Amer. Chem. Soc., (this volume).
88. Bassett, R. L., Kharaka, Y. K., and Langmuir, D. Critical review of the equilibrium constants for clay minerals, in Jenne, E. A., ed., "Chemical Modeling in Aqueous Systems, Speciation, Sorption, Solubility, and Kinetics," Amer. Chem. Soc., (this volume).

Disclaimer: The reviews expressed and/ or the products mentioned in this article represent the opinions of the author(s) only and do not necessarily represent the opinions of the U.S. Geological Survey.

RECEIVED November 16, 1978.

Chemical Modeling of Trace Metal Equilibria in Contaminated Soil Solutions Using the Computer Program GEOCHEM

SHAS V. MATTIGOD and GARRISON SPOSITO

Department of Soil and Environmental Sciences, University of California, Riverside, CA 92521

The study of trace metal chemistry in soils has assumed added importance recently because of the possible deleterious effects on soil organisms, plants, and human beings that could result from the accumulation of these metals to toxic levels in soils contaminated, e.g., by the land disposal of sewage sludges or by the seepage of geothermal brines lost accidently from storage lagoons. Although the soil solution is an open, dynamic, natural water system whose composition reflects many reactions occurring simultaneously among its aqueous constituents and between those constituents and the assembly of mineral and organic solid phases present, a knowledge of the general features of soil trace metal equilibria is expected to be a useful guide to predicting what will occur in nature if contamination takes place. These general features cannot be assessed conveniently by laboratory experiments because of the complexity of soil solutions. A more viable alternative is assessment by a computer model that is based on thermodynamic association and solubility product constants.

The computer program GEOCHEM is adapted and being developed for soil solutions from the REDEQL2 program created originally by Morel and Morgan and their coworkers at Caltech (1). The principal ways in which GEOCHEM differs from REDEQL2 are: 1) it includes thermodynamic data for a few hundred additional soluble complexes and solids that are particularly relevant to trace metal equilibria in soil, 2) it contains a subroutine for cation exchange on constant charge surfaces that is based on a thermodynamic model, and 3) it contains a subroutine for the estimation of single-ion activity coefficients at ionic strengths up to 3 M.

A number of important theoretical problems have arisen in connection with the development of GEOCHEM. These problems may be classified broadly into four categories: (a) stability constants for trace metal complexes with inorganic ligands; (b) stability constants for trace metal complexes with organic ligands; (c) solubility product constants for soil clay minerals, and (d) thermodynamic cation exchange constants and exchanger

0-8412-0479-9/79/47-093-837$05.00/0
© 1979 American Chemical Society

phase activity coefficients. Any reasonably accurate computation of trace metal equilibria in a soil solution must be preceded by a successful attack on these four problem areas.

Stability Constants for Soluble Inorganic Complexes of Trace Metals

Studies involving trace metal speciation in soil solutions require the values of stability or association constants of complexes of the trace metals with a number of inorganic and organic ligands which exist in these environments. Much of these data can be obtained readily from published compilations (2-7). However, experimental data on association constants for complexes of the trace metals with ligands such as CO_3^{2-}, HCO_3^-, PO_4^{3-}, HPO_4^{2-}, $H_2PO_4^-$ are not often available. In view of this lack of experimental data, modeling studies of chemical equilibrium in natural water systems have typically followed two approaches. One of the approaches, represented by computer programs such as SOLMNEQ and WATEQ (8, 9), is to include only those complexes for which the values of experimental association constants are available. The second approach consists of including in the equilibrium calculations estimates of the values of the association constants of complexes that are believed to form, but for which experimental data do not yet exist, along with experimentally derived values of association constants (10, 11). The latter approach seems preferable because it tends to include in the calculations all known interactions of metals and ligands through complex formation and thus provides results which are expected to be qualitatively more reliable than the results obtained through the former approach.

The methods available for calculating ion association constants consist of theoretical methods based on electrostatics (12, 13) and empirical methods based on correlations of association constants with properties of ions such as electronegativity, charge, radii, coordination number, etc. (14, 15, 16, 17, 18). Recently, Mattigod and Sposito (19) estimated the association constants for complexes of a number of cations of the first transition metal series with many of the inorganic ligands considered to be of importance in the trace metal chemistry of soil solutions. The association constants calculated by Mattigod and Sposito (19) with the method of Kester and Pytkowicz (13), for carbonate and phosphate complexes appear to be systematically different from those estimated recently with the correlation method of Nieboer and McBryde (17). The latter method has a firmer empirical basis and, therefore, has been preferred. Values of the revised estimated and measured association constants for some trace metal-inorganic ligand complexes are listed in Table I. These and other estimated values have been incorporated into GEOCHEM. They will be replaced when reliable experimental values become available.

There is some experimental evidence which indicates that mixed-ligand complexes of trace metals may be very stable and dominant in soils (20). In modeling the speciation of elements in sea water, Dyrssen and Wedborg (10) found that mixed-ligand complexes of certain trace metals were significant. At present, experimental data on mixed-ligand complexes are rather sparse. However, stability constants can be estimated by a statistical method suggested by Dyrssen, Jagner and Wengelin (21). Currently, GEOCHEM includes estimated stability constants of hydroxy-chloro complexes of a few trace metals. Lack of data on many mixed-ligand complexes is clearly one of the significant shortcomings in modeling studies at present.

Table I.

Estimated and measured common logarithms of association constants for some trace metal-inorganic ligand ion pairs at 25°C, 1 atm

Ligand	Mn^{2+}	Fe^{2+}	Co^{2+}	Ni^{2+}	Cu^{2+}	Zn^{2+}
CO_3^{2-}	4.52	5.31	5.53	5.78	6.73[1]	4.76[2]
HCO_3^-	1.95[3]	2.72	2.89	3.08	4.29	2.79
PO_4^{3-}	7.19	7.93	8.13	8.37	9.85	8.02
HPO_4^{2-}	3.58[4]	3.60[5]	3.04	2.93[6]	3.20[6]	3.30[7]
$H_2PO_4^-$	1.35	2.70[5]	1.49	1.53	1.76	1.60[7]

[1]Schindler et al (22), [2]Bilinski, Huston and Stumm (23)

[3]Morgan (24), [4]Smith and Alberty (25), [5]Nriagu (26),

[6]Sigel et al (27), [7]Nriagu (28)

Stability Constants for Soluble Organic Complexes of Trace Metals

The soluble, metal-complexing, organic fraction of sewage sludge is a heterogeneous assembly of molecules that is characterized by wide ranges of chemical composition, molecular weight, and functional group acidity (29). This complexity poses a difficult problem for the chemical modeling of soil solutions contaminated by the incorporation of trace metal-bearing sludges. A succesful model will require, in particular, information about the many possible reactions of trace metals with the fulvic acid fraction of sludges.

The dissociation of protons from fulvic acid (FA) extracted from anaerobically-digested sewage sludge has been investigated by potentiometric titration (29). In that study, a number of important facts was brought to light which suggest that the behavior of fulvic material in a soil solution will indeed be complex. According to Sposito and Holtzclaw (29), there appear to be four separate classes of dissociable functional groups that range in acidity from very strong (ionized at $pH < 2$) to very weak (ionized at $pH > 10$). (These classes are just four out of a continuum of classes of functional group acidity in sludge-derived FA, since no acid or alkaline final end points appear in the titration curve between pH 1 and 11.) Based on this evidence, Sposito et al. (30), suggested that the acidic functional group classes in sludge-derived FA be designated I, II, III and IV for those groups that titrate approximately in the pH ranges < 3, 3-5, 5-8, > 8, respectively.

The reactions between sludge-derived fulvic material and trace metals are poorly understood and, at present, no reliable thermodynamic stability constants are available. A provisional approach to resolving this difficulty, which has been employed often in modeling studies, is to identify carefully certain classes of known organic acids whose proton dissociation constants fall into the ranges observed for the soil solution organics and which can be expected to be present or to simulate closely the organic acids present (whether "natural" or from contamination) in a soil solution. Table II lists a set of aromatic, aliphatic, and amino acids that are frequently observed in soil leachates, together with the relative concentrations of each in a mixture of model water-soluble soil organics that provides the total of 2.2 meq of H^+ found per gram of sludge-derived FA (30). It has been observed that the functional groups in these acids are similar to the functional groups present in sludge fulvic material. The measured stability constants for trace metal complexes with these model organic acids are assumed to be good approximations to the unknown stability constants for the assembly of soil solution organics, in the sense that a mixture of the model organic acids whose proton titration curve semiquantitatively simulates that of sludge-derived FA (Figure 1) will also produce a comparable distribution of various trace metals among organic and inorganic species in a soil solution. The several obvious limitations and dangers in this provisional approach, presently used in GEOCHEM, are not considered to be as serious as completely neglecting the organic speciation of trace metals in a soil

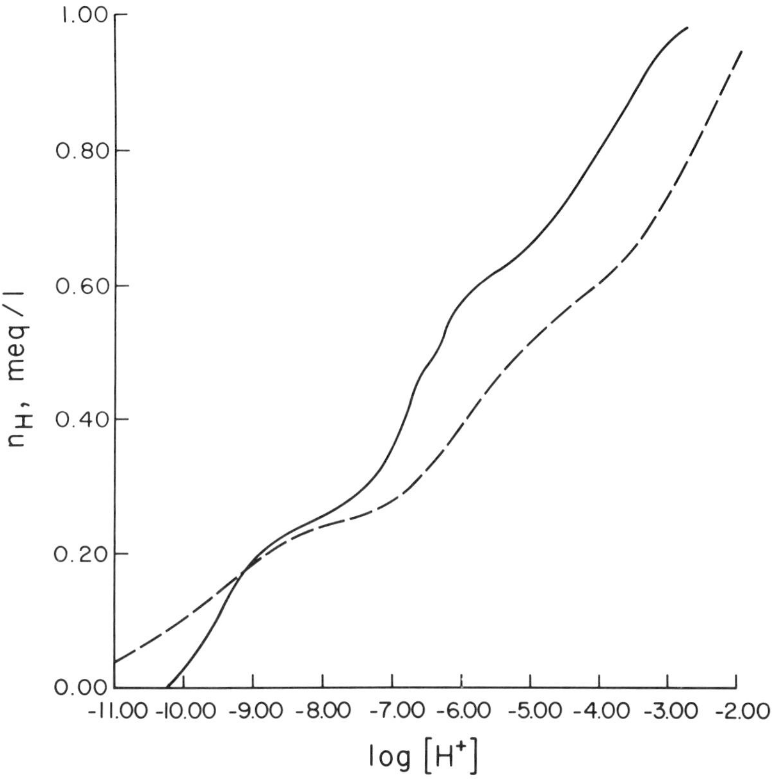

Figure 1. Comparison between the potentiometric titration curve for a sewage-sludge-derived fulvic acid (——) and the simulated titration curve for a model fulvic acid (– – –)

solution whose organic constituents may approach 10^{-3} M in total concentration (e.g., a soil solution affected by sewage sludge application).

Table II.

Organic acids used as a model for sludge-derived fulvic acid

Acid	Concentration (pM)
Benzenesulfonic	4.27
Salicylic	4.27
Phthalic	3.97
Citric	4.14
Maleic	3.97
Ornithine	4.36
Lysine	4.36
Valine	4.36
Arginine	4.49

Solubility Product Constants for Soil Clay Minerals

Solid phases comprise the dominant fraction of the soil volume and, as such, the dissolution, precipitation and ion exchange properties of the soil minerals have a profound influence on the composition of soil solutions. Therefore, it is essential to include the reactions these solids may undergo in soils in simulation studies of chemical equilibrium. Reliable thermodynamic data on the stability of many of the important oxides, hydroxides, carbonates, phosphates, and silicates are available and can be incorporated into a computer program such as GEOCHEM. In many soils, phyllosilicates dominate the clay-size fraction. Smectites as a group of the phyllosilicates are ubiquitous in soils and sediments. Because of their large surface areas and high cation exchange capacities, smectites, when present, play a significant part in influencing the composition of soil solutions. Unlike other clay minerals, smectites also exhibit a broad range of chemical compositions and, therefore, it is unlikely that the standard free energy of formation of every naturally-occurring smectite will be determined experimentally. This difficulty has prompted a number of attempts to estimate $\Delta G^{o}_{f,298.15}$ for silicates in general (31, 32, 33) and smectites in particular (34, 35). These methods are based on simple physical models which are related primarily to classical electrostatics. The success of these methods is based on the observation that the short-range coordination environment of a cation that is found in silicate minerals does not tend to vary much from one mineral to another. Since covalency plays its most important role in nearest neighbor interactions, it follows that its contribution to the structural energy of a mineral will be about the same among silicates and, therefore, that the relative stabilities of the minerals largely will be determined by longer-ranged ionic interactions.

Nriagu (34), in comparing his method with the method of Tardy and Garrels (35), concluded that the accuracy of estimation of $\Delta G^{o}_{f,298.15}$ values of phyllosilicates in either method was about the same. But, each of these methods contains a critical, ad hoc assumption that is difficult to justify on geochemical grounds. Tardy and Garrels (35) had to assume that the hydroxyl ions in Mg-bearing layer silicates are associated only with magnesium ions, insofar as thermochemical calculations are concerned. Nriagu (34) was forced to decrease the values of $\Delta G^{o}_{f,298.15}$ for solid KOH and NaOH by about 15 percent from the measured values in order to obtain agreement between the results of his method and experimental data for Na- and K-containing clay minerals. These problems with the methods of Tardy and Garrels (35) and Nriagu (34) prompted the development of a method by Mattigod and Sposito (36) for estimating the $\Delta G^{o}_{f,298.15}$ values of smectites which not only shows a slight improvement in prediction, but also is free of the arbitrary assumptions inherent in some of the other methods. This method employs the concept of ionic bonding as applied to clay minerals, mentioned earlier, along with a statistical equation to relate the charge on a smectite due to isomorphous substitution and the valence and radius of the interlayer cation to the free energy changes brought about by changes in coordination environment. These changes in free energy occur in transferring a cation from relatively high-potential sites in a hydroxide to a relatively low-potential interlayer site in a smectite. The relationship developed by Mattigod and Sposito (36) is:

$$\Delta G^{o}_{f} = \Sigma n_i \, \Delta G^{o}_{f} (n_i) - (\Sigma n_i Z_i - 12) \, \Delta G^{o}_{f} \, H_2O_{(1)} - |\delta| \qquad (1)$$

and $\quad \ln |\delta| = 1.9283\ C + 0.3501\ R - 0.2819\ Z + 3.5427$

where, ΔG^{o}_{f} = standard free energy of formation of a smectite
n_i = reaction coefficient of the ith hydroxide
Z_i = charge on the ith cation (including Si)

ΔG_f^o = standard free energy of formation of the ith hydroxide component
C = cation exchange capacity (due to isomorphous substitution) of the smectite per formula unit, in eq/fw
R = Pauling radius (Å) of the exchangeable (interlayer) cation
Z = valence of the exchangeable cation

Equation 1 provided estimates of $\Delta G^o_{f,298.15}$ for smectites which showed better agreement with the experimentally derived values than the other two methods (34, 35). A comparison among the methods is given in Table III.

Cation Exchange Phenomena

The task of calculating the composition of a soil exchanger phase in equilibrium with a soil solution has two distinct parts: the thermodynamic exchange equilibrium constants must be determined and the activity coefficients of the components of the exchanger phase must be estimated. The first part of the problem, that of obtaining exchange equilibrium constants, is not particularly difficult since a number of measurements of ΔG^o_{ex} for the common clay minerals (e.g., smectites, vermiculites, and kaolinites) has been published. Alternatively, a semi-empirical approach such as the James-Healy model included in REDEQL2 (1) may be employed to estimate a standard free energy of adsorption for a cationic species.

Smectites are one of the most important soil clay minerals as regards cation exchange. The reversible exchange reactions of these minerals with metal cations may be pictured as kinetically-favored (i.e., rapid) precipitation-dissolution reactions (44). From this point of view, it is meaningful to write the "exchange half-reaction":

$$MX_m(s,aq) = M^{+m}(aq) + mX^{-1}(s,aq) \qquad (2)$$

where X refers to one equivalent of the anionic portion of a smectite exchanger and M^{+m} is an exchangeable cation. The typical exchange reaction then is pictured as a set of pairs of reactions such as Eq. 2, with each reaction pair involving two different metal cations. The analogy between Eq. 2 and the dissolution reaction for a solid is evident.

However, there are some important differences between the equilibrium constant for Eq. 2 and the usual K_{SO}. First, the compound on the left-hand side of Eq. 2 is not a dry solid at standard temperature and pressure, but instead is a homoionic smectite in contact with an aqueous solution. The standard state for $MX_m(s,aq)$ accordingly is the homoionic clay mineral at standard temperature and pressure in equilibrium with an infin-

Table III. Comparison between estimated and experimental $\Delta G^{o}_{f,298.15}$ values (kJ/mol) of some smectites.

Mineral	Composition	Estimated values: TARDY and GARRELS method (35)	Estimated values: NRIAGU method (34)	Estimated values: Eq. 1 This paper	Experimental values	(Ref.)
1. Aberdeen	$Mg_{0.2075}(Al_{0.18}Si_{3.82})(Al_{1.29}Fe^{3+}_{0.335}Mg_{0.445})O_{10}(OH)_2$	-5220.6	-5173.0	-5211.1	-5218.7	(37)
2. Aberdeen	$Al_{0.1383}(Al_{0.18}Si_{3.82})(Al_{1.29}Fe^{3+}_{0.335}Mg_{0.445})O_{10}(OH)_2$	-5193.3	-5159.7	-5182.8	-5200.0	(38)
3. Belle Fourche	$Mg_{0.1325}(Al_{0.065}Si_{3.935})(Al_{1.515}Fe^{3+}_{0.225}Mg_{0.29})O_{10}(OH)_2$	-5223.6	-5184.6	-5209.4	-5222.8	(37)
4. Belle Fourche	$Al_{0.0883}(Al_{0.065}Si_{3.935})(Al_{1.515}Fe^{3+}_{0.225}Mg_{0.29})O_{10}(OH)_2$	-5206.2	-5175.3	-5188.5	-5213.4	(38)
5. Smectite from Houston Black Clay	$Mg_{0.225}(Al_{0.30}Si_{3.70})(Al_{1.345}Fe^{3+}_{0.405}Mg_{0.27})O_{10}(OH)_2$	-5209.4	-5158.8	-5208.5	-5214.7	(39)
6. Beidellite (Mg-sat)	$Mg_{0.135}Ca_{0.01}Na_{0.07}K_{0.095}(Al_{0.45}Si_{3.55})$ $(Al_{1.41}Fe^{3+}_{0.415}Fe^{2+}_{0.055}Mg_{0.205})O_{10}(OH)_2$	-5220.2	-5180.5	-5185.0	-5200.1	(40)
7. Beidellite (K-sat)	$K_{0.37}Ca_{0.01}Na_{0.07}(Al_{0.45}Si_{3.55})$ $(Al_{1.41}Fe^{3+}_{0.415}Fe^{2+}_{0.055}Mg_{0.205})O_{10}(OH)_2$	-5238.4	-5211.8	-5205.7	-5215.4	(40)
8. Colony	$Mg_{0.195}(Al_{0.19}Si_{3.81})(Al_{1.52}Fe^{3+}_{0.22}Mg_{0.29})O_{10}(OH)_2$	-5258.3	-5217.2	-5253.3	-5262.4	(41)
9. Colony I	$Mg_{0.185}(Al_{0.19}Si_{3.80})(Al_{1.58}Fe^{3+}_{0.19}Mg_{0.26})O_{10}(OH)_2$	-5261.4	-5219.0	-5253.1	-5267.6	(42)
10. Colony II	$Mg_{0.21}(Al_{0.19}Si_{3.81})(Al_{1.52}Fe^{3+}_{0.21}Mg_{0.29})O_{10}(OH)_2$	-5264.6	-5222.7	-5261.9	-5262.1	(42)
11. Castle Rock	$Mg_{0.21}(Al_{0.32}Si_{3.68})(Al_{1.52}Fe^{3+}_{0.14}Mg_{0.46})O_{10}(OH)_2$	-5329.8	-5293.8	-5333.0	-5336.8	(42)
12. Upton	$Mg_{0.17}(Al_{0.07}Si_{3.93})(Al_{1.55}Fe^{3+}_{0.20}Mg_{0.24})O_{10}(OH)_2$	-5236.9	-5195.6	-5226.8	-5218.0	(42)
13. Clay Spur (Na-sat)	$Na_{0.27}Ca_{0.1}K_{0.02}(Al_{0.06}Si_{3.94})(Al_{1.52}Fe^{3+}_{0.19}Mg_{0.22})O_{10}(OH)_2$	-5268.4	-5227.4	-5242.1	-5226.4	(43)
14. Cheto (Ca-sat)	$Ca_{0.185}Na_{0.02}K_{0.02}(Al_{0.07}Si_{3.93})(Al_{1.52}Fe^{3+}_{0.14}Mg_{0.33})O_{10}(OH)_2$	-5287.5	-5227.2	-5262.7	-5245.3	(43)

itely dilute solution containing the cation M^{+m} (45). In addition, the species X^{-1}(s,aq) is the anionic part of the smectite exchanger, considered as a (charged) solid at standard temperature and pressure that is in contact with an infinitely dilute aqueous solution. Thus X^{-1} is not a dissolved species, as it would be if Eq. 2 referred to an ordinary dissolution reaction. Lastly, it should be noted that the equilibrium constant for Eq. 2 cannot be determined easily because of the difficulty of measuring the activity of X^{-1}(s,aq). This fact need not undermine the utility of Eq. 2, however, because one may write

$$\mu^{o}(X^{-1}(s,aq)) = \frac{1}{m}\,[\mu^{o}(MX_{m}(s,aq)) - \mu^{o}(M^{+m}(aq))] \quad (3)$$

for any M and X for which the equilibrium constant in Eq. 2 may be taken to be unity. With the free energy of formation of $\mu^{o}(X^{-1}(s,aq))$ so determined, the values of the equilibrium constants of all other exchange half-reactions can be found, given the available values of the free energies of formation of one mole of the aqueous metal cations and one equivalent of the homoionic clay minerals.

There is evidence (46, 47) to suggest that the half-reaction

$$HX(s,aq) = H^{+}(aq) + X^{-1}(s,aq) \quad (4)$$

may be assigned an equilibrium constant near unity, as may be the exchange reaction between Na^{+} and H^{+} on a smectite. The lack of selectivity between H^{+} and Na^{+} is the more firmly established hypothesis of the two (46). If Eq. 4 truly had an equilibrium constant of unity, then H-smectites would behave as very strong electrolytes in aqueous solutions (e.g., a 1% suspension of H-montmorillonite with a CEC equal to 1 meq/g would exhibit a pH value of 2). The available published data on H-smectite suspensions are not conclusive, but they suggest strong electrolyte behavior. Given this hypothesis, it follows that

$$\mu^{o}(X^{-1}(s,aq)) = \mu^{o}(HX(s,aq)) \quad (5)$$

for any smectite.

The direct measurement of $\mu^{o}(HX(s,aq))$ is difficult because pure H-smectites are not easy to prepare. But this free energy of formation can be estimated using the accurate correlation technique discussed previously (36) and the hypothesis of no selectivity in H^{+}-Na^{+} exchange. Thus

$$\mu^{o}(HX(s,aq)) = \mu^{o}(NaX(s,aq)) + \mu^{o}(Na\ (aq)) \quad (6)$$

where $\mu^{o}(NaX(s,aq)) = \Delta G^{o}_{f,298.15}$ for one equivalent of Na-smectite, estimated according to Eq. 1. It then follows that the equilibrium constant for Eq. 2 is given by the expression

$$\log K_{Mex} = \frac{1}{5.707} [\mu^{o}(MX_m(s,aq)) - \mu^{o}(M^{+m}(aq)) - m\mu^{o}(HX(s,aq))] \quad (7)$$

at 298.15 K, where $\mu^{o}(MX_m(s,aq)) = \Delta G^{o}_{f,298.15}$ for $\underline{m}$ equivalents of M-smectite. The values of K_{Mex}/K_{Naex} for two montmorillonites and several exchangeable cations of interest in soils are listed in Table IV. These equilibrium constants may be treated in much the same way as are ordinary solubility product constants in the GEOCHEM program. Aside from understanding that the K_{Mex} values are relative, there is a need to modify the equilibrium calculation for a soil solution so as not to permit $X^{-1}(s,aq)$ to contribute to the ionic strength of a solution in contact with the exchanger phase.

Table IV.

$\log K_{Mex}/K_{Naex}$ values for the dissociation of various cations from two smectite surfaces

Cation	Smectite: Camp Berteau	Smectite: Chambers
Mg^{2+}	- 1.81	- 3.51
Ca^{2+}	- 2.12	- 3.37
Na^{+}	0	0
K^{+}	- 1.41	- 1.04
Ni^{2+}	- 6.52	- 8.18
Cu^{2+}	-11.61	-13.26
Zn^{2+}	- 8.27	- 9.85
Cd^{2+}	-10.67	-11.95
Pb^{2+}	-21.09	-22.01

The problem of estimating the activity coefficients of the species MX_m which make up the exchanger phase is quite formidable. Fundamentally, what is needed is an equation for the activity coefficient that is comparable in simplicity and generality with the Davies equation. Unfortunately, the measurements of exchanger activity coefficients that are available apply only to two-component systems and exhibit no parameters of the same generality as is the ionic strength in the case of aqueous solutions. On the other hand, there is evidence that two-component smectite exchangers containing cations such as Na^+, K^+, Ca^{2+}, Mg^{2+}, and the trace metals are nearly ideal solutions (48, 49). This fact and the absence of information concerning multicomponent exchangers suggest that a good, working first approximation is to set all activity coefficients in a smectite exchanger equal to unity. This approach assumes that exchange selectivity will in most cases of interest be dominated by the interaction between the exchangeable cations and the exchanger instead of by interactions among the exchangeable cations themselves. Clearly, much work remains to be done on this important problem. This ion exchange model is being tested presently for its reliability in predicting the exchangeable ion composition on smectites.

Ionic Strength Corrections

In the computer program REDEQL2 (and GEOCHEM), the activity coefficients for aqueous species are calculated from the Davies equation. Generally, this equation is used for the computation of activity coefficients of ions in aqueous solutions up to an ionic strength of 0.5 M, even though Dyrssen et al. (21) indicate that this equation can be used for ionic strength up to 1.0 M. In saline soils and soils contaminated with geothermal brines, the ionic strengths of the soil solution may exceed 0.5 M. This fact poses the necessity of using equations which have been developed to describe the activity coefficients of ions in concentrated, multicomponent electrolyte solutions. As part of a study on the chemistry of ore-forming fluids, Helgeson (50) has proposed that the true individual ion activity coefficients for ions present in small concentrations in multicomponent electrolyte solution having sodium chloride as the dominant component be approximated by a modified form of the Stokes-Robinson equation. The equation proposed is:

$$\log \gamma_i = -\frac{AZ_i^2 I^{\frac{1}{2}}}{1+a_i B I^{\frac{1}{2}}} + B^{o} I \tag{8}$$

where, γ_i = individual activity coefficient of the ith trace ion
Z_i = charge on the ith trace ion
I = true ionic strength

A and B = the Debye-Hückel Parameters
a_i = "Distance of closest approach" of the ith ion (cm)
B^o = Deviation function (0.041 at 25°C)

According to Helgeson (50), equation 8 can be used to estimate the individual ion activity coefficient for ions present in small concentrations in sodium chloride solutions of true ionic strength up to 3.0 M. Since saline soils and geothermal brines are often dominated by sodium chloride, it will be appropriate to use the equation proposed by Helgeson (50). Therefore, in GEOCHEM, ionic activity coefficient calculations for such systems are performed by equation 8.

The "distance of closest approach" parameters for many of the ions and complexes included in GEOCHEM are not available. This lack of data has been circumvented by assigning arbitrarily 4, 5, and 6 A for mono-, di- and tri- (and higher) valent ions respectively. A similar assumption has been made previously by Truesdell and Jones (9).

Applications

At present, there is a serious effort being made to assess the impact of geothermal power development on agriculture in the Imperial Valley of California. One aspect of the problem involves the possible contamination of agricultural soils from accidental spills of geothermal brine. A good semi-quantitative guide to trace metal solubility and mobility in contaminated soils can be compiled from results obtained from GEOCHEM.

As an example, one of the possible means of contamination of soils is from irrigating them with water accidentally contaminated with brines. GEOCHEM can be used to simulate the change in trace metal speciation when varying quantities of brine mix with irrigation water. One example of a computed metal speciation in a mixture of brine and irrigation water is shown in Table V. The computations predict that the alkali metals will be present mainly as free ions. Among the alkaline earth metals, Ca, Mg, and Sr are present principally as free ions; a minor part of these metals is predicted to be present as the $CaSO_4^o$, $MgSO_4^o$, and $SrSO_4^o$ species. Also, a minor portion of Mg is predicted to be in the form of a solid borate. However, the bulk of Ba is predicted to precipitate as barite ($BaSO_4$), with some as solid $BaAsO_4$. The metals present in trace quantities such as Cu, Ni, Zn, and Pb, are predicted to be present principally as $MHCO_3^+$ and MCO_3^o complexes, the exception being the metal Cd, which is present mainly as Cd^{2+} and $CdCl^+$. Regarding boron and arsenic, the calculations predict that the solution concentrations of these elements would be controlled by the solubility of solid phases such as $Mg_2B_2O_5 \cdot H_2O$ and $BaAsO_4$. (In simulating the trace metal speciation when soils

Table V.
Calculated metal and ligand speciation in a mixture of brine and irrigation water

Component	Total Analytical Conc. (-log M)	Dominant Species*
Ca	2.63	Ca^{2+}, $CaSO_4^{o}$
Mg	3.08	Mg^{2+}, $Mg_2B_2O_5 \cdot H_2O$ (Solid), $MgSO_4^{o}$
Sr	5.18	Sr^{2+}, $SrSO_4^{o}$
Ba	6.08	$BaSO_4$(Solid), $BaAsO_4$(Solid)
K	3.28	K^{+}
Na	1.82	Na^{+}
Cs	6.59	Cs^{+}
Li	4.18	Li^{+}
Fe	6.19	$Fe(CO_3)_3^{3-}$
Mn	6.59	Mn^{2+}, $MnHCO_3^{+}$, $MnSO_4^{o}$
Cu	6.76	$CuHCO_3^{+}$, $CuCO_3^{o}$
Cd	7.92	Cd^{2+}, $CdSO_4^{o}$, $CdCl^{+}$
Zn	6.82	Zn^{2+}, $ZnHCO_3^{+}$, $ZnSO_4^{+}$
Ni	6.64	Ni^{2+}, $NiCO_3^{o}$, $NiHCO_3^{+}$
Pb	6.63	Pb^{2+}, $PbCO_3^{o}$, $PbHCO_3^{+}$
Al	7.00	$Al_2Si_2O_5(OH)_4$(Solid), $Al(OH)_3^{o}$
CO_3	2.49	HCO_3^{-}, $H_2CO_3^{o}$
SO_4	2.48	SO_4^{-}, $CaSO_4^{o}$, $NaSO_4^{-}$
Cl	1.97	Cl^{-}
F	4.66	F^{-}
PO_4	6.72	$CaHPO_4^{o}$, $MgHPO_4^{o}$, $NaHPO_4^{-}$, HPO_4^{2-}, $H_2PO_4^{-}$
Si	3.59	$Si(OH)_4^{o}$
B	4.05	$Mg_2B_2O_5 \cdot H_2O$ (Solid)
As	6.80	$BaAsO_4$(Solid)
Se	7.10	$CaSeO_3^{o}$, $MgSeO_3^{o}$, $HSeO_3^{-}$
NO_3	4.61	NO_3^{-}

* only those species which constitute greater than 5% of the respective total analytical concentrations are listed.

are irrigated with these contaminated waters, additional reactions such as ion exchange and adsorption would have to be included, of course.) It is envisioned that these types of computations of the speciation of metals and ligands would be helpful in predicting the effects of brine spillage on the trace metal solubility and mobility in the soils of the Imperial Valley.

There is also an urgent need to know the possible deleterious effects of sewage sludge application on agricultural land. One of the potential problems arising out of the disposal of sewage sludge is the likelihood of accumulation of trace metals to toxic levels in soils. This concern is reflected in a recent report by Page (51), which examines the multiplicity of fates for trace elements introduced into agricultural soils by the land application of municipal wastes. One possible means of assessing the effects of sewage sludge application would be to simulate the probable speciation of the various elements with a computer program such as GEOCHEM. Table VI lists a part of the result of a simulation, using GEOCHEM, regarding the probable trace metal distribution in soil solutions extracted from soils

Table VI.
Distribution of metals and ligands (% of total) in saturation extracts of some sewage sludge amended soils

	-log Total conc.	Free metal	CO_3	SO_4	Cl	"Fulvate"
			San Miguel*			
Cu	5.69	19	14	5	<1	62
Cd	3.73	57	-	11	29	2
Zn	4.67	78	2	19	<1	<1
			Holtville†			
Cu	5.36	-	-	-	-	100
Cd	7.01	56	<1	8	27	8
Zn	4.67	70	9	13	-	5

Note: Other components included in the simulations were: Ca, K, Na, Fe, Mn, PO_4, NO_3, OH. These computations are based on "equilibrium" soil solutions and therefore, ion exchange and adsorption modeling are not included in the simulation.

*pH = 5.1 † pH = 7.6

amended with sewage sludge. These results subsequently were used in interpreting the observed patterns of Cd uptake by plants which were growing in these soils. It is seen from Table VI that Cu is present predominantly as a soluble organic complex in both soils, but roughly a fifth of total soluble copper in the San Miguel soil (pH=5.1) is present as Cu^{2+} ion. The computations indicate that, in the same soil, about half the soluble Cd would be free, whereas a third would be present as complexes with Cl. Even though the total soluble Cd in the Holtville Soil is very much smaller than the total soluble Cd concentration found in the San Miguel soil, the predicted relative proportion of Cd^{2+}, Cd bound to SO_4 and Cd bound to Cl are similar for both soils. In either soils, it is predicted that soluble Cd-organic complexes would not be significant. Similarly, soluble Zn-organic complexes are insignificant in either soil. It is also predicted that approximately 78% of the soluble Zn in the San Miguel soil would be in the form of Zn^{2+} ion and the rest would be present as $ZnSO_4^o$. But, in the Holtville soil, about 70% of the soluble Zn is in the free ionic form and about 9% and 13% of soluble Zn is predicted to be in the form of $ZnHCO_3^+$ and $ZnSO_4^o$, respectively.

An interesting aspect of these simulations is that the speciation of Cu, Cd and Zn with respect to inorganic and organic ("fulvate") ligands were qualitatively similar to what was observed in gel filtration studies (52). In these studies it was noticed that Cu was predominately bound to organics whereas Zn and Cd were observed to be predominately bound to inorganics.

The two examples of simulation of metal-ligand speciation in soil systems suggests that such studies will be a useful adjunct to the examination of trace metal chemistry in soils.

Acknowledgements

This research was supported in part by a grant from Kearney Foundation of Soil Science, in part by a contract from The Lawrence Livermore Laboratory as a contribution to its Imperial Valley Environmental Project, and in part from the Environmental Protection Agency under grant number R804516010. The contents of this paper do not necessarily reflect the views and policies of EPA. GEOCHEM will be available from the authors, on request, after January 1979.

Abstract

The computer program GEOCHEM is adapted and being developed for soil solutions from the REDEQL2 program originally created by Morel and Morgan at Caltech. GEOCHEM differs from REDEQL2 in its inclusion of 1) data for a few hundred additional soluble complexes and solids that are relevant to trace metal studies, 2) a subroutine for cation exchange that is based in thermodynamics, and 3) a subroutine for the estimation of single-ion activity coefficients at ionic strengths up to 3 M. Four categories of important theoretical problems were confronted in connection with the development of GEOCHEM. These problems were the lack of data regarding (a) stability constants of trace metal complexes with many important inorganic and mixed ligands and (b) stability constants of trace metal complexes with naturally occurring organic ligands; (c) solubility product constants for soil clay minerals, and (d) thermodynamic exchange constants and exchanger phase activity coefficients. The resolution of these problems has been discussed. Two representative applications of GEOCHEM in its current form to the calculation of trace metal equilibria in a mixture of irrigation water and a geothermal brine and in the aqueous phase of a sewage sludge amended soil are presented and discussed.

Literature Cited

1. McDuff, R.E., and Morel, F.M.M. Description and use of the chemical equilibrium program REDEQL2. Tech. Rpt. EQ-73-02. W.M. Keck Lab, Caltech 75 p. (1974).
2. Baes, C.F., Jr., and Mesmer, R.E. "The Hydrolysis of Cations." 489 p. John Wiley and Sons, New York, 1976.
3. Bjerrum, J., Schwarzenbach, G., and Sillen, L.G. "Stability Constants of Metal-Ion Complexes, with Solubility Products of Inorganic Substances." 131 p. The Chem. Soc., No. 17, London, 1958.
4. Martell, A.E., and Smith, R.N. "Critical Stability Constants," 257 p. Plenum Press, New York 1976.
5. Sillen, L.G., and Martell, A.E. "Stability Constants of Metal-Ion Complexes." 754 p., The Chem. Soc. London, 1964.
6. Sillen, L.G., and Martell, A.E. "Stability Constants of Metal-Ion Complexes." 865 p. The Chem. Soc. London, 1971.
7. Yatsimirskii, K.B., and Vasilev, V.P. "Instability Constants of Complex Compounds. 214 p. Consultants Bureau, New York, 1960.
8. Kharaka, Y.K., and Barnes, I. SOLMNEQ: Solution-mineral equilibrium computations, NTIS #PB215899, 75 p. 1973.
9. Truesdell, A.H., and Jones, B.F. WATEQ, A computer program for calculating chemical equilibria of natural waters, NTIS #PB-220464, 73 p. 1973.

10. Dyrssen, D., and Wedborg, M. Equilibrium calculations of the speciation of elements in sea water, p. 181-195, in Goldberg, E.D., ed., "The Sea," John Wiley and Sons, New York, 1974.
11. Zirino, A., and Yamamoto, S. A pH-dependent model for the chemical speciation of copper, zinc, cadmium, and lead in sea water. Limnol. Oceangr. 17, 661-671 (1972).
12. Fuoss, R.M. Ionic association III. The equilibrium between ion-pairs and free ions. J. Amer. Chem. Soc. 80, 5059-5061 (1958).
13. Kester, D.R., and Pytkowicz, R.M. Theoretical model for the formation of ion-pairs in sea water. Mar. Chem. 3, 365-374 (1975).
14. Garrels, R.M., and Christ, C.L. "Solution, Minerals, and Equilibria." 450 p., Harper and Row, New York, 1965.
15. Nieboer, E., and McBryde, W.A.E. Free energy relationships in coordination chemistry I. Linear relationships among equilibrium constants. Can. J. Chem. 48, 2549-2564 (1970).
16. Nieboer, E., and McBryde, W.A.E. Free energy relationships in coordination chemistry II. Requirements for linear relationships. Can. J. Chem. 48, 2565-2573 (1970).
17. Nieboer, E., and McBryde, W.A.E. Free energy relationships in coordination chemistry III. A comprehensive index to complex stability. Can. J. Chem. 51, 2512-2524 (1973).
18. Van Panthaleon Van Eck, C.L. On the mutual stability relationship of MeA complexes consisting of a metal ion Me and a ligand A in aqueous solution, for consequetive values of n. Recl. Trav. Chim. Pays-Bas. 72, 529-537 (1953).
19. Mattigod, S.V., and Sposito, G. Estimated association constants for some complexes of trace metals with inorganic ligands. Soil Sci. Soc. Am. J. 41, 1092-1097 (1977).
20. Ramamoorthy, S., and Manning, P.G. Equilibrium studies of metal-ion complexes of interest to natural waters - VIII. J. Inorg. Nucl. Chem. 36, 695-698 (1974).
21. Dyrssen, D., Jagner, D., and Wengelin, F. "Computer Calculation of Ionic Equilibria and Titration Procedures." 250 p. Almquist and Wiksell, Stockholm, 1968.
22. Schindler, P., Reinert, M., and Gamsjager, H. "Loslichkeitkonstanten und Breie Bildungenthalpie von $Cu_2(OH)_2$ CO_3 (Malachit) and $Cu_3(OH)_2(CO_3)_2$ (Azurit) bei 25°C. Helv. Chim. Acta. 51, 1845-1856 (1968).
23. Bilinski, H., Huston, R., and Stumm, W. Determination of the stability constants of some hydroxo and carbonate complexes of Pb (II), Cu (II), Cd (II) and Zn (II) in dilute solutions by anodic stripping voltammetry and differential pulse polarography. Anal. Chim. Acta. 84, 157-164 (1976).
24. Morgan, J.J. Chemical equilibria and kinetic properties of manganese in natural waters. p. 561-624, in Faust, S.D., Hunter, J.V., eds. "Principles and Application of Water Chemistry." John Wiley and Sons, New York, N.Y., 1965.

25. Smith, R.M., and Alberty, R.A. The apparent stability constants of ionic complexes of various adenosine phosphates with divalent cations. J. Amer. Chem. Soc. 78, 2376-2380 (1956).
26. Nriagu, J.O. Stability of vivianite and ion-pair formation in the system $Fe_3(PO_4)_2$-H_3PO_4-H_2O. Geochim. Cosmochim. Acta. 36, 459-470 (1972).
27. Sigel, H., Becker, K., and McCormick, D.B. Ternary complexes in solution. Biochem. Biophys. Acta. 148, 655-664 (1967).
28. Nriagu, J.O. Solubility equilibrium constant of α-Hopeite Geochim. Cosmochim. Acta. 37, 2357-2361 (1973).
29. Sposito, G., and Holtzclaw, K.M. Titration studies on the polynuclear, polyacidic nature of fulvic acid extracted from sewage sludge-soil mixtures. Soil Sci. Soc. Amer. J. 41, 330-336 (1977).
30. Sposito, G., Holtzclaw, K.M., and Keech, D.A. Proton binding in fulvic acid extracted from sewage sludge-soil mixtures. Soil Sci. Soc. Amer. J. 41, 1119-1125 (1977).
31. Chen, Chao-Hsia. A method for estimation of standard free energies of formation of silicate minerals at 298.15 K. Amer. J. Sci. 275, 801-807 (1975).
32. Karpov, I.K., and Kashik, S.A. Computer calculation of standard isoparic-isothermal potentials of silicates by multiple regression from a crystallochemical classification. Geochem. Int. 5, 706-713 (1968).
33. Tardy, Y., and Garrels, R.M. Prediction of Gibbs energies of formation of compounds from the elements. II. Monovalent and divalent metal silicates. Geochim. Cosmochim. Acta. 41, 87-92 (1977).
34. Nriagu, J.O. Thermochemical approximations for clay minerals. Amer. Mineral. 60, 834 (1975).
35. Tardy, Y., and Garrels, R.M. A method for estimating the Gibbs energies of formation of layer silicates. Geochim. Cosmochim. Acta. 38, 1101-1116 (1974).
36. Mattigod, S.V., and Sposito, G. Improved method for estimating the standard free energies of formation of smectites. Geochim. Cosmochim. Acta. (In press).
37. Kittrick, J.A. Montmorillonite equilibria and the weathering environment. Soil Sci. Soc. Amer. Proc. 35, 815-820 (1971).
38. Kittrick, J.A. Stability of monmorillonites: II. Aberdeen montmorillonite. Soil Sci. Soc. Amer. Proc. 35, 820-823 (1971).
39. Carson, C.D., Kittrick, J.A., Dixon, J.B., and McKee, T.R. Stability of soil smectite from a Houston black clay. Clays and Clay Min. 24, 151-155 (1976).
40. Misra, V.K., and Upchurch, W.J. Free energy of formation of beidellite from apparent solubility measurements. Clays and Clay Min. 24, 327-331 (1976).
41. Weaver, R.M., Jackson, M.L., and Syers, J.K. Magnesium and silicon activities in matrix solutions of montmorillonite containing soils in relation to clay mineral stability. Soil Sci. Soc. Amer. Proc. 35, 823-830 (1971).

42. Weaver, R.M., Jackson, M.L., and Syers, J.K. Clay mineral stability as related to activities of aluminum, silicon and magnesium in matrix solution of montmorillonite-containing soils. Clay and Clay Min. 24, 246-252 (1976).
43. Huang, W.H., and Keller, W.D. Gibbs free energies of formation calculated from dissolution data using specific mineral analysis. III. Clay minerals. Amer. Mineral. 58, 1023-1028 (1973).
44. Sposito, G., and Mattigod, S.V. On the chemical foundation of the sodium adsorption ratio. Soil Sci. Soc. Amer. J. 41, 323-329 (1977).
45. Gaines, G.L., Jr., and Thomas, H.C. Adsorption studies on clay minerals. II. A formulation of the thermodynamics of exchange adsorption. J. Chem. Phy. 21, 714-718 (1953).
46. Foscolos, A.E., and Barshad, I. Equilibrium constants between both freshly prepared and aged H montmorillonites and chloride salt solutions. Soil Sci. Soc. Amer. Proc. 33, 242-247 (1969).
47. Peigneur, P., Maes, A., and Cremers, A. Heterogeneity of charge density distribution in montmorillonite as inferred from cobalt adsorption. Clays and Clay Min. 23, 71-75 (1975).
48. Gast, R.G. Alkali metal cation exchange on Chambers montmorillonite. Soil Sci. Soc. Amer. Proc. 36, 14-19 (1972).
49. Sposito, G., and Mattigod, S.V. Ideal behavior in Na - trace metal cation exchange on Camp Berteau montmorillonite. Clays and Clay Min. (In press).
50. Helgeson, H.C. Thermodynamics of hydrothermal systems at elevated temperatures and pressures. Amer. J. Sci. 267, 729-804 (1969).
51. Page, A.L. Fate and effects of trace elements in sewage sludge when applied to agricultural lands. A literature review study. Rept. Office Res. Dev. U.S. Environ. Protection Agency, Cincinnati, Ohio, 97 p., 1974.
52. Sposito, G., Holtzclaw, K.M., and Baham, J. Analytical properties of the soluble, metal-complexing fractions in sludge-soil mixtures: II. Comparative structural chemistry of fulvic acid. Soil Sci. Soc. Amer. J. 40, 691-697 (1976).

RECEIVED November 16, 1978.

38

A Comparison of Computerized Chemical Models for Equilibrium Calculations in Aqueous Systems[1]

D. K. NORDSTROM[2], L. N. PLUMMER[3], T. M. L. WIGLEY[4], T. J. WOLERY[5], J. W. BALL[6], E. A. JENNE[6], R. L. BASSETT[7], D. A. CRERAR[8], T. M. FLORENCE[9], B. FRITZ[10], M. HOFFMAN[11], G. R. HOLDREN, JR.[12], G. M. LAFON[13], S. V. MATTIGOD[14], R. E. McDUFF[15], F. MOREL[15], M. M. REDDY[16], G. SPOSITO[14], and J. THRAILKILL[17]

The interpretation of chemical processes in aqueous systems requires the use of modern electronic computers, particularly in the calculation of multicomponent, multiphase equilibria. Commonly, the first concern of solution chemists and aqueous geochemists is to calculate the distribution and activities of species on the assumption that equilibrium exists throughout the aqueous phase. Species distribution can then be used in several areas of analytical and applied chemistry, e.g. to examine the availability of free and reactive ions, to test solubility hypotheses, and to determine the potential bioavailability of nutrients or toxic substances. Species distribution also forms the basis for more complex computations involving solutions which change composition by reaction with other solutions and with gases and solids. Equilibrium calculations of this type are particularly helpful in solving interpretive problems encountered in such fields as chemical and environmental engineering, geochemistry, biochemistry and aquatic ecology.

This symposium demonstrates quite clearly that we depend heavily on chemical models, especially computerized models, to interpret aqueous chemical processes. Several computer programs which solve problems of simultaneous chemical equilibria are being used by a rapidly increasing number of investigators and it is necessary to review the inherent assumptions and limitations of these aqueous models. There is a temptation to use these models as ready-made interpretations

[1] Current addresses for authors can be found on page 892.

0-8412-0479-9/79/47-093-857$09.00/0
This chapter not subject to U.S. copyright
Published 1979 American Chemical Society

of reality without a clear understanding of their weaknesses. These programs and models are usually developed for specific purposes and taken together they represent a wide range of capabilities and features. This paper reviews the state-of-the-art of equilibrium computations by providing a compilation and description of aqueous models in current use and by examining their consistency through a comparison of species distributions and saturation indices for two hypothetical test cases: a river water sample and a seawater sample. This review will also serve to complement the review on machine computation efficiency by Leggett (1) and the review by Perrin (2) on the applications of digital computers to analytical chemistry.

The Chemical Equilibrium Problem and Its Thermodynamic Basis

An "aqueous chemical model" needs to be defined separately from the computer program which executes calculations based on the model. Such a model can be defined as a theoretical construction which allows us to predict the thermodynamic properties of electrolyte solutions. There are several ways of constructing an aqueous model, e.g. from the Bjerrum ion association theory (3), or the Fuoss ion association theory (4,5) or the Reilly, Wood and Robinson mixed electrolyte theory (6). The choice of model must be made on the basis of the problem to be solved and each model carries with it its own set of assumptions and restrictions. Nearly all computerized models are based on the ion association theory and within this framework the species distribution problem can be formulated in two distinct but thermodynamically related ways: the equilibrium constant approach and the Gibbs free energy approach. Both approaches are subject to the conditions of 1) mass balance and of 2) chemical equilibrium. The mass balance condition requires that the computed sum of the free and derived (complexes) species be equal to the given total concentration. Chemical equilibrium requires that the most stable arrangement for a given system be found, as defined by the equilibrium constants for all mass action expressions of the system, or through the use of Gibbs free energies for all of the components and derived species. In the equilibrium constant approach the mass action expressions are substituted into the mass balance conditions resulting in a set of nonlinear equations which must be solved simultaneously. The Gibbs free energy approach is simply a transformation of variables through the thermodynamic relation:

$$\Delta G_r = \Delta G_r^\circ + RT\ln K = 0$$

which allows a different numerical approach. The total Gibbs free energy function is then minimized for a given set of species and their mole numbers subject to the mass balance requirement. In the former approach equilibrium constants are needed for the data base whereas in the latter approach free energy values are needed. This difference in the data base can often be an important limitation. At the present time there are more reliable and available equilibrium constants than free energy values.

By either thermodynamic approach the problem can be stated numerically as one of finding a solution to a set of nonlinear equations. It is usually not feasible to simultaneously solve these equations in exact form for a multicomponent, multiphase system and therefore an iteration procedure must be utilized. The standard method of solving the problem by the equilibrium constant approach is to use linearized matrix inversion. Convergence assumes, of course, that the solution not only exists but that it is unique. If a system can have several thermodynamically metastable states (local minima in the Gibbs function) then several nonunique solutions are possible. Recent papers by Othmer (7) and Caram and Scriven (8) have pointed out that uniqueness is characteristic of ideal systems whereas for non-ideal systems a solution may occur at the global minimum (most stable equilibrium point) but it also may occur at a nonunique local minimum. For applications in aquatic chemistry the problem of nonuniqueness is particularly important in the interpretation of solid precipitation and dissolution processes.

The choice of the thermodynamic approach dictates the general category of numerical techniques to be used. Optimization techniques such as pattern search, linear programming, steepest descent and gradient methods are all appropriate to the Gibbs function approach whereas Newton-Raphson, successive approximations and nested iterations are best suited to the equilibrium constant approach. These techniques and several other mathematical methods have been documented along with the thermodynamic formalism in the excellent reviews by Zeleznik and Gordon (9) and Van Zeggeren and Storey (10). In this report we prefer to describe the main features and capabilities of current models rather than digress on the mathematical details of each since "... any method of calculation that can be made reliable is a good method when equilibrium compositions are the only concern" (9).

Chemical models can be further characterized by their application. The intended utilization of a model usually directs the developmental stages of choosing the type of model and the mathematical method. We have grouped current computerized models into "major schools" according to their point of origin and their application. Since the second generation models frequently used the same basic numerical approach as their predecessors, the classification into schools also tends to separate different mathematical formulations.

Major Schools of Computerized Models

Generalized Method of the Equilibrium Constant Approach. The basic mathematics of the equilibrium constant approach was derived in general form by Brinkley (11,12) and Kandiner and Brinkley (13). The development of the equations was well suited for adaptation to digital computers and Feldman, et al. (14) utilized the approach to calculate the equilibrium composition of high temperature gaseous mixtures. This approach has been strongly favored by geochemists. For example, Crerar (15) has revised this method so that arbitrary independent equilibrium constants can be used. He employed Newton-Raphson iteration with curve-crawler techniques (16) for rapid convergence and applied his routine to problems of hydrothermal chemistry. It should be pointed out that these methods are completely general, in the mathematical sense, so that only the reactions taking place in the system under investigation need be coded for computation. This routine can be useful for laboratory and experimental systems where most of the species and associated data base of non-general programs are unnecessary.

Generalized Method of the Gibbs Free Energy Minimization. The mathematical formulation of the free energy approach was pioneered by White et al. (17) who pointed out the advantages of using alternative numerical techniques such as steepest descent and linear programming. Dayhoff et al. (18) computed equilibrium compositions for prebiological and planetary atmospheres with this approach and showed how these compositions changed at different temperatures and pressures for systems containing C, H, O, N, P, S and Cl. Holloway and Reese (19) have solved the equilibrium composition of the system C-O-H-N at high temperatures by a computerized free energy minimization model and Karpov and Kaz'min (20) have computed the distribution of species in seawater using a dual algorithm technique. For relatively simple systems where the free energies are available and reliable, the Gibbs minimization approach is convenient and dependable. For large complex systems, however, the equilibrium constant approach is preferred. If and when an accurate and internally consistent set of thermodynamic data becomes available, the Gibbs minimization will likely find greater use. We now present a brief review of non-general computerized models which uses the equilibrium constant approach and incorporates individual reactions as part of the program.

Specific Programs in Analytical Chemistry. Following the lead of Brinkley and others, several programs were independently developed to solve problems involving aqueous equilibria in analytical and physical chemistry. One of the earliest programs, HALTAFALL, came from the work of Sillen and his colleagues (21,22). This program was designed to calculate species distribution in an aqueous phase from a knowledge of the appropriate stability constants and total concentrations. The program is general enough to handle mixing problems such as titrations involving separation of gaseous or solid phases and organic solvent extractions. The computations are carried out at constant ionic strengh and constant temperature although these restrictions can be modified by the programmer. When apparent stability constants at different ionic strengths than the equilibrium problem are used, corrections are made by the Davies equation (80). The general method of calculation is by successive approximation. A companion program, LETAGROP VRID, was written to calculate stability constants for aqueous complexes from various measurements (23). Both programs are written in ALGOL. Perrin (24) and Perrin and Sayce (25) developed the COMICS program for the calculation of the equilibrium distribution of species using a type of "brute force" successive approximation (see below). COMICS has gone through several stages of modification which have been discussed by Leggett (1) and will not be further elaborated on here, except to mention a more recent version called SIAS (26). A similar program named EQBRAT is described by Detar (27) and it complements other programs (all in FORTRAN) which can handle a wide range of chemical problems. Bos and Meershoek (28) introduced Newton's method to titration calculations with the program EQUIL, written in PL/1, which increased the computation efficiency and was protected against non-convergence problems.

Successive Approximation Programs For Natural Water Equilibria. Garrels and Thompson (29) were the first to use the method of successive approximation in a hand calculation to solve an equilibrium problem in aqueous geochemistry. Their analysis of the species distribution in seawater in terms of ion association influenced the development of several models. Barnes and Clarke (30) found this approach useful in the investigation of the corrosion properties of wells and developed the WATCHEM program to interpret iron corrosion processes. This approach was also used to initialize PATHI (31) by a subroutine called SOLSAT which existed as a separate program. A recent version of this program, called EQUIL, has evolved from the work of Fritz (32) and Droubi (33) and it is used with the programs DISSOL and EVAPOR mentioned below. The most general first-generation programs of this type are WATEQ (34,35), SOLMNEQ (36), and EQ3 (37). All three programs have been designed to accept water analyses with on site values for pH, Eh and temperature. There is no proton mass balance condition, only mass balances on cations and anions are carried out. SOLMNEQ carries a data base in the form of a table of equilibrium constants for the range 0-350°C and EQ3 contains a similar data base for 0-300°C whereas WATEQ uses the Vant

Hoff equation or analytical expessions for equilibrium constants as a function of temperature and is considered reliable for the range 0-100°C. SOLMNEQ has additional features such as the silica and Na-K-Ca geothermometers for evaluating geothermal reservoir temperatures and includes several more trace element species than WATEQ. SOLMNEQ has been expanded to include organic complexes and ion exchange equilibria and uses the pressure dependence of the equilibrium constants. EQ3 also uses pressure dependent equilibrium constants. WATEQ was originally written in PL/1 and has been revised and translated into FORTRAN by Plummer et al. (38) in a program called WATEQF. Manganese speciation has been added to WATEQF and the successive approximation procedure was revised to give much faster convergence. 21 aqueous species and 17 minerals of uranium have been added to WATEQF in a recent modification (39). The advantages of using PL/1 optimizing code with reorganization into several subprogram blocks along with the expansion of the data base to include several trace elements and the rapid convergence of WATEQF have been incorporated into WATEQ2 (40). A shorter version, called WATSPEC, which is preferrable for handling routine water analyses has been published by Wigley (41).

The method of successive approximations has been conveniently described by Wigley (41) where either a "brute force" method or a "continued fraction" method can be used. The brute force method is the classical approach where mass action expressions are substituted directly into the mass balance conditions and solved for total concentrations which are then compared to the analytical values. In the continued fraction method, the non-linear equations are rearranged to solve for free ion concentrations which are initially assumed to be equal to the total concentrations, as detailed by Wigley (42). These two methods are best illustrated by a simple example. Assume a solution which contains free Ca^{2+} ions, free CO_3^{2-} ions, and only one ion pair: $CaCO_3^\circ$. The mass balance conditions are given by

$$mCa(total) = mCa^{2+} + mCaCO_3^\circ \quad (1)$$

$$mCO_3(total) = mCO_3^{2-} + mCaCO_3^\circ \quad (2)$$

with the restriction that (assuming an ideal solution)

$$K = \frac{mCaCO_3^\circ}{(mCa^{2+})(mCO_3^{2+})} \quad \text{or } mCaCO_3^\circ = K(mCa^{2+})(mCO_3^{2-}) \quad (3)$$

Substituting equation 3 into equations 1 and 2 gives

$$mCa(total) = mCa^{2+} + K(mCa^{2+})(mCO_3^{2-}) \quad (4)$$

and $mCO_3(total) = mCO_3^{2-} + K(mCa^{2+})(mCO_3^{2-})$. (5)

Using the brute force method,

$$mCa^{2+} = mCa(total) - K(mCa^{2+})(mCO_3^{2-}) \quad (6)$$

$$mCO_3^{2-} = mCO_3(total) - K(mCa^{2+})(mCO_3^{2-}) \quad (7)$$

where mCa^{2+} and mCO_3^{2-} are assumed equal to mCa(total) and mCO_3(total) for the first estimate. Equations 6 and 7 give new values for the free ion concentrations which are then used in the next iteration. Using the continued fraction method equations 4 and 5 are rearranged to

$$mCa^{2+} = \frac{mCa(total)}{1 + KmCO_3^{2-}} \quad (8)$$

$$mCO_3^{2-} = \frac{mCO_3(total)}{1 + KmCa^{2+}} \quad . \quad (9)$$

As before, the first estimate is made by assuming that free ion concentrations are equal to total concentrations but the second estimate comes from equations 8 and 9 which vary less than equations 6 and 7 because of the form of the equation.

This alternate form of the equations produces a faster convergence as shown in an example given by Wigley (41) and also converges more rapidly than Newton-Raphson. EQ3 employs an additional control on the continued fraction method which generates monotone sequences (43,44). Its chief virtues are strict error bounds and increased stability with respect to a wide range of analyses of aqueous solutions used as input.

Other programs of this general type include SEAWAT (45) which was specifically designed for seawater calculations, MIRE (46) which was specifically designed for anoxic marine pore waters, IONPAIR and NOPAIR (47) and CALCITE (48) which were designed for freshwaters in carbonate terrains and KATKLE 1 (49) which has been used for soil water geochemistry.

Newton-Raphson Programs for Experimental and Natural Water Equilibria. Morel and Morgan (49) developed the FORTRAN program REDEQL for the calculation of multicomponent metal-ligand equilibria with considerable flexibility and includes a large number of metal-ligand complexes. The program is based on the equilibrium constant approach and uses Newton-Raphson iteration to find the solution to a function which compares the difference between the total calculated component concentration and the total analytical component concentration. REDEQL has the capability of imposing mineral saturation to allow dissolution and/or precipitation of various solids. Another option is the calculation of interaction intensities and capacities (50,51,52). This program and its subsequent modifications have led to several "second generation" programs which are widely used by environmental engineers: REDEQL2, MINEQL and GEOCHEM. This family of programs can simulate adsorption behavior using the James-Healy approach, the Schindler-Stumm complexation approach or a combination of surface complexation and electric double layer (53-59). These programs also contain a data base for a constant temperature of 25°C, other temperatures cannot be calculated without changing all the equilibrium constants to those temperatures. MINEQL (60) has greater clarity and flexibility in a more compact program which utilizes Gaussian elimination to solve the matrix equation. GEOCHEM (61, unpublished data) has a greatly expanded data base covering over 800 organic species and more than 2000 inorganic species and it includes ion exchange reactions for simulating soil water reactions. Convergence problems are occasionally encountered with Newton-Raphson methods which are commonly caused by poor estimates of initial concentrations supplied by the programmer. However, several numerical techniques including under-relaxation, curve-crawling and pre-iteration optimization of starting estimates can greatly increase the reliability of the Newton-Raphson method (15,37,62). A good example is provided by the program EQUIL (62) which utilizes matrix scaling, eigen vector analysis, matrix modifications and a convergence forcer to achieve rapid and reliable convergence.

Reaction Path Simulation. The first application of computer techniques to problems of mass transfer in geochemistry began with the work of Helgeson and colleagues (63,64). The general approach of the program PATHI (31) is to describe a partial equilibrium reaction path (e.g. mineral dissolution and rock weathering) in terms of ordinary differential equations which are linear and can be solved by matrix algebra. The initial condition, which consists of an aqueous fluid, must be calculated by a species distribution iteration scheme such as the successive approximation method. Progressive reaction states can be computed by incrementing the progress variable, ξ (65), iteratively and checking the aqueous phase for saturation at each step and then dissolving or precipitating the required mass(es) of the appropriate mineral phase(s) to retain equilibrium. Thus, by integrating a set of differential equations a reaction path can be followed progressively until overall chemical equilibrium is reached by the specified system. This method has been applied to the study of weathering reactions, diagenesis, metamorphism and hydrothermal metasomatism (64), in the formation of ore deposits (66) and scaling of conduits caused by precipitation from geothermal brines (67). PATHI has been revised and modified to simulate granite weathering (32)and evaporation of closed basin lakes

(36) with the programs DISSOL and EVAPOR which use a related program, EQUIL, to calculate species distribution. These programs have been applied to the prediction of soil salinity and optimal irrigation doses as well as the general evolution of soil water during evaporation (68,69).

Several difficulties with PATHI led to very long execution times and occasional abnormal program termination. Claude Herrick and others (67) greatly improved the efficiency by adopting high-order Gear integration techniques (70,71). However, drift error incurred in the integration remains an annoyance, and the differential equation approach cannot be applied if the starting solution is supersaturated with any solids.

Wolery (37) has written a PATHI-like program, EQ6, which uses the Newton-Raphson method to solve the system of algebraic equations instead of their differential counterparts at each stage of reaction progress. The drift problem is thereby avoided. Taylor's series expressions, based on finite differences, are utilized to follow the course of the simulation and to predict starting estimates at each new point of reaction progress. EQ6 has some affinities in its numerical approach to the REDEQL school and Crerar's (15) program and it can precipitate an equilibrium assemblage of precipitates from an initially supersaturated solution.

Another program well suited for reaction path simulation is MIX2 (72) which uses a regression technique on the charge balance condition for the aqueous phase in computing pH after a reaction step. MIX2 can solve problems in mixing and titration of aqueous solutions (73,74), evaporation, heterogeneous equilibrium, and non-equilibrium heterogeneous reactions (75). Although MIX2 avoids the drift problems associated with PATHI, the aqueous model is limited to the major species in the system CaO-MgO-Na_2O-K_2O-H_2SO_4-H_2CO_3-HCl-H_2O and considers only one phase boundary at a time. More advanced mass transfer programs using a log linearization technique for simultaneous solution of mass action, mass balance and charge balance equation are in preparation (76) which have broad applications to mass transfer problems.

Test Case Results

Collectively, the programs mentioned above represent the "state of the art" in the calculation of the equilibrium distribution of species in aqueous systems. As a means of examining the consistency of these programs, two test cases (a dilute river water and an average seawater analysis) were compiled and mailed to more than fifty researchers who have been active in the field of chemical modeling. These test cases may overlook many of the features of specific programs, but they provide a common basis by which most of the programs can be compared. One approach to the comparison of aqueous models is to tabulate and examine the thermodynamic data, activity coefficients, choice of complexes, etc. Alternatively the approach taken in this survey has been to examine the results predicted by the aqueous models, an approach that integrates all the aspects of each model. In this manner the differences between models can be seen in terms of the actual results which are of value in applications to specific problems. Differences will always be apparent between the thermodynamic data base used in different models but it is difficult to ascertain whether a certain difference has any effect on the final result unless the results are all compared. It should be emphasized, however, that any type of comparison will always be inadequate because any criterion that is chosen as a basis for comparison is usually not given the same priority by another model. For example, models that are developed for reaction path simulation may sacrifice computation efficiency on a species distribution calculation if more efficiency is gained in the reaction progress calculation. In this instance, as in many others, the computerized model is developed for a specific purpose and design priorities are assigned accordingly.

Each researcher was asked to complete a questionnaire describing the details of their program and to return the computed results on the two test cases. The responses received include representatives of most of the major aqueous models known. Table I identifies the programs and researchers who supplied the information for the particular program and includes general information on types of computers used, number of cards in the source deck, primary and secondary references and availability.

Table II gives a general description of the program features such as total number of elements, aqueous species, gases, organic species, redox species, solid species, pressure and temperature ranges over which calculations can be made, an indication of the types of equations used for computing activity coefficients, numerical method used for calculating distribution of species and the total number of iterations required by these models for each of the two test cases. The chemical analyses for the two test cases are summarized in Table III. The seawater compilation was prepared in several units to assure consistency between concentrations for proper entry into the aqueous models.

The results of the river water and seawater test cases computed by the aqueous models listed in Table I are summarized in Tables IV-X. Tables IV and V compare selected major and minor species computed for the river water test case, and Tables VI and VII make a similar comparison for the seawater test case. Table VIII compares activity coefficients computed for the major species in seawater and Table IX and X tabulate saturation indices for selected minerals in the river water and seawater test cases. The saturation index, SI, is defined as log IAP/Ksp where IAP is the ion activity product for the mineral and Ksp is the thermodynamic solubility product constant.

Discussion

The remainder of this paper is a plea for caution and restraint in interpreting the results of the test cases shown here, and in the use of computerized chemical models in general. Tables IV-X show both remarkable agreement and disagreement. In general, there is better agreement between the major species concentrations than the minors and the results for the river water tend to agree better than those for seawater. We would expect better agreement in the river water test case because of the smaller amount of complexing in the more dilute solution and the more consistent and reliable activity coefficients which can be obtained at low ionic strength.

There are several limitations which lead to the discrepancies in Tables IV-X. First of all, no model will be better than the assumptions upon which it is based. The models compiled in this survey are based on the ion association approach whose general reliability rests on several non-thermodynamic assumptions. For example, the use of activity coefficients to describe the non-ideal behavior of aqueous electrolytes reflects our uncertain knowledge of ionic interactions and as a consequence we must approximate activity coefficients with semi-empirical equations. In addition, the assumption of ion association may be a naive representation of the true interactions of "ions" in aqueous solutions. If a consistent and comprehensive theory of electrolyte solutions were available along with a consistent set of thermodynamic data then our aqueous models should be in excellent agreement for most systems. Until such a theory is provided we should expect the type of results shown in Tables IV-X. No degree of computational or numerical sophistication can improve upon the basic chemical model which is utilized.

The second limitation, almost a corollary of the first, is the reliability of the equilibrium constants (or free energies) used in the model. It is quite common for solubility product constants and complex stability constants to vary by 1 to 3 orders of magnitude and have been observed to vary by that much in these models. Quite clearly this amount of uncertainty can cause large differences in the computed results and probably contributes the largest single source of error. The thermodynamic properties of substances are currently in a state of refinement by many researchers and continual updating of the aqueous models is needed. This situation is quite frustating for investigators who are involved in solute transport modeling since they need a general chemical model with a fixed data base. An internally consistent data base can be obtained

Table I
General Information

Program Name	EQUIL	EQ3	GEOCHEM	IONPAIR	MINEQL2	MIRE	MINEQL/ REDEQL2	REDEQL2	SEAWAT	SOLMNEQ	WATEQF	WATEQ2	WATSPEC	SIAS/ COMICS
Language	FORTRAN	FORTRAN	FORTRAN	FORTRAN	FORTRAN	FORTRAN	FORTRAN	FORTRAN	FORTRAN	PL/1	FORTRAN	PL/1 (optimizing)	FORTRAN	FORTRAN
Computers	UNIVAC 1110	CDC 6400/6600 7600	IBM 360/50	IBM 370/ 165II	CDC CYBER 74	IBM 360	IBM 370	CDC CYBER 74	CDC 6400 IBM 360 IBM 370 IBM 7094 DEC 10	IBM 370	IBM 370[1]	IBM 370 Honeywell 60/68/80	ICL 1903 IBM 370	IBM 360
Number of Cards	3300	3048	3630	606	1500	420	1500	4000	184	2000	1857	3000	423	253
Primary Reference(s)	(32)	(37)	(50)	(47)	(60)	(46)	(50)	(50)	(45)	(36)	(34,35)	(34,35)	(41)	(25)
Secondary Reference(s)	(33)	(43,44)	(85)	(87,88)	(85)	(86)	(85)	(85)			(38)	(40)	(42)	(26)
Available[2] From	B. Fritz*	T.J. Wolery*	S.V. Mattigod* G. Sposito*	J. Thrail-kill*	M.R. Hoff-mann*	G.R. Holdren Jr.*	F.M.M. Morel R.E. McDuff*	M.R. Hoff-man*	G.M. Lafon*	R.L. Bassett* Y.K. Kharaka[3]	L. N. Plummer*	J.W. Ball* E.A. Jenne*	Geo Ab-stracts[4]	T. M. Florence*

[1]WATEQF is also available for use on an UNIVAC 1110 computer by contacting M.M. Reddy. [2]Coauthors who contributed data given in this and subsequent tables are identified by an asterisk,*. [3]U.S. Geological Survey, 345 Middlefield Road, Menlo Park, CA 94025. [4]Geo Abstracts Limited, University of East Anglia, Norwich NR4 7TJ, England; price $1.70. WATSPEC calculations on subsequent tables were contributed by T.M.L. Wigley.

Table II
Descriptive Program Features

Program Name	EQUIL	EQ3	GEOCHEM	IONPAIR	MINEQL2	MIRE	MINEQL/ REDEQL2	REDEQL2	SEAWAT	SOLMNEQ	WATEQF	WATEQ2	WATSPEC	SIAS/ COMICS
Elements[1]	20	18	44	8	38	11	*	44	7	24	19	29	16	*
Aqueous Sp.	93	140	2000	26	*	49	*	266	23	181	105	220	69	*
Gases	3	8	2	1	*	0	*	2	1	3	3	3	2	*
Organics	0	0	889	0	*	1	*	35	0	10	0	12	0	*
Redox Sp.	9	14	20	0	*	0	*	24	0	12	8	12	6	*
Minerals	120	130	185	5	*	variable	*	154	variable	158	101	309	40	*
Activity Coefficients[2]	B-dot	B-dot	Davies or B-dot	EDH	Davies	EDH	Davies	Davies	B-dot	B-dot	B-dot or Davies	B-dot or Davies	B-dot or Davies	**
Temperature Range °C	0-300	0-300	25	0-25	25	25	25	25	0-100	0-350	0-100	0-100	0-100	25
Pressure range bars	1	to steam saturation or 500 bars	1	1	1	1	1	1	1-1000	1-1000	1	1	1	1
Iteration and convergence method	explicit equations	Monotone sequences and secant method	Newton-Raphson	Free ion molalities by difference	Newton-Raphson	continued fraction	Newton-Raphson	Newton-Raphson	continued fraction	continued fraction for anions only	continued fraction	continued fraction	continued fraction	brute force
Fresh Water[3]	<10	5	-	7	†	3	†	32†	-	7	1	2	1	†
Seawater[3]	<30	6	156	-	†	5	†	57†	4	11	4	4	5	29

[1]excluding those of water. [2]EDH - extended Debye Huckel, B-dot - extended Debye Huckel with added B˙I term, Davies - Davies equation.
[3]Number of iterations where one iteration is defined as one pass through the mass action equations of the aqueous model. *depends on data base.
**Constant ionic medium. †depends on initial guess by user.

Table III
Test Case Data

River Water		Seawater*					
Species	mg/l	species	mg/kg soln	mmol/kg H_2O	mmol/l	mg/l	GFW
Na	12.	Ca	412.3	10.6617	10.5272	421.931	40.08
K	1.4	Mg	1291.8	55.08565	54.39113	1321.976	24.305
Ca	12.2	Na	10768.	485.4435	479.3230	11019.54	22.9898
Mg	7.5	K	399.1	10.5794	10.4461	408.423	39.0983
Si	8.52	Cl	19353.	565.7625	558.6293	19805.09	35.453
HCO_3^-**	75.2	SO_4	2712.	29.2615	28.8926	2775.35	96.0576
Cl	9.9	Alkal.†	141.682	2.40659	2.37625	144.992	61.0171
SO_4	7.7	Br	67.3	0.87294	0.86194	68.872	79.904
B	0.050	Sr	8.14	0.096285	0.095071	8.3302	87.62
Br	0.006	B	4.45	0.42665	0.42127	4.5540	10.81
I	0.0018	SiO_2	4.28	0.073828	0.072897	4.3800	60.0843
F	0.10	F	1.39	0.075829	0.074873	1.4225	18.9984
PO_4	0.210	Ba	0.02	0.000151	0.000149	0.0205	137.33
NO_3	0.898	I	0.062	0.0005064	0.0005000	0.06345	126.9045
NO_2	0.019	PO_4	0.06	0.000655	0.000647	0.0614	94.9714
NH_4	0.144	NO_3	0.29	0.004847	0.004786	0.2968	62.0049
Fe(II)	0.015	NO_2	0.02	0.000451	0.000445	0.0205	46.0055
Fe(III)	0.0007	NH_4	0.03	0.00172	0.00170	0.0307	18.0383
Mn	0.0044	Fe	0.002	0.0000371	0.0000366	0.00205	55.847
Al	0.005	Mn	0.0002	0.00000377	0.00000373	0.000205	54.9380
Zn	0.00049	Al	0.002	0.0000768	0.0000759	0.00205	26.9815
Cd	0.0001	Zn	0.0049	0.00007768	0.00007670	0.005014	65.38
Hg	0.00001	Cd	0.0001	0.000000922	0.000000910	0.000102	112.41
Pb	0.00003	Hg	0.00003	0.000000155	0.000000153	0.0000307	200.59
Cu	0.0005	Pb	0.00005	0.000000250	0.000000247	0.0000512	207.2
Co	0.0005	Cu	0.0007	0.0000114	0.0000112	0.000716	63.546
Ni	0.0018	Co	0.00005	0.000000879	0.000000868	0.0000512	58.9332
Cr	0.0005	Ni	0.0017	0.00002056	0.00002030	0.001740	58.71
Ag	0.00004	Cr	0.0003	0.00000598	0.00000590	0.000307	51.996
Mo	0.0005	Ag	0.00004	0.000000384	0.000000379	0.0000409	107.868
As	0.002	Sb	0.00033	0.000002809	0.000002774	0.0003377	121.75
H_2S	0.002	Mo	0.005	0.0000540	0.0000533	0.00512	95.94
DO	10.94	As	0.004	0.0000553	0.0000546	0.000409	74.9216
Eh (V)	0.440	Li	0.181	0.027027	0.026686	0.18523	6.941
DOC	2.5	Rb	0.117	0.0014188	0.0014009	0.11973	85.4678
T°C	9.5	Cs	0.0004	0.00000312	0.00000318	0.00409	132.9054
pH	8.01	DO				6.6	
Density	1.00	Eh (V)	0.500				
		T°C	25.0				
		pH	8.22				
		Density	1.02336				

*Original data are in ppm, except total titration alkalinity which is 2.322 meq/kg soln. All other units have been derived from these values using: density = g/cc, salinity = 35. ‰, kg soln/kgH_2O = 1.03642731, and the values of Gram Formula Weight given. The derived units are given to two additional significant figures. **Titration alkalinity as HCO_3^-. † Total titration alkalinity is 2.322 meq/kg soln, expressed here as HCO_3^-. The total titration alkalinity includes carbonate as well as non-carbonate alkalinity. Instead of total alkalinity, the input may be expressed as total inorganic carbon which is 2.022 mmol/kg soln, or 123.377 ppm, 126.259 mg/l, 2.06924 mmol/l, 2.09566 mmol/kgH_2O, as HCO_3^-. Alternatively total inorganic carbon may be expressed as $H_2CO_3^*$ ($H_2CO_3^* \doteq H_2CO_3^\circ + CO_2^\circ$) + HCO_3^-(total) + CO_3^{2-}(total). In this case, $H_2CO_3^*$ is 0.742030 ppm, 0.759364 mg/l, 0.0122429 mmol/l, 0.0123992 mmol/kgH_2O; HCO_3^-(total) is 109.469 ppm, 112.026 mg/l, 1.83598 mmol/l, 1.85942 mmol/kgH_2O; and CO_3^{2-}(total) is 12.9598 ppm, 13.2625 mg/l, 0.221228 mmol/l, 0.223831 mmol/kgH_2O.

Table IV. p(m), -Log Molality, of Selected Major Species in River Water Test Case

PROGRAM	EQUIL	EQ3	IONPAIR	MINEQL2*	MIRE	MINEQL/ REDEQL2	REDEQL2	SOLMNEQ	WATEQF	WATEQ2	WATSPEC
Ca^{2+}	3.532	3.529	3.522	3.54†	3.530	3.54†	3.539	3.527	3.525	3.525	3.525
$CaSO_4^\circ$	5.537	5.578	5.6	5.58†	5.533	5.55†	5.58	5.545	5.578	5.577	5.545
$CaHCO_3^+$	5.288	5.343	-	5.17†	5.409	5.22†	5.27	5.518	5.722	5.723	5.714
$CaCO_3^\circ$	5.681	5.933	6.	5.76†	5.732	5.82†	5.27	5.959	6.000	6.001	5.992
Mg^{2+}	3.523	3.519	3.518	3.53	3.521	3.54	3.53	3.519	3.519	3.519	3.520
$MgSO_4^\circ$	5.586	5.750	5.7	5.47	5.593	5.45	5.47	5.726	5.767	5.756	5.600
$MgHCO_3^+$	5.636	5.491	5.6	5.16	5.568	5.21	5.25	6.027	5.495	5.496	5.486
$MgCO_3^\circ$	5.470	6.133	6.	5.55	5.988	5.61	5.65	5.622	6.156	6.157	6.147
Na^+	3.283	3.282	3.283	3.28	3.284	3.28	3.28	3.283	3.283	3.283	3.283
$NaSO_4^-$	6.447	6.819	7.	6.82	7.280	6.81	6.82	6.617	6.819	6.820	6.793
$NaHCO_3^\circ$	6.505	-	-	-	6.500	-	-	6.496	6.495	6.496	6.494
K^+	4.446	4.446	4.45	4.45	4.444	4.45	4.45	4.446	4.446	4.446	4.446
KSO_4^-	7.830	7.896	-	7.59	7.819	7.58	7.60	7.942	7.935	7.936	7.910
SO_4^{2-}	4.129	4.122	4.12	4.14	4.128	4.15	4.14	4.122	4.121	4.121	4.164
Cl^-	3.401	3.554	3.554	3.55	3.556	3.55	3.55	3.558	3.554	3.554	3.554
HCO_3^-	2.924	2.920	2.915	2.93	2.919	2.91	2.93	2.918	2.917	2.917	2.917
CO_3^{2-}	5.164	5.346	5.3	5.01	5.168	5.11	5.12	5.328	5.334	5.333	5.333
$B(OH)_3^\circ$	5.361	-	-	5.37	-	-	-	5.379	5.354	5.355	5.355
$B(OH)_4^-$	6.570	-	-	6.41	-	-	6.42	6.726	6.690	6.711	6.688
Br^-	-	-	-	7.12	-	-	7.12	-	7.124	7.124	7.120
F^-	5.281	-	-	5.29	-	-	5.29	5.282	5.284	5.284	-
$H_4SiO_4^\circ$	3.528	-	-	3.53†	-	-	3.53†	3.523	3.851	3.520	3.520
H^+	7.988	7.989	-	-	-	8.01	-	7.990	7.987	7.987	7.987
OH^-	5.966	6.518	6.	5.94	5.967	-	5.95	6.525	6.502	6.501	6.502
Ionic Strength	.00238	.00239	.00240	.00300	.0024	-	.00300	.00240	.00240	.00241	.00239

*Calculated at 25°C. †MINEQL2 and REDEQL2 results were calculated in such a way that the solution was equilibrated with supersaturated phases. P, Fe, Al and Si species are not exactly comparable to the same species from the other programs because of this equilibrium process and, of course, all species have been affected to some degree.

Table V

p(m), - Log Molality, of Selected Minor Species in River Water Test Case

PROGRAM	EQUIL	EQ3	IONPAIR	MINEQL2*	MIRE	REDEQL2	SOLMNEQ	WATEQF	WATEQ2	WATSPEC
Cr^{3+}	-	-	-	20.02	-	16.15	-	-	-	-
Mn^{2+}	7.100	8.205	-	7.14	7.138	7.13	7.099	7.119	7.119	-
$MnOH^{+}$	-	11.403	-	9.28	9.804	9.28	-	10.264	10.347	-
$MnSO_4^{\circ}$	9.166	10.383	-	9.18	9.213	9.18	9.289	9.864	9.864	-
Fe^{2+}	6.711	6.639	-	15.18†	6.896	15.18†	8.389	12.167	6.579	11.953
$FeOH^{+}$	7.075	7.428	-	15.82†	6.855	15.83†	26.711	14.257	8.669	13.860
Fe^{3+}	19.04	16.212	-	20.88†	-	20.640†	25.811	17.635	18.596	17.408
$Fe(OH)_4^{-}$	-	8.072	-	10.44†	-	10.35†	18.172	6.989	8.340	7.028
Ni^{2+}	-	-	-	7.59	-	7.93	-	-	8.892	-
$NiCO_3^{\circ}$	-	-	-	-	-	7.76	-	-	7.539	-
Cu^{2+}	10.15	8.763	-	9.67	-	10.42	14.772	-	11.293	-
$Cu(OH)_2^{\circ}$	-	-	-	13.68	-	13.22	-	-	9.045	-
Ag^{+}	-	9.660	-	-	-	15.51	9.651	-	15.891	-
$AgHS^{\circ}$	-	-	-	-	-	-	-	-	9.432	-
Zn^{2+}	8.131	8.129	-	8.15	-	8.48	7.129	-	8.504	-
$ZnCO_3^{\circ}$	-	-	-	-	-	8.51	-	-	8.720	-
$Zn(HS)_2^{\circ}$	-	-	-	-	-	-	-	-	8.790	-
Cd^{2+}	-	-	-	9.10	-	9.41	-	-	11.612	-
$CdOH^{+}$	-	-	-	10.15	-	11.46	-	-	12.651	-

$CdHS^+$	-	-	-	-	-	-	-	-	9.078	-
Hg^{2+}	-	16.875	-	21.07	-	20.15	41.872	-	-	-
Al^{3+}	15.69	13.697	-	17.49†	-	17.164†	13.598	13.908	15.565	14.778
$Al(OH)_4^-$	7.090	6.732	-	8.66	-	8.37†	6.765	6.741	8.406	6.733
Pb^{2+}	9.845	14.380	-	12.07	-	11.93	9.843	-	11.749	-
$PbOH^+$	-	-	-	11.21	-	11.68	-	-	10.378	-
$PbCO_3^\circ$	-	9.839	-	9.89	-	9.86	-	-	10.025	-
NO_3^-	4.839	-	4.84	-	-	-	4.839	4.839	4.839	4.839
$NH+$	5.098	-	-	5.14	-	5.13	5.105	5.106	5.106	10.392
PO_4^{3-}	10.03	-	-	11.41†	-	11.40†	10.189	10.156	10.156	-
HPO_4^{2-}	5.818	-	-	7.07†	5.773	7.16†	5.813	5.793	5.793	-
$H_2PO_4^-$	6.694	-	-	7.78†	6.654	8.06†	6.622	6.625	6.625	-
$HAsO_4^{2-}$	-	-	-	-	-	-	13.990	-	7.595	-
$H_2AsO_4^-$	-	-	-	-	-	-	15.046	-	8.885	-
S^{2-}	-	17.140	-	-	-	-	16.489	12.613	12.870	12.615
HS^-	-	7.268	-	-	-	-	7.299	7.288	7.544	7.288
H_2S°	-	8.076	-	-	-	-	8.071	8.146	8.403	8.146
I^-	-	-	-	7.785	-	7.86	-	-	-	-

*Calculated at 25°C. †MINEQL2 and REDEQL2 results were calculated in such a way that the solution was equilibrated with supersaturated phases. P, Ca, Fe, Al and Si species are not exactly comparable to the same species from the other programs because of this equilibration process, and, of course, all species have been affected to some degree.

Table VI

p(m), -Log Molality, of Selected Major Species in Sea Water Test Case

PROGRAM	EQUIL	EQ3	GEOCHEM	K+K*	MINEQL2	MIRE	MINEQL/ REDEQL2	REDEQL2	SEAWAT	SIAS	SOLMNEQ	WATEQF	WATEQ2	WATSPEC
Ca^{2+}	2.013	2.027	2.21	2.024	2.12†	2.074	2.03	2.11†	2.053	2.058	2.013	2.024	2.025	2.023
$CaSO_4^\circ$	3.062	2.928	3.12	3.046	3.00†	2.744	2.97	3.00†	2.764	2.781	3.056	2.940	2.938	2.947
$CaHCO_3^+$	4.261	4.234	4.48	4.504	4.92†	4.366	4.26	4.84†	4.242	4.285	4.261	4.481	4.449	4.465
$CaCO_3^\circ$	4.636	4.540	4.59	4.755	5.30†	4.734	4.63	5.30†	4.715	4.365	4.635	4.649	4.616	4.665
$CaCl^+$	-	-	2.46	-	-	-	-	-	-	-	-	-	-	-
Mg^{2+}	1.304	1.319	1.46	1.328	1.34	1.322	1.33	1.34	1.347	1.357	1.299	1.311	1.312	1.310
$MgSO_4^\circ$	2.305	2.175	2.47	2.190	2.11	2.299	2.17	2.13	2.008	2.080	2.360	2.235	2.222	2.241
$MgHCO_3^+$	3.777	3.554	3.83	3.405	4.14	3.676	3.60	4.06	3.588	2.684	3.834	2.655	3.624	3.597
$MgCO_3^\circ$	3.619	3.940	4.14	3.745	4.32	3.684	3.67	4.32	4.121	4.664	3.689	4.045	4.014	4.062
$MgCl^+$	-	-	1.79	-	-	-	-	-	-	-	-	-	-	-
Na^+	0.336	0.320	0.41	0.328	0.34	0.335	0.33	0.34	0.321	0.314	0.336	0.320	0.320	0.323
$NaSO_4^-$	1.964	2.208	1.94	2.286	2.18	2.716	2.20	2.19	2.107	-	1.965	2.178	2.195	2.146
$NaHCO_3^\circ$	3.873	-	3.40	3.466	-	3.851	3.84	-	-	-	3.860	3.785	3.752	3.807
K^+	1.982	1.983	2.04	2.007	2.01	2.000	1.99	2.01	1.981	1.976	1.982	1.983	1.983	1.985
KSO_4^-	3.839	3.748	3.76	3.802	3.45	3.789	3.67	3.46	3.916	-	3.871	3.768	3.766	3.736
Cl^-	0.257	0.247	0.33	0.25	0.25	0.265	0.25	0.25	0.247	0.253	0.256	0.247	0.247	0.249
SO_4^{2-}	1.906	1.823	2.11	1.804	1.91†	1.836	1.84	1.89†	2.009	1.723	1.881	1.811	1.808	1.822
HCO_3^-	2.884	2.816	2.87	2.792	3.56	2.815	2.90	3.39	2.808	-	2.885	2.825	2.836	2.845
CO_3^{2-}	4.419	4.371	4.48	4.622	4.91	4.415	4.46	4.89	4.369	5.507	4.484	4.422	4.566	4.443
$B(OH)_3^\circ$	3.435	-	3.51	-	3.49	-	3.45	3.48	-	-	3.506	3.430	3.431	3.481
$B(OH)_4^-$	4.226	-	4.23	-	4.06	-	4.20	4.10	-	-	4.234	4.257	4.255	4.015
Br^-	-	-	3.06	-	3.07	-	-	3.07	3.059	3.065	-	3.059	3.059	3.059

F^-	4.228	-	4.36	-	4.39	-	-	4.38	-	4.411	4.379	4.410	4.410	-
MgF^+	4.792	-	4.53	-	4.56	-	-	4.58	-	-	4.464	4.442	4.443	-
Sr^{2+}	4.016	-	4.28	-	4.13†	-	-	4.119†	-	-	4.045	4.016	4.016	4.016
$H_4SiO_4^\circ$	4.155	4.159	4.16	-	4.19†	-	-	4.17†	-	-	4.161	4.145	4.145	4.147
H^+	8.125	8.126	8.15	7.966	-	-	-	-	8.095	-	8.125	8.093	8.093	8.098
OH^-	5.603	5.606	5.65	6.038	5.46	5.552	-	5.54	5.613	-	5.601	5.660	5.660	5.571
Ionic Strengh	0.663	0.6772	0.5298	-	0.65	0.655	-	0.65	0.6595	-	0.6662	0.6801	0.6799	0.6770

*Karpov and Kaz'min (1972) using Gibbs free energy minimization on a similar seawater. †MINEQL2 and REDEQL2 results were calculated in such a way that the solution was equilibrated with supersaturated phases. Si, Al, Fe, Ca, Sr, Ba, Zn SO_4, and PO_4 species are not exactly comparable to the same species from the other programs because of this process, and, of course, all species have been affected to some degree.

Table VII

p(m), -Log Molality, of Selected Minor Species in Seawater Test Case

PROGRAM	EQUIL	EQ3	GEOCHEM	MINEQL2	MIRE	REDEQL2	SIAS	SOLMNEQ	WATEQF	WATEQ2	WATSPEC
Li^{+}	4.572	-	4.66	-	-	-	-	4.568	4.573	4.573	4.573
Rb^{+}	-	-	5.87	-	-	-	-	-	-	5.848	-
Cs^{+}	-	-	8.56	-	-	-	-	-	-	8.506	-
Ba^{2+}	-	-	6.93	8.03†	-	8.01†	-	6.857	6.821	6.821	6.821
Cr^{3+}	-	-	25.63	28.34	-	24.55	-	-	-	-	-
Mn^{2+}	8.456	9.287	9.33	8.96	8.592	8.94	-	8.478	8.654	8.654	-
$MnCl^{+}$	-	-	10.26	8.74	9.067	8.75	-	29.376	8.880	8.881	-
Fe^{3+}	18.740	17.466	22.94	20.71†	-	20.26†	-	-	17.897	17.897	17.714
$Fe(OH)_3^{\circ}$	-	-	11.28	17.10†	-	24.14†	-	-	8.071	8.071	7.805
$Fe(OH)_4^{-}$	-	7.439	12.41	9.96†	-	9.94†	-	-	7.664	7.664	7.678
Ni^{2+}	-	-	9.23	7.77	-	7.92	7.834	-	-	8.813	-
$NiCl_2^{\circ}$	-	-	9.74	-	-	8.10	-	-	-	9.335	-
$NiCO_3^{\circ}$	-	-	7.91	5.75	-	8.61	-	-	-	7.590	-
Cu^{2+}	13.152	9.056	11.21	11.35	-	11.32	9.579	-	-	10.153	-
$Cu(OH)_3^{\circ}$	-	-	12.40	3.22	-	9.06	-	-	-	7.985	-
$CuCO_3^{\circ}$	-	-	10.01	10.82	-	10.80	-	-	-	9.070	-
Ag^{+}	-	14.797	16.79	16.82	-	16.79	-	14.477	-	14.394	-
$AgCl_4^{3-}$	-	9.517	12.48	12.08	-	12.05	-	9.617	-	9.684	-
Zn^{2+}	7.323	7.346	8.35	7.74†	-	8.06†	7.690	7.321	-	7.547	-
$ZnCO_3^{\circ}$	-	-	7.33	-	-	8.95†	-	-	-	7.894	-
Cd^{2+}	-	-	11.11	10.59	-	10.70	10.606	-	-	11.257	-
$CdCl^{+}$	-	-	9.99	9.47	-	9.41	-	-	-	9.371	-

$CdCl_2^\circ$	-	-	9.94	9.34	-	9.48	-	-	-	9.369	-
Hg^{2+}	-	23.310	-	-	-	-	-	23.426	-	-	-
$HgCL_3^-$	-	9.316	-	-	-	-	-	9.965	-	-	-
Al^{3+}	15.963	16.076	16.37	16.67†		16.15†		19.277	15.956	16.091	16.996
$Al(OH)_3^\circ$	7.600	-	7.72	-	-	7.58†	-	-	-	7.665	-
$Al(OH)_4^-$	7.287	7.114	7.25	7.53†	-	7.32†	-	10.611	7.115	7.259	7.114
Pb^{2+}	10.446	13.746	11.68	-	-	-	-	10.442	-	11.335	-
$PbCl_2^\circ$	10.380	13.591	11.41	-	-	-	-	0.372	-	11.017	-
$PbCO_3^\circ$	-	9.602	10.09	-	-	-	-	9.743	-	9.743	-
NO_3^-	5.315	-	5.30	-	-	-	-	-	5.315	5.314	5.315
PO_4^{3-}	9.979	-	10.48	11.81†	-	11.76†	-	10.126	10.482	10.480	-
HPO_4^{2-}	6.858	-	7.02	8.48†	6.673	8.48†	-	6.897	6.989	6.988	-
$H_2PO_4^-$	8.474	-	8.49	9.93†	8.295	10.06†	-	8.449	8.382	8.381	-
AsO_4^{3-}	-	-	11.01	-	-	-	-	-	-	10.007	-
$HAsO_4^{3-}$	-	-	8.45	-	-	-	-	-	-	7.264	-
I^-	-	-	6.30	6.30	-	6.31	-	-	-	6.296	-

†MINEQL2 and REDEQL2 results were calcualted in such a way that the solution was equilibrated with supersaturated phases. Si, Al, Fe, Ca, Sr, Ba, Zn, SO_4, and PO_4 species are not exactly comparable to the same species from the other programs because of this process, and of course, all species have been affected to some degree.

Table VIII

Activity Coefficients of Selected Major Species in Sea Water

PROGRAM	EQUIL	EQ3	MIRE	SEAWAT	SOLMNEQ	WATEQF	WATEQ2
Ca^{2+}	0.245	0.245	0.231	0.245	0.244	0.249	0.249
$CaSO_4^\circ$	1.172	1.000	0.507	1.150	1.180	1.170	1.170
$CaHCO_3^+$	0.714	0.714	0.589	0.669	0.737	0.747	0.747
$CaCO_3^\circ$	1.172	1.000	1.123	1.150	1.180	1.170	1.170
Mg^{2+}	0.315	0.315	0.296	0.314	0.263	0.288	0.288
$MgSO_4^\circ$	1.172	1.000	1.123	1.150	1.180	1.170	1.170
$MgHCO_3^+$	0.671	0.670	0.589	0.669	0.670	0.747	0.747
$MgCO_3^\circ$	1.172	1.000	0.387	1.150	1.180	1.170	1.170
Na^+	0.671	0.670	0.631	0.688	0.670	0.706	0.706
$NaSO_4^-$	0.683	0.682	0.589	0.669	0.720	0.747	0.747
$NaHCO_3^\circ$	1.172	-	1.123	1.150	1.180	1.170	1.170
K^+	0.626	0.626	0.589	0.625	0.626	0.622	0.622
KSO_4^-	0.650	0.647	0.589	0.669	0.720	0.747	0.747
Cl^-	0.627	0.626	0.589	0.625	0.626	0.622	0.622
SO_4^{2-}	0.618	0.167	0.158	0.167	0.167	0.181	0.181
HCO_3^-	0.690	0.690	0.631	0.669	0.720	0.675	0.747
CO_3^{2-}	0.188	0.187	0.195	0.190	0.222	0.207	0.311
$B(OH)_3^\circ$	1.172	-	-	-	1.180	1.170	1.170
$B(OH)_4^-$	0.671	-	-	-	0.605	0.747	0.747
Br^-	-	-	-	0.625	-	0.747	0.747
F^-	0.650	-	-	-	0.649	0.747	0.747
MgF^+	-	-	-	-	0.689	0.747	0.747
Sr^{2+}	0.207	-	-	-	0.207	0.311	0.311
$H_4SiO_4^\circ$	1.000	-	1.123	-	1.180	1.170	1.170
H^+	0.804	0.805	-	0.750	0.804	0.747	0.747
OH^-	0.650	0.649	0.589	0.685	0.649	0.747	0.747
Ionic Strength	0.663	0.677	0.655	0.660	0.666	0.680	0.680

Table IX

Saturation Index for Selected Minerals in River Water Test Case

MINERAL	FORMULA	EQUIL	EQ3	IONPAIR	MIRE	SOLMNEQ	WATEQF	WATEQ2	WATSPEC
Calcite	$CaCO_3$	-0.51	-0.585	-0.673	-0.461	-0.765	-0.634	-0.634	-0.63
Dolomite	$CaMg(CO_3)_2$	-0.73	-0.248	-1.340	-0.730	-1.329	-1.384	-1.386	-1.38
Siderite	$FeCO_3$	-1.37	+0.456	-	-2.329	-3.377	-7.347	-1.760	-7.13
Rhodochrosite	$MnCO_3$	-1.81	-3.225	-	-2.097	-2.136	-2.180	-2.180	-
Gypsum	$CaSO_4 \cdot 2H_2O$	-3.00	-2.962	-	-3.081	-2.942	-3.057	-2.969	-3.25
Celestite	$SrSO_4$	-	-	-	-	-	-	-	-
Hydroxyapatite	$Ca_5(PO_4)_3OH$	+2.82	-	-	+5.891	+5.046	-1.784	-1.722	-
Fluorite	CaF_2	-4.39	-	-	-	-3.338	-3.079	-3.074	-
Ferric Hydrox.(Am)	$Fe(OH)_3$	-	-	-	-	-7.584	+1.304	-	-
Goethite	$FeO(OH)$	+5.59	-	-	-	-1.484	+7.810	+6.843	+5.06
Hematite	Fe_2O_3	+8.11	+18.332	-	-	-3.252	+15.144	+13.223	+13.52
Gibbsite (crypt.)	$Al(OH)_3$	-0.08	+1.948	-	-	-0.058	-0.336	-1.989	-1.19
Birnessite	MnO_2	-	-	-	-	-	-4.114	-4.114	-
Chalcedony	SiO_2	-	+0.490	-	-	+0.217	-0.142	+0.189	-
Quartz	SiO_2	+0.47	+0.776	-	-	+0.697	+0.405	+0.736	+0.74
Kaolinite	$Al_2Si_2O_5(OH)_4$	+1.83	+5.826	-	-	-	+1.638	-1.021	+2.32
Sepiolite	$Mg_2SiO_{7.5}(OH) \cdot 3H_2O$	-4.55	-	-	-	-5.734	-3.699	-	-
FeS Amorphous	FeS	-	-	-	-	-	-7.644	-2.313	-
Mackinawite	FeS	-	-	-	-	-	-6.928	-1.580	-

Table X

Saturation Index for Selected Minerals in Seawater Test Case

MINERAL	FORMULA	EQUIL	EQ3	GEOCHEM	MIRE	SEAWAT	SOLMNEQ	WATEQF	WATEQ2	WATSPEC	ACM[1]	RWR[2]
Calcite	$CaCO_3$	+0.60	+0.806	+0.67	+0.621	+0.631	+0.597	+0.742	+0.774	+0.72	+0.568 to +0.674	-
Dolomite	$CaMg(CO_3)_2$	+2.30	+3.439	+1.79	+2.277	+2.305	+2.219	+2.330	+2.394	+2.30	-	-
Siderite	$FeCO_3$	-2.65	-10.276	-12.42	-	-	-6.691	-9.006	-8.973	-8.83	-	-
Rhodochrosite	$MnCO_3$	-3.57	-4.444	-4.45	-4.077	-	-3.709	-3.727	-3.695	-	-	-
Gypsum	$CaSO_4 \cdot 2H_2O$	-0.47	-0.399	-0.76	-0.487	-0.840	-0.441	-0.439	-0.348	-0.60	-	-0.582
Celestite	$SrSO_4$	-	-	-0.13	-	-	-0.988	-0.610	-0.609	-1.32	-	-
Barite	$BaSO_4$	-	-	+1.12	-	-	-0.501	+0.097	+0.097	-0.34	-	-
Hydroxyapatite	$Ca_5(PO_4)_3OH$	+4.16	-	+3.53	+7.642	-	+7.140	+0.605	+0.670	-	-	-
Fluorite	CaF_2	-1.61	-	-2.61	-	-	-1.048	-0.742	-0.751	-	-	-
Ferric Hydr.(Am)	$Fe(OH)_3$	-	-	-2.93	-	-	-	+0.712	+0.706	-	-	-
Goethite	$FeO(OH)$	+5.64	-	+2.58	-	-	-	+7.809	+7.803	+5.04	-	-
Hematite	Fe_2O_3	+8.21	+16.518	+5.65	-	-	-	+15.228	+15.229	+13.54	-	-
Gibbsite Crypt.	$Al(OH)_3$	-0.63	+0.216	-0.57	-	-	-4.954	-1.685	-1.817	-2.69	-	-
Birnessite	MnO_2	-	-	-1.01	-	-	-	-2.993	-2.993	-	-	-
Manganite	$MnO(OH)$	-	-	-2.03	-	-	-	-1.335	-1.336	-	-	-
Chalcedony	SiO_2	-	-0.415	-1.41	-	-	-0.522	-0.537	-0.537	-0.54	-	-
Quartz	SiO_2	-0.14	-0.143	-0.04	-	-	-0.092	-0.055	-0.054	-0.06	-	-
Kaolinite	$Al_2Si_2O_5(OH)_4$	-0.47	+0.645	-0.67	-	-	-2.108	-2.108	-2.384	-2.24	-	-
Sepiolite Cryst.	$Mg_2Si_3O_{7.5}(OH) \cdot 3H_2O$	-1.96	-	+1.09	-	-	-1.908	+1.059	+1.034	-	-	-

[1]Apparent constant method (see text for explanation). [2]Calculated from the equations of Reilly, Wood and Robinson (6) by M. Lafon.

through the approach offered by Haas and Fisher (77) but it will take some time to compile and evaluate the appropriate data. In the meantime the lack of carefully evaluated thermodynamic data will continue to be the most serious limitation for any type of chemical modeling.

Thirdly, another corollary of the first limitation, is the inconsistency and inadequacy of activity coefficient equations. Some models use the extended Delbye-Huckel equation (EDH), others the extended Debye-Huckel with an additional linear term (B-dot, 78, 79) and others the Davies equation (some with the constant 0.2 and some with 0.3, 80). The activity coefficients given in Table VIII for seawater show fair agreement because seawater ionic strength is not far from the range of applicability of the equations. However, the accumulation of errors from the consideration of several ions and complexes could lead to serious discrepancies. Another related problem is the calculation of activity coefficients for neutral complexes. Very little reliable information is available on the activity of neutral ion pairs and since these often comprise the dominant species in aqueous systems their activity coefficients can be an important source of uncertainty.

The fourth limitation is the assumption made about the redox state of aqueous systems. The distribution of redox species depends on what redox potential is assumed to dominate the chemical equilibrium. The possible alternatives include the Fe^{2+}/Fe^{3+} couple, the S^{2-}/SO_4^{2-} couple, the O_2/H_2O couple, the O_2/H_2O_2 couple, the NO_3^-/NH_4^+ couple and the redox potential measured with a platinum electrode. The wide range of values for iron, manganese, chromium and arsenic species is partly due to the inherent redox assumptions. To examine how different redox controls affect the distribution of species while other factors are kept constant, several redox options were computed on WATEQ2 for both test cases. These results are shown in Table XI. The redox elements, iron and arsenic, can be distributed according to several imposed redox potentials given total concentrations of iron and arsenic. Ferrous and ferric ions along with the two dominant forms of oxidized and reduced arsenic were computed by all the possible redox options. The range of concentrations of these species is several orders of magnitude and includes most of the values listed in Tables V and VII. If the application of a chemical model is to interpret natural water chemistry, including redox reactions, then individual redox elements such as total ferrous and total ferric should be analyzed separately when possible rather than assuming that they can be distributed according to some other equilibrium. Homogeneous redox equilibrium may not be often obtained in real systems (81) and imposed redox equilibria may not represent a realistic distribution of species.

A fifth factor is the total number of complexes considered by an aqueous model. If one model has 2 or 3 times as

Table XI
Variations in the distribution of selected redox species (in p(m)) caused by changing the imposed redox potential.

Imposed Redox Potential	Implied Eh(volts)	Fe^{2+}	Fe^{3+}	$H_2AsO_3^{\circ}$	$HAsO_4^{2-}$
River Water Test Case					
Pt Eh	0.440	11.78	17.25	25.80	7.60
O_2/H_2O	0.783	17.89	17.25	38.03	7.60
O_2/H_2O_2	0.141	6.81	17.61	15.14	7.60
NO_3^-/NO_2^-	0.445	11.87	17.25	25.98	7.60
NO_3^-/NH_4^+	0.328	9.78	17.25	21.81	7.60
S^{2-}/SO_4^{2-}	-0.532	6.56	29.36	7.59	24.04
Fe^{2+}/Fe^{3+}	0.073	6.58	18.50	12.72	7.60
Seawater Test Case					
Pt Eh	0.550	13.94	17.90	29.17	7.26
O_2/H_2O	0.731	17.85	17.90	36.98	7.26
O_2/H_2O_2	0.133	7.94	18.09	16.76	7.26
NO_3^-/NO_2^-	0.390	12.08	17.90	25.45	7.26
NO_3^-/NH_4^+	0.269	10.04	17.90	21.36	7.26

many metal-ligand complexes as another model then it will have a lower concentration of free ligand, assuming negligible differences in the data base. For example, the inclusion of the calcium and magnesium chloride complexes by GEOCHEM tends to lower the free calcium, magnesium and chloride ion concentrations below those of the other models.

A sixth source of difference, which was unavoidable in preparation of the test cases, is the various ways each program handles the carbonate system. From a practical standpoint, the inorganic carbon system of natural waters is usually determined from the titration alkalinity. Because the titration alkalinity includes both carbonate and non-carbonate alkalinity, the titration alkalinity must be corrected for non-carbonate alkalinity. Most models correct for the presence of $H_2BO_3^-$ and $H_3SiO_4^-$ but many other minor species should be considered and there is no general agreement as to the precise correction for non-carbonate alkalinity. In devising the correction used originally in WATEQ, Truesdell and Jones also considered the possibility that some non-carbonate alkalinity species were kinetically slow to react in the titration and thus were not included in the titration alkalinity. Even if we could agree on which species to subtract from the titration alkalinity, the computed carbonate system is still dependent on the equilibrium constants and activity coefficients used by the model to compute the actual concentrations of non-carbonate alkalinity species.

Some aqueous models accept only total inorganic carbon rather than titration alkalinity or carbonate alkalinity. For this reason, the sea water analysis of Table III includes total inorganic carbon which was calculated from pH, total alkalinity and salinity using the apparent sea water constants of Mehrbach et al.(82) for the dissociation of carbonic acid and the boric acid dissociation constant of Lyman (83), as expressed by Li et al. (84).

Certainly, there are differences in aqueous models in current use and the carbonate calculations will depend in part on whether the source of carbon data was total alkalinity or total inorganic carbon from Table III. For example, using the program WATEQF and data in ppm from Table III, Table XII compares some computed parameters of the carbonate system of seawater using total alkalinity and total inorganic carbon from Table III.

Table XII

Comparison of carbonate parameters computed by WATEQF in Seawater Test Case

Computed Parameter	Inorganic Carbon Data Source: Total Alkalinity (= 141.682 ppm as HCO_3^-)	Inorganic Carbon Data Source: Total Inorganic Carbon (= 2.09566×10^{-3} m)
Total carbon (m)	2.1421×10^{-3}	2.0957×10^{-3}
$mHCO_3^-$	1.4961×10^{-3}	1.4636×10^{-3}
mCO_3^{2-}	3.7880×10^{-5}	3.7057×10^{-5}
$\log P_{CO_2}$	-3.399	-3.409
SI calcite	0.7418	0.7323

The differences shown in Table XII are really quite small and well within the uncertainties of the thermodynamic data of the aqueous model. The close agreement shown in Table XII indicates that the carbonate system of WATEQF is reasonably compatible with the apparent constants of Mehrbach et al. (82) and Lyman (83). However, in comparing results in the carbonate system computed by other aqueous models, there is a potential for differences depending on the compatibility of the equilibrium constants and activity coefficients used with the apparent constant approach for seawater.

A seventh limitation is the fact that not all of the aqueous models compute temperature corrections. The programs of the REDEQL school and those used in analytical chemistry contain a data base of equilibrium constants at 25°C and are not as reliable at other temperatures. Since the river water test case was given a temperature of 9.5°C there will be some differences between programs that correct for temperature and those that do not. The temperature correction can cause substantial changes in computed results when interpreting the chemistry of natural waters because these systems commonly vary both diurnally and seasonally over a large range in temperature and because many equilibrium constants are strongly temperature dependent.

If all of these limitations were overcome, the aqueous models described here should give consistent results. To reemphasize, the largest single source of discrepancy is the thermodynamic data base used by each model. This limitation is not apparent among the major species in a dilute solution but as the ionic strenth increases and/or the concentration of the constituent decreases, the discrepancies markedly increase. This problem becomes particularly acute for trace elements where apparently small changes in equilibrium constants or pH or redox potential or temperature may produce very large changes in trace element speciation.

The effect of the thermodynamic data is particularly striking from a comparison of saturation indices in Tables IX and X. The mineral showing the best agreement in both test cases is calcite which would be expected since its properties have been extensively studied, especially in the marine environment. Using the apparent constant method the SI of calcite in seawater varies from 0.568 to 0.674 depending on the choice of data. The low value is that of Berner (90) which is compatible with the apparent dissociation constants for carbonic acid in seawater of Lyman (83). The high value is from Ingle et al. (88) which is compatible with the dissociation constants of Mehrbach et al. (82). In the river water test case gypsum SI values show very good agreement but for many other minerals there is considerable disagreement of one order of magnitude or more. SI values for several minerals show both supersaturation and undersaturation and these conditions are particularly striking for hydroxyapatite, goethite, hematite and kaolinite. In the seawater test case gypsum does not show as good agreement as in river water with values ranging from -0.35 to -0.84. Many minerals again show more than an order of magnitude difference in SI values. From an examination of these tables it should be clear that any interpretation of mineral saturation states in an aquatic environment depends greatly on the chosen chemical model which, in turn, depends upon the reliability of the thermodynamic data base and any inherent assumptions of the behavior of electrolyte solutions.

Summary

In this review over 30 computerized chemical models have been described which can calculate the distribution of species in an aqueous system at equilibrium. Every computerized model was developed for somewhat different purposes and there is no general purpose model which can be used for all of the applications described in this report. However, an attempt has been made to point out the major differences between models as a guide to researchers interested in chemical modeling. The applications include titration simulation, solubility testing, adsorption modeling, ion exchange modeling, and progressive mass transfer reactions in heterogeneous systems. Two hypothetical test cases: a dilute river water analysis and a seawater analysis were run on a total of 13 different programs to determine the distribution of species and the saturation indices for several minerals. A comparison of these results demonstrate generally good agreement for the major species and rather poor agreement for the minor species. The major source of discrepancy is the thermodynamic data base used by the various models. Other important limitations include the number of complexes in each model, the form of the activity coefficient equation, the redox assumptions, the form of the alkalinity input and the non-carbonate alkalinity correction, and temperature and pressure corrections. The discrepancies in the test case results indicate that a great deal of caution must be exercised when interpreting aqueous chemical equilibria by a chemical modeling approach. More attention should be paid to the assumptions of a model and a great deal more work is needed on the evaluation of thermodynamic data in order to provide a consistent set of values.

Abstract

A survey of computer programs which are currently being used to calculate the distribution of species in aqueous solutions, especially natural waters, has been made in order to 1) provide an inventory of available programs with a short description of their uses, 2) compare the consistency of their output for two given test solutions and 3) identify major weaknesses or problems encountered from their use. More than a dozen active programs which can be used for distribution of species and activity calculations for homogeneos equilibria among the major anions and cations of natural waters have been inventoried. Half of these programs can also accept several trace elements including Fe, Al, Mn, Cu, Ni, Zn, Cd, Pb, Ag, Hg, As,

Ba, Sr, and B. Consistency between programs was evaluated by comparing the log of the molal concentrations of free ions and complexes for two test solutions: a hypothetical seawater analysis and a hypothetical river water analysis. Comparison of the free major ion concentrations in the river water test case shows excellent agreement for the major species. In the seawater test case there is less agreement and for both test cases the minor species commonly show orders of magnitude differences in concentrations. These differences primarily reflect differences in the thermodynamic data base of each chemical model although other factors such as activity coefficient calculations, redox assumptions, temperature corrections, alkalinity corrections and the number of complexes used all have an affect on the output.

Literature Cited

1. Leggett, D. J. Machine computation of equilibrium concentrations - some practical considerations, Talanta 24, 535-542 (1977).
2. Perrin, D.D. Recent applications of digital computers in analytical chemistry, Talanta 24, 339-345 (1977).
3. Bjerrum, N. Ionic association I. Influence of ionic association on the activity of ions at moderate degrees of association, Kgl. Danske Videnskab. Selskab. Math-fys. Medd. 7, 1-48 (1926).
4. Fuoss, R. M. Properties of electrolyte solutions, Chem. Rev. 17, 27-42 (1935).
5. Bockris, J. O'M. and Reddy, A. K. N. "Modern Electrochemistry," 622 p. Plenum Press, New York, 1970.
6. Reilly, P. J., Wood, R. H. and Robinson, R. A. Prediction of osmotic and activity coefficients in mixed-electrolyte solutions, J. Phys. Chem. 75, 1305-1315 (1971).
7. Othmer, H. G. Nonuniqueness of equilibria in closed reacting systems, Chem. Eng. Sci. 31, 993-1003 (1976).
8. Caram, H. S. and Scriven, L. E. Nonunique reaction equilibria in non-ideal systems, Chem. Eng. Sci. 31, 163-168 (1976).
9. Zeleznik, F. J. and Gordon, S. Calculation of complex chemical equilibria, Ind. Eng. Chem. 60, 27-57 (1968).
10. Van Zeggeren, F. and Storey, S. H. "The Computation of Chemical Equilibria," 176 p. Cambridge University Press, London, England, 1970.
11. Brinkley, S. R. Notes on the condition of equilibrium for systems of many constituents, J. Chem. Phys. 14, 563-564 (1946).
12. Brinkley,, S. R., Calculation of the equilibrium composition of systems of many constituents, J. Chem. Phys. 15, 107-110 (1947).

13. Kandiner, H. J. and Brinkley, S. R. Calculation of complex equilibrium relations, Ind. Eng. Chem. 42, 850-855 (1950).
14. Feldman, H. F., Simons, W. H. and Bienstock, D. Calculating equilibrium compositions of multicomponent, multiphase, chemical reacting systems, U.S. Bur. Mines Rep. Invest. 7257, 22 p. (1969).
15. Crear, D. A method for computing multicomponent chemical equilibria based on equilibrium constants, Geochim. Cosmochim. Acta 39, 1375-1384 (1975).
16. Acton, F. S. "Numerical Methods that Work," 541 p. Harper and Row, New York, 1970.
17. White, W. B., Johnson, S. M. and Dantzig, G. B. Chemical equilibrium in complex mixtures, J. Chem. Phys. 28, 751-755 (1958).
18. Dayhoff, M.O., Lippincott, E. R., Eck, R. V. and Nagarajan, G. Thermodynamic equilibrium in prebiological atmospheres of C, H, O, N, P, S and Cl, NASA SP-3040, Washington, D.C., 260 p. 1967.
19. Holloway, J. R. and Reese, F. O. The generation of N_2-CO_2-H_2O fluids for use in hydrothermal experimentation I. Experimental method and equilibrium calculations in the C-O-H-N system, Amer. Mineral. 59, 587-597 (1974).
20. Karpov, I. K. and Kaz'min, L. A. Calculation of geochemical equilibria in heterogeneous multicomponent systems, Geochem. Int. 9, 252-265 (1972).
21. Ingri, N., Kakolowicz, W., Sillen, L. G. and Warnquist, B. High-speed computers as a supplement of graphical methods - V. HALTAFALL, a general program for calculating the composition of equilibrium mixtues, Talanta 14, 1261-1286 (1967).
22. Dyrssen, D., Jagner, D. and Wengelin, F. "Computer Calculation of Ionic Equilibria and Titration Procedures," 250 p. John Wiley , New York, 1968.
23. Ingri, N. and Sillen, L. G. High-speed computers as a supplement of graphical methods IV. An ALGOL version of LETAGROP VRID, Arkiv. Kemi 23, 97-121 (1965).
24. Perrin, D. D. Multiple equilibria in assemblages of metal ions and commplexing species: a model for biological systems, Nature 206, 170-171 (1965).
25. Perrin, D. D. and Sayce, I. G. Computer calculation of equilibrium concentrations in mixtures of metal ions and complexing species, Talanta 14, 833-842 (1967).
26. Fardy, J. J. and Sylva, R. N. SIAS, a computer program for the generalized calculation of speciation in mixed metal-ligand aqueous systems, AAEC/E445, Lucas Heights, Australia, 20 p. (1978).
27. Detar, D. F. "Computer Programs for Chemistry," Vol. II, 260 p. W. A. Benjamin, New York, 1969.

28. Bos, M. and Meershoek, H. Q. J. A computer program for the calculation of equilibrium concentrations in complex systems, Anal. Chim. Acta 61, 185-194 (1972).
29. Garrels, R. M. and Thompson, M. E., A chemical model for seawater at 25°C and one atmosphere total pressure, Amer. J. Sci. 260, 57-66 (1962).
30. Barnes, I. and Clarke, F. E., Chemical properties of ground water and their encrustation effects on wells, U.S. Geol. Survey Prof. Paper 498-D, 58 p. (1969).
31. Helgeson, H. C., Brown, T. H., Nigrini, A. and Jones, T. A., Calculation of mass transfer in geochemical processes involving aqueous solutions, Geochim. Cosmochim. Acta. 34, 569-592 (1970).
32. Fritz, B. "Etude thermodynamique et simulation des reactions entre mineraux et solutions application a la geochemie des alterations et des eaux continentales." PhD. Thesis, Univ. Louis Pasteur, Strasbourg, France, 152 p., 1975.
33. Droubi, A. Geochimie des sels et des solutions concentrees pars evaporation. Modele thermodynamique de simulation. Application aux sols sales du Tchad. Ph.D. Thesis, Univ. Louis Pasteur, Strasbourg, France, 177 p., 1976.
34. Truesdell, A. H. and Jones, B. F. WATEQ, a computer program for calculating chemical equilibria of natural waters, NTIS Tech. Rept. PB2-20464 Springfield,VA 77 p. (1973).
35. Truesdell, A. H. and Jones, B. F. WATEQ, a computer program for calculating chemical equilibria of natural waters, J. Res. U. S. Geol. Survey 2, 233-274 (1974).
36. Kharaka, Y. K. and Barnes, I., SOLMNEQ: Solution-mineral equilibrium computations, NTIS Tech. Rept. PB214-899, Springfield, VA 82 p. (1973).
37. Wolery, T. J., Some chemical aspects of hydrothermal processes at mid-oceanic ridges - a theoretical study. I. Basalt - sea water reaction and chemical cycling between the coeanic crust and the oceans. II. Calculation of chemical equilibrium between aqueous solutions and minerals, Ph.D. Thesis, Northwestern Univ., Evanston, IL, 1978.
38. Plummer, L. N., Jones, B. F. and Truesdell, A. H., WATEQF - a FORTRAN IV version of WATEQ, a computer program for calculating chemical equilibrium of natural waters, U.S. Geol. Survey Water Resour. Invest. 76-13, 61 p. (1976).
39. Lueck, S. L., Runnells, D. D. and Markos, G., Computer modelling of urnaium species in natural waters: applications to exploration, Geol. Soc. Amer. Ann. Mtg. Abstracts, 1978.
40. Ball, J. W., Jenne, E. A. and Nordstrom, D. K., WATEQ2 - a computerized chemical model for trace and major element speciation and mineral equilibria of natural waters, in

Jenne, E. A., ed., "Chemical Modeling in Aqueous Systems. Speciation, Sorption, Solubility, and Kinetics," Amer. Chem. Soc., 1978 (This volume).
41. Wigley, T.M.L. WATSPEC: a computer program for determining the equilibrium speciation of aqueous solutions, Brit. Geomorph. Res. Group Tech. Bull. 20, 48 p. (1977).
42. Wigley, T. M. L., Ion pairing and water quality measurements, Can. J. Earth Sci. 8, 468-476 (1971).
43. Wolery, T. J. and Walters, L. J., Jr., Calculation of equilibrium distributions of chemical species in aqueous solutions by means of monotone sequences, Math Geol. 7, 99-115 (1975).
44. Walters, L. J., Jr. and Wolery, T. J., A monotone-sequences algorithm and FORTRAN IV program for calculation of equilibrium distributions of chemical species, Comp. Geosci. 1, 57-63 (1975).
45. Lafon, G. M., Some quantitative aspects of the chemical evolution of the oceans, Ph.D. Thesis, Northwestern Univ., Evanston, IL., 136 p., 1969.
46. Holdren, G. R., Jr., Distribution and behavior of manganese in the interstitial waters of Chesapeake Bay sediments during early diagenesis, Ph.D. Thesis, The Johns Hopkins Univ., Baltimore, MD, 191 p., 1977.
47. Thrailkill, J., Solution geochemistry of the water of limestone terrains, Univ. Kentucky Water Resour. Inst. Res. Rept. 19, 125 p (1970).
48. Van Beek, C. G. E. M., personal communication.
49. Van Breeman, N. Calculation of ionic activities in natural waters, Geochim. Cosmochim. Acta 37, 101-107 (1973). (1973).
50. Morel, F. and Morgan, J. J. A numerical method for computing equilibria in aqueous chemical systems, Env. Sci. Tech. 6, 58-67 (1972).
51. Morel, F., McDuff, R. E. and Morgan, J. J. Interactions and chemostasis in aquatic chemical systems: Role of pH, pE, solubility and complexation, p. 157-200, in Singer, P.C., ed., "Trace Metals and Metal-Organic Interactions in Natural Waters," Ann Arbor Science Publishers, Ann Arbor, Michigan, 1973.
52. Morel, F., McDuff, R. E. and Morgan, J. J. Theory of interaction intensities, buffer capacities, and pH stability in aqueous systems, with application to the pH of seawater and a heterogeneous model ocean system, Mar. Chem. 4, 1001-1028 (1976).
53. James, R. O. and Healy, T. W. Adsorption of hydrolyzable metal ions at the oxide-water interface. I. Cobalt (II) adsorption on silicon dioxide and titanium dioxide as model systems II. Charge reversal of silicon dioxide and titanium dioxide colloids by adsorbed cobalt (II), lanthanum (III) and thorium (IV) as model systems III.

Thermodynamic model of adsorption, J. Colloid Interface Sci. 40, 42-81 (1972).

54. Schindler, P. W. and Gamsjaeger, H. Acid-base reactions of the titanium dioxide (anatase) - water interface and the point of zero charge of titanium dioxide suspension, Kolloid - Z. Z. Polym. 250, 759-763 (1972).

55. Schindler, P. W., Fuerst, B., Dick, R., Wolfe, P. U. Ligand properties of surface silanol groups. I. Surface complex formation with iron (3+), copper (2+), cadmium (2+) and lead (2+), J. Colloid Interface Sci. 55, 469-475 (1976).

56. Schindler, P. W., Waelti, E. and Fuerst, B. The role of surface hydroxyl roups in the surface chemistry of metal oxides, Chimia 30, 107-109 (1976).

57. Hohl, H. and Stumm, W. Interaction of lead (2+) with hydrous-alumina, J. Colloid Interface Sci. 55, 281-288 (1976).

58. Westall, J. and Hohl, H. A general method for the computation of equilibria and determination of equilibrium constants for adsorption at hydrous oxide surfaces, in "Abstracts of Papers," Amer. Chem. Soc. Meeting, Miami Beach, FL., 1978.

59. Davis, J. A., III and Leckie, J.O. Surface ionization and complexation at the oxide/water interface, in Jenne, E. A., ed. "Chemical Modeling in Aqueous Systems. Speciation, Sorption, Solubility, and Kinetics." Amer. Chem. Soc., 1978 (This volume).

60. Westall, J. C., Zachary, J. L. and Morel F. M. M. MINEQL, a computer program for the calculation of chemical equilibrium composition of aqueous system, Tech. Note 18, Dept.Civil Eng. Mass. Inst. Tech., Cambridge, MA., 91 p., (1976).

61. Mattigod, S. V. and Sposito, G. Chemical modeling of trace metal equilibria in contaminated soil solutions using the computer program GEOCHEM, in Jenne, E.A., ed., "Chemical Modeling in Aqueous Systems. Speciation, Sorption, Solubility, and Kinetics." Amer. Chem. Soc., 1978 (This volume).

62. I, T.-P. and Nancollas, G.H. EQUIL - a computational method for the calculation of solution equilibria, Anal. Chem. 44, 1940-1950 (1972).

63. Helgeson, H. C. Evaluation of irreversible reactions in geochemical processes involving minerals and aqueous solutions - I. Thermodynamic relations, Geochim. Cosmochim. Acta 32, 853-877 (1968).

64. Helgeson, H. C., Garrels, R. M. and Mackenzie, F. T. Evaluation of irreversible reactions in geochemical processes involving minerals and aqueous solutions - II. Applications, Geochim. Cosmochim. Acta 33, 455-481 (1969).

65. DeDonder, Th. and Van Rysselberghe, P. "The Thermodynamic Theory of Affinity," Stanford University Press, Stanford, California, 1936.

66. Helgeson, H. C. A chemical and thermodynamic model of ore deposition in hydrothermal systems, Mineral Soc. Amer. Spec. Paper 3, 155-186 (1970).
67. Miller, D. G., Piwinskii, A. J. and Yamauchi, R. The use of geochemical-equilibrium computer calculations to estimate precipitation from geothermal brines, UCRL-52197 Livermore, Calif., 34 p., (1977).
68. Droubi, A., Cheverry, C., Fritz, B. and Tardy, Y. Geochimie des eaux et des sels dans les sols des polders du Lac Tchad: Application d'un modele thermodynamique de simulation de l'evaporation, Chem. Geol. 17, 165-177 (1976).
69. Droubi, A., Fritz, B. and Tardy, Y. Equilibres entre mineraux et solutions; Programmes de calcul appliques a la prediction de la sahire des sols et des doses optimales d'irrigation, Cah. ORSTOM, ser. Pedol XIV, 13-38 (1976).
70. Gear, C. W. The automatic integration of ordinary differential equations, Communications of the ACM 14, 176-179 (1971).
71. Gear, C. W. Algorithm 407-DIFSUB for solution of ordinary differential equations, Communications of the ACM 14, 185-190 (1971).
72. Plummer, L. N., Parkhurst, D. L. and Kosiur, D. R. MIX2: A computer program for modeling chemical reactions in natural waters, U.S. Geol. Survey Water Resour. Inv. Rept. 75-61, 73 p. (1975).
73. Plummer, L. N. Mixing of sea water and calcium carbonate ground water, Geol. Soc. Amer. Mem. 142, 219-236 (1975).
74. Wigley, T. M. L. and Plummer, L. N. Mixing of carbonate waters, Geochim. Cosmochim. Acta 40, 989-995 (1976).
75. Plummer, L. N. Defining reactions and mass transfer in part of the Floridan aquifer, Water Resour. Res. 13, 801-812 (1977).
76. Parkhurst, D. L. Plummer, L. N. and Thorstenson, D. C. Chemical models in ground-water systems, Geol. Soc. Amer. Ann. Mtg. Abstracts 1978.
77. Haas, J. L., Jr. and Fisher, J. R. Simultaneous evaluation and correlation of thermodynamic data, Amer. J. Sci. 276, 525-545 (1976).
78. Helgeson, H. C. Thermodynamics of hydrothermal systems at elevated temperatures and pressures, Amer. J. Sci. 267, 729-804 (1969).
79. Helgeson, H. C. and Kirkham, D. H. Theoretical prediction of the thermodynamic behavior of aqueous electrolytes at high pressures and temperatures II. Debye-Huckel parameters for activity coefficients and relative partial molal properties, Amer. J. Sci. 274, 1199-1261 (1974).
80. Davies, C. W. "Ion Association," 190 p. Butterworths, Washington, D.C., 1962.

81. Morris, J. C. and Stumm, W. Redox equilibria and measurements of potentials in the aquatic environment, p. 270-285, in Gould, R. F., ed., "Equilibrium Concepts in Natural Water Systems," Adv. Chem. Ser. 67, 1967.
82. Mehrbach, C., Culberson, C. H., Hawley, J. E. and Pytkowicz, R. M. Measurement of the aparent dissociation constants of carbonic acid in seawater at atmospheric pressure, Limnol. Oceanogr. 18, 897-907 (1973).
83. Lyman, J. Buffer mechanism of seawater, Ph.D. Thesis, Univ. Calif., Los Angeles, CA., 196 p., 1956.
84. Li, Y. H., Takahashi, T. and Broecker, W. S., Degree of saturation of $CaCO_3$ in the oceans, J. Geophys. Res. 74, 5507-5525 (1969).
85. McDuff, R. E. and Morel, F. M., Description and use of the chemical equilibrium program REDEQL2, Keck Lab. Tech. Rept. EQ-73-02, Calif. Inst. Tech., Pasadena, CA, 75., (1973).
86. Holdren, G. R., Jr. and Bricker, O.P., Distribution and control of dissolved iron and manganese in the interstitial waters of the Chesapeake Bay, p. 178-196, in Drucker, H. and Wildung, R. E., ed., "Biological Implications of Metals in the Environment," ERDA Symposium Series 42, 1977.
87. Thrailkill, J., Carbonate chemistry of aquifer and stream water in Kentucky, J. Hydrol. 16, 93-104 (1972).
88. Thrailkill, J., Carbonate equilibria in karst waters, p. 745-771, in Yevjevich, V., ed., "Karst Hydrology and Water Resources," Vol. 2, Water Resources Pub., Ft. Collins, CO., 1976.
89. Wigley, T. M. L., Plummer, L. N., and Pearson, F. J., Jr., Mass transfer and carbon isotope evolution in natural water system, Geochim. Cosmochim. Acta 42, 1117-1139 (1978).
90. Berner, R. A., The solubility of calcite and aragonite in seawater at atmospheric pressure and 34.5°/oo salinity, Amer. J. Sci. 276, 713-730 (1976).

RECEIVED November 16, 1978.

[2] Department of Environmental Sciences, University of Virginia, Charlottesville, VA 22903.
[3] U.S. Geological Survey, National Center, MS 432, Reston, VA 22092.
[4] Climatic Research Unit, University of East Anglia, Norwich, NR4 7TJ, England.
[5] Lawrence Livermore Laboratory, University of California, P.O. Box 808, Livermore, CA 94550.
[6] U.S. Geological Survey, WRD, 345 Middlefield Rd., MS 21, Menlo Park, CA 94025.
[7] U.S. Geological Survey, Denver Federal Center, MS 416, Box 25046, Lakewood, CO 80225.
[8] Department of Geological & Geophysical Sciences, Princeton University, Princeton, NJ 08540.
[9] Australian Atomic Energy Commission Research Establishment, Lucas Heights, N.S.W., Australia.
[10] Institut de Geologie, 1, rue Blessig, 67084 Strasbourg, France.
[11] Department of Civil and Mineral Engineering, University of Minnesota, Minneapolis, MN 55455.
[12] Department of Geological Sciences, University of Rochester, Rochester, NY 14627.
[13] Department of Earth & Planetary Sciences, The Johns Hopkins University, Baltimore, MD 21218.
[14] Department of Soil & Environmental Sciences, University of California, Riverside, CA 92521.
[15] Department Civil Engineering, Massachusetts Institute of Technology, Cambridge, MA 02139.
[16] State of New York Dept. of Health, New Scotland Avenue, Albany, NY 12201.
[17] Department of Geology, University of Kentucky, Lexington, KY 40506.

INDEX

INDEX

B

E

I

J

K

N

T

U